GW01606162

GAS CONDITIONING AND PROCESSING

Volume 2: The Equipment Modules

By:
John M. Campbell

In Collaboration with: **Larry L. Lilly, Chapter 11**
Robert N. Maddox, Energy Data

8th Edition
Edited By:
Robert A. Hubbard

This book is based upon – and is the successor to – the classic work of the same title authored through seven editions and many printings by **Dr. John M. Campbell, Sr.** from 1966-1998.

GAS CONDITIONING AND PROCESSING

Volume 2: The Equipment Modules

Eighth Edition

Library of Congress Catalog Card No.: 73-157183

3rd Printing, February 2004

Printed and Bound in the U.S.A.

ISBN 0-9703449-1-0

DISCLAIMER

Published by:

John M. Campbell and Company
1215 Crossroads Blvd.
Norman, Oklahoma U.S.A. 73072
Phone: 1-405-321-1383
Fax: 1-405-321-4533
Website: www.jmcampbell.com
E-mail: jmc@jmcampbell.com

Special Thanks To:

Daniel Stowe for his tireless effort in bringing this book to publication.
Martina Dreyer for the much improved graphics in the book.
Betty Dalton for typing the new text and proof reading of this book.
John Morgan for review of the manuscript and editing.

LIST OF CHAPTERS

TABLE OF CONTENTS

Volume 2 of "Gas Conditioning and Processing" is a continuation of Volume 1, "The Basic Principles." These two volumes are a single book published in two parts.

LIST OF FIGURES

LIST OF TABLES

ABOUT "John M. Campbell and Company (JMC)"...

JMC was formed in 1968 to provide professional services to the oil and gas industry. To date, JMC has provided training and consulting to virtually every major oil company worldwide. Our traditional area of focus continues to be production facilities, gas processing, and commercial issues.

Training. Since 1968 over 30,000 professionals have attended our training programs which combine a strong theoretical foundation with outstanding practical experience. Our programs are not just technical entertainment, but relevant, focused and results-oriented.

Our ***Standard Foundation Course*** titles are listed below. Please visit our website (**www.jmcampbell.com**) to find out more about each course and available dates.

- ❒ Overview of Gas Processing (G-2)
- ❒ Gas Conditioning and Processing (G-4)
- ❒ Gas Conditioning and Processing (LNG Emphasis) (G-4 LNG)
- ❒ Accelerated Selected Topics in Gas Processing (G-5)
- ❒ Gas Treating and Sulfur Recovery (G-6 EP)
- ❒ Refinery Gas Treating and Sulfur Recovery (G-6 REF)
- ❒ Process Simulation in Gas Conditioning and Processing (G-7)
- ❒ CO_2 Surface Facilities (G-8)
- ❒ Gas Measurement for Engineering Personnel (M-1)
- ❒ Gas Measurement for Field Personnel (M-2)
- ❒ Production/Processing Facilities (P-2)
- ❒ Applied Water Technology (W-1)
- ❒ Corrosion Management in Production/Processing Operations (W-2)
- ❒ Oilfield Corrosion and Water Treatment (W-3)

We also offer "***Expanded and Special Topic***" courses. These courses are shorter in length from our "Standard Foundation Courses", but no less intense.

- ❒ Process Engineering Fundamentals (G-41)
- ❒ Gas Dehydration and Hydrate Inhibition (G-42)
- ❒ Refrigeration and NGL Extraction (G-43)
- ❒ Distillation: Design and Operation (G-44)

Our "***Equipment Courses***" are also shorter in length than our "Foundation" courses. They are:

- ❒ Piping Design and Specification (P-21)
- ❒ Process Vessel Specification and Design (P-22)
- ❒ Heat Transfer Equipment (P-23)
- ❒ Pumps and Compressors (P-24)

Consulting. John M. Campbell and Company's consulting activities range from conceptual design and process selection to troubleshooting and debottlenecking. Our worldwide exposure allows us to develop innovative solutions which may not be readily apparent to others. Our philosophy is to maintain a small, highly skilled staff augmented by exceptional internal consultants.

Publishing. John M. Campbell & Co. (JMC) publishes reference books for the oil and gas industry. Many of these books have been the standard reference material in their industry for over 25 years. Volumes 1 and 2 of "Gas Conditioning and Processing", used in the Campbell Gas Course®, are the most "borrowed" books from many engineers' shelves. These books are updated frequently to reflect changing industry technology and demands.

To find out how to order our books, please visit our website at www.jmcampbell.com or contact us at 1-405-321-1383 for availability and ordering information. A price list for all books is available upon request. Adult training courses using these books also are available exclusively through John M. Campbell and Company.

GAS CONDITIONING AND PROCESSING "SERIES"
Volume 1: ***The Basic Principles***, one of a four volume series, has been published for the natural gas processing industry for over 30 years. This edition has been edited to reflect continuing changes in technology and the manner in which it is practiced. This book addresses: overview of gas processing; material and energy balances; phase behavior; physical properties; water-hydrocarbon equilibrium; hydrates; applied thermodynamics; process control; and flow of fluids.
Volume 2: ***The Equipment Modules*** is a continuation of the material presented in Volume 1. This edition includes information for applying today's technology and the current business requirements to selecting and operating gas processing and production surface facilities. This book aids in decisions relating to separation; heat transfer; pumps; compressors; refrigeration; fractionation and absorption; glycol and solid bed dehydration.
Volume 3: ***Computer Applications for Production/Processing Facilities*** is an extension of the information in Volumes 1 and 2 with emphasis on the more detailed calculations required for computer modeling and simulation. Equations of state, heavy component characterization, rotating equipment modeling, fractionation, fluid flow and separation are covered. All topics are addressed from a practical application of computer modeling techniques applicable to any simulator.
Volume 4: ***Gas Treating and Sulfur Recovery*** concentrates on problems associated with treating and removing H_2S, CO_2 and other sulfur compounds often associated with natural gas production. A detailed view of commercial amine type processes; carbonate processes; physical absorption methods; liquid product treating; solid bed sweetening; sulfur production; and tail gas conditioning is presented.
APPLIED WATER TECHNOLOGY
This book focuses on water handling and disposal problems for produced water associated with natural gas and oil production. The subjects covered include: water sampling and analysis; water formed scales; corrosion control; microbiology; water processing equipment; water injection system; water treatment for EOR; boiler water and cooling water treatment.

COMPUTER PROGRAMS
GCAP® 8th Edition. The key equations and correlations in Volumes 1 and 2 of Gas Conditioning and Processing have been programmed for use with personal computers. The books serve as your manuals, providing an explanation of the assumptions and limitations of the calculations.
Example problems in the book can be used as a guide to familiarize the user with these computer programs.

11

SEPARATION EQUIPMENT

The design of equipment for the separation of vapors and liquids is essential to almost all processes. The design concepts of a simple separator may be extended to several other processes such as fractionation towers, glycol dehydrators, two-phase flow lines, slug catcher design, desalters, etc. The purpose of this chapter is to review the principles governing the basic separation process and set forth some criteria for use in the planning and operation of the equipment involved.

The basic equipment for separating liquid from vapor uses both *gravitational* and *centrifugal* force. The gravitational force is used by reducing velocity so the liquid can settle out in the space provided. Centrifugal force is used by changing the direction of flow. A true separator, as defined herein, depends on gravitational force to a substantial degree and has sufficient liquid retention time to allow for effective vapor-liquid disengagement.

FABRICATION SPECIFICATIONS

Most process vessels are fabricated to applicable codes established by government agencies, professional and trade groups, and/or by individual companies. These codes specify fabrication standards, inspection requirements, over-pressure protection as well as numerous other design and operational issues. The standards require that pressure vessels are manufactured to a consistent and recognized standard thus ensuring a safe working environment for personnel. In most states and countries, manufacture of pressure vessels to an acceptable code is mandated by law.

Current addresses for detailed information about some of the commonly used codes are:

ASME Pressure Vessel Code American Society of Mechanical Engineers 22 Law Drive, P.O. Box 2900 Fairfield, New Jersey USA 07007-2900, *phone* (800) 843-2763 *fax* (973) 882-1717 *website*: www.asme.org	**BS 5500 – Specification for Unfired Fusion Welded Pressure Vessels** British Standard Institution, BSI Standards 389 Chiswick High Road London W4 4AL, UNITED KINGDOM *phone* 0181-996-7000 *fax* 0181-996-7001
CSA B51-97, Part 1 – Boiler, Pressure Vessel, and Pressure Piping Code (1997 Edition) Canadian Standards Association (CSA) 178 Rexdale Boulevard, Rexdale (Toronto) Ontario M9W 1R3 CANADA	**Australian Pressure Vessel Code, AS 1210 Unfired Pressure Vessels** Standards Australia GPO Pox 5420, Sydney, NSW 2001 AUSTRALIA *phone* 61 2 8206 6010 *fax* 61 8206 6020, *E-mail* sales@standards.com.au *website* www.standards.com.au

CODAP 95 – French Code for Construction of Unfired Pressure Vessels SNCT-Publications, 39-41 Rue Louis Blanc 92400 Courlendie, FRANCE	**The Dutch Pressure Vessel Code**, published by Sdu Uitgevers Postbus 20014 2500 EA Den Haag, THE NETHERLANDS *phone* 31 70 378 9880 *fax* 31 70 378 9783
A. D. Merkblatt Code Published by Carl Heymanns Publishing, Luxemburger Street 449, 50939 Cologne, GERMANY *phone* 0221 97373 901 *fax* 0221 94373 901 *website* www.heymanns.com *e-mail* marketing@heymanns.com	

Two codes — ASME Section VIII Division 1 and 2 and BS 5500 are probably the two most widely recognized pressure vessel codes, worldwide.

Vessel Shell Thickness

The basic formula for calculating the required wall thickness for a cylindrical vessel under internal pressure given by the two codes mentioned above are shown below:

ASME

$$t = \frac{P D_i}{2\,SE - 1.2\,P} + C$$

$$t = \frac{P D_o}{2\,SE + 0.8\,P} + C$$

BS 5500

$$t = \frac{P D_i}{2\,S - P} + C$$

$$t = \frac{P D_o}{2\,S + P} + C \qquad (11.1)$$

Where:

			SI	FPS
t	=	wall thickness	mm	in
P	=	design pressure	MPa(g)	psig
D_i	=	inside vessel diameter	mm	in
D_o	=	outside vessel diameter	mm	in
E	=	joint efficiency	(see Table 11.1)	
S	=	maximum allowable stress (see table below)	MPa	psi
C	=	corrosion allowance		

ASME Sec. VIII Div. 1	S = (1/3.5)(Tensile Strength)[1]
ASME Sec. VIII Div. 2	S = (1/3)(Tensile Strength)
BS 5500	S = (2/3)(Specified Minimum Yield Strength, SMYS)

TABLE 11.1

Joint Efficiency

Double-Welded Butt Joints		Single-Welded Butt Joints (backing strip left in place)	
Fully radiographed	1.00	Fully radiographed	0.90
Spot radiographed	0.85	Spot radiographed	0.80
No radiograph	0.70	No radiograph	0.65

1 Prior to 1998 the factor applied to the tensile strength was 1/4, or 25%.

For a commonly used carbon steel plate (A-515, Gr 70) the tensile strength is 483 MPa [70 000 psi] and the SMYS is 262 MPa [38 000 psi]. Based on these, the maximum allowable stress for this particular steel under the above mentioned codes are:

ASME Div. 1	138 MPa [20 000 psi]
ASME Div. 2	161 MPa [23 300 psi]
BS 5500	175 MPa [25 300 psi]

These allowable stresses have a direct impact on the vessel wall thickness – hence weight and cost. In general, the ASME code is the most conservative. There is a movement to consolidate the European codes in the future.

For spherical shells,

$$t = \frac{PR}{2\,SE - 0.2\,P} \tag{11.2}$$

When seamless pipe is used for vessel shells, one can use Equation 11.1 by substituting the proper tensile strength. It should be noted, though, that seamless pipe wall thicknesses have an allowable variation of 12.5% from the nominal. Always use 87.5% of the nominal thickness in design calculations.

Estimation of Vessel Weight and Footprint

Equation 11.1 gives the wall thickness for the shell for any vessel under internal pressure. Pressure vessel codes include equations for calculating the thicknesses of various parts of vessels. In general, Equation 11.1 will give the largest thickness for the vessel and may be used for preliminary weight calculations. The pressure used is the design pressure of the vessel. Many companies will specify the design pressure to be 1.1 times the maximum operating pressure. In turn, the maximum operating pressure is usually 5-10% above normal operating pressure. These guidelines are widely used in the industry but are not mandatory. In setting vessel design pressures, one must consider several factors. If, for example, the calculated design pressure is slightly above the maximum working pressure of an ANSI flange class, the design pressure might be reduced to avoid the additional cost of the higher pressure flanges. Conversely, equipment whose operating pressure is set by the vapor pressure of a liquid, e.g. fractionation columns, may require a larger margin, say 20-30% between operating and design pressure.

The weight of an empty vessel (including heads) may be estimated from the following equations:

SI	FPS
$W_b = 0.032\,dt$	$W_b = 13.8\,dt$

(11.3)

Where:		SI	FPS
W_b =	mass per unit length	kg/m	lbm/ft
d =	internal diameter	mm	in
t =	wall thickness, including corrosion allowance	mm	in

Equation 11.3 is based on typical separator type vessels where L/D ≈ 3-5. For longer (taller) vessels such as contactors and fractionators, Equation 11.3 will over estimate the weight by about 10-20%.

The weight of the internals (W_I) also may be estimated from Table 11.2. The weight of the external nozzles (W_N) also may be estimated from Table 11.2. The weight of the vessel can therefore be estimated as

$$W_v = W_b L + W_I + W_N \tag{11.4}$$

Where: L = seam to seam length in m or ft

Note that this is the weight of the vessel and does not include fluids from operation or hydrostatic test.

TABLE 11.2

Weight of Pressure Vessel Accessories

Vessel Internal Weight in Pounds (W_I) Manways										
		Mist Eliminators				Distillation Trays				
Vessel Diameter		Vane		Mist Mat		Normal		Light Wt		
mm	ft	kg	lbm	kg	lbm	kg	lbm	kg	lbm	
616	2.0	6	14	5	12	32	70	23	50	
770	2.5	8	17	7	15	48	105	34	75	
924	3.0	10	22	9	19	73	160	50	110	
1078	3.5	13	28	10	23	95	210	68	150	
1232	4.0	15	33	12	27	127	280	91	200	
1386	4.5	18	40	15	32	159	350	113	250	
1540	5.0	21	46	16	36	200	440	141	310	
1694	5.5	25	53	19	41	236	520	168	370	
1848	6.0	27	60	21	46	284	625	200	440	
2002	6.5	31	68	23	51	331	730	234	515	
2156	7.0	34	75	25	56	386	850	272	600	
2310	7.5	38	84	28	62	440	970	311	685	
2464	8.0	42	93	31	68	504	1110	354	780	
2618	8.5	47	103	34	74	563	1240	397	875	
2772	9.0	53	116	36	80	635	1400	445	980	
2926	9.5	57	126	39	86	703	1550	499	1100	
3080	10.0	62	137	42	93	794	1750	553	1220	
3234	10.5	65	143	45	99	862	1900	608	1340	

DN 500 [20 in] Manways*		
ANSI Class	kg	lbm
150	317	700
300	544	1200
600	952	2100
900	1269	2800

* Includes blind flange, nuts and bolts

External Nozzle Weights in kg (W_N)												
ANSI Class	Nominal Nozzle Sizes (DN)											
	50	75	100	150	200	250	300	350	400	450	500	600
150	4	7	11	20	29	43	61	75	97	150	194	267
300	5	11	18	32	50	66	100	129	168	277	321	513
600	7	18	27	54	79	129	165	233	315	424	564	823
900	13	20	34	70	118	170	249	351	437	625	767	1379

External Nozzle Weights in lbm (W_N)												
ANSI Class	Nominal Nozzle Sizes (inches)											
	2	3	4	6	8	10	12	14	16	18	20	24
150	9	16	25	45	65	95	135	165	215	331	428	589
300	12	25	40	70	110	145	220	285	370	610	708	1131
600	15	40	60	120	175	285	365	515	695	935	1245	1815
900	28	45	75	155	260	375	550	775	965	1379	1693	3041

For skid-mounted, modular construction the following factors have been found satisfactory for preliminary estimates:

Piping: $W_p = 40\%$ of W_v Structural steel $W_s = 10\%$ of W_v

Electrical & instrument: $W_e = 8\%$ of W_v $W_{skid} = W_v + W_p + W_e + W_s$

Care should be used when applying these factors as an excessive number of block valves or optional flow paths may increase the piping weight. For more accurate estimates, a preliminary piping layout should be made for all major piping (4 inch and larger) and the total piping weight taken as 150% of the weight of the major piping. Experience has shown that as a project progresses the "additional" piping required (simply piping not originally thought of) increases dramatically so all preliminary estimates must be given a healthy safety factor to account for the additional piping and equipment certain to be added. The following preliminary estimates for plot area of modular constructed process vessels can be made:

	Horizontal Vessels	**Vertical Vessels**
Module Width	I.D. × 2	I.D. × 2
Module Length	S/S × 1.5	I.D. × 2.5
Module Height	I.D. × 2 + 1 meter	S/S × 1.5 + 1 meter

Note these are approximations and assume a nominal amount of piping within the module. If pumps are included in the module, care should be taken to ensure adequate skirt height is provided for NPSH requirements. Additional piping space should also be provided for bypassing pumps, minimum flow requirements, etc. For better estimates, a preliminary "loose" piping layout should be made. Experience has shown that if a piping layout is "tight" in the preliminary stages, it will not fit in the detailed stage. Additional plot plan and deck loading estimates for specific pieces of equipment are given in Appendix 11A.

Vessel Fabrication

Most fabricators, as a matter of simple economics, must stock a limited amount and size of shell plate and heads. Furthermore, the major fabrication cost is often labor. Having to cut and chamfer shell plate to produce an "odd-length" vessel may be more costly than to "give away" extra plate. It also might be less expensive to use thicker steel plate in a vessel requiring thinner plate (from Code calculations) if the latter requires special order and handling. The moral: It often is unrealistic to hold the fabricator to an exact length or wall thickness specification; both delivery and cost could be affected. Many fabricators, for example, stock shell plate in 760 mm [30 in] length increments. Head seam to head seam (S-S) dimensions for the increment stocked often result in the most economical vessel. One should establish minimum or maximum dimensions, but some latitude should be given the vendor since it may affect both price and delivery time. Standard practice should be to insist that the fabricator quote to your minimum sizes and then offer an alternate quotation for consideration if he desires.

VAPOR-LIQUID SEPARATION EQUIPMENT

Efficient separation of vapor and liquid is a critical processing operation. Undersized or inefficient separators can result in numerous processing problems. Separators are used upstream of compressors, glycol and amine contactors, mol sieve dehydrators and the like. Carryover of hydrocarbons, water and solid particulates can cause mechanical damage to compressors and contaminate solvents. The operating costs associated with the repair or replacement of equipment/solvents often far exceeds the initial cost of the separator.

In addition, in those applications where gas is processed to meet a hydrocarbon dewpoint, entrainment of liquid hydrocarbons in the sales gas leaving the cold separator can have a significant effect on the dewpoint of the gas. This often requires that the cold separator be operated at lower temperatures to compensate for the carryover effect. This in turn increases refrigeration costs.

The above discussion deals with entrainment of liquid in the gas phase, but entrainment of gas in the liquid (often called carryunder) occurs as well. If carryunder is severe it can load up low and intermediate flash gas compressors with excess gas not planned for in the initial design.

A “complete” separator must have the following:

1. A primary separation section to remove the bulk of the liquid from the gas.
2. Sufficient liquid capacity to handle surges of liquid and to adequately degas the liquid.
3. Sufficient diameter and length (height) to allow the small droplets to settle out by gravity (to prevent undue entrainment).
4. A means of reducing turbulence in the main body of the separator so that proper settling may take place.
5. A mist extractor to capture entrained droplets and those too small to settle by gravity.
6. An inlet device to absorb the momentum (kinetic energy) of the entering fluids.
7. Vortex breaker on liquid outlet nozzle(s).
8. A method of solids removal, e.g. sand jets.
9. Manways or hand-holes to access the vessel for inspection and cleaning.
10. Proper pressure and level controls, alarms and shutdowns.

The two most common types of separators are vertical and horizontal. Relative advantages of each are listed below:

Vertical	Horizontal
Generally used in gas dominated service where liquid quantity is low.	Generally used in liquid dominated service, i.e. crude oil systems where gas flowrate is low.
Smaller footprint.	More interfacial area – better for 3-phase separation and foaming fluids.
Gas handling capacity is not a function of liquid level.	Better at handling surges and slugs.
Level vs. liquid inventory relationship is linear.	Less headroom

In general, one selects the separator with the lower installed cost and which meets the separation objectives. There are no hard and fast rules regarding separator type. Economics favor horizontal separators in liquid dominated systems (low to moderate GOR) and vertical separators in vapor dominated systems (high GOR).

Some separators are classified as scrubbers. Scrubbers are usually vertical vessels and are designed for systems where little or no liquid is expected. They are often used upstream of compressors, contactors, fuel systems, etc. as secondary separators, downstream of the primary separator.

Scrubbers typically include proprietary, high-efficiency internals, such as centrifugal elements to improve separation efficiency. They seldom contain liquid surge capacity and often employ snap-acting liquid dump valves.

The installation of scrubbers upstream of equipment sensitive to liquid entrainment is good practice, but it must be remembered that physics of separation do not change just because a vessel is called a scrubber rather than a separator. The moral – many scrubbers are undersized. Scrubbers should not be used as primary separation vessels.

Separator Components

Figure 11.1 is an example schematic of a 3-phase separator. Figures 11.2 and 11.3 are typical cutaway views showing various separator parts.

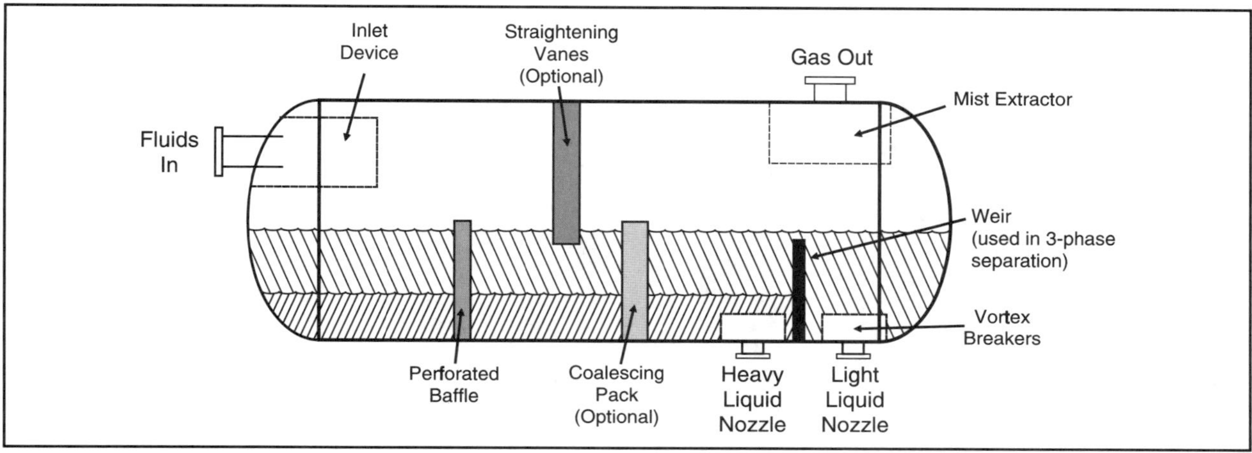

Figure 11.1 Schematic of an Example 3-Phase Separator

Inlet devices. Entering fluids can have a high velocity; 6-10 m/s [20-33 ft/sec] is not unusual. In cases where the inlet piping is undersized 20-30 m/s [66-100 ft/sec] have been experienced. The kinetic energy (often represented by the term ρv^2) of the entering fluids is high and must be dissipated before the fluids enter the gravity separation portion of the separator. Inlet devices, used to dissipate the energy, come in several designs from simple deflection plates to proprietary vane or centrifugal units. These will be discussed in more detail later in this chapter.

Gravity separation section. The bulk of the vapor and liquid will have (hopefully) segregated at this point. In this section, liquid droplets large enough to settle by gravity (typically 150-300 μm) will do so. In horizontal separators, some type of vane unit may be used here to reduce turbulence. In the liquid section of the separator, coalescing packs may be used to enhance oil-water separation.

Mist extraction. Droplets smaller than about 150-300 μm will normally not settle in the gravity separation section. It is seldom economic to increase the vessel diameter, to lessen the gas velocity in order to collect these droplets. Mist extractors are used to improve separation efficiency and depending on their design, will typically collect most droplets down to about 20-30 μm. Mist extractors will be discussed in more detail later in this chapter.

Separator sizing correlations are based on steady flow. However, experience dictates that surge capacity be built in. There is no way of determining the necessary surge capacity, as the degree of surging can never be known exactly in advance. Consequently, the design capacity chosen has to be a

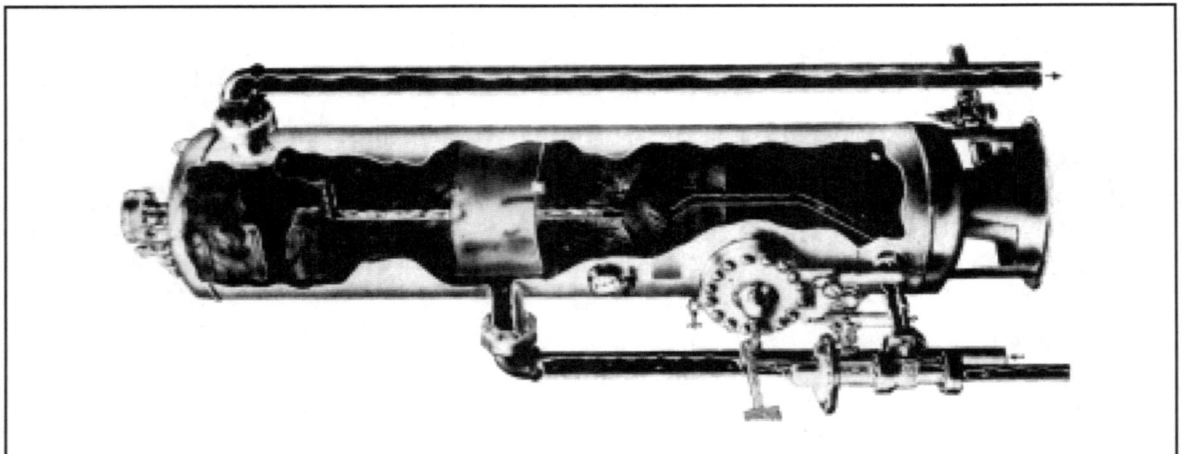

Figure 11.3 Cut-Away View of Vertical Separator

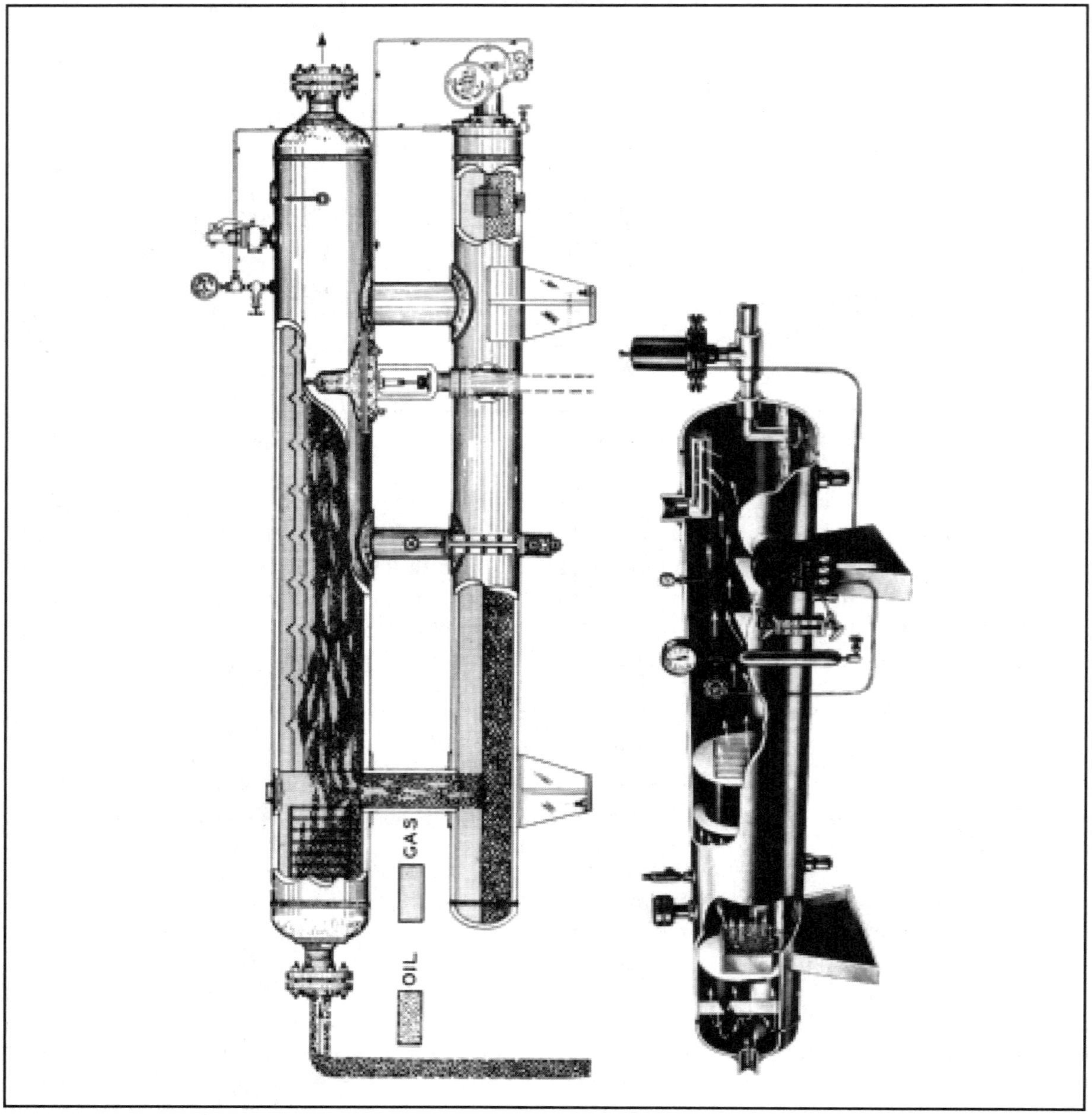

Figure 11.2 Cut-Away View of Horizontal Separators

compromise between cost and the process requirements. Typical design margins for throughput or capacity range from about 1.2 to 1.5 with the lower end of the range used for normal production systems and the high end used for production in hilly terrain, riser flow, and gas-lift systems.

In the gravity separation section, the length of a horizontal separator has a greater effect on capacity than the height of a vertical type. In the horizontal vessel, the path of any droplet ideally has a trajectory similar to that of a bullet from a gun. Therefore, the length necessary depends on:

1. Droplet size
2. Gas velocity
3. Droplet density
4. Vessel diameter
5. Degree of turbulence
6. Gas density

In the ideal case, turbulence may be neglected and effect of gravity assumed constant. Droplet velocity affects length – as it increases the droplets will travel farther before settling out. Although corrections have to be made for the nonideal conditions that exist, an analysis shows that the capacity of horizontal separators of a given diameter may be increased by increasing length.

Application of the same principles to vertical separators shows slightly different results. There, the droplet velocity is working against gravity, which makes separation more difficult. In order to keep separator height reasonable, allowable velocities are often lower than in horizontal vessels. Sufficient length is needed only for the droplet velocity to become zero and for the droplet to start falling. Increasing the length above this point does not improve separation efficiency.

If too much liquid is carried into the gravity separation section, the falling of the small droplets is hindered and entrainment becomes more prevalent because the mist extractor floods. This is one reason internal design is critical. Theory alone does not give the designer sufficient insight into separator sizing. Feedback from actual separator performance is vital in refining separator design. Alternatively, computer based computational fluid dynamics (CFD) software allows the designer to visualize fluid flow patterns through the separator without experimental testing. This approach is widely used by separator designers today.

Principles of Separation

Separation equipment employs one or more of the following mechanisms: (1) gravity settling, (2) centrifugal force, (3) impingement, (4) electrostatic precipitation, (5) sonic precipitation, (6) filtration, (7) adhesive separation, (8) adsorption, or (9) thermal. The primary mechanisms in oil-gas separation are 1, 2, and 3.

The problem is complicated because particles of varying size and characteristics, both liquid and solid, must be removed. Furthermore, size and cost of the equipment required are always practical considerations.

Particle Size. This is normally defined by its diameter in microns, $1\mu = 1 \times 10^{-6}$ m $= 3.3 \times 10^{-6}$ ft. Particles larger than about 20-30 μm may be separated by properly designed equipment. Particles smaller than this represent a problem. Settling, impingement and centrifugal force exhibit limited effectiveness in removing these small particles. Properly sized coalescing filter separators can remove droplets to about 1 μm.

There are several ways of describing the average particle diameter:

1. Diameter of a sphere having average volume:

$$D_p = \left[\frac{\sum(n\,D^3)}{\sum n}\right]^{1/3}$$

2. Diameter of sphere having average area:

$$D_p = \left[\frac{\sum(n\,D^2)}{\sum n}\right]^{1/2}$$

3. Diameter of sphere having geometric mean diameter:

$$D_p = e^B$$

Where:

e = natural logarithm base
B = Σ(n ln D)/Σn
D_p = average particle diameter
D = diameter of given size particle
n = number of particles of given sized (dimensionless)
(D and D_p in consistent units)

The usual probability of particle sizes normally takes the form shown in Curve a, Figure 11.4. Often the curve will be skewed to the right or left as with Curve b, or several peaks may be encountered as with curve c.

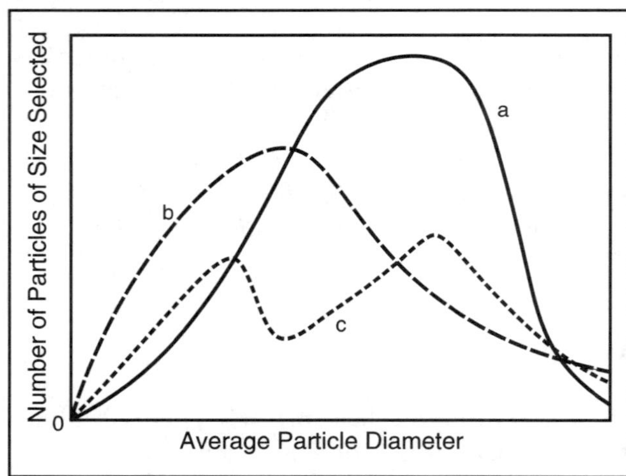

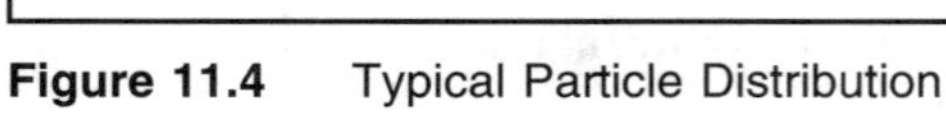
Figure 11.4 Typical Particle Distribution

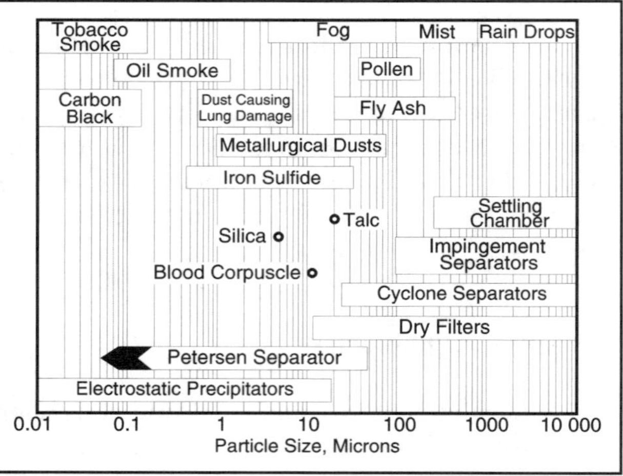

Figure 11.5 Typical Particle Distribution[11.1]

Figure 11.5 gives examples of particle sizes for comparison purposes. The general break point between fog and mist is approximately 100 microns.

The determination of average particle size is relatively easy with solids, but few methods have yet been devised for liquids. This determination of particle sizes may be considered somewhat academic because size changes continually in an actual line – yet it is by necessity the logical starting point in design.

Equation 11.5[11.6, 11.7] can be used to estimate the approximate size of the largest droplets, $d_{p_{max}}$ formed in the inlet pipe to the vessel.

$$\frac{d_{p_{max}}}{d_{pipe}} = 4.5 \left[\frac{\sigma}{\rho_g \, v_g^2 \, d_{pipe}} \right]^{0.6} \left(\frac{\rho_g}{\rho_L} \right)^{0.4} \qquad (11.5)$$

Where:

			SI	FPS
$d_{p_{max}}$	=	largest droplet size	m	ft
d_{pipe}	=	inlet pipe inside diameter	m	ft
σ	=	liquid surface tension	N/m	poundal/ft
ρ_g	=	gas density	kg/m^3	lbm/ft^3
v_g	=	gas velocity	m/s	ft/sec
ρ_L	=	liquid density	kg/m^3	lbm/ft^3

Some typical surface tensions are shown below:

	Surface Tension, dyne/cm @ 38°C [100°F]
HP Oil/Condensate	10-20
LP Oil/Condensate	20-30
NGL	5-15
Water	70
TEG	45*
1 dyne/cm = 0.001 N/m = 0.0022 poundal/ft * @ 25°C [77°F]	

Example 11.1: Estimate the maximum droplet size in the inlet piping to a separator. The velocity is 7 m/s [23 ft/sec]. The gas and liquid densities are 70 kg/m^3 and 700 kg/m^3 [4.4 lbm/ft^3 and 44 lbm/ft^3] respectively. The liquid surface tension is 20 dynes/cm and the inlet pipe diameter is 0.406 m [16 in].

SI Solution:

$$\frac{d_{p_{max}}}{d_{pipe}} = 4.5 \left[\frac{(20)(0.001)}{(70)(7)^2 (0.406)} \right]^{0.6} \left(\frac{70}{700} \right)^{0.4} = 0.0022$$

$$d_{p_{max}} = (0.0022)(0.406) = 0.000\,89 \text{ m} = 890 \; \mu\text{m}$$

FPS Solution:

$$\frac{d_{p_{max}}}{d_{pipe}} = 4.5 \left[\frac{(20)(2.20 \times 10^{-3})}{(4.4)(23)^2 (1.33)} \right]^{0.6} \left(\frac{4.4}{44} \right)^{0.4} = 0.0022$$

$$d_{p_{max}} = (0.0022)(1.33) = 0.00029 \text{ ft} = 890 \; \mu\text{m}$$

As an estimate for flow in pipes, the smallest droplets will generally have a size 5 to 10 times smaller than the largest. For flow across chokes or pressure reduction valves, the smallest droplets can be much smaller than this.

Most particles have no electrical charges of any magnitude; therefore, if a charge is desired (electrostatic precipitation), this normally must be imposed by artificial means. For all practical purposes, the particles may be considered neutral.

Mist eliminator designs are essentially redistributors of the particle distribution curve. An ordinary horizontal separator might have the particle distributions throughout its length as shown in Figure 11.6.

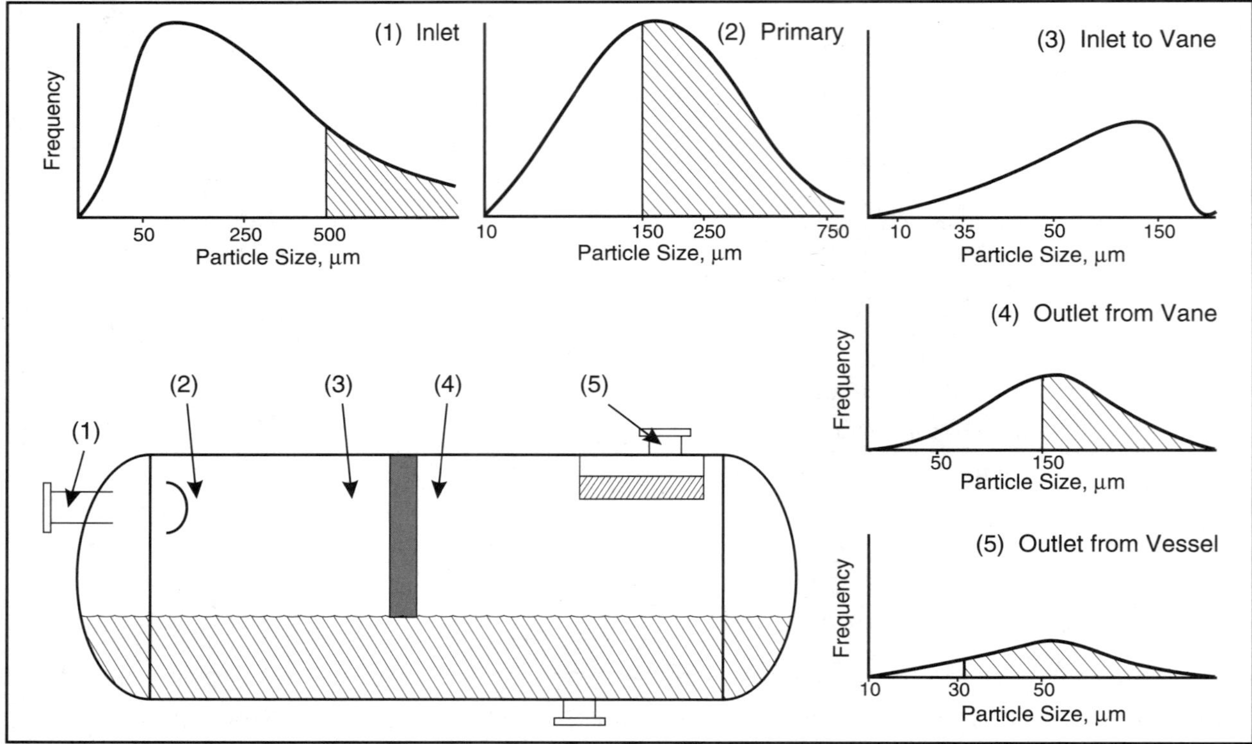

Figure 11.6 Example of Particle Distribution in Separator

The distribution depends on internal design, the character of the fluids involved and the equipment immediately upstream. Taking a large pressure drop across a valve, particularly with high gas-oil ratio condensates, can atomize the liquid and create small particles smaller than 2 µm in diameter.

A. An inlet particle distribution that is skewed and with several distinct particle sizes enters the vessel as shown in (1).

B. The stream hits the inlet momentum device where bulk liquid (particles larger than 500 microns) are removed.

C. The distribution shown in (2) enters the primary separation area where gravity effectively separates all particles 150 microns and larger.

D. The inlet to the vane mist eliminator will have been changed to the distribution in (3). At this point the "vane" type mist eliminator through inertial force, centrifugal force, and reduction of turbulence, coalesces the small droplets into larger droplets. Some of these droplets would be large enough to fall to the liquid by paths provided in the vane section. A significant number of large droplets would pass through the vane as shown in (4).

E. The outlet from the vane would enter the secondary separation section where gravity would again separate all those particles larger than 150 microns.

F. This new particle distribution would enter the mist eliminator where an action similar to Step D would occur so that the final outlet from the vessel would have a distribution of particles 30 microns and less.

There are several points to be made about the above description. Many vendors claim that they remove all the particles down to 10 microns. This is not commercially viable in an entrainment type separator. In oil and gas production, most separators are sized to remove liquid droplets larger than about 150 m by gravity settling. Smaller droplets must coalesce to form larger droplets (D_p > 150 μm) which may be separated by gravity. Mist extractors are commonly used to provide coalescence. Under ideal conditions, a 100 mm [4 in] thick wire mesh demister pad will allow 30-40% of the droplets larger than 20 μm to pass through, although the efficiency rises to near 100% for droplets larger than 60-70 μm. As you can see, separation efficiency is dependent on particle size distribution.

Experience has indicated that minimum particle size is related to inlet velocity, number of restrictions and amount of pressure drop in the inlet piping. These factors lead to inlet nozzle guidelines as discussed later, but the vendor usually has no control over the inlet piping. Therefore, although performance guarantees are nice to have, the ultimate responsibility for a particular design is shared between the design engineer and the vendor.

Gravity Separation

Two forces, gravity and drag, act upon a liquid droplet in a gas-liquid separator.

Equation 11.6 shows the relationship between terminal velocity and separation parameters and fluid properties. The terminal velocity is the gas velocity where the liquid droplet is suspended in the gas flow, moving neither up nor down.

$$v_t = \left(\frac{4\,g\,D_p}{3\,C_d}\right)^{0.5}\left(\frac{\rho_L - \rho_g}{\rho_g}\right)^{0.5} \qquad (11.6)$$

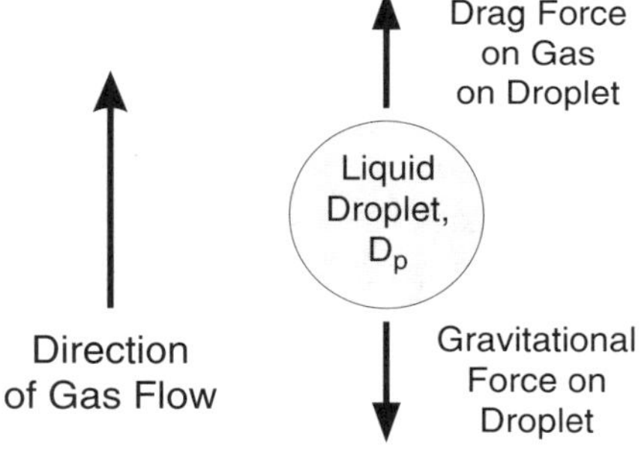

The drag coefficient, C_d, is a function of particle diameter, shape, terminal velocity, gas density and viscosity. Solution for C_d is iterative, much like the solution for the friction factor in fluid flow.

One approach is to evaluate C_d for three separate flow regimes - laminar (or Stokes' Law), intermediate and turbulent (or Newton's Law). This approach assumes that the plot of drag coefficient versus Reynold's number can be approximated by three straight lines. This does not result in a serious loss of accuracy at typical oil-gas separator conditions.

Using this approach, the terminal velocity of a particle falling through a fluid by the pull of gravity may be represented by

$$v_t = \left[\frac{4 g D_p^{N+1} (\rho_p - \rho_f)}{3 A \mu^N \rho_f^{(1-N)}}\right]^{\left(\frac{1}{2-N}\right)} \qquad (11.7)$$

"A" and "N" are constants related to the flow regime and drag coefficient on the system as determined by

$$K = D_p \left[\frac{g \rho_f (\rho_p - \rho_f)}{\mu^2} \right]^{1/3}$$

The values of "A" and "N" for use with Equation 11.7 are found from the table

Flow Regime	K	A	N
Laminar (Stokes' Law)	K < 3.3	24.0	1.0
Intermediate	3.3 < K < 43.6	18.5	0.6
Turbulent (Newton's Law)	K > 43.6	0.44	0

When liquid is being separated from gas in laminar flow (Stokes' Law), Equation 11.7 becomes

$$v_t = \frac{g D_p^2 (\rho_L - \rho_g)}{18 \mu} \tag{11.8}$$

Centrifugal Settling. Stokes' Law may be applied to this process if the effect of gravity, g, is replaced by a, the acceleration due to centrifugal force. The terminal velocity then becomes

$$v_t = \frac{a D_p^2 (\rho_L - \rho_g)}{18 \mu} \tag{11.9}$$

Where:

Symbol		(applies to Equations 11.6 to 11.9)	SI	FPS
a	=	acceleration due to centrifugal force	m/s²	ft/sec²
D_p	=	particle diameter	m	ft
g	=	acceleration due to gravity	9.81 m/s²	32.2 ft/sec²
v_t	=	particle terminal velocity	m/s	ft/sec
ρ_L	=	density of liquid	kg/m³	lbm/ft³
ρ_g	=	density of gas	kg/m³	lbm/ft³
ρ_p	=	density of dispersed particle	kg/m³	lbm/ft³
ρ_f	=	density of continuous fluid	kg/m³	lbm/ft³
μ	=	viscosity of gas	kg/m·s	lbm/ft-sec

Note: 1 cp = 0.001 kg/m·s = 6.72×10^{-4} lbm/ft-sec

SEPARATOR SIZING

The determination of separation vessel diameter is based on the above principles. It is a semi-empirical approach since one cannot measure things like droplet size and other variables. Furthermore, many of the assumptions in the previous equations are not satisfied in actual practice.

One form of the equation assumes that Newton's law is valid for the gravity-settling portion of the separator. Equation 11.6 then becomes,

$$v_t = \left(\frac{4\,g\,D_p}{(3)(0.44)}\right)^{0.5}\left(\frac{\rho_L-\rho_g}{\rho_g}\right)^{0.5} \tag{11.10}$$

For a particular droplet size, the first term on the right-hand side of Equation 11.10 is fixed. For example, if D_p = 200 μm this term is equal to 0.077 m/s [0.25 ft/sec]. This term is often referred to as the sizing coefficient, K_s and the general model for sizing gas-liquid separators becomes,

$$v = K_s\left(\frac{\rho_L-\rho_g}{\rho_g}\right)^{0.5} \tag{11.11}$$

Where:			SI	FPS
	ρ_L	= liquid density	kg/m³	lb/ft³
	ρ_g	= gas density	kg/m³	lb/ft³
	v	= allowable gas velocity	m/s	ft/sec
	K_s	= an empirical constant	m/s	ft/sec

All density terms are at the pressure and temperature of separation.

The value of K_s depends on all factors that affect separation other than density – particle size, types of internals, vortex action, foaming, pulsating flow, presence of solids, separation length, varying gas-liquid ratios, and the like. It is not surprising that K_s varies widely in different applications. How do you predict it? From experience!

The following values of K_s are taken from API 12J.

Separator Type	Height or Length, m [ft]	K_s Factor	
		SI	FPS
Vertical	3.0 [10], or taller	0.055-0.107	0.18-0.35
Horizontal	3.0 [10]	0.122-0.152	0.40-0.50
	Other	$K_3(L/3)^{0.56}$	$K_{10}(L/10)^{0.56}$

Notice that K_s values for horizontal are cited as a function of length. This length correction for horizontal separators must be used carefully. The values shown were obtained on separators with a length-to-diameter ratio averaging about 5:1. In effect, K_s only increases significantly with length as this L/D increases. A separator 10 ft by 50 ft may not have higher value of K_s than one 4 ft by 20 ft (the same L/D ratio). The former may tend to possess a slightly higher value of K_s but it would not be as much greater as the API correlation would indicate. Unless L/D is greater than 5:1, it is difficult to justify a value of K_s greater than 0.15 [0.50] regardless of length.

If a horizontal separator is to be used with an L/D other than 5, multiply the value of K_s by the term

$$\left[\frac{L/D}{5}\right]^{0.56}$$

to correct for extra length in the absence of foaming. If severe foaming is expected you may wish to use a minimum value of K_s and add extra length, based on experience.

Calculation of the separator diameter proceeds from Equation 11.11 using the following relationships.

$$v = \frac{q_a}{A} = \frac{4\,q_a}{\pi d^2 F_g} \quad ; \quad d = \sqrt{\frac{4\,q_a}{\pi\, v F_g}} \tag{11.12}$$

Where:

			SI	FPS
v	=	allowable gas velocity	m/s	ft/sec
q_a	=	actual volumetric rate	m^3/s	ft^3/sec
d	=	separator diameter	m	ft
F_g	=	fraction of cross sectional area available for gas flow (See Table 11.3 or Figure 11A.1 in Appendix 11A)		

The actual volumetric rate can be calculated from Equations 11.13 or 11.14.

$$q_a = \frac{m}{3600\,\rho_g} \tag{11.13}$$

$$q_a = \left(\frac{q_s}{86\,400}\right)\left(\frac{P_s}{P}\right)\left(\frac{T}{T_s}\right)(z) \tag{11.14}$$

Where:

			SI	FPS
m	=	mass flowrate	kg/h	lbm/hr
ρ_g	=	gas density	kg/m^3	lbm/ft^3
q_s	=	standard gas flowrate	std m^3/d	scf/day
P_s	=	standard pressure	kPa	psia
P	=	actual separator pressure	kPa	psia
T	=	actual separator temperature	K	°R
T_s	=	standard temperature	K	°R
z	=	gas compressibility factor @ separator P and T		

Combining Equations 11.11, 11.12, and 11.14 yields Equation 11.15 which gives the gas handling capacity of the separator as a function of d, T, P, and K_s.

$$q_s = 67\,824(K_s)(d^2)F_g\left(\frac{P}{P_s}\right)\left(\frac{T_s}{T}\right)\left(\frac{1}{z}\right)\left(\frac{\rho_L - \rho_g}{\rho_g}\right)^{0.5} \tag{11.15}$$

Use of Mass Flowrate

It is sometimes more convenient to use mass flow rate for sizing purposes. This is often done in sizing fractionators and absorbers. The mass velocity (w) is related to linear velocity by the equation

$$w = 3600\, v\, \rho_g$$

If this relationship is substituted into Equation 11.11, the result is

$$w = 3600\, K_s\, [(\rho_L - \rho_g)(\rho_g)]^{0.5} \qquad (11.16)$$

Also,

$$w = \frac{4\, m}{\pi\, d^2\, F_g} \qquad (11.17)$$

Where:			SI	FPS
	w =	mass velocity	$kg/m^2{\cdot}h$	lbm/ft^2-hr
	m =	mass flow	kg/h	lbm/hr
	v =	linear velocity	m/s	ft/sec
	ρ_g =	gas density (actual)	kg/m^3	lbm/ft^3
	ρ_L =	liquid density	kg/m^3	lbm/ft^3
	d =	separator I.D.	m	ft
	F_g =	fraction of area available to gas		
	K_s =	gas sizing parameter	m/s	ft/sec

Combining Equations 11.16 and 11.17 enables one to solve for diameter

$$d = \frac{0.0188\left(\dfrac{m}{F_g\, K_s}\right)^{0.5}}{[(\rho_L - \rho_g)(\rho_g)]^{0.25}} \qquad (11.18)$$

The value of "m" is related to standard volume rate of flow as follows:

SI (q_{sc} at 15°C and 101.325 **kPa**)	**FPS** (q_{sc} at 60°F and 14.7 psia)
$m = (1762)(10^6 \text{ std m}^3/\text{d})(\text{MW gas})$	$m = 110(\text{MMscfd})(\text{MW gas})$
$= (51\ 060)(10^6 \text{ std m}^3/\text{d})(\gamma_{gas})$	$= 3180(\text{MMscfd})(\gamma_{gas})$

The calculation of vessel diameter for a given flowrate (or the determination of allowable flowrate for a given diameter) depends primarily on the choice of K_s. The API sets forth a range of possible values. But, this range is large, varying in some cases by a factor of two. What is a suitable value within this range?

In the final analysis the choice must be made from experience and data from comparable separation facilities and the efficiency of separation required. In some field applications, a nominal amount of carryover may be tolerable and a larger K_s (smaller diameter separator) may be specified than in some process applications. Separation equipment ahead of dehydrators, amine contactors, compressors, etc. is critical.

Our experience is that for vertical separators with standard wire mesh or vane type mist extractors, K_s values vary from about 0.07 to 0.105 m/s [0.23 to 0.35 ft/sec]. For proprietary inlet devices and centrifugal mist extractors, K_s values as high as 0.25 m/s [0.82 ft/sec] have been specified by some companies. These would generally be used in scrubbing applications with low liquid loadings, with no slugging or surging flows.

For horizontal separators with vane pack or wire mesh mist extractors, K_s values vary from about 0.07 to 0.15 m/s [0.23 to 0.5 ft/sec].

Recognizing that design flowrates are usually based on reservoir engineering estimates and that transient surges and slugs can often result in instantaneous flows which are higher than average flow, it

Example 11.2: Calculate the diameter of the vertical separator needed for a gas flow rate of one million std m^3 per day [35.4 MMscfd] if gas density is 80 kg/m^3 [5.0 lbm/ft^3] and the liquid density is 800 kg/m^3 [50 lbm/ft^3]. Use a value of K_s of 0.07 m/s [0.23 ft/sec]. The gas relative density is 0.7. The design factor for this application is 1.2.

SI Solution:

$$m = (51\,060)\,(1.0)\,(0.7) = 35\,700 \text{ kg/h}$$

Use a design factor of 1.2

$$q_a = \frac{1.2 \text{ m}}{3600\ \rho_g} = \frac{(1.2)(35\,700)}{(3600)(80)} = 0.149 \text{ m}^3/\text{s}$$

$$v = K_s\left(\frac{\rho_L - \rho_g}{\rho_g}\right)^{0.5} = 0.07\left(\frac{800-80}{80}\right)^{0.5} = 0.21 \text{ m/s}$$

$$d = \sqrt{\frac{4\,q_a}{\pi\, v\, F_g}}$$ for a vertical separator, $F_g = 1.0$

$$d = \sqrt{\frac{(4)(0.149)}{(\pi)(0.21)(1.0)}} = 0.95 \text{ m or } 950 \text{ mm}$$

FPS Solution:

$$m = (3180)\,(35.4)\,(0.7) = 78\,800 \text{ lbm/hr}$$

Use a design factor of 1.2

$$q_a = \frac{1.2 \text{ m}}{3600\ \rho_g} = \frac{(1.2)(78\,800)}{(3600)(5.0)} = 5.25 \text{ ft}^3/\text{sec}$$

$$v = K_s\left(\frac{\rho_L - \rho_g}{\rho_g}\right)^{0.5} = 0.23\left(\frac{50-5}{5}\right)^{0.5} = 0.69 \text{ ft/sec}$$

$$d = \sqrt{\frac{4\,q_a}{\pi\, v\, F_g}}$$ for a vertical separator, $F_g = 1.0$

$$d = \sqrt{\frac{(4)(5.25)}{(\pi)(0.69)(1.0)}} = 3.1 \text{ ft or } 37 \text{ in}$$

is common practice to adjust upwards the expected average flow using a design factor. Typical design factors range from about 1.2 to 1.5 depending on the location and nature of the inlet stream. The higher end of the range should be used for gas-lifted production, flow in hilly terrain and riser systems on offshore platforms.

All separators experience carryover or entrainment. Even "so-called" high efficiency separators using various proprietary internals can experience liquid entrainment of 2-3%. Carryover is not necessarily intolerable. Liquid entrainment from primary production separators may not be a major operating problem if secondary separation is provided downstream. However, in critical separation applications, the cost of the problems created by the entrainment often far exceed the incremental cost of a properly sized vessel.

Foaming and *emulsified* liquids also affect capacity. Foam must be broken to obtain a good vapor-liquid separation. It takes contact surface and time (length) to break it physically. Anti-foam chemicals have been successfully employed in many locations. In some instances, horizontal separators with an L/D greater than 10:1 have been used to help destabilize the foam.

The severity of the foaming problem is made worse by the design of the control system ahead of the separator when a substantial pressure drop may occur. When foaming is anticipated this should be considered in vessel sizing.

In the sizing of vertical vapor-liquid vessels like absorbers and fractionators, it is customary to use values of K_s = 0.043-0.067 m/s [0.14-0.22 ft/sec] for trayed towers. This is designed to give a rather efficient vapor-liquid separation over the short distance between the trays. This is consistent with short separator experience. Sizing of towers is discussed in more detail in Chapter 17.

Values of "F_g." The fraction of the total area available for gas flow is shown in Table 11.3. One can extrapolate this table beyond 0.55. The value for h/d = 0.60 would be one minus the F_g value at h/d = 0.4. Or, $F_g = 1 - 0.626 = 0.374$. Appendix 11A contains a figure for the same purpose.

TABLE 11.3

Values of "F" as a Function of Liquid Depth in a Horizontal Separator

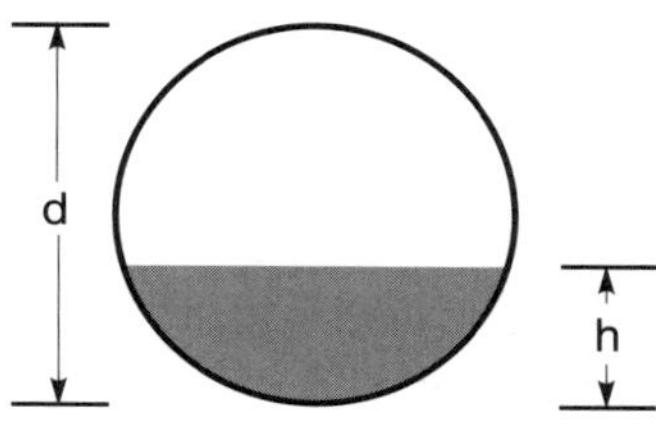

h/d	F_g	h/d	F_g
0.00	1.000	0.30	0.748
0.05	0.981	0.35	0.688
0.10	0.948	0.40	0.626
0.15	0.906	0.45	0.564
0.20	0.858	0.50	0.500
0.25	0.804	0.55	0.436

Liquid Retention Time

One factor not covered by the gas sizing equations is liquid retention time. It takes a finite time for gas to break out of liquid.

Retention time is an indirect way of fixing the volume of a separator necessary to handle the liquid flow rate. Separator liquid volume equals liquid flow rate times retention time. For some separator designs, liquid volume requirements (set by retention time) have a greater effect on size than gas flow rate. This is true particularly of large crude oil separators where the gas-oil ratio is low. The liquid is the controlling factor.

The equation for calculating the required separator liquid volume is

$$V_L = \frac{(q_L)(t)}{1440} \tag{11.19}$$

Where:		SI	FPS
V_L	= required separator liquid volume	m^3	bbl
q_L	= liquid throughput	m^3/d	bbl/day
t	= design retention time	min	min

Typical retention times are as follows:

Natural gas-oil	1-3 minutes
Reflux accumulators	5-10 minutes
Fractionation feed surge tanks	8-15 minutes
Refrigerant surge tanks	4-7 minutes
Refrigerant economizers	2-3 minutes

API 12J gives the following guidelines for gas-oil separation.

Oil Relative Density	Minutes
Below 0.85	1
0.85-0.93	1 to 2
0.93-1.0	2 to 4

Retention or residence time is affected also by composition, foaming, the presence of solids and emulsions.

Retention time is a very important factor in separation vessels where a chemical reaction might be occurring. A good example is water deoxygenation. A retention time of 5 minutes is preferred, with 3 minutes being marginal. This time is necessary for the chemical reaction to proceed and to obtain good chemical mixing.

As noted later in this chapter, liquid retention volume must also be provided for control purposes, particularly when alarm and shut-down controls are involved.

A second issue in sizing the liquid section is providing sufficient cross sectional area to degas the liquid. For low viscosity liquids this is normally not the limiting criterion in separator design, but for viscous liquids particularly in systems where the liquid loading is high, degassing may be limiting.

In general it is assumed that if bubbles larger than 200-300 μm are allowed to escape to the vapor phase, carryunder of vapor should be minimal. Stokes' law may be used to estimate the bubble rising velocity.

$$v = \frac{g D_p^2 (\rho_L - \rho_g)}{18 \mu_L} \qquad (11.20)$$

Where:

Symbol		Description	SI	FPS
v	=	bubble rising velocity*	m/s	ft/sec
D_p	=	bubble diameter	m	ft
ρ_L	=	liquid density	kg/m^3	lbm/ft^3
ρ_g	=	gas density	kg/m^3	lbm/ft^3
μ_L	=	liquid viscosity	kg/m·s	lbm/ft-sec
g	=	gravitational acceleration	$9.81\ m/s^2$	$32.17\ ft/sec^2$

* this is therefore maximum liquid downward velocity

THREE-PHASE OR LIQUID-LIQUID SEPARATIONS

Three-phase separators handle gas plus two immiscible liquid phases. The two liquid phases might be oil and water, glycol and condensate, etc. Do not anticipate perfect separation between the liquid phases, particularly if the gas-oil ratio is high or the oil has any emulsifying or foaming tendencies. For primary separation, carryover of 5-10% oil into water or water into oil is not uncommon for three-phase separators operating at design conditions.

Equipment used downstream of a three-phase separator or free water knockout to meet crude and sales specifications and/or water discharge regulations include:

1. Wash tanks and settling tanks
2. Gun barrels (type of settling tank)
3. Heater treaters
4. Coalescers
5. Electrostatic precipitators
6. Gas floatation cells
7. Hydrocyclones

The equipment actually selected depends on several factors, including crude oil density and viscosity, water cut, stability of emulsion, specifications, etc.

A three-phase separator must be sized to provide effective vapor/liquid separation and also to provide adequate liquid/liquid separation. Because of the difficulty in separating liquid phases, this criterion often controls the separator size.

The separation of two immiscible liquid phases (typically water and oil) would be routine except for the presence of emulsifying agents in the produced fluids. Emulsifiers may be naturally present, finely divided solids from the reservoir, corrosion products, asphaltenes, etc. or they may be artificially introduced into the crude-water mixture such as surfactants or other treating chemicals such as corrosion and hydrate inhibitors.

Regardless of how introduced, an emulsifier stabilizes the emulsions which form during intense mixing and shearing of the produced fluids in the wellbore, flowline and across chokes. A stable emulsion may consist of oil or water droplets as small as 1 μm. An emulsion is said to be "stable" when these tiny droplets will not coalesce due to the presence of the emulsifier on the surface of the droplet. This impedes coalescence by making the interfacial film more durable or by imparting a charge to the droplet. From Stokes' Law, settling velocity is proportional to the droplet diameter squared, so tiny droplets remain virtually suspended in the continuous phase.

In addition, the viscosity of an emulsion can be several times higher than the viscosity of either phase. Figure 11.7 shows the relationship between the viscosity of an oil-water emulsion and water cut.

This highly viscous emulsified layer, often referred to as an emulsion pad, acts as a barrier to good oil-water separation.

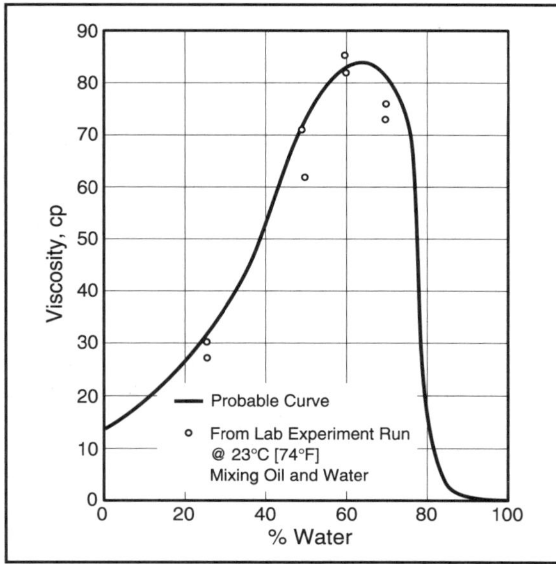

Figure 11.7 Viscosity of an Oil-Water Emulsion

Example 11.3: Estimate the cross-sectional area necessary to degas a crude oil. The oil viscosity is 10 cp. The oil and gas densities are 800 kg/m^3 [50 lbm/ft^3] and 10 kg/m^3 [0.625 lbm/ft^3] respectively. The oil rate is 8000 m^3/d [50 300 bbl/day]. Assume a bubble diameter of 250 µm.

SI Solution:

$D_p = 250\ \mu m = 250 \times 10^{-6}$ m $\qquad \rho_L = 800\ kg/m^3$

$\mu_L = 10$ cp $= 0.010$ kg/m·s $\qquad \rho_g = 10\ kg/m^3$

$$v = \frac{(9.81)(250 \times 10^{-6})^2 (800 - 10)}{(18)(0.010)} = 0.002\ 69\ m/s$$

$$q_L = (8000\ m^3/d)\left(\frac{1\ d}{86\ 400\ s}\right) = 0.0926\ m^3/s$$

$$A = q/v = (0.0926\ m^3/s)\left(\frac{1\ s}{0.002\ 69\ m}\right) = 34.4\ m^2$$

Assume a horizontal separator with L/D = 4 and level at 50%.
The interface is a rectangle, so its dimensions must be

$$(L)(D) = 4(D)^2 = 34.4\ m^2$$

D = 2.9 m $\qquad$ L = 11.7 m

FPS Solution:

$D_p = 250\ \mu m = 0.000\ 82$ ft $\qquad \rho_L = 50\ lbm/ft^3$

$\mu_L = 10$ cp $= 6.72 \times 10^{-3}$ $\qquad \rho_g = 0.625\ lbm/ft^3$

$$v = \frac{(32.17)(0.00082)^2 (50 - 0.625)}{(18)(6.72 \times 10^{-3})} = 0.008\ 84\ ft/sec$$

$$q_L = (50\ 300\ bbl/day)\left(\frac{5.61\ ft^3}{bbl}\right)\left(\frac{1}{86\ 400 sec}\right) = 3.27\ ft^3/sec$$

$$A = q/v = (3.27\ ft^3/sec)\left(\frac{1\ sec}{0.008\ 84\ ft}\right) = 369\ ft^2$$

Assume a horizontal separator with L/D = 4 and level at 50%.
The interface is a rectangle, so its dimensions must be

$$(L)(D) = 4(D)^2 = 369\ ft^2$$

D = 9.6 ft $\qquad$ L = 38.4 ft

Oil-water separation proceeds in three steps:

1. **Destabilization.** Emulsions can be destabilized through the use of chemicals (demulsifiers) and gentle mixing. Heat may also help destabilize the emulsion by reducing the effectiveness of the emulsifying agents.

2. **Coalescence.** Coalescence is the key step in separation. Droplet settling velocity is proportional to D_p^2. Coalescence rates can be increased by gentle mixing and mechanical

coalescers such as plate packs and fiber mats. In crude oil dehydrators, where the water concentration is low (< 5 vol%) electrostatic coalescers are sometimes used.

3. **Settling.** The final step in oil-water separation is settling. Higher settling rates are favored by larger droplet sizes, reduced viscosity of the continuous phase and reduced turbulence. Settling vessels are sometimes heated especially in the case of heavier, viscous crude oils. Adequate residence time and interfacial area are critical aspects of design. Centrifugal separation is used in some applications. This increases settling velocity by increasing the gravitational term, g, in Stokes' law, Equation 11.8.

Retention Time

The usual approach in design is to allow equal retention time for the two liquid phases. The following are recommendations from API 12J.

Oil Relative Density	Retention Time, min
Below 0.85	3-5
Above 0.85 and >100°F	5-10
80-100°F	10-20
60-80°F	20-30

For glycol/hydrocarbon separation in a cold separator, a minimum retention time of 30 minutes in the glycol phase is recommended.

Settling Velocity

Settling velocity sets the cross-sectional area of the oil-water separation section.

The basic settling characteristics of two immiscible phases are represented by Stokes' Law (Equation 11.8). Typically, a value of D_p = 150-300 μm is suitable for designing liquid-liquid separators. Stokes Law assumes unhindered settling which seldom, if ever, occurs in practice. Stokes' Law provides a reasonable estimate of settling velocity when the dispersed phase is dilute (less than 5%) in relation to the continuous phase.

Design of oil-water separation equipment should ideally be based on physical tests where possible. Unfortunately this is not always practical. One correlation[11.2] based on limited testing with four different oils is presented in Figure 11.8.

The upper end of the design range is based on relatively unstable emulsions, the lower end on stable (or “tight”) emulsions. Ultimately inlet droplet size distribution and emulsion stability are the primary drivers in sizing oil-water separation equipment. Since it is not practical to accurately measure droplet size distribution, the design of oil-water separators relies to a large extent on a Stokes' law type model modified with empirical data.

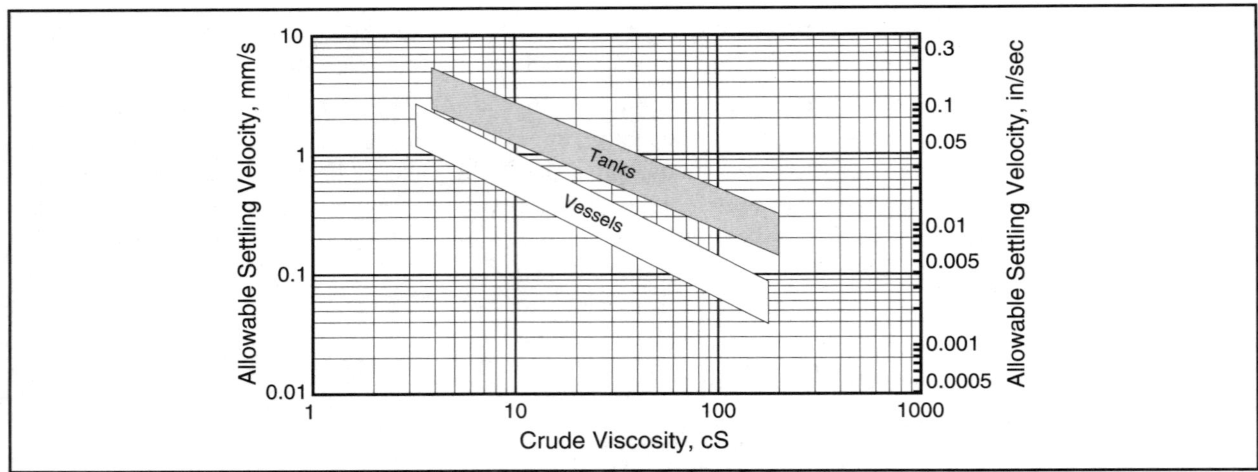

Figure 11.8 Generalized Design Windows for Primary Separators(11.2)

Campbell has put forward an empirical correlation, Equation 11.21, based on unstable emulsions which did not require heating or chemicals to resolve.

$$v = \frac{(A)(C)(\rho_w - \rho_o)}{\mu}(L_c) \tag{11.21}$$

Where:		SI	FPS
v	= settling velocity	m/s	ft/sec
ρ_w	= water density	kg/m^3	lb/ft^3
ρ_o	= oil density	kg/m^3	lb/ft^3
μ	= oil viscosity	kg/m·s	lbm/ft-sec
L_c	= length correction	$0.52(L)^{0.2}$	$1.35(L)^{0.2}$
	where L =	m	ft
A	= coefficient	0.167	1.79
C	= correlation constant found from Fig. 11.9		

The L_c in Equation 11.21 is the effective length of the liquid separation section.

Viscosity is often measured in centipoise (cP). For use in Equation 11.21, remember that

$$1 \text{ cP} = 0.000\,672 \text{ lb/ft-sec} = 0.001 \text{ kg/m·s}$$

The velocity predicted by Stokes' Law may be 2-3 times that observed. But remember, Stokes' Law assumes unhindered settling. Notice in Figure 11.9 that the settling velocity decreases with amount of water. This is an indirect indication of the effect of hindrance.

The actual size of the separation section must meet both the retention time and settling velocity criteria. Several vessel geometries can be used. For three-phase separation, horizontal separators are generally preferred because increasing the vessel length decreases settling velocity. Length-to-diameter ratios typically vary from 3 to 5.

Once the settling velocity has been determined, calculation of the settling area proceeds as follows.

$$A = \frac{q}{v} \qquad (11.22)$$

Where:			SI	FPS
	A =	settling area	m^2	ft^2
	q =	flowrate of continuous phase	m^3/s	ft^3/sec
	v =	settling velocity	m/s	ft/sec

Settling velocity, v, can be calculated from Stokes' Law, Figure 11.8, Equation 11.21 or empirical data, if available. For horizontal separators, the settling area will be a rectangle, for vertical separators, a circle.

The liquid volume of the separator is determined by the retention time.

Example 11.4: Calculate the settling area required in a 3-phase separator. Oil and water properties are shown below.

	Oil	Water
ρ	840 kg/m^3 [52.5 lbm/ft^3]	980 kg/m^3 [61.2 lbm/ft^3]
μ	2.5 cp	0.8 cp
rate	4800 m^3/d [30 200 bpd]	700 m^3/d [4400 bpd]

1) From Figure 11.8, at a viscosity of 2.5/0.84 = 3 cS, the settling velocity varies from 1.2 to 2.8 mm/s [0.0039 to 0.0092 ft/sec].

2) From Stokes' Law, using a 300 μm droplet size

$$v = \frac{(9.81)(300 \times 10^{-6})^2(980-840)}{(18)(0.0025)} = 2.7 \times 10^{-3} \text{ m/s} = 2.7 \text{ mm/s } [0.009 \text{ ft/sec}]$$

3) From Equation 11.21*, the settling velocity is

$$v = \frac{(0.167)(0.5 \times 10^{-6})(980-840)}{(0.0025)}(0.52)(6)^{0.2} = 0.0035 \text{ m/s} = 3.5 \text{ mm/s } [0.011 \text{ ft/sec}]$$

* from Figure 11.9, at Water cut ≈ 13%, oil density = 37°API, and C = 0.5 × 10^{-6}.
Assume L = 6 m [19.7 ft].

Calculate area based on 3.0 mm/s (0.0030 m/s) [0.0098 ft/sec]

$$q = 4800 \text{ m}^3/\text{d} = 0.0556 \text{ m}^3/\text{s}$$

$$A = \frac{q}{v} = \left(\frac{0.0556 \text{ m}^3}{\text{s}}\right)\left(\frac{1 \text{ s}}{0.0030 \text{ m}}\right) = 18.5 \text{ m}^2$$

For horizontal separator (L/d = 4)

$4 d^2 = 18.5 \text{ m}^2$ d = 2.15 m [7.1 ft]   L = 8.6 m [28 ft]

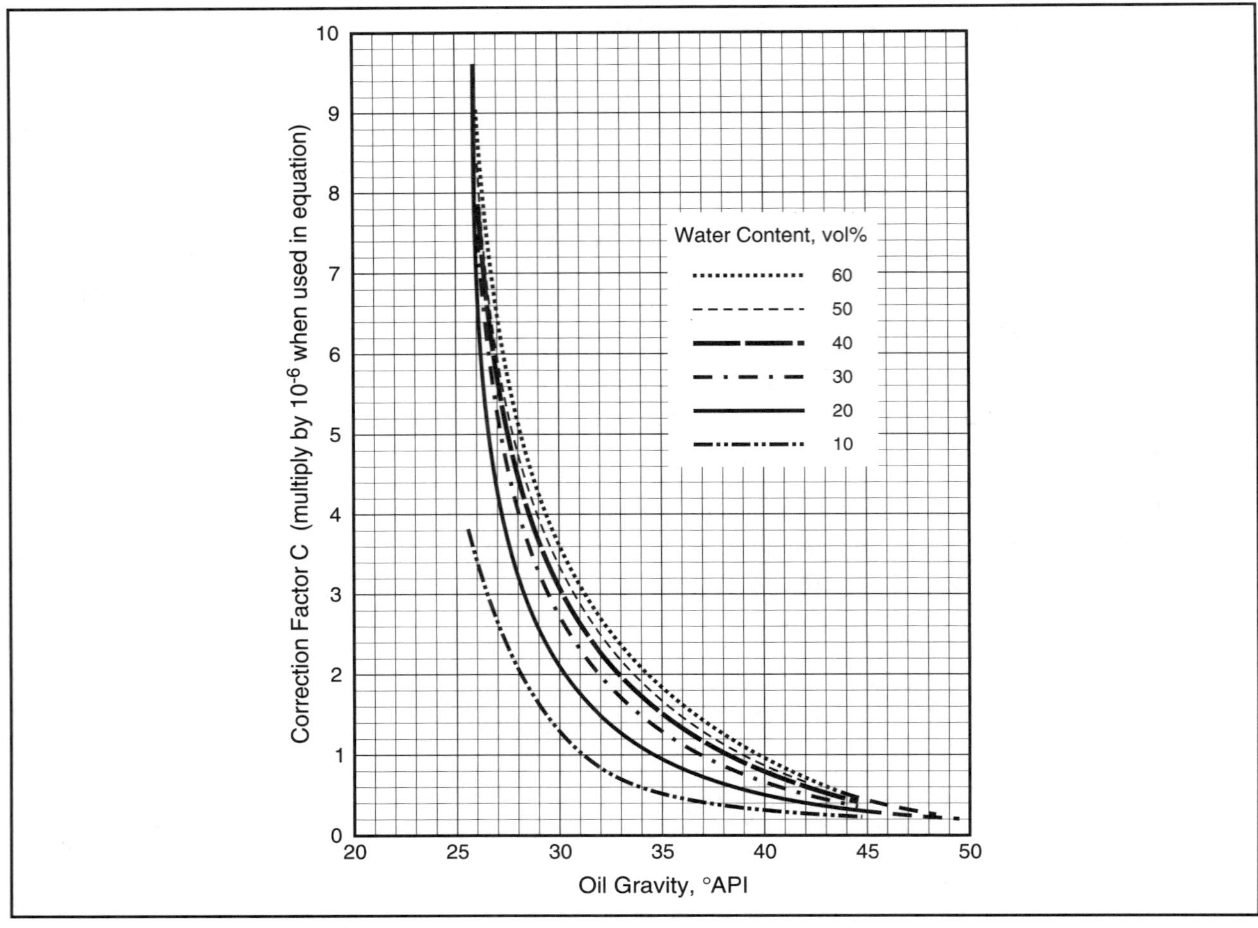

Figure 11.9 Correction Factor for Liquid-Liquid Separation

VESSEL INTERNALS

The proper selection of internals can significantly enhance the operation of separators. Proprietary internals often are helpful in improving separator efficiency at design conditions, but they usually cannot overcome an improper design or operation well away from design conditions.

Mist Extraction

Equipment involving the separation of liquid hydrocarbons and natural gas often uses *impingement-type* or centrifugal mist-extraction elements. Impingement devices are usually vane type or knitted wire.

The *vane type* consists of a labyrinth formed with parallel metal sheets with suitable liquid collection "pockets." The gas, in passing between plates, has to change direction a number of times. Obviously, some degree of centrifugal force is introduced, for as the gas changes direction the heavier particles tend to be thrown to the outside and are caught in the pockets provided.

The surface of the element is usually wet, and small particles striking it are absorbed into the liquid film. Inasmuch as the pockets are perpendicular to the gas flow, the liquid thus formed does not have to flow against the gas. Consequently, small compact units have a large capacity.

Vane units are manufactured by several vendors. They may be installed vertically (for horizontal flow) or horizontally (for vertical flow). Horizontal units are not widely used in oil-gas separation.

The collection pockets may be single or double. Single pocket units are simpler and less prone to fouling but can experience some liquid reentrainment, particularly at higher velocities. Double-pocket units better protect the collected liquid from reentrainment and can operate at higher velocities.

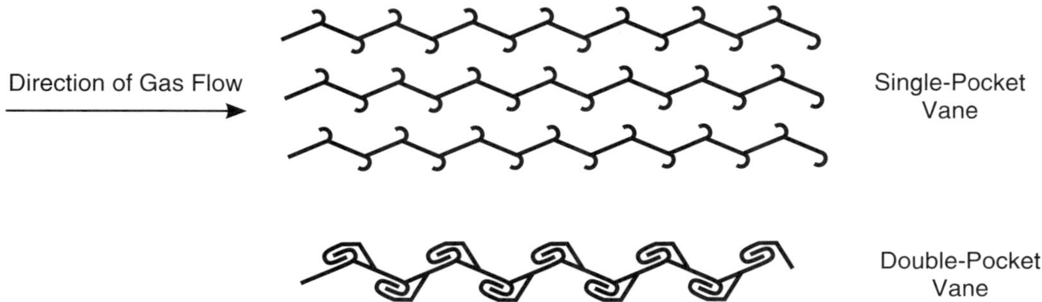

The effective capacity of a double-pocket vane is somewhat higher than a single pocket, but they should be used in clean service only.

As the plates are placed closer together and more pockets are provided, greater agitation, centrifugal force, and collection surface are provided, but the pressure drop is increased correspondingly. Thus, for a given flow rate, the collection efficiency is normally some function of the pressure drop.

In the average application this pressure drop varies from 25-250 mm H_2O [1-10 in H_2O] of water. Because of this pressure drop and to prevent gas bypassing the extractor, a liquid-collection pan incorporating a liquid seal is necessary for the liquid to drain properly.

When using the vane type mist extractor one must be careful that the pressure drop across it does not exceed its height above the liquid level, if a downcomer pipe is used. Otherwise, liquid will be "sucked out" overhead. In addition, plugging in the downcomer pipe can restrict drainage.

The cross-sectional area of vane type mist extractors may be estimated by using Equation 11.11 and the sizing criteria below.

Direction of Gas Flow	K_s	
	SI	FPS
Vertical	0.12 m/s	0.4 ft/sec
Horizontal	0.20-0.30 m/s	0.65-1.0 ft/sec

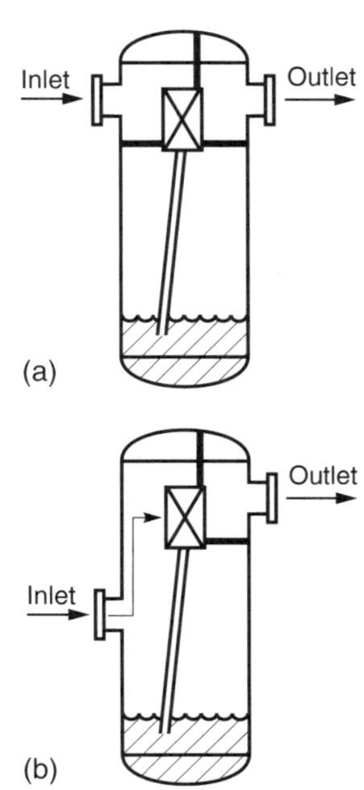

Vane type units are installed both in vertical and horizontal separators. Their use in horizontal separators is widespread because the gas flow is horizontal and the bottom of the vane can be sealed in the separator liquid. The use of vane packs in vertical separators is less common. In vertical separators the vane pack may be installed "in-line" (a) between the inlet and outlet nozzles or it may be installed in front of the outlet nozzle (b). The in-line configuration should only be used in clean service where the liquid loading is low and there is no slugging or surging.

Historically, the most common type of mist extractor for vertical separators is the wire mesh.

The element consists of wire knitted into a pad having a number of unaligned, asymmetrical openings. Although similar in appearance to filter media, its action is somewhat different. Filter media are rather dense and have small openings. This knitted wire, on the other hand, has about 97 to 98% void space and collects the particles primarily by impingement.

The material is available in single wound units of varying thickness in diameters up to 900 mm [3 ft], or in laminated strips for insertion through manholes in large process vessels.

The principle of separation is similar to that of the vane-type unit. The gas flowing through the pad is forced to change direction a number of times, although centrifugal action is not so pronounced as in vane packs. Impingement is the primary mechanism.

A liquid particle striking the metal surface which it does not "wet," flows downward where adjacent wires provide some capillary space. At these points, liquid collects and continues to flow downward. Surface tension tends to hold these drops on the lower face of the pad until they are large enough for the downward force of gravity to exceed that of the upward gas velocity and surface tension.

Efficiency is a function of the number of targets presented. This may be accomplished by increasing the pad thickness, changing wire diameter, or the closeness of the weave.

The wire mesh normally used falls within the following range:

Wire diameter:	0.076-0.28 mm [0.003-0.011 in]
Void volume:	92 to 99.4 percent
Density:	48-529 kg/m^3 [3-33 lb/ft^3]
Surface area:	164-1970 m^2/m^3 [50-600 ft^2/ft^3]

The most commonly used wire mesh has a void volume of 97 to 98%, a bulk density of approximately 192 kg/m^3 [12 lb/ft^3], a surface area of 328-410 m^2/m^3 [100 to 125 ft^2/ft^3], with a wire diameter of 0.28 mm [0.011 in]. A pad thickness of 100-150 mm [4-6 in] is sufficient for most separator applications, although thicknesses up to 450 mm [18 in] have been used in glycol dehydrators. In routine separator service, 100-150 mm [4-6 in] will normally suffice.

Any common metal may be used in these units, including carbon steel, stainless steel, aluminum, monel, etc. The pressure drop is a function of the entrainment load, the pad design, and gas velocity, but will typically not exceed 30 mm [1.2 in] H_2O in the average installation. Because of this small pressure drop, the elements do not have to be "held down" and are normally only wired to the support grid to prevent shifting unless surging flow is anticipated.

Experience has shown that the support grid should contain at least 90% free area to eliminate any restrictions to liquid drainage. The pads are light weight so that a light angle-iron support is often adequate.

When both liquid and solids are present, a portion of the latter obviously will be scrubbed out. When only dry solids are present, the removal efficiency is substantially less. At the present time, though, wire mesh type units are only considered for clean service.

The capacity of a wire mesh unit is represented by Equation 11.11. A maximum value of K_s = 0.107 [0.35] generally is satisfactory for non-viscous liquids and assuming that the gravity separation section of the separator has removed the bulk of the liquids.

At high liquid loadings mist extractors can "flood" resulting in liquid entrainment. Maximum liquid loadings on wire mesh pads are about 2.4 m^3/h per m^2 of flow area [1 US gpm/ft^2]. Maximum loadings on vane pack units are about twice that of wire mesh. With materials like glycol and amine, which wet metal very well, a Teflon® or Dacron®-stainless steel co-knit pad may prove desirable. Remember, the liquid must be nonwetting in order to stay as droplets that can run down the wires and coalesce into bigger droplets. A wetting fluid will tend to "run up" the wires.

One additional caution regarding the use of wire mesh mist extractors is that viscous fluids, such as glycol, do not drain as readily as non-viscous fluids such as water and light hydrocarbons. Some have proposed derating the K_s factor for wire mesh pads by the ratio $[1.0/(\mu_L)]^{0.04}$, where μ is in cp, for liquids with a viscosity greater than 1cp.

Some operators have successfully used a wire-mesh and vane pack unit in series. The vane pack is installed downstream of the wire mesh. The vane pack has a higher capacity than the wire mesh and at design rates the wire mesh unit will flood. In this "flooding" mode it acts as a coalescer for the vane pack improving separation efficiency. It also improves turndown. At lower vapor rates the wire mesh unit becomes the primary mist eliminator device.

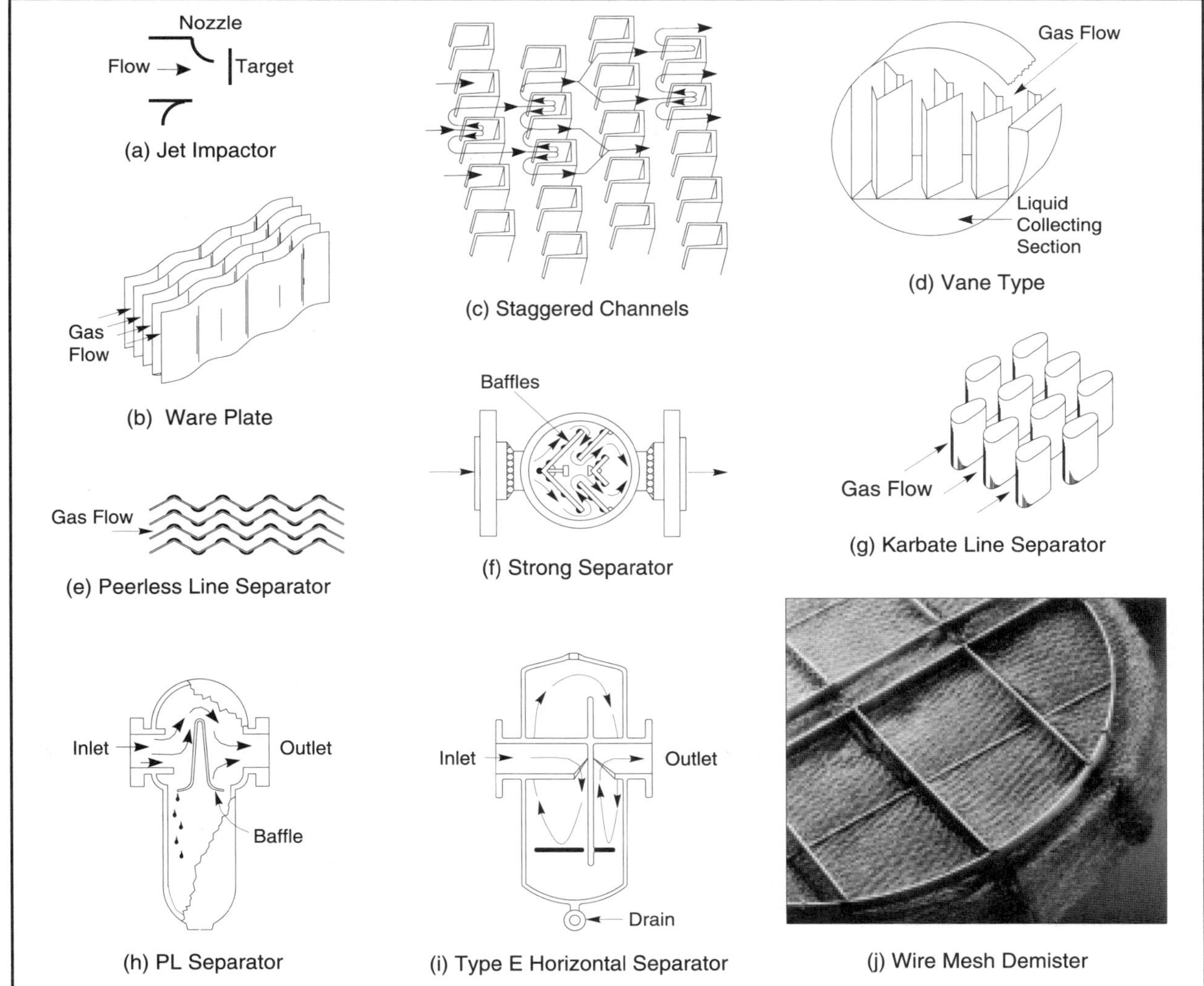

Figure 11.10 Examples of Typical Impingement Separators(11.3)

Centrifugal Separation

It has been noted that centrifugal force is an integral part of separation processes. The standard oil and gas separator may have an inlet that utilizes centrifugal force to separate the larger droplets.

The same principle is used in some mist-extractor elements except that higher velocities are needed in order to separate the smaller droplets. The velocity needed for separation is a function of the particle diameter, particle and gas densities, and the gas viscosity.

With a given system, the size of particle collected is inversely proportional to the square root of the velocity. Consequently, the success of a cyclonic mist extractor is dependent on the velocity attained. Furthermore, the velocity needed to separate a given size of particle must increase as the density of the particle becomes less. In addition to producing the necessary velocity, the mist extractor must provide an efficient means of collecting and removing the particles collected to prevent re-entrainment.

Several types of centrifugal elements are used.

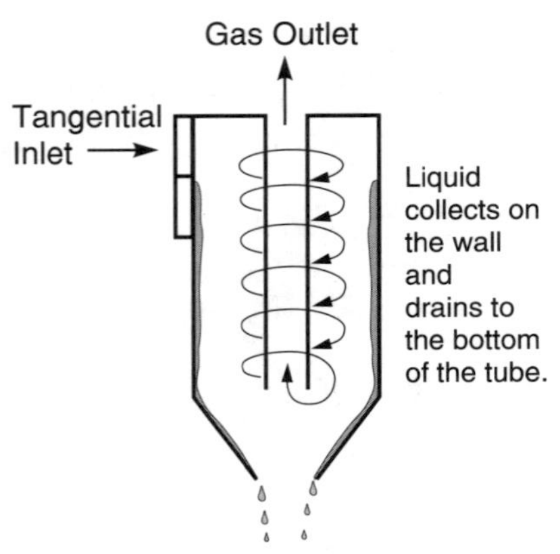

Reverse Flow Cyclone

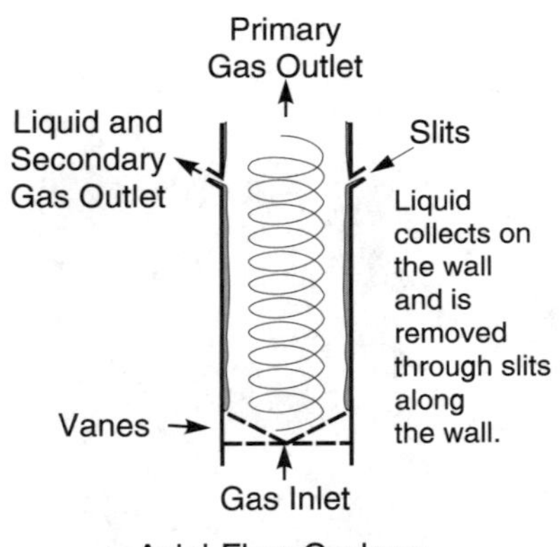

Axial Flow Cyclone

Historically the most common type is referred to as the reverse flow cyclone. Its primary advantage is its robustness. Disadvantages include higher pressure drop and larger size relative to other designs.

Another type is the axial-flow cyclone. In the axial-flow cyclone the centrifugal "spin" is created by a vane installed in the flow pipe. The clean gas flows out of the unit without reversing direction. The primary advantage of the axial flow cyclone is the lower pressure drop and smaller size – allowing several cyclones to be bundled together in a unit.

The main disadvantage of the axial flow cyclone is that a portion of the gas flow (~ 5-10%) exits the slits with the liquid. This gas is heavily liquid loaded and should undergo some type of secondary separation before exiting from the vessel. Some manufacturers use an impingement device, such as wire mesh to remove liquid from the secondary gas in axial flow cyclones.

This leads to the third type of cyclone – recycle axial cyclone. This type is similar to the axial-flow type, but the secondary gas is recycled back into the primary axial cyclone. This is done by installing ports on the cyclone tube which connect the low pressure area in the center of the cyclone with the outside cyclone compartment. This is the basis of the popular Portatest separator which was introduced in Canada in the 1970s.

Reference 11.4 provides a review of centrifugal devices as well as other separator internals.

The most common application of centrifugal elements in separators is the use of multicyclone bundles. Here several cyclone tubes are installed as one internal unit, although monocyclones have been proposed as well.[11.5]

Figure 11.11 shows the overall collection efficiency of a cyclone or centrifugal mist extractor when processing pipeline dust. The recommended flow rate for such devices is usually proportional to ΔP to the 0.5 power. So, efficiency is obtained at the expense of pressure drop.

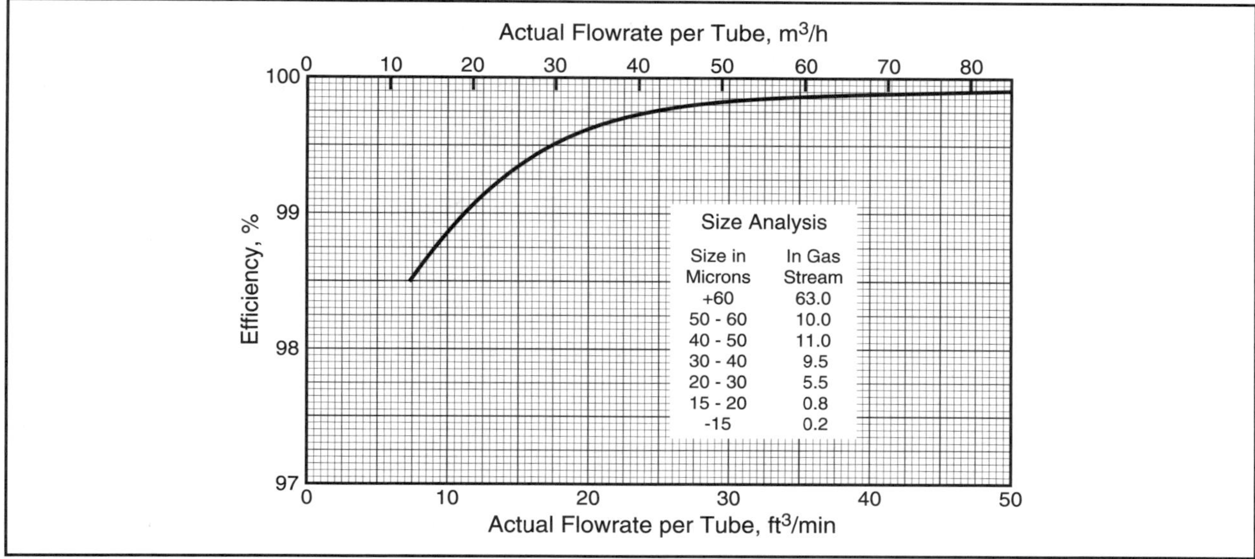

Figure 11.11 Effective Efficiency of a Centrifugal Mist Extractor

The multicyclone separator has been used efficiently in the separation of dusts as well as liquid particles. It is substantially self-cleaning and can handle relatively large quantities of both. In some cases an impingement-type element, e.g. Figure 11.10, has been effectively used ahead of such a unit. The first acts as an agglomeration agent for the smaller particles which may then be separated in the centrifugal unit. This configuration also increases turndown as the impingement device is more efficient at lower flowrates.

The chief disadvantages of any centrifugal unit are higher pressure drop and turndown, i.e. the efficiency falls off rapidly as the velocity decreases below a certain point. Therefore, when widely varying loads are encountered, this type of unit shows decreased advantage. However, if a higher pressure drop is tolerable the unit may be designed for the lowest flow rate anticipated since it gives efficient separation at greater than rated capacity.

Inlet Devices

As mentioned earlier, inlet nozzle velocities in separators can range from 10-30 m/s [33-100 ft/sec].

Some type of inlet device is required to dissipate the energy of the inlet fluids. Further, this energy dissipation should occur over a distance (rather than instantaneously) to avoid shearing of tiny droplets which makes subsequent separation more difficult.

The simplest device is a flat or dished plate placed in the fluid path downstream of the inlet nozzle. It is cheap and simple but not very effective at higher velocities. A half-open inlet pipe (a) is commonly used as well. When the fluid enters the vessel it flows through a pipe which has the bottom cut out of it as shown at right.

Fluid In

(a) Half-open inlet pipe

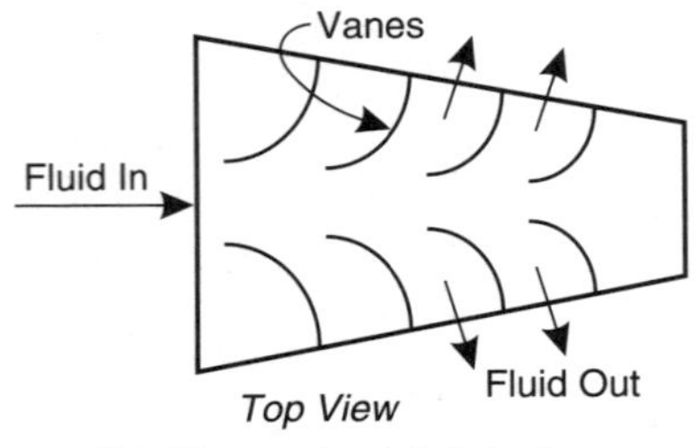

(b) Vane-type inlet device

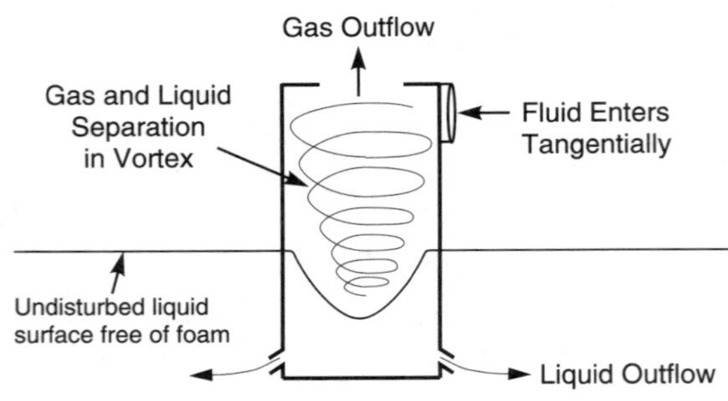

(c) Portatest (cyclone-type inlet device)

The liquid drops to the bottom of the separator and the vapor flows up and around the pipe to the mist extractor. Although widely used, this type of inlet device requires relatively low inlet nozzle velocities to be effective.

Others have used staggered rows of angle iron or channel iron similar to Figure 11.10(c).

Proprietary designs include vane devices similar to the one shown at left (b). Examples include Shell's Schoepentoeter and Zeta Dynamics ZPV inlet device.

Cyclone devices have also been successfully employed. The Portatest vortex tube (c) has been used in foaming applications to break the foam at the vessel inlet before the liquid enters the liquid collection section of the vessel.

In addition to foam breaking, the cyclone devices can also operate at high inlet nozzle velocities which may be useful in debottlenecking applications.

Nozzles

Nozzles must be of a size to minimize erosion/corrosion, pressure drop, entrainment, etc. The equations following may be used to estimate nozzle sizes.

Inlet
$$v_i = \frac{A}{(\rho_m)^{0.5}} \tag{11.23}$$

Gas Outlet
$$v_g = \frac{B}{(\rho_g)^{0.5}} \tag{11.24}$$

Liquid Outlet
$$v_L = C \tag{11.25}$$

Where:

	SI	FPS
v_i, v_g, v_L =	m/s	ft/sec
ρ_m, ρ_g =	kg/m^3	lbm/ft^3
A =	60	50
B =	75	60
C =	1.0	3.3

Sizing correlations for inlet nozzles depend to a large extent on the inlet device. Equation 11.23 is based on a simple half-open inlet pipe. Proprietary inlet devices may allow the use of higher inlet velocities.

Liquid outlets should be equipped with anti-vortex devices (vortex breakers) to prevent gas from going out with the liquid. Several types are shown in Figure 11.12.

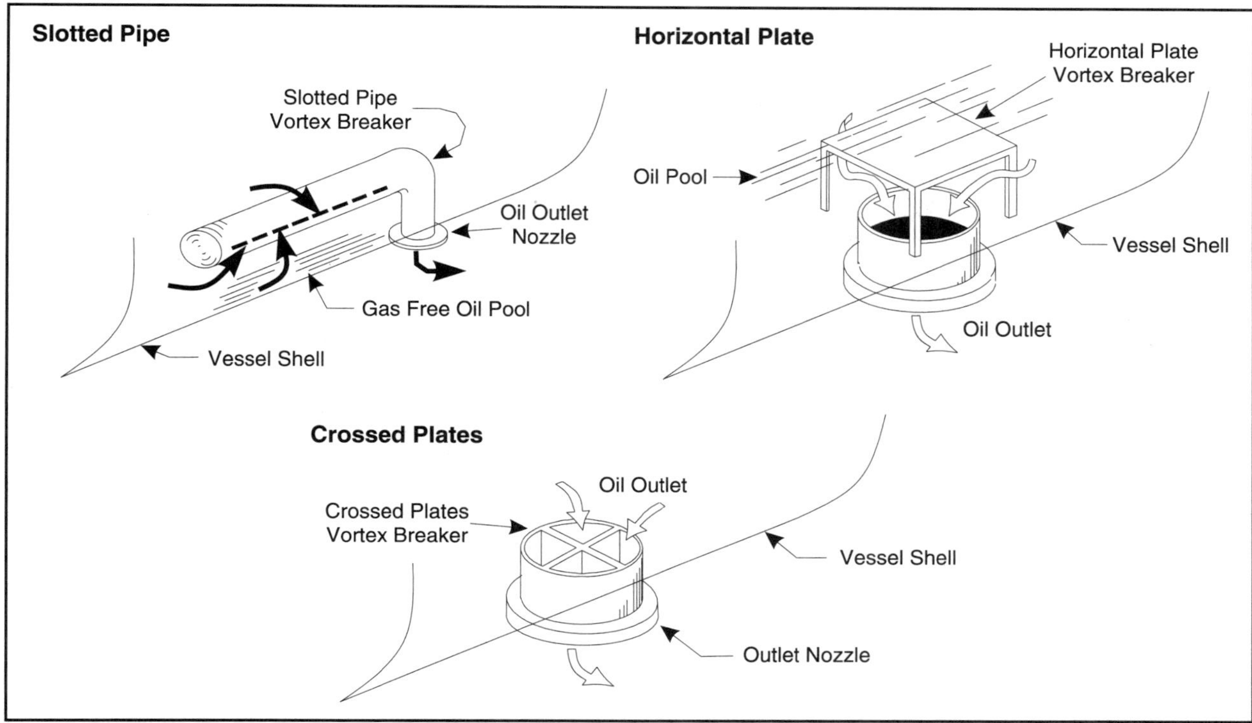

Figure 11.12 Several Types of Anti-Vortex Devices

GAS CLEANING

Cleaning normally refers to the removal of small solid particles and liquid mist entrainment from a gas stream. There are six basic types of gas cleaning equipment – coalescing filter separators, dust filters, gas-liquid separators, centrifugal scrubbers, electrostatic precipitators, and oil-bath scrubbers. All are, in effect, two-section devices – bulk removal followed by some method of removing the "fines."

Filter separators are probably the most common type used today. Electrostatic precipitators and oil bath scrubbers are not widely used in the gas industry.

Filter separators are used in two applications: liquid coalescing and dust removal. In the first application, tiny liquid droplets ($< 1\ \mu m$) which cannot be removed from the gas in a standard impingement separator are coalesced across the filter element. The larger droplets, now several hundred microns in diameter, exit the filter and are removed from the gas by gravity settling or mist extractor. The filter elements are usually compacted glass fibers. When properly sized, coalescing filter separators can be very effective in the removal of fine aerosols from the gas.

Coalescing filter separators are commonly used upstream of dry desiccant dehydrators, glycol dehydrators and reciprocating compressors. They are also sometimes used downstream of cold separators in hydrocarbon dewpoint facilities to remove entrained droplets from the gas as well as in many fuel gas systems.

Dust filters are similar to coalescing filter separators, except the primary purpose of the filter is to remove particulates. The most common application of this filter is removal of fines downstream of dry desiccant dehydrators.

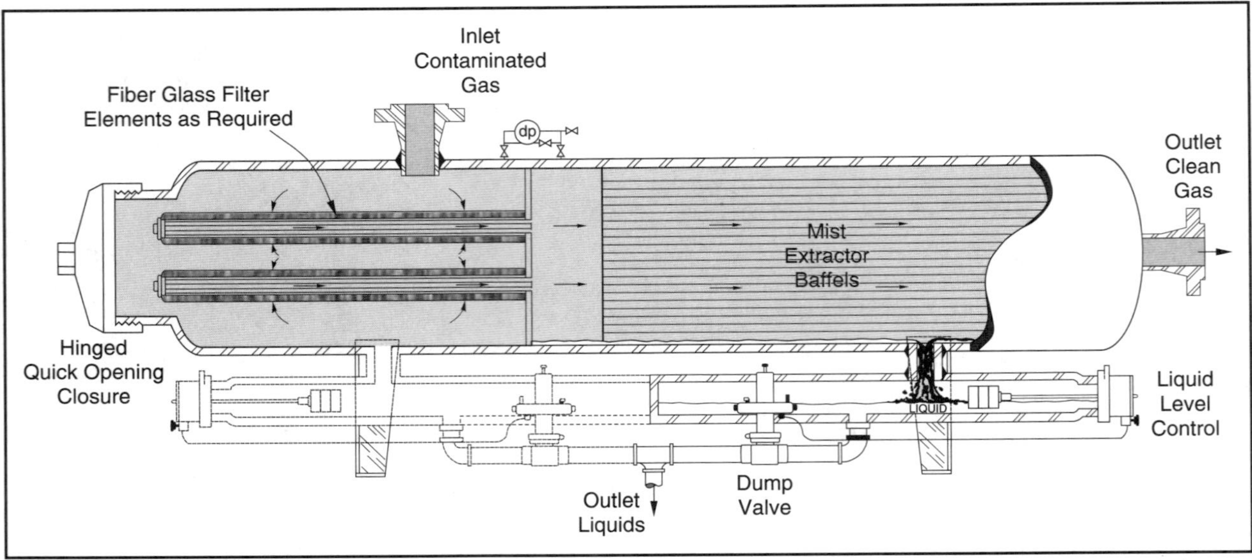

Figure 11.13 A Typical Horizontal Filter Separator-Scrubber

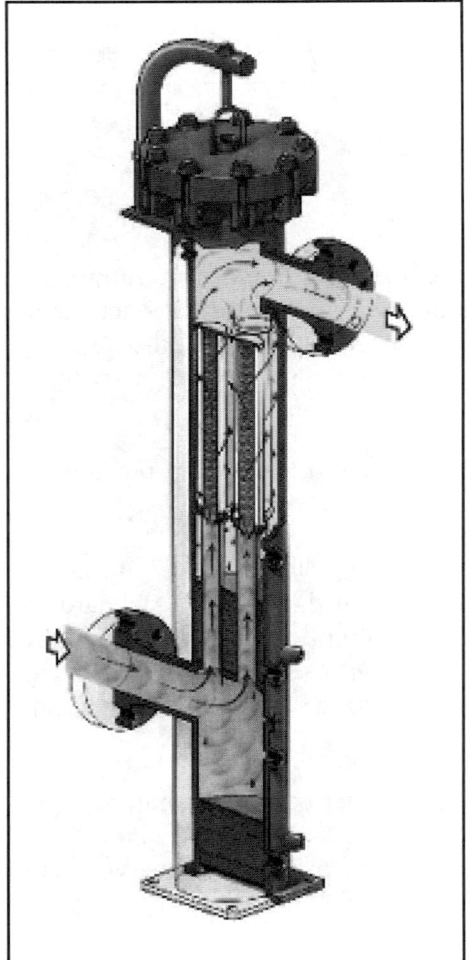

Figure 11.14 Vertical Filter Separator
(Courtesy Pall Filter)

Filter separators can be constructed in the vertical or horizontal orientation. Figure 11.13 shows a double barrel horizontal filter separator. Feed gas enters from the left and flows radially inward through the filters in the first chamber. The filters are mounted on perforated tubes and the gas flows down the tubes through a tubesheet and into a second chamber. This chamber contains a mist extractor, usually a vane-type, which removes the liquid droplets from the gas. Liquid collected in either chamber flows through downcomers to the liquid collection chamber (lower barrel) below.

A common vertical orientation is shown in Figure 11.14. In this design the gas enters the bottom portion of the separator and flows upward through the tubesheet and into the filter tubes. The gas flows radially outward through the filter elements. Liquid, coalesced on the glass fibers drains to the bottom portion of the filter element and drips off of the filters accumulating in a liquid collection chamber. Some manufacturers treat the surface of the glass fibers with a chemical that reduces wetting and promotes droplet formation. Other vertical orientations are used, i.e. gas enters the top chamber, flows radially inward through the filters and out of the vessel in the bottom chamber.

A properly sized filter separator can be an important and effective component of a gas processing facility. Unfortunately, they have acquired a poor reputation because many filter separators are undersized. Most suppliers claim removal efficiencies of 99-99.9% of droplets larger than 1 μm. This is technically feasible, but is often not achieved in practice.

Evaluation of vendor proposals is difficult because the sizing calculations used by the vendor are proprietary. But, sizing is critical. Remember that the "back-end" of a filter separator is a mist extractor and/or a gravity settling section. The diameter of this section should be based on sizing criteria presented earlier in this Chapter. Some vendors propose smaller vessels based on the efficiency of filter separation. It does no good to coalesce small droplets if they are then re-entrained in the collection chamber.

CONTROLS

Control of pressure and level are basic for good separator operation. The choice of control modes, sensitivity characteristics, control hardware, etc. depends on the purpose of the separator and the process modules immediately preceding and following it.

As a general rule, the pressure of the separator must be held rather constant, independent of the operation of adjacent equipment. Usually this means pressure control on the gas outlet. The controller mode will typically be proportional or proportional plus integral (P+I). Offset due to flow rate changes normally presents no problem because proportional band settings are usually narrow. However, the vessel design pressure and high pressure alarm or shutdown controls must be consistent with the range of pressure expected for the setpoint and and offset anticipated.

A common practice is to set the high pressure shutdown at the maximum operating pressure. This is typically 5-10% above the normal operating pressure.

The separator off-gas often flows to a compressor. Depending on the type of compressor and driver, pressure control may be achieved with a pressure control valve or by directly manipulating a compressor variable, e.g. speed, valve unloaders, slide valve, etc. In general, constant speed centrifugal compressors will likely require a control valve. For variable speed compressors and reciprocating compressors equipped with valve unloaders and/or clearance pockets, pressure control can often be implemented by manipulating the compressor. A more detailed discussion of compressor control can be found in Chapter 15.

If efficient vapor-liquid separation is the primary purpose of the vessel, the liquid level should be held relatively constant. As per the principles outlined in Chapter 9, proportional control or proportional plus integral is adequate. If the vessel is in surging or slugging service, the primary objective may be to stabilize the outlet liquid flowrate. In this case some variation in liquid level is required.

Figure 11.15 shows a split-range approach sometimes used for slugging/surging applications. So long as liquid input is rather steady only Valve A is operating. It is fully closed at a controller output of 4 mA [3 psi] and fully open at 12 mA [9 psi]. It would typically have a narrow proportional band to minimize offset.

What happens if a large slug of liquid hits the separator? The level rises, controller output increases toward 12 mA [9 psi], Valve A is fully open, but the level continues to rise. At 12 mA [9 psi], Valve B starts to open to relieve the surge. How fast it opens depends on the setting (proportional band). Once the slug has passed and the level is back within normal range, Valve B closes and waits for the next such upset.

This system provides sensitive routine level control plus the added capability for relieving surges not possible with a single valve system. Notice that the line to Valve B may come off the same liquid nozzle as Valve A or from a separate one at another point in the vessel. In a horizontal separators greater than 6 m [20 ft] in length, two liquid nozzles may be a justifiable expense. In a vertical separator one would always use a single liquid nozzle.

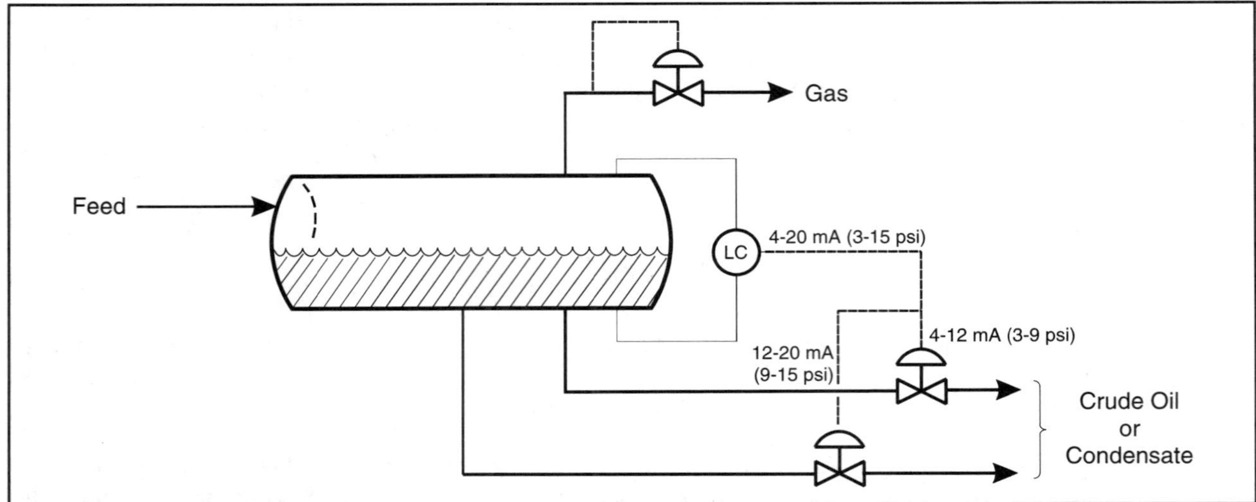

Figure 11.15 Example of Split-Range Level Control

Check the size of the liquid nozzles. They tend to be too small (for competitive cost reasons) even though they may technically meet specifications.

Suppose that the liquid is being pumped out of the separator. Figure 11.16 shows three different arrangements. In (a) the level control valve simply changes the pump discharge pressure so that the pump capacity matches the rate of liquid accumulating in the vessel. A minimum flow by-pass circuit is provided to protect the pump from minimum flow. A second alternative (b) is to provide a rate controlled by-pass valve. This is more expensive than (a) because of the extra valves and controls needed, but with this method the pump does not continuously run in recycle.

Throttling and bypassing waste power. In some installations it is desirable to use a variable speed drive. The level controller then adjusts pump speed to maintain level.

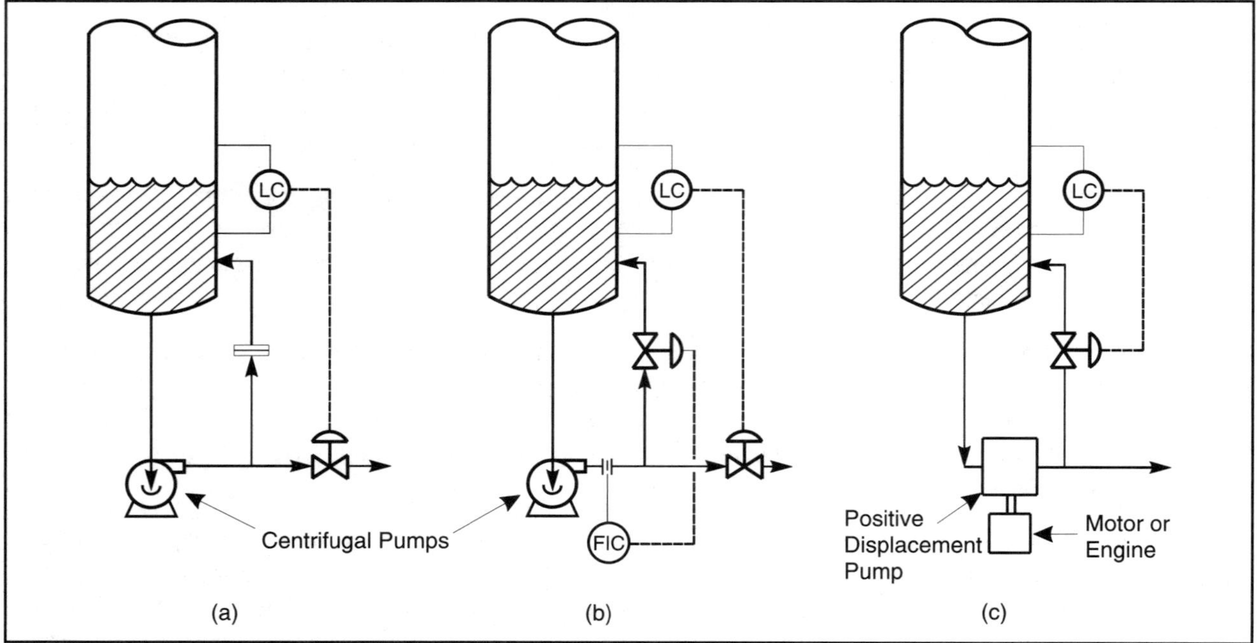

Figure 11.16 Example of Level Control with a Pump

As shown in (c), it is not possible to throttle a positive displacement pump. The level will control by-pass rate or pump speed.

If the pump capacity is significantly greater than the liquid rate, the pump will continuously run in a recycle mode. This can heat the liquid in the separator reducing liquid recovery and possibly NPSHA.

Liquid Residence Time and Controls

Final separator design must conform to residence time requirements that are compatible with the control system. Figure 11.17 shows a horizontal separator equipped with high and low level shut-down (LSDH and LSDL) as well as high and low level alarms (LAH and LAL). These are shown in relation to the normal liquid level (NLL).

If surging or slugging flow is expected, sufficient volume must be provided on both the low and high side so that the alarms do not trigger routinely. Once an alarm has been activated, adequate time must be provided for the operator to take corrective action before shutdown occurs. If the vessel is installed in an unmanned facility, alarms are not required.

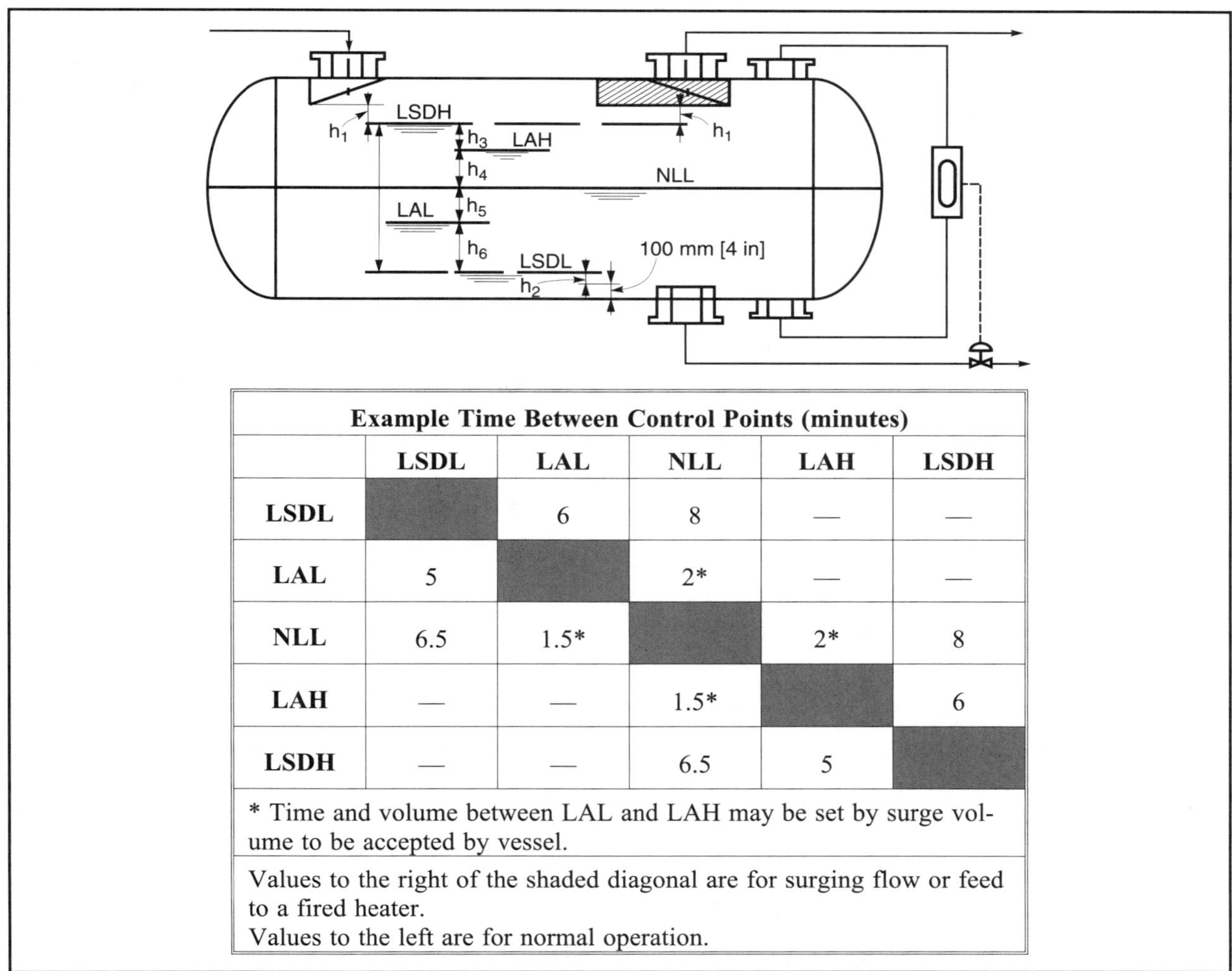

Example Time Between Control Points (minutes)					
	LSDL	LAL	NLL	LAH	LSDH
LSDL		6	8	—	—
LAL	5		2*	—	—
NLL	6.5	1.5*		2*	8
LAH	—	—	1.5*		6
LSDH	—	—	6.5	5	

* Time and volume between LAL and LAH may be set by surge volume to be accepted by vessel.

Values to the right of the shaded diagonal are for surging flow or feed to a fired heater.
Values to the left are for normal operation.

Figure 11.17 Guidelines for Separator Control Volumes

For severe slugging, additional liquid volume must be added to the separator to accommodate the slug. This volume is typically added between the NLL and the LAH. In some slug catchers, this volume may represent as much as 50-60% of the separator capacity. Slug sizes can be estimated from two-phase simulation models discussed in Chapter 10. In the absence of a model, the slug size can be assumed to be 3-5 seconds of liquid-full flow at feed pipe velocity.

The table in Figure 11.17 provides guidelines for control and operator intervention times. Most control volumes (LAL to LAH) range from 2 to 4 minutes of liquid flow depending on the application. In surge vessels and accumulators the control time may be greater because the objective is not so much to control level, but to stabilize the flow.

Operator intervention times are more difficult. For locally mounted level controllers at least five minutes should be provided to allow the operator to access the vessel and troubleshoot the problem. For controllers located in the control room intervention times are often less, in some cases as low as one minute. Operator intervention times are frequently set by "what can we afford". For low liquid rates, required liquid holdup for control and intervention purposes is inexpensive. For high liquid rates this is not true and large holdup volumes can have a significant impact on vessel size.

It should be apparent that separator and control design should be compatible. A substantial portion of separator volume may be required for control purposes. Failure to properly coordinate process and control functions results in too frequent alarms and shut-downs and/or inefficient separation.

Three-Phase Separation Control

Figure 11.18 shows various ways of controlling the high pressure separation of gas, oil and water. Shown are the use of interface controls, buckets and weirs. In *interface control* a displacement float reacts to the different density of the two adjacent liquid phases. Only this difference affects buoyancy change with level. Thus, it may be very unstable. *Buckets* are chambers within the vessel where one or more liquid phases are segregated. *Weirs* are used to aid in segregation to eliminate the need for interface controls.

Methods (3) and (4) of Figure 11.18 provide maximum control stability since all control interfaces are gas-liquid. However, these designs are complex and the buckets are prone to filling with sand.

Another type of three phase separator design involves a boot-type separator as shown in Figure 11.19. These separators are frequently used in glycol hydrocarbon separation service in LTS and refrigeration type plants. The level of the glycol phase is controlled with an interphase controller, while the hydrocarbon level is controlled with a normal level controller. This separator configuration is also occasionally used in oil/water service.

Boot-type separators are preferred when the flowrate of the heavier, aqueous phase is small compared to the lighter, hydrocarbon phase. The boot facilitates level control and provides a separate chamber where a heating coil can be included to reduce the viscosity of the glycol phase.

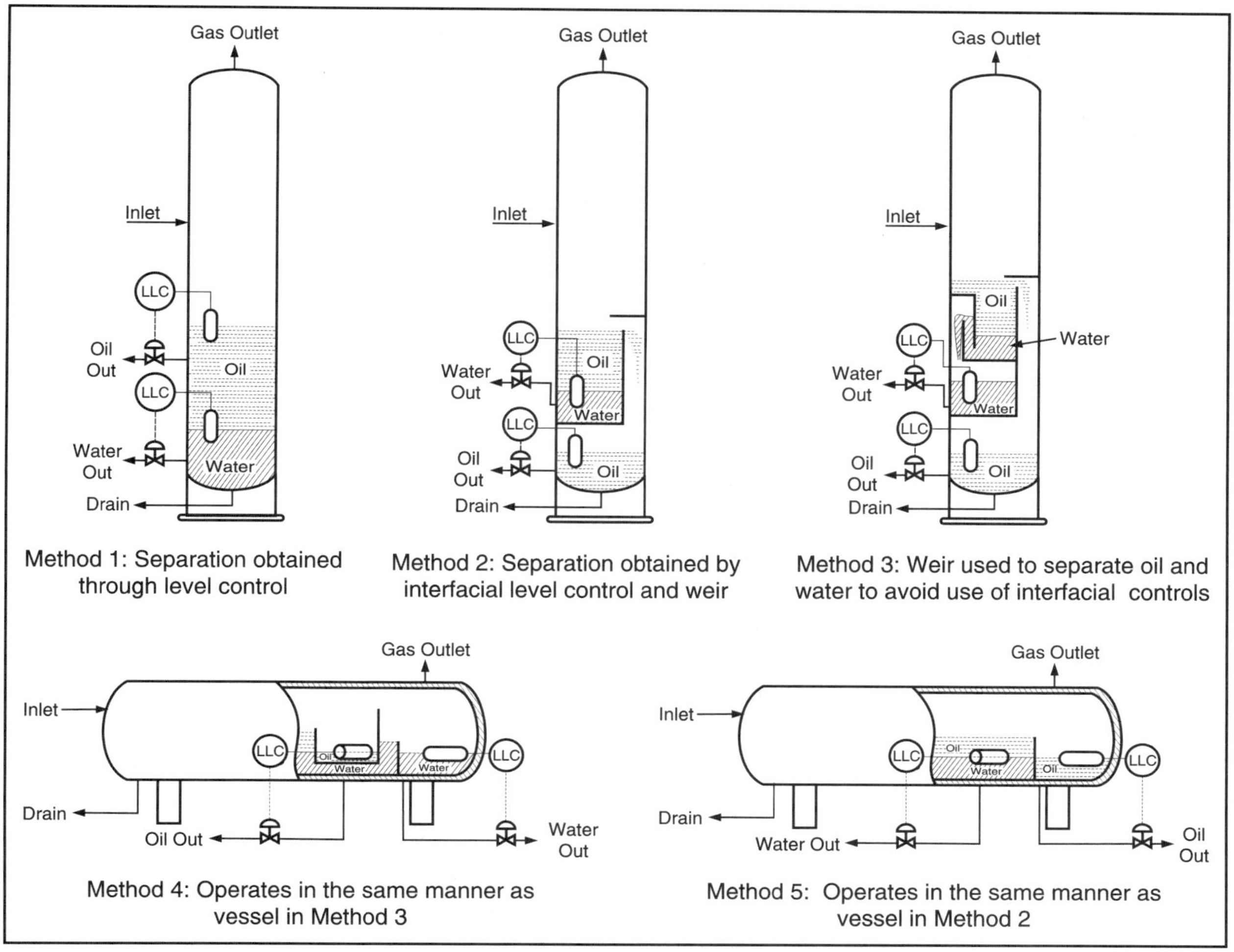

Method 1: Separation obtained through level control

Method 2: Separation obtained by interfacial level control and weir

Method 3: Weir used to separate oil and water to avoid use of interfacial controls

Method 4: Operates in the same manner as vessel in Method 3

Method 5: Operates in the same manner as vessel in Method 2

Figure 11.18 Various Methods for Controlling Three-Phase Separation

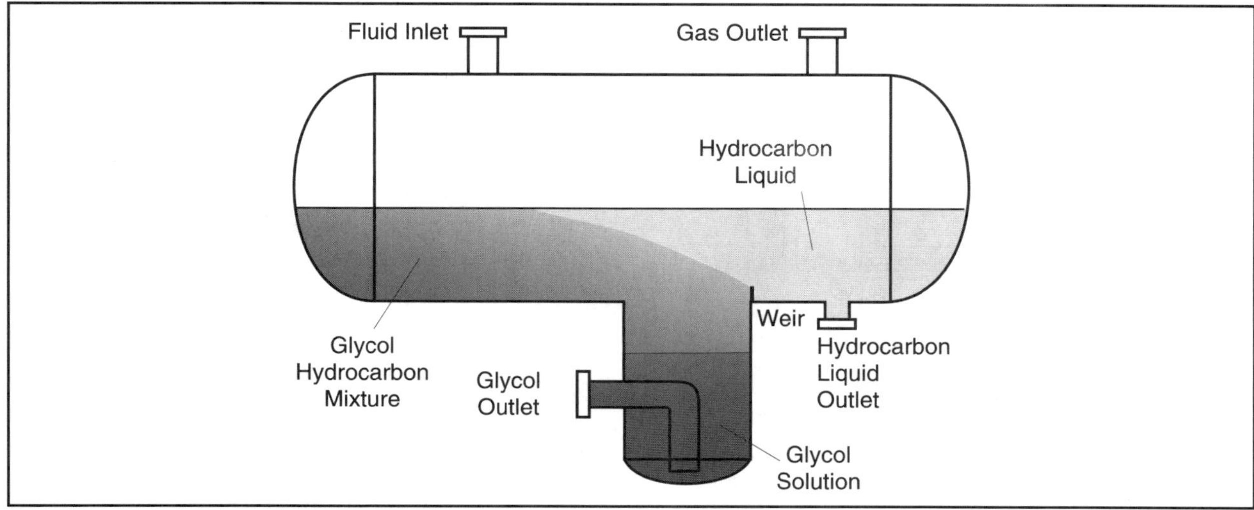

Figure 11.19 Horizontal Three-Phase Glycol Hydrocarbon Separator

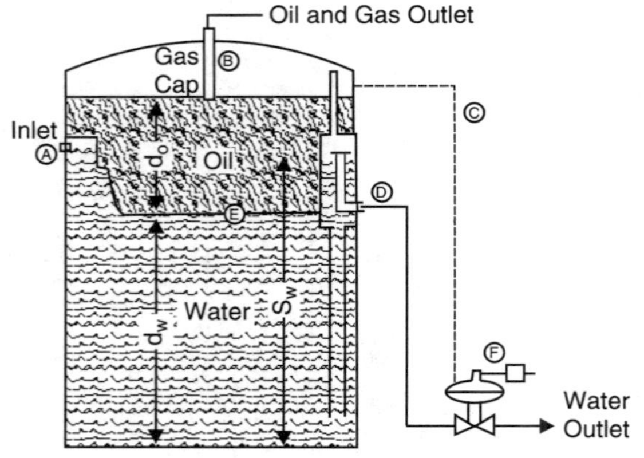

In atmospheric pressure, free-water knock-out applications (see figure on the left), the free-water settles out to the bottom and the oil and gas pass overhead – usually to a treater. The outlet B is extended into the vessel to ensure the presence at all times of a gas cap and to control the oil level.

The location of the oil-water interface E is fixed by the position of adjustable nozzle D. Since the water outlet box is vented into the gas cap, rotation of this nozzle changes the interface level for:

$$d_o\rho_o + d_w\rho_w = S_w\rho_w + \Delta P_f \tag{11.26}$$

Where:

d_o = thickness of oil in vessel
d_w = height of water in vessel proper
S_w = height of water in outlet box
ρ_o, ρ_w = density of oil and water respectively
ΔP_f = pressure drop in downcomer pipe and outlet box

In a well designed system operating at normal capacity, ΔP_f may be neglected. Equation 11.26 may then be rearranged to read:

$$d_w = S_w - d_o\left(\frac{\rho_o}{\rho_w}\right) \tag{11.27}$$

Consequently, any change in S_w will cause a corresponding change in d_w and d_o. As S_w is decreased, d_o gets larger, so that small changes in S_w give correspondingly larger changes in d_w.

In applications where no instrument air or gas is available water is removed from the vessel with a simple weight-loaded, self-acting diaphragm valve. With the top of the diaphragm vented into the vessel gas cap, the only force acting on the underside of the diaphragm is the head of liquid. The weight is then simply adjusted on the arm to compensate for the desired liquid head operating the valve.

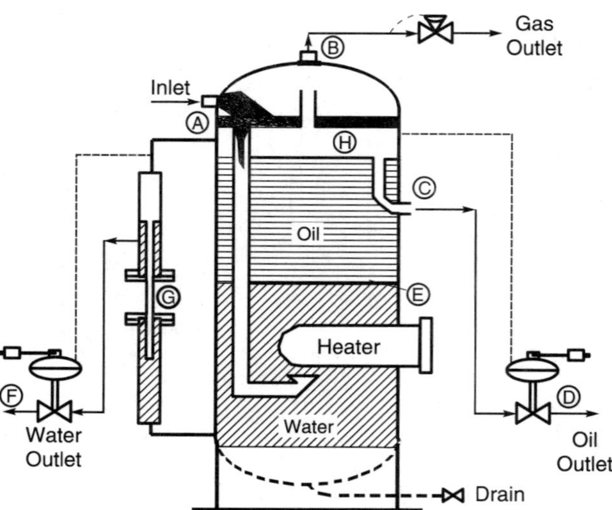

Shown at left is a similar system for a heater treater. In this case, though, an additional valve is necessary to remove the oil which is separated from the gas in the top section. This valve operates in the same manner as the water valve.

The water syphon Ⓖ is a type commonly used and the principle is the same as the rotating nozzle. It also must be connected to the gas space with a pressure equalizing line. The oil outlet is a fixed opening so located that a gas cap Ⓗ may be preserved below the top plate.

The control system shown with the high pressure free-water knockout to the right is similar to that for the low pressure unit. In this case, though, a conventional liquid-level controller is incorporated into the water-outlet compartment.

The oil simply overflows into the gas-outlet line with the combined stream passing through the gas back-pressure valve. Because of liquid surging, the controlling connection should be upstream from where streams comingle.

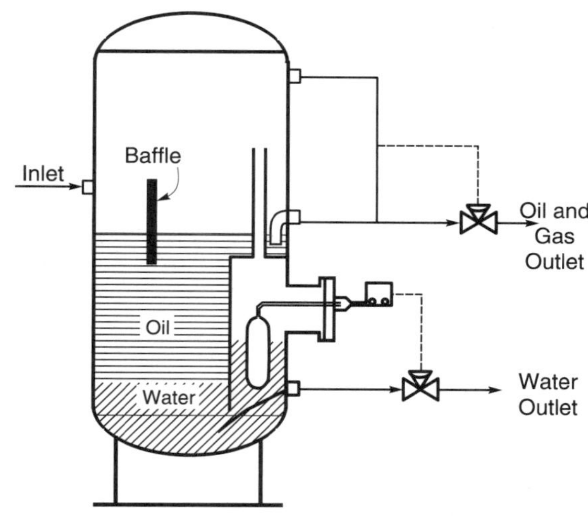

SUMMARY

The foregoing should make it clear that choice of separation vessels is never routine. As a general rule, the quality and reliability of your production/processing system will be determined to a large degree by the quality of your separation decisions.

REFERENCES

11.1 Barker, W. F., *Oil and Gas J.* (Dec. 27, 1982), p. 186.

11.2 Polderman, H. G., Bouma, J. S., and H. van der Poel, "Design Rules for Dehydration Tanks and Separator Vessels," SPE 38816, San Antonio, Texas (5-8 Oct., 1997).

11.3 Perry, J. H., *et al.*, *Chemical Engrs. Handbook*, 4th Ed., McGraw-Hill, New York (1963).

11.4 Swanborne, R. A., *et al.*, "New Separator Internals Cut Revamping Costs," *Jour. Pet. Tech.* (Aug. 1995), p. 688.

11.5 Orange, L., "Cyclone-Type Separators Score High in Comparative Tests," *Oil and Gas J.* (22 Jan., 1990), p. 54.

11.6 Davies, J. T., "Turbulence Phenomena," Academic Press, London (1972), p. 355.

11.7 Hinze, J. O., "Fundamentals of the Hydrodynamic Mechanism of Splitting in Dispersion Processes," *AICHE Jour.*, Vol. 1 No. 3 (Sept. 1955), p. 289-295.

NOTES:

APPENDIX 11A

This appendix contains some general correlations from the references shown below for estimating weight, size and relative cost of various equipment. Costs are a function of time and size. Indices are published regularly which show price changes with time. The *Oil and Gas Journal* publishes one of particular use in the petroleum industry.

The change of cost with size or capacity is quite often a straight line on a log-log plot within a given size range which can be represented by the equation

$$C_1 = C_2 \left(\frac{x_1}{x_2} \right)^a$$

Where: C_1 and C_2 = the cost of two different units
x_1 and x_2 = capacity (size) of these units

Typical values of "a" are shown below.

Equipment	Value - "a"	Limits of Applicability
Steel pressure tanks, 15-100 psia	0.54	$1\text{-}100 \times 10^6$ USgal
Liquid pumps, centrifugal	0.15	0.1-10 hp
Liquid pumps, centrifugal	0.60	10-1000 hp
Liquid pumps, triplex reciprocating	0.60	1-1000 hp
Fractionation towers	1.00	10-100 in diam. vs. price/ft
Low pressure storage tanks (steel)	0.54	$1\text{-}100 \times 10^6$ USgal
Heat exchangers	0.59	10-20 000 ft^2 surface
Fans, blowers, compressors	0.66	10-10 000 hp
Refrigeration units	0.66	10-500 tons rating
Liquid filters	0.56	10-1000 ft^2 filtering area
Furnaces, tubular	0.54	0.1-100 MMBtu/hr duty
Furnaces, Dowtherm	0.35	0.1-3.0 MMBtu/hr duty

The tables and figures which follow are useful approximations for planning purposes only. Please note that some of the weights are dry weights and some include normal liquid weight during operation.

REFERENCES

11A.1 Lilly, L. L., private communication.

11A.2 Mayer, J. L. and R. W. Coggins, *World Oil* (Aug. 1969), p. 29.

11A.3 McMinn, R. E., *Ibid.* (Oct. 1969), p. 100.

11A.4 Philbeck, W. T., *Ibid.* (Nov. 1969), p. 98.

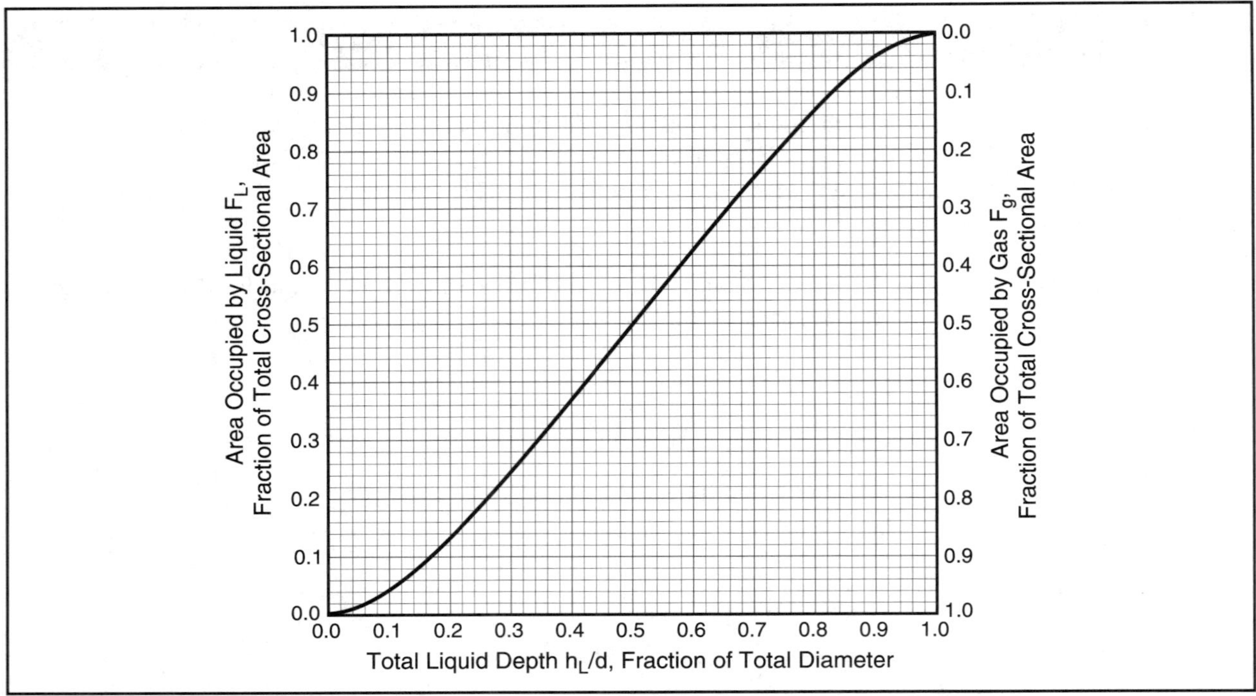

Figure 11A.1

The above figure is for a horizontal, circular vessel. F_L or F_g (on the ordinate) is the area occupied by the liquid as a fraction of total cross-sectional area. h_L is the total depth of liquid as a fraction of the total diameter.

For three-phase separation with two liquid phases, the F_L for the lightest liquid phase can be found by subtracting F_L for the heaviest phase from the total F_L.

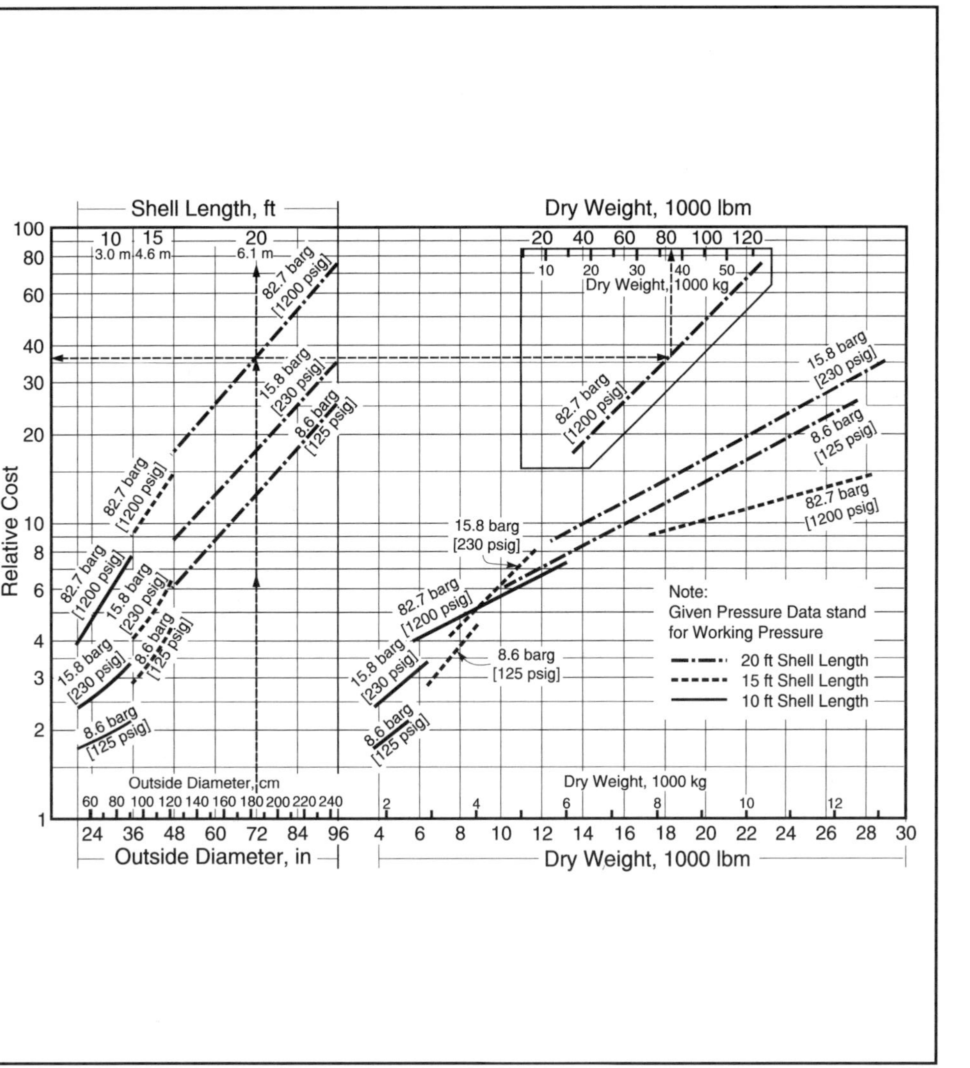

Figure 11A.2 Relative Cost and Weight of Horizontal Separators

Cost and weight of various sizes of horizontal separators. Price includes skid-mounted ASME vessel complete with controls and external corrosion protection

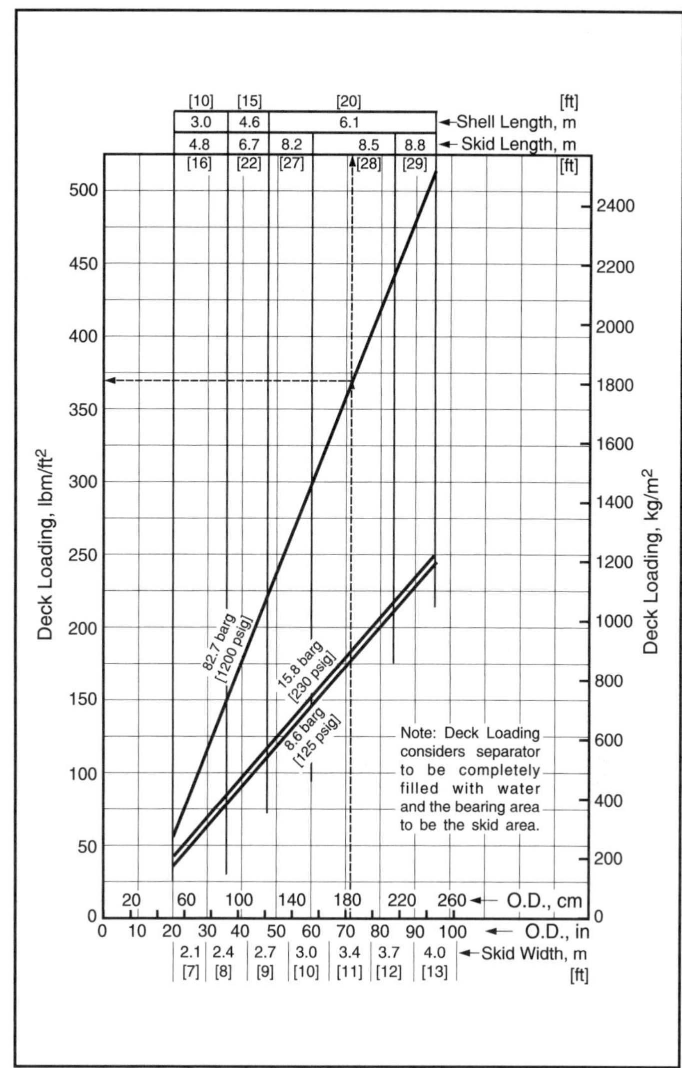

Figure 11A.3 Relative Cost and Weight of Horizontal Separators

Deck loading characteristics and space required for horizontal separators. Units are assumed to be filled with water which can occur during testing, or if controls malfunction.

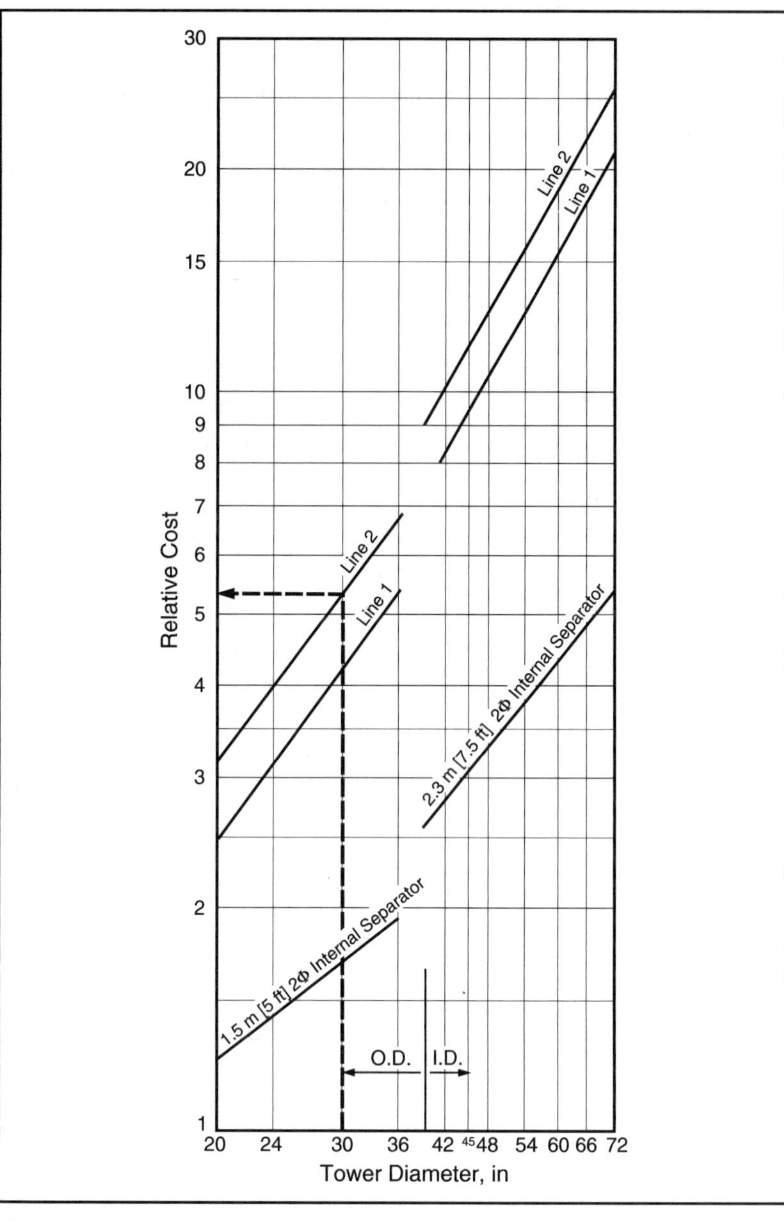

Figure 11A.4 Relative Cost of Glycol Contactors

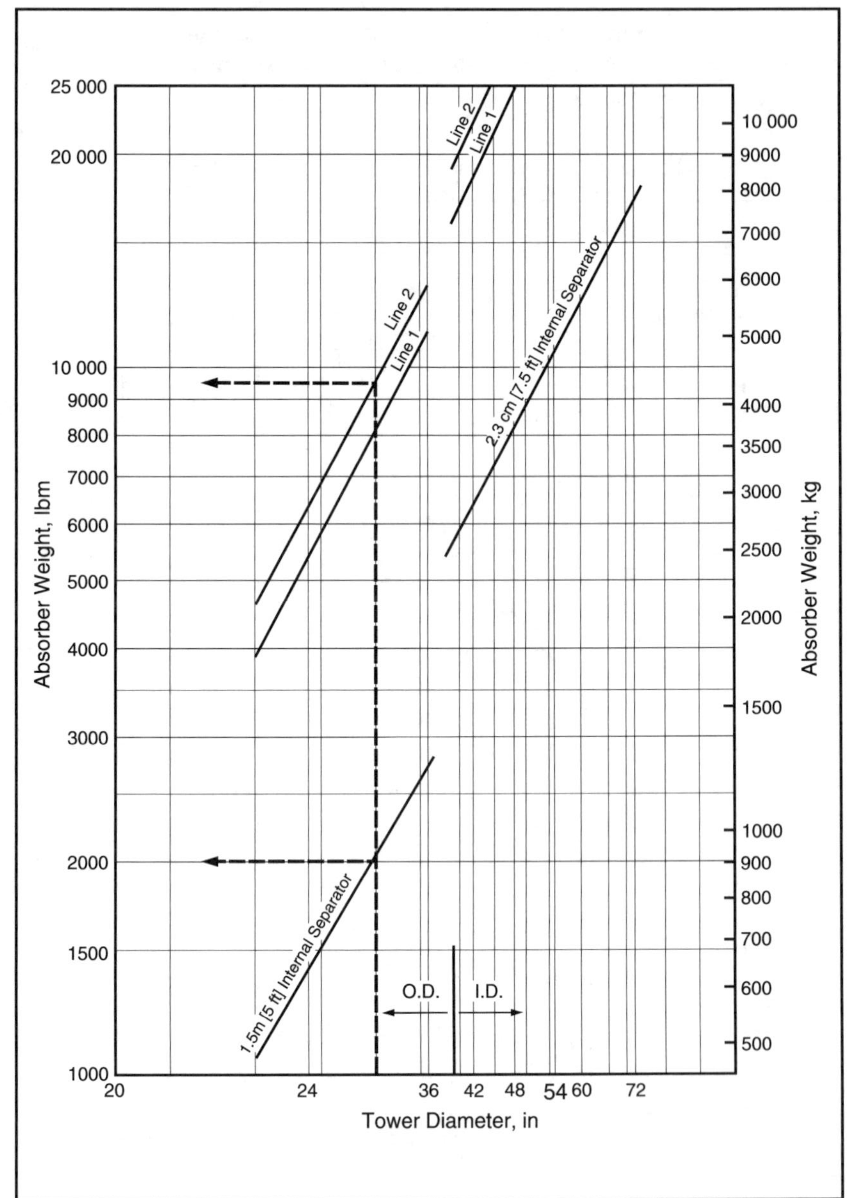

Figure 11A.5 Weight of Glycol Contactors

Approximate Deck Areas and Weights of Glycol Regeneration Units, lbm				
Inlet Gas Flow, MMscfd	Approximate Plot Size, ft × ft		Approximate Weights	
	65°F Depression	80-100°F Depression	65°F Depression	80-100°F Depression
5	4 × 6	4 × 8	2100	2700
10	4 × 9	4 × 10	2400	3400
20	5 × 12	5 × 14	3700	4900
30	6 × 14	6 × 16	4800	9500
40	6 × 16	6 × 19	6900	10 400
50	6 × 18	6.5 × 18	10 100	12 500
60	6 × 18	6.5 × 18	10 100	12 500
70	6.5 × 17	7 × 18	12 900	16 500
80	6.5 × 17	7 × 21	12 900	18 500
90	7 × 17	7 × 21	23 800	25 900
100	7 × 17	7 × 21	23 800	25 900

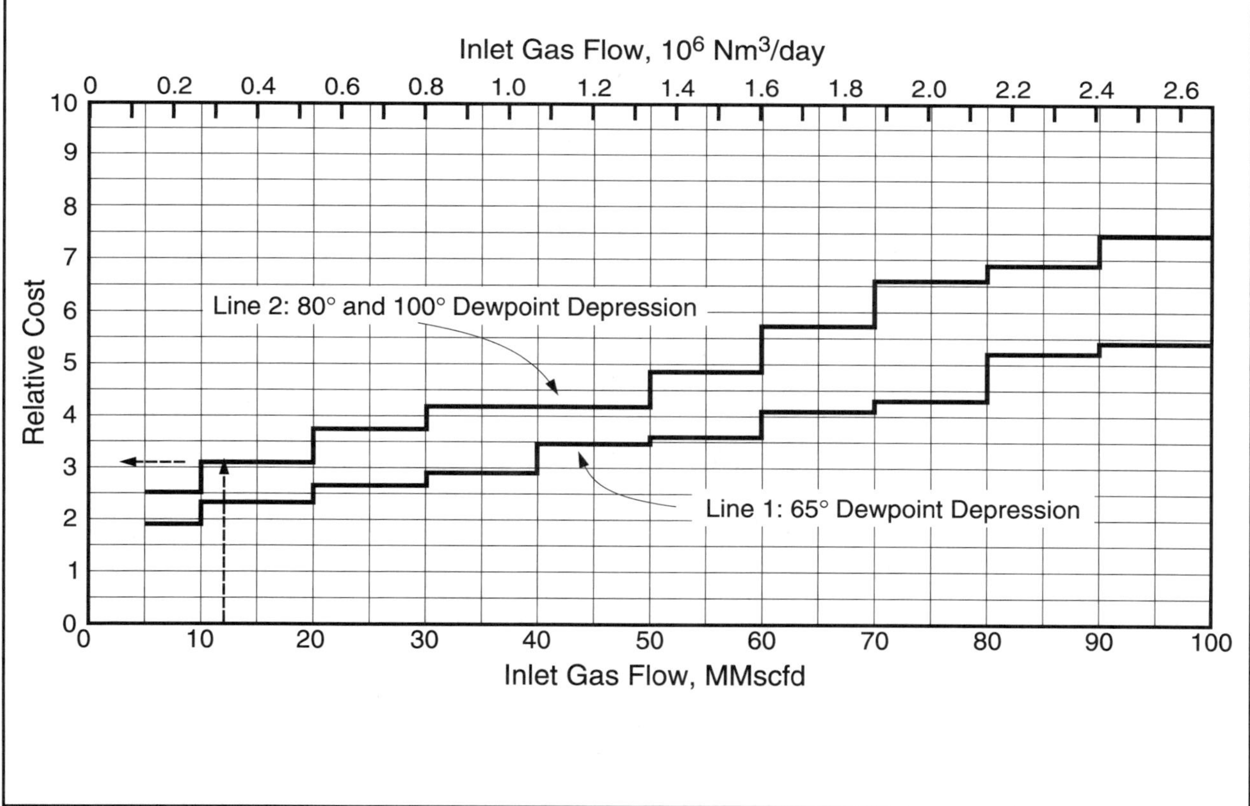

Figure 11A.6 Relative Cost, Weight and Space Requirements for Glycol Regeneration Units

Figure 11A.7 Relative Cost and Weight of an Amine Absorber

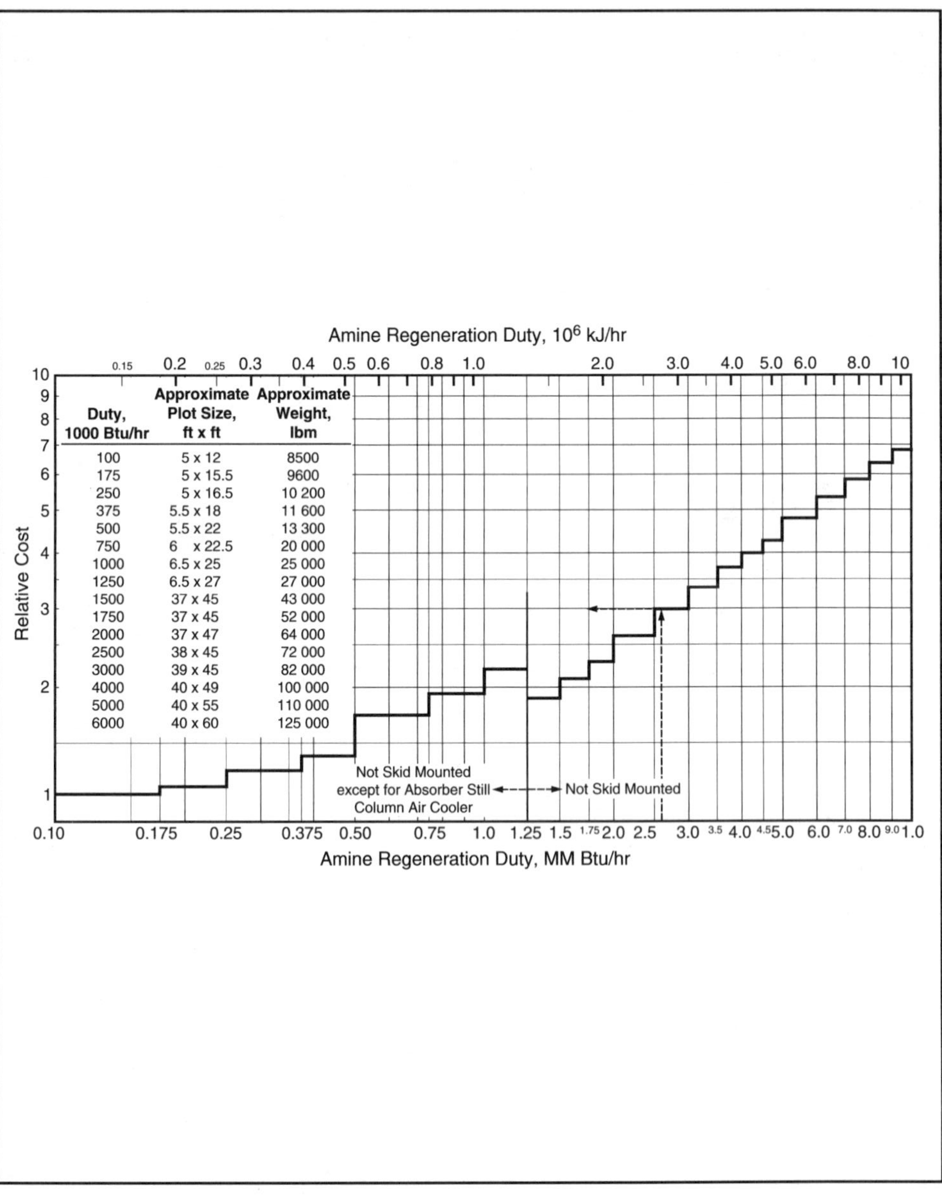

Duty, 1000 Btu/hr	Approximate Plot Size, ft x ft	Approximate Weight, lbm
100	5 x 12	8500
175	5 x 15.5	9600
250	5 x 16.5	10 200
375	5.5 x 18	11 600
500	5.5 x 22	13 300
750	6 x 22.5	20 000
1000	6.5 x 25	25 000
1250	6.5 x 27	27 000
1500	37 x 45	43 000
1750	37 x 45	52 000
2000	37 x 47	64 000
2500	38 x 45	72 000
3000	39 x 45	82 000
4000	40 x 49	100 000
5000	40 x 55	110 000
6000	40 x 60	125 000

Figure 11A.8 Relative Cost, Weight and Space Requirements of an Amine Regeneration Unit

Figure 11A.9 Relative Cost and Weight of Horizontal Oil Dehydrators

Cost and weight of various sizes of horizontal filter media type treaters. ASME vessels are skid-mounted, equipped with all controls and provided with external corrosion protection.

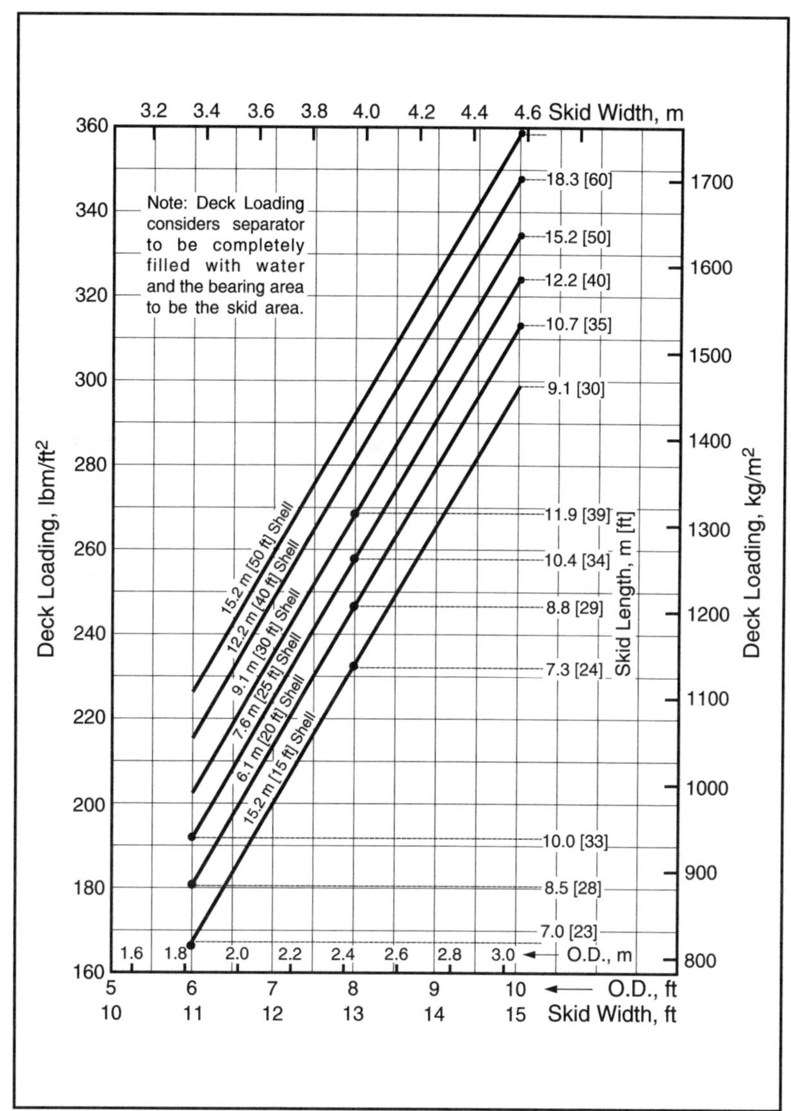

Figure 11A.10 Relative Cost and Weight of Horizontal Oil Dehydrators

Space requirements and deck loading data for horizontal treaters. Information from this chart is helpful when locating vessels on platform foundations in marsh or water.

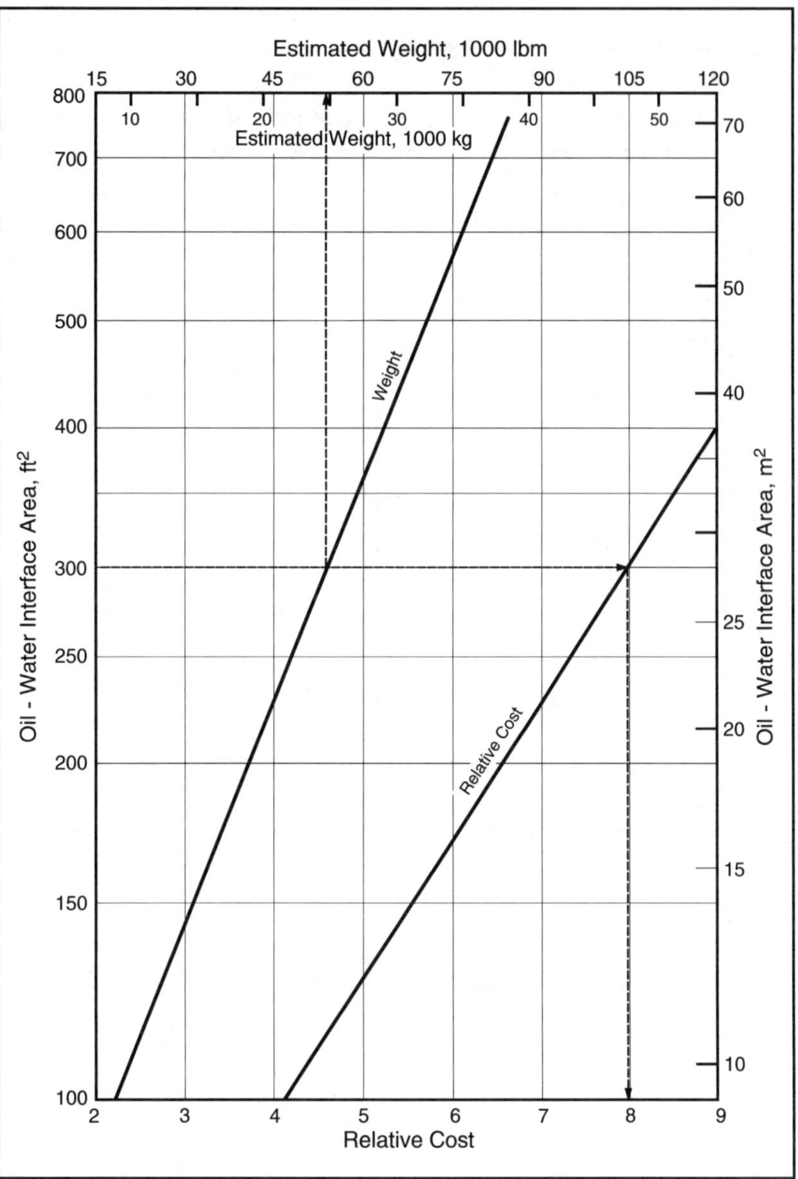

Figure 11A.11 Estimated Costs and Weights for Crude Oil Dehydrators with Electrically Assisted Coalescence

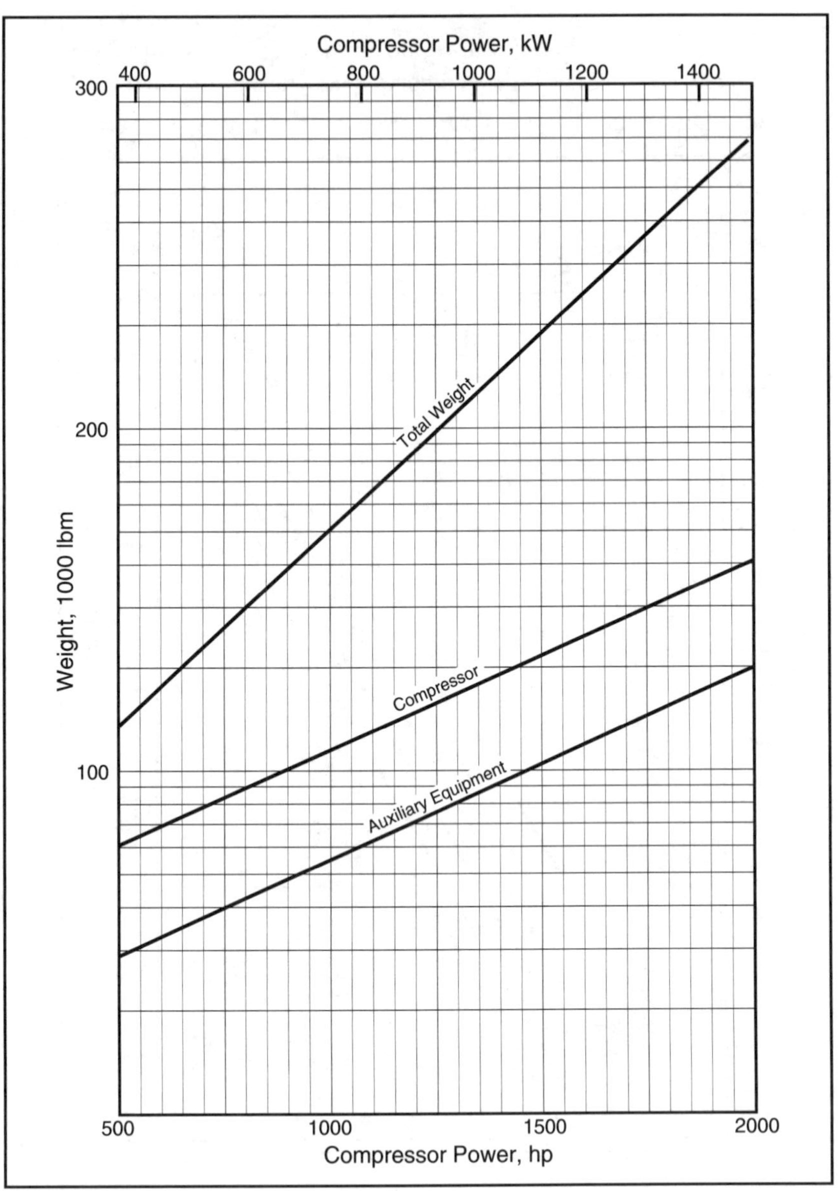

Figure 11A.12 Reciprocating Compressor Weight Estimate

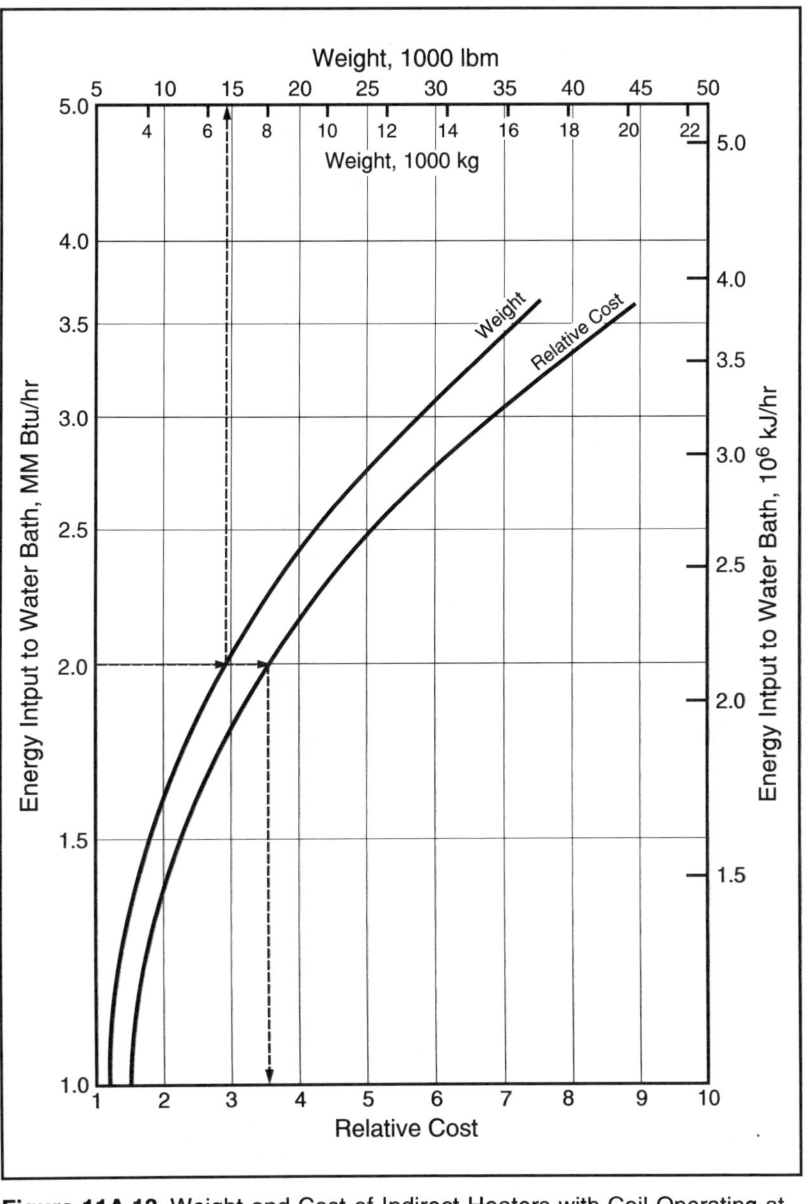

Figure 11A.13 Weight and Cost of Indirect Heaters with Coil Operating at 500 psi

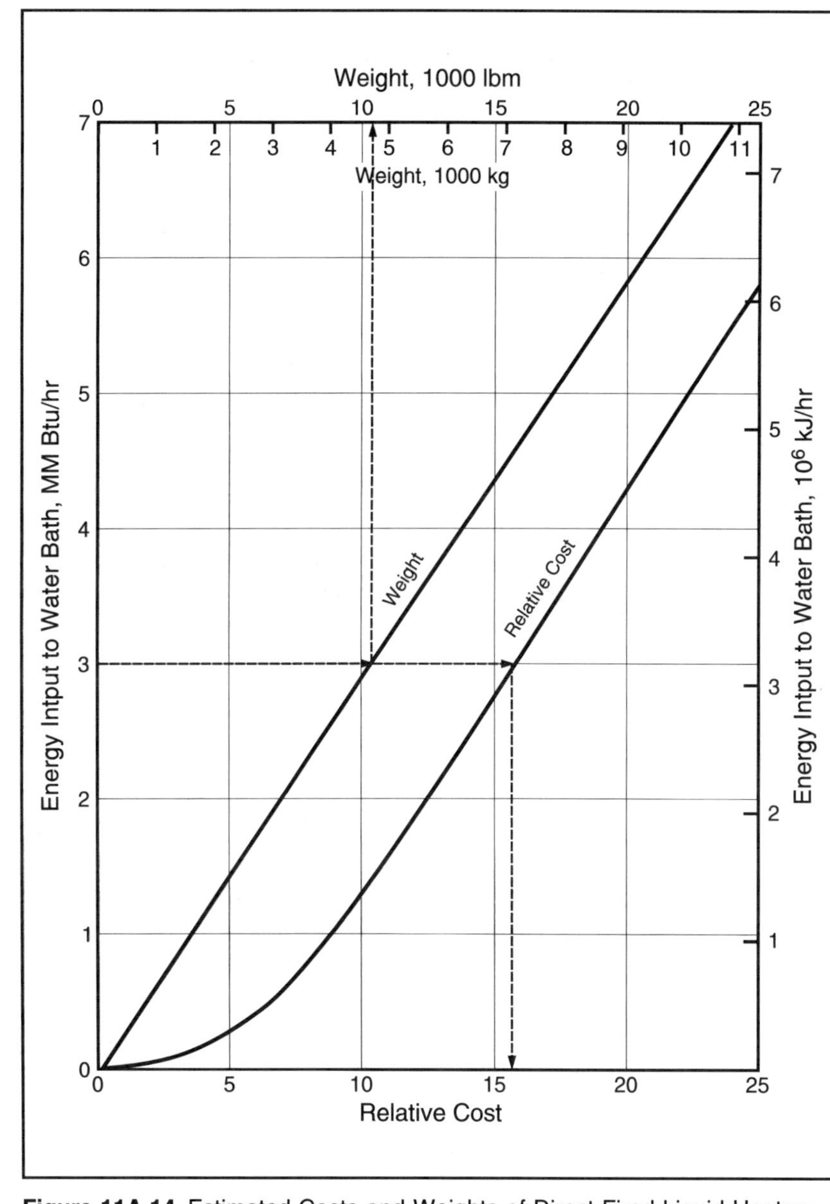

Figure 11A.14 Estimated Costs and Weights of Direct-Fired Liquid Heaters, 25 psi Maximum Operating Pressure

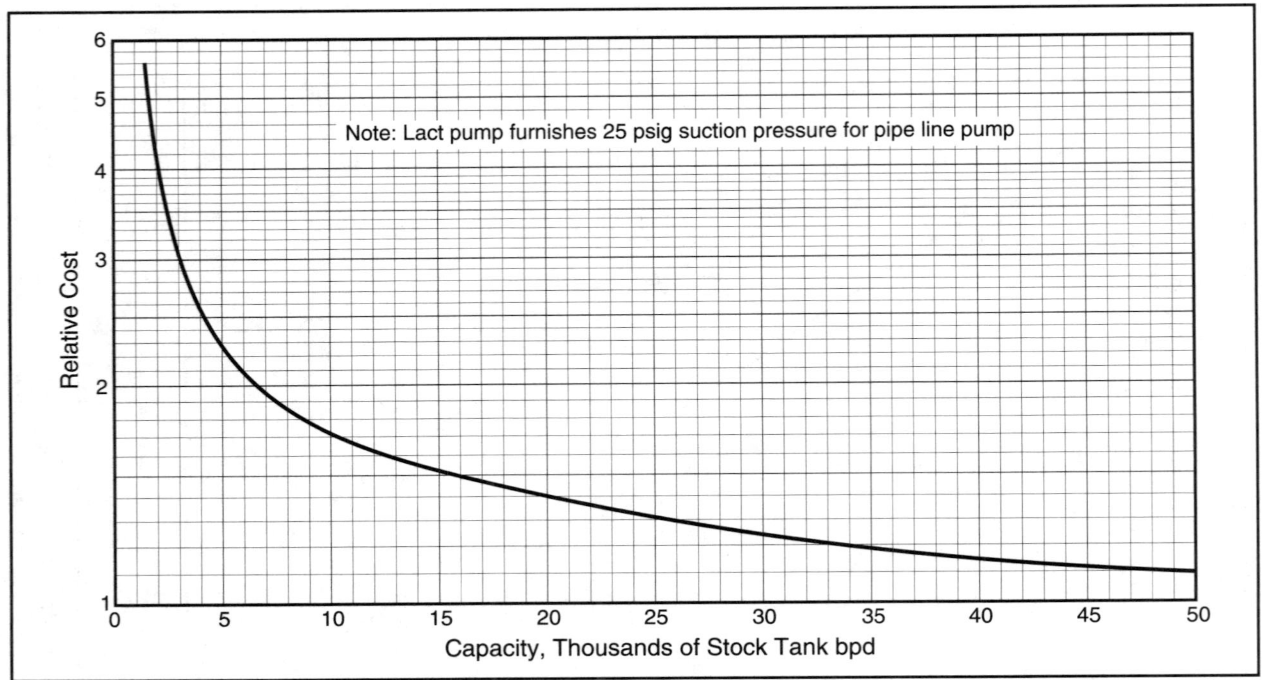

Figure 11A.15 Cost of Normal LACT Units

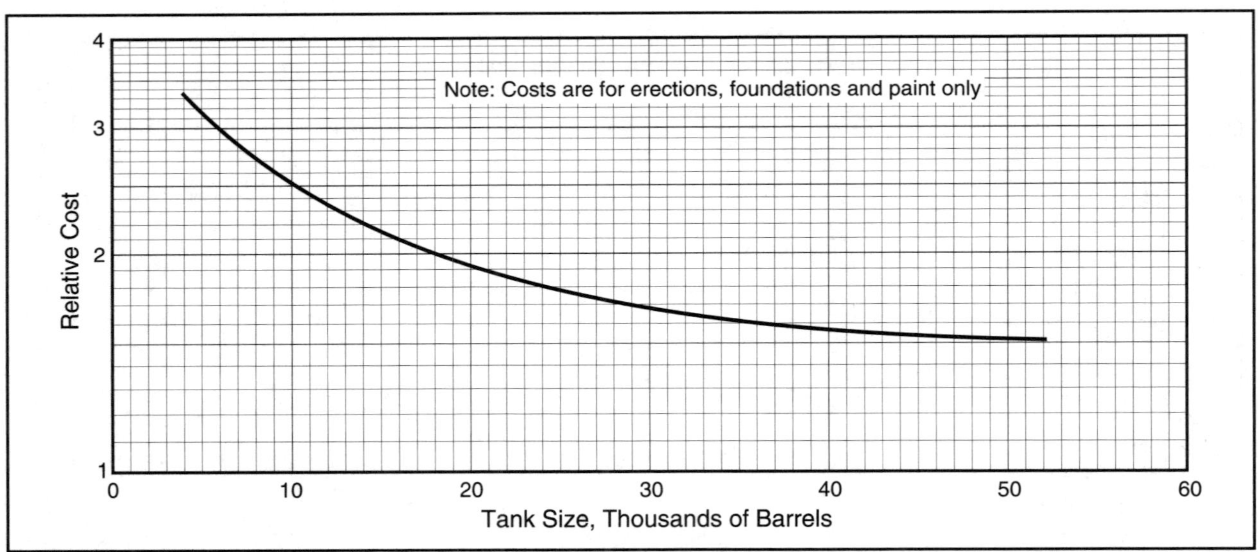

Figure 11A.16 Relative Cost of Oil Storage

12

FUNDAMENTALS OF RATE PROCESSES

Engineers are concerned about rate – the rate of heat transfer, the rate of doing work (power), the rate of flow, the rate of mass transfer, etc. The form of the rate equation used is the same for all engineering disciplines. The basic form is

$$\text{Rate} \propto \frac{\text{Driving Force}}{\text{Resisting Force}} \tag{12.1}$$

The proportionality sign may be replaced by an equal sign if we add a proportionality constant to the right-hand side of the equation.

$$\text{Rate} = \frac{(\text{k})(\text{Driving Force})}{\text{Resisting Force}} \tag{12.2}$$

Constant "k" is the general proportionality constant. For each specific application it is given a name like heat transfer coefficient, mass transfer coefficient, etc.

Driving Force

The table below shows the common driving forces.

Process	Driving Force	Rate
Fluid flow – piping, vessels, porous media	Pressure	mass/unit time
Heat transfer – heating, cooling, mass transfer	Temperature	energy/unit time
Mass transfer – absorbers, fractionators, stabilizers	Concentration	mass/unit time
Electricity	Voltage	amperes

Some mass transfer processes like absorption and fractionation also involve heat transfer. A change of phase is involved. So…a given piece of equipment must be designed so that both the mass and the equivalent energy change involved may be transferred between phases. This is why design demands both a heat and material balance.

Resisting Force

Two variables affecting resistance may be defined easily – length (L) and area (A). Resistance to flow is proportional to the distance of flow. The longer the flow path the greater is the resistance, on a relative basis.

Resistance is inversely proportional to area. The larger the area available for flow, the less the relative resistance will be per unit of flow.

Equation 12.2 may then be written generally as:

$$\text{Rate} = \left(\frac{k}{L}\right)(A)(\text{Driving Force}) \tag{12.3}$$

If no factors which affect resistance other than L or A are included in the working equation, the proportionality constant must contain in its numerical value the effect of all these factors.

In some cases L is not included because it is small and not easily, or accurately, measurable. If L is not shown in the equation it is included in the proportionality constant.

The proportionality constant must always be an empirical number. In a test one can measure rate, driving force, area, and L (if feasible). An equation of the form of 12.3 is then solved for the proportionality constant. The "k" thus found is applicable only to the system being tested. What is done is to make a series of controlled tests using various combinations of the known variables affecting the process. A correlation for "k" is then prepared.

So...k is like z, K, H, S and a large number of like quantities. Each is an empirical number which can be used reliably only if it was obtained on a system comparable to one on which it is applied. No matter how complex the equation, or how many numbers are used to the right of the decimal place, this is still true. This is why no one correlation can be used indiscriminately for all applications.

Some Examples of Linear Flow

Linear flow is flow in only one direction. If we further have only one resisting force, some very simple rate equations can be written which illustrate the rate equation concept.

Electricity

Basic equation: $$I = \frac{E}{R}$$

Where:

I = rate in amperes
E = driving force in volts
R = resistance in ohms

Note that the units have been defined so that "k" = 1.0.

Conduction of Heat

Basic equation: $$Q = \left(\frac{k}{L}\right)(A)(\Delta T)$$

Where:

Q = rate of heat transfer
k = thermal conductivity (proportionality constant)
L = length of flow
A = area through which transfer occurs
ΔT = temperature driving force

Heat Across a Thin Fluid Film

Basic equation: $$Q=(h)(A)(\Delta T)$$

Where: h = film coefficient (proportionality constant)

The value of "h" depends on all factors which affect film thickness as well as any other factors affecting resistance.

Mass Transfer

Basic equation:

$$Q_m=(k_G)(A)(c_1-c_2)$$

or

$$Q_m=(k_{GA})(c_1-c_2)$$

Where:

Q_m = mass rate

k_G or k_{GA} = mass transfer coefficient

c_1 and c_2 = concentration change causing mass transfer

When area is not measurable directly it is incorporated into the coefficient k_{GA}. In some cases area is made a function of volume, as noted for packings in Chapter 8.

Porous Media Fluid Flow

Basic equation: $$Q=\left(\frac{k}{\mu}\right)(A)\left(\frac{\Delta P}{L}\right)$$

Where:

Q = mass flow rate

k = permeability (proportionality constant)

A = area through which flow occurs

μ = viscosity of fluid

ΔP = pressure driving force

L = length of flow section

Notice that viscosity has been added to the resisting forces. Permeability now represents all resistances except A, L and μ.

The above examples show how the rate equations are set up. We have shown them using "delta" notation. They also may be written in differential form and then integrated.

RESISTANCES IN SERIES

Each of the previous equations has been written for the calculation of rate across a single resistance. In a given system or piece of equipment there may be more than one resistance in series. Series flow is where the total flow passes through a sequence of resistances.

Let us illustrate this case using heat transfer between two fluids, across a solid surface.

There are three basic resistances in series – two fluid films and the solid wall separating the fluids. The fluid films are very thin so it is not practical to include "L" as a resistance.

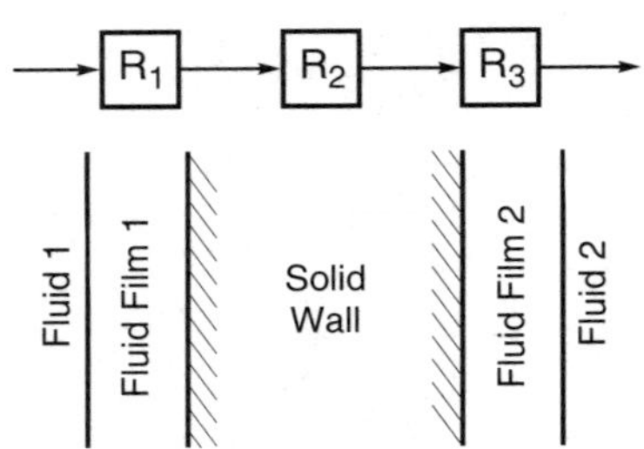

Total $R = R_1 + R_w + R_2$. But since the same amount of heat must flow through each resistance, $Q = Q_1 = Q_w = Q_2$. Or,

$$Q_1 = \frac{\Delta T_1}{R_1} = (h_1)(A_1)(\Delta T_1)$$

$$Q_w = \frac{\Delta T_w}{R_w} = \left(\frac{k_w}{L_w}\right)(A_w)(\Delta T_w)$$

$$Q_2 = \frac{\Delta T_2}{R_2} = (h_2)(A_2)(\Delta T_2)$$

So,

$$R_1 = \left(\frac{1}{h_1}\right)(A_1)$$

$$R_w = \left(\frac{L_w}{K_w}\right)(A_w)$$

$$R_2 = \left(\frac{1}{h_2}\right)(A_2)$$

Now, $\Delta T = \Delta T_1 + \Delta T_w + \Delta T_2$. So, we can combine terms to arrive at a total equation

$$Q = \frac{\Delta T}{R} = \frac{\Delta T}{\frac{1}{h_1 A_1} + \frac{L_w}{k_w A_w} + \frac{1}{h_2 A_2}}$$

This equation will be used in Chapter 13. The equivalent for fluid flow will be used in Chapter 10.

RESISTANCES IN PARALLEL

If the resistances are in parallel the total flow rate is divided among these resistances. The flow through each resistance is proportional to that resistance. The driving force is the same across each parallel resistance.

To illustrate this, assume that the figure below applies to an electrical circuit with three wires in parallel.

For this case,

$$I = I_1 + I_2 + I_3 \quad , \quad E = E_1 + E_3 = E_2 + E_3$$

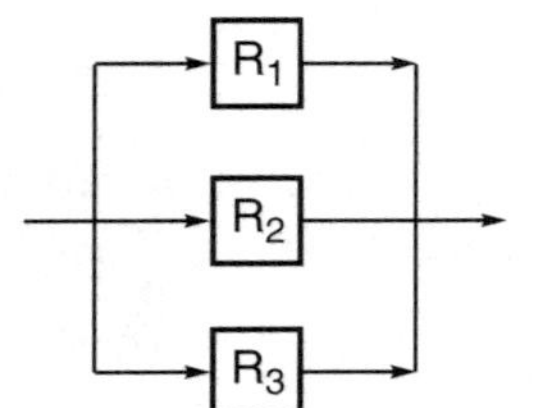

and

$$I_1 = \frac{E}{R_1} \quad , \quad I_2 = \frac{E}{R_2} \quad , \quad I_3 = \frac{E}{R_3}$$

The same basic relationships apply for all flow systems containing resistances in parallel.

COMPOUND RESISTANCES

A given system may have resistances in both series and parallel. Consider the simple electrical system shown below.

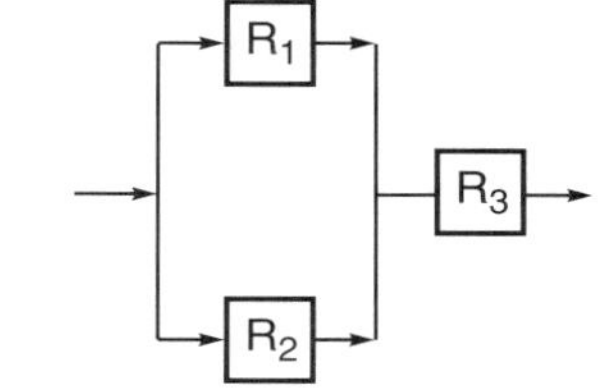

$$I = I_1 + I_2 = I_3 \quad \textit{and} \quad E = E_1 + E_3 = E_2 + E_3$$

$$E_1 = E_2$$

If one can calculate the value of R_1 and R_2, the relative flow rate through each parallel line can be determined.

A similar example is found in the well bore. It is customary to divide a reservoir into a series of layers for calculation purposes. Each layer (1, 2, 3, etc.) has a given permeability and height assigned from logs and cores. The reservoir proper is therefore characterized as a series of parallel resistances.

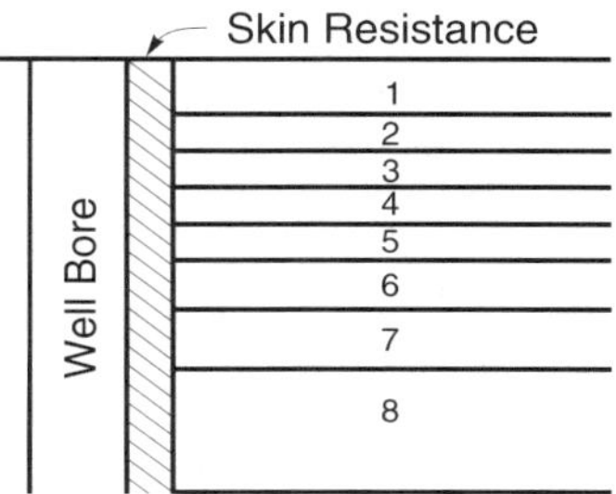

In the process of drilling a reservoir some damage may result around the well bore. This "skin" is a resistance in series.

In Chapter 10, the looping of a line is another example of compound resistances.

GENERAL APPLICATIONS

The rate equations, when corrected for the effect of resistance patterns, serve as a foundation for performance along with thermodynamic concepts, equilibrium, phase behavior and other principles discussed in the first ten chapters. Keep these things in mind as we begin consideration of specific equipment and processes.

In absorption and fractionation, where vapor and liquid phases are involved, the rate of transfer of any component from one phase to another is governed by its relative concentration in the respective phases. If its concentration is less in the vapor phase than in the liquid phase, the component will vaporize at some rate. With reverse relative concentration, condensation will occur.

At low pressures, concentration is often expressed in terms of partial pressures. At higher pressures, outside the ideal gas range, partial fugacities may be used. As noted elsewhere, concentration may be expressed as a mole fraction or any other unit that makes the calculation convenient.

The area of contact between phases is a critical variable that may be controlled to some degree. In adsorption processes the active surface area of the solid is a factor. Since most of this area occurs in the small capillaries that honeycomb the solid, capillary size can also be a factor. Obviously, a molecule larger than the capillary diameter will be unable to even get to the internal surface and effective mass transfer would be negligible. As capillary size increases, the surface area per unit volume of adsorbent decreases. Consequently, in some adsorption processes, the effective area available is limited by the size of molecules being processed.

The rate at which a molecule can move down a capillary large enough to accommodate it will depend on a number of factors such as wet-ability of the fluid, diffusion in the fluid phase, etc. Such variables are necessarily included in the mass transfer coefficient since they are not formally shown in the equation.

In the transfer of mass between vapor and liquid, the area of contact may be increased by breaking up the mass of gas or liquids into droplets – the smaller the droplet, the greater the contact area per unit of mass. A number of mechanical devices have been developed for this purpose. The velocity of the gas flowing upward through the contacting device is used to agitate the liquid into a froth. By restricting the gas flow, it passes through the liquid as smaller bubbles. This combination action supplies the area of contact. The amount of frothing, as controlled by gas velocity, is limited by the fact that entrainment of such liquid into the gas stream reduces the effective contactor efficiency. The combination of gas velocity and liquid droplet size must be such that the liquid will drop back into the mass of liquid from whence it came.

In quiescent contact between vapor and liquid the area of contact is very limited. Furthermore, the concentration of molecules is not uniform throughout each phase. Contacting the gas with the liquid as small bubbles not only increases surface area but also induces turbulence. The latter promotes better diffusion throughout the liquid phase.

Area of contact may also be increased by flowing the liquid along a large surface, in a thin sheet, countercurrent to the gas. The use of columns packed with ceramic or metal objects having a large surface area per unit volume is commonly done to achieve this.

Regardless of the mechanical device used to promote a large area of contact, the design of a mass transfer unit will always involve at least some elements of the following:

1. Assurance that the design is compatible with the conservation of mass and energy as represented by the first law of thermodynamics.
2. Some method for predicting the rate of transfer as a function of driving and resisting forces.
3. Knowledge of the equilibrium relationships applicable.
4. Fixing the allowable rates of fluid through the contacting device.

Design of fractionators and absorbers involves all four factors.

Finally, there is a philosophical reason for discussing the basic rate equation. Engineers tend to "hide" in their chosen areas of expertise and fail to "seek out" knowledge in other areas or perform in them. This is neither self-serving nor fulfilling the employer's needs. Principles are not bounded by geography, language or disciplinary areas. They apply to all systems. A basic understanding of the principles enables one to interact in a meaningful manner with "experts" in other areas – make suggestions, review work and generally communicate, all necessary ingredients in project development. One of my goals has been to remove fear of the unknown in the hope it will encourage you to continue to expand your technological horizons.

13

HEAT TRANSFER

The transfer of heat is necessary for control of: (1) a fluid temperature and/or its composition and phase; (2) the rate of mass transfer between phases; (3) the rate of chemical reactions and (4) suitable temperatures to prevent failure or reduced service life of the equipment. Provision for heat transfer is incorporated into most equipment. However, this chapter is devoted only to that equipment whose primary purpose is transfer of heat.

Heat transfer equipment can be divided into the following basic types.

Fluid-Fluid:	Pipe-in-Pipe, Shell-and-Tube, Plate, Printed Circuit, Spiral, Coil and Special Types for Specific Services
Fired Heaters:	Direct and Indirect
Coolers Utilizing Air:	Fin-Fan® (Aerial), Cooling Towers, Combination Air-Water

Each will be discussed to some degree in this chapter. The emphasis is on fluid-fluid types.

HEAT TRANSFER MECHANISMS

By definition, heat is that energy transferred solely as a result of a temperature difference, that is independent of mass transfer. There are three mechanisms of heat transfer – *conduction*, *convection* and *radiation*.

Conduction of heat occurs by the excitation of adjacent molecules where said molecules have little or no movement. Conduction thus is the primary mechanism in solids and may be an important component mechanism with some liquids at low flowrates.

Convection is that mechanism where heat energy is transferred by the physical movement of molecules from place to place. Any factor which enhances or hinders this movement affects the rate of heat transfer by convection. In most commercial fluid-fluid exchangers, convection is the most important mechanism.

As discussed in Chapter 12 and later herein, the usual heat transfer process is governed by a group of resistances in series. There are two fluid films governed primarily by convection, the solid separating the fluid governed by conduction and possibly some other corrosion, scale or deposition films also governed by conduction.

Radiation is the process whereby a body emits *heat waves* that may be absorbed, reflected or transmitted through a colder body. The sun heats the earth by means of electromagnetic waves.

A hot body emits a whole spectrum of wave lengths. Radiation which affects the eye as light extends roughly from 0.4-0.8 μm in wave length. To the right of this visual spectrum is the *infrared* region; to the left is the *ultraviolet* region. Heat is transferred throughout the full wave length range. As temperature increases the predominant wave lengths become shorter.

BASIC CONDUCTION/CONVECTION EQUATIONS

In most fluid-fluid exchangers the temperatures are not high enough for radiation to be a significant mechanism. Since the coefficients used to calculate performance are empirical, they incorporate any radiant effects that might have been present in the test system.

One may calculate the heat transfer process by the equation:

$$Q = Q_1 = Q_w = Q_2 = (h_1)(A_1)(\Delta T_1) = (k/L)(A_w)(\Delta T_w) = (h_2)(A_2)(\Delta T_2) \qquad (13.1)$$

The values of "h" are proportionality constants used to characterize the liquid film resistance determined from experimental data or general correlations. The value "k" is the thermal conductivity of the solid separating the two fluids – a measurable property of that solid.

It is convenient to show total heat transfer per unit time in terms of an overall heat transfer coefficient "U."

$$Q = U A \Delta T_m \qquad (13.2)$$

Where U is determined based on a reference area, A. For shell and tube exchangers A is the outside tube wall area, for plate exchangers it is the plate area. For finned tubes, A can be based on the extended area (including the area of the fins) or it can be the bare tube area.

In shell and tube exchangers, the overall coefficient "U" is related to the film coefficients and thermal conductivity by the equations:

$$\frac{1}{U_o} = \frac{1}{h_o} + \frac{A_o L}{k A_w} + \frac{A_o}{h_i A_i} + F_i + F_o \qquad (13.3)$$

Where:

U_o = overall heat transfer coefficient based on the outside tube wall area
h_o = film coefficient on the outside of the tube
h_i = film coefficient on the inside of the tube
k = thermal conductivity of solid wall
A_o = outside tube wall area
A_i = inside tube wall area
A_w = average wall area of piping or tubing
L = wall thickness of pipe or tubing
F_o = fouling factor on the outside tube wall
F_i = fouling factor on the inside tube wall

The fouling factor (F) accounts for scale, rust, and other deposits which form on the surface with use and are an additional resistance to heat flow. The fouling factor will vary widely with conditions.

Values of the overall coefficient "U" may be predicted from Equation 13.3 or from actual performance. Most heat exchanger quotations show the overall "U" used in their sizing or rating. These, plus plant operating data, are a valuable source of information for future planning.

EFFECTIVE ΔT

Equation 13.2 is the basic equation used for sizing heat exchangers. It contains the term ΔT_m. This is the mean ΔT because the ΔT across the wall surface varies with heat transferred as shown below.

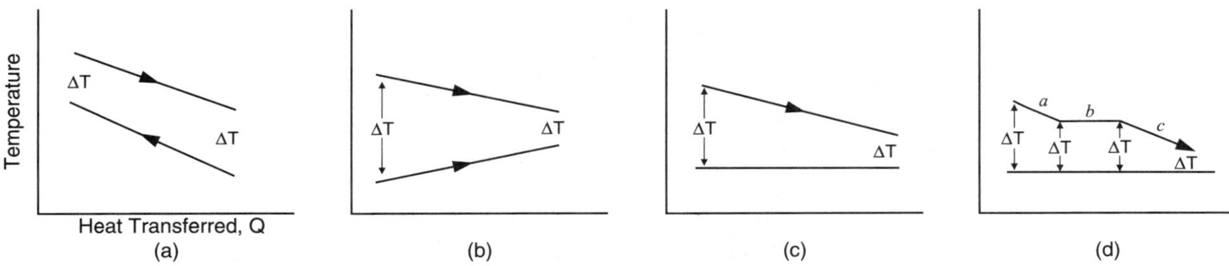

Legend for above figure.
- (a) Two fluids flowing countercurrent, no phase change.
- (b) Two fluids flowing concurrent, no phase change.
- (c) One fluid flowing and one boiling (or condensing).
- (d) Superheated vapor being cooled to saturation (*a*) condensing (*b*) and being subcooled as liquid *(c)*. The other fluid is boiling or condensing.

The only temperatures that we can measure conveniently are at the inlet and outlet ends of the exchanger. Thus, we can measure two ΔTs. The larger we will call ΔT_1, the smaller ΔT_2. ΔT_2 is also called the *approach*. It designates how close the temperatures of the two fluids approach each other in the exchanger.

In concurrent flow the fluids flow in the same direction. In countercurrent flow they flow in opposite directions. Most exchangers use countercurrent flow, or as close to it as possible, since it is more efficient.

The basic equation for estimating ΔT_m is:

$$\Delta T_{m_{corr}} = (F)(\Delta T_m) = (F)\left[\frac{\Delta T_1 - \Delta T_2}{\ln\left(\Delta T_1/\Delta T_2\right)}\right] \tag{13.4}$$

Where:

ΔT_m	=	mean temperature difference
$\Delta T_{m_{corr}}$	=	mean temperature difference corrected for heat exchanger configuration
F	=	ΔT_m correction factor
ΔT_1	=	largest ΔT (at one end of the heat exchanger)
ΔT_2	=	smallest ΔT (at one end of the heat exchanger)
ln	=	logarithm to the base *e*

The value of F depends on the geometry of the fluid flow in the exchanger and will be discussed later for each exchanger type. F = 1.0 for counter-flow exchangers, but will be less than 1.0 for cross flow exchangers and multipass exchangers where both concurrent and counter flow occur.

Equation 13.4 is based on several assumptions, the primary one being that the heating and cooling curves (T vs. Q) for the exchanger fluids are straight lines. This is true only when there is no

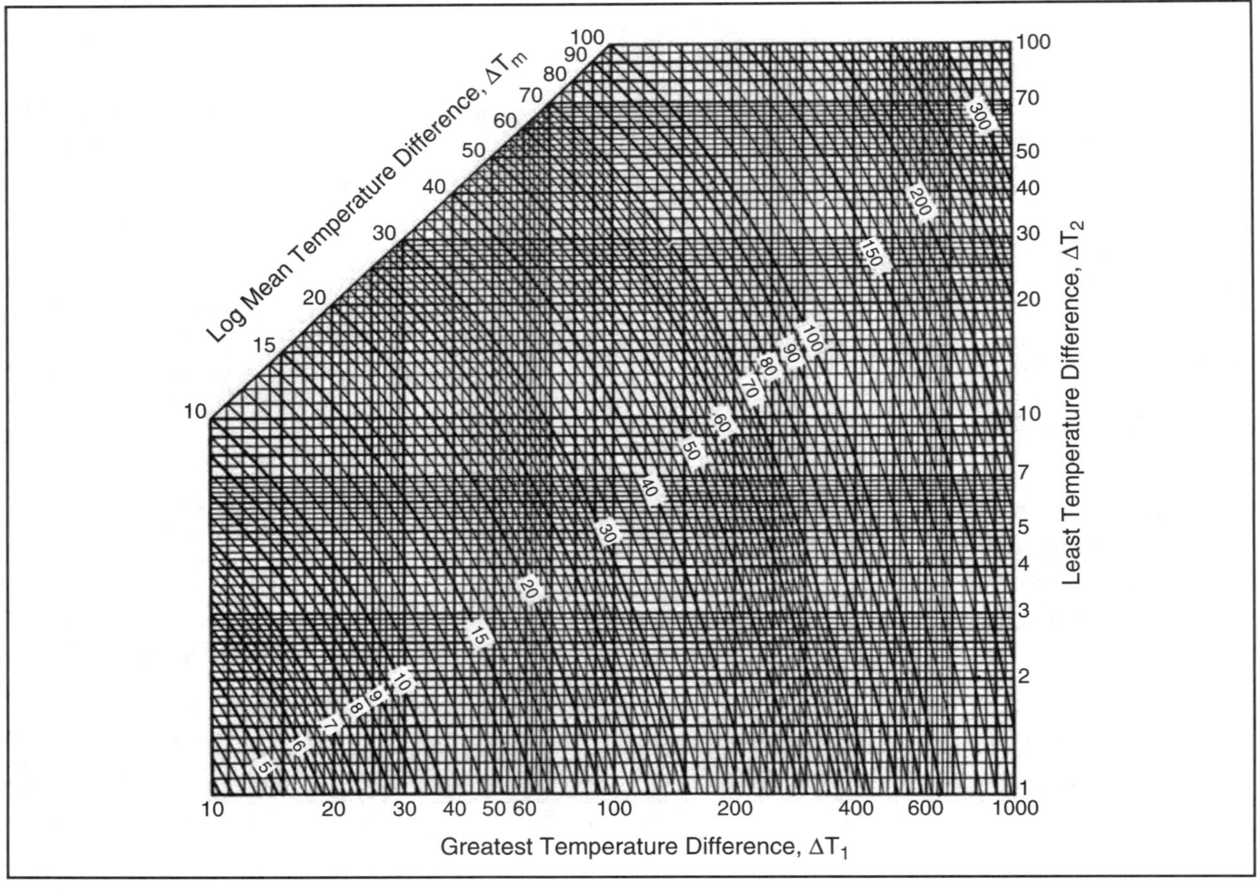

Figure 13.1 Nomograph for ΔT_m

phase change in the exchanger and/or the physical properties of the fluids do not change significantly. ΔT_m may also be determined from Figure 13.1.

Approach

The smaller temperature approach, ΔT_2, is an economic choice. Its specification governs heat exchanger cost. As ΔT_2 gets smaller, ΔT_m becomes smaller and required area becomes larger. As ΔT_m approaches zero, area approaches infinity. Since the cost of the heat exchanger is depends on the area, specification of approach has a direct effect on cost.

For shell and tube exchangers, the approach used will often be in the following range:

Aerial coolers	10-25°C [18-45°F]
Water cooling of hydrocarbon liquids and gases	8-12°C [14-22°F]
Liquid-liquid heat exchange	11-25°C [20-45°F]
Refrigeration chillers on gas-liquid streams	4-6°C [7-11°F]

For compact exchangers closer approaches are usually economically justifiable.

When specifying heat exchangers, it often is desirable to specify a maximum or minimum approach to the vendor. This does not fix the actual approach. It merely establishes an upper or lower limit, below or above which the actual approach must occur.

Vaporizing (Boiling) Liquids

There is a special concern when one of the heat exchanger fluids is vaporizing. This occurs in refrigeration chillers and fractionator reboilers, as two examples.

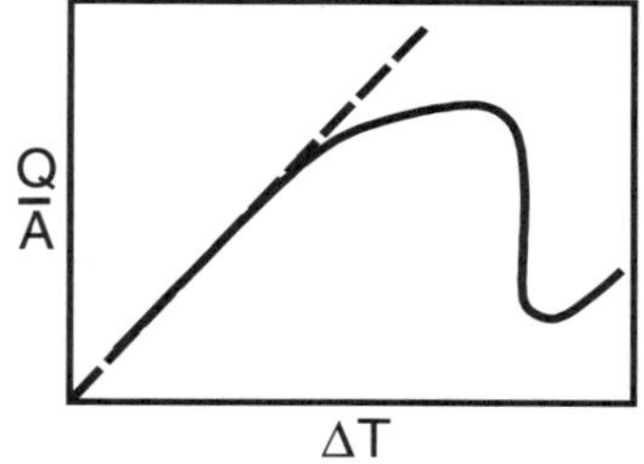

From Equation 13.2, you would expect a plot of Q/A versus ΔT to yield a line like the dashed line at right. It does except for boiling liquids. The solid curve at right is what really occurs with boiling liquids. At some value of ΔT the curve changes direction and Q/A decreases rapidly to a minimum, after which it begins to rise again. Why?

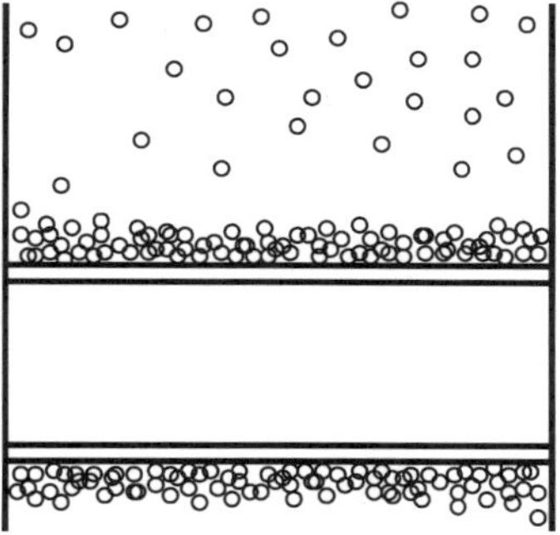

As shown in the sketch at right, a layer of gas bubbles can build up around a tube if vaporization occurs at the tube wall faster than the vapor can disengage and rise through the liquid. This layer of bubbles forms an extra resistance in series and is a type of fouling factor.

When ΔT across the tube reaches a critical point, the bubble layer forms and Q/A decreases. If ΔT continues to increase, the layer resistance stabilizes and Q/A begins to increase again.

The critical ΔT depends on the liquid and the character of the tube surface. The critical ΔT may occur as low as 20-35°C [36-63°F]. Special tube surfaces are marketed which are designed to minimize bubble layer formation.

There are two basic mechanical factors which affect vapor disengagement – spacing and arrangement of the exchanger, and the area available between the liquid and vapor phases. As vapor forms it must leave the surface quickly. There also must be enough surface area so that the vapor-liquid interface does not limit vapor disengagement.

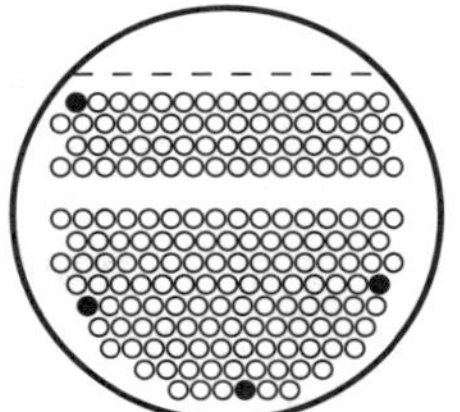

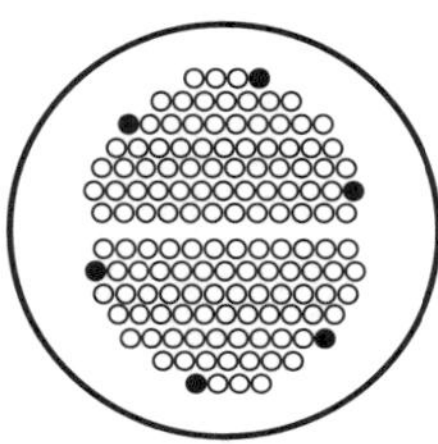

In the figure at right are shown two tube configurations among the many available. This is known as *triangular layout* since the tubes in adjacent rows are not directly above or below each other. To improve vapor disengagement between tubes, the tube pitch is sometimes increased to 1.5 to 2 times the tube diameter when shell and tube exchangers are used in boiling service.

Another alternative is the *square layout* where tubes in adjacent rows are directly above or below each other. Although not as common as triangular layout, square layout has been used in corrosive service such as amine regeneration.

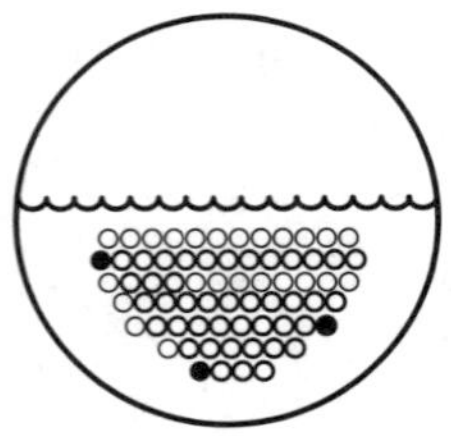

Notice in the figure at left that there is room above the tubes for vapor. The arrangement shown is typical for chillers and reboilers where the liquid covers all tubes. The area of the liquid surface must be sufficient to allow the vapor to escape from the liquid without significant entrainment. The top half of this exchanger is a separator. This means that the shell diameter must be larger than that needed to merely hold the tubes.

As the tubes are farther apart there is more room for vapor to rise. But, the cost of the exchanger increases. Be sure the low bid on your reboiler or chiller has sufficient vapor space.

Also, be sure the vapor outlet nozzles and piping have sufficient area. If not, vapor can back up, "choke" the exchanger, and limit capacity even though the tube area is adequate.

BASIC HEAT BALANCE

Heat Units

The rate of heat transfer (Q) normally is expressed in kW or Btu/hr.

$$1\ \text{Btu/hr} = 0.293\ \text{W}$$
$$1\ \text{kW} = 3413\ \text{Btu/hr}$$

Overall and film coefficients (U and h) are proportionality coefficients whose units depend on the rest of the equation. The most common units are shown below.

$$1\ \text{Btu}/(\text{hr-ft}^2\text{-°F}) = 5.678\ \text{W}/(\text{m}^2\cdot\text{K})$$

Thermal conductivity (k) is also a proportionality coefficient but is expressed per unit length of the conduction flow path.

$$1(\text{Btu-ft})/(\text{hr-ft}^2\text{-°F}) = 1.73\ \text{W}/(\text{m}\cdot\text{K})$$
$$1\ (\text{Btu-in})/(\text{hr-ft}^2\text{-°F}) = 0.144\ \text{W}/(\text{m}\cdot\text{K})$$

It should be noted that as used here, Kelvin (K) refers to a temperature difference not an absolute temperature, so °C and K have the same value.

In a fluid-fluid heat exchanger the overall balance reduces to $\Delta H = Q$. Furthermore, the ΔH of one fluid equals the ΔH of the other fluid if one ignores heat losses to, or heat gains from, the atmospheric air. Heat loss or gain is normally considered to be zero in exchanger heat balances. If the vessel is properly insulated the heat loss or gain should be less than 5% of Q. This is compensated for when choosing the actual equipment.

Thus, for any two-fluid heat exchanger, three equations can be written.

1) $Q = m_{12}\Delta h_{12} = m_{12}C_{p_{12}}(T_2 - T_1)$ *
2) $Q = m_{34}\Delta h_{34} = m_{34}C_{p_{34}}(T_4 - T_3)$ *
3) $Q = U\,A\,\Delta T_m$

* C_p used when no phase change occurs

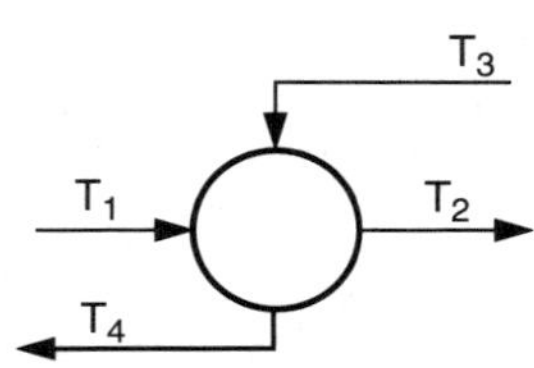

Example 13.1: 6500 m^3/d [40 900 bbl/day] of a 28°API [0.89 rel. ρ] crude oil is cooled from 84°C [183°F] to 62°C [143°F] in a counterflow shell and tube exchanger with 254 000 kg/h [560 000 lb/hr] of a 60% MEG-40% H_2O cooling medium. The cooling medium enters the exchanger at 38°C [100°F]. The heat capacity of the oil is 2.11 kJ/kg·°C [0.505 Btu/lbm-°F] and the heat capacity of the cooling medium is 3.19 kJ/kg·°C [0.762 Btu/lbm-°F]. The overall heat transfer coefficient is 330 W/m^2·°C [58.2 Btu/hr-ft^2-°F].

Calculate the exchanger duty, the temperature of the cooling medium leaving the exchanger, and the exchanger surface area.

SI Solution:

1) Calculate exchanger duty, Q

$$m = \left(\frac{6500\ \text{m}^3}{\text{d}}\right)\left(\frac{890\ \text{kg}}{\text{m}^3}\right)\left(\frac{1\ \text{d}}{86\ 400\ \text{s}}\right) = 67.0\ \text{kg/s}$$

$$Q = mC_p\Delta T = \left(\frac{67.0\ \text{kg}}{\text{s}}\right)\left(\frac{2.11\ \text{kJ}}{\text{kg·°C}}\right)(84 - 62°\text{C}) = 3110\ \text{kW}$$

2) Calculate temperature of cooling medium out

$$Q = mC_p\Delta T\ \text{(cooling medium)} \quad m = \left(\frac{254\ 000\ \text{kg}}{\text{h}}\right)\left(\frac{1\ \text{h}}{3600\ \text{s}}\right) = 70.6\ \text{kg/s}$$

$$\Delta T = \frac{Q}{mC_p} = \frac{3110\ \text{kW}}{\left(\frac{70.6\ \text{kg}}{\text{s}}\right)\left(\frac{3.19\ \text{kJ}}{\text{kg·°C}}\right)} = 13.8°\text{C}$$

$$T_{out} = 51.8°\text{C}\ \text{(use 52°C)}$$

3) Calculate Area, A

$$\Delta T_m = \frac{32 - 24}{\ln\left(\frac{32}{24}\right)} = 27.8°\text{C}\ \text{(use 28°C)}$$

84°C
ΔT_1 = 32°C
62°C
52°C
ΔT_2 = 24°C
38°C

$$A = \frac{Q}{U\Delta T_m} = \frac{3\ 110\ 000\ \text{W}}{\left(\frac{330\ \text{W}}{\text{m}^2·°\text{C}}\right)(28°\text{C})} = \underline{\underline{337\ \text{m}^2}}$$

FPS Solution:

1) Calculate exchanger duty, Q

$$m = (40\ 900)(14.6)(0.89) = 531\ 000\ \text{lbm/hr}$$

$$Q = mC_p\Delta T = \left(\frac{531\ 000\ \text{lbm}}{\text{hr}}\right)\left(\frac{0.505\ \text{Btu}}{\text{lbm-°F}}\right)(183 - 143°\text{F}) = 10.7 \times 10^6\ \text{Btu/hr}$$

2) Calculate temperature of cooling medium out

$$Q = mC_p\Delta T\ \text{(cooling medium)}$$

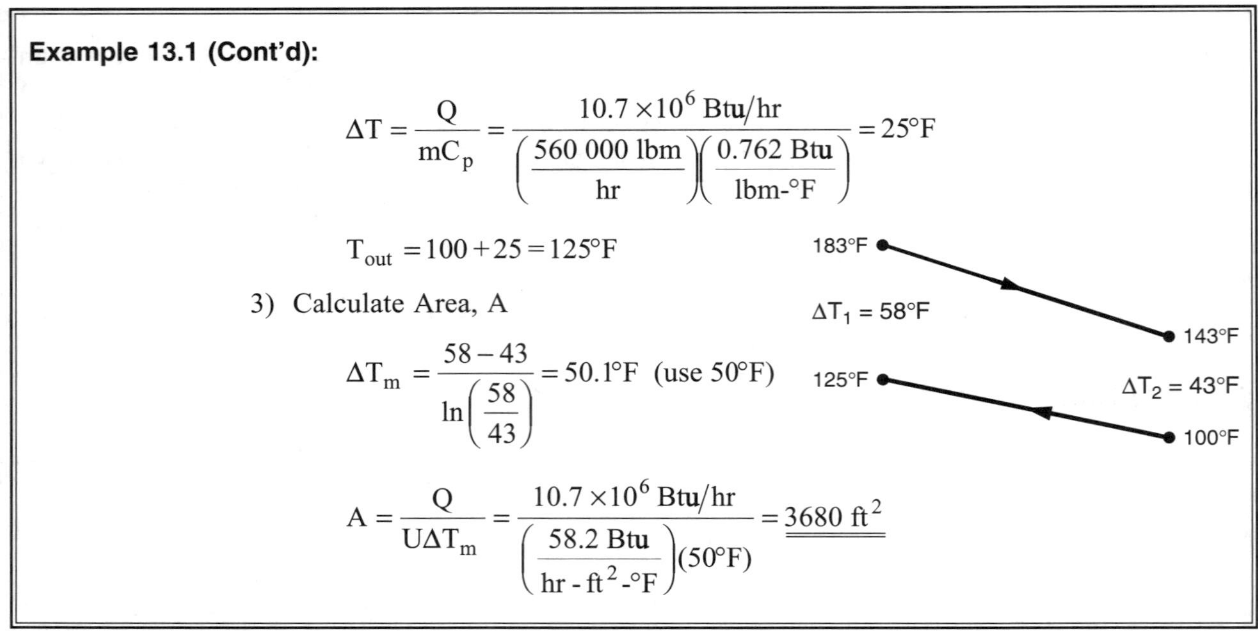

Example 13.1 (Cont'd):

$$\Delta T = \frac{Q}{mC_p} = \frac{10.7 \times 10^6 \text{ Btu/hr}}{\left(\frac{560\ 000 \text{ lbm}}{\text{hr}}\right)\left(\frac{0.762 \text{ Btu}}{\text{lbm-°F}}\right)} = 25°\text{F}$$

$$T_{out} = 100 + 25 = 125°\text{F}$$

3) Calculate Area, A

$$\Delta T_m = \frac{58 - 43}{\ln\left(\frac{58}{43}\right)} = 50.1°\text{F (use 50°F)}$$

$$A = \frac{Q}{U\Delta T_m} = \frac{10.7 \times 10^6 \text{ Btu/hr}}{\left(\frac{58.2 \text{ Btu}}{\text{hr - ft}^2\text{-°F}}\right)(50°\text{F})} = \underline{\underline{3680 \text{ ft}^2}}$$

In design calculations, both mass flowrates will be known as well as three temperature, say T_1, T_2, and T_3. The unknowns are Q, A, and the remaining temperature, T_4. The above equations can be used as follows.

- Use Equation 1 to calculate Q
- Use Equation 2 to calculate T_4
- Use Equation 3 to calculate A

For performance calculations, the exchanger is existing and the objective is to calculate the exchanger performance for a set of conditions. In this calculation, A is known as well as m_{12}, m_{34}, T_1, and T_3. The unknowns are T_2, T_4, and Q. This is often a trial and error solution due to the calculation of ΔT_m.

Step 1: Estimate T_2, and use Equation 1 to calculate Q

Step 2: Use Equation 2 to calculate T_4

Step 3: Use Equation 3 to calculate Q and compare it to the Q calculated in step 1. If not equal, repeat steps 1-3.

Performance calculations are most often done on a computer, but TEMA has developed a set of graphs which can be used to facilitate a performance calculation. Figure 13.2 shows the graph for a counterflow exchanger. The nomenclature has been adapted to match that of the discussion in this section. This graph is only valid for counterflow exchangers with no phase change. In those exchangers where a phase change occurs, the graph can still be used but the heat capacity, C_p, should include the latent heat. In other words, C_p is a pseudo C_p which includes the latent heat effect.

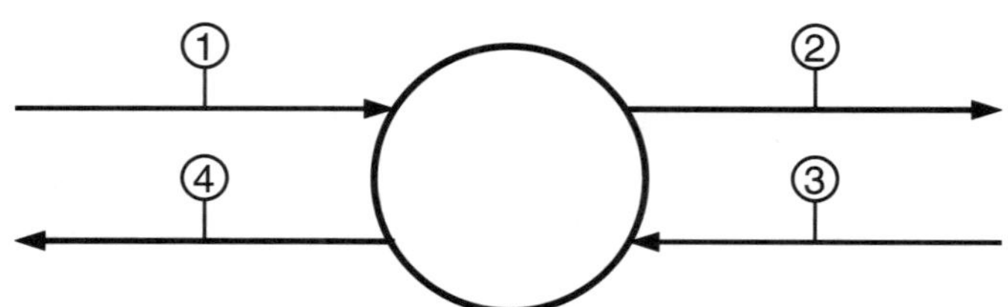

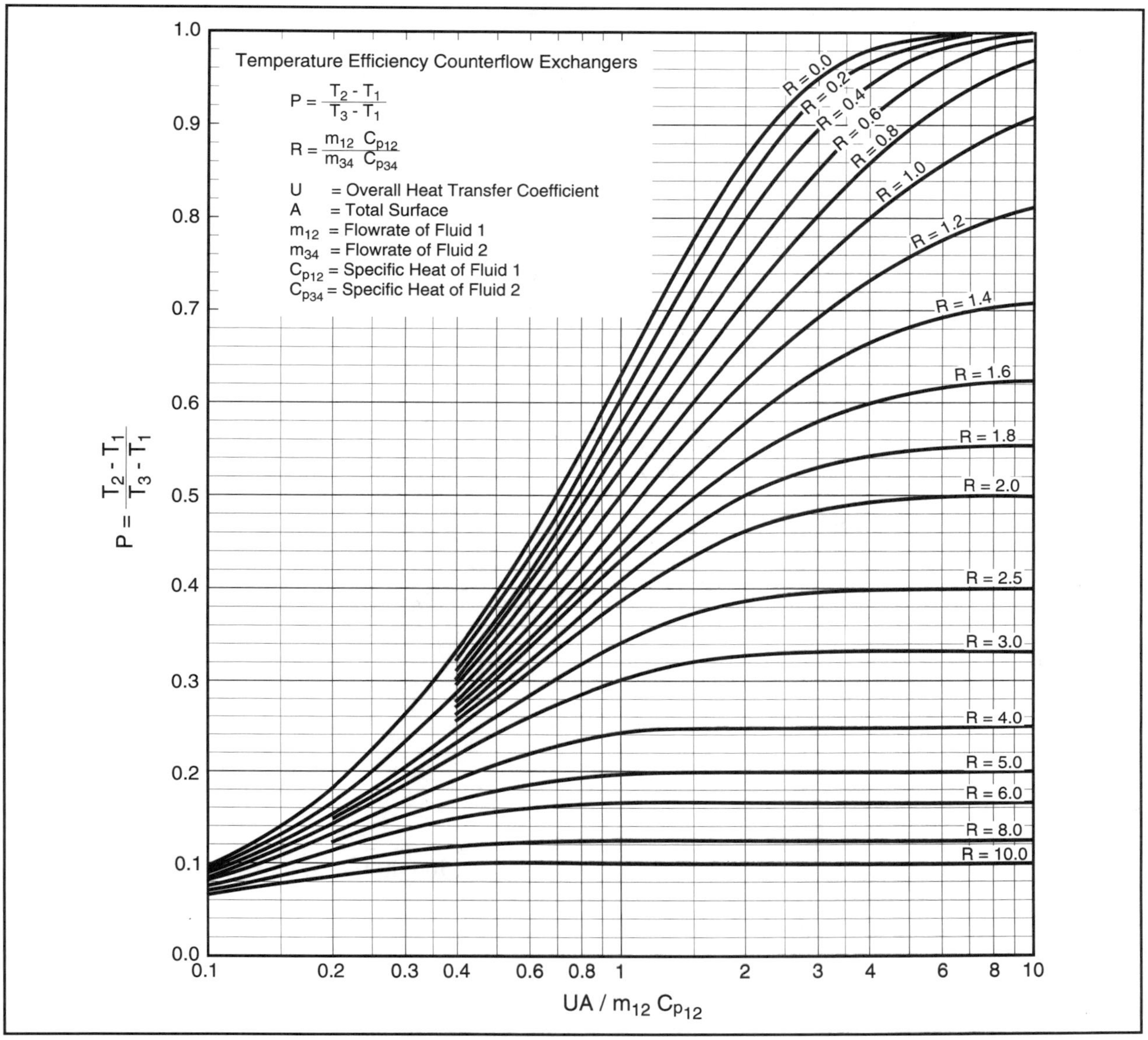

Figure 13.2 Temperature Efficiency for Counterflow Exchangers *(Reproduced with permission from TEMA)*

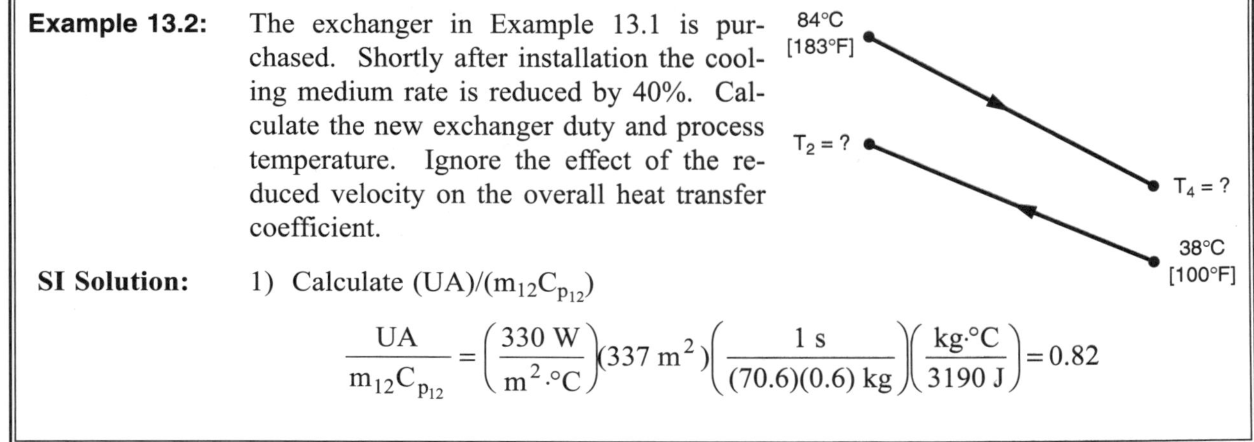

Example 13.2: The exchanger in Example 13.1 is purchased. Shortly after installation the cooling medium rate is reduced by 40%. Calculate the new exchanger duty and process temperature. Ignore the effect of the reduced velocity on the overall heat transfer coefficient.

SI Solution: 1) Calculate $(UA)/(m_{12}C_{p_{12}})$

$$\frac{UA}{m_{12}C_{p_{12}}} = \left(\frac{330\ \text{W}}{\text{m}^2\cdot{}^\circ\text{C}}\right)(337\ \text{m}^2)\left(\frac{1\ \text{s}}{(70.6)(0.6)\ \text{kg}}\right)\left(\frac{\text{kg}\cdot{}^\circ\text{C}}{3190\ \text{J}}\right) = 0.82$$

Example 13.2 (Cont'd):

2) Calculate R

$$R = \frac{(70.6)(0.6)(3.19)}{(67)(2.11)} = 0.96$$

From Figure 13.2, P ≈ 0.45

$$P = 0.45 = \frac{T_2 - T_1}{T_3 - T_1} = \frac{T_2 - 38}{84 - 38}\text{ , solve for } T_2 \qquad T_2 = 59°C$$

Calculate Q

$$Q = mC_p\Delta T = \left(\frac{(70.6)(0.6)\text{ kg}}{\text{s}}\right)\left(\frac{3.19\text{ kJ}}{\text{kg·°C}}\right)(59 - 38°C) = 2840\text{ kW}$$

Calculate T_4

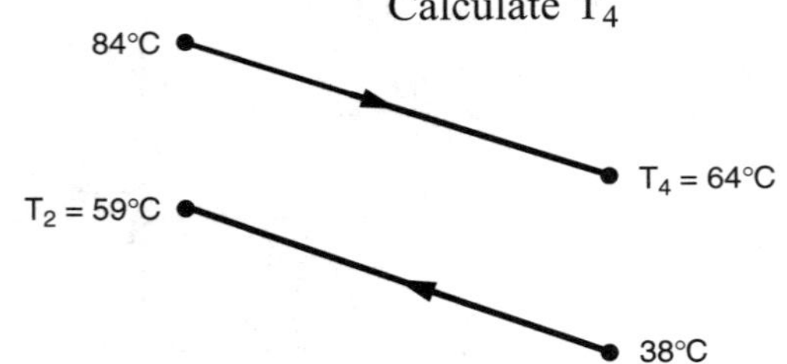

$$\Delta T = \frac{Q}{mC_p} = \frac{2840\text{ kW}}{\left(\frac{67\text{ kg}}{\text{s}}\right)\left(\frac{2.11\text{ kJ}}{\text{kg·°C}}\right)} = 20°C$$

$$T_4 = 84 - 20 = 64°C$$

FPS Solution: 1) Calculate $(UA)/(m_{12}C_{p_{12}})$

$$\frac{UA}{m_{12}C_{p_{12}}} = \left(\frac{58.2\text{ Btu}}{\text{hr-ft-°F}}\right)(3680\text{ ft}^2)\left(\frac{1\text{ hr}}{(560\ 000)(0.6)\text{ lbm}}\right)\left(\frac{\text{lbm-°F}}{0.762\text{ Btu}}\right) = 0.84$$

2) Calculate R

$$R = \frac{(560\ 000)(0.6)(0.762)}{(531\ 000)(0.505)} = 0.95$$

From Figure 13.2, P ≈ 0.46

$$P = 0.46 = \frac{T_2 - T_1}{T_3 - T_1} = \frac{T_2 - 100}{183 - 100}\text{ , solve for } T_2 \qquad T_2 = 138°F$$

Calculate Q

$$Q = mC_p\Delta T = \left(\frac{(560\ 000)(0.6)\text{ lbm}}{\text{hr}}\right)\left(\frac{0.762\text{ Btu}}{\text{lbm-°F}}\right)(138 - 100°F)$$

$$= 9.73 \times 10^6\text{ Btu/hr}$$

Example 13.2 (Cont'd):

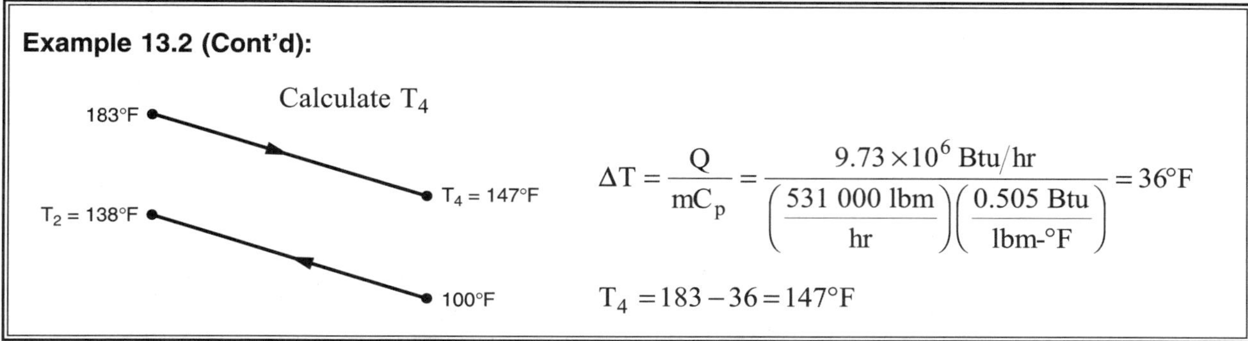

Calculate T_4

$$\Delta T = \frac{Q}{mC_p} = \frac{9.73 \times 10^6 \text{ Btu/hr}}{\left(\frac{531\ 000 \text{ lbm}}{\text{hr}}\right)\left(\frac{0.505 \text{ Btu}}{\text{lbm-°F}}\right)} = 36°\text{F}$$

$$T_4 = 183 - 36 = 147°\text{F}$$

OVERALL HEAT TRANSFER COEFFICIENTS

The best source of "U" values is from actual tests on comparable equipment. Continued tests over a period of time build up a data base for various conditions of operation, fluids and types of exchangers. A second good source of "U" values is data from reliable vendors.

Figure 13.3 is a general summary of "U" values for shell-and-tube type exchangers. The upper nomograph is for exchangers with only nominal fouling that operate at acceptable fluid velocity levels.[(13.1)] The lower table is another compilation of such data. Correlations like these are not recommended for design calculations but are useful for general estimation of behavior and/or the evaluation of alternative systems.

Calculation of "U" from Equation 13.3 may be necessary. If so, values of "k" and "h" are required.

Example 13.3: High pressure natural gas is to be cooled with cooling water in a compressor after-cooler. The outside tube diameter is 19 mm [0.75 in] and the inside diameter is 14.8 mm [0.584 in]. The estimated film coefficients are 454 W/m²·K [80 Btu/hr-ft²-°F] for the gas and 1703 W/m²·K [300 Btu/hr-ft²-°F] for the cooling water. The thermal conductivity of the tube wall is 52 W/m·K [30 Btu/hr-ft-°F]. The fouling factor on both the gas and water side is estimated to be 0.000 18 m²·K/W [0.001 ft²-hr/Btu]. Estimate the overall heat transfer coefficient. Assume natural gas is on the tube side.

A_o = 0.06 m²/m [0.196 ft²/ft] $\quad$ A_i = 0.0465 m²/m [0.153 ft²/ft]

A_w = 0.0529 m²/m [0.174 ft²/ft] $\quad$ L = 0.0021 m [0.00689 ft]

$$\frac{1}{U} = \frac{1}{h_o} + \frac{A_o L}{A_w k} + \frac{A_o}{A_i h_i} + F_i + F_o$$

$$= \frac{1}{1703} + \frac{(0.06)(0.0021)}{(0.0529)(52)} + \frac{0.06}{(0.0465)(454)} + 0.00018 + 0.00018$$

$$= 0.003\,835 \qquad U = 261\,\text{W/m}^2\text{·K}$$

Similarly, for FPS units U = 45.9 Btu/hr-ft²-°F

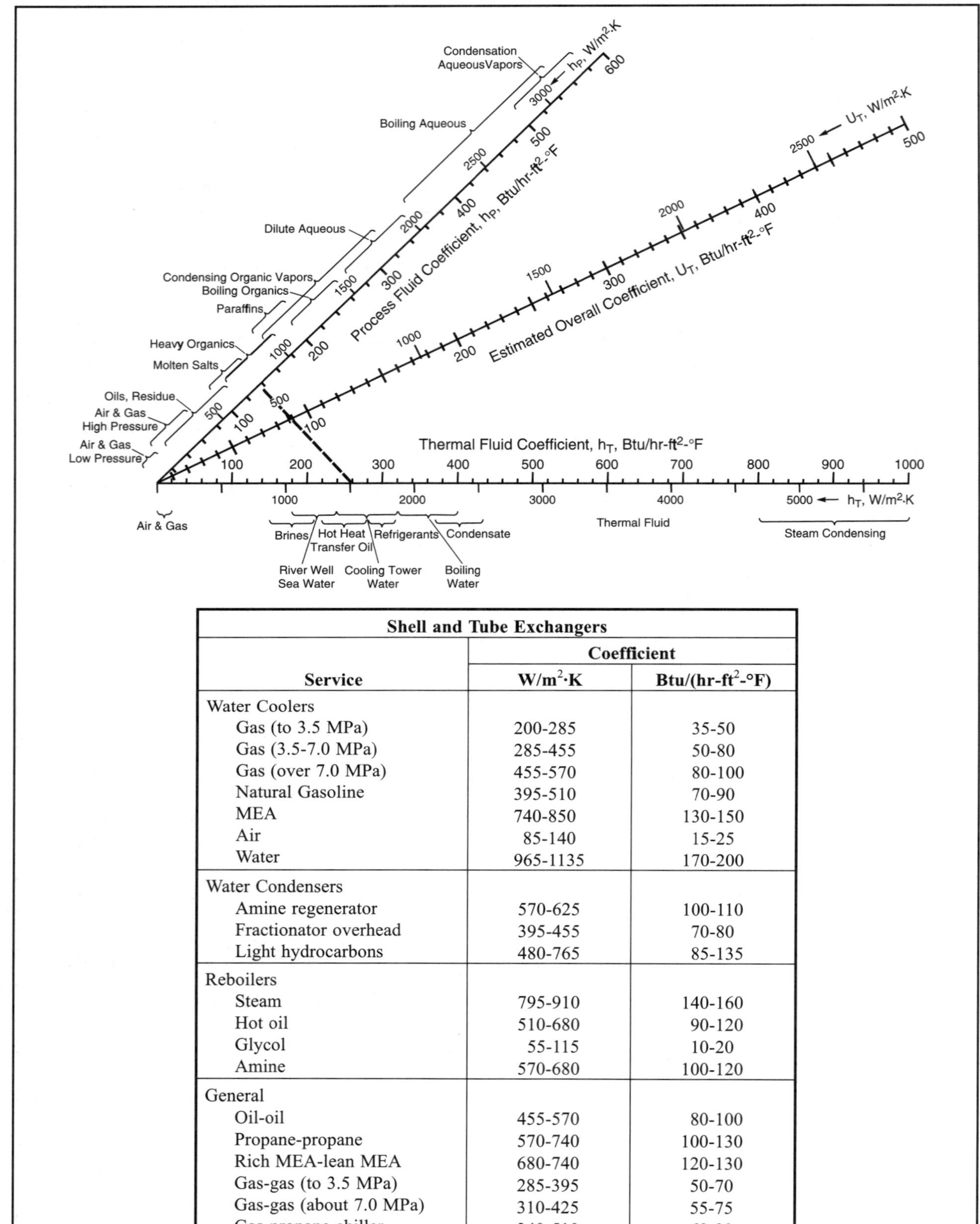

Shell and Tube Exchangers		
	Coefficient	
Service	$W/m^2{\cdot}K$	$Btu/(hr\text{-}ft^2\text{-}°F)$
Water Coolers		
Gas (to 3.5 MPa)	200-285	35-50
Gas (3.5-7.0 MPa)	285-455	50-80
Gas (over 7.0 MPa)	455-570	80-100
Natural Gasoline	395-510	70-90
MEA	740-850	130-150
Air	85-140	15-25
Water	965-1135	170-200
Water Condensers		
Amine regenerator	570-625	100-110
Fractionator overhead	395-455	70-80
Light hydrocarbons	480-765	85-135
Reboilers		
Steam	795-910	140-160
Hot oil	510-680	90-120
Glycol	55-115	10-20
Amine	570-680	100-120
General		
Oil-oil	455-570	80-100
Propane-propane	570-740	100-130
Rich MEA-lean MEA	680-740	120-130
Gas-gas (to 3.5 MPa)	285-395	50-70
Gas-gas (about 7.0 MPa)	310-425	55-75
Gas-propane chiller	340-510	60-90

Figure 13.3 Typical Values of Overall Heat Transfer Coefficients

Thermal Conductivity of Solids

The table following shows the thermal conductivity of common materials at about 93°C [200°F].

Material	Thermal Conductivity - k	
	W/(m·K)	Btu/(hr-ft-°F)
Copper	386	223
Aluminum	173	100
Admiralty Brass	121	70
Mild Steel	43	25
Silicon Bronze	26	15
Stainless Steel (18 Cr-8 Ni)	14	8
Inconel	14	8
90-10 Cu-Ni	52	30
70-30 Cu-Ni	31	18
Monel	26	15
Titanium	17	10

The choice of metal depends on thermal conductivity, cost, temperature of operation and the presence of corrosion/erosion problems. With sea water cooling, for example, the use of titanium may be indicated to minimize chloride cracking, even though it possesses low thermal conductivity and is more expensive to buy and fabricate.

Fluid Film Coefficients

The basic equation for calculating fluid films is obtained by dimensionless analysis.[13.2] The correlation of data shows a proportionality relationship among the Nusselt, Reynolds and Prandtl dimensionless numbers. In a solvable form this relationship can be written as follows.

$$h = A\left(\frac{k}{d}\right)\left(\frac{d\,v\,\rho}{\mu}\right)^{a}\left(\frac{C_p\,\mu}{k}\right)^{b} \tag{13.5}$$

Where:		SI	FPS
h	= film coefficient	W/(m²·K)	Btu/(hr-ft²-°F)
A	= proportionality constant		
k	= fluid thermal conductivity	W/(m·K)	Btu/(hr-ft-°F)
d	= effective diameter	m	ft
v	= fluid velocity	m/s	ft/hr
ρ	= fluid density	kg/m³	lbm/ft³
C_p	= fluid specific heat	kJ/(kg·K)	Btu/(lbm-°F)
μ	= fluid viscosity	kg/(m·s)	lbm/(ft-hr)
a	= coefficient on Reynolds No.		
b	= coefficient on Prandtl No.		

In Equation 13.5 the effective diameter depends on the circumstances:

- Tube inside diameter for fluid flowing inside tube
- Tube outside diameter for fluid flowing normal to one or more tubes
- For annular flow between concentric pipes use an equivalent diameter defined by the equation

$$d_e = \frac{d_2^2 - d_1^2}{d_1} \tag{13.6}$$

Where:

d_e = equivalent diameter
d_1 = inside diameter of outside pipe
d_2 = outside diameter of inside pipe

The velocity term is the volumetric flowrate divided by the cross-sectional area of flow. When flow is normal to a series of tubes, the cross-sectional area is the area available for flow between the first row of tubes.

Equation 13.5 applies only for Newtonian fluids where viscosity is a proper measure of the rate of shear versus shear stress (See Chapter 10). Gases and ordinary hydrocarbon liquids are Newtonian in nature. Correlations for estimating the viscosity and density of fluids are shown in Chapter 3.

Table 13.1 summarizes the values of A, a and b in Equation 13.5 that apply for the *turbulent flow* of fluids in and around cylindrical tubing.(13.3) The values shown are based on data from heat exchangers in general chemical service. However, they can prove useful for estimating the performance of a specific heat exchanger. One very important use of Equation 13.5 is prediction of the change in performance of a heat exchanger with change in fluid or flow conditions.

The following conversion factors are useful in solving Equation 13.5.

1 centipoise (cp) = 2.419 lbm/(ft-hr) = 0.001 kg/m·s
1 m = 3.281 ft, 1 ft = 0.305 m
1 kg/m^3 = 0.0624 lbm/ft^3, 1 lbm/ft^3 = 16.02 kg/m^3
1 Btu/(lbm-°F) = 4.19 kJ/(kg·K)

Fluid Properties

Prediction of the density and viscosity terms is possible from correlations like those shown in Chapter 3. Many references contain thermal conductivity data for gases and liquids.(13.4, 13.5) Shown following are some representative thermal conductivities for gases and liquids at atmospheric pressure and about 38°C [100°F].

The "k" of liquids does not vary much with pressure. There is a change with temperature, but it is not large over the usual temperature range of fluid-fluid exchangers. An equation for estimating the "k" of crude oil and condensates is:

$$\textbf{SI:} \quad k = 0.131 - (0.000142)(°C)$$
$$\textbf{FPS:} \quad k = 0.0773 - (0.0000456)(°F) \tag{13.7}$$

Where:

k = W/(m·K) or Btu/(hr-ft-°F)

TABLE 13.1

Coefficients for Equation 13.5 for Turbulent Flow in Shell-and-Tube Type Heat Exchangers (Fluid-Fluid)

Service	Coefficient A	Exponent a	Exponent b
Inside Vertical Tubes			
Cooling gas	0.0265	0.8	0.3
Cooling liquid (high visc.)	0.027	0.8	0.33
Heating gas	0.0243	0.8	0.4
Heating liquid (low visc.)	0.023	0.8	0.4
Heating liquid (high visc.)	0.027	0.8	0.33
Boiling liquid	0.029	0.8	0.4
Inside Horizontal Tubes			
Cooling gas	0.023	0.8	0.3
Cooling liquid (low visc.)	0.0265	0.8	0.3
Cooling liquid (high visc.)	0.027	0.8	0.33
Heating gas	0.0225	0.8	0.4
Heating liquid (low visc.)	0.0255	0.8	0.4
Heating liquid (high visc.)	0.027	0.8	0.33
Flow normal to a single tube, cooling or heating			
Gas	0.26	0.6	0.3
Liquid	0.38	0.56	0.3
Gas or liquid flow normal to banks of tubes			
Staggered	0.33	0.6	0.33
In-Line	0.26	0.6	0.33

Material	Thermal Conductivity	
	W/(m·K)	Btu/(hr-ft-°F)
Liquid		
Freon 12	0.083	0.048
C_3-C_{10}(n-paraffins)	0.14-0.15	0.079-0.083
Water	0.62	0.36
Ammonia	0.50	0.29
Benzene	0.16	0.091
Crude oil	0.13	0.076
Ethylene Glycol	0.26	0.15
Diethylene Glycol	0.21	0.12
Triethylene Glycol	0.19	0.11
Monoethanolamine	0.24	0.14
Diethanolamine	0.22	0.13
Gas		
Air	0.026	0.015
Ammonia	0.028	0.016
Benzene	0.012	0.0069
Butane	0.017	0.010
Carbon Dioxide	0.019	0.011
Methane	0.035	0.020
Propane	0.021	0.012
Water Vapor	0.021	0.012
Freon	0.011	0.0064

Data on the "k" values of gases at high pressure are sparse. From the kinetic theory of gases "k" is theoretically proportional to gas viscosity. One can correct the "k" for pressure using corresponding viscosity correction as an approximation.

The Prandtl number ($C_p\mu/k$) is essentially constant with pressure and temperature. For natural gas a value of 0.79 is suitable for this number.

Effect of Velocity

The velocity of the fluid, inside or outside the tubes, has a significant effect on exchanger performance. For flow inside tubes,

$$h \propto v^{0.8}$$

Within reasonable limits, h increases as velocity increases. The reason? The film is thinner and there is less resistance to heat flow.

But, as velocity increases so does pressure drop. For flow inside tubes,

$$\Delta P \propto v^{1.8}$$

If pressure drop is critical and must be recovered by pumping or compression, an economic compromise is necessary. So ... specification of the heat exchanger must consider pressure drop considerations. A good rule-of-thumb is that the cost of the pressure drop over the life of the exchanger should be approximately 1/3 the cost of the surface area.[13.6] This gives a design near the economic optimum.

One must also consider varying throughput. If the exchanger is too large at the lowest flowrates expected, you can expect "U" to be lower than at rated capacity. For developing fields it might be more efficient to plan on two or more exchangers in parallel. The manifold is constructed initially with exchangers being added as needed to maintain efficient heat transfer. If block valves are added initially, no shutdown is needed for future additions. Sometimes some of the tubes are plugged during early development or near the end of the facility life to maintain desirable velocity.

When using parallel equipment of any kind, positive controls must be added to ensure proper flow distribution. You cannot rely on header design alone.

There is a special consideration in cases where a corrosion or hydrate inhibitor is being injected. In order to achieve good distribution in all tubes, a certain minimum tube velocity must be maintained. From test, it has been established that

$$v_m = \frac{A}{\sqrt{\rho}} \tag{13.8}$$

Where:			SI	FPS
	ρ	= fluid density	kg/m³	lb/ft³
	V_m	= min. tube velocity	m/s	ft/sec
	A	= correlation constant	26	21

For gases, ρ is found from methods comparable to those discussed in Chapter 3, at average exchanger pressure and temperature.

Fouling Factors

The fouling factors (F_i and F_o) shown in Equation 13.3 must be estimated from experience. The figure at right shows a temperature gradient across a wall, including the possible corrosion or depositional scales. If scale forms, how fast will it form? This is a question that requires a detailed analysis.

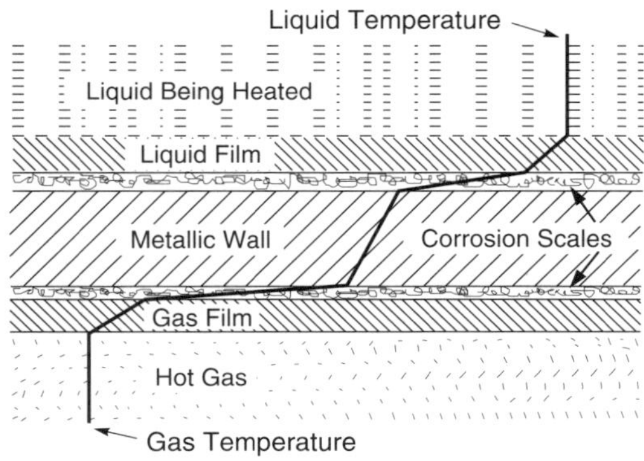

Some erroneously use a fouling factor as an arbitrary safety factor. Use of a fouling factor is all right so long as the number used is a realistic one compatible with expected performance. If too large a number is used it controls "U" and invalidates the calculation.

In Equation 13.3 the units of F are the reciprocal of those for "U" or "h." For oil and gas systems the table below gives a range of fouling factors typically encountered.

The lower end of the range applies to clean natural gas and sea water. The high end applies to dirty water and heavy crude oils. Fouling factors are a strong function of velocity. Higher velocities result in lower values for F. Deposition of solids such as salt, paraffins, asphaltenes can have a significant effect on F. The TEMA standards[(13.7)] provide an excellent compilation of typical fouling factors.

Units of "h"	F
W/(m^2·°C)	0.0001-0.0005
Btu/hr-ft^2-°F	0.0005-0.003

SHELL-AND-TUBE EXCHANGERS

Figures 13.4-13.7 show the basic characteristics of shell-and-tube exchangers. The major manufacturers of such equipment have a trade association (TEMA) which has a set of standards. They are not a code but are used commonly in bid specifications. Class R exchangers are used most commonly in the petroleum industry,[(13.7)] although Class C exchangers have been specified in some E & P applications.

Shell and tube exchangers are the most versatile of the exchanger types. They can handle a wide range of fluids, both single phase and two phase. They are robust and can be manufactured from a variety of materials. They are also heavy, have a relatively large footprint, and are expensive. In several applications in oil and gas processing they have been superceded by compact exchangers (discussed later in this Chapter).

Figure 13.4 shows the general head and shell types. The other figures show the details of various types of exchangers using the nomenclature shown in Table 13.2.

The choice of configuration depends on a number of considerations – fluids involved, corrosion potential, problems of cleaning, allowable pressure drop, heat transfer efficiency. Heat exchanger selection is not routine.[(13.8)]

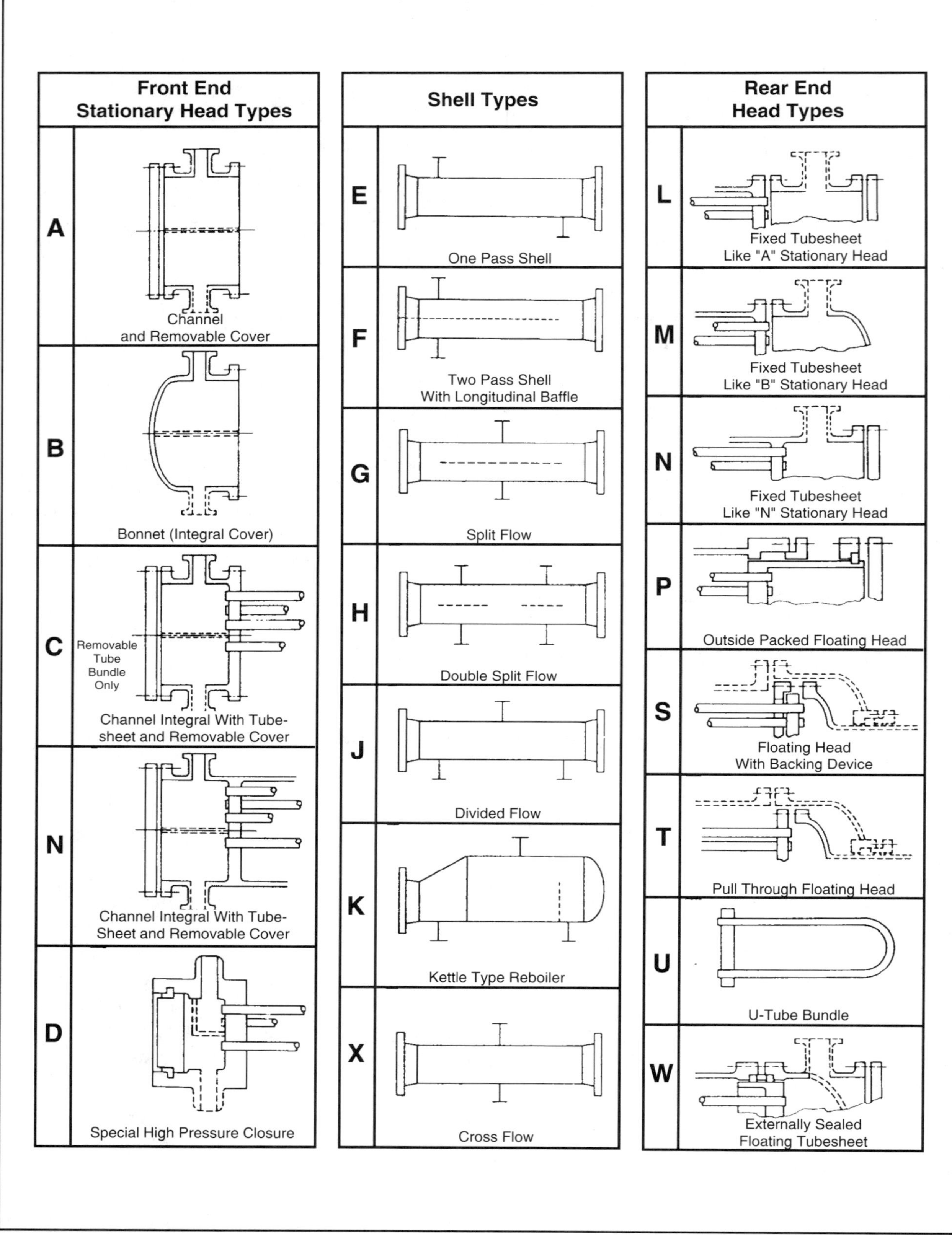

Figure 13.4 Basic Mechanical TEMA Characteristics *(Courtesy Tubular Exchanger Mfgrs. Assn. -- TEMA)*

TABLE 13.2

Number Guide for Use with Figures 13.5, 13.6, and 13.7

1. Stationary Head-Channel	21. Floating Head Cover-External
2. Stationary Head-Bonnet	22. Floating Tubesheet Skirt
3. Stationary Head Flange-Channel or Bonnet	23. Packing Box Flange
4. Channel Cover	24. Packing
5. Stationary Head Nozzle	25. Packing Follower Ring
6. Stationary Tubesheet	26. Lantern Ring
7. Tubes	27. Tie Rods and Spacers
8. Shell	28. Transverse Baffles or Support Plates
9. Shell Cover	29. Impingement Baffle
10. Shell Flange-Stationary Head End	30. Longitudinal Baffle
11. Shell Flange-Rear Head End	31. Pass Partition
12. Shell Nozzle	32. Vent Connection
13. Shell Cover Flange	33. Drain Connection
14. Expansion Joint	34. Instrument Connection
15. Floating Tubesheet	35. Support Saddle
16. Floating Head Cover	36. Lifting Lug
17. Floating Head Flange	37. Support Bracket
18. Floating Head Backing Device	38. Weir
19. Split Shear Ring	39. Liquid Level Connection
20. Slip-on Backing Flange	

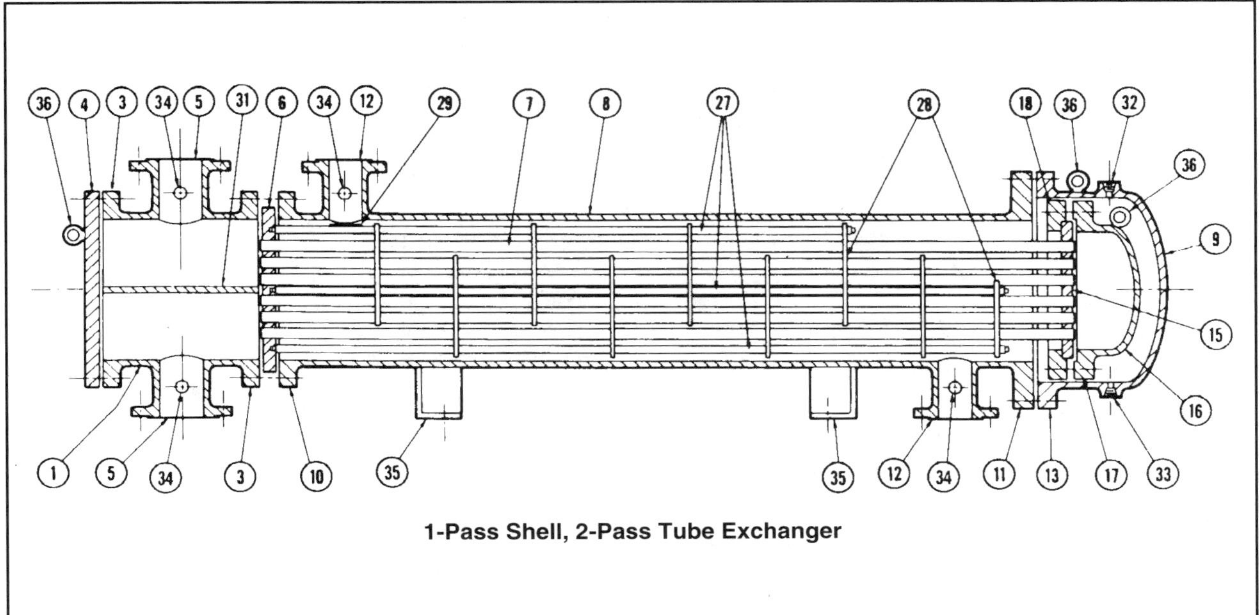

1-Pass Shell, 2-Pass Tube Exchanger

Figure 13.5 One Type of Tubular Exchanger *(Courtesy Tubular Exchanger Mfgrs. Assn. -- TEMA)*

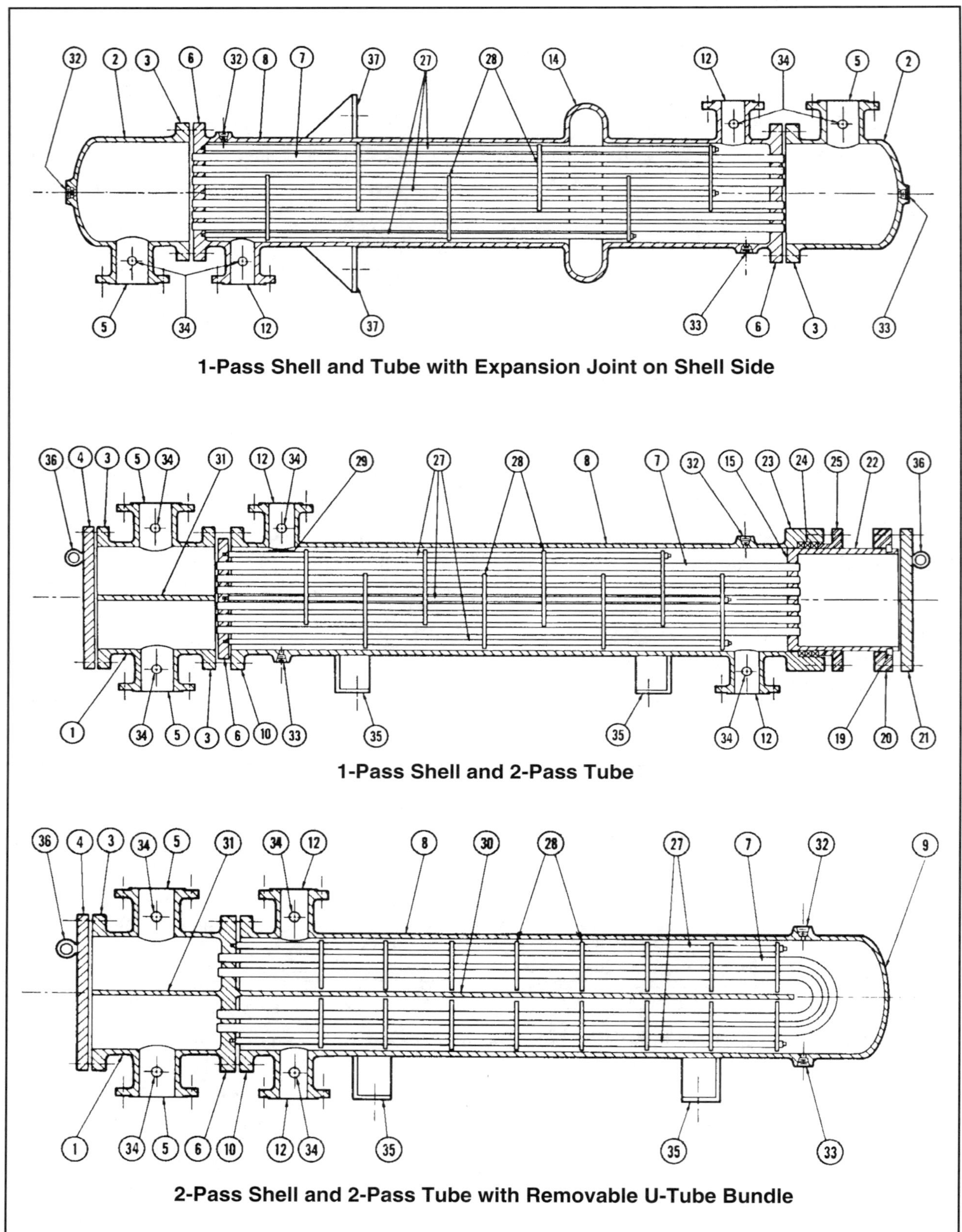

Figure 13.6 Three Other Examples of Tubular Exchangers *(Courtesy Tubular Exchanger Mfgrs. Assn. -- TEMA)*

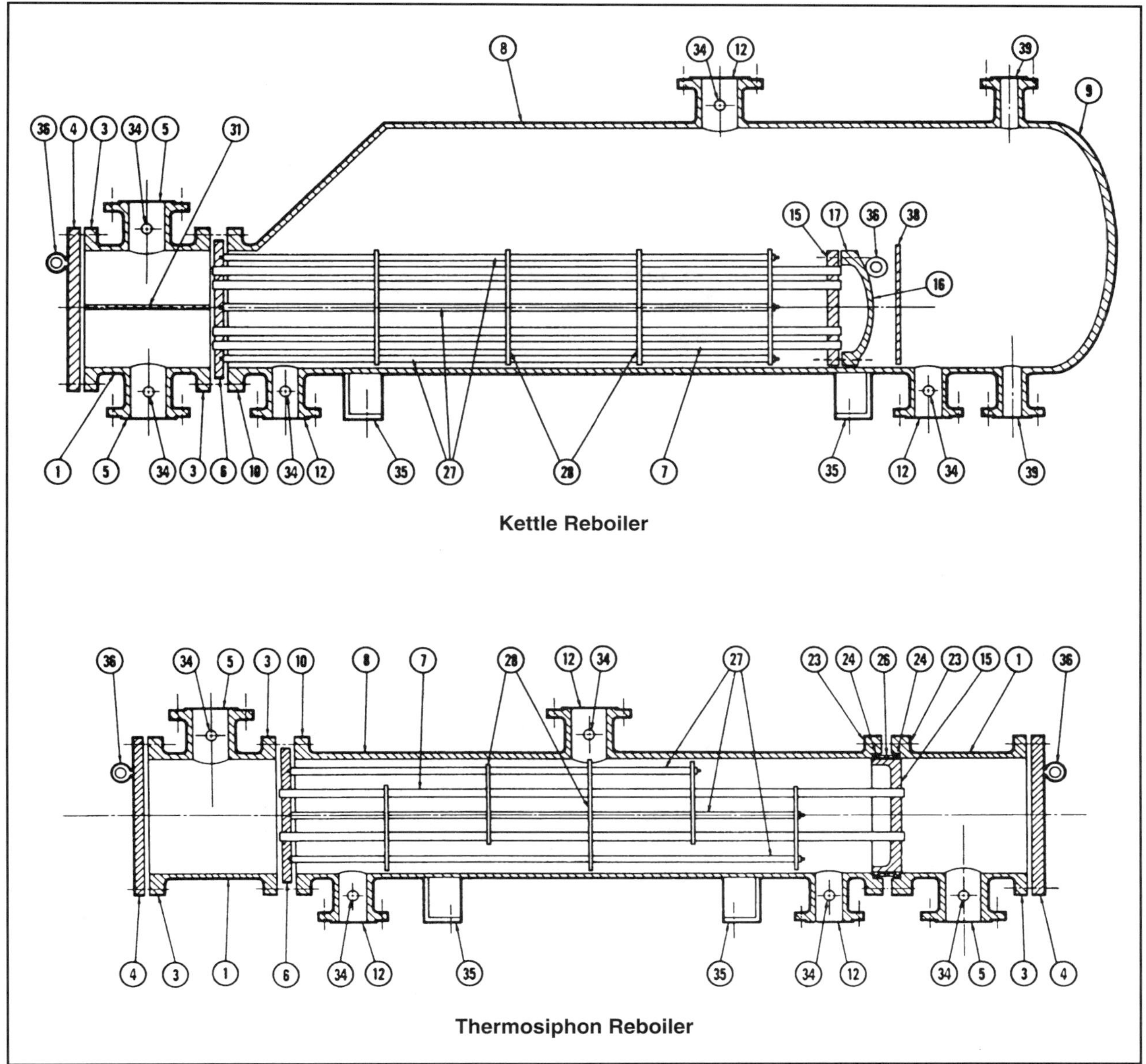

Figure 13.7 Two Common Types of Reboilers *(Courtesy Tubular Exchanger Mfgrs. Assn. -- TEMA)*

Do you need removable or nonremovable tube bundles? The latter are relatively inexpensive and provide maximum protection against shell-side leakage but they are not accessible for mechanical shell-side cleaning. A type of expansion joint is sometimes needed to relieve differential thermal expansion stresses.

Removable tube bundles consist of U tubes (hairpin type) or straight tubes with a floating head. The former is the least expensive, can be used with very high pressures on the tube-side. But, mechanical cleaning is difficult, it is very difficult to replace tubes, and leakage around the longitudinal baffle can impair performance.

The floating head exchanger is the most versatile and most expensive. Obtaining a positive seal between tube-side and shell-side fluids is critical in many cases. The pressure differential between shell- and tube-sides is limited by the seal.

Although there are exceptions, most tubes used are 15-25 mm [5/8-1.0 in] diameter, 19 mm [3/4 in] is the most common. The larger size normally is used when fouling is anticipated, to facilitate mechanical cleaning. The tube length may be as large as 12 m [40 ft] but tubes about half this length are more commonly employed.

The tube pitch is the distance between the centerline of the tubes. The minimum tube pitch is 1.25 times the tube O.D. For a 19 mm [3/4 in] tube this is 23.8 mm [15/16 in]. A larger tube pitch is often used in fouling applications.

The tube bundle can be arrayed in a triangular, square or rotated-square layout. Triangular usually gives better shell-side "h" values and more heat transfer area for a given shell diameter. However, the other arrangements are easier to clean, have a lower pressure drop, and may be preferred in boiling applications to provide a clearer path for vapor to exit the tube bundle.

There are several shell configurations. E type shells are by far and away the most common. The F shell is often used on U-tube exchangers. The tube bundle has a longitudinal baffle which must seal against the shell wall (see the bottom of Figure 13.6). This creates a potential leakage path in the exchanger for the shell-side fluid, and if the leakage is large, heat transfer efficiency is compromised.

Type G and H shells are often used in low pressure drop applications such as thermosyphon reboilers. There are no cross baffles in this exchanger between the feed nozzle(s) and the tube sheets. For 19 mm [3/4 in] tubes, maximum unsupported tube length is limited to 1.5 m [5 ft], so the G-type shell has a maximum length of 3 m [10 ft] and the H-type 6 m [20 ft].

J-type (divided flow) shells shorten the liquid flow path. These are often used in low pressure-drop applications. X-type (crossflow) shells are also used in very low pressure-drop services such as condensers. Multiple inlet and outlet nozzles can be used.

The baffles serve two purposes: to increase the heat transfer coefficient by forcing the shell-side fluid to traverse the tube bundle several times (thus maintaining high velocity on the shell side) and to support the tube bundle.

Several baffle designs are available but the most common is the single segmental baffle shown to the below.

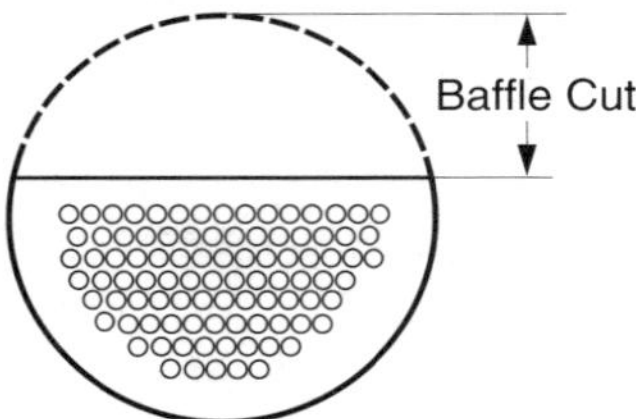

The baffle cut sets the area open for flow around the baffle. It is expressed as a fraction of inside shell diameter. Typical baffle cuts range from 0.2 to 0.35. This opening is often called the baffle window, and it should provide roughly the same flow area as the crossflow area between the baffles.

The distance between the baffles is termed the baffle pitch. It typically ranges from 20-100% of the shell diameter.

The baffle cut and pitch are the primary factors affecting shell-side pressure drop and heat transfer coefficient. For low pressure drop applications double and triple segmental baffles are sometimes used. Rod baffles, which consist of a series of plastic rods wedged in the tube bundle, are reported to give superior heat transfer and tube support.

LMTD Correction Factor

The factor "F" in Equation 13.4 is essentially one in a pipe-in-pipe exchanger and in counterflow shell-and-tube exchangers with an equal number of shell and tube passes. If the number of shell and tube passes differ, or if the exchanger is a crossflow type (e.g., TEMA "J or X," and aerial coolers), F will have a value less than 1.0, depending on the flow configuration used. As a practical economic consideration, it is seldom that a configuration should be chosen where F is less than about 0.80.

Figures 13.8-13.13 provide a means to estimate the factor F shown on the left ordinate. Values of P and R on these figures are found by the equations. In shell and tube applications "t" refers to tube-side temperatures and "T" refers to the shell side. For aerial coolers the terminology is just the opposite.

$$P = \frac{t_2 - t_1}{T_1 - t_1} \qquad R = \frac{T_1 - T_2}{t_2 - t_1}$$

For a given value of P and R, find the corresponding value of F. If the values of P and R do not intersect within the grid, simply record F as less than 0.5.

This correlation is a modified version that has been in use for a many years.*(13.9)* An alternate calculation to that shown has been developed.*(13.10)* The net results are essentially the same.

Figures 13.12 and 13.13 are suitable for an aerial cooler or for any case where one fluid flows normal to a bank of tubes.

In Example 13.4, notice that the AES configuration requires about 7% more area than a counterflow fixed tubesheet type. This would increase the exchanger cost, not only because the area is greater but also because the exchanger construction is more complex. What is the offsetting benefit of this configuration? – a removable tube bundle for cleaning and a floating tubesheet for less stress due to differential temperature expansion.

Example 13.4: The exchanger in Example 13.1 is actually a type AES exchanger with four-tube passes and one shell pass. Recalculate ΔT_m and A, using Figure 13.8. The crude oil is on the tube side of the exchanger.

T_1 t_2 t_1 T_2

$$\left.\begin{aligned} P &= \frac{62-84}{38-84} = 0.48 \\ R &= \frac{38-52}{62-84} = 0.64 \end{aligned}\right\} \text{From Figure 13.8} \quad F \approx 0.93$$

$$\Delta T_{m_{corr}} = (0.93)(27.8°C) = 25.9°C \text{ (use 26°C)}$$

$$A = \frac{Q}{U\Delta T_m} = \frac{3\,110\,000}{(330)(26)} = \underline{\underline{362 \text{ m}^2}}$$

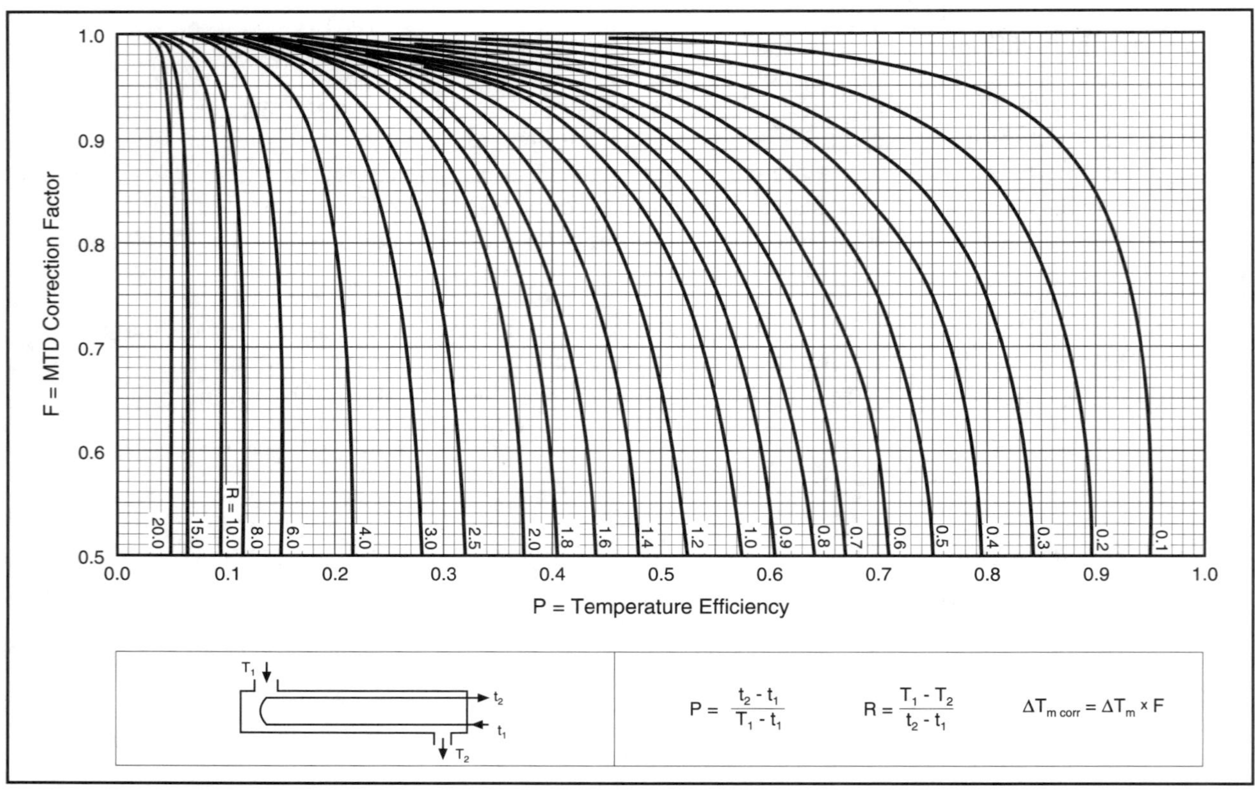

Figure 13.8 MTD Correction Factor (1 shell pass, 2 or more tube passes)

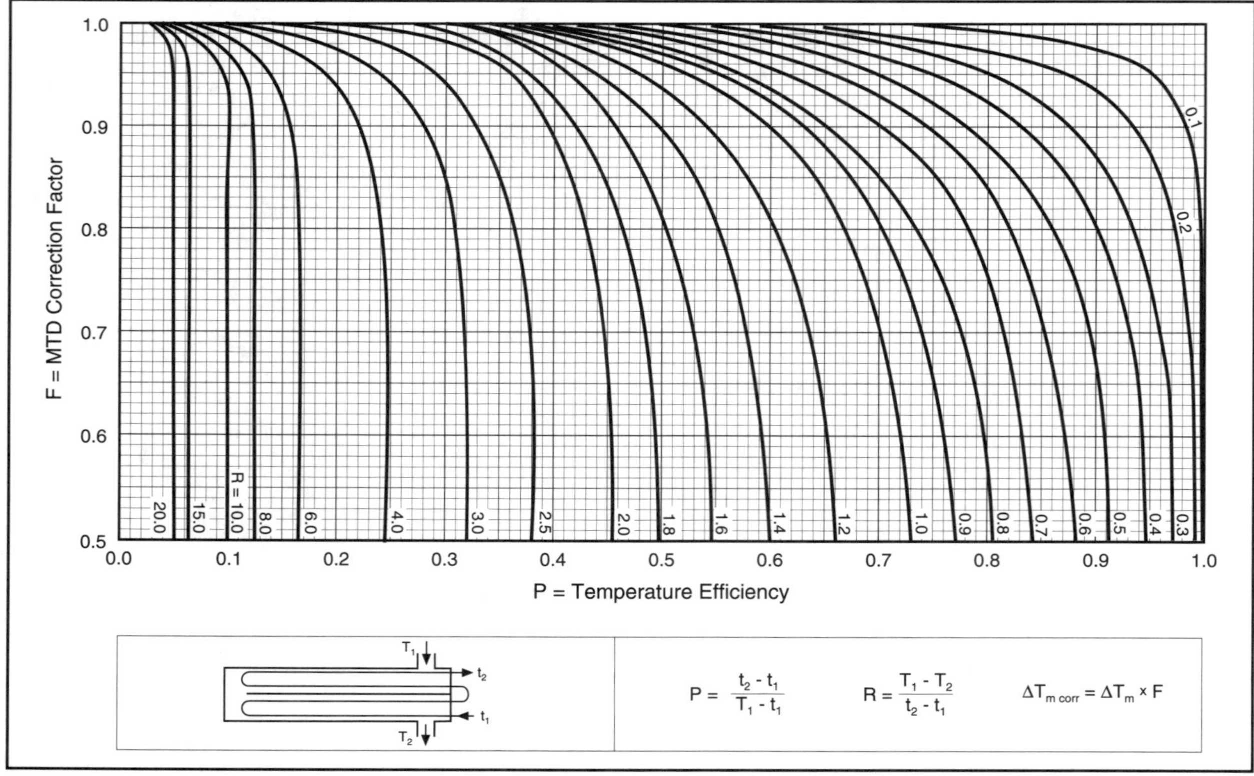

Figure 13.9 MTD Correction Factor (2 shell pass, 4 or more tube passes)

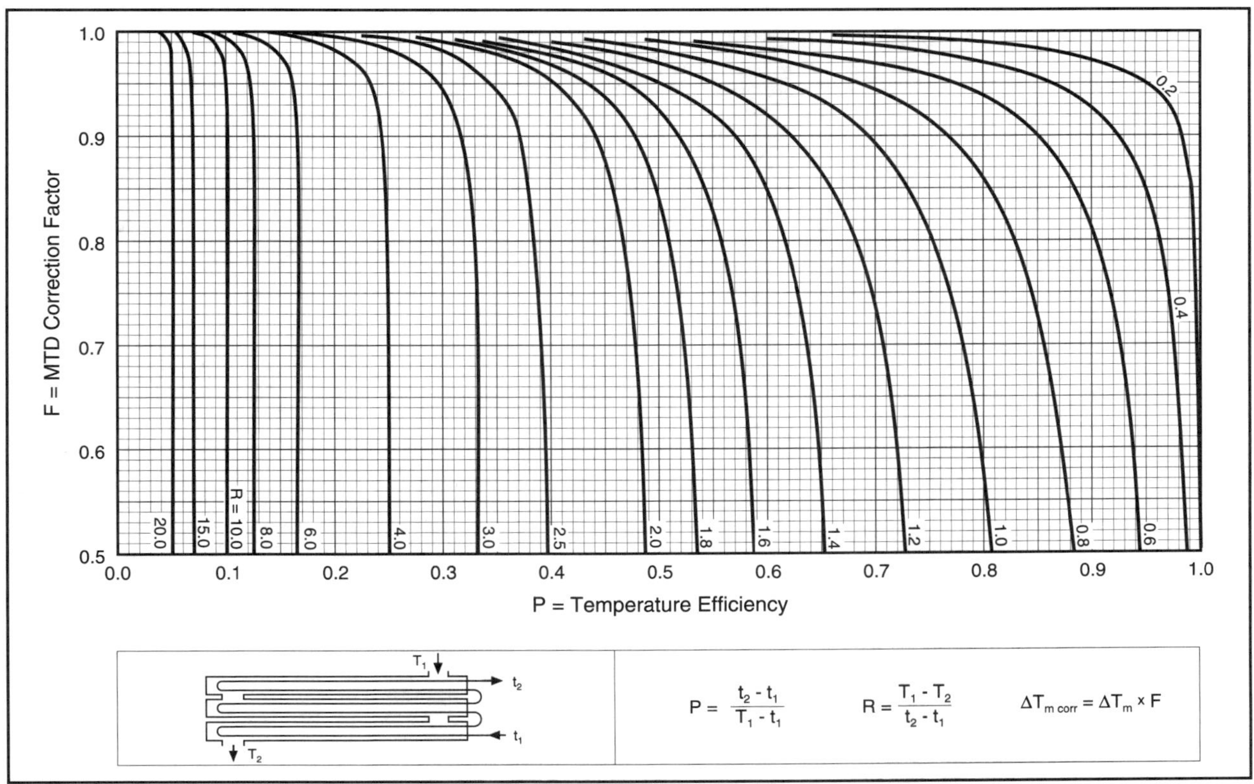

Figure 13.10 MTD Correction Factor (3 shell pass, 6 or more tube passes)

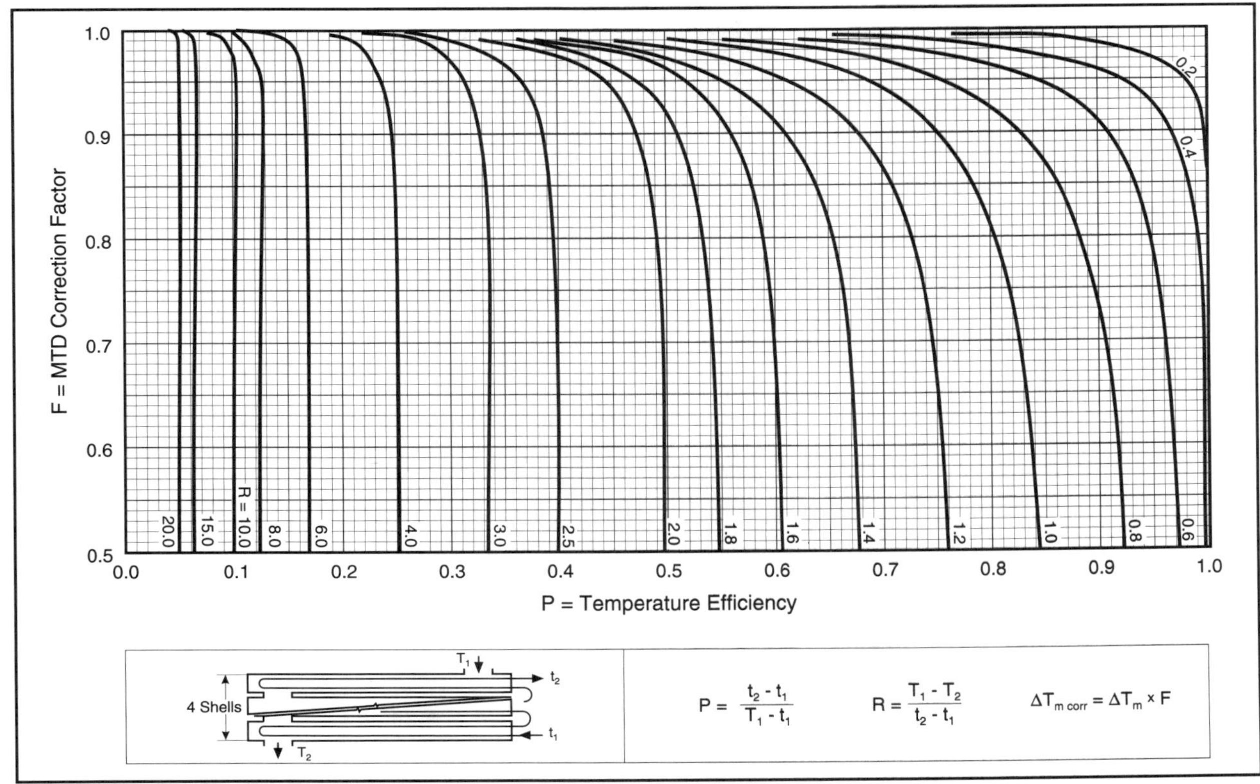

Figure 13.11 MTD Correction Factor (4 shell pass, 8 or more tube passes)

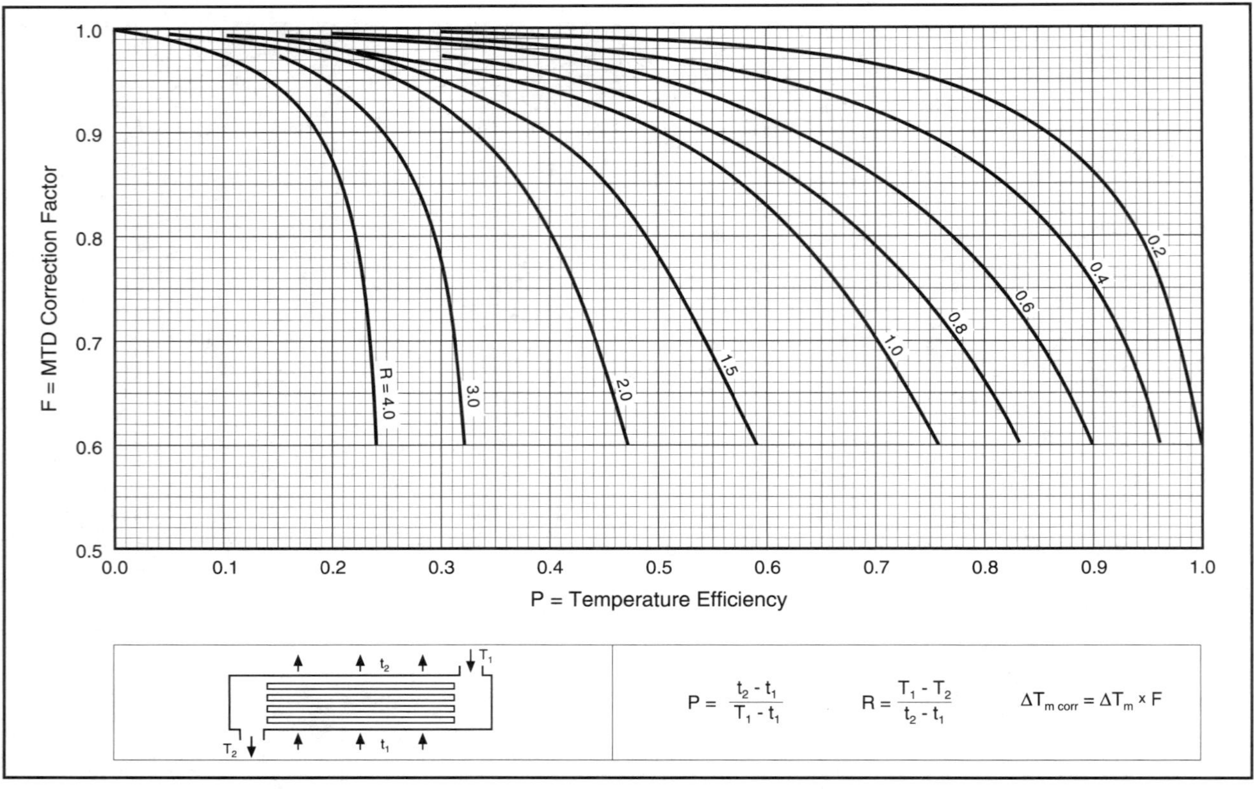

Figure 13.12 MTD Correction Factor (1 pass, 1 or more parallel rows of tubes)

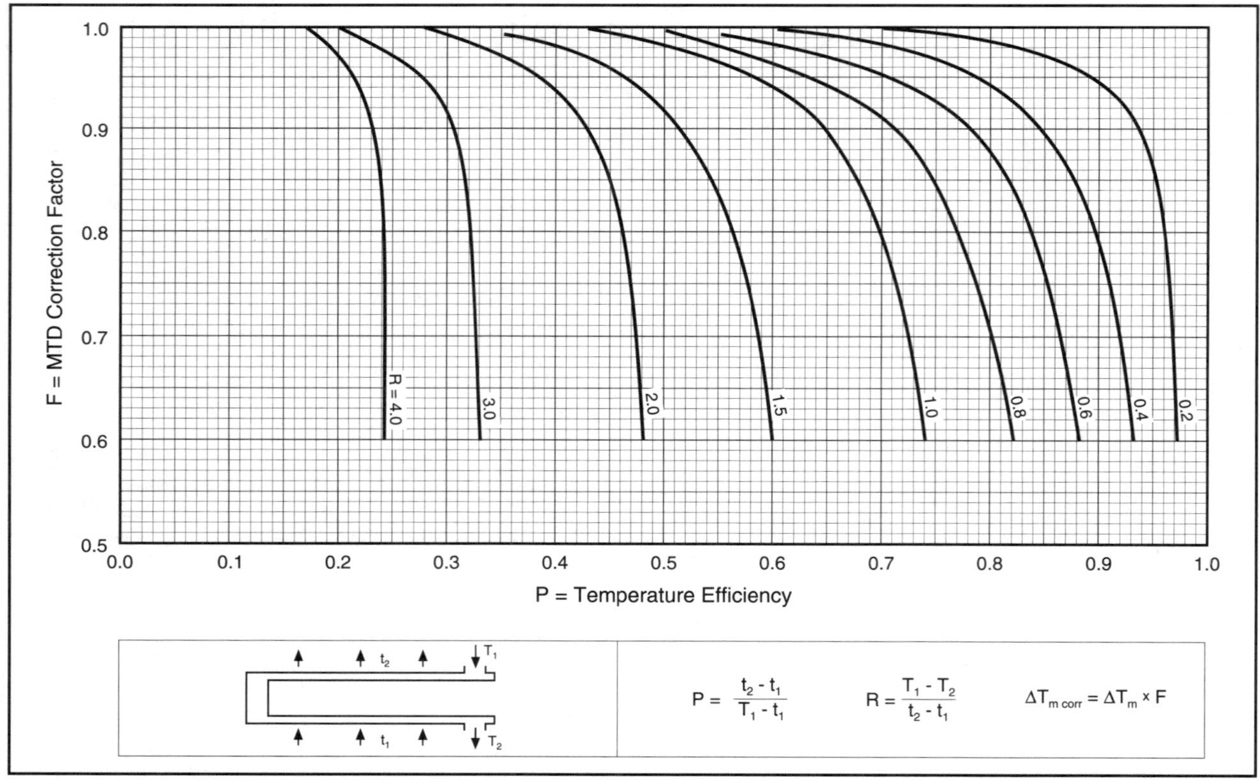

Figure 13.13 MTD Correction Factor (2 passes, 2 rows of tubes – for more than 2 passes, use F=1.0)

What about a configuration employing two shell passes and four or more tube passes? This configuration is actually two exchangers in series. The correction factor for this configuration is essentially 1.0. This would result in less surface area but would likely be more expensive due to the construction of two exchanger shells rather than one.

Fluid Placement

This obviously affects the value of F in the LMTD calculation. But, the major consideration may be the character of the fluid itself. The following general guidelines are useful.

A. Shell-Side
   1. Viscous fluid to increase (generally) the value of "U"
   2. Fluid having the lower flowrate
   3. Condensing or boiling fluid

B. Tube-Side
   1. Toxic and lethal fluids to minimize leakage
   2. Corrosive fluids
   3. Fouling fluids; increased velocity minimizes fouling but enhances erosion
   4. High temperature fluids requiring alloy materials
   5. High pressure fluids to minimize cost
   6. Fluid for which pressure drop is most critical

These are not mutually exclusive considerations. Some priorities must be established; some compromises are necessary. For example, condensing may be done on the tube side when special metallurgy is required. In this case, vertical tubes may be a better choice than horizontal tubes.

In some cases a series of exchangers (train) is required. One then must divide the total heat transfer duty to optimize the number and size of each unit.[13.11, 13.12]

Factors governing testing, troubleshooting, etc. are covered in References 13.13-13.19.

Estimation of Mechanical Design

As part of the early planning function, it may be desirable to estimate the physical size of the shell and tube exchanger being considered. Figure 13.14 provides an easy method to accomplish this.[13.12] The equation for use with this figure is:

$$A = A_0 F_1 F_2 F_3 \qquad (13.9)$$

Where:

A = area on left-hand ordinate of figure

A_0 = area calculated from heat transfer equation

F_1, F_2, F_3 = correction factors shown in Figure 13.14

F_1, F_2 and F_3 are equal to one for 19 mm [3/4 in] tubes on a 23.8 mm [15/16 in] triangular pitch, one tube pass and a fixed tube sheet exchanger, respectively.

For a given value of A, various diameter/length combinations are suitable. These would need to be checked for fluid velocities. L/D ratios less than 3:1 may suffer from poor fluid distribution. Ratios in range 6:1-10:1 generally are a good compromise. These L/D ratios are shown as dashed lines in Figure 13.14.

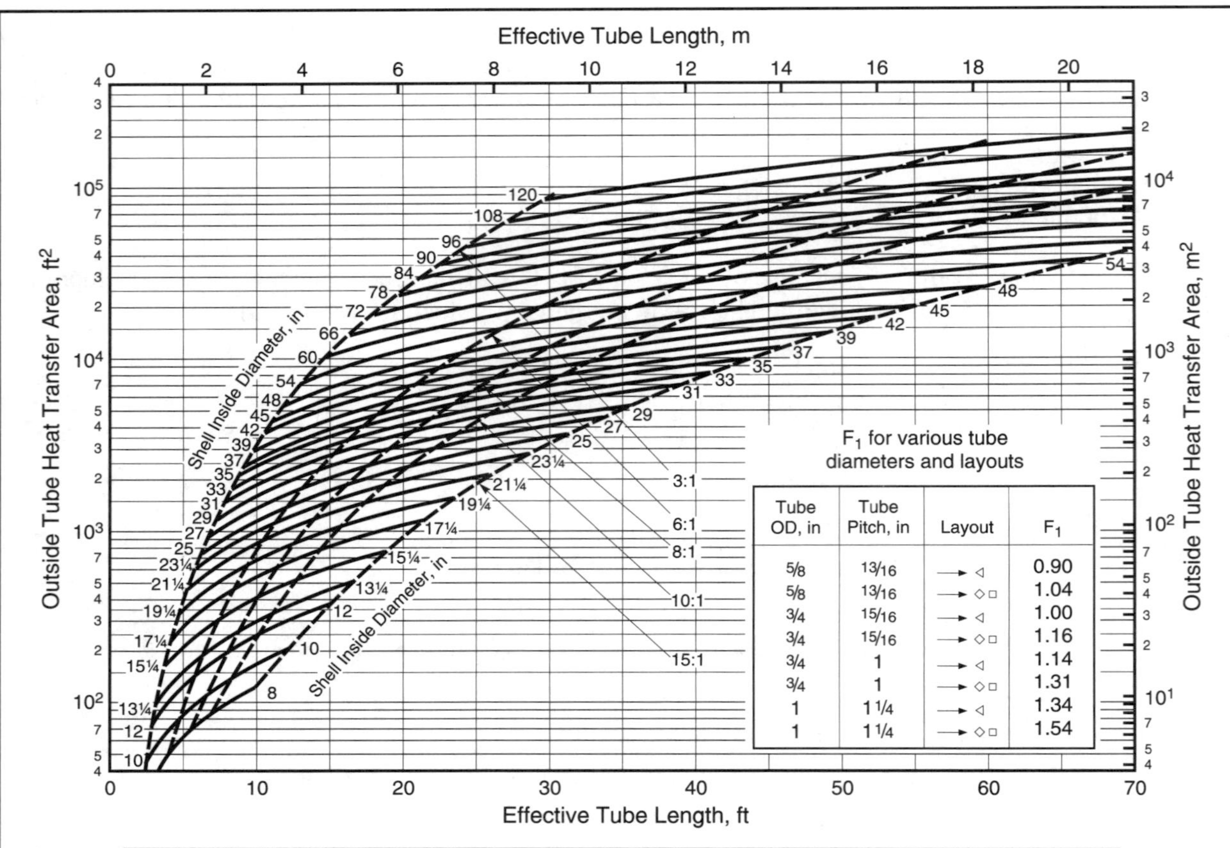

Tube OD, in	Tube Pitch, in	Layout	F_1
5/8	13/16	→ ◁	0.90
5/8	13/16	→ ◇□	1.04
3/4	15/16	→ ◁	1.00
3/4	15/16	→ ◇□	1.16
3/4	1	→ ◁	1.14
3/4	1	→ ◇□	1.31
1	1 1/4	→ ◁	1.34
1	1 1/4	→ ◇□	1.54

F_2 for Various Numbers of Tube-Side Passes*

Inside shell diameter, in	Number of tube-side passes: 2	4	6	8
Up to 12	1.20	1.40	1.80	–
13¼ to 17¼	1.06	1.18	1.25	1.50
19¼ to 23¼	1.04	1.14	1.19	1.35
25 to 33	1.03	1.12	1.16	1.20
35 to 45	1.02	1.08	1.12	1.16
48 to 60	1.02	1.05	1.08	1.12

*Since U-tube bundles must always have at least two passes, use of this table is essential for U-tube bundle estimation. Most floating-head bundles also require an even number of passes.

F_3 for Various Tube-Bundle Constructions

Type of tube bundle construction	Inside shell diameter, in: Up to 12	13 to 21	23 to 35	37 to 48	Above 48
Split backing ring (TEMA S)	1.30	1.15	1.09	1.06	1.04
Outside packed floating head (TEMA P)	1.30	1.15	1.09	1.06	1.04
U-Tube* (TEMA U)	1.12	1.08	1.03	1.01	1.01
Pull-through floating head (TEMA T)	–	1.40	1.25	1.18	1.15

*Since U-tube bundles must always have at least two tube side passes, it is essential to also use the above table for this configuration.

Figure 13.14 Estimation of Shell and Tube Exchanger Size

COMPACT EXCHANGERS

Plate and Frame Exchangers

Where applicable, the plate and frame exchanger (PHE) has become a viable alternative for heat exchange. In many services it is lighter, more compact, less expensive, and offers better overall performance than more traditional types. It is very competitive in many services in frontier and offshore applications. Offshore, the use of plate exchangers for sea water cooling has become almost universal. References 13.20-13.24 provide useful additional information and are the source of illustrations in this section.

Figure 13.15 summarizes some basic characteristics of this exchanger. Part (a) shows the basic construction. A series of corrugated, pressed metal plates are clamped together and held in place by bolts through the end plates and a pressure plate. Each such plate is gasketed to prevent leakage. There are four flow channels on each plate which can be blanked off or combined in different ways to form different flow patterns. An assembled plate exchanger is shown in (b) of Figure 13.15.

Shown in (c) are plates with chevron-type grooves. A low chevron angle plate at left is called a high theta (θ) plate; a high angle is a low theta plate. Different manufacturers use different corrugated or embossed patterns for competitive advantage.

Also shown in Part (c) of Figure 13.15 is the fact that plates may be mixed and matched to provide more efficient heat transfer. One of the "tricks" of the trade is providing the best combination. Remember, the heat transfer plates are grooved on both sides. Since the plates are symmetrical, the direction of the grooves and their angle can be reversed on alternate plates.

The term theta is defined by Equation 13.10 later in this section. It is a measure of heat transfer effectiveness. As in all heat exchange, effectiveness is purchased with pressure drop. A high theta plate has a higher pressure drop (and heat transfer coefficient) than a low theta plate.

Part (d) is an example of mixing plates for cooling a large amount of process water with cooling water.[13.20] The optimum mixture of plate types reduces the number of plates (and thus cost) significantly for each specified pressure drop.

The basic idea is to form a series of interlocking flow channels which produce high velocity. The metal-to-metal contact also helps plate rigidity. Any metal that can be cold worked can be used. Titanium is a common plate material for sea water cooling. Stainless steels, Monel, nickel, Incoloy, etc., also are used in some services.

The size and thickness of plates depend on the metal used, system pressure, and process specifications. Pressure deflection must be limited to prevent leaking.

Plate thickness ranges from 0.5-3.0 mm [1/50-1/8 in] with an average gap between plates of 1.5-5.0 mm [1/16-1/5 in]. Most plates have an area less than 1.5 m^2 [16 ft^2]. The pressure limitation is obvious. A pressure of about 2.0 MPa [290 psia] is a typical maximum although slightly higher pressures have been used. Where possible, a maximum pressure of about 1.0 MPa [150 psia] is preferred.

Temperature is limited by metallurgy and gasket materials. Some installations are operating up to about 250°C [480°F]. But, the utility of the plate exchanger diminishes rapidly above 150°C [300°F].

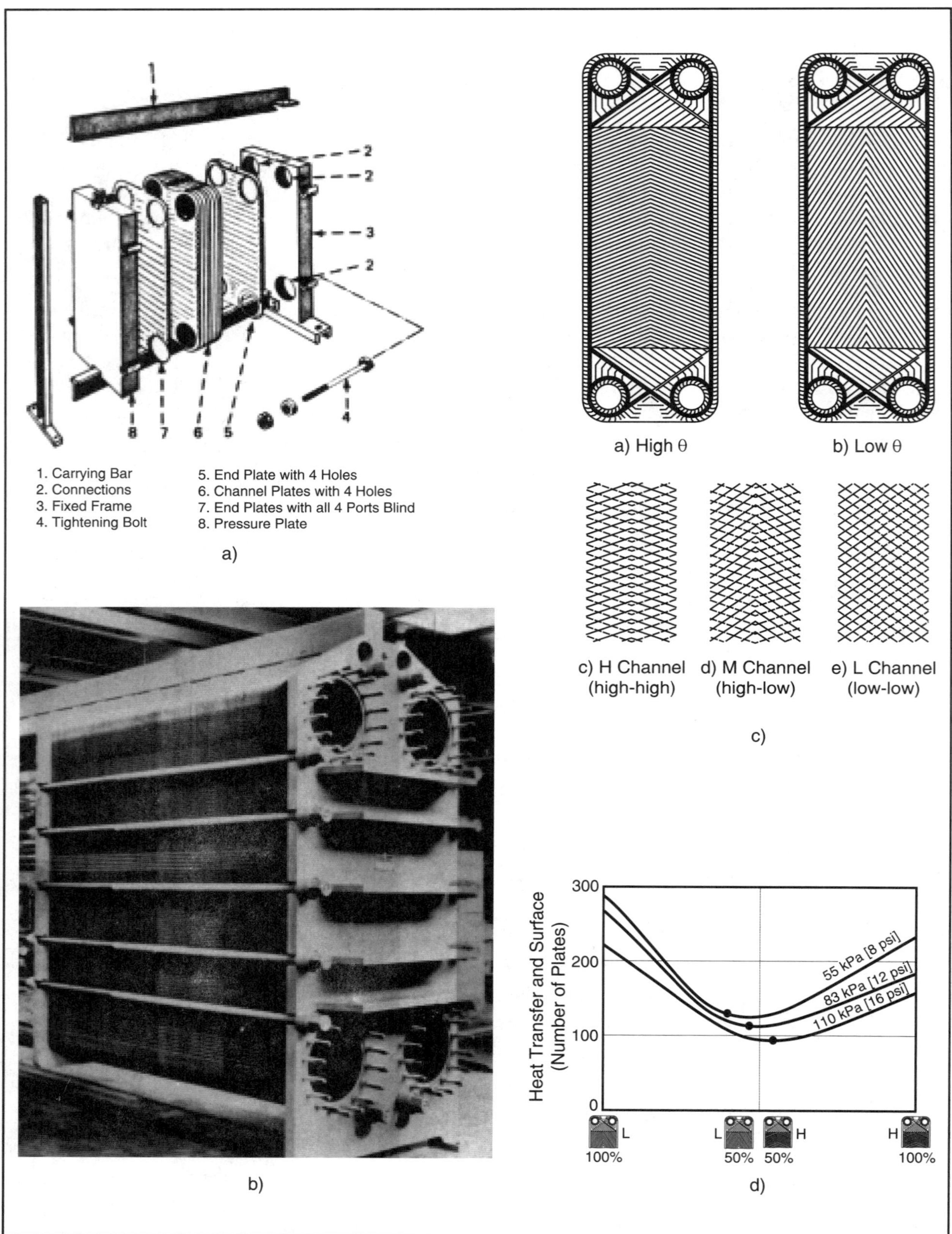

Figure 13.15 General Characteristics of Plate Exchangers

Theta Factor

This factor also is called NTU (number of transfer units). This theta can be stated in terms of the required duty or the enthalpy change of one or both fluids. A consistent set of units must be used to make theta a dimensionless number. The total area is twice the area of one side of the thermal plates. The end plates are not involved in heat transfer.

$$\theta = \frac{t_i - t_o}{\Delta T_m} = \frac{UA}{mC_p} \tag{13.10}$$

Where:

t_i = inlet temperature of fluid to channel
t_o = outlet temperature of fluid to channel
U = overall heat transfer coefficient
A = total area of thermal plates
m = mass flowrate of fluid per hour
C_p = specific heat of fluid

Determination of LMTD

LMTD is determined in the same manner as before, the only difference is in "F." The figure at right is an approximate correlation for F in plate exchangers. For an equal number of passes on the hot and cold fluids and for a low NTU (or θ), F approaches one. An unequal number of passes usually occurs only when the two fluids involved have widely different flowrates. Where a different number of passes is required, the plate exchanger may not be the best choice.

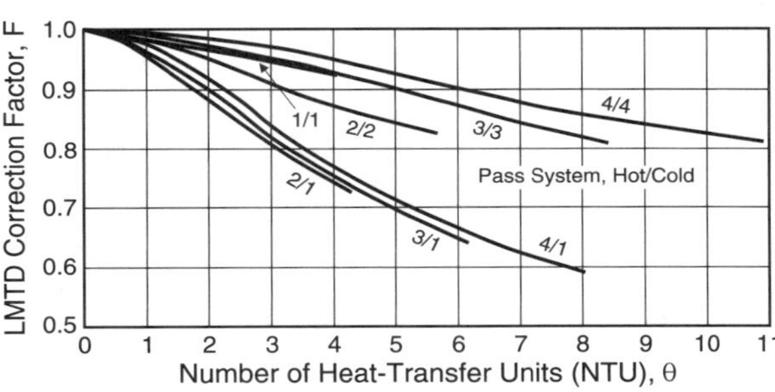

Heat Transfer Coefficients

The film coefficients use the same form as Equation 13.5. They may be estimated by the equations:

For turbulent flow[(13.23)]

$$h = A\left(\frac{k}{d_e}\right)\left(\frac{d_e w}{\mu}\right)^{0.65}\left(\frac{C_p \mu}{k}\right)^{0.4} \tag{13.11}$$

For laminar flow[(13.25)]

$$h = A' C_p w \left(\frac{d_e w}{\mu}\right)^{-0.62}\left(\frac{C_p \mu}{k}\right)^{-0.67}\left(\frac{\mu}{\mu_w}\right)^{0.14} \tag{13.12}$$

Where:			SI	FPS
k	=	thermal conductivity	W/(m·K)	Btu/hr-ft-°F
d_e	=	(4Wb)/(2W + 2b)	m	ft
W	=	plate width	m	ft
w	=	mass flowrate	kg/(s·m^2)	lbm/(hr-ft^2)
μ	=	fluid viscosity	kg/(m·s)	lbm/(ft-hr)
C_p	=	specific heat	kJ/(kg·K)	Btu/(lbm-°F)
μ_w	=	fluid viscosity at wall	kg/(m·s)	lbm/(ft-hr)
A	=	proportionality constant	0.2536	0.2536
A'	=	proportionality constant	0.742	0.742
h	=	film coefficient	W/(m^2·K)	Btu/(hr-ft^2-°F)
b	=	mean distance between plates	m	ft

The overall heat transfer coefficient is

$$\frac{1}{U} = \frac{1}{h_1} + \frac{L}{k_w} + \frac{1}{h_2} + F_f \qquad (13.13)$$

which is the same form as Equation 13.3. For a flat plate all heat transfer surface areas are equal and cancel out.

One can calculate area using Equation 13.2. An alternate approach designed primarily for computer usage has been developed by Jackson and Troup.[(13.24)] It is based on theta and does not use the LMTD equation. Comparable results should be obtained.

Pressure Drop

The pressure drop depends on plate design, arrangement and flow pattern. Reference to manufacturer's specifications is recommended although Reference 13.26 shows an estimation method. Pressure drop is about the same as (or less than) that in a comparable duty shell and tube exchanger.

Corrosion/Fouling

In general, fouling is minimized by high velocity and a clean surface. So, a plate exchanger will have less fouling tendency than a shell and tube. Since the plates are thin, corrosion standards are critical. A corrosion rate of over 0.05 mm [2 mils] per year would be unacceptable in most instances. Because of erosion/corrosion considerations, the choice of metallurgy is more critical in plate than in tubular exchangers.

General Considerations

One must compare all factors to choose the proper heat exchanger. Within its range of applicability, the plate exchanger is frequently less expensive than the tubular, particularly where alloy construction is needed. One comparison is shown below based on the crude oil cooler in Example 13.1. Notice the material upgrade for the plate exchanger. Although it cost twice as much per unit area, it had only about 1/6 the area of the tubular. It ended up being smaller, lighter and less expensive.

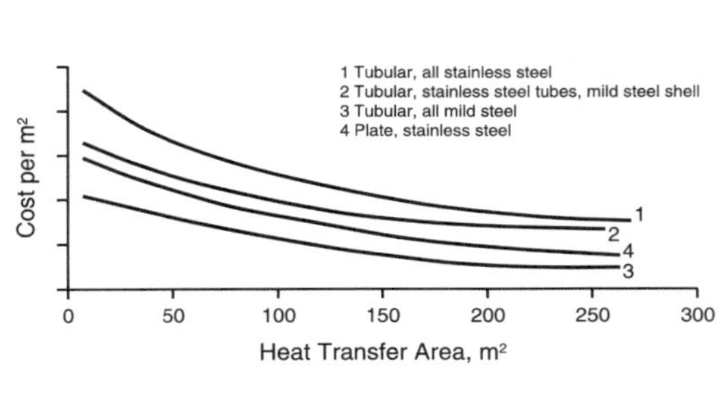

	Shell and Tube	Plate Heat Exchanger
Design pressure	100 psi/250°F	100 psi/250°F
Passes	One shell side/four tube side	One/one
Overall "U"	58 Btu/hr-ft^2-°F	336 Btu/hr-ft^2-°F
Required surface	3940 ft^2	635 ft^2
Materials	All carbon steel	Carbon steel frame/316 SS plates
Pressure drops	Hot side: 21 psi Cold side: 5 psi	Hot side: 9.7 psi Cold side: 8.2 psi
Size	3 ft diameter/24 ft length	5 ft length/3 ft width
Weight	23 800 lbm	3837 lbm
Plot space required	4 ft × 60 ft	5 ft × 6 ft
Cost/ft^2	$10.60	$21.42
Quoted price	$41,700	$13,607

In spite of possessing many favorable features, the plate exchanger is not an automatic choice. With high pressure gas and/or condensing or boiling fluid, the tubular may be superior even in the P and T range of the plate type.

BRAZED ALUMINUM EXCHANGERS

Plate-Fin Exchangers

Brazed aluminum plate-fin heat exchangers (BAHX) are frequently used in low temperature gas processing service. Often installed in a "cold-box," they can be designed to handle up to 10 fluids in a single exchanger and can operate at temperatures as low as –269°C [–452°F].

A brazed aluminum heat exchanger is composed of alternating layers of corrugated fins and flat separator sheets called parting sheets (Figure 13.16). A stack of fins and parting sheets comprise

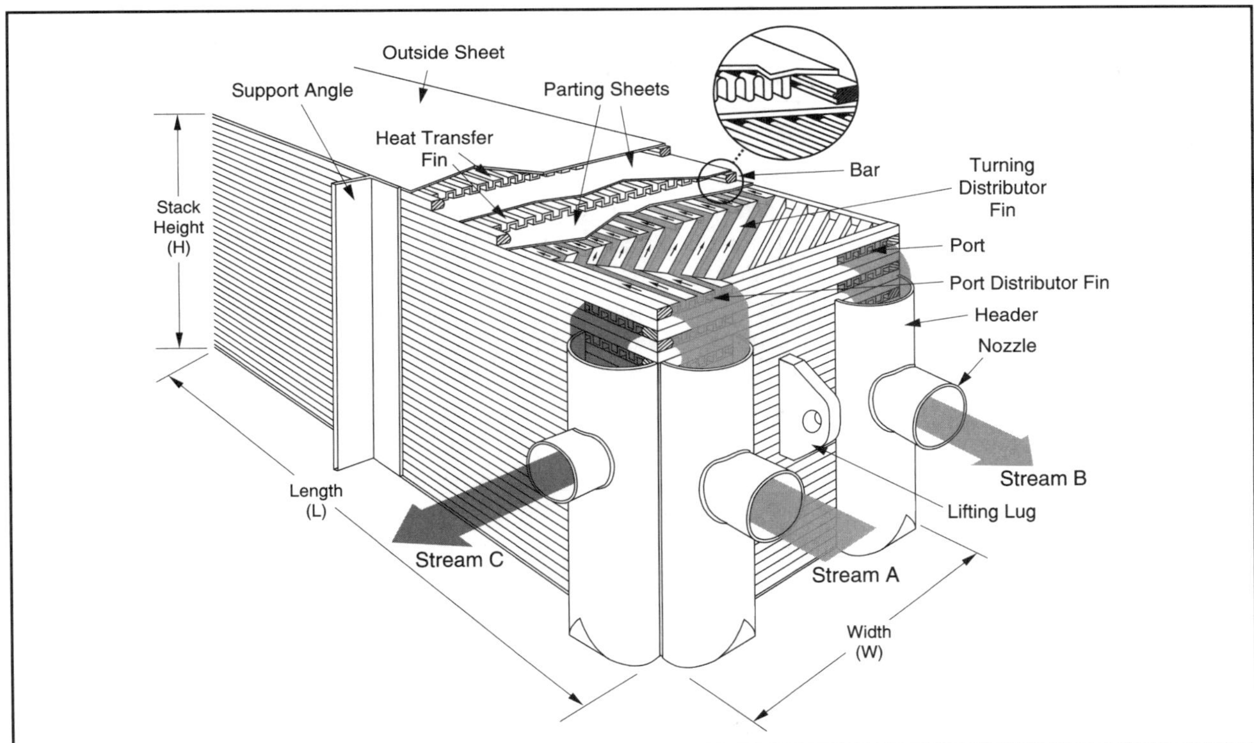

Figure 13.16 Basic Components of a Brazed Aluminum Plate-Fin Heat Exchanger *(Courtesy Chart Heat Exchangers)*

the heat exchanger, sometimes referred to as the "core." Each fluid pass in a core has the appearance of a section of the wall of a cardboard box. The inside and outside panels represent the parting sheets and the corrugations represent the fins. The number of layers, type of fins, stacking arrangement, and stream circuiting will vary depending on the application requirements. A typical plate fin exchanger is shown below.

Fluid configuration may be counterflow, crossflow, and cross-counterflow circuiting. Temperature approaches of 1.7°C [3°F] on single-phase fluids and 2.8°C [5°F] on two-phase fluids can be achieved. Often, corrected mean temperature differences of 2.8-5.6°C [5-10°F] are employed in brazed aluminum heat exchanger applications.

Brazed aluminum heat exchangers are compact and light-weight. A typical high pressure brazed aluminum heat exchanger with a design pressure of 4100-9650 kPa [600-1400 psig] will provide 90-120 m^2/m^3 [300-400 ft^2/ft^3] of heat transfer area per exchanger volume. This is six to eight times the surface density of comparable shell and tube exchangers. Additionally, a typical high pressure brazed aluminum heat exchanger will have a density of 1200-1450 kg/m^3 [75-90 lbm/ft^3] versus 4000 kg/m^3 [250 lbm/ft^3] for comparable shell and tube exchangers. The net effect of these differences is that a brazed aluminum heat exchanger will provide approximately 25 times more surface per weight of equipment than comparable shell and tube exchangers. This decrease in exchanger weight and volume reduces foundation, support, plot plan, and insulation requirements.

Brazed aluminum heat exchangers are designed and constructed to comply with the "ASME Boiler and Pressure Vessel Code," Section VIII, Division I, or other applicable standards. The aluminum alloys used comply with ASME Section II, Part B, "Nonferrous Materials," or the requirements of the specified code authority.

Aluminum alloy 3003 is generally used for the parting sheets, corrugated fins, and bars which form the rectangular heat exchanger block. These parts are metallurgically bonded by a brazing process at temperatures of about 593°C [1100°F]. The brazing alloy is an aluminum silicon metal and is provided on or with the parting sheets.

Brazed aluminum heat exchangers should be used with clean fluids since they are more susceptible to plugging than other types of heat exchanger equipment; however, proper filters or strainers will prevent heat exchanger fouling. Brazed aluminum should not be used with fluids which are corrosive to aluminum. Mercury and caustic soda are extremely corrosive to aluminum and should not be introduced into the exchanger. Hydrogen sulfide and carbon dioxide are not a corrosion problem in streams with water dewpoint temperatures below the cold end temperature of the exchanger.

Design of plate-fin exchangers is not routine, particularly when two-phase streams are present. Cooling and heating curves (T vs. Q) must be developed for each fluid present. The heat exchanger is then broken into "zones" where the cooling and heating curves are essentially linear. The corrected LMTD and heat transfer coefficients are determined for each zone and the zone area is calculated from Equation 13.2. The areas of each "zone" are then summed to give the total exchanger area.

This procedure becomes increasingly complex as more streams are included in the exchanger. Accurate thermodynamic and physical property data are essential. The first and second laws of thermodynamics cannot be violated. Check that total duties of all fluids sum to zero and no temperature crosses can occur throughout the exchanger.

Heat leak in cryogenic heat exchangers is another factor which will affect the cooling curve. It acts as an unwanted heat flow into the heat exchange fluids and will reduce the effective LMTD. For well insulated exchangers, heat leak normally has a negligible effect on the LMTD. However, the amount of heat leak should always be checked and combined as another warm stream on the cooling curve to determine its effect on the LMTD.

Figure 13.17 shows an BAHX application in a simple mechanical refrigeration process. Inlet gas is cooled by heat exchange with the cold processed gas, cold separator liquid, and propane refrigerant. All of the exchanger surface area is contained in one unit but there are actually three separate exchanger services — gas-gas, gas-liquid, and gas-refrigerant. The cooling (heating) curves for these services are shown in Figure 13.18.

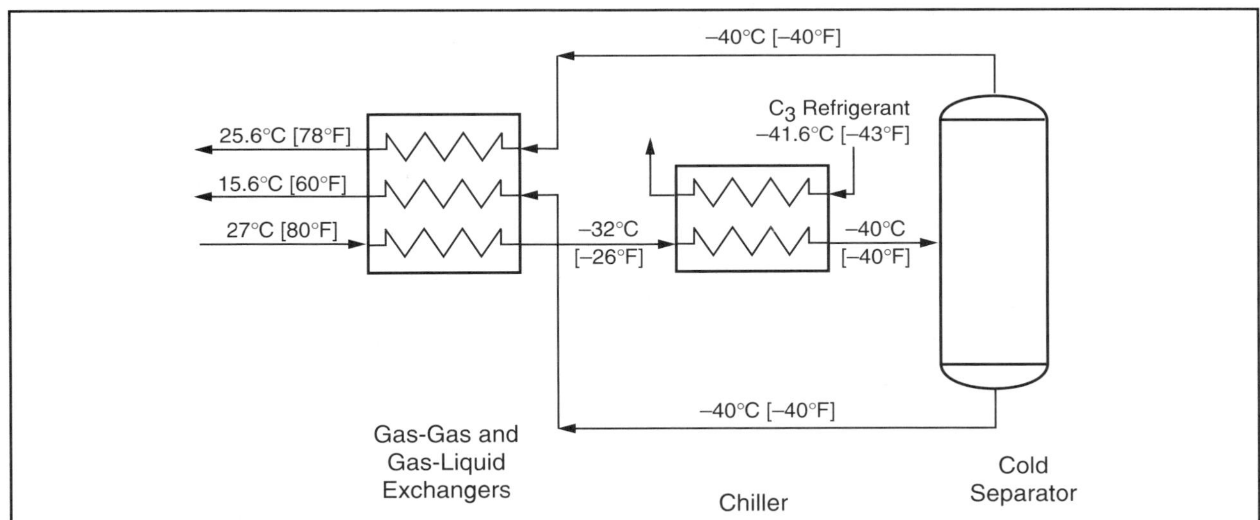

Figure 13.17 BAHX Application in Mechanical Refrigeration Service

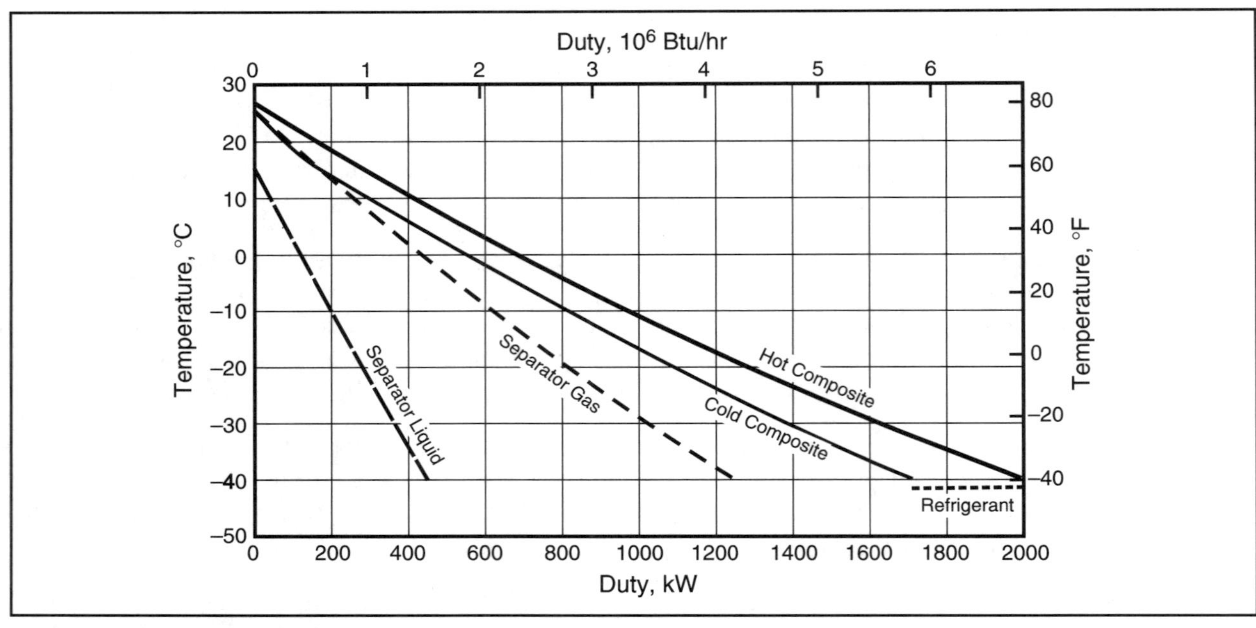

Figure 13.18 Heating/Cooling Curves for BAHX in Figure 13.17

For purposes of exchanger sizing the heating curves for the cold separator liquid and gas are combined and the result is the "hot composite" curve in Figure 13.17. ΔT_m and area are calculated for this section of the exchanger using the feed gas cooling curve (cold composite) and the hot composite curve. The remainder of the exchanger is for feed gas-propane refrigerant service.

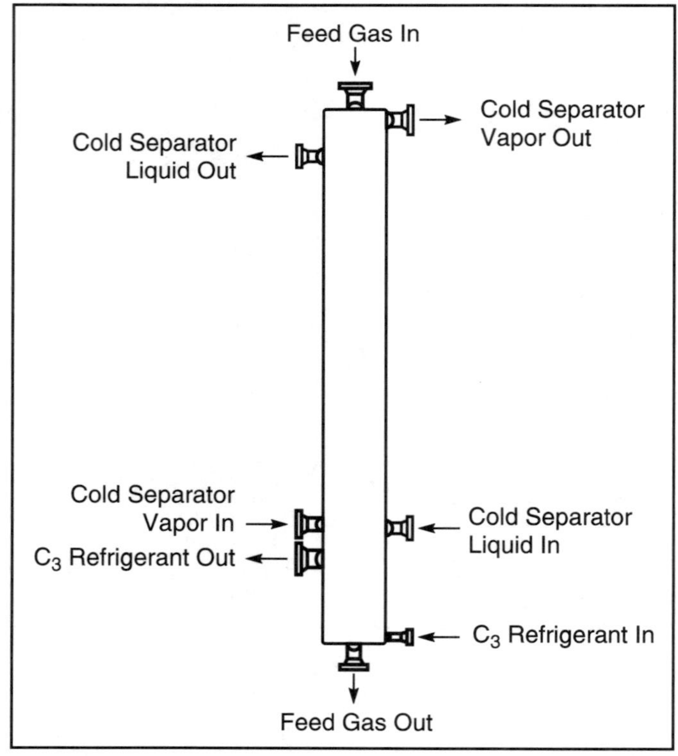

Figure 13.19 Nozzle Orientation for BAHX in Figure 13.17
(Courtesy Chart Heat Exchangers)

The exchanger configuration is shown in Figure 13.19. Approximately 15% of the exchanger area is allocated to chiller service, the remainder to gas-gas and gas-liquid exchange. The overall exchanger length is 6.4 m [21 ft], width 0.63 m [25 in] and height 0.59 m [23.2 in]. Space and weight requirements for shell and tube exchangers (3 required) would be significantly greater than this. The BAHX exchanger weight (empty) is approximately 4000 kg [8800 lbm].

Printed Circuit Heat Exchangers

Printed circuit exchangers (PCHEs) were introduced in the oil and gas industry in the early 1980s. PCHEs are constructed from flat metal plates into which flow channels have been chemically milled or etched. The figure below shows one such plate. Flow patterns vary depending on the exchanger specification. Passages are typically 1-2 mm [0.04-0.08 in] deep.

The plates are stacked and diffusion bonded together to form a heat exchanger core. Two or more fluids can be accomodated in the core. Diffusion bonding is a welding process in which the plates are compressed together and heated to just below the melting temperature of the material. At these conditions the plates fuse together forming a solid block of material containing thousands of flow passages.

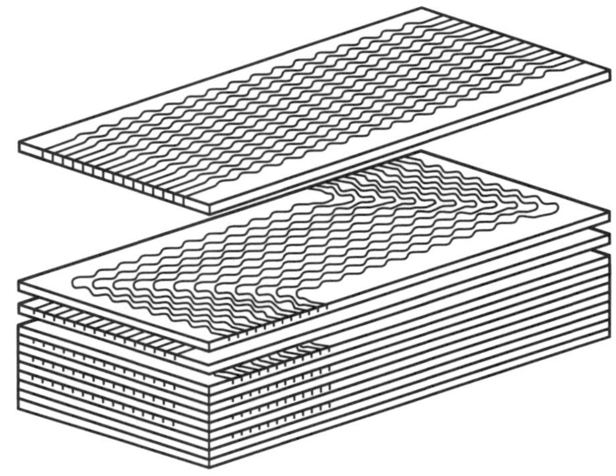

To complete the exchanger construction, fluid headers and nozzles are welded to the core to direct the fluids to the appropriate passages. An example of a PCHE is shown in Figure 13.20.

Figure 13.20 Printed Circuit Heat Exchanger *(Courtesy Heatric)*

Because the exchanger core is essentially a fused block of material, the design pressures are very high, up to 50 000 kPa(g) [7280 psig]. The material of construction is typically 316L stainless, but other materials can be used.

The primary advantages of the PCHEs is similar to other compact exchangers. The size and weight is often less than 25% of a comparable shell and tube exchanger. In addition, the high design pressure precludes the need for pressure relief on the exchanger. In offshore applications the installed cost of a PCHE is almost always less than a shell and tube, particularly if the shell and tube must be constructed of materials other than carbon steel. PCHEs do not require gaskets and are not subject to mercury induced corrosion.

One disadvantage of the PCHE is narrow flow passages which may be subject to plugging in dirty service, and the inability to mechanically clean the exchanger. Deposits such as hydrates or paraffins can often be removed by periodic warming of the exchanger or by injection of a solvent or inhibitor. Solid deposits may be removed by back-puffing or offline chemical cleaning. In general it is standard practice to install an inlet strainer (~ 300 μm [0.012 in] opening).

PCHEs are subject to failure due to thermal fatigue if operated in a service with frequent and severe (> 50°C [90°F]) temperature fluctuations.[(13.27)] Some failures of these exchangers have occurred in compressor aftercooler applications where fluctuating gas flows coupled with unstable temperature control led to several thousand temperature variation cycles. In applications where the temperature difference in the exchanger is large and one of the flowrates frequently varies, it is extremely important that the temperature control loop be tuned to avoid these temperature fluctuation cycles. In addition, PCHEs should not be used in on/off applications where wide temperature differences exist between the fluid, e.g., mol sieve regeneration heater.

Thermal fatigue is of little concern in high efficiency counterflow applications such as gas-gas and gas-liquid exchangers in gas processing facilities.

PIPE-IN-PIPE EXCHANGERS

This type of exchanger may be advantageous for relatively low heat loads where one stream is a gas or viscous liquid or for relatively small exchangers operating at high pressure. Figure 13.21 shows various details of an exchanger employing longitudinal fins.

In this exchanger a piece of pipe serves as a shell. Inside is a single concentric pipe or a group of pipes. A commercial type used commonly consists of a single U-tube unit which may be manifolded in series and parallel to satisfy the heat duty.

This is very efficient. The cost is competitive so long as the total area needed for heat exchange is not too large. For very large areas, weight and/or volume may prove to be unfeasible for a given application.

Figure 13.21 also shows finned tubes. Those shown are all on the outside of the tubes, but they also can be used inside. The shape and style vary widely.

The purpose of the fin is to increase the surface area to the fluid. The area of the bare pipe plus fins is called the *extended area*. By extending the area with fins the resistance per unit heat transferred is decreased. Thus, fins are used with that fluid whose film coefficient is so low that areas must be increased to give an economical rate of heat transfer.

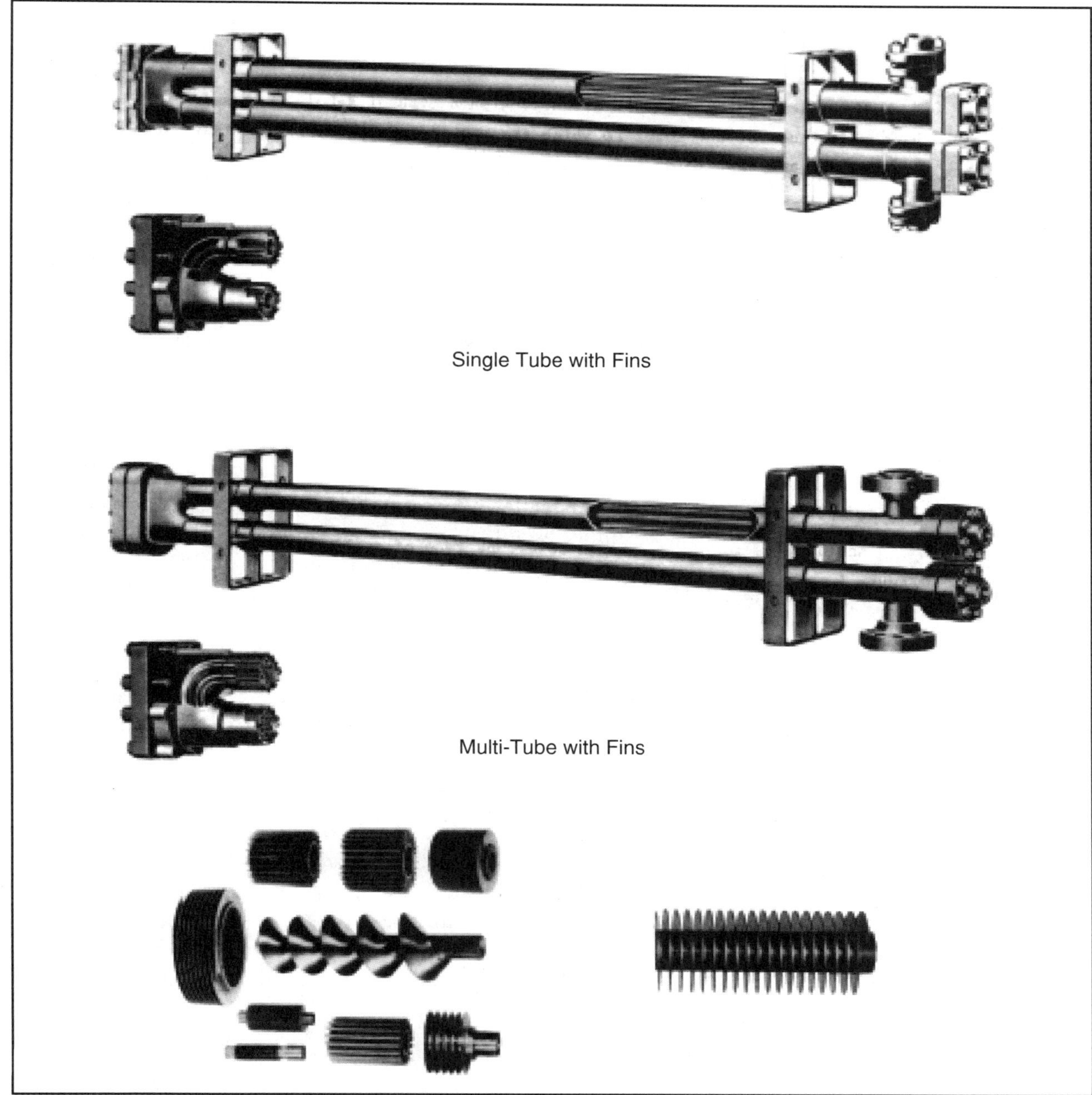

Figure 13.21 Some Details of Finned Tubes and Pipe-in-Pipe Heat Exchangers

Fins are almost always used with aerial coolers because the air film heat transfer coefficients are low. Finned tubes also may be used in all other types of heat exchangers. They should not be used arbitrarily, however. Fins increase pressure drop and are difficult to clean of scale and solids which become imbedded in them. With corrosive fluids, erosion-corrosion may be enhanced due to impingement and turbulence problems. Unless the fins are firmly attached, vibration may cause mechanical separation of the fin from the tube.

The design of these units is similar to other exchangers. Film coefficients are calculated from a Nusselt number correlation like Equation 13.5, often called the Seider-Tate correlation.[13.28, 13.29]

Maximum velocity is limited by erosion, vibration and pressure drop. The maximum velocity desirable can be estimated.

$$v_{max} = \frac{A}{\rho^{0.5}} \tag{13.14}$$

Where:		SI	FPS
	v = velocity	m/s	ft/sec
	ρ = density	kg/m^3	lbm/ft^3
	A = factor	122	100

A velocity higher than this may be used sometimes when employing an erosion resistant material with sweet fluid that is also free of solids.

PIPE COILS

Pipe coils come in many forms. One example is the combination heat exchanger-surge tank used in small glycol dehydrators. This is simply a coil of pipe or tubing wrapped around a mandrel and inserted into a vessel. Other vessels contain pipe running longitudinally with "U" bends at the end. Banks of pipes may be placed in the water basin of a cooling tower in units called atmospheric sections.

In all cases like these, the liquid on the outside is flowing at a low rate, or is essentially stationary. The heat flux (heat transferred per unit time per unit area) is relatively low, but this type of exchanger is inexpensive and is suitable for some limited applications. The calculation is equivalent to that shown for other fluid exchangers.[13.30]

SPECIAL EXCHANGERS

The need for close approaches and high surface areas may require exchangers possessing less weight and higher surface area per unit volume than more traditional types. At very low temperatures, such as LNG exchangers, a 1°C increase in approach may require an extra energy input equivalent to 55-97 $kW/10^6$ std m^3 of gas processed. Thus exchangers are needed with minimum resistance to flow, and maximum surface per unit weight and volume.

Figure 13.22 shows several common types used.[13.31] The Hampson model uses tubing wound on a mandrel. This minimizes thermal stresses, gives a large area per unit volume, and minimizes channeling on the shell side. The Trane exchanger consists of corrugated aluminum sheets brazed between flat aluminum plates in layers and sealed with aluminum channels to form the flow passages. This exchanger is the BAHX discussed earlier. The Joy-Collins exchanger has two (or more) concentric tubes joined together by soldering spiral-wound metal helices between their walls.

The Ramens Lamella design is a shell-and-tube type using a cylindrical bundle of flattened tubes to obtain more area per unit volume. These lamellas (flattened tubes) are welded and thus are not easy to repair or remove. Depending on metallurgy, this exchanger is available for pressure to 3.6 MPa [525 psia] and temperatures to 500-600°C [932-1112°F].

All of these may be used so that the streams involved may periodically change channels (reversing process). This is frequently used to revaporize solids frozen out of high pressure feed (such as water and CO_2) by using a lower pressure residue stream.

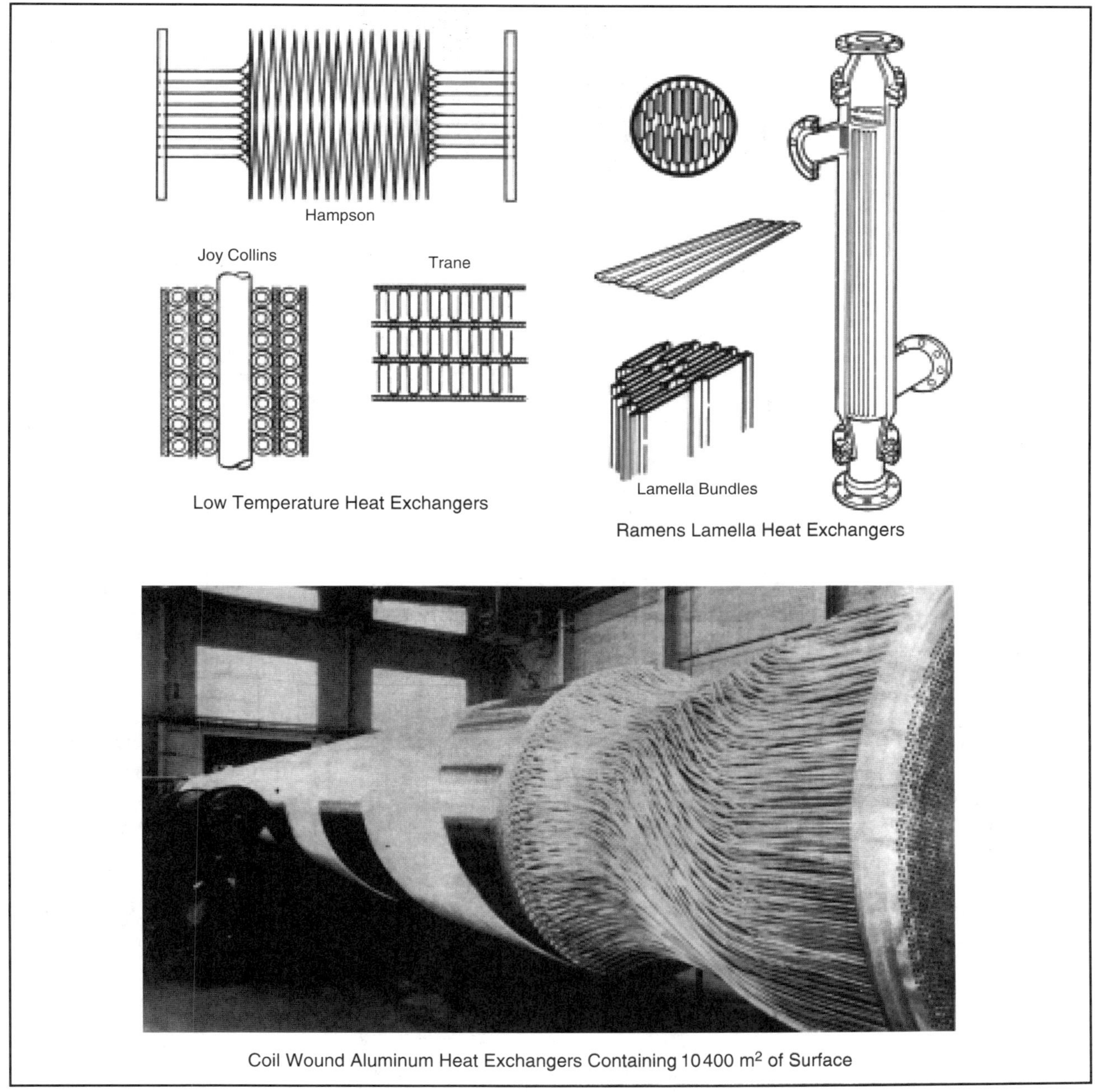

Figure 13.22 Examples of Special Service Exchangers

The coil-wound, aluminum heat exchanger also shown in Figure 13.22 has become very popular for LNG service. One main problem has been plugging from solid debris which has sometimes necessitated back-flushing with dry nitrogen or a similar material. Small tube size is the price one pays for maximum area per unit volume.

The regenerative heat exchanger of the type used in other processes also shows some promise. This may consist of one or two vessels containing a high heat capacity packing. Two vessels are required for continuous service. One is sufficient for use in intermittent processes. In operation, one stream gives up heat to the packing, which heat is, in turn, picked up by the other stream. To be effective, the two streams must have about the same enthalpy requirements and be fairly close in average temperature. Regenerative heat exchangers are fairly cheap to manufacture and may have about 6500

m^2 [70 000 ft^2] of surface per cubic meter of volume. Several types of aluminum packing are commonly used.

RADIANT HEAT TRANSFER

This becomes a significant factor as the temperature increases. It is the major factor in most fired heaters.

The basic equation relating the radiation variables is known as the Stefan-Boltzmann Law:

$$Q = k A (T_1^4 - T_2^4) F_e F_a \tag{13.15}$$

Where:			SI	FPS
k	=	constant	5.72×10^{-11}	1.73(E-09)
A	=	surface area	m^2	ft^2
T_1	=	higher temperature	K	°R
T_2	=	lower temperature	K	°R
F_e	=	emissivity factor	–	–
F_a	=	geometry factor	–	–
Q	=	heat gained or lost	kW	Btu/hr

The geometric factor accounts for relative size and shape, and distance between the two bodies exchanging radiant heat.

A given body can absorb or emit radiant energy. At thermal equilibrium the ratio of the emissive power of a surface to its absorptivity is the same for all bodies. A perfect radiator has an emissivity of one. Such a surface must have an absorptivity of one and reflectivity of zero. This hypothetical surface is commonly referred to as a *black body*. The ratio of the emissive power of an actual surface to that of a "black body" is known as the *emissivity*. At thermal equilibrium the emissivity and absorptivity of a body are identical.

Below, typical emissivity values are listed for the types of surface commonly used in processing.

Surface	Emissivity
Aluminum	0.040-0.055
Iron and Steel	
Rolled sheet steel	0.66
Oxidized iron	0.74
Iron oxide (rusted surface)	0.87
Galvanized sheet iron	0.28
Brick (red)	0.93
Glass	0.94
Roofing paper (black)	0.91
Paints	
Black lacquer	0.80-0.95
Flat black	0.97
Aluminum	0.40-0.60
White enamel	0.90
Oil paints (all colors)	0.92-0.96

The above shows that aluminum surfaces will reflect more radiant heat from solar radiation. Generally, darker, less glossy surfaces are more efficient emitters of heat. The type of surface is more important than color.

We paint primarily to protect and beautify, but in doing so it should be kept in mind that the type of surface can have a noticeable effect on heat gain or loss, particularly outdoors where rapid changes can occur between a clear sunny day and a clear black night.

Lauer has prepared a series of correlations for estimating radiant heat transfer using a pseudo-film coefficient. He equated ($hA\Delta t$) to Equation 13.17 to solve for an equivalent "h."

INDIRECT FIRED HEATERS

There are different configurations, two of which are shown in Figure 13.23. The hot combustion gas and flame heat an intermediate liquid which, in turn, heats a fluid flowing through a coil or a series of tubes. The intermediate liquid must be stable at atmospheric pressure and the maximum temperature involved. It is typically water, heat transfer oil or a eutectic salt, depending on the temperature level. Indirect heaters have proven safe, reliable and convenient to use. Both radiation and convection are involved.

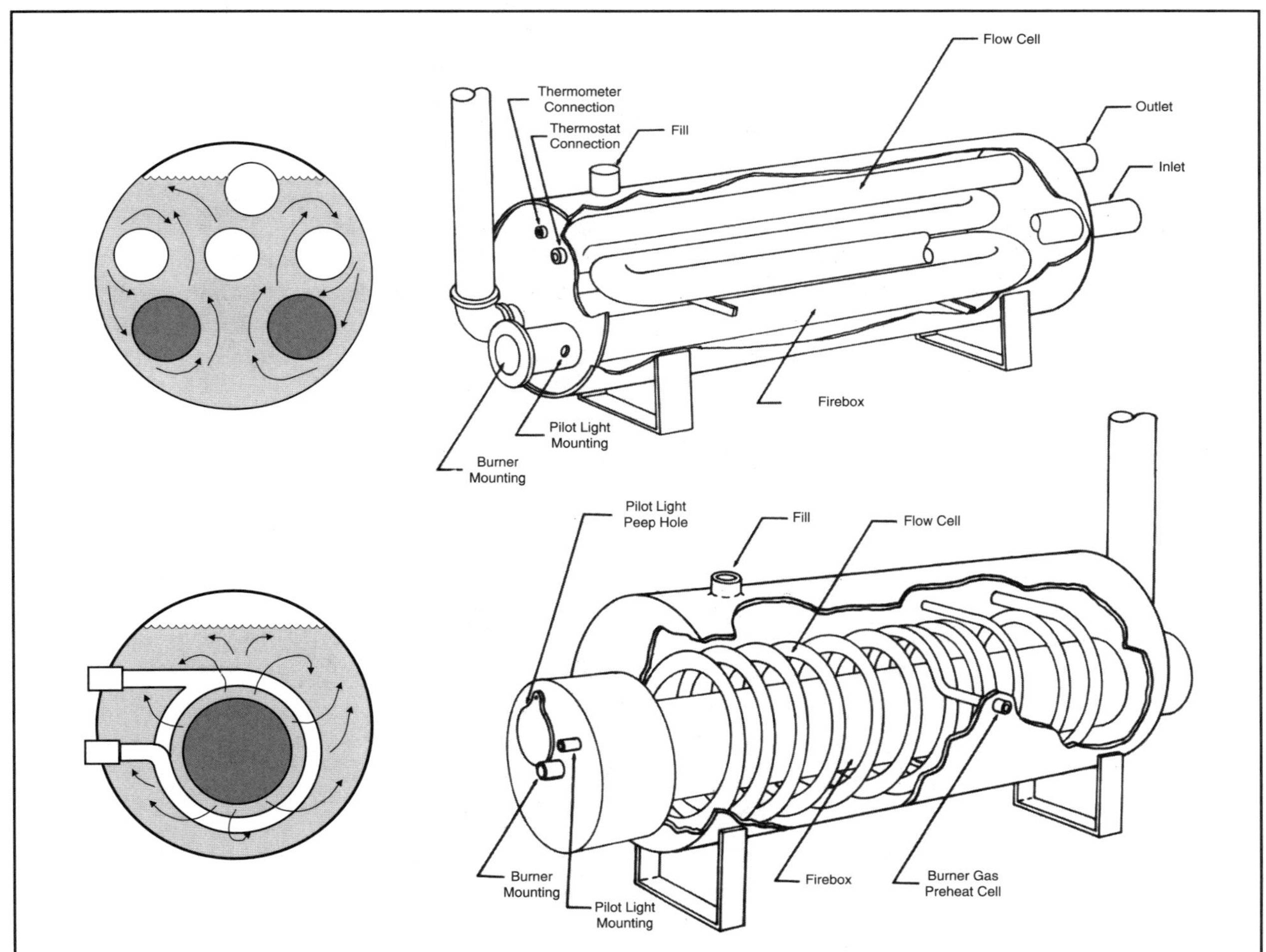

Figure 13.23 Basic Types of Indirect Fired Heaters

The intermediate liquid transfers heat between the fire tube and the fluid being heated by natural convection. This limits the rate of heat flux per unit area. Indirect heaters are seldom used to produce outlet fluid temperatures above 260°C [500°F]. The primary use is to heat oil and gas in production operations where the heat loads are not large. This configuration is also used in amine, glycol, and oil stabilization reboilers. In these cases, there is no intermediate heat transfer fluid, heat is transferred directly from the firetube to the process fluid.

For line heaters in natural gas service (water-bath heaters) typical sizing criteria are:

Fluid coil: Overall "U" (in-service) 250-300 $W/m^2 \cdot °C$ [44-53 $Btu/hr\text{-}ft^2\text{-}°F$]

Fire tube: Heat flux 28-38 kW/m^2 [9000-12 000 $Btu/hr\text{-}ft^2$]

For maximum firetube life it is recommended that heat flux not exceed 28 kW/m^2 [9000 $Btu/hr\text{-}ft^2$].

DIRECT FIRED HEATERS

Direct fired heaters are the heat exchangers where the heat is liberated by the combustion of fuel (usually natural gas) within an internally insulated enclosure. The heat is transferred to process fluid which flows through the heater in a series of tubes. Typically the tubes are installed along the walls and roof of the combustion chamber where heat transfer occurs primarily by radiation. Most heaters today also include a separate tube bank section where the heat transfer is primarily by convection.

Although fired heater duties sizes can vary from as small as 150 kW [500 MBtu/hr] up to 150 MW [500 MMBtu/hr] the vast majority of heater applications fall between 3 MW [10 MMBtu/hr] and 100 MW [350 MMBtu/hr]. The most common fired heater applications in E & P applications include still bottom heaters in lean oil plants, regeneration gas heaters in dry desiccant dehydration systems, hot oil (or other heat transfer fluid) heaters, oil heaters upstream of oil dehydration units, and boilers.

A wide variety of models and heating configurations are employed. The choice depends on fuel cost, thermal efficiency, temperatures desired, size of the heat load and the fluid being heated. Figures 13.24 and 13.25 show various types of commonly available direct fired heaters.[13.32]

In Figure 13.24 the fluid tubes are horizontal. The characteristics can be summarized as follows.

A. *Cabin* — The radiant section normally lines the walls with burners in the floor. An economical and high efficiency unit that currently is the most popular of horizontal tube units. Normal duty range: 3-30 MW [10-100 MMBtu/hr].

B. *Two-Cell Box* — Only two boxes are shown but three or four can be used. Vertically fired from floor to give an economical, high efficiency design. Normal duty rating: 30-75 MW [100-250 MMBtu/hr].

C. *Cabin with Bridgewall* — This divider provides two sections which can be fired individually. Can be fired horizontally or vertically. Normal duty range: 6-30 MW [20-100 MMBtu/hr].

D. *End Fired Box* — Horizontally fired as name implies. Normal duty range: 1-15 MW [5-50 MMBtu/hr].

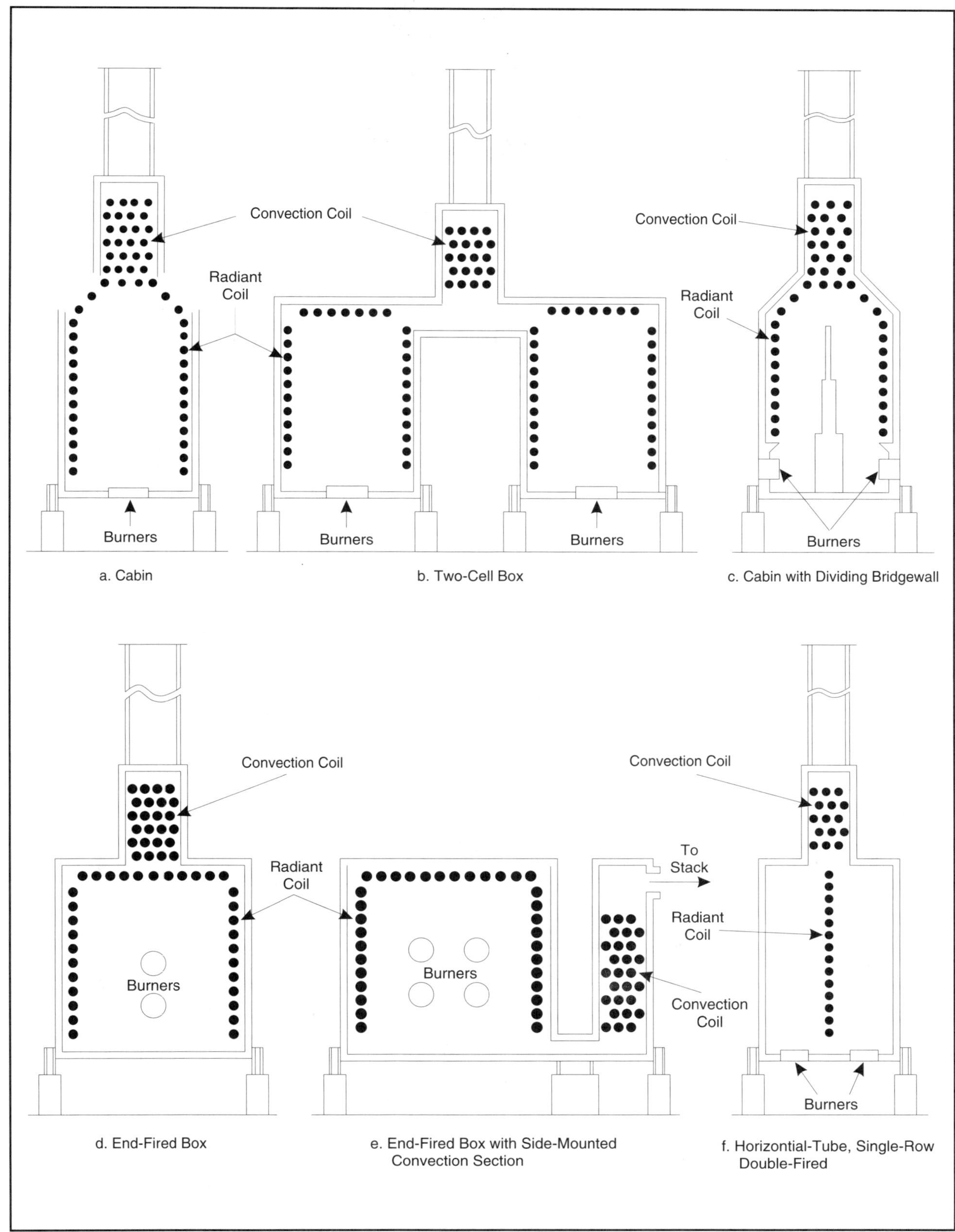

Figure 13.24 Basic Types of Direct Fired Heaters with Horizontal Tubes

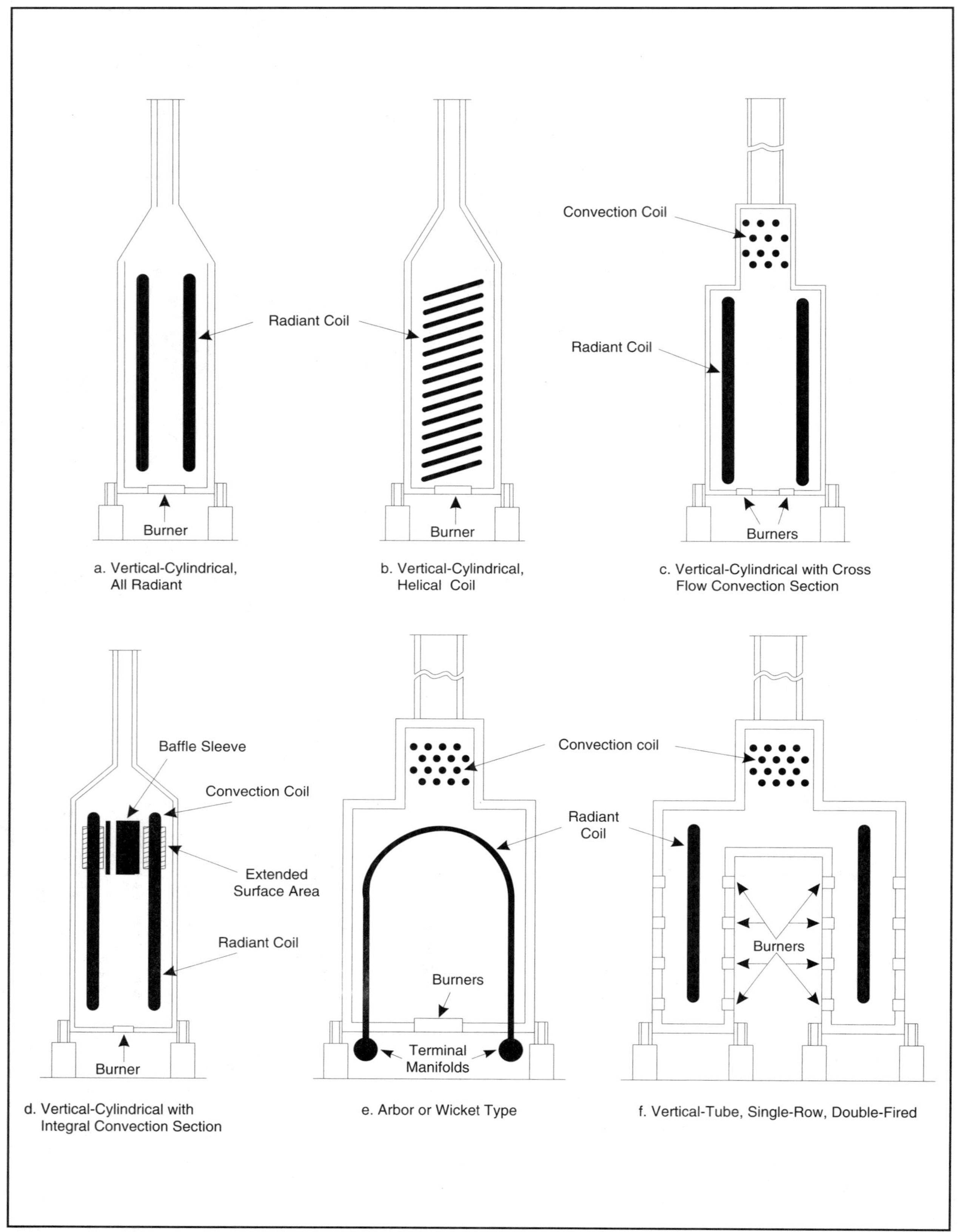

Figure 13.25 Basic Types of Direct Fired Heaters with Vertical Tubes

E. *End Fired Box with Side Mounted Convection Section* — An older type unit that may be used in new installations with high ash, poor grade fuels. More expensive design. Normal duty range: 15-60 MW [50-200 MMBtu/hr].

F. *Single Row, Double Fired* — Consists of one or more cells shown. Often used for reactor-feed heating services. Normal duty range: 6-15 MW [20-50 MMBtu/hr].

Figure 13.25 shows some comparable vertical tube heaters. The characteristics can be summarized briefly as follows.

A. *All Radiant* — Low cost, low efficiency design that is compact. Normal duty range 0.1-6 MW [0.5-20 MMBtu/hr].

B. *Cylindrical, Helical Coil* — A basic low cost, low efficiency alternate to (A). Not feasible to have parallel flow coils for fluid. Normal duty range: 0.1-6 MW [0.5-20 MMBtu/hr].

C. *Cylindrical, with Cross-Flow Convection* — Probably the most popular of new vertical flow units. It is an economical, high efficiency, compact unit. Normal duty range: 3-60 MW [10-200 MMBtu/hr].

D. *Cylindrical, with Integral Convection* — There are a large number of existing units of this type. Not often purchased now for new installation because of the limited thermal efficiency. Normal duty range: 3-30 MW [10-100 MMBtu/hr].

E. *Arbor or Wicket* — Used most commonly for heating large quantities of gas where low pressure drop is desired. Several arbor coils may be used in one heater unit. Normal duty range: 15-30 MW [50-100 MMBtu/hr].

F. *Single Row, Double Fired* — Most expensive configuration but provides high and rather uniform heat flux. Normal duty range: 6-35 MW [20-125 MMBtu/hr].

Direct fired heaters have evolved over the years to reflect the need for greater efficiency and more reliable performance. With high flame temperatures and low convection film coefficients, the development of "hot spots" and tube failure always has been a problem. Tube metallurgy selection normally is a compromise between initial cost and service life. The choice of material, method of welding, configuration used, etc. must be based on experience.

Common Heater Configurations

One of the most common heater designs in E & P operations is the horizontal-tube cabin design as illustrated in Figure 13.26. The structure consists of refractory covered side walls, end walls, roof, and floor. Heating capacity determines the cross-section of the combustion chamber and length of the heater. Horizontal tubes extend along the length of the side walls surrounding the combustion chamber. Usually tubes are omitted from the end-walls in order to simplify construction. Burners can be located in the floor for vertical firing, or in the end-walls for horizontal firing against a refractory target-wall located in the center of the heater. Hip tubes are arranged in a sloping pattern with roof tubes in a horizontal pattern. Shock tubes are located in the first row above the roof tubes. The side-wall tubes, hip tubes, and shock tubes are bare without extended surface because they receive radiant heat from the combustion chamber which subjects them to high heat fluxes.

Hot combustion gases flow from the combustion chamber through the roof tubes and shock tubes into a convection section which usually contains finned tubes arranged in a staggered (triangular) pattern. Heat is transferred by convection from the combustion gases to the extended surface (fins) on the tubes and then by conduction along the fins to the inner surface of the tube where it enters the process fluid.

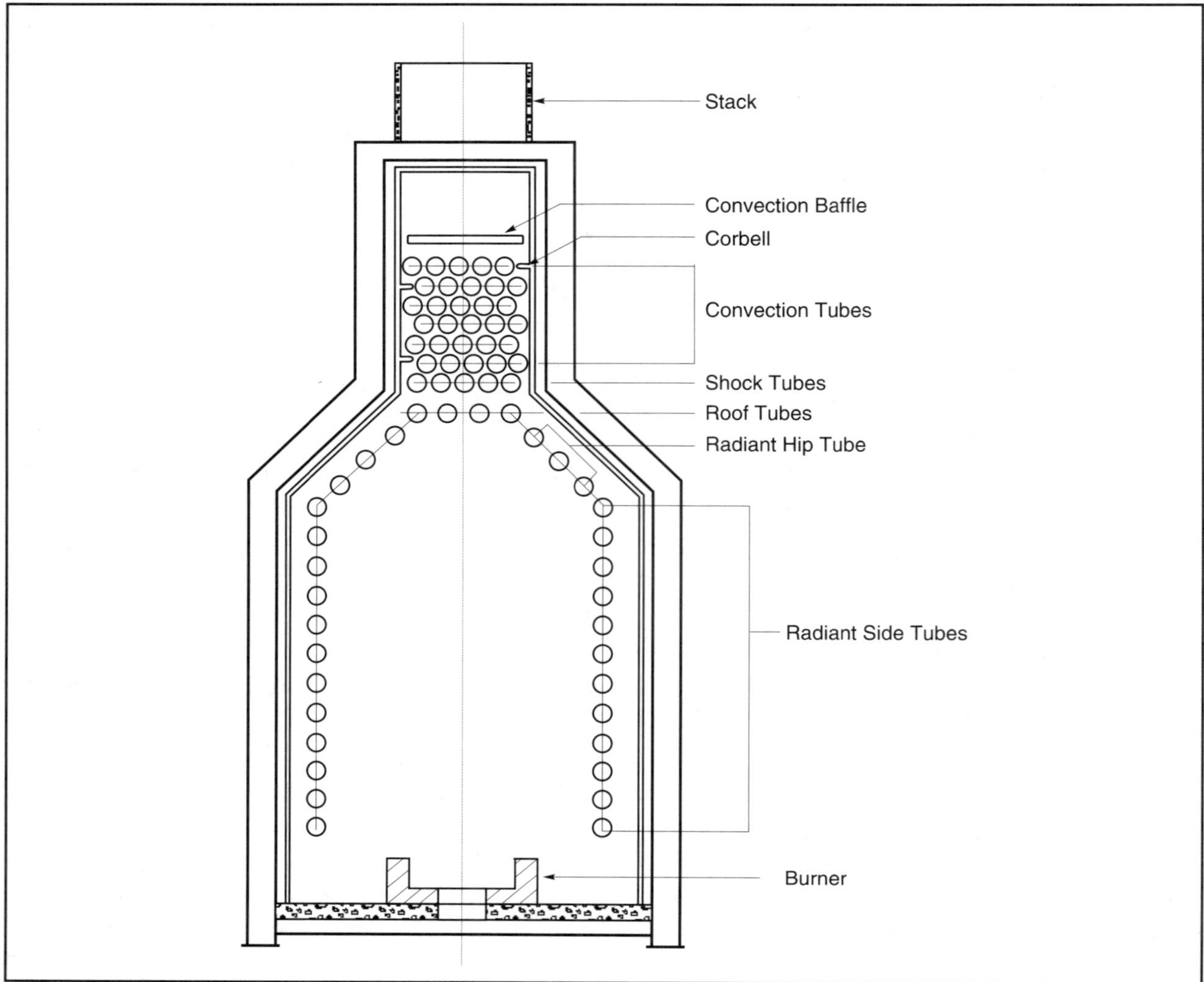

Figure 13.26 Cabin-Horizontal Tube Heater

Countercurrent flow is used except for some special services. The process fluid normally enters the convection tubes and flows downward through one or more parallel sections. From the convection section it flows through the shock tubes, and the roof tubes, and the side wall tubes for final heating before it leaves the heater.

Another popular heater design, particularly in smaller heaters, is the vertical cylindrical heaters. These heaters are usually selected because of their low initial cost or because of limited space available for installation. The necessary heat transfer surface can be provided in the cylindrical structure, but space available for burners limits the design flexibility. Although the vertical cylindrical design produces a simple and compact heater, it has fewer desirable features than the horizontal-tube cabin unit. Repair or replacement of tubes often require the use of special hoisting equipment.

A straight tube coil heater is shown in Figure 13.27. The heater shell is a vertical refractory covered cylinder which provides an elongated circular combustion chamber. The wall tubes are vertical and are arranged around the circumference of the combustion chamber. A convection section is usually provided to maximize thermal efficiency. In many cases the convection sections have a rectan-

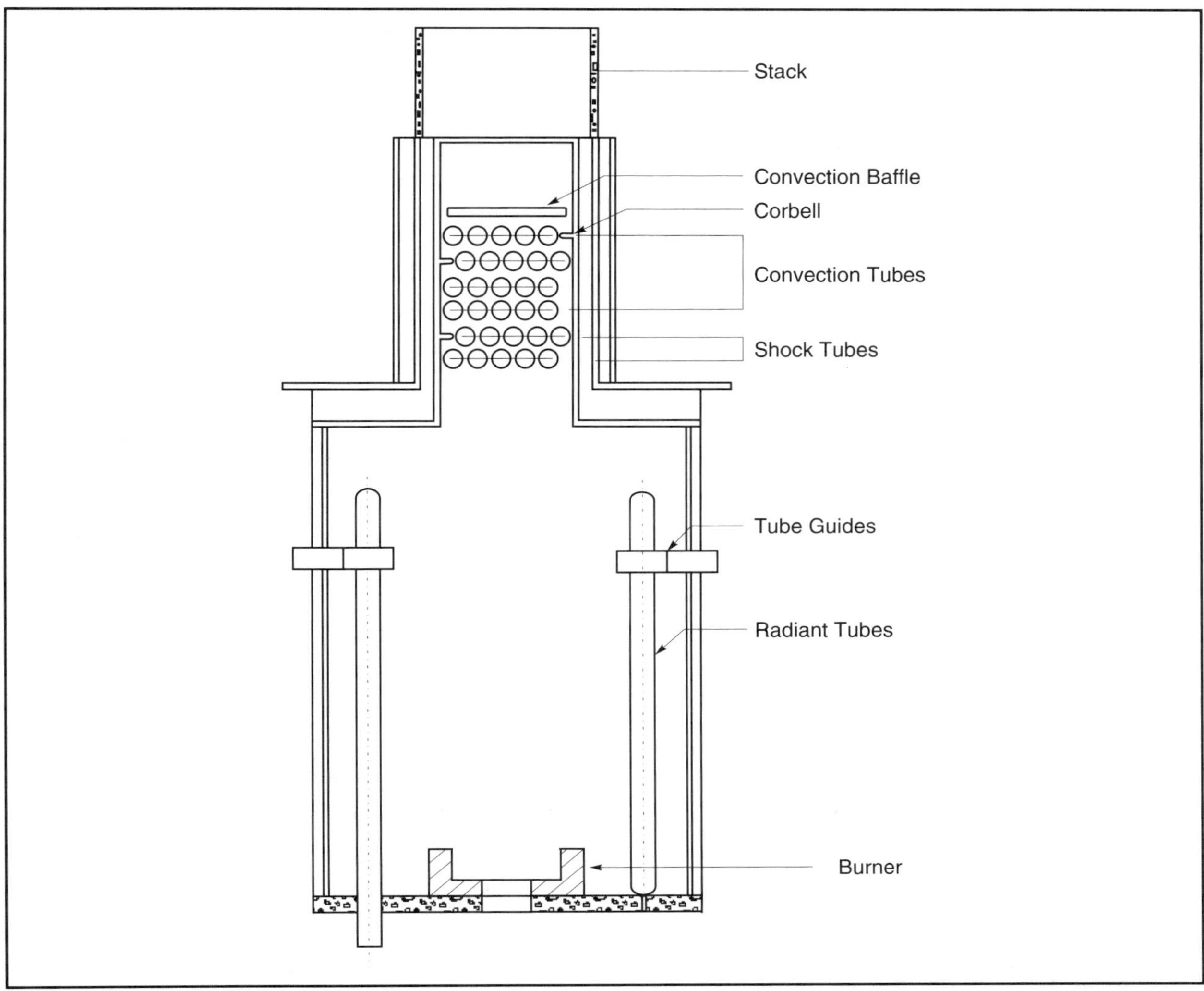

Figure 13.27 Vertical Straight Tube Heater

gular cross-section and is arranged like the convection section of the cabin-horizontal tube heater, complete with shock tubes.

Air Supply and Flue-Gas Removal

Aside from the major classifications according to service and configuration, fired heaters can also be grouped according to their methods of combustion-air supply and flue-gas removal.

Combustion air flow is induced or drawn into a fired heater when hot flue gas of relatively low density is confined in a structure and isolated from higher-density air at ambient temperature. The buoyancy of the hot flue gas contained in the fired heater creates "draft" (a local pressure less than atmospheric pressure), which induces flow of air into the combustion chamber. Since this draft results from a natural stack effect, it is termed natural draft. The majority of fired-heater installations are such natural-draft types, where the stack effect both introduces the combustion air and removes the flue gas.

Obstruction to the flow of flue gas through a fired heater can result in a local positive pressure somewhere in the structure. It is the function of the stack in a natural-draft heater to generate draft sufficient to overcome such obstructions and to maintain a negative pressure throughout.

An induced-draft fired heater incorporates an induced-draft fan, in lieu of a stack, to maintain a negative pressure and to induce the flow of combustion air and the removal of flue gas.

A forced-draft fired heater is one wherein the combustion air is supplied under positive pressure by means of a forced-draft fan. Even with air supplied under positive pressure, the combustion chamber and all other parts of the fired heater are maintained under negative pressure, and the flue gas is removed by the stack effect.

Radiant and Convection Sections

The bulk of the heat transfer in a fired heater occurs in the radiant section, accounting for about 2/3 of total heat transferred.

The actual radiant heat transfer depends on the temperature and emissivity of the gas, tube layout and spacing. Higher radiant duty heaters with a lower proportion of convective heat transfer are less expensive heaters, but also incur higher maintenance expenses since the tubes, clips and refractories are exposed to higher temperatures. In addition the higher tube wall temperatures raise the potential for coke deposition (hot spots) and product degration.

Typical radiant heat fluxes for heaters in E & P operations range from 30-40 kW/m^2 [10 000-12 000 Btu/hr-ft^2].

The shock tubes are considered part of the radiant section because they "see" the flame. The tubes above the shock tubes make up the convection section of the heater. The convection tubes are typically finned. Increasing the surface area of the convection section results in lower stack gas temperatures and a higher thermal efficiency but increases heater cost and flue gas pressure drop.

Typical heat fluxes in the convection section are 60-80 kW/m^2 [20 000-24 000 Btu/ft^2-hr]. This is based on bare tube area but takes into account the effect of the fins.

Burners

The fundamental criteria for selecting a burner include (1) the ability to handle fuels having a reasonable variation in calorific value, (2) provision for safe ignition and easy maintenance, (3) a reasonable turndown ratio between maximum and minimum firing rates, (4) predictable flame patterns for all fuels and firing rates, and (5) noise.

Burners designed for gaseous fuel only are classified into two basic categories: premix inspirating and rawgas burning.

Premix Inspirating. The premix burner relies on the kinetic energy from expansion of the fuel gas through an orifice to inspirate (draw in) and mix combustion air prior to ignition at the burner tip. Approximately 50 to 60% of the combustion air is inspirated as primary air into the burner ahead of the ignition point.

Some of the advantages of this type of burner are:

1. Operating flexibility is good. Premix burners can operate at low excess-air rates and are not significantly affected by changes in wind velocity and direction.
2. Flame length is short, and flame pattern sharply defined at high heat-release rates.
3. Burner orifices are fairly large and located in the cold zone of the burner. They are less prone to plugging than the smaller openings on noninspirating gas burners.

Some of the disadvantages of inspirating burners are:

1. Relatively high fuel gas pressures must be available. Below about 70 kPa(g) [10 psig] the percentage of inspirated air falls rapidly and flexibility is reduced.
2. Flashback of the flame from the burner tip to the mixing orifice may occur at low gas pressures.
3. The noise level of premix inspirating burners is higher than that of noninspirating types.

Raw-gas burning. The nozzle-mixing, raw-gas burner receives fuel gas from the gas manifold without any premixing of combustion air. The gas is then burned at a tip equipped with a series of small ports.

Some of the advantages of this type of burner are:

1. Best turndown ratio.
2. It can operate at very low gas pressures on a wide variety of fuels and without flashback.
3. Noise level is reasonably low.

Some of the disadvantages of raw-gas burners are:

1. Heater efficiency varies over its wide turndown range. Because no primary air is inspirated, combustion-air adjustments must be made over the full operating range of the burner.
2. Burner ports diameter is very sensitive. Enlargement of the port opening will often result in unsatisfactory flame conditions.
3. Flames tend to lengthen, and flame conditions may become unsatisfactory if the burner is operated beyond its design level.
4. The burner ports are exposed to the hot zone and are subject to plugging at low velocities and high temperatures.

Heater emissions, particularly NO_x, are a critical issue in heater specification. The formation of NO_x can occur due to the oxidation of nitrogen in the fuel (fuel NO_x) or from the combination of the nitrogen and oxygen in the air (thermal NO_x). The amount of NO_x formed depends on several factors but the primary ones are flame temperature and burner residence time.

Maximum burner efficiency is favored at low excess air operation, typically less than 15-20%. Unfortunately, this operating region also corresponds to maximum NO_x formation. Low NO_x burners are designed with a two-stage combustion process. The initial combustion takes place in a fuel-rich zone, followed by a second air-rich zone. NO_x formation is minimized in the first zone because the oxygen is fully consumed by the fuel. The secondary zone oxidizes any CO formed due to incomplete combustion in the first zone.

Some operators have used water or steam injection at the burner to lower combustion temperatures. Recirculation of flue gas back to the burner can also lower NO_x by reducing available oxygen and combustion temperature.[13.33-13.35]

Stack Design

The main functions of a stack are to induce the flow of combustion air into the heater, and to produce draft sufficient to overcome all obstructions to the flow of flue gas while maintaining a negative pressure throughout the heater. A heater should not be operated with greater than atmospheric pressure at any point within the structure. Positive pressure within a heater creates a driving force for the outward movement of hot gases, which can lead to serious overheating and corrosion of the steel structure. A simple plot showing the draft in a typical fired heater is shown in Figure 13.28.

The draft produced by a column of hot flue gas depends on the density difference between the hot gas and the ambient air. It can be estimated from Equation 13.16.

$$\text{Draft} = (C)(L_s)(P)\left(\frac{1}{T_a} - \frac{1}{T_g}\right) \tag{13.16}$$

Where:			SI	FPS
		Draft	mbar	in H_2O
L_s	=	stack height	m	ft
P	=	atmospheric pressure	kPa	psia
T_a	=	ambient temperature	K	°R
T_g	=	flue-gas temperature	K	°R
C	=	conversion constant	0.34	0.52

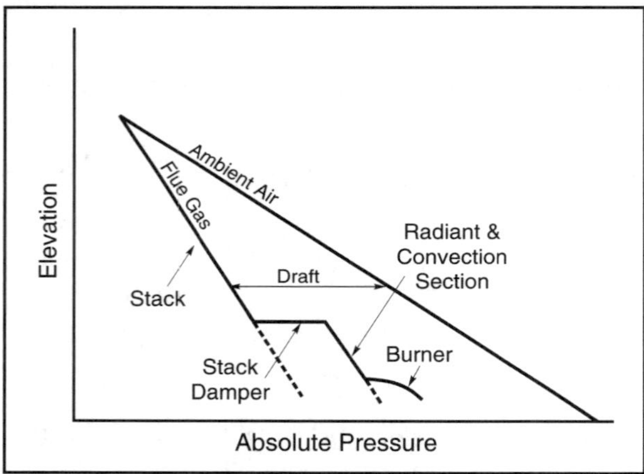

Figure 13.28 Variation of Stack Draft with Elevation for a Fired Heater

Because of heat loss through the stack casing, the flue-gas temperature at the top of the stack is substantially lower than at the inlet at the stack base. The magnitude of the differential depends on several factors, including the stack dimensions and the effectiveness of the stack insulation.

For most applications, the average flue-gas temperature in the stack may be conservatively estimated at 42°C [75°F] less than the inlet flue-gas temperature.

As noted above, a major objective in stack design is to ensure negative pressure throughout the heater. The most critical location in this respect occurs where the flue gas enters the convection section. It is recommended that the stack design be based on a negative pressure of 0.12 mbar [0.05 in H_2O] at this point.

It should be noted that in most heaters the convection section exerts a stack effect due to its physical height and serves to reduce the overall draft requirement.

To assure good flue-gas distribution throughout the convection section, it is usual for a flue-gas opening to be provided at every 12 m [40 ft] of convection section length.

Assuming that there are no requirements dictating a minimum stack height, a reasonable basis for stack diameter sizing would result in a flue-gas mass velocity in the range of 3.6-4.9 $kg/m^2{\cdot}s$ [0.75-1.0 lb/ft^2-sec].

The determination of flue-gas flow is an approximation. Operation at excess air levels greater than design produces increased quantities of flue gas. Furthermore, efficiency decline due to fouling of heat-transfer surfaces also results in greater flue-gas quantities. It is good design to add 25% to the flue-gas design flow when estimating the flue-gas frictional losses.

PROCESS COOLING

In all processes, heat must be rejected to ambient (heat sink). Examples include compressor aftercoolers, refrigeration and reflux condensers, steam condenser, to name a few. In general, from Carnot's Law we know that the process efficiency increases as the heat sink temperature decreases. Selection of the heat rejection (process cooling) method therefore has economic implication to the process. There are three primary methods of process cooling summarized below:

Once-Through Cooling Water		
		Heat Sink Temperature: water ambient temperature Heat Transfer Fluid: water Approach*: 5-10°C [9-18°F]
Advantages:	1)	Simplicity, low capital cost
	2)	Typically gives lowest process temperatures
	3)	Water is heat transfer fluid
	4)	Exchangers are small, minimizing footprint
	5)	Water is less susceptible to ambient temperature fluctuations
Disadvantages:	1)	Limited availability, e.g. desert applications
	2)	Environmental limits on temperature of water returned to environment
	3)	Water is usually corrosive and fouling
	4)	Freezing
	5)	Water can be contaminated by process fluid

Cooling Towers		
		Heat Sink Temperature: wet-bulb air temperature Heat Transfer Fluid: water Approach*: 15-20°C [27-36°F]
Advantages:	1)	Water is heat transfer fluid
	2)	Exchangers are small, minimizing footprint
	3)	Large supply of ambient water not necessary
Disadvantages:	1)	High capital and operating cost
	2)	Make-up water supply required
	3)	Chemicals are necessary to treat water for corrosion, scaling, algae, etc.
	4)	Cooling water blowdown disposal
	5)	Freezing

Aerial Coolers		
		Heat Sink Temperature: dry bulb air temperature Heat Transfer Fluid: air Approach*: 10-20°C [18-36°F]
Advantages:	1)	Air readily available everywhere
	2)	Low environmental impact
	3)	Lower maintenance than cooling towers
	4)	Less fouling than water
Disadvantages:	1)	Poor heat transfer coefficients lead to large footprint and equipment size
	2)	Large variation in ambient temperature day-night, winter-summer, rainy-sunny, etc.
	3)	Low ambient temperatures can lead to freezing of process fluids
	4)	Air flow must be free of obstruction
	5)	Fan noise an issue when installed in populous areas
* process fluid to heat sink temperature		

Cooling towers are seldom used in most oil and gas processing applications. Aerial coolers are the most popular in onshore facilities. Offshore, either aerial coolers or once-through cooling water is typically used depending on water temperature, water depth, and platform cost.

COOLING TOWERS

The sketch below summarizes the process involved. It is only possible to cool the water in the tower when the entering air is unsaturated (above its dewpoint temperature). In the process of passing unsaturated air countercurrent to the warm water entering, some of the water evaporates. The latent heat of that water evaporating must be supplied from somewhere. Most comes from the sensible cooling of water, which does not evaporate. A further cooling is supplied by the air flowing through the cooling tower. This is the mechanism by which water is cooled – evaporating a portion to cool the remainder. The maximum amount evaporated is limited by the air's capacity to hold water vapor. The actual amount of water evaporated will depend on the effectiveness of the mass transfer (efficiency of air-water contact, area of contact, distribution, etc.). The driving force for evaporation is the water concentration difference between the actual water content of the air and its saturation value, i.e. 100% relative humidity. The resisting force is a function of effective area of air-water contact.

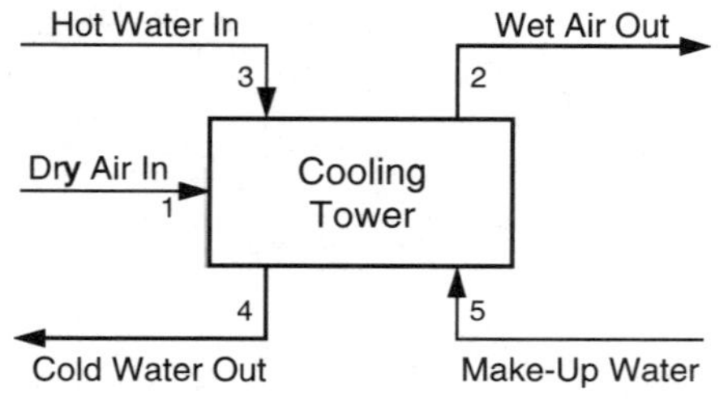

The following quantities must be known or specified for cooling tower design:

Dry Bulb Temperature. The maximum design temperature of the air as measured by a temperature recording device.

Wet Bulb Temperature. The design temperature established by passing air over a thermometer whose bulb is covered with wicking, e.g. cotton cloth, saturated with water. This may be established using a *sling psychrometer* or a device of the type shown at right. Some of the water on the wick evaporates as the unsaturated air passes over it, cooling the bulb. However, following cooling, heat is transferred to the bulb by the thermometer stem and the air. If the process is continued for several minutes, at high enough air rates, a temperature equilibrium is established. This is the wet bulb, the minimum temperature to which the water could be cooled in a cooling tower.

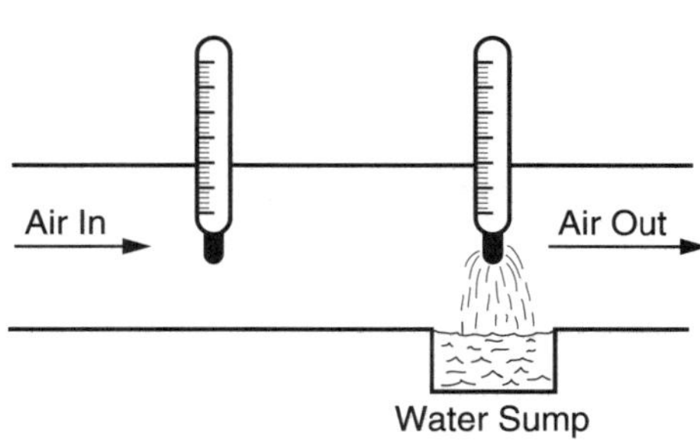

Dewpoint. The temperature at which the air is saturated with water at the pressure involved (atmospheric pressure), i.e 100% relative humidity.

Humidity Ratio (ω). Ratio of the mass of water vapor (M_v) to the mass of dry air (M_a) in an air-water vapor mixture.

Relative Humidity (R.H.). The ratio of the partial pressure of the water vapor in the air mixture (P_p) to the saturation (vapor) pressure (P_v) at the same temperature. The relative humidity (R.H.) is equal to 1.0 at saturation (dewpoint) conditions. "R.H." is related to "ω" by the equation

$$R.H. = \frac{P_p}{P_v} = \frac{\omega P_a}{0.622 P_v} \tag{13.17}$$

Where:

ω = $(0.622\ P_p)/P_a$ (for ideal gases)

P_a = partial pressure of the dry air at total pressure P

Range. The temperature difference between the hot cooling water entering the cooling tower and the cool water leaving. Established by the tower heat load and the water circulation rate. It is a measure of tower loading. The most common range is 5-20°C [9-36°F]. Both range and water circulation rate have a significant effect on tower size.

The equation for calculating range is

$$\text{Range} = \frac{\text{Heat Load}}{(\text{Water Rate})(\text{Water } C_p)} \tag{13.18}$$

Approach. The temperature difference between the cold water leaving the tower and the wet bulb temperature. Approach usually has the greatest effect on tower size and cost. An 8-9°C [14-16°F] approach is typical.

Tower cost depends on the approach and the wet bulb temperature. The table below is based on an approach of 9°C [16°F] which means that the outlet water temperature shown is 9°C [16°F] above the wet bulb.

Outlet Water Temperature	**Relative Cost**
32°C	1.0
30	1.1
25	1.35
20	2.0

Thus, as the wet bulb decreases, to keep the approach the same results in higher cost. The wet bulb seldom exceeds 30°C [86°F]. For most areas of the world it is often 24°C [75°F] or less.

It is customary to specify a dry bulb and wet bulb that will not be exceeded over 5 percent of the time. Designing for a higher temperature is very expensive compared to the benefit.

Energy Balance

In a tower operating at steady state conditions, enthalpy input = enthalpy output. The following overall energy balance may then be written.

$$Q = (m_L)(4.19)(T_3 - T_4) = m_a\left[(0.92)(T_2 - T_1) + (\omega_2 h_{w2} - \omega_1 h_{w1})\right] \tag{13.19}$$

Where:

	ω	=	kg water vapor per kg dry air
	h	=	enthalpy, kJ/kg
	m	=	kg per hour
	T	=	Temperature, °C
Subscripts:	1	=	inlet conditions for air
	2	=	outlet conditions for air
	3	=	hot water in
	4	=	cool water out
	a	=	dry air
	w	=	water vapor
	L	=	liquid water

Equation 13.19 assumes the air is at, or near, atmospheric pressure. For a fixed set of temperature conditions, this equation may be solved for "m_a," the necessary flowrate to produce these temperatures. This equation is possible because there is no change in the mass of dry air in or out, or process water in or out. The make-up water preserves the latter equality. The make-up water requirements are found by a water balance, which reduces to

$$m(\text{kg/h}) = m_a(\omega_2 - \omega_1) \tag{13.20}$$

In an adiabatic tower, the energy lost by the water equals the energy gained by the air stream. The water makeup (including driftage loss) is typically 2-5% of the circulation rate.

Psychrometric Chart

In solving some of the previous equations, a psychrometric chart like Figure 13.29 is useful. The saturation curve is the dewpoint temperature at 100 kPa pressure of saturated air.

The wet bulb and dry bulb temperatures obtained from a psychrometer test fix a point on this chart. From this point one can find both the moisture content of the air and the relative humidity.

The dewpoint of unsaturated air is found by reading horizontally from the point fixed by the wet bulb and dry bulb to the dewpoint (saturation) curve.

If we use a 9°C [16°F] approach to a wet bulb temperature of 21°C [70°F], the cold water should leave the tower at 30°C [86°F]. If this cold water is used in a heat exchanger where the approach is 10°C [19°F] to cool a process fluid, the process fluid will leave at 40°C [104°F]. So ... knowing the wet bulb and the two approaches involved, one can predict what process temperature can be achieved in the cooling tower.

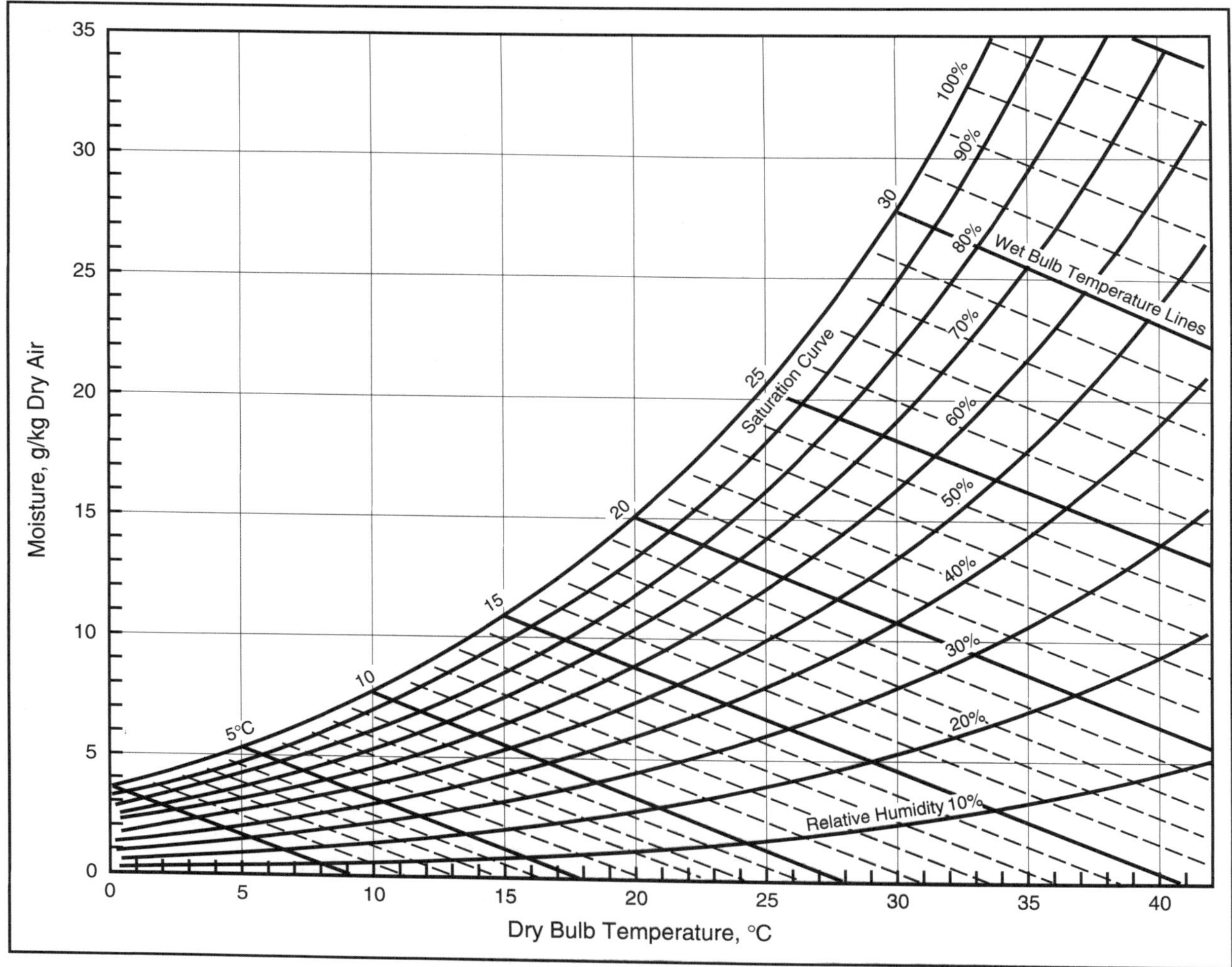

Figure 13.29 Psychrometric Chart Barometric Pressure: 100 kPa

The values in Figure 13.29 are dependent on pressure and only apply at 100 kPa [14.5 psia]. One can correct for pressure by using ideal gas laws. In most cases this correction is negligible.

Example 13.5: From a test, the Dry Bulb temperature = 25°C and the Wet Bulb temperature = 21°C. What is the relative humidity and moisture content of the air?

From Figure 13.29, Rel. Hum. (R.H.) = 70%, Moisture content = 14 g/kg air

Practical Tower Choice

There are five basic types of cooling towers – atmospheric spray, counter-flow, cross-flow, hyperbolic natural draft, and coil-shed. Air is moved by either natural air movement, induced draft (fans suck air through), or forced draft (fans blow air through).

The atmospheric spray tower is normally used only for small installations. It is seldom recommended for gas processing applications. A counter-flow tower has a small footprint and might be preferred for oil-contaminated or heavily sedimented waters. Except for these cases, the cross-flow tower is usually preferred. It gives low draft loss, may use a higher water loading, a longer "range," and achieve closer approaches with lower fan power requirements. A cross-flow tower also requires a lower pumping head and does not require nozzles (using a fill material instead).

Hyperbolic towers are used primarily in the electric power industry in generator stations. They offer favorable economics in very large installations.

A coil-shed tower is simply a cross-flow tower with coils in the base. Rather than circulating water from a basin to external coolers, some of the external process cooling is done in the tower proper. They are very flexible and may be used in combination with external coolers for multiple cooling functions.

Induced draft is usually preferred over forced draft because of a tendency for the latter to recirculate wet exit gas back into the inlet air stream. This adversely affects tower performance by increasing the inlet air wet bulb. The fan power requirements may be estimated for both types.

There are many mechanical considerations too numerous to discuss in detail here. One is the choice of materials. Redwood has been the traditional material but fiberglass reinforced plastic has gained wide acceptance. It is lighter than redwood and the cost is generally competitive. Fire retardant plastics cost about 50-60% more. Even stainless steel has been used successfully. The initial cost is higher but it can show real advantage if long service life, with minimum maintenance cost, enters the picture.

Data are readily available from manufacturers and in the literature for the general sizing and specification of cooling towers. Short-cut analyses provide convenient input for planning purposes.[13.36, 13.37]

AERIAL COOLERS

The aerial cooler is very popular even where water for cooling is available. It is mechanically simple, flexible and eliminates the nuisance and cost of water treating. However, it also has several disadvantages. In warm climates it is not capable of producing as low a temperature as water. This can increase the capital cost of the system substantially by lowering thermal efficiency. If water is available, a very careful study should be made of the cooling alternatives. This study should not merely compare the cost of the cooling system but should examine the cost of the alternative on the total system.

Figure 13.30 is a schematic view of the two basic types of aerial coolers. The forced draft type is cheaper. However, many prefer the induced draft type on the basis that it is more efficient.

With the forced draft type, there is a possibility that hot air leaving the top will flow around the unit and be drawn through again. Any by-passed air will be hotter than the ambient air and will lower the cooling efficiency. This by-passing of hot air is less likely with the induced draft type because air is exhausted at higher velocity through the reduced area of the fan ring and is more likely to be dissipated by normal winds.

Air is abundant but has a heat transfer effectiveness of about 1-2% compared to water. The use of fins on the air side makes air about 20% as effective as water, if the fouling effect of water on exchange is ignored. With the poor quality of water often used, air may be 50-70% as effective as water.

Historically, many vendors have compromised on design to minimize initial cost. One study of aerial coolers (purchased as quoted) showed that two-thirds performed at less than design capacity.

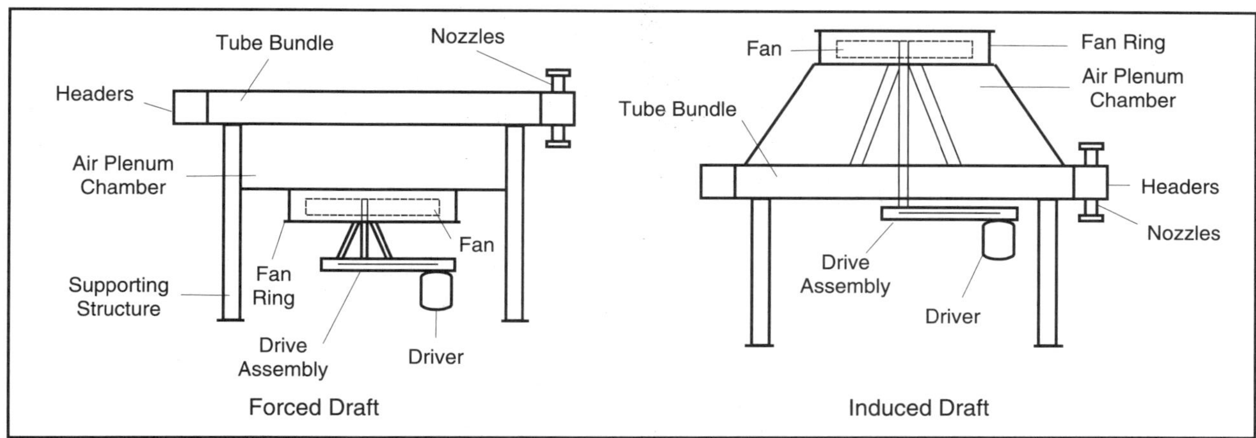

Figure 13.30 The Two Basic Types of Aerial Coolers *(Courtesy the Rainey Corp.)*

About 75% of these had insufficient air flow. About one-half of the units were within 10% of quoted performance. These results may not be typical of all units, but these units were purchased by a well engineered, major oil company. I mention this to point out that a "guarantee" is no substitute for careful customer appraisal.

The following data sheet illustrates the variables that are involved in designing and specifying an aerial cooler. One must consider the cooling tube design, fan arrangement, number of transfer bays, control of cooled fluid temperature as the ambient air temperature changes and the usual mechanical considerations.(13.38-13.42) API 661 is used extensively to define air cooler specifications.

DATA SHEET – AIR COOLED HEAT EXCHANGER Sheet of

No.			
1	CUSTOMER		EQUIPMENT TAG NO.
2	PLANT LOCATION		MANUFACTURER:
3	SERVICE		
4	SIZE	TYPE: INDUCED/FORCED DRAFT	NO. OF BAYS
5	SURFACE PER UNIT - FINNED TUBE	FT^2; BARE TUBE	FT.^{2}
6	HEAT EXCHANGED BTU/HR	MTD (EFF.)	°F
7	TRANSFER RATE - FINNED TUBE	; BARE TUBE, SERVICE	CLEAN BTU/HR. FT.^{2}°F
8	PERFORMANCE DATA		
9	TUBE SIDE		
10	FLUID CIRCULATED		GRAVITY, LIQ. °API/SG@ 60°F
11	TOTAL FLUID ENTERING LBS/HR		POUR POINT °F
12		IN / OUT	BUBBLE POINT °F
13	TEMPERATURE °F		DEW POINT °F
14	LIQUID LBS/HR		SPECIFIC HEAT BTU/LB °F @ °F
15	VAPOR LBS/HR, MW		LATENT HEAT BTU/LB @ °F
16	NONCOND. LBS/HR, MW		COND. (vap./liq) BTU-FT/HR · FT.^{2}°F @ °F
17	STEAM LBS/HR		OUTLET PRESSURE PSIG/PSIA
18	WATER LBS/HR		ALLOWABLE PRESSURE DROP PSI
19	VISCOSITY (LIQ/VAP) cP		CALC. PRESSURE DROP PSI
20			FOULING RESISTANCE, INSIDE
21	AIR SIDE		
22	AIR QUANTITY (LBS/HR) (SCFM)		ALTITUDE FT.
23	AIR QUANTITY/FAN ACFM		TEMPERATURE IN °F
24	STATIC PRESSURE INCHES OF WATER		TEMPERATURE OUT °F
25	FACE VEL. FPM (STD)	MASS VEL. LBS/HR FT.^{2}	MINIMUM AMBIENT °F
26	DESIGN - MATERIALS - CONSTRUCTION		
27	DESIGN PRESSURE PSIG	TEST PRESSURE PSIG	DESIGN TEMP. °F
28	TUBE BUNDLE	HEADER	TUBE
29	SIZE NO. TUBE ROWS	TYPE	MATERIAL
30	NO./BAY	MATERIAL	ASTM SEAMLESS/WELDED
31	ARRANGEMENT-	PASSES-NO. ARR'GMENT*	OD. IN. MIN. THKNS. IN.
32	BUNDLES IN PARALLEL IN SER.	PLUG-DESIGN	NO. BUNDLE LENGTH FT.
33	BAYS IN PARALLEL IN SER.	MATERIAL	PITCH IN.△
34	BUNDLE FRAME	GASKET MATERIAL	FIN
35	MISCELLANEOUS	CORROSION ALLOW. IN.	TYPE
36	STRUCTURE MTG-PIPERACK/GRADE	SIZE INLET NOZZLE IN.	MATERIAL
37	SURFACE PREPARATION	SIZE OUTLET NOZZLE IN.	OD. IN. STOCK THKNS. IN.
38	LOUVERS AUTO/MANUAL	RATING & FACING	NO./IN.
39	CODE - ASME VIII STAMP - YES/NO		
40	MECHANICAL EQUIPMENT		
41	FAN	DRIVER	SPEED REDUCER
42	MFR. & MODEL	TYPE	TYPE
43	NO./BAY HP/FAN	NO./BAY HP/DRIVER	NO./BAY
44	DIAMETER, FT. RPM	RPM	MODEL
45	NO. BLADES PITCH ADJ. AUTO	ENCLOSURE	AGMA HP RATING
46	BLADE MATERIAL ANGLE	V/PHASE/Hz	RATIO /1
47	HUB MATERIAL	MFR.	MFR.
48	CONTROL ACTION ON AIR FAILURE -	FAN PITCH MIN./MAX.;	LOUVERS OPEN/CLOSED
49	NOTES: *GIVE TUBE COUNT OF EACH PASS WHEN IRREGULAR		
50			
51			
52	PLOT AREA	PROPOSAL DWG. NO.	
53	Prepared: Approved:	Date: Rev.	Rev. Rev.

ADVANTAGES OF INDUCED DRAFT DESIGN

1. Easier to shop assemble, ship and install.
2. The hoods offer protection from weather.
3. Easier to clean underside when covered with lint, bugs, debris.
4. More efficient air distribution over the bundle.
5. Less likely to be affected by hot air recirculation.

DISADVANTAGES OF INDUCED DRAFT DESIGN

1. More difficult to remove bundles for maintenance.
2. High temperature service limited due to effect of hot air on the fans.
3. More difficult to work on fan assembly, i.e. adjust blades due to heat from bundle, and their location.
4. Fan power is higher due to larger inlet air volumes.

ADVANTAGES OF FORCED DRAFT DESIGN

1. Easy to remove and replace bundles.
2. Easier to mount motors or other drivers with short shafts.
3. Lubrication, maintenance, etc. more accessible.
4. With reinforced straight side panels to form a rectangular box-type plenum, shipping and mounting is greatly simplified, permitting complete preassembled shop-tested units.
5. Best adapted for cold climate operation with warm air recirculation.

DISADVANTAGES of the force draft design are the list of "advantages of the induced draft design."

Design of an optimum aerial cooler is a trial-and-error calculation that should not be undertaken by a novice. There are many combinations of air flowrate, tube design, type of fins, and fan type and speed. For a given cooling service, the designer must assume an air flowrate. Increased rate increases U, LMTD, pressure drop and fan power requirements. For each rate there is a further change with tube diameter, pitch and fin type. So, multiple calculations are required to truly optimize design.

Heat Transfer Calculations

The basic calculation approach is the same as other exchangers. Table 13.3 shows a group of overall heat transfer coefficients based on bare tube area.*(13.38)* These are useful as a first step in planning before choosing a particular fin type on the outside of the tube.

TABLE 13.3

Typical Overall Heat Transfer Coefficients (based on bare tube area)

Liquid Coolers			
Material	**Heat-transfer coefficient, Btu/hr-ft²-°F**	**Material**	**Heat-transfer coefficient, Btu/hr-ft²-°F**
Oils, 20° API:		Heavy oils, 8-14° API:	
200°F avg. temp.	10-16	300°F avg. temp.	6-10
300°F avg. temp.	13-22	400°F avg. temp.	10-16
400°F avg. temp.	30-40	Diesel oil	45-55
Oils, 30° API:		Kerosene	55-60
150°F avg. temp.	12-23	Heavy naphtha	60-65
200°F avg. temp.	25-35	Light naphtha	65-70
300°F avg. temp.	45-55	Gasoline	70-75
400°F avg. temp.	50-60	Light hydrocarbons	75-80
		Alcohols & most organic solvents	70-75
		Ammonia	100-120
		Brine, 75% water	90-110
		Water	120-140
Oils, 40° API:		50% ethylene glycol & water	100-120
150°F avg. temp.	25-35		
200°F avg. temp.	50-60		
300°F avg. temp.	55-65		
400°F avg. temp.	60-70		

Condensers	
Material	**Heat-transfer coefficient, Btu/hr-ft²-°F**
Steam	140-150
Steam -	
10% noncondensibles	100-110
20% noncondensibles	95-100
40% noncondensibles	70-75
Pure light hydrocarbons	80-85
Mixed light hydrocarbons	65-75
Gasoline	60-75
Gasoline-steam mixtures	70-75
Medium hydrocarbons	45-50
Medium hydrocarbons w/ steam	55-60
Pure organic solvents	75-80
Ammonia	100-110

Vapor Coolers (vapor only, no condensation)					
	Heat-transfer coefficient, Btu/hr-ft²-°F				
Material	**10 psig**	**50 psig**	**100 psig**	**300 psig**	**500 psig**
Light hydrocarbons	15-20	30-35	45-50	65-70	70-75
Medium hydrocarbons & organic solvents	15-20	35-40	45-50	65-70	70-75
Light inorganic vapors	10-15	15-20	30-35	45-50	50-55
Air	8-10	15-20	25-30	40-45	45-50
Ammonia	10-15	15-20	30-35	45-50	50-55
Steam	10-15	15-20	25-30	45-50	55-60
Hydrogen - 100%	20-30	45-50	65-70	85-95	95-100
- 75% vol.	17-28	40-45	60-65	80-85	85-90
- 50% vol.	15-25	35-40	55-60	75-80	85-90
- 25% vol.	12-23	30-35	45-50	65-70	80-85

The optimum air temperature rise across the tubes may be estimated by the equation

$$(t_2 - t_1) = (0.005)(U)\left[\frac{(T_2 + T_1)}{2} - t_1\right] \tag{13.21}$$

Where:				SI	FPS
	t_2	=	outlet air temperature	°C	°F
	t_1	=	inlet air temperature	°C	°F
	T_2	=	temperature of process fluid out	°C	°F
	T_1	=	temperature of process fluid in	°C	°F
	U	=	value from Table 13.3		Btu/hr-ft²-°F

The optimum air temperature rise is also a function of the range $(T_1 - T_2)$ of the process fluid. The value of $(t_2 - t_1)$ calculated from Equation 13.21 should be corrected using the equation

$$CF = 0.89 + A(T_1 - T_2) \tag{13.22}$$

Where:				SI	FPS
	CF	=	correction factor	dimensionless	
	A	=	constant	0.0025	0.0014

For a specified cooling load and conditions, the outlet air temperature can be estimated. From this $\Delta T_{m_{corr}}$ can be found to calculate bare tube area. Figures 13.12 and 13.13 give "F," the ΔT_m correction factor.

Tube sizes can vary from 15.9-38.1 m [5/8-1 1/2 in), but the standard size is 25.4 mm [1 in). Tube layout is triangular. Tube pitch is the minimum distance between the tube centerlines which avoids fin contact or overlap.

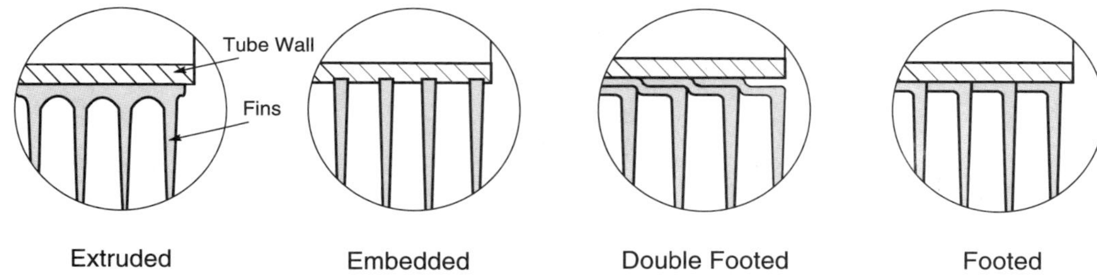

The types of fins vary with the service. They are either tension wrapped, embedded or extruded. The latter are the most expensive. Fin height varies from 12-15 mm [1/2-5/8 in] and normally there are 8-11 fins per inch.

The air-side film coefficient for a typical fin tube (based on extended area) can be estimated from the equation

$$h_a = \frac{A(v_g)^{0.6}}{d^{0.3}} \tag{13.23}$$

Where:			SI	FPS
h_a	=	air-side film coefficient	kJ/h·m²·°C	Btu/hr-ft²-°F
v_g	=	air velocity by tubes	m/s	ft/sec
d	=	outside diameter of bare tubes	cm	in
A	=	constant	92.1	1.67

This "h" would be used with other data in Equation 13.3 to find an overall "U" for comparison with values in Table 13.3.

Fans

The fan power requirements can be estimated from the equation

$$kW = \frac{(\Delta P_a)(q_{air})}{(A)(Efficiency)} \tag{13.24}$$

Where:			SI	FPS
q_{air}	=	air flowrate	m³/s	ft³/sec
ΔP_a	=	air pressure drop in cooler	kPa	in H_2O
A	=	constant	1.0	142

Efficiency varies from 0.4-0.75; 0.7 is a useful planning number.

The volumetric air flowrate in Equation 13.24 can be calculated from Equation 13.25.

$$q_{air} = \frac{m_{air}}{\rho_{air}} \tag{13.25}$$

Where:			SI	FPS
q_{air}	=	volumetric air flow	m³/s	ft³/sec
m_{air}	=	mass air flow	kg/s	lbm/sec
ρ_{air}	=	air density at fan inlet conditions	kg/m³	lbm/ft³

The air density, ρ_{air}, is calculated from the gas density in Equation 3.12 in Chapter 3. The compressibility factor, z, is 1.0. For forced-draft coolers, the density is calculated at fan inlet conditions and should account for elevation and relative humidity. Table 13.4 shows average atmospheric pressures at various elevations. The molecular weight of air can be calculated from Equation 13.26 and is a function of relative humidity and pressure.

$$MW = \frac{\left(\frac{RH}{100}\right)(P_w^s)(18) + \left[P - \left(\frac{RH}{100}\right)(P_w^s)\right](29)}{P} \tag{13.26}$$

			SI	FPS
Where:	R.H.	= relative humidity	–	–
	P_w^s	= water vapor pressure at air temperature*	kPa	psia
	P	= air pressure	kPa	psia

* see steam tables in Appendix B, pages 452 and 465

TABLE 13.4

Altitude and Atmospheric Pressures

Altitude above Sea Level		Barometer		Atmospheric Pressure	
meters	feet	mm Hg abs.	in Hg abs.	kPa	psia
0	0	760.0	29.92	101.325	14.696
153	500	746.3	29.38	99.49	14.43
305	1000	733.0	28.86	97.63	14.16
458	1500	719.6	28.33	95.91	13.91
610	2000	706.6	27.82	94.18	13.66
763	2500	693.9	27.32	92.46	13.41
915	3000	681.2	26.82	90.80	13.17
1068	3500	668.8	26.33	89.15	12.93
1220	4000	656.3	25.84	87.49	12.69
1373	4500	644.4	25.37	85.91	12.46
1526	5000	632.5	24.90	84.32	12.23
1831	6000	609.3	23.99	81.22	11.78
2136	7000	586.7	23.10	78.19	11.34
2441	8000	564.6	22.23	75.22	10.91
2746	9000	543.3	21.39	72.39	10.50
3050	10 000	522.7	20.58	69.64	10.10

Cooler pressure drop varies with air rate, tube diameter, pitch and number of tube rows. For planning purposes a pressure drop of 25 to 35 Pa [0.10 to 0.15 in H_2O] per tube row can be used. For most gas processing applications the number of tube rows varies from 3-6.

Figure 13.31 is a rough estimate of horsepower based on the value of "U" from Table 13.3 for bare tubes. The number of fans and heat transfer bays will vary with the installation. Table 13.5 can be used to estimate exchanger size. Two fans/unit are generally preferred because of the additional flexibility in controlling air flow.

Noise control is a concern in populated areas. A common specification is that fan and motor noise shall not exceed 85-90

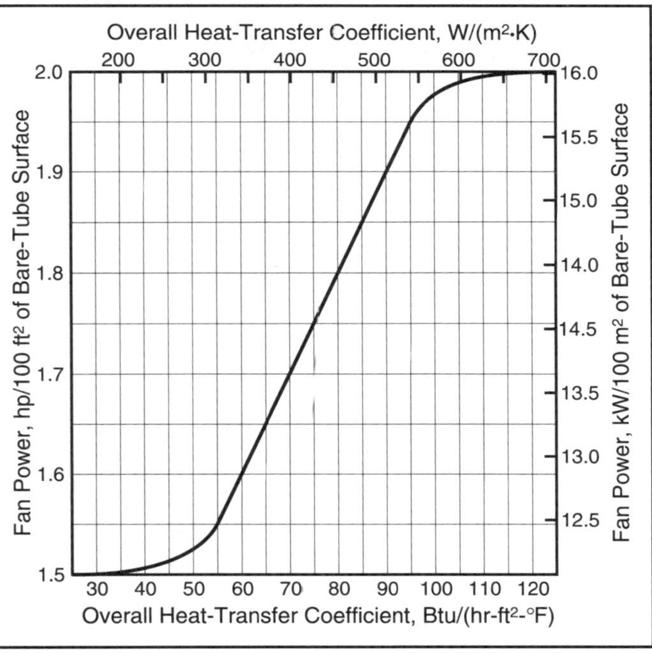

Figure 13.31 Approximate Fan Power Correlation

TABLE 13.5

Approximate Bare Tube Area Versus Unit Size(13.38)

1-in O.D. bare tube on $2^{3/8}$-in pitch						
Approximate unit width, ft	Tube length, ft	Fans per unit	No. of tube rows in depth			
			3	4	5	6
4	4	1	49	64	81	97
	6	1	73	97	122	146
	8	2	98	129	163	194
	10	2	123	162	204	243
6	6	1	121	160	201	240
	8	1	161	213	268	320
	12	2	242	320	402	481
	14	2	282	374	469	561
8	8	1	224	297	373	446
	10	1	280	372	466	558
	12	1	336	446	559	669
	14	1	392	520	652	781
	16	2	448	595	746	892
	20	2	560	744	932	1116
	24	2	672	892	1119	1339
10	10	1	351	466	584	699
	12	1	421	559	701	839
	14	1	491	652	817	979
	16	1	561	746	934	1119
	20	2	702	932	1168	1399
	24	2	842	1119	1402	1678
	30	2	1053	1399	1752	2098
	32	2	1123	1492	1869	2238
12	12	1	515	685	858	1028
	14	1	601	799	1001	1199
	16	1	687	913	1144	1370
	20	1	859	1142	1430	1713
	24	2	1031	1370	1716	2056
	30	2	1289	1713	2145	2570
	32	2	1374	1827	2288	2741
	36	2	1546	2056	2574	3084
	40	2	1718	2284	2861	3426
14	14	1	700	931	1166	1397
	16	1	800	1064	1333	1597
	20	1	1000	1330	1666	1996
	24	2	1201	1597	1999	2395
	30	2	1501	1996	2499	2994
	32	2	1601	2129	2666	3194
	36	2	1801	2395	2999	3593
	40	2	2001	2661	3332	3992
16	16	1	897	1190	1492	1785
	20	1	1121	1488	1865	2232
	24	1	1345	1785	2238	2678
	30	2	1682	2232	2798	3348
	32	2	1794	2381	2984	3571
	36	2	2018	2678	3357	4018
	40	2	2242	2976	3730	4464
18	20	1	1247	1655	2075	2483
	24	1	1496	1987	2490	2980
	30	2	1870	2483	3112	3725
	32	2	1995	2649	3320	3974
	36	2	2244	2980	3735	4470
	40	2	2494	3311	4150	4967
20	20	1	1404	1865	2337	2798
	24	1	1685	2238	2804	3357
	30	2	2106	2798	3505	4197
	32	2	2246	2984	3739	4477
	36	2	2527	3357	4206	5036
	40	2	2808	3730	4674	5596

Notes:

1. Assume 4 rows of tubes in depth except for the following condition:
   a. If the temperature range on the process side is 10°F or less, assume 3 rows.
   b. If the temperature range of the process fluid falls between 10°F and 20°F, and special materials of construction are required, assume 3 rows.
   c. If the temperature range of the process fluid is between 100°F and 200°F and/or the assumed overall heat-transfer rate is less than 60, assume 5 rows.
   d. If the temperature range of the process fluid is between 200°F and 300°F and/or the overall heat-transfer rate is less than 40, assume 6 rows.
   e. If the temperature range of the process fluid is greater than 300°F and/or overall heat-transfer rate is less than 30, assume 8 rows.

2. Relative to 14 BWG, the effect of tube-wall thickness on cost is:

Average gage	Cost factor
12 Bwg	1.025
14 Bwg	1.0
16 Bwg	0.99

3. Relative to 6 rows of tubes, the effect of the number of tube rows on cost is:

Rows	Cost factor
4	1.10
5	1.05
6	1.00
8	0.95

4. Relative to length of 24 ft, the effect of tube length on cost is:

Tube length, ft	Cost factor
10	1.15
12	1.13
14	1.11
16	1.08
18	1.06
20	1.05
24	1.00
30	0.95
32	0.93
36	0.89
40	0.85

5. Because of shipping limitations the widest tube bundle that can be shop fabricated and shipped to a plantsite is 12 ft. Wider bundles must be field fabricated.

dBA at a distance of three feet from the fan ring. One can estimate the sound pressure level by the equation

$$dBA = 65 + 3(\log V) + 10 \log(hp) + 20 (\log d) \tag{13.27}$$

Where:

dBA	=	relative sound level in decibel
log	=	logarithm to base 10
V	=	fan tip speed. (0.001)(ft/min)
hp	=	fan horsepower
d	=	fan diameter, ft

As noted previously, various methods of fan control are used. The primary criteria are temperature control of the process fluid and power consumption. It is feasible to drive *variable speed* fans with standard induction motors using some type of a.c. adjustable frequency drives (AFD) – a variable voltage inverter, a pulse width modulator or current source types.

One alternative is to use a *variable pitch* fan. It offers rather precise temperature control, provides energy savings and is convenient for cold weather operations.[13.43] It also tends to cost more and may involve more routine maintenance. The choice between variable speed and variable pitch depends on local circumstances and the biases of the purchaser. If power costs are large and temperature control is critical, one or the other normally will be chosen.

The other control alternatives are fluid by-pass, on-off operation (with possibly several fans per cooling bay) and the use of louvers or shutters. By-pass and louvers may be effective in some cases but they are energy inefficient. On-off fan control is simple and may be used if there are a lot of fans in the same service. Winter protection is required in cold climates. In this case, the use of louvers plus some form of variable air rate control is desirable.[13.42] This is one case where a variable pitch fan plus louvers may be the best system to control internal air circulation.

Outlet temperature is controlled primarily by air rate. Louvers, variable pitch fan blades, and variable speed motors are all used to control temperature. Louvers may be manually adjustable for seasonal or night-day air temperature changes, or controlled automatically. We have found automatic louver control less than satisfactory in those cases where a close tolerance is required on outlet fluid temperatures and the louvers are operating almost closed (where a small change in position causes a large change in air flowrate). In those cases where large air temperature changes are encountered, a variable pitch fan may prove efficient. Some report trouble with the pitch control, but this has not been a problem in my experience. Pitch and speed controls are expensive but in this era of high energy costs they can prove profitable.

Fan power is an important operating cost consideration. A ten percent change in air flowrate will cause about a 35 percent change in power used, assuming efficiency stays constant. Actual power consumption required for a given heat transfer depends on many factors. One is the clearance between the fan and the fan ring. Close clearances are more expensive to fabricate. Consider this in comparing capital cost from different vendors.

There are several rules-of-thumb used by some for fan specification. One is that the distance between tubes and fan should be 0.4-0.5 of fan diameter to allow good air mixing. Another is that the ratio of fan ring area to tube area should not be less than 0.4.

The total cost of air cooling is very sensitive to the temperature approach between entering air and the cooled fluid leaving. It is seldom that an approach less than 10-11°C [18-20°F] can be justified. Most coolers use a 15-25°C [27-45°F] approach to ambient air temperature. Anything larger than this is clearly uneconomical.

The total cost (operating plus amortized capital cost) increases sharply as approach decreases. An approach of 11°C [20°F] has a total cost about 1.6 times that of a 22°C [40°F] approach.

This approach limitation adversely affects the economics of aerial cooling in many cases. Using a 38°C [100°F] D.B. and a 24°C [75°F] W.B. affords a reasonable comparison between air and water cooling. The fluid leaving an air cooler with a 16°C [29°F] approach will be at 54°C [129°F]. If a cooling tower exchanger is used, the temperature leaving likely will be about 42°C [108°F]. This 12°C [22°F] difference can have a significant effect on the process efficiency as well as the size (and cost) of equipment affected by the temperature level achieved in the coolers.

Operators usually prefer air coolers, but they sometimes pay a high price for the convenience. Examine the alternatives carefully before making a choice.

COMBINATION COOLERS

Figure 13.32 shows one type of combination cooler that may offer advantage where water is expensive to treat or is in short supply. It is an aerial cooler preceded by an evaporative section. The entering air is cooled with water as necessary. When the air temperature is low, the water may be shut off. Water rate may be decreased at intermediate temperatures.

This unit has a higher capital cost but may offer a total cost saving in some applications. When equipped with controls to vary fan power output, this unit offers flexibility at minimum operating cost.

Some have accomplished a similar result by using an air cooler for water. This cooling is supplemented as needed with a cooling tower. The combination unit is usually preferred.

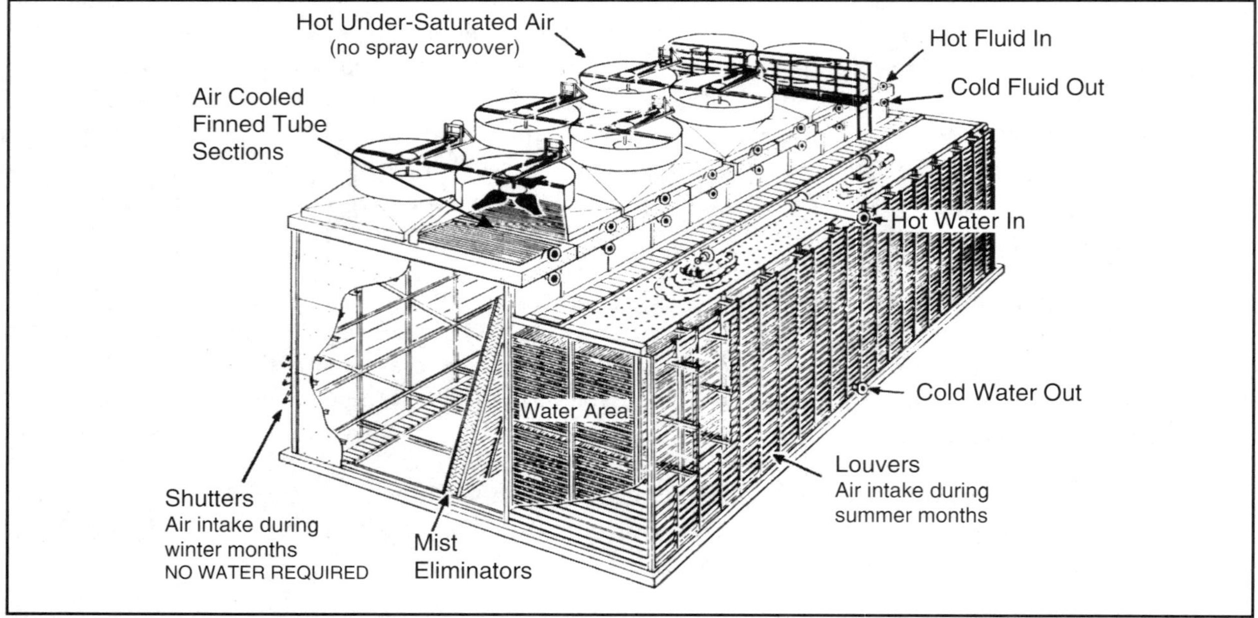

Figure 13.32 Combination Air-Water Cooler

CHOICE OF HEAT EXCHANGERS

It should be apparent from preceding discussions that choice of heat exchangers involves many factors. It is relatively easy to choose one that will work. An intelligent choice involves choosing equipment that optimizes cost of the total system without compromising operating reliability.

Heat exchangers normally cost less per unit of energy transferred than any other type of energy equipment. If you compromise on exchanger size, you must pay dearly for this in the cost of companion equipment in many instances. Since heat loads vary with flowrates, some flexibility must be provided. If done wisely, a little extra heat exchange capacity is cheap "insurance."

There are some "rules" one should follow.

1. Do not specify or purchase a HEX without consideration of its effect on the total process.
2. Do not make the capital cost of the HEX alone a sole criterion for purchase.
3. Acquaint the vendor with details of service and point out the choice will be made on both initial and operating cost, not initial capital cost alone.
4. Use realistic pressure drop specifications since this affects size and cost. Allow as much pressure loss as economics dictates for the actual system and not merely reproduce a standard spec that might not apply.

Remember: the vendor presumably understands his product, but he knows only as much about the application of his product as the customer conveys to him. The majority of less-than-satisfactory exchanger installations is as much the fault of the customer as it is the vendor.

REFERENCES

13.1 Frank, O., *Chem. Eng.* (May 13, 1974), p. 126.

13.2 Campbell, J. and G. L. Farrar, *Effective Communication for the Technical Man*, Campbell Petr. Series, Norman, Oklahoma (1980).

13.3 Lauer, B. E., Reprint "hf," *Oil and Gas J.* (circa 1953).

13.4 Li, *AICHE Jour.*, Vol. 22 (1976), p. 927.

13.5 Lenoir, *Pet. Ref.*, Vol. 36, No. 8 (1957), p. 162.

13.6 Steinmeyer, D., "Understanding ΔP and ΔT in Turbulent Flow Heat Exchangers," *Chem. Eng. Prog.* (June 1996), p. 49.

13.7 *Standard of the Tubular Exchanger Mft. Assoc.*, 6th Ed., New York (1978).

13.8 Poddar, T. K. and G. T. Polley, "Optimize Shell-and-Tube Heat Exchanger Design," CEP (Sept. 2000), p. 41.

13.9 Bowman, R. A., *et al., Trans. ASME* (May 1940), p. 283.

13.10 Wales, R. E., *Chem. Engr.* (Feb. 1981), p. 77.

13.11 Chen, C. C., *Ibid.* (Mar. 1984), p. 155.

13.12 Bell, K. J., *Oil and Gas J.* (Dec. 4, 1978), p. 59.

13.13 Caglayan, A. N. and P. Buthod, *Ibid.* (Sept. 6, 1976), p. 91.

13.14 Strickland, J. R., *Ibid.*, p. 100.

13.15 Kern, R., *Chem. Eng.* (Sept.12, 1977), p. 169.

13.16 Roebuck, A. H., *Oil and Gas J.* (Dec. 4, 1978), p. 70.

13.17 Mehra, D. K., *Chem. Eng.* (July 25, 1983), p. 47.

13.18 Yokell, S., *Ibid.*, p. 57.

13.19 Crane, R. and R. Gregg,, *Ibid.*, p. 76.

13.20 Hammond, R. H., *Oil and Gas J.* (May 17, 1982), p. 78.

13.21 Raju, K. and J. Chand, *Chem. Eng.* (Aug. 11, 1980), p. 133.

13.22 Cross, P. H., *Ibid.* (Jan. 1, I979), p. 87.

13.23 Buonopane, R.A., *et al., Chem. Eng. Prog.*, Vol. 59, No. 7 (1963), p. 57.

13.24 Jackson, B. W. and R. A. Troup, *Ibid.*, Symp. Series, Vol. 62, No. 64 (1966), p. 185.

13.25 *Ibid., Chem. Eng. Prog.*, Vol. 60, No. 7 (1964), p. 62.

13.26 Cooper, A., *Chem. Eng.*, Vol. 285 (May 1974), p.280.

13.27 Burn, J., Johnston, A. M., and N. M. Johnston, "Experience with Printed Circuit Heat Exchangers," 1999 European GPA Continental Meeting," Budapest (Sept. 1999).

13.28 Purohit, G. P., *Chem. Eng.* (May 16, 1983), p. 62.

13.29 Sieder, E. N. and G. E. Tate, *Ind. Eng. Chem.*, Vol. 28, No. 12 (1936), p. 1429.

13.30 Patil, R. K., et al., *Chem. Eng.* (Dec. 13, 1982). p. 85.

13.31 Rorschach, R. L., *Oil and Gas J.* (June 13, 1966). p. 90.

13.32 Berman, H. L., *Chem. Eng.* in four parts (1978): June 19, p. 99; July 31, p. 87; Aug. 14, p. 129 and Sept. 11, p. 165.

13.33 Wood, S., "Select the Right NO_x Control Technology," *Chem. Eng. Prog.* (Jan 1994), p. 32.

13.34 Colannino, J., "Low-Cost Techniques Reduce Boiler NO_x," *Chem. Eng. Prog.* (Feb. 1993), p. 100.

13.35 Kunz, R. G., Smith, D. D., and E. M. Adams, "Predict NO_x from Gas-Fired Furnaces," *Hydr. Proc.* (Nov. 1996), p. 65.

13.36 Meytsar, J., *Hydr. Proc.* (Nov. 1978), p. 238.

13.37 Ucheyama, T., *Ibid.* (Dec. 1976), p. 93.

13.38 Brown, R., *Chem. Eng.* (Mar. 27, 1978), p. 108.

13.39 Glass, J., *Ibid.*, p. 120.

13.40 Baker, W. J., *Hydr. Proc.* (May 1980), p. 173.

13.41 Ganapathy, V., *Oil and Gas J.* (Dec. 3, 1979), p. 74.

13.42 Rubin, F. L., *Hydr. Proc.* (Oct. 1980), p. 147.

13.43 Monroe, R. C., *Hydr. Proc.* (Dec. 1980), p. 122.

14

PUMPS

The choice of a pump is not easy. There are thousands of manufacturers of the dozen or so basic models of pumps. Within each model there are numerous combinations of metallurgy, bearings and seals, packing, mechanical configurations, etc. The final choice involves input from one experienced in pump applications and a careful appraisal of pump needs.

However, there are some basic principles and criteria that apply to all pumps. These will be reviewed in this chapter. The emphasis is on process type pumps used in surface facilities although a brief overview is provided for downhole (wellbore) production pumps. The Standards of the Hydraulic Institute[14.1] and API 610, Centrifugal Pumps for General Refinery Services, provide the basis of the specification of process pumps. Downhole pump standards are set forth by the American Petroleum Institute and individual companies.

Basic Process Pumps

These may be divided into two categories: *positive displacement* and *kinetic* or *dynamic*. The former increase pressure by direct mechanical action. Kinetic pumps impart kinetic energy to the fluid, a portion of which is converted to pressure by reduction in velocity. It is convenient to subdivide these categories.

Positive Displacement

A. Reciprocating
   1. Piston
   2. Plunger

B. Rotary
   1. Gear
   2. Screw
   3. Vane

Kinetic (or Dynamic)

A. Centrifugal
   1. Radial flow
   2. Axial flow
   3. Mixed flow
   4. Special high head, low flow

B. Peripheral (regenerative turbine)

The choice of type depends on liquid density and viscosity, the presence of solids or entrained gases, liquid corrosivity, head (pressure differential across pump) and liquid flowrate.

The table below outlines advantages and disadvantages of centrifugal and positive displacement pumps.

Centrifugal	Positive Displacement
Most common choice for majority of pump applications Lower cost per unit of flow Less maintenance Better for slurries and liquids containing solids	Better for low flow/high head applications Better for viscous fluids More efficient
Not suitable for low flow/high head applications Performance declines on viscous liquids Performance adversely affected by entrained gases	More expensive per unit of flow Higher maintenance Higher NPSH requirements (plunger pumps)

Viscosity can have a profound effect on performance, particularly of centrifugal pumps. In pump literature you may encounter the viscosity unit SSU (Saybolt Seconds Universal). SSU is the time in seconds that it takes 60 cm^3 of liquid to flow out of a standard Saybolt viscosimeter. The conversion formula is

$$\text{For SSU} < 100, \text{ viscosity } \mu, \text{ in cS} = (0.226 \text{ SSU}) - (195/\text{SSU})$$

$$\text{For SSU} > 100, \text{ viscosity } \mu, \text{ in cS} = (0.220 \text{ SSU}) - (130/\text{SSU})$$

Almost all of the crude oils, condensates and process fluids in the petroleum industry possess a viscosity less than 500 SSU [110 cS]. From the Hydraulic Institute Standards, a centrifugal pump with a capacity of 227 m^3/h [1000 US gpm] designed to produce 60 m [200 ft] of head will have the following approximate change in performance parameters (relative to water) when pumping a 500 SSU [110 cS] viscosity liquid.

Efficiency	82%
Capacity	99%
Head	96%
Power	116%

Solid-liquid slurries present a particular problem for all rotating pumps. If used at all in this service, slower speeds must be used. When using a pump for reservoir liquids, remember that consideration must be given to the possibility that solids also may be produced. A special "slurry" pump may be necessary if solids separation is not provided, particularly early in well life.

If the liquid contains over 4-5% entrained gas, the use of a positive displacement pump is indicated.

A corrosive liquid usually requires special linings or metallurgy. Low velocity normally is desired, since erosion is the natural companion of corrosion.

Subject to the above limitations, the pump type ultimately chosen usually depends to some degree on the combination of head and liquid flowrate considerations. Figure 14.1 shows the typical range of performance of common pump types.

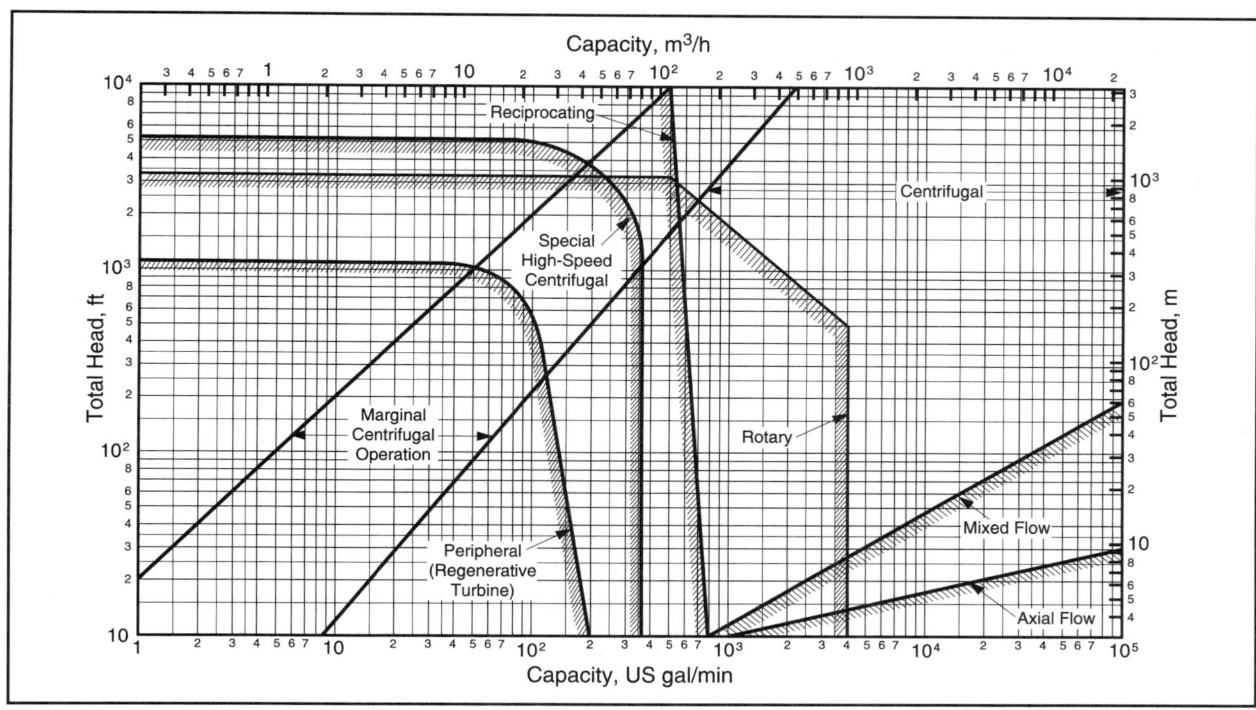

Figure 14.1 Typical Application Range of Common Pumps[(14.2)]

Units for Head and Flowrate

Centrifugal pumps impart energy to the fluid. This energy is termed the pump head and is expressed in meters [feet] of fluid pumped. The diagram at right illustrates head by using a manometer which connects the suction and discharge piping.

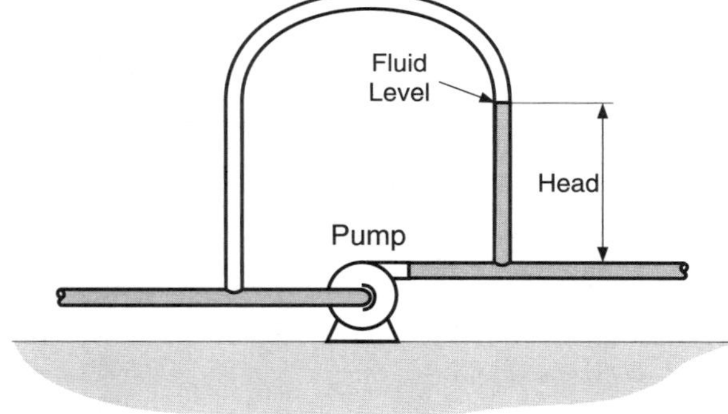

The pressure rise across the pump depends on fluid density.

The relationship between head and ΔP is represented by the equation

$$\Delta P = (A)(\gamma)(H) \tag{14.1}$$

Where:			SI	FPS
	ΔP =	pressure rise across the pump	kPa	psi
	γ =	liquid specific gravity		
	H =	pump head	m	ft
	A =	unit conversion factor	9.81	0.433

The head in feet is 3.281 times the head in meters.

In America it has been traditional to use "gpm" as an abbreviation for U.S. gallons per minute. The use of cubic feet or cubic meters, per hour or per second, is sometimes more convenient for calculation purposes. The following conversion factors may prove useful:

$$1 \text{ m}^3 = 35.31 \text{ ft}^3 = 264 \text{ U.S. gal} = 220 \text{ U.K. gal}$$

Pump Power Requirements

The amount of power required to pump a given amount of liquid to the desired pressure is an application of the First and Second Laws of Thermodynamics discussed in Chapter 7 (Volume 1). The procedure is the same for all pumps.

The basic approach involves establishing the energy gained by the fluid, assuming that this energy change is accomplished adiabatically and reversibly (isentropically). This theoretical energy is corrected to an actual energy by means of a thermodynamic efficiency term which has been established by test. This efficiency encompasses the error in the isentropic assumption as well as true mechanical energy losses in the machinery.

The energy balance around the equipment may be reduced to the form

$$\int_{P_1}^{P_2} V dP = \Delta H_{isen} = W_{theor} \qquad (14.2)$$

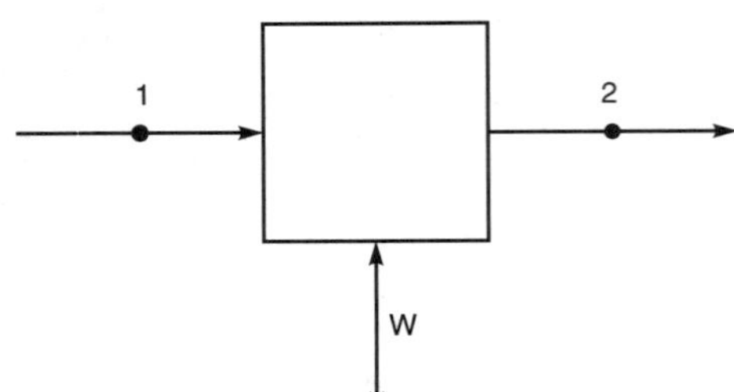

The use of ΔH to find work theoretical is limited primarily to compressors and gas expanders. Liquid systems use the integral to find work.

The work done by a pump is determined on the assumption that the liquid is incompressible. Since H is small for most pumps, we usually apply Equation 14.2 as follows

$$V(\Delta P) = W_{theor} \qquad (14.3)$$

where "V" can be expressed as a specific volume or as a volume rate of flow.

The theoretical equations use the term "work" but we actually are concerned with power, the work per unit time. So, W is power in the actual case.

In countries using FPS units the traditional unit for power is horsepower (hp). In the SI metric system the watt (W), with an appropriate prefix, is the standard power unit.

1 horsepower (hp) = 0.746 kW

Equation 14.3 may be expressed in working form as

$$\text{Power} = \frac{(A)(q)(P_2 - P_1)}{E} \qquad (14.4)$$

Where:		SI	FPS
Power =		kW	hp
A =	constant	1.0	5.83×10^{-4}
q =	volume rate of liquid flow	m^3/s	gpm (US)
P_2 =	outlet pressure of pump	kPa	psia
P_1 =	inlet pressure of pump	kPa	psia
E =	pump efficiency, fraction		

The value of q may be found from the equation

$$q = \frac{m}{\rho} \tag{14.5}$$

Where: m = mass rate/unit time

ρ = fluid density

The value of "E" will depend on the pump type and the service. The values vary widely, particularly for centrifugal pumps.

The value of ΔP from Equation 14.1 may be substituted for $(P_2 - P_1)$ in Equation 14.4 to calculate power directly from fluid head.

Net Positive Suction Head (NPSH)

NPSH is a critical factor in pump selection. It is the positive head required at the suction flange for proper pump performance.

Sufficient NPSH must be provided to prevent formation of small gas bubbles which then collapse, releasing energy that can damage the pump. This is called *cavitation.*

From a phase behavior viewpoint, these bubbles can only occur when the pressure is at or below the bubblepoint. So, the pressure at the suction flange must be sufficiently above this bubblepoint to prevent gas from forming.

The Net Positive Suction Head Required (NPSHR) in a centrifugal pump is a function of pump design and is governed by rotation speed (N), inlet (eye) area of the impeller, type and number of vanes, and like factors. With reciprocating pumps, NPSHR depends on speed and suction valve design.

In pump selection one must provide an available NPSH (NPSHA) that is equal to, or preferably greater than, the NPSHR of the pump.

The NPSHA available for pumps may be found by the following equation:

$$\text{NPSHA} = \frac{A\left(P_s - P_v - \Delta P_f\right)}{\gamma} + H_s - H_m - H_{ac} \tag{14.6}$$

Where:			SI	FPS
	P_s =	absolute pressure of suction vessel	kPa	psia
	ΔP_f =	fitting and friction loss in suction line	kPa	psi
	H_s =	height between lowest drawdown line of suction vessel and centerline of pump suction	m	ft
	H_{ac} =	acceleration head (for reciprocating pumps only)	m	ft
	H_m =	safety margin	m	ft
	P_v =	vapor pressure of liquid at pump suction flange	kPa	psia

The value of A will depend on the units used for H and P, as shown in the table below.

P	H	A	NPSHA
bar	m	10.22	m
kPa	m	0.102	m
psi	ft	2.31	ft

For bubblepoint liquids, $P_s = P_v$ because of equilibrium in a closed tank. In these instances NPSHA can only be increased by adjusting H_s. This is done by raising the suction vessel, lowering the pump or using vertical "deep well" pumps. Alternatively the NPSHR of the pump may be decreased by installing an inducer in the pump suction or using a booster pump. Remember NPSHA ≥ NPSHR.

The safety margin, H_m, (usually a company or personal specification) is advisable because NPSHR values can vary from pump manufacturer's specifications and the pump can operate at higher than design flowrates which increases NPSHR.

The ΔP_f term is calculated from the equations in Chapter 10 of Volume 1. The length, L, must be increased by the equivalent length of all fittings placed on the suction side. This length addition may be as high as three times the actual length of suction pipe! For centrifugal pumps, suction velocities should be limited to a maximum 1 m/s [3 ft/sec] and short radius elbows, long suction lines and numerous fittings should be avoided.

Reciprocating pumps require a correction called *acceleration head* (H_{ac}). When designing reciprocating pump inlet piping with no pneumatic pulsation control device, one must provide for the continual acceleration and deceleration the fluid undergoes with plunger movement. The following equation reasonably defines the acceleration head for suction pipes less than 15 m [50 feet].

$$H_{ac} = \frac{(A)(C)(K)(N)(L)(q)}{d^2} \tag{14.7}$$

Where:

		SI	FPS
H_{ac} =	acceleration head	m	ft
N =	rotational velocity of crankshaft	rev/min	rev/min
L =	length of suction pipe	m	ft
q =	pump rate	m^3/s	ft^3/sec
d =	suction pipe diameter	m	ft
A =	coefficient	3.32	1.0

C = fluid velocity factor:
- 0.0158 for single acting, simplex power pump
- 0.0079 for double acting, simplex power pump
- 0.0079 for single acting, duplex power pump
- 0.0046 for double acting, duplex power pump
- 0.0026 for single acting, triplex power pump
- 0.0016 for single acting, quintuplex power pump
- 0.0011 for single acting, septuplex power pump

K = fluid compressibility factor:
- 0.4 compressible liquids (C_2, etc.)
- 0.5 most hydrocarbons
- 0.67 amine, glycol, water
- 0.71 deaerated, hot water

The value determined must be subtracted from the NPSHA already calculated and still maintain a reasonable margin of available head. Line size is inversely proportional to acceleration head, i.e., the larger the diameter, the smaller the acceleration head loss.

In suction lines longer than 15 m [50 ft], internal disturbances and pressure wave velocities become significant, and the acceleration head found by the equation probably would be more than actually existed.

The introduction of properly-sized suction side pneumatic or gas pulsation bottles adjacent to the pump practically eliminates acceleration head. Installation of these reduces the size of suction pipe significantly.

One fitting usually forgotten in design is the startup screen normally placed in the inlet line. Destructive cavitation may occur during startup caused by this oversight, so include it in the list of fittings. Unfortunately, published data is limited, but a judgmental allowance should be made.

Pump vendors seldom take such inlet screens into consideration for their NPSHR values. Some specifically prohibit their use after startup. To be safe, question the vendor about screen use with his equipment.

Pump suctions, then, should be examined carefully, using line sizes for pipe and fittings which provide adequate suction head margin at the pump, during and after startup procedures.

Specific Speed

This is a type of index that can be applied to all types of pumps. It is sometimes used to aid in pump selection and use. The equation is:

$$N_s = \frac{(A)(N)(q)^{0.5}}{H^{0.75}} \tag{14.8}$$

Where:			SI	FPS
	A	= conversion constant	1.0	0.0194
	N	= pump speed	rpm	rpm
	q	= liquid flowrate	m^3/s	gpm (U.S.)
	H	= head	m	ft

The primary use of specific speed is in the classification of centrifugal pumps.

CENTRIFUGAL PUMPS

Part (a) of Figure 14.2 shows a schematic of an overhung impeller, single-stage centrifugal pump. The liquid enters the impeller eye, at the left, is accelerated through the impeller, and leaves the impeller tip at high velocity. In a single-stage pump the fluid enters the discharge volute where the velocity is decreased to discharge nozzle velocity.

Wear rings limit the fluid leakage from the discharge to the suction between the impeller and the pump case. The shaft is sealed by ring packing installed in a stuffing box (as shown in the drawing) or by mechanical seals. Mechanical seals are preferred in most hydrocarbon applications.

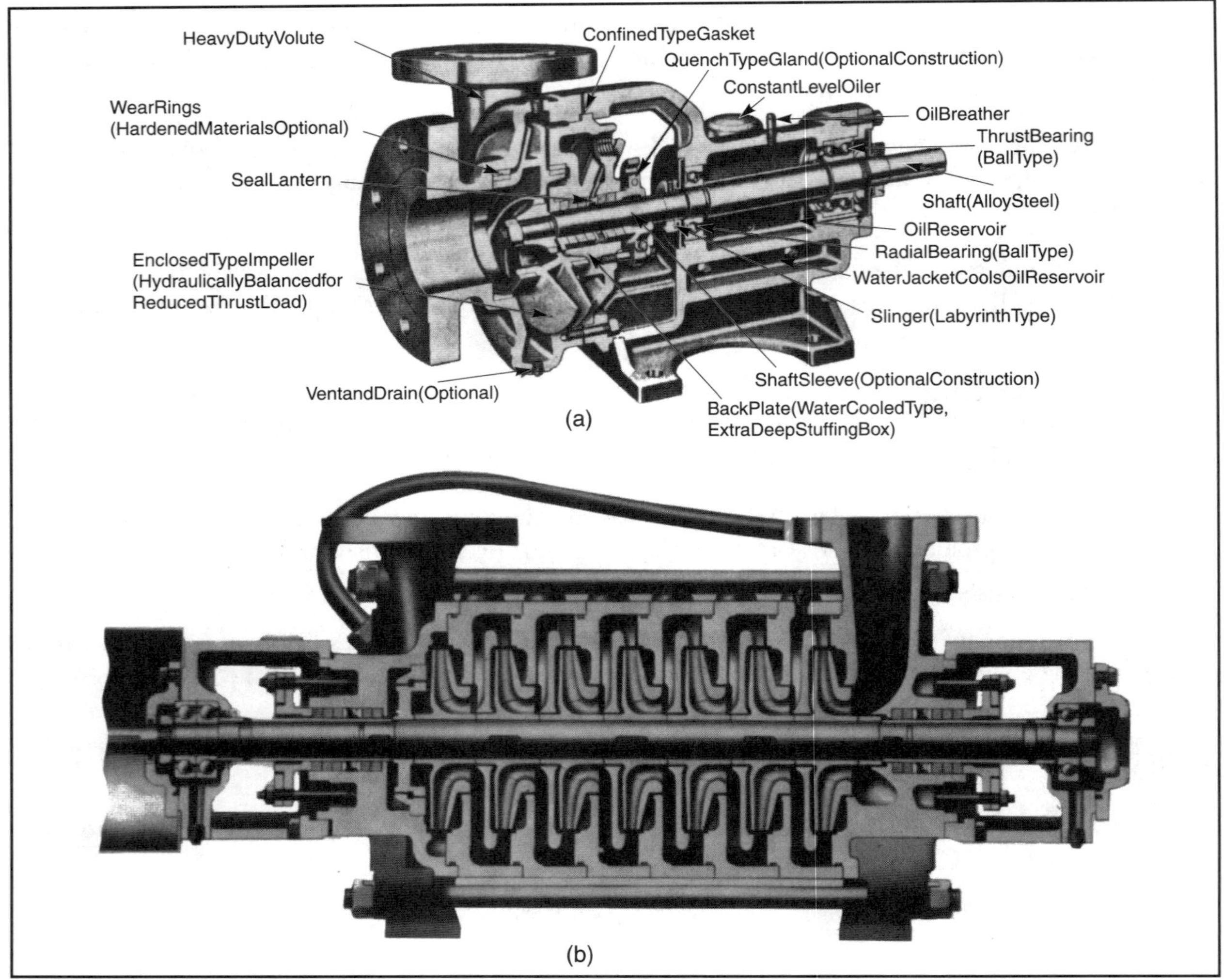

Figure 14.2 Cutaway of Centrifugal Pumps

Both radial and thrust bearings are used. In smaller, single stage pumps these are typically ball or roller bearings. In larger multistage pumps sleeve bearings are frequently used for radial bearings.

In many cases, the head produced by a single pump stage is less than that required. Several impellers can be placed in series on the same shaft to achieve multi-stage performance. This is shown in the lower portion of Figure 14.2. Note that the discharge from a stage is the inlet to the next one. Whether the pump is horizontal (as shown) or vertical, the number of stages is limited only by mechanical factors. Vertical pumps can accommodate more stages than horizontal pumps. A view of a multi-stage, vertical centrifugal pump is shown later in Figure 14.15.

In multistage pumps, the axial thrust across each impeller is additive. This can result in a total axial thrust which is too high for the thrust bearing. It is often necessary to balance the axial thrust. This is done two ways. The first is to use a balance piston as shown in Figure 14.2(b). The pressure on the "inboard" side of the balance piston is full discharge pressure. The pressure on the "outboard" side is suction pressure. The balance piston creates an axial thrust which balances the thrust created by the impellers.

Alternatively, the impellers in a pump can be installed "back-to-back" with half the impellers facing one direction and the other half the opposite direction. This also balances the axial thrust.

A number of impeller types, sizes and mechanical configurations are used in centrifugal pumps. Each pump has a *characteristic* curve of the types shown in Figure 14.3.

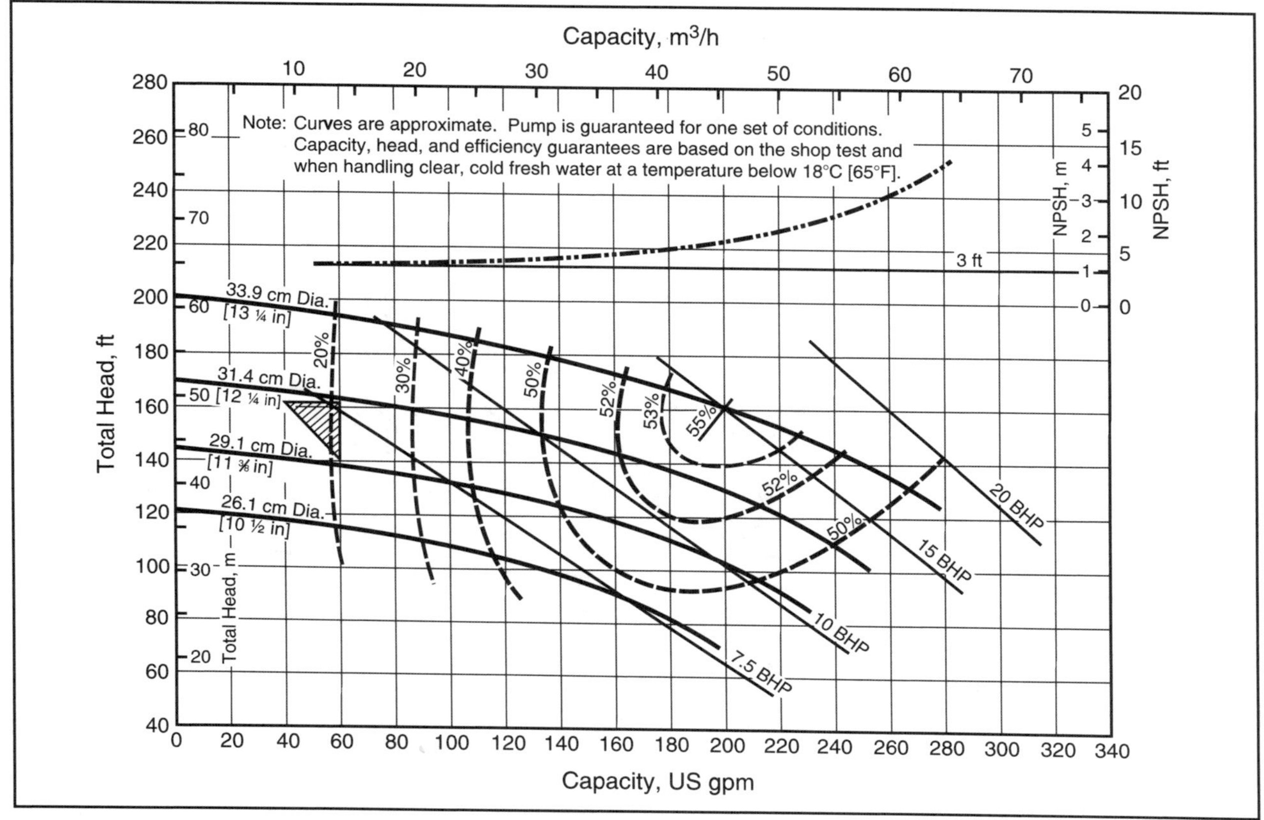

Figure 14.3 Typical Characteristic Curves for a Centrifugal Pump

Shown as a function of flowrate and impeller diameter are head, NPSHR, power and overall efficiency.

All parameters, except for power, are independent of the fluid being pumped (assuming a non-viscous fluid). The values quoted for power are based on water. The power required for other fluids can be determined by multiplying the water power times the rel. ρ, γ, of the fluid.

The maximum efficiency point (MEP) is the point where the specific speed and suction specific speed are calculated. It is widely used as a reference point when evaluating pump performance at an off-design condition.

The NPSHR shown on a pump curve is sometimes confusing. The units on NPSHR are m [ft] of liquid pumped. However, the NPSHR test is run on water. NPSHR represents the NPSHA when the pump experiences a 3% head loss as the NPSHA is reduced. Cold water is the most severe test of NPSHR. Water has a very high latent heat of vaporization and is essentially incompressible. Pump cavitation in cold water service can be very destructive.

Cavitation is less of an issue for fluids with a lower latent head of vaporization and which are somewhat compressible. Most hydrocarbons, particularly NGLs, fall into this category. The Hydraulic

Standards Institute allow a reduction in NPSHR (measured by water test) when dealing with higher vapor pressure fluids. This correction is shown in Figure 14.4.

Figure 14.4 was created based on data for the fluids shown. It is expected that other high vapor pressure fluids would exhibit similar characteristics, but that is not confirmed by experimental data. The total NPSHR correction cannot exceed 50% of the NPSHR for cold water.

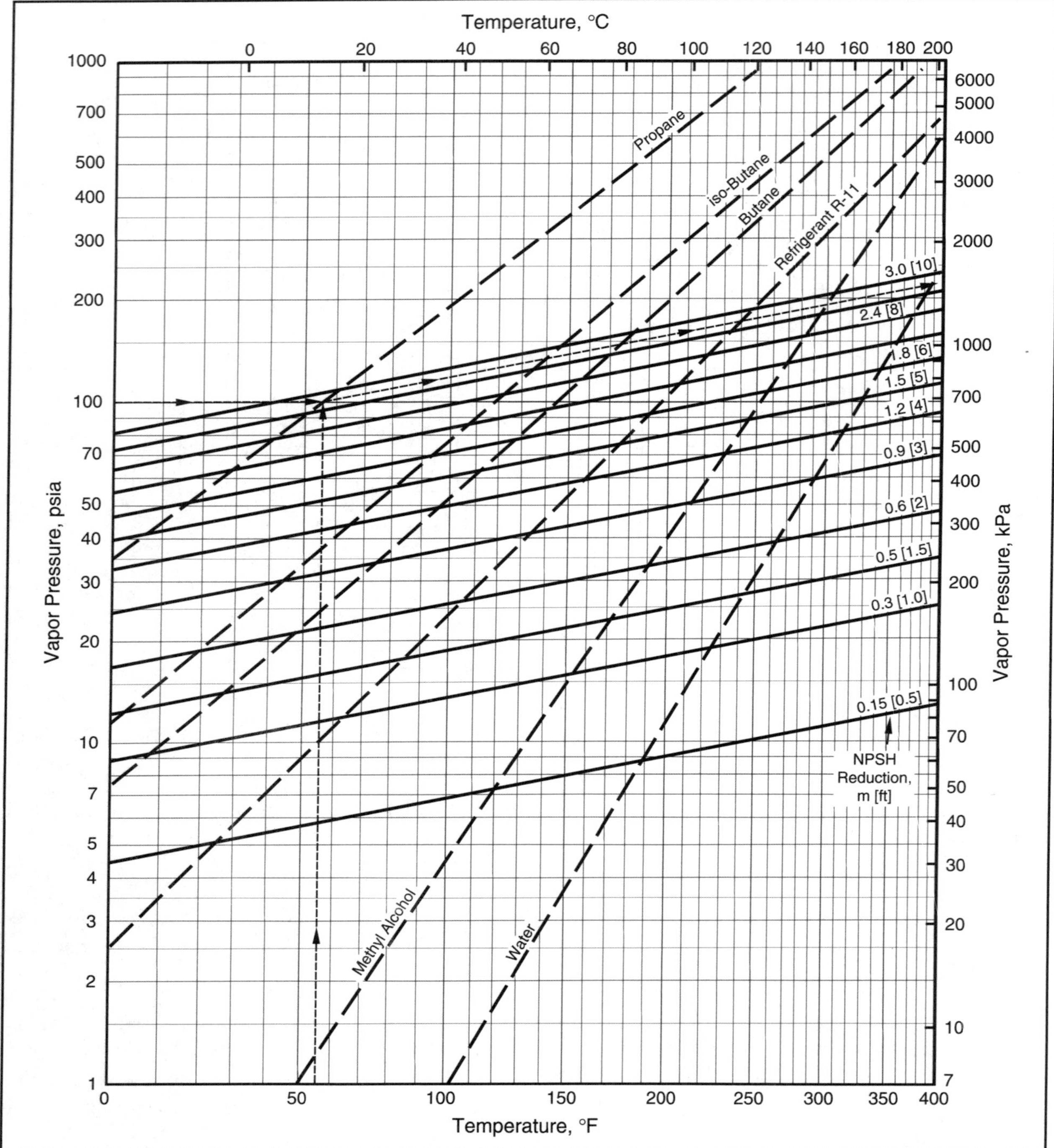

Figure 14.4 NPSH Reduction for Pumps Handling Hydrocarbon Liquids and High Temperature Water
(Reproduced with permission from Hydraulic Standards Inst.)

Note that the pump capacity and pump head depend on the impeller diameter. Larger impeller diameters increase head and flow because the tip speed of the impeller increases. The characteristic relationships are know as the fan or affinity laws and are shown below:

$$q \propto N \text{ or } D$$

$$H \propto (N \text{ or } D)^2$$

$$\text{Power} \propto (N \text{ or } D)^3$$

Where:

N = pump rotational speed
D = impeller diameter

Once a pump has been selected, and the impeller diameter fixed, its performance curves can be plotted as shown in Figure 14.5. The characteristic head, efficiency, power and NPSH curves depend on the impeller type. Figure 14.5 depicts a characteristic curve for a typical process pump with $N_s \approx 40$.

Figure 14.6 shows the effect of specific speed on the likely maximum efficiency of centrifugal pumps. Also shown is the impeller shape and characteristic performance curve consistent with a particular specific speed. Each efficiency curve is for a similar family of pumps with each point being a different pump. It is apparent that peak efficiency occurs for values of N_s between 30 and 60 in Figure 14.6. The use of multiple stages is indicated when too high a head results in a low specific speed, hence low efficiency pump.

In multistage pumps, the specific speed is based on individual impellers, not the entire pump.

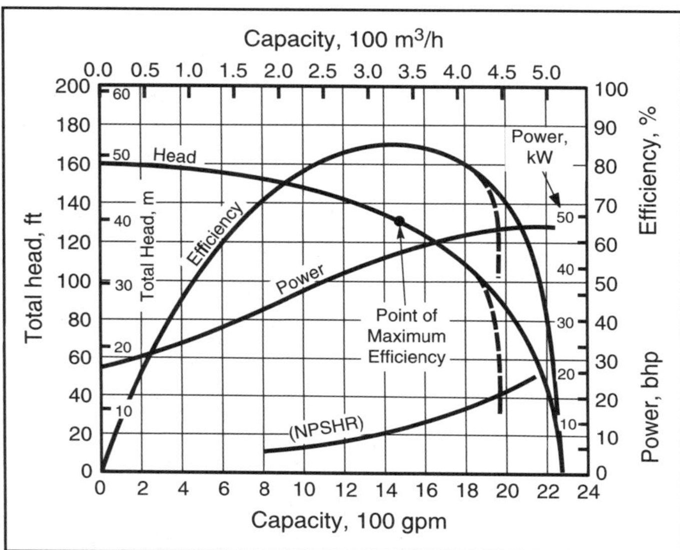

Figure 14.5 Performance Characteristics for a Centrifugal Pump[(14.3)]

Suction Specific Speed

In some pumps cavitation can occur even though NPSHA exceeds NPSHR. This has led to the development of the *suction specific speed* concept.[(14.4)]

$$N_{ss} = \frac{(A)(N)(q)^{0.5}}{(NPSH)^{0.75}} \tag{14.9}$$

Where:

			SI	FPS
N	=	rpm	–	–
q	=	liquid flowrate	m³/s	gpm
NPSH	=	required NPSH	m	ft
A	=	unit conversion factor	1.0	0.0194

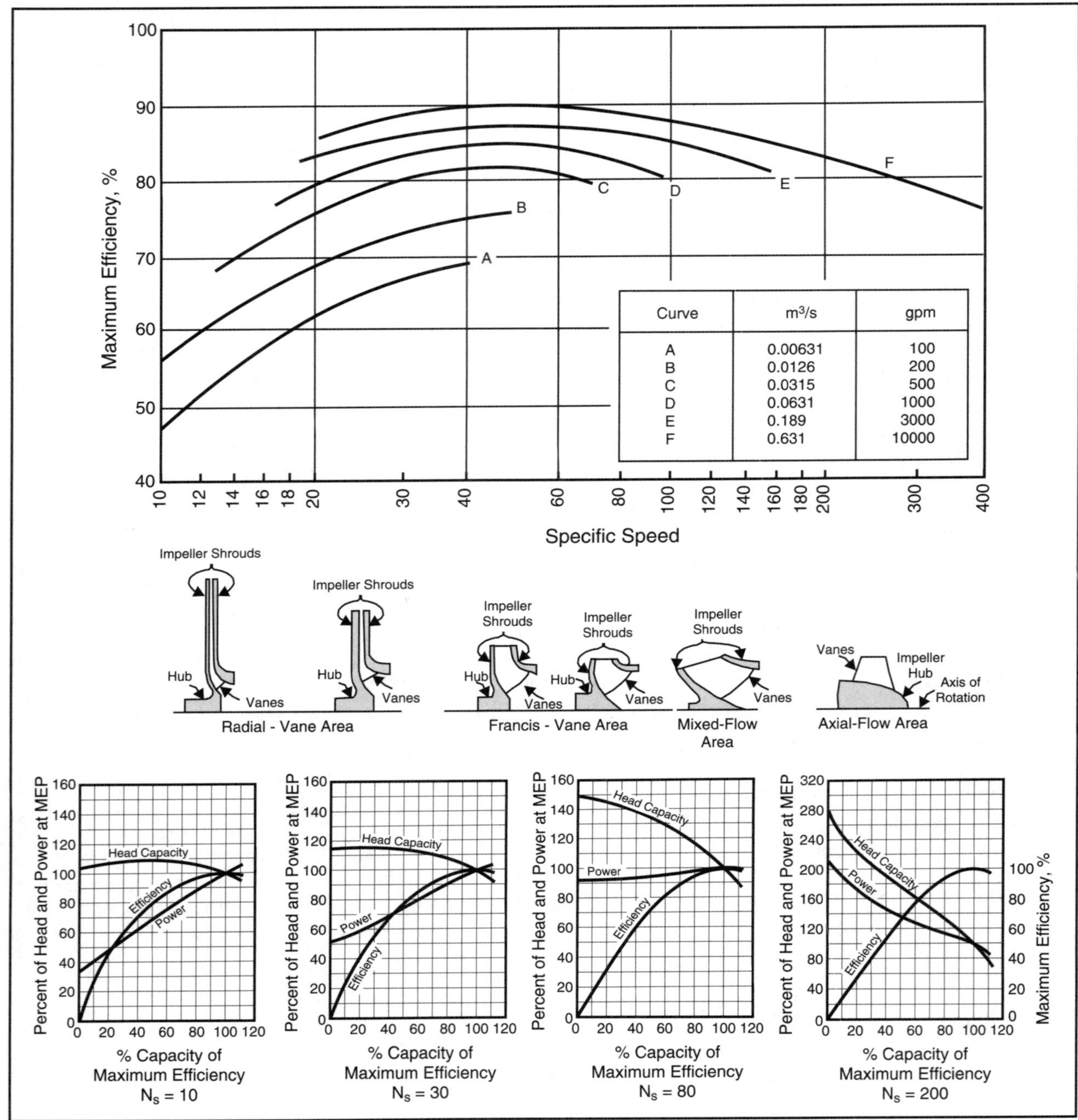

Figure 14.6 Overall Efficiency vs. Specific Speed for Centrifugal Pumps[14.4]

Cavitation may occur at low flow conditions if N_{SS} is greater than about 120. It has been shown that in water service, cavitation can occur at 30-40% of best efficiency flow even though the NPSHA is 2.0 to 2.5 times greater than the NPSHR from the manufacturer's curves. This is due to internal recirculation in the impeller. In hydrocarbon service this phenomena is less troublesome, but in any case it is recommended that pumps with an N_{SS} greater than 200 be avoided. Once a pump has been chosen, it is advisable to make this calculation, particularly for pumps expected to operate at well below their design rate for part of the time. Minimum flow recirculation piping and controls are recommended for high energy centrifugal pumps. Reference 14.5 provides an excellent summary of pumping conditions which warrant minimum flow installation.

Pump Selection and System Performance

Selection of a centrifugal pump requires knowledge of the hydraulic characteristics of system in which the pump will be installed. The "system head" is the amount of energy which must be imparted to the liquid at the pump to achieve the required flowrate. System head is made up of the two components: static head and dynamic head. Static head arises due to a difference in elevation or pressure level. It is independent of system flow. Examples of static head include the elevation difference between sea level and platform deck in a sea water supply pump or the difference between the surge tank and contactor pressure in an amine circulation pump. Dynamic head is due to friction losses in the system. It is the ΔP_f term (expressed in head) discussed in Chapter 10.

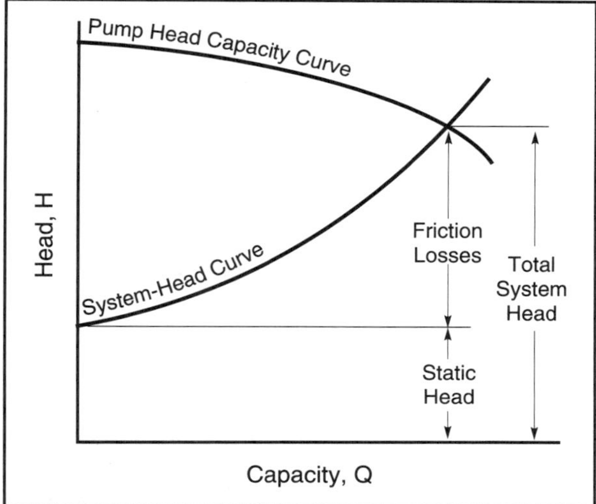

Figure 14.7 System Requirements vs. Pump Capacity

Figure 14.7 shows a system head curve and a typical pump head curve. The intersection point of the two head curves represents the flowrate which will be delivered by the pump at a particular speed.

Control

In process applications it is unusual for a system to operate at a single fixed capacity. The system demand frequently varies. Remember that a given pump operating in a system will deliver the flowrate which corresponds to the intersection of the pump head and system head curves. In order to control the flow through the pump it is necessary to change the shape of either one or both of the curves.

The traditional method for controlling the flow through any dynamic pump like a centrifugal has been to use a valve on the outlet. Liquid is throttled through this valve so that the system head is artificially increased to give the desired flow in accordance with the pump's characteristic curves. This has been a satisfactory arrangement, even though it possesses some basic mechanical faults. From an economic viewpoint, the primary fault is that it wastes energy. For this reason (among several), variable speed drive systems are being increasingly employed. With a variable speed drive the pump head curve is adjusted in accordance with the affinity laws. Figure 14.8 illustrates the use of these two control strategies.

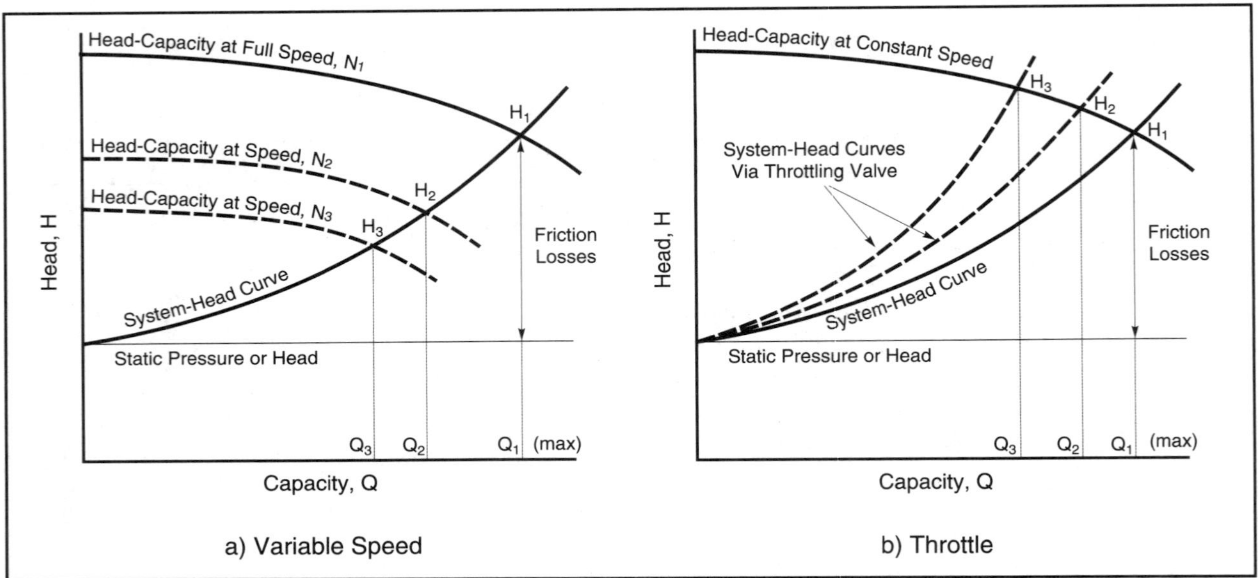

Figure 14.8 Capacity Control of a Centrifugal Pump[14.3]

Considerable power savings can be realized in some systems through the use of variable speed drivers. Figure 14.9 illustrates the magnitude of these savings in one system.

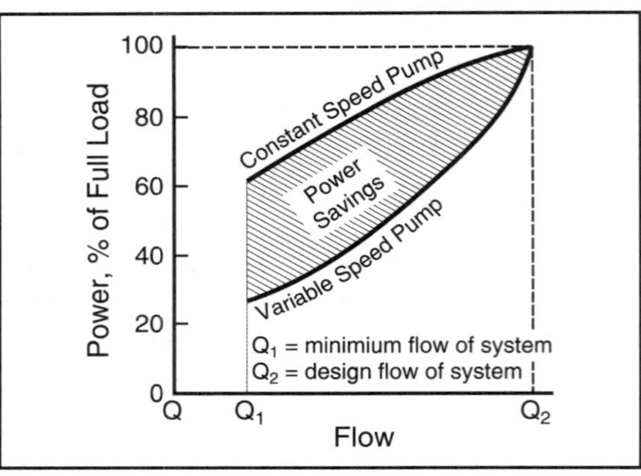

Figure 14.9 Potential Power Savings from Variable Speed Drives

There also are some other advantages of variable speed systems. Soft startups are possible without drawing a large amount of current, since there is no closed discharge valve. Pump wear is reduced as the pump does not operate with artificially high discharge pressures. At constant speed, shaft deflection usually increases with decreasing flow. With variable speed, this deflection is less. Hydraulic noise also is decreased. Last, but not necessarily least, throttling valve maintenance is eliminated.

There are five basic methods for varying pump speed: solid state a.c. and d.c. motor control, mechanical, electro-mechanical and fluid speed changers between the driver and pump. There are discussions of these in the process literature.[14.6, 14.7]

Pumps in Series/Parallel

Many applications require pumps to be installed in series or parallel operation. Series installation is frequently encountered when booster pumps are used. Parallel configuration is often used when several pumps are required to maximize availability. When two pumps are operated in series, the pump head of the two curves is added together to give the combined performance curve for both pumps. Figure 14.10 is an example of such a curve for two pumps in series.

Figure 14.11 shows the combined performance curve of two centrifugal pumps in parallel. The head capacity curves of each pump are additive in flow.

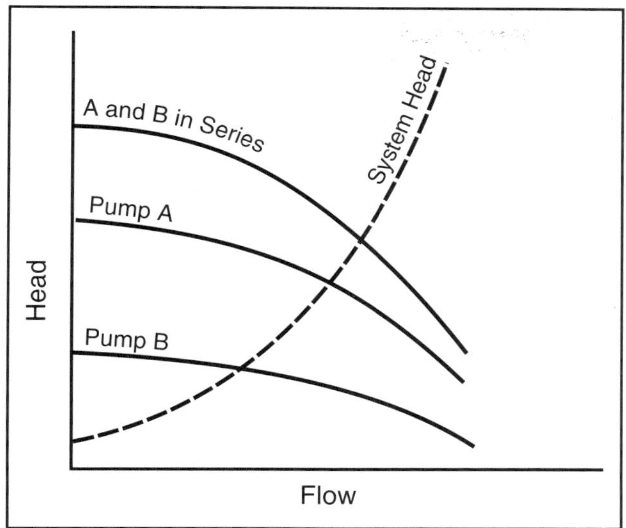

Figure 14.10 Effective Centrifugal Pump Characteristic Curve, Pumps in Series

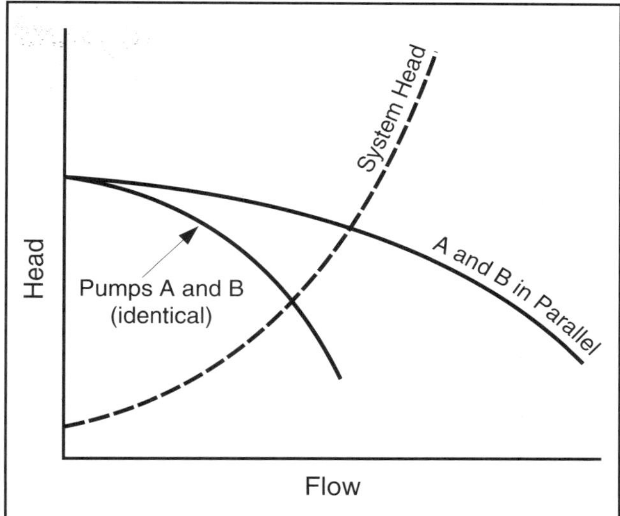

Figure 14.11 Effective Centrifugal Pump Characteristic Curve, Pumps in Parallel

A system head curve has been superimposed on Figure 14.11. It shows that the flowrate through the system does not double when two pumps are operated. In fact, this example shows that the flowrate increases only about 20% for two pumps vs. one pump. It is important to note that if the two pumps are truly identical they will share the new flow equally ie, each pump operates at 60% of the single pump flow.

If the pumps are not identical due to slight dissimilarities in the impeller or due to different maintenance schedules, the load sharing will be more imbalanced, possibly to the point of reducing the flow below the minimum flow shut down of one of the pumps. Individual flow control may be required in these applications.

General

Centrifugal pumps are very flexible and offer a high power output per unit weight and volume. However, like all high speed devices, they can offer unsatisfactory reliability if proper attention is not given to a host of mechanical details. Table 14.1 lists 89 sources of problems. In perusing this list, one can recognize the mechanical considerations that need attention in pump selection, specification, and operation.

RECIPROCATING PUMPS

These are positive displacement devices that move a quantity of liquid fixed by the piston, plunger or diaphragm displacement on the fluid end. They offer particular advantage with slurries, highly viscous liquids and in low flow applications when a high fluid head is required.

There are two basic types of drives. A *power pump* is driven by a driver like a motor, turbine, etc. A *direct acting pump* is driven by steam, gas, air, etc.

Figure 14.12 shows the fluid end of a plunger pump and one type of diaphragm pump. These are the most common types employed in petroleum applications. The plunger pump (as illustrated) is

TABLE 14.1

"Checklist" of Centrifugal Pump Problems[14.8]

No.	Problem
1.	Measuring instruments not correctly calibrated and/or incorrectly mounted.
2.	Air entering the pump during operation, or the pumping system not completely deaerated before starting.
3.	Insufficient speed.
4.	Wrong direction of rotation.
5.	Discharge pressure required by the system is greater than that for which the pump was designed.
6.	Available NPSH too low (including suction lift too high).
7.	Excessive amount of vapor entrained in liquid.
8.	Excessive leakage through wearing surfaces.
9.	Viscosity of liquid higher than viscosity of liquid for which the pump was originally designed.
10.	Impeller and/or casing partially (or fully) clogged with solid matter.
11.	Waterways of impeller and/or of casing very rough.
12.	Fins, burrs, sharp edges, etc., in the path of the liquid.
13.	Impeller damaged.
14.	Outer diameter of impeller machined to a lower diameter than specified.
15.	Faulty casting of impeller and/or casing.
16.	Impeller incorrectly assembled in casing.
17.	System requirements too far out on the head/capacity curve.
18.	Obstruction in suction and/or discharge piping.
19.	Foot valve jammed or clogged up.
20.	Suction strainer filled with solid matter.
21.	Suction strainer covered with fibrous matter.
22.	Incorrect layout of suction and/or discharge piping.
23.	Incorrect layout of suction sump.
24.	The operation of one pump (in a system having two or more pumps operating either in parallel or in series, or in a combination of these) seriously affected by the operation of the other pumps.
25.	Water level in suction sump (or tank, or well) below pump intake.
26.	Speed too high.
27.	Pumped liquid of higher specific gravity than anticipated.
28.	Oversize impeller.
29.	Total head of system either higher or lower than expected.
30.	Misalignment between pump and driver.
31.	Rotating parts rubbing on stationary parts.
32.	Worn bearings.
33.	Packing improperly installed.
34.	Incorrect type of packing.
35.	Mechanical seal exerts excessive pressure on seat.
36.	Gland too tight.
37.	Improper lubrication of bearings.
38.	Piping imposing strain on pump.
39.	Pump running at critical speed.
40.	Rotating elements not balanced.
41.	Excessive lateral forces on rotating parts.
42.	Insufficient distance between outer diameter of impeller and volute tongue.
43.	Faulty shape of volute tongue.
44.	Undersize suction and/or discharge piping and fittings (sometimes causing cavitation).
45.	Loose valve or disk in the system, causing premature cavitation in the pump.
46.	Bent shaft.
47.	Bore of impeller not concentric with its outer diameter and/or not square with its face.
48.	Misalignment of parts.
49.	Pump operates at a very low capacity.
50.	Improperly designed baseplate and/or foundation.
51.	Resonance between operating speed of pump and natural frequency of foundation and/or other structural elements of pumping station.
52.	Rotating parts running off-center because of worn bearings or damaged parts.
53.	Improper installation of bearings.
54.	Damaged bearings.
55.	Water-seal pipe plugged.
56.	Seal cage improperly located in stuffing box, preventing sealing fluid from entering space to form seal.
57.	Shaft or shaft sleeves worn or scored at the packing.
58.	Failure to provide cooling liquid to water-cooled stuffing boxes.
59.	Excessive clearance at bottom of stuffing box, between shaft and casing.
60.	Dirt of grit in sealing liquid.
61.	Stuffing in box eccentric in relation to shaft.
62.	Mechanical seal improperly installed.
63.	Incorrect type of mechanical seal for given operating conditions.
64.	Internal misalignment of parts, preventing seal washer and seal from mating properly.
65.	Sealing face not perpendicular to shaft axis.
66.	Mechanical seal that has been run dry.
67.	Abrasive solids in liquids that come into contact with seal.
68.	Leakage under sleeve due to gasket and O-ring failure.
69.	Bearing-housing bores not concentric with water end.
70.	Damaged or cracked bearing housing.
71.	Excessive grease in bearings.
72.	Faulty lubrication system.
73.	Improper installation of bearings (damage during assembly, incorrect assembly, wrong type of bearings, etc.).
74.	Bearings not lubricated.
75.	Dirt getting into bearings.
76.	Water entering the bearing housing.
77.	Balancing holes clogged up.
78.	Failure of balancing device.
79.	Too-high suction pressure.
80.	Tight fit between line bearing and its seats (which may prevent it from sliding under axial load).
81.	Pump not primed and allowed to run dry.
82.	Vapor or air pockets inside the pump.
83.	Operation at too-low capacity.
84.	Parallel operation of poorly matched pumps.
85.	Internal misalignment due to too much pipe strain, poor foundations or improper repair.
86.	Internal rubbing of rotating parts on stationary parts.
87.	Worn bearings.
88.	Lack of lubrication.
89.	Rotating and stationary wearing rings made of identical materials with identical physical properties.

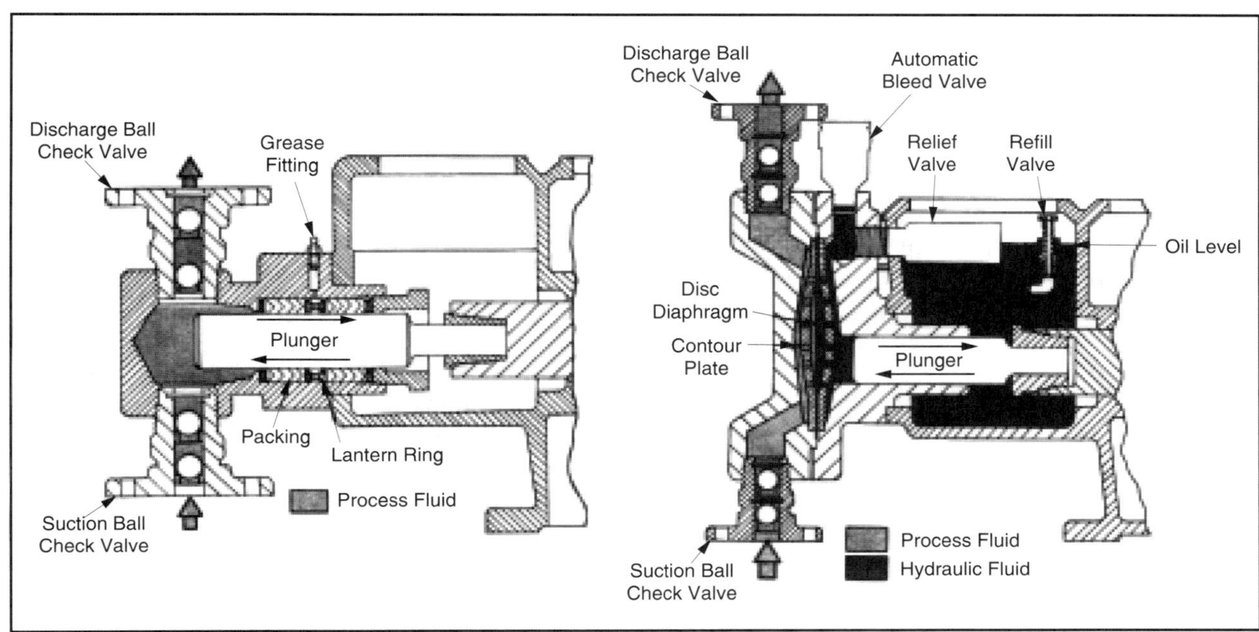

Figure 14.12 View of Fluid End of Reciprocating Pump(14.9)

preferred over the piston pump (not shown) because the plunger's fixed packing provides a better seal than piston rings.

The diaphragm pump is usually employed in corrosive or erosive situations, or where even a small amount of leakage is detrimental. In Figure 14.12, there is a plunger that moves the diaphragm with hydraulic fluid. As it reciprocates it hydraulically moves the diaphragm back-and-forth. Generally, this is preferred over direct coupling of the plunger to the diaphragm.

Ball check valves are shown. These are common in low capacity pumps like metering or chemical feed pumps. Larger pumps may use wing-guided or disk-type valves.

It is common to apply some sort of hard coating to the plunger to minimize scoring and prolong life. Chrome plating, ceramics and nickel alloys are common. In glycol dehydration and sweetening plants (a common application), chrome plating is often used.

The fluid end may be single-acting or double-acting. The pumps shown in Figure 14.12 are single-acting. There is only one inlet and outlet. If a second inlet and outlet are placed on the right side of the plunger, suction will occur at one end while discharge occurs at the other, i.e., the pump is double-acting. In very high head, low flowrate applications the single-acting pump is preferred. In applications where a piston is suitable, a double-acting pump may be preferred.

One basic disadvantage of a reciprocating pump is the velocity surge that occurs during the stroke. One solution is to use more than one pumping element in parallel. A simplex pump possesses only one plunger or piston, a duplex has two, and so on. Shown following are the velocity variations for various pump types, using single-acting pumping elements.

Number	Pump Type	Velocity Variation, %
2	Duplex	160
3	Triplex	25
5	Quintuplex	7
7	Septuplex	4

The velocity variation is the range of velocities related to the average. In a triplex pump the minimum velocity is 82% of the average, the maximum being 107% – a difference of 25%.

A triplex, single-acting reciprocating pump is a common choice in glycol dehydration and liquid sweetening units. It is an effective compromise between pump complexity and flow surge control.

Plunger diameter will vary from 19-63 mm [0.75-2.5 in] with the stroke varying from 25-200 mm [1-8 in]. For reliable performance, the allowable pump speed should depend on stroke length, varying from about 600 rpm for a 1 inch stroke to 200 rpm for 8 inches. For a typical 75 mm [3 in] stroke triplex pump, a maximum rpm of about 400 rpm is recommended.

The overall efficiency for a powered reciprocating pump typically will be 85-92% when operating above a 50% load factor. The efficiency of direct-acting pumps will be lower, depending primarily on their speed. Figure 14.13 shows the correct hook-up of a reciprocating pump.

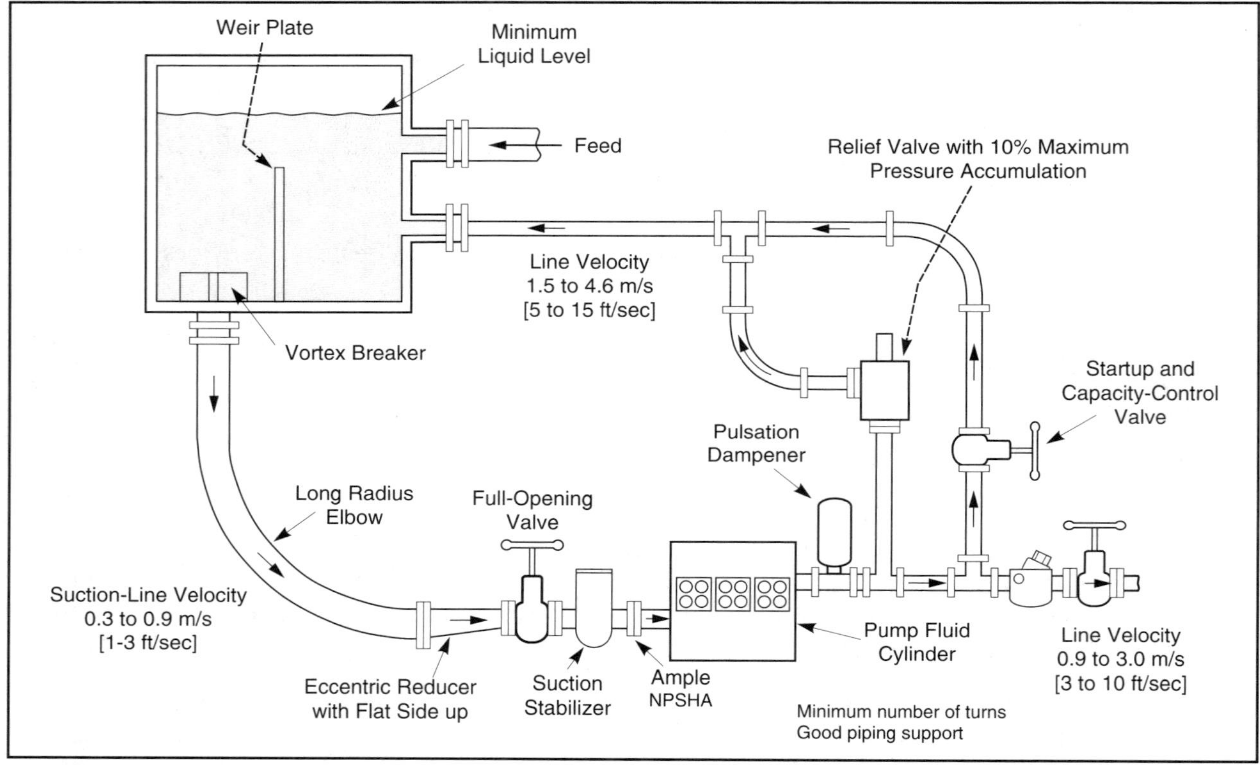

Figure 14.13 Example Installation for a Plunger Pump

Since it is a positive displacement device, a by-pass must be provided for startup and capacity control. Speed and by-passing liquid are the only effective flow control devices available. Reference 14.10, from which this figure is taken, is an excellent review of reciprocating pumps.

Reciprocating pumps have a significantly lower power density than centrifugal pumps. The mass and cost per unit of flow is much higher than for a centrigual pump. Foundations tend to be heavier and more robust and maintenance costs are higher due to valve and packing wear.

ROTARY POSITIVE DISPLACEMENT PUMPS

The rotary pump is a very flexible unit that can handle large quantities of viscous liquid. It is used widely in fluid power applications. Figure 14.14 shows several types of rotary pumps. Of these, the three-screw pump is used most commonly in the petroleum industry. It is simple, rugged, has no valves or parts to foul, is relatively quiet and produces pulsation-free liquid.

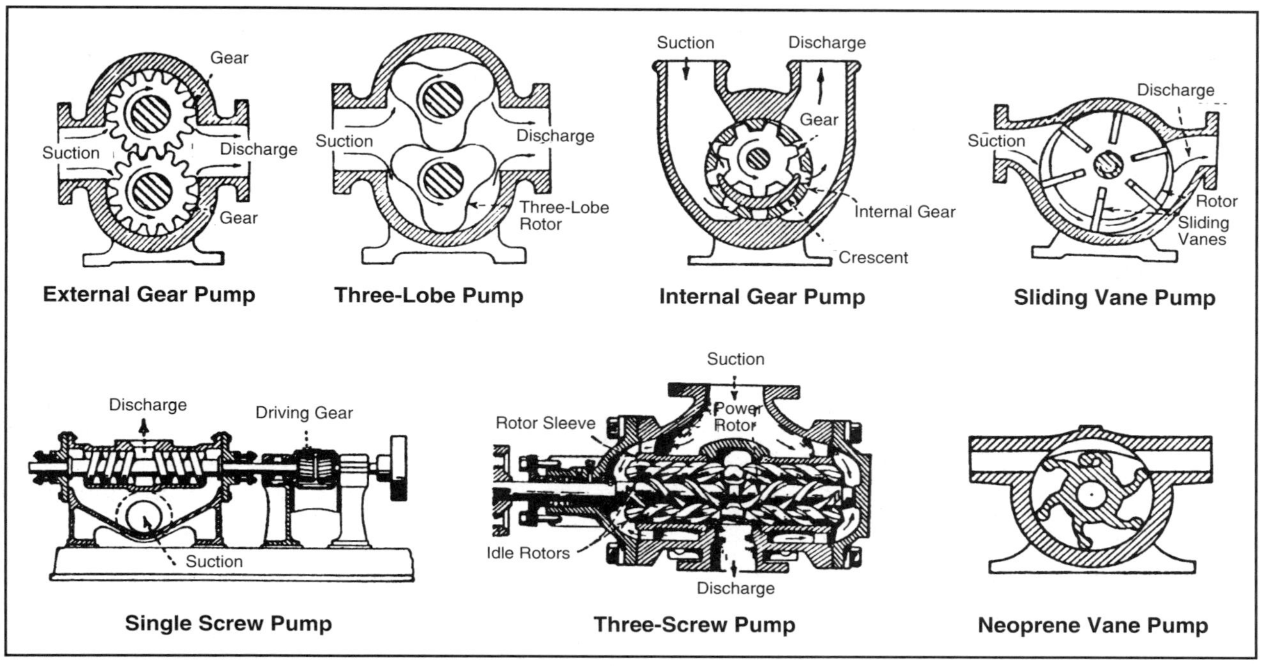

Figure 14.14 Various Types of Common Rotary Pumps

Gear pumps are useful in transfer pump applications. They may be the best choice when differential pressure exceeds 1030 kPa [150 psi] and viscosity exceeds about 70-80 cp, self-priming is desired and/or entrained vapor compromises centrifugal pump viability. Shown at right is a cutaway view of a double-end gear pump.

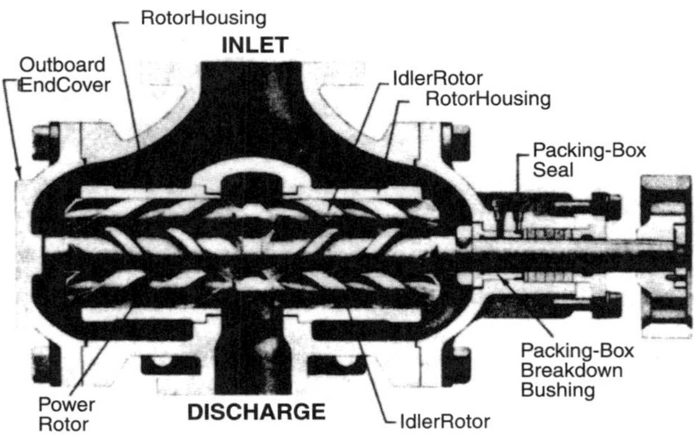

Vacuum Pumps

A vacuum pump is really a compressor. It compresses gas, possibly with entrained liquid, from low pressure to atmospheric pressure. The choice depends on service.

The simplest vacuum pump is an ejector whereby a motive fluid is passed through an orifice or nozzle to increase velocity. This produces an artificially low pressure in the "throat" capable of "pumping" gas.

The motive fluid can be any stream on which a pressure drop is feasible and economical. Steam has been used but rising energy and steam costs have reduced the number of applications. In some cases, natural gas is used when a pressure drop is needed anyway. One application is found on glycol regeneration units when a vacuum is needed.

Most vacuum pumps are some version of the vane or lobe type rotary pumps. As with the types shown in Figure 14.14, oil lubrication is required. In service like sea water deoxygenation, some of the water components contaminate the oil and a treating system is required. Thus, "oil ring" pumps offer limited application in this service.

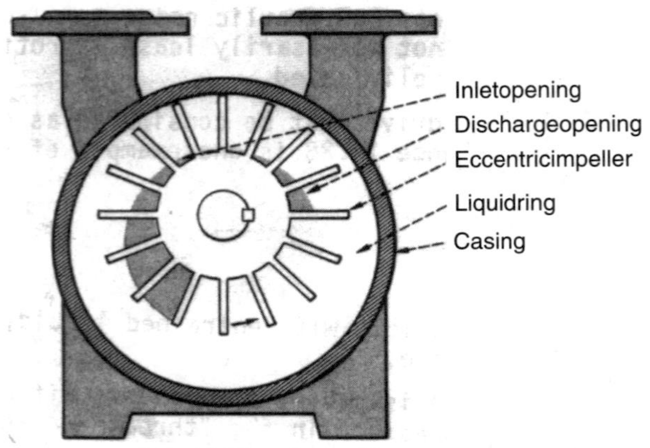

Liquid Ring Pump/Compressor

The "water ring" pump may be the choice when pumping vapors incompatible with lube oil. Sea water deoxygenation is one good example of this. A water-ring pump consumes more energy than oil sealed pumps and requires large amounts of seal liquid. With water, this seal is a once-through process.

Roots blowers are dry lobe type pumps whose lubrication system is isolated from process vapors. They will handle large volumes of vapor. In some cases they are operated in series with mechanical pumps to produce the best results. One concern is to keep the compression ratio low enough in the roots blower so that overheating does not occur.

The type of pump depends also on factors like capacity need and the pressure required. In vacuum technology the pressure unit of the *Torr* is commonly used,

$$1 \text{ Torr} = 1 \text{ mm Hg} = 0.133 \text{ kPa}$$

Often it is convenient to express absolute pressure in Torrs instead of kPa or psia when working below atmospheric pressure.

VERTICAL LIFT PUMPS

There are many applications where liquid is to be pumped from a sump, a body of water, a basin, etc. A common pump used is a vertical, multi-stage centrifugal of the type represented by Figure 14.15. The number of stages will vary with the service. A pump much like that shown usually is used as a seawater supply pump in offshore production systems. It can be located at an appropriate depth below the sea water surface to reduce intake of surface debris. Vertical pumps are frequently used in process applications where space or NPSHA is limited and/or high head requirements exist.

Four basic systems are used to lift liquids in a wellbore. One of these, gas-lift, is outside the scope of this book. The other three are shown in Figure 14.16. Part (a) shows five different rod pumps. All are similar in operation. The upper portion moves up and down. By using two ball valves, the pump barrel is alternately filled and emptied. The amount of liquid pumped is regulated by barrel size, stroke and speed. The pump shown is connected to the surface by "sucker rods."

Part (b) of Figure 14.16 shows a multi-stage centrifugal pump small enough in diameter to be lowered in a wellbore. The basic operating characteristics are the same as regular centrifugal pumps. Most pumps in this service are in the mixed flow range wherein the impeller imparts both centrifugal and axial force on the liquid. Details of this type of pump are found in Reference 14.11.

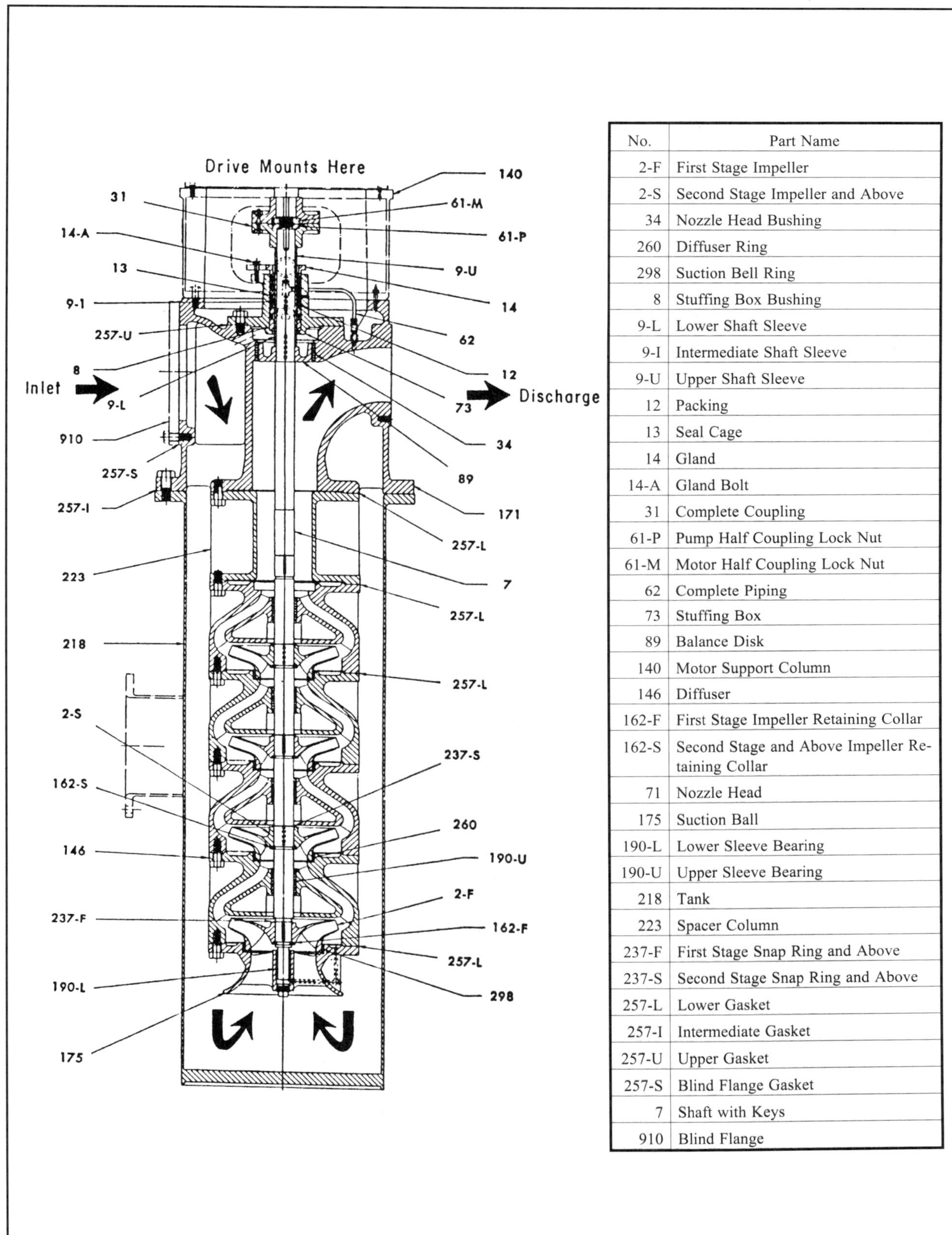

No.	Part Name
2-F	First Stage Impeller
2-S	Second Stage Impeller and Above
34	Nozzle Head Bushing
260	Diffuser Ring
298	Suction Bell Ring
8	Stuffing Box Bushing
9-L	Lower Shaft Sleeve
9-I	Intermediate Shaft Sleeve
9-U	Upper Shaft Sleeve
12	Packing
13	Seal Cage
14	Gland
14-A	Gland Bolt
31	Complete Coupling
61-P	Pump Half Coupling Lock Nut
61-M	Motor Half Coupling Lock Nut
62	Complete Piping
73	Stuffing Box
89	Balance Disk
140	Motor Support Column
146	Diffuser
162-F	First Stage Impeller Retaining Collar
162-S	Second Stage and Above Impeller Re-taining Collar
71	Nozzle Head
175	Suction Ball
190-L	Lower Sleeve Bearing
190-U	Upper Sleeve Bearing
218	Tank
223	Spacer Column
237-F	First Stage Snap Ring and Above
237-S	Second Stage Snap Ring and Above
257-L	Lower Gasket
257-I	Intermediate Gasket
257-U	Upper Gasket
257-S	Blind Flange Gasket
7	Shaft with Keys
910	Blind Flange

Figure 14.15 Four Stage Vertical Centrifugal Pump

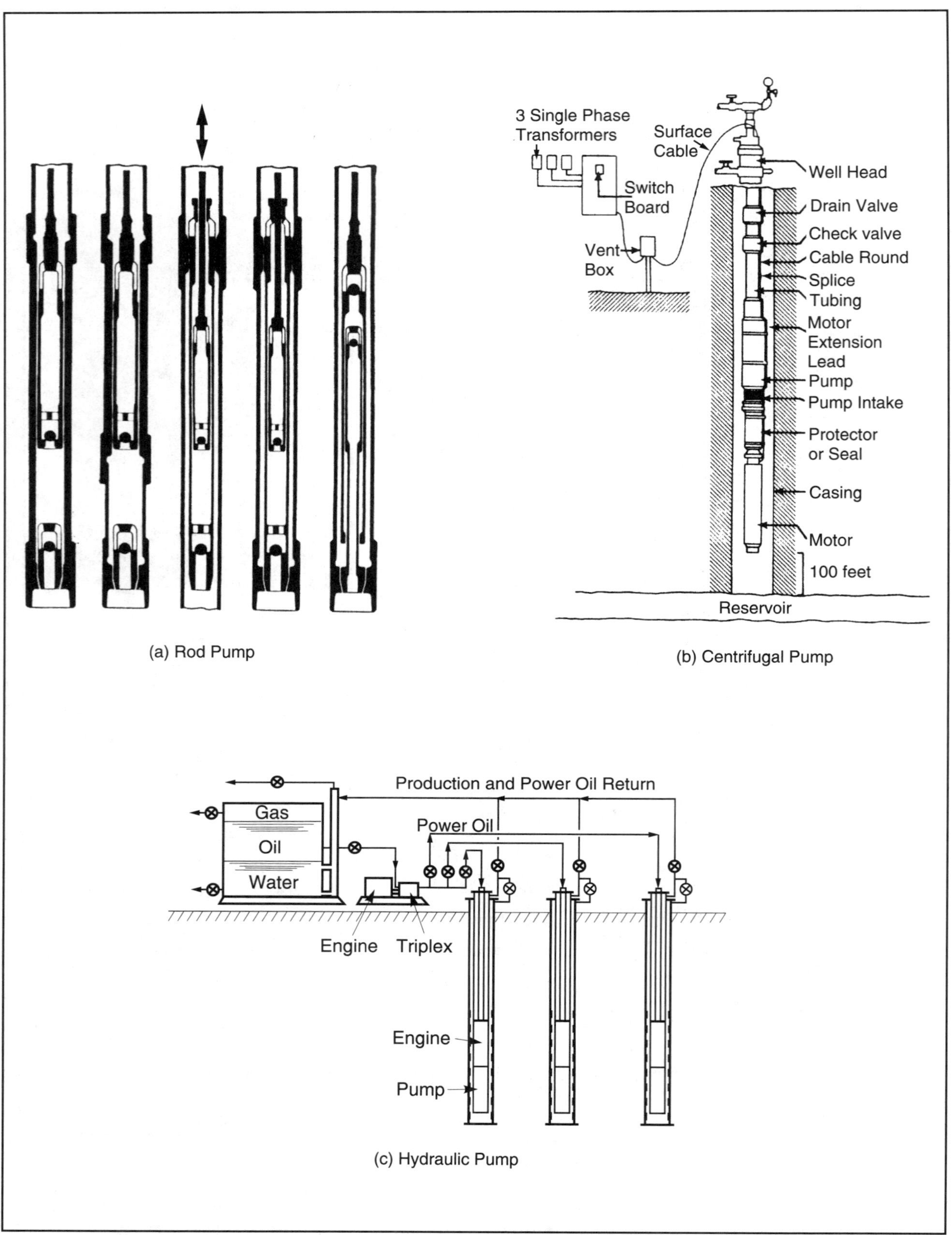

(a) Rod Pump

(b) Centrifugal Pump

(c) Hydraulic Pump

Figure 14.16 Example Well-Lift Pump Applications[(14.9)]

Part (c) is a hydraulic system. Power oil under pressure is used to drive the pump. Generally, power oil is pumped down the inner pipe string and the mixture of power oil and liquid production return together in the tubing annulus. These are separated and a portion of the oil is used for power oil, to complete the cycle.

The net pressure on the pump is the column pressure minus the formation pressure; this also is equal to the differential pressure on the engine when the engine and pump pistons have the same diameter.

Each of these basic pump types has an optimum operating range but the final choice often is based on economics. Figure 14.17 shows a summary of relative cost versus pump depth and liquid production rate.(14.12) Part (d) is a composite of Parts (a)-(c) and shows the optimum investment cost range for each type. The specific results shown apply only to one study, but the general pattern shown is typical.

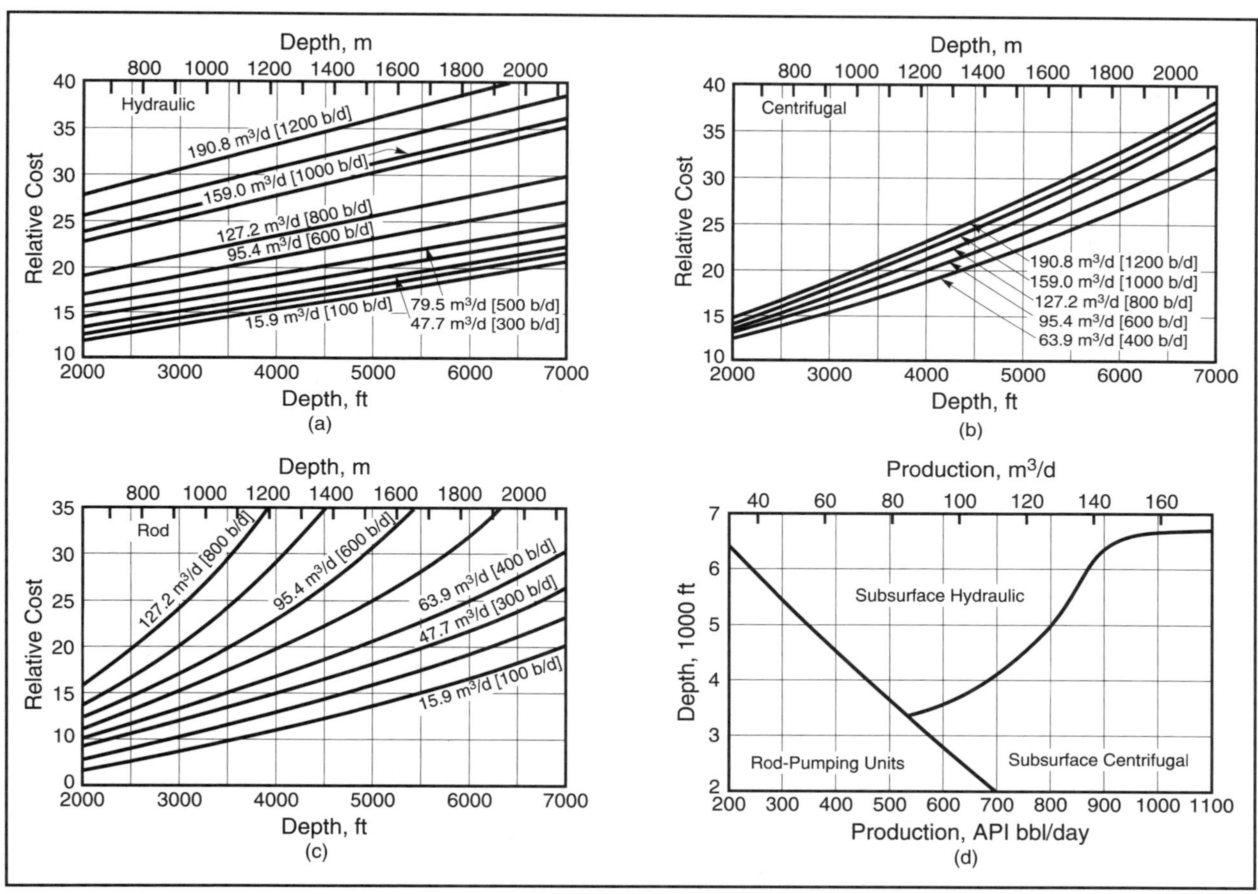

Figure 14.17 Relative Cost and Application of Wellbore Pumps

The relative investment cost shown is for comparison purposes. A given value on the ordinate represents a comparable cost for the units. Absolute cost will vary with many factors.

HYDRAULIC POWER RECOVERY EXPANDERS

In some instances power can be recovered by depressurizing liquid across hydraulic power recovery turbine (HPRT), which looks very much like a centrifugal pump, instead of a valve. The HPRT is basically a pump running backwards to produce power or to drive other equipment.

All rotating pumps are "free wheeling" and can run backward. Centrifugal pumps have proven satisfactory as HPRTs. In addition, there are several types of hydraulic turbines manufactured specifically for that service.

In a centrifugal pump the liquid enters at what normally would be the outlet and leaves at the "eye." A backward running pump has several critical limitations. Its efficiency may be about 10% less than a special power recovery turbine, even though its first cost is less. Also, control is more complicated, for you must contend with some independent control elements. With a flashing liquid, efficiency can be lower because the impeller tends to shed the gas which bypasses the liquid.

The performance curves of a centrifugal pump are different when used as a turbine. Efficiency is several percent lower and occurs at a higher capacity. Manufacturers have developed correction factors to predict pump performance as a turbine.

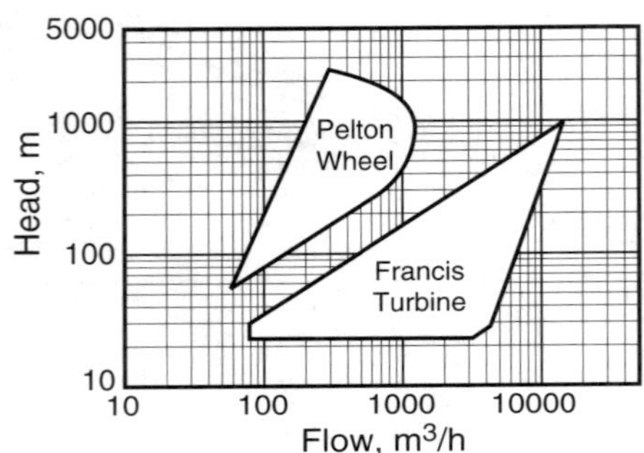

There are two types of regular hydraulic turbines – a Pelton wheel and a Francis runner. The Pelton wheel has a series of jets which impinge directly on a rotating impulse wheel. As shown at left, it is used primarily for high head, low flow applications. The Francis runner uses wheels shown earlier for centrifugal pumps.

The figure below shows performance of one turbine. The curve shapes are for example purposes and actual values will vary with the specific model.

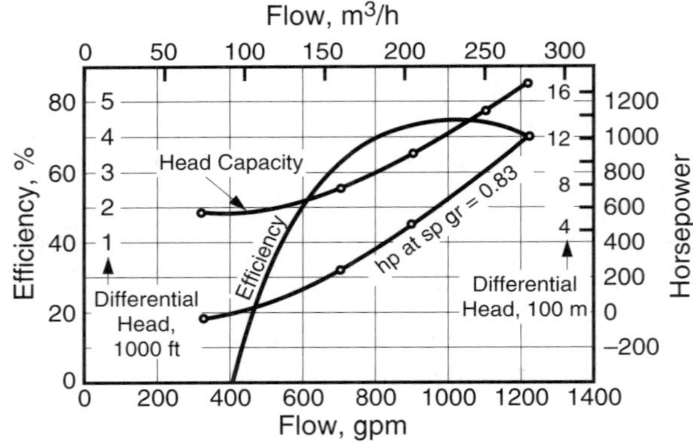

If the HPRT is run by liquid from a vessel where a level is being maintained, a level controller is used.

References 14.13-14.16 review hydraulic expanders in more detail.

REFERENCES

14.1 Hydraulic Institute, "Standards for Centrifugal, Rotary and Reciprocating Pumps," Cleveland, Ohio.
14.2 Stindt, W. H., *Chem. Eng.* (Oct. 11, 1971).
14.3 Karassik, I. J., *Chem. Eng.* (Oct. 4, 1982), p. 84.
14.4 Sutton, G. P., *Hydr. Proc.* (June 1978), p. 103.
14.5 Taylor, I., *Chem. Eng.* (May 11, 1987), p. 53.
14.6 Doll, T. R., *Chem. Eng.* (Aug. 9, 1982), p. 46.
14.7 Chauprade, R., *Hydr. Proc.* (Oct. 1981), p. 128.
14.8 Hedidiah, S., *Chem. Eng.* (Oct. 24, 1977), p. 124.
14.9 Poynton, J. P., *Hydr. Proc.* (Nov. 1981), p. 279.
14.10 Henshaw, T. L., *Chem. Eng.* (Sept. 21, 1981), p. 105.
14.11 Coltharp, E. D., *Jour. Pet. Tech.* (April 1984), p. 645.
14.12 Johnson, L. D., *Oil and Gas J.* (Aug. 26, 1968), p. 110.
14.13 Braun S. S., *Ibid.* (May 21, 1973), p. 128.
14.14 Swearingen, J. S. and B. G. Schulz, *Ibid.* (July 5, 1976), p. 70.
14.15 Buse, F., *Chem. Eng.* (Jan. 26, 1981), p. 113.
14.16 Brennan, J. R., *Hydr. Proc.* (July 1981), p. 72.

NOTES:

15

COMPRESSORS

Compressors can be generally classified in two categories:

1. Positive Displacement: this type of compressor includes reciprocating, rotary screw, sliding vane, liquid ring and rotary lobe. The compression principle is volumetric displacement – reducing the gas volume increases pressure.
2. Kinetic or Dynamic: this type of compressor includes centrifugal and axial compressors. The compression principle is acceleration and deceleration of the gas – kinetic energy is converted into pressure rise.

Reciprocating and centrifugal compressors are by far and away the most popular compressors used in E & P applications. Rotary screw compressors are gaining in popularity in low to moderate pressure gas boosting service, refrigeration systems and fuel gas compression for gas turbines.

Rotary lobe and vane type compressors are used primarily in specialized services. Lobe and vane compressors may prove useful, particularly when discharge pressures do not exceed about 200 kPa [30 psig] and the flowrate is not too large. Both possess favorable weight and vibration characteristics. Primary uses include vapor recovery service, low pressure gas gathering, and closed cycle solid bed dehydration units.

Table 15.1 is a comparison of various compressors and Table 15.2 various drivers. Notice that three different listings are used for reciprocating compressors. Two are for separable units using high speed (H.S.) and low speed (L.S.) drivers. A separable compressor is one where the gas compressor is

TABLE 15.1

Comparative Guidelines to Selection of Compressors(15.1)

Compressor Type	Reliability	Initial Cost	Installation Cost	Efficiency	Maintenance Cost	Weight/ Space	Run Length	Ease Movement	Remove Adapt-ability	Adaptability to Change of Conditions
Low Pressure Screw or Lobe	E	E	E	G	E	E	E	E	E	G
Low Pressure Sliding Vane	G-P	E	E	G	F	E	F	E	E	G
H.S. Recip. Separable	G	E	E	G	G	E	G	E	G	E
L.S. Recip. Separable	E	G-F	G-F	E	E	F	G	P	G	E
Integral Recip Compressor	E	G-P	G-P	E	E	F	E	G-P	G	E
Centrifugal	E	E-F	E	E-G	E	E	E	G-P	E	F-P

Edit Copy 2 — 8/31/2000

TABLE 15.2

Comparative Guidelines to Selection of Gas Compressor Drivers(15.1)

Driver Type	Reliability	Initial Cost	Installation Cost	Efficiency	Maintenance Cost	Weight/ Space	Run Length	Ease Movement	Remove Adaptability	Adaptability to Reciprocating	Adaptibility to Centrifugal Compressor
High Speed Gas Engine	G	E	E	G	G	E	G	E	G	E	F
Low Speed Integral Gas Engine	E	G-P	G-P	E	E	F	E	P	G	E	P
Gas Turbine	E	G-P	E-F	G*	E	E	E	E	E	F	E
Steam Turbine	G	G	G	G-P	E	E	E	P	F	F	E
Electric Motor	E	E-G	E	E	E	E	E	E-P	E	G	E

Rating Symbols	
E - Excellent	F - Fair
G - Good	P - Poor

*Thermal Efficiency can be increased by use of Recuperator and/or Waste Exhaust Heat Recovery

Exhaust Heat Recovery can also be utilized sometimes on Reciprocating Engines.

a separate unit directly coupled to, or belt driven by, a separate driver (often on a common skid). The difference between "high speed" and "low speed " is somewhat arbitrary, but low speed can be considered anything less than about 500-600 rpm. The integral compressor has an internal combustion engine driver on the same frame, integral with the gas compression cylinders. Integral compressors are seldom used in new applications today due to their significant weight and space requirements. In addition, the gas engine drivers have difficulty complying with emission standards.

Figure 15.1 is the general representation of the application range of most of the compressors shown. The inlet flowrate is actual flow at suction P and T. There is a broad overlap shown. Frequently the choice is made on the economic and the mechanical factors involved.

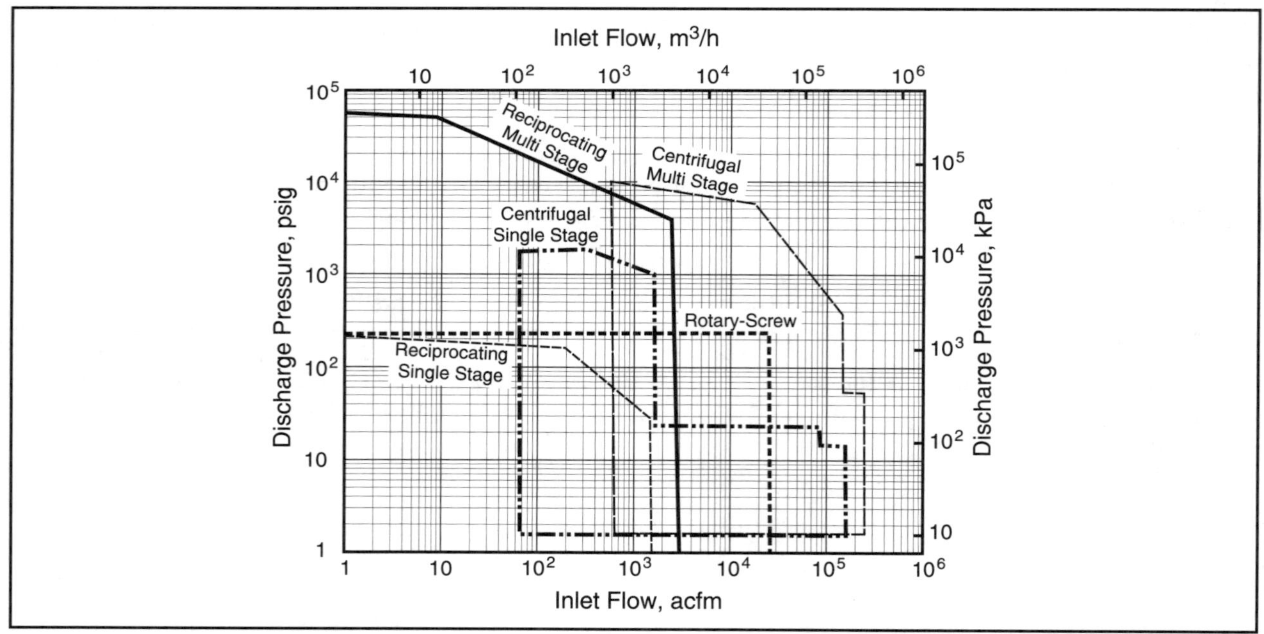

Figure 15.1 General Range of Application of Compressors *(Adapted with permission from GPSA)*

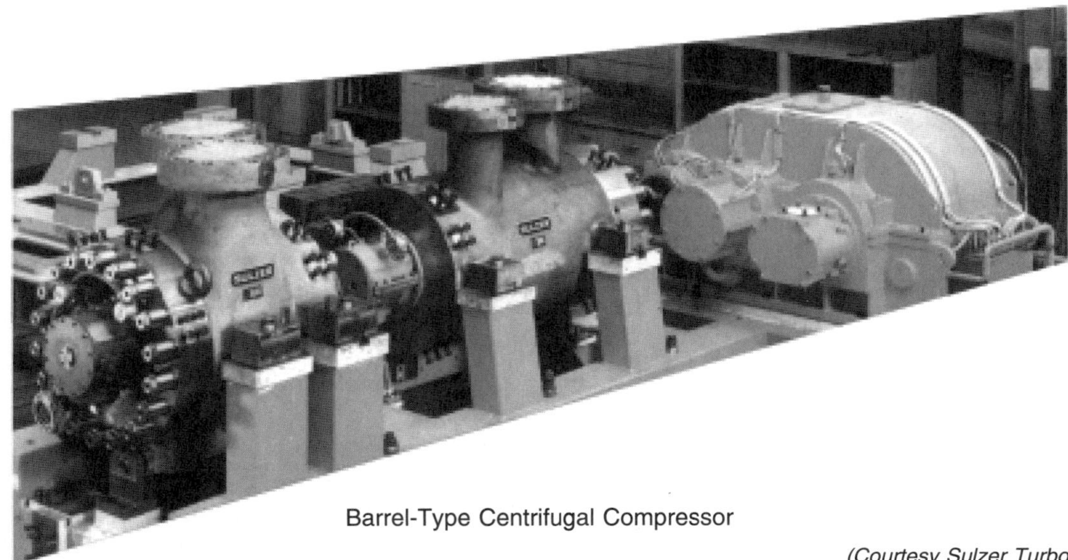

Barrel-Type Centrifugal Compressor

(Courtesy Sulzer Turbosystems)

If the inlet flow exceeds about 850-1700 m^3/h [500-1000 acfm], the centrifugal compressor is the most common choice. It produces the most power per unit weight and volume, and is less expensive per unit of power. It is ideal for high flowrate and low to high head situations. In offshore and frontier area applications, the centrifugal is the usual choice. However, for all of these advantages, a centrifugal is not always the best choice. Its overall efficiency is often lower than a reciprocating compressor and fuel consumption is higher. It also is less "forgiving" of errors in specification and changes in operating conditions.

Unless the changes in gas rate and gas composition have been accurately forecasted, the centrifugal unit may be forced to operate (if at all) outside its satisfactory range. Liquids, solids and other contaminants can prove troublesome. Centrifugal compressors are relatively simple and provide satisfactory service when carefully chosen and applied with proper controls. The majority of the problems are due to poor specification and installation.

If you are considering a reciprocating unit, what type may be most desirable? In production operations when boosting small quantities of gas, a high-speed separable with driver speeds up to 1800 rpm may be chosen. In small sizes, to several hundred kW, a small belt-driven unit may be suitable. For larger units a direct coupled, horizontally-opposed unit might be used. High speed units operating above 900 rpm are available to about 1500 kW [2000 hp]. Units operating below 900 rpm are

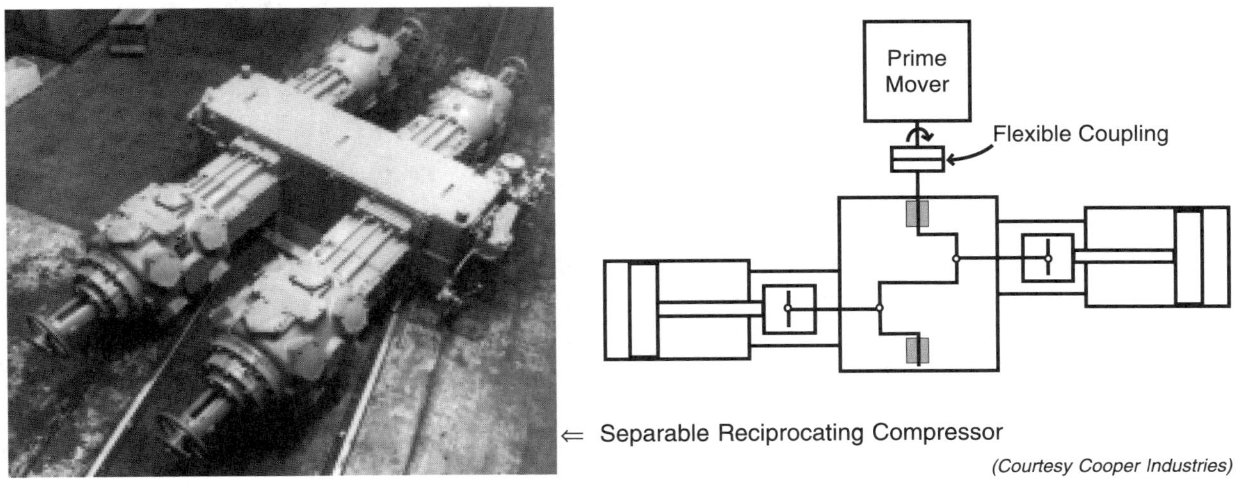

⇐ Separable Reciprocating Compressor

(Courtesy Cooper Industries)

available to 12 000 kW. Slow speed separables can be driven by a gas engine or electric motor. In large frame sizes an electric motor drive is often preferred because a gearbox is not required.

Large integral compressors operating in a speed range of 300-600 rpm have the highest efficiency and lowest maintenance. They are seldom purchased in new applications. In spite of this, there are many existing applications (on land) where they are still used. Long life, low maintenance and ease of operation are all advantages of integral compressors.

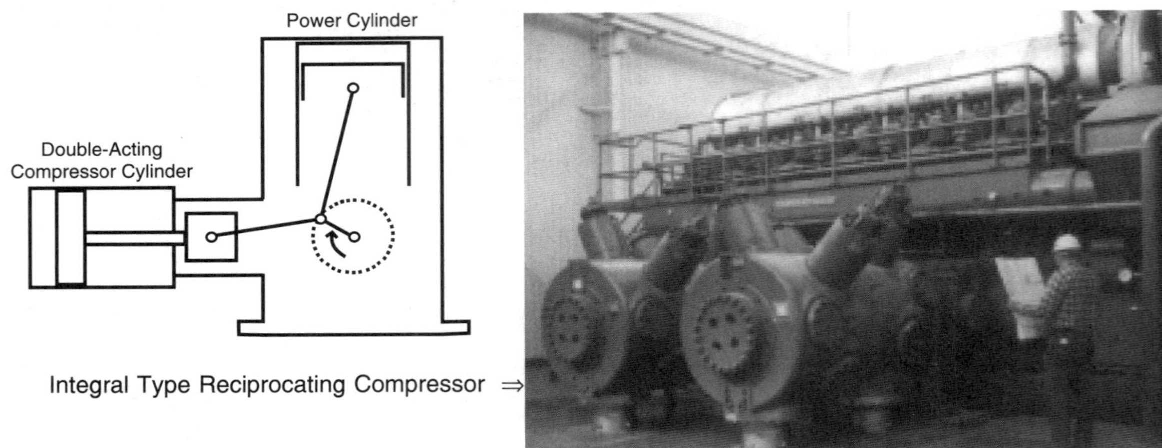

Integral Type Reciprocating Compressor ⇒

Rotary screw compressors have become increasingly popular in the oil and gas industry. The primary applications are in gas boosting, refrigeration, and fuel gas compression for gas turbine installations. Maximum discharge pressure is usually limited to about 2000-4000 kPa [300-600 psig], but higher discharge pressures are possible. Most screw compressors are twin rotor and the gas is compressed by positive displacement in the ever diminishing spaces between the convolutions of the rotors. Frame sizes are available up to 10000 m^3/h [6000 acfm] displacement. Two configurations are available — oil-free or oil-flooded. The most common drivers are electric motor and gas engines.

(Courtesy Aerzen)
Rotary Screw Compressor

(Courtesy Aerzen)
Twin Rotor Screws

COMPRESSOR POWER REQUIREMENTS

The *theoretical* amount of energy needed to compress a given amount of a gas between specified suction and discharge conditions is independent of the compressor unit. The *actual* amount of energy used depends on the efficiency of the compressor and its driver.

The basic thermodynamic equation is written

$$\Delta H = \int VdP = -W_{theor} \quad (15.1)$$

on the basis that the process is reversible ($\Delta S = 0$) and adiabatic ($Q = 0$). The subscript "theor" signifies these are the assumptions being used.

The error in these assumptions is corrected by an overall efficiency represented by the symbol "E."

$$W_{act} = \frac{W_{theor}}{E} \tag{15.2}$$

The value of "E" is used as a decimal in Equation 15.2 although it often is expressed as a percentage. It includes the effect of the thermodynamic (isentropic) efficiency and the mechanical efficiency. *Mechanical efficiency* refers only to frictional and other mechanical losses.

Another efficiency, called *polytropic efficiency* is sometimes used in lieu of an isentropic efficiency. It will be defined and discussed in a later section.

The first step is the calculation of theoretical work (power). Using an efficiency from an acceptable correlation, one can then obtain actual power needs. It is usual to calculate power per stage and then multiply by the number of stages to find the total for that unit.

Use of Enthalpy Correlations to Obtain Power Requirements

Most computer solutions use the equation:

$$\text{Theoretical Power} = m(h_{2_{isen}} - h_1) \tag{15.3}$$

Where:

m = gas flowrate

$h_{2_{isen}}$ = outlet enthalpy per unit mass for an isentropic path

h_1 = inlet enthalpy per unit mass

Entropy and enthalpy values are calculated from an equation of state. Necessary data include inlet gas rate, pressure, temperature and composition plus outlet pressure. The calculation proceeds as follows:

Step 1: Determine h_1 and s_1 from T_1 and P_1

Step 2: Assume isentropic path, so $s_2 = s_1$

Step 3: From s_2 and P_2 determine $h_{2_{isen}}$

Step 4: $\Delta h_{isen} = h_{2_{isen}} - h_1$, this is the isentropic or adiabatic head

The figure at right illustrates the procedure. Line AB is for an isentropic compression. With a reciprocating compressor it will be one stage; with a centrifugal it can be one stage or that number of stages in series before gas cooling is required. Line BC is a gas cooling step. Line CD represents a subsequent stage or stages, if any.

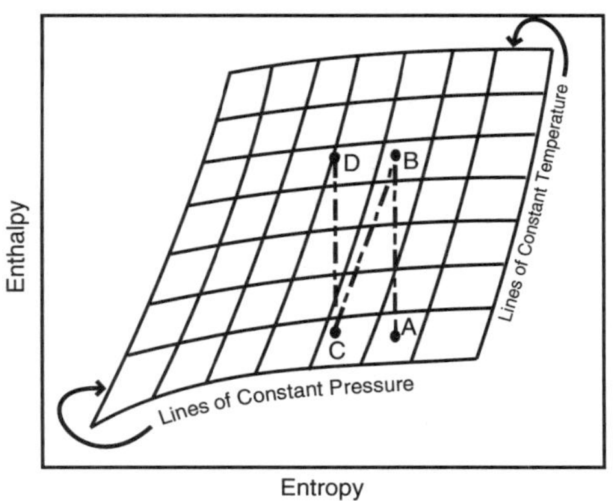

Figure 15.2 (a & b) can be used to find an approximate Δh_{isen} in a manual calculation for a 0.65-0.75 relative density hydrocarbon gas. This figure also can be used to estimate the outlet temperature for a specified set of conditions.

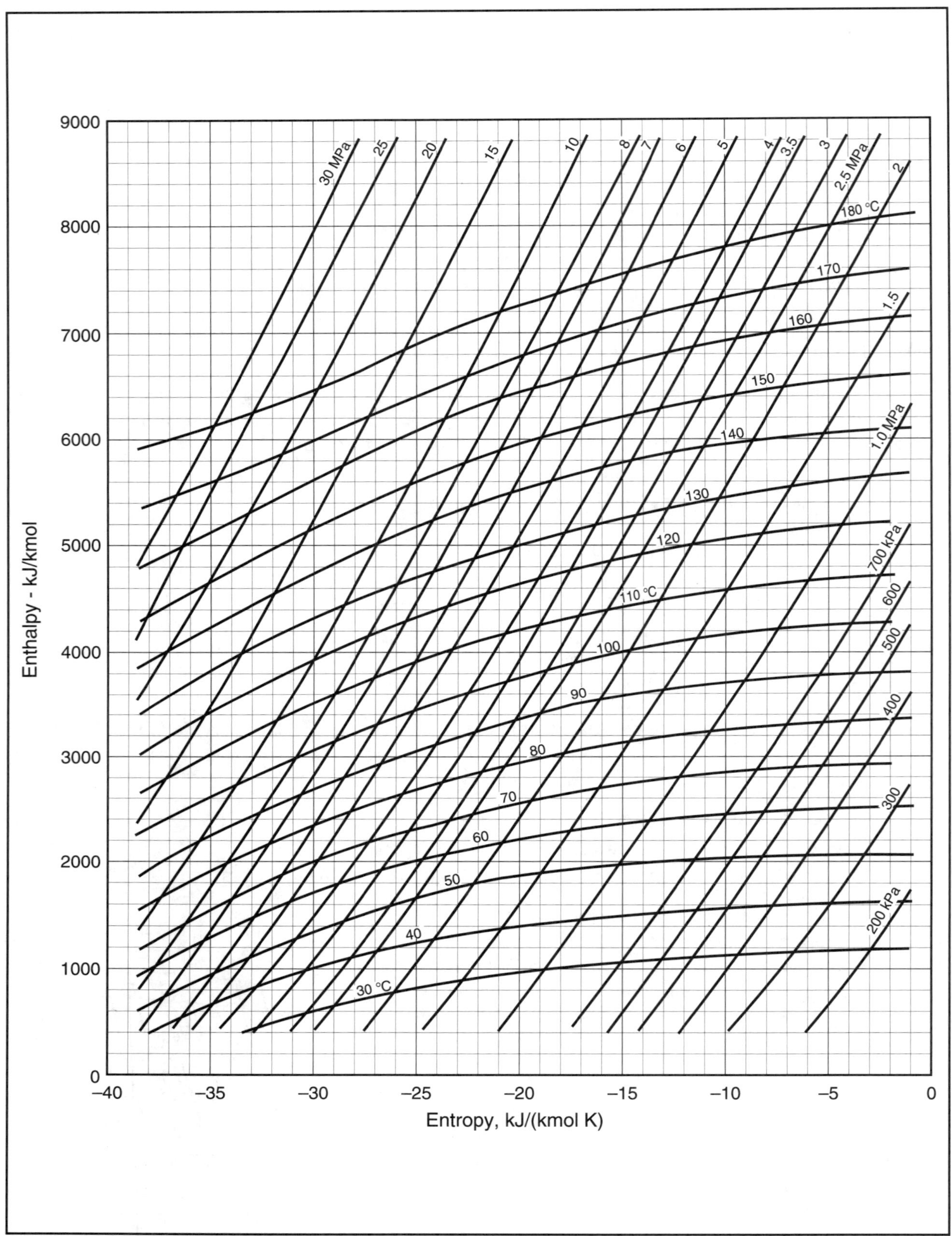

Figure 15.2(a) Enthalpy-Entropy Diagram for a 0.65-0.75 Relative Density Sweet Natural Gas

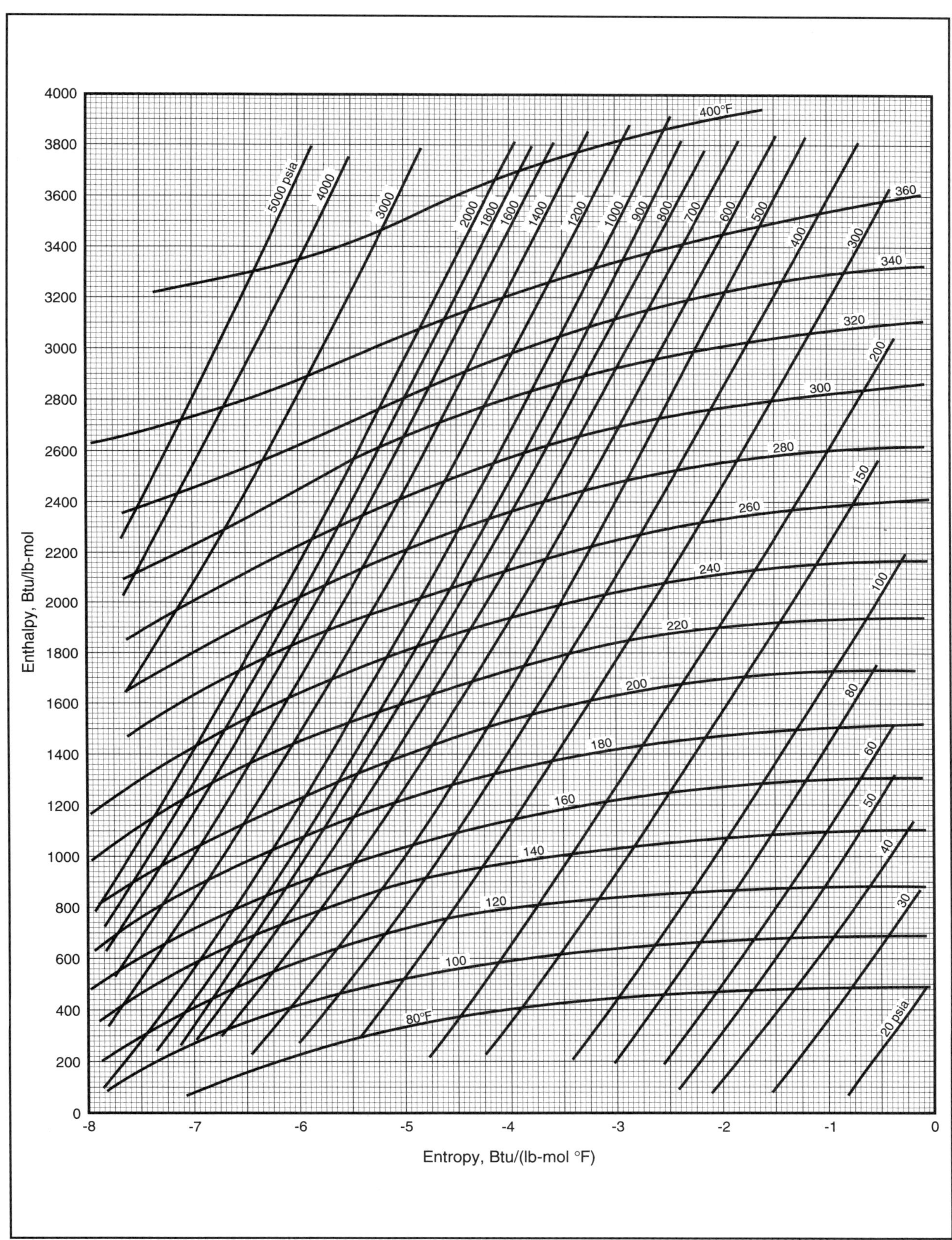

Figure 15.2(b) Enthalpy-Entropy Diagram for a 0.65-0.75 Relative Density Sweet Natural Gas

The general equation is:

$$\text{Power} = \frac{(\text{Mass of Gas per Unit Time})(\Delta h_{theor} \text{ per Unit Mass})}{(\text{Efficiency})(\text{Energy Conversion Factor})} \tag{15.4}$$

The common units employed are shown in the table below.

Power	Mass	Δh	Energy Conv. Factor
kW	(kg or kmol)/s	kJ/kg or kJ/kmol	1.0
kW	(kg or kmol)/h	kJ/kg or kJ/kmol	3600
kW	(lbm or lb-mol)/hr	Btu/lb or Btu/lb-mol	3413
bhp	(lbm or lb-mol)/hr	Btu/lb or Btu/lb-mol	2545
bhp	lbm/min	ft-lbf/lbm	33000

Compressor efficiencies vary with compressor type, size, and throughput. They can only be determined (after-the-fact) by compressor test although compressor manufacturers can usually provide good estimates. For planning purposes the following values may be used.

	Efficiency (E)
Centrifugal	0.65-0.80
H.S. Reciprocating	0.65-0.75
L.S. Reciprocating	0.75-0.85
Rotary Screw	0.65-0.75

The above values represent overall efficiencies. Overall efficiencies include the mechanical losses (bearings, seals, gearbox, etc.). The mechanical efficiency varies with compressor size and type but 95% is a useful planning number.

$$\text{Overall Efficiency} = (\text{Isentropic or Polytropic Efficiency})(\text{Mechanical Efficiency})$$

Useful Conversion Factors

SI:

$$\text{kmol/h} = 1762\,(10^6 \text{ std m}^3/\text{d})$$

$$\text{kg/h} = 51060\,(10^6 \text{ std m}^3/\text{d})(\gamma) = 1762\,(10^6 \text{ std m}^3/\text{d})(\text{MW of gas})$$

FPS:

$$\text{lb-mol/hr} = 110\,(\text{MMscfd})$$

$$\text{lb/hr} = 3180\,(\text{MMscfd})(\gamma) = 110\,(\text{MMscfd})(\text{MW of gas})$$

The previous equations are based on the following conversion factors:

SI:

$$\text{Density of Air} = 1.225 \text{ kg/m}^3$$

$$1 \text{ kmol} = 23.64 \text{ std m}^3$$

(standard conditions are 101.325 kPa and 15°C)

$$10^6 \text{ std m}^3 = 42\ 300 \text{ kmol}$$

FPS:

$$\text{Density of Air} = 0.0764 \text{ lb/ft}^3$$

$$1 \text{ lbmol} = 379.5 \text{ scf}$$

(standard conditions are 14.7 psia and 60°F)

$$1 \text{ MMscf} = 2636 \text{ lbmol}$$

In converting from Δh to power the following *conversion factors* are convenient:

$$1\,\text{hp} = 0.746\,\text{kW} = 2685\,\text{kJ/h} = 2545\,\text{Btu/hr}$$
$$1\,\text{kW} = 1.34\,\text{hp} = 3600\,\text{kJ/h} = 3413\,\text{Btu/hr}$$

Example 15.1: A compressor is to compress 1×10^6 std m^3/d [35.4 MMscf/day] from 1.0 MPa [145 psia] to 3.0 MPa [435 psia]. The suction temperature is 30°C [86°F]. Selected gas properties are shown below:

$\gamma_g = 0.7$ $\qquad z_a = 0.98$

The adiabatic (isentropic) efficiency is 78% and the overall efficiency is 75%.

Calculate the power and discharge temperature.

SI Solution: From Figure 15.2(a) at 30°C and 1.0 MPa, h_1 = 1000 kJ/kmol, s_1 = -18.8 kJ/kmol

$s_1 = s_2$, at s_2 = -18.8 kJ/kmol and P_2 = 3.0 MPa,

$h_{2_{isen}}$ = 3800 kJ/kmol $\qquad T_{2_{isen}} \approx 100°C$

$$m = \left(\frac{1\,000\,000 \text{ std m}^3}{\text{d}}\right)\left(\frac{1 \text{ kmol}}{23.64 \text{ std m}^3}\right)\left(\frac{1 \text{ d}}{24 \text{ h}}\right) = 1762 \text{ kmol/h}$$

(Note: can also use conversion factor on page 160)

m = (1.0)(1762) = 1762 kmol/h

From Equation 15.4,

$$\text{Power} = \frac{(1762)(3800-1000)}{(3600)(0.75)} = 1800 \text{ kW}$$

$$h_{2_{act}} = h_1 + \frac{(h_{2_{isen}} - h_1)}{E_{isen}} = 1000 + \left(\frac{3800-1000}{0.78}\right) = 4600 \text{ kJ/kmol}$$

at $h_{2_{act}}$ = 4600 kJ/kmol and P_2 = 3.0 MPa, $T_{2_{act}} \approx$ 114°C

FPS Solution: From Figure 15.2(b) at 86°F and 145 psia,

h_1 = 470 Btu/lbmol, s_1 = -3.7 Btu/lbmol-°F

$s_1 = s_2$, at s_2 = -3.7 Btu/lbmol-°F and P_2 = 435 psia,

$h_{2_{isen}}$ = 1670 Btu/lbmol-°F $\qquad T_{2_{isen}} \approx 212°F$

$$m = \left(\frac{35\,400\,000 \text{ scf}}{\text{day}}\right)\left(\frac{1 \text{ lbmol}}{379.5 \text{ scf}}\right)\left(\frac{1 \text{ day}}{24 \text{ hr}}\right) = 3890 \text{ lbmol/hr}$$

(Note: can also use conversion factor on page 160)

m = (110)(35.4) = 3890 lbmol/hr

Example 15.1 (Cont'd.):

From Equation 15.4,

$$\text{Power} = \frac{(3890)(1670-470)}{(2545)(0.75)} = 2450 \text{ hp}$$

$$h_{2_{act}} = h_1 + \frac{(h_{2_{isen}} - h_1)}{E_{isen}} = 470 + \left(\frac{1670-470}{0.78}\right) = 2010 \text{ Btu/lbmol}$$

at $h_{2_{act}}$ = 2010 Btu/lbmol and P_2 = 435 psia, $T_{2_{act}} \approx$ 238°F

Calculation of Power from the P-V Integral

For adiabatic compression of an *ideal gas* one can use the expression below to solve the term $\int VdP$.

$$PV^k = \text{constant}$$

where k is the ratio of specific heats, C_p/C_v. This ratio is convenient, for it is relatively constant with temperature for an ideal gas.

If one substitutes the above solution for $\int VdP$ into Equation 15.1, and inserts a compressibility term to correct the ideal gas assumption, an approximation for Δh_{isen} (adiabatic or isentropic head) results.

$$\Delta h_{isen} = \frac{T_1 z_a R}{\left(\frac{k-1}{k}\right)(MW)}\left[\left(\frac{P_2}{P_1}\right)^{\frac{k-1}{k}} - 1\right] \qquad (15.5)$$

Where:

		SI	FPS
Δh_{isen}	= isentropic enthalpy change (head)	(see Table 15.3)	
T_1	= suction temperature	K	°R
z_a	= average compressibility factor $(z_1 + z_2)/2$	–	–
R	= gas law constant	(see Table 15.3)	
k	= ratio of heat capacities (C_p/C_v)	–	–
MW	= gas molecular weight	kg/kmol	lb/lbmol
(P_2/P_1)	= compression ratio	–	–

TABLE 15.3
Values of Δh and R for Equation 15.5

Head	R
kJ/kg	8.314 kJ/kmol·K
m	848 kg·m/kmol·K
ft-lbf/lbm	1545 ft-lbf/lbmol-°R
Btu/lbm	1.99 Btu/lbmol-°R

The value of z_a can be estimated from $(z_1 + z_2)/2$. The values of z may be found by one of the methods in Chapter 3 or from Figure 15.3. As noted, this figure is an approximation but is useful for natural gas manual compression calculations.

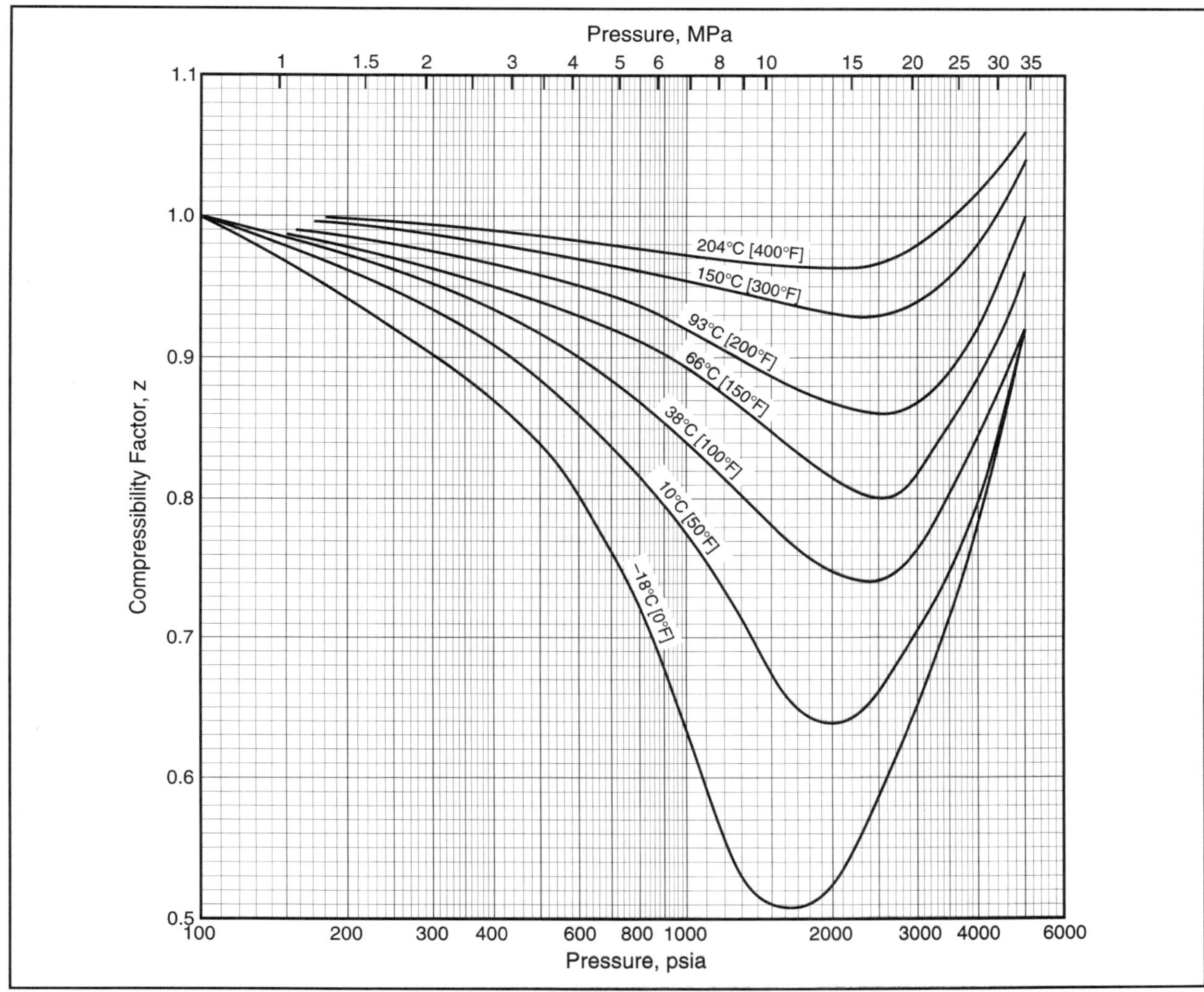

Figure 15.3 Approximate z Chart for a 0.65-0.75 Relative Density Sweet Natural Gas for Use in Compressor Calculations

The value of "k" can be found from Equation 15.6 when gas composition is known.

$$k = \frac{\sum (y_i)(C_{p_i})}{\sum (y_i)(C_{p_i}) - R} \qquad (15.6)$$

Where:			SI	FPS
y_i	=	mol. fr. of each component		
C_{p_i}	=	molar heat capacity from Table 15.4		
R	=	ideal gas law constant	8.314 kJ/kmol·K	1.99 Btu/lbmol-°R

Table 15.4 presents molar heat capacities of several natural gas components at various temperatures.

TABLE 15.4(a)

Molar Heat Capacity of Hydrocarbons, kJ/kmol•°C

Component	Temperature °C								
	-25	0	10	25	50	75	100	125	150
Methane	34.30	34.93	35.20	35.72	36.74	37.87	39.20	40.53	41.99
Ethane	47.13	49.88	50.90	52.67	55.72	58.82	62.11	65.29	65.56
Propane	64.18	68.78	70.61	73.52	78.56	83.59	88.82	93.82	98.84
i-Butane	83.48	90.08	92.69	96.82	103.62	110.41	117.34	123.93	130.52
n-Butane	85.28	91.27	93.69	97.45	105.33	110.33	117.02	123.33	130.40
i-Pentane	101.90	110.37	113.68	118.79	127.34	135.58	144.03	152.01	160.00
n-Pentane	105.13	112.60	115.57	120.21	130.69	136.16	144.45	152.18	161.45
Hexane	123.40	133.30	137.14	143.11	152.71	162.31	171.88	181.08	190.19
Heptane	142.94	154.54	159.01	165.99	177.14	188.29	199.40	210.05	220.59
Nitrogen	29.08	29.11	29.09	29.11	29.12	29.14	29.20	29.22	29.28
Hydrogen Sulfide	33.31	33.67	33.82	34.03	34.38	34.73	35.08	35.43	35.79
Carbon Dioxide	34.70	35.96	36.41	37.12	38.21	39.26	40.29	41.20	42.10

TABLE 15.4(b)

Molar Heat Capacity of Hydrocarbons, Btu/(lbmol-°R)

Component	Temperature °F						
	0	50	100	150	200	250	300
Methane	8.23	8.42	8.65	8.95	9.28	9.64	10.01
Ethane	11.44	12.17	12.95	13.78	14.63	15.49	16.34
Propane	15.65	16.88	18.17	19.52	20.89	22.25	23.56
i-Butane	20.40	22.15	23.95	25.77	27.59	29.39	31.11
n-Butane	20.80	22.38	24.08	25.81	27.55	29.23	30.90
i-Pentane	24.94	27.17	29.42	31.66	33.87	36.03	38.14
n-Pentane	25.64	27.61	29.71	31.86	33.99	36.08	38.13
Hexane	30.17	32.78	35.37	37.93	40.45	42.94	45.36
Heptane	34.96	38.00	41.01	44.00	46.94	49.81	52.61
Nitrogen	6.95	6.95	6.96	6.96	6.97	6.98	7.00
Hydrogen Sulfide	8.00	8.09	8.18	8.27	8.36	8.46	8.55
Carbon Dioxide	8.38	8.70	9.00	9.29	9.56	9.81	10.05

The value of “k” may also be approximated from the natural gas relative density.

$$k = 1.3 - (0.31)(\gamma_g - 0.55) \tag{15.7}$$

Equation 15.6 should be used when gas composition is known. Equation 15.7 is purely empirical, based on typical natural gases that contain no substantial quantities of non-hydrocarbons and whose relative density does not exceed one.

Example 15.2: Rework Example 15.1 using Equation 15.5.

SI Solution: From Equation 15.7, k = 1.3 – (0.31)(0.7 – 0.55) = 1.25

$$\Delta h_{isen} = \frac{(303)(0.98)(8.314)}{\left(\frac{0.25}{1.25}\right)(0.7)(28.97)}\left[\left(\frac{3.0}{1.0}\right)^{0.2} - 1\right] = 150 \text{ kJ/kg}$$

$$m = \left(\frac{1\,000\,000 \text{ std m}^3}{\text{d}}\right)\left(\frac{1 \text{ kmol}}{23.64 \text{ std m}^3}\right)\left(\frac{(0.7)(28.97)\text{ kg}}{\text{kmol}}\right)\left(\frac{1 \text{ d}}{86\,400 \text{ s}}\right)$$
$$= 9.93 \text{ kg/s}$$

(Note: can also use conversion factor on page 160)

$$m = (51\,060)(1.0)(0.7) = 35\,740 \text{ kg/h} = 9.93 \text{ kg/s}$$

From Equation 15.4

$$\text{Power} = \frac{(9.93)(150)}{(0.75)} = \underline{\underline{1980 \text{ kW}}}$$

FPS Solution: From Equation 15.7, k = 1.3 – (0.31)(0.7 – 0.55) = 1.25

$$\Delta h_{isen} = \frac{(546)(0.98)(1545)}{\left(\frac{0.25}{1.25}\right)(0.7)(28.97)}\left[\left(\frac{435}{145}\right)^{0.2} - 1\right] = 50\,100 \text{ ft-lbf/lbm}$$

$$m = \left(\frac{35\,400\,000 \text{ scf}}{\text{day}}\right)\left(\frac{1 \text{ lbmol}}{379.5 \text{ scf}}\right)\left(\frac{(0.7)(28.97)\text{ lbm}}{\text{lbmol}}\right)\left(\frac{1 \text{ day}}{1440 \text{ min}}\right)$$
$$= 1310 \text{ lbm/min}$$

(Note: can also use conversion factor on page 160)

$$m = (3178)(35.4)(0.7) = 78\,800 \text{ lbm/hr} = 1310 \text{ lbm/min}$$

From Equation 15.4

$$\text{Power} = \frac{(1310)(50\,100)}{(33\,000)(0.75)} = \underline{\underline{2650 \text{ hp}}}$$

Combination of Equations 15.4 and 15.5 yields a convenient working equation for compression power.

$$\text{Power/Stage} = \left(\frac{A}{E}\right)\left(\frac{k}{k-1}\right)(q)\left(\frac{P_s}{T_s}\right)(T_1)\left[\left(\frac{P_2}{P_1}\right)^{\frac{k-1}{k}} - 1\right](z_a) \tag{15.8}$$

Where:		SI	FPS
Power/Stage =		kW	hp
A =	conversion factor	11.57	3.03
T_1 =	suction temperature	K	°R
P_s =	standard pressure	kPa	psia
T_s =	standard temperature	K	°R
q =	gas flowrate	10^6 std m^3/d	MMscfd
k =	C_p/C_v	–	–
z_a =	$(z_1 + z_2)/2$	–	–
E =	overall efficiency (decimal)	–	–

Example 15.3:

SI Solution:

$$\text{Power} = \left(\frac{11.57}{0.75}\right)\left(\frac{1.25}{0.25}\right)(1.0)\left(\frac{101.3}{288}\right)(303)\left[\left(\frac{3.0}{1.0}\right)^{0.2} - 1\right](0.98)$$

$$= \underline{\underline{1980 \text{ kW}}}$$

FPS Solution:

$$\text{Power} = \left(\frac{3.03}{0.75}\right)\left(\frac{1.25}{0.25}\right)(35.4)\left(\frac{14.7}{520}\right)(546)\left[\left(\frac{435}{145}\right)^{0.2} - 1\right](0.98)$$

$$= \underline{\underline{2660 \text{ hp}}}$$

When applied to a centrifugal compressor, Equation 15.8 may be applied for any number of stages between coolers. There is a practical limit to the gas temperature allowable in any compressor. Metallurgical, lubrication and efficiency factors are involved. If Equation 15.8 or an equivalent method is used, a separate determination of outlet temperature is required. Equation 15.9 below, can be used to estimate the discharge temperature.

$$T_D = T_1\left[1 + \frac{\left[(P_2/P_1)^{(k-1)/k} - 1\right]}{E_{isen}}\right] \tag{15.9}$$

The efficiency term in the denominator of Equation 15.9 is the isentropic efficiency. For centrifugal compressors a value of 0.65 to 0.80 is suitable. For reciprocating compressors use 0.7 to 0.75 for high speed units and 0.83 to 0.90 for low speed.

The suction temperature to the next stage depends on the effectiveness of the interstage cooler, if one is used. For aerial cooling this temperature will likely be about 14-16°C [25-30°F] above the air

Example 15.4: Estimate the discharge temperature of the compressor in Examples 15.1 and 15.2, using Equation 15.9.

SI Solution:

$$T_D = 303\left[1 + \frac{(3.0)^{0.2} - 1}{0.78}\right] = 398\ K = 125°C$$

FPS Solution:

$$T_D = 546\left[1 + \frac{(3.0)^{0.2} - 1}{0.78}\right] = 718°R = 258°F$$

dry bulb temperature used in the design basis. For water cooling it will be about 5-10°C [9-18°F] above the water temperature.

Power from "Quickie" Charts

The above two basic calculation methods have been used to prepare some correlations to satisfy the need for a quick, easy approximation of power requirements. One of these is shown in Figure 15.4.

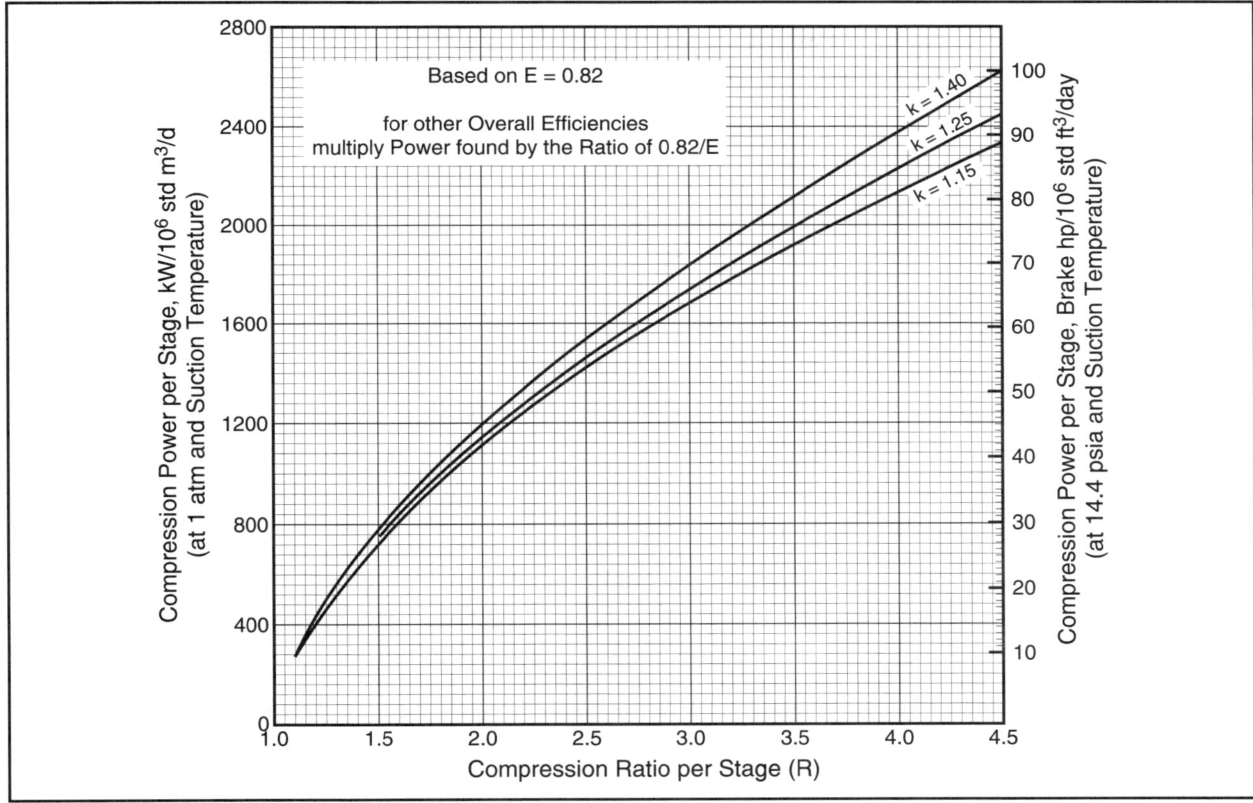

Figure 15.4 Simple Correlation for Estimating Compressor Power

Figure 15.4 is an example of a simple correlation that can be used for estimating compressor power. As a general rule, this figure will give slightly higher results than more exact methods.

The volumetric flowrate shown on the ordinate is at suction temperature, not standard temperature. So, the volumetric flowrate of gas at standard conditions must be converted to the conditions specified by the following equation.

SI:

$$q\ (\text{multiplier for chart}) = 10^6\ \text{std m}^3/\text{d}\left(\frac{P_s}{101.3}\right)\left(\frac{T_1}{T_s}\right) \tag{15.10}$$

Where:

std m^3/d = flowrate at 101.3 kPa and 15°C
T_s = standard temperature, K
T_1 = suction temperature, K
P_s = standard pressure, kPa

FPS:

$$q\ (\text{multiplier for chart}) = \text{MMscf/day}\left(\frac{P_s}{14.4}\right)\left(\frac{T_1}{T_s}\right) \tag{15.11}$$

Where:

MMscf/day = flowrate at P_s and T_s
T_s = standard temperature, °R
T_1 = suction temperature, °R
P_1 = standard pressure, psia

The power from Figure 15.4 is multiplied by the adjusted "q" to find power per stage. The charts are for E = 0.82. They may be used for a different compressor efficiency by adjusting for the efficiency of the unit involved.

An even simpler correlation, which can be derived from Equation 15.8, is shown below:

$$\text{Power} = (q_{std})\left(\frac{P_2}{P_1}\right)(\text{Factor}) \tag{15.12}$$

Where:

			SI	FPS
Power	=	compressor power	kW	hp
q_{std}	=	standard volumetric flow	10^6 std m^3/d	MMscfd
P_2/P_1	=	compression ratio per stage		
Factor	=	see guidelines at right		

	SI	FPS
Single Stage	530-630	20-24
Two Stage	580-680	22-26
Three stage	605-710	23-27

Simple correlations like those shown in Figure 15.4 and Equation 15.12 are useful for early planning purposes. Flowrates and conditions are more uncertain than the calculation. In reality, the calculation of compressor power requirements is the most routine of all the considerations involved. The selection of the compressor type, size, driver, control system, and auxiliary equipment (such as scrubbers) are usually more critical to obtain a satisfactory unit.

Example 15.5: Estimate the power of the compressor in Example 15.1 using Equation 15.12.

SI Solution: use factor at the top of range, factor = 630

$$\text{Power} = (1.0)(3.0)(630) = \underline{\underline{1900\ \text{kW}}}$$

FPS Solution: use factor at the top of range, factor = 24

$$\text{Power} = (35.4)(3.0)(24) = \underline{\underline{2550\ \text{hp}}}$$

For planning purposes, before a specific compressor and driver has been chosen, E = 0.82 for low speed reciprocating units and E = 0.73 for centrifugal units are suitable, safe values. In dealing with a vendor, be certain which efficiency is being quoted. It is not so important which efficiency is being used so long as the communication is clear.

However you determine it, the power requirement is used to pick a standard, discrete size of unit. Normally, the compressor will be equipped with a driver capable of producing more power than that calculated under the most adverse compression conditions anticipated. Because of size limitations, flowrate variations or availability requirements, units in parallel should be considered. The power calculation provides a guideline for considerations like these.

Isentropic Versus Polytropic Efficiency

Traditionally compressor vendors have used polytropic rather than the isentropic efficiency when quoting compressor performance. This is because polytropic efficiency [E_{poly}] is essentially independent of compression ratio and gas composition. Polytropic efficiency is based on an imaginary "polytropic path" which is reversible and non-adiabatic (isentropic is reversible and adiabatic). For a polytropic path, the exponent "n" is substituted for "k" in Equations 15.5 and 15.8. In Equation 15.8 the "overall" efficiency would then be based on E_{poly} rather than E_{isen}. Polytropic efficiency can only be determined by compressor test. Some useful approximate relationships between polytropic and isentropic equations are shown below.

$$\frac{n-1}{n} = \left(\frac{k-1}{k}\right)\left(\frac{1}{E_{poly}}\right) \qquad (15.13)$$

$$E_{isen} = \frac{\left[\left(\frac{P_2}{P_1}\right)^{\frac{k-1}{k}} - 1\right]}{\left[\left(\frac{P_2}{P_1}\right)^{\frac{n-1}{n}} - 1\right]} \qquad (15.14)$$

Equation 15.14 is shown graphically in Figure 15.5.

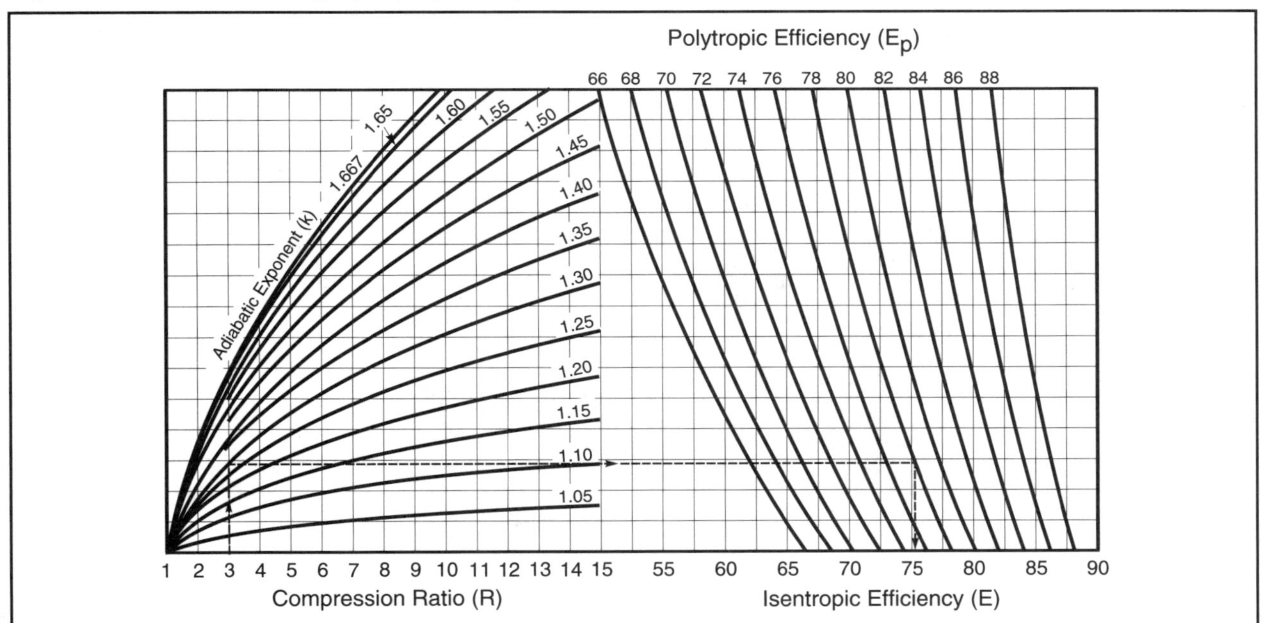

Figure 15.5 Relationship Between Isentropic and Polytropic Efficiency

Example 15.6: A compressor manufacturer has quoted a polytropic efficiency of 78% for a centrifugal compressor. If the compression ratio is 3.0 and k = 1.3, estimate the adiabatic efficiency from Figure 15.5.

Dashed lines on graph show solution, E_{ad} = 75%

Both approaches produce the same basic result. Process engineers tend to favor the isentropic method while compressor specialists favor polytropic.

CENTRIFUGAL AND AXIAL COMPRESSORS

Centrifugal compressors should be considered when actual inlet flows exceed about 850 m^3/h [500 acfm]. They offer the advantages of more power per unit weight and they operate essentially vibration free. This makes them particularly attractive for offshore locations or where air transportation to remote locations is necessary. Their initial cost normally is less than for reciprocating compressors but the efficiency is often less and utility costs may be higher. In addition, the ability of a centrifugal compressor to handle changing conditions (pressures, temperature, molecular weight) may be limited.

Centrifugal compressors are manufactured in three configurations:

1. overhung impeller
2. horizontally split
3. vertically split (barrel-type)

Overhung impellers are typically used in single stage service. The booster compressor in a turboexpander unit is a good example of this type of compressor. The impeller is usually open, backward-bladed.

Figure 15.6 Horizontally Split Case Centrifugal Compressor *(Courtesy of Sulzer Turbosystems)*

Horizontally split cases are often used in high volume, lower pressure applications. Figure 15.6 shows a horizontally split case. These compressors have casings split horizontally at the mid-section. The halves are bolted and doweled together. The internal parts such as the rotor, bearings and seals are accessible for repair and inspection by removing the top half. To facilitate access, compressor nozzles are sometimes oriented downwards from the bottom half of the case.

In vertically split or barrel type compressors the compressor "barrel" contains the rotor and diaphragms and is removed from the end of the compressor. These compressors are typically used in higher pressure, lower volume applications. If a spare barrel is available they can be repaired more quickly than horizontally split compressors. They require additional plot area since the space must be provided to remove the barrel from the case (similar to a shell and tube heat exchanger).

Shown below is a cut-away portion of a vertically-split centrifugal compressor using a closed, backward-bladed impeller, the most common type in natural gas operations.

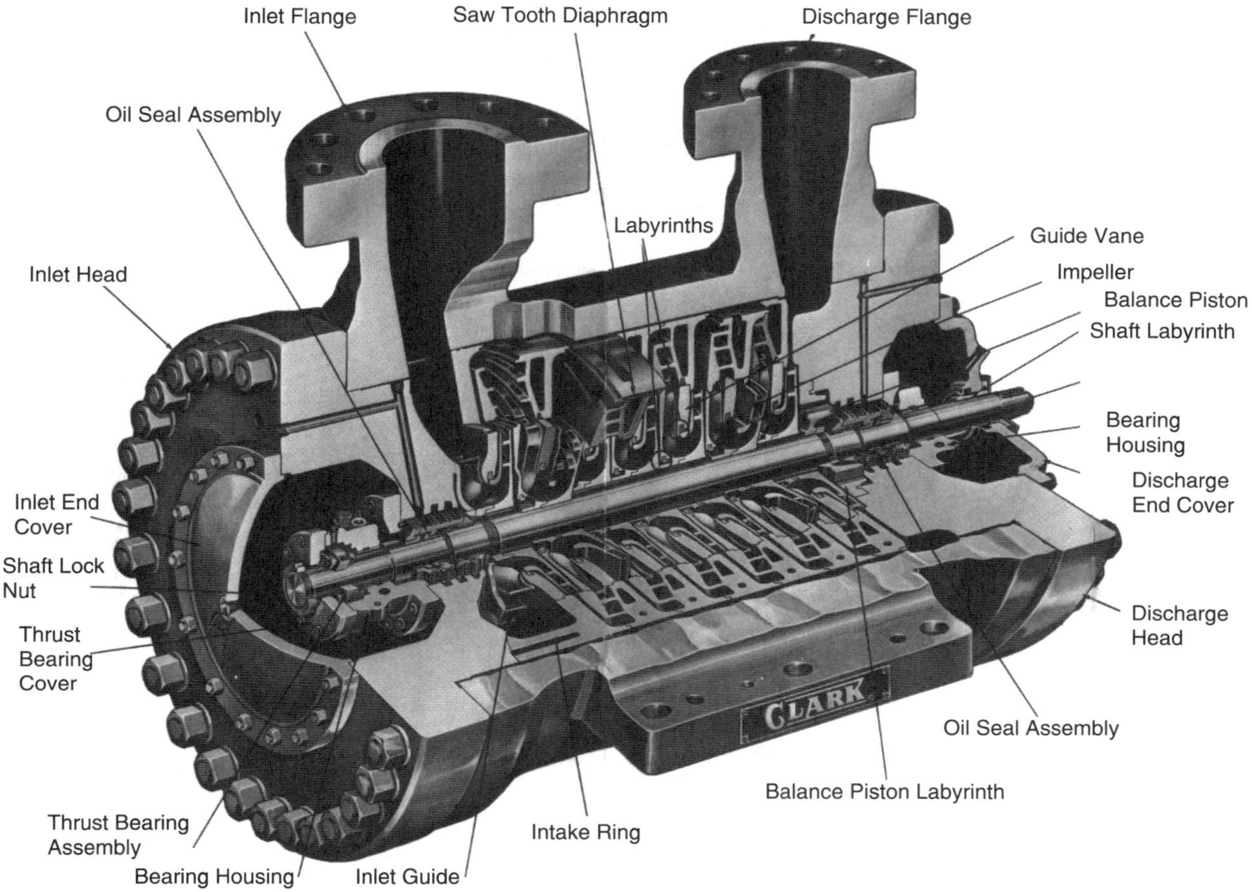

The form of impeller blade affects the characteristic performance curve, the operating range and the efficiency of a centrifugal compressor. Shown below are three types.

Open Radial-Bladed Impeller **Closed Backward-Bladed Impeller** **Open Backward-Bladed Impeller**

The same basic specific speed and diameter relationships used for pumps in Chapter 14 apply also for compressors, particularly centrifugal and axial. Figure 15.7 is based on the equations:

$$d_s = \frac{(A)(d)(H)^{0.25}}{q^{0.5}} \qquad N_s = \frac{(B)(N)(q)^{0.5}}{H^{0.75}} \tag{15.15}$$

Where:

		SI	FPS
d_s =	specific diameter	m	ft
N_s =	specific speed	–	–
d =	diameter	m	ft
q =	gas flowrate	m^3/s	ft^3/sec
H =	head per stage	m	ft
N =	rpm	–	–
A =		0.74	1.00
B =		2.44	1.00

Figure 15.7 shows the relationship between these quantities for centrifugal compressors. Normally one would not use a centrifugal when the efficiency is less than about 0.7 unless weight or vibration limitations dictate it. The number of stages and impeller diameter combinations must be chosen to achieve a satisfactory efficiency.

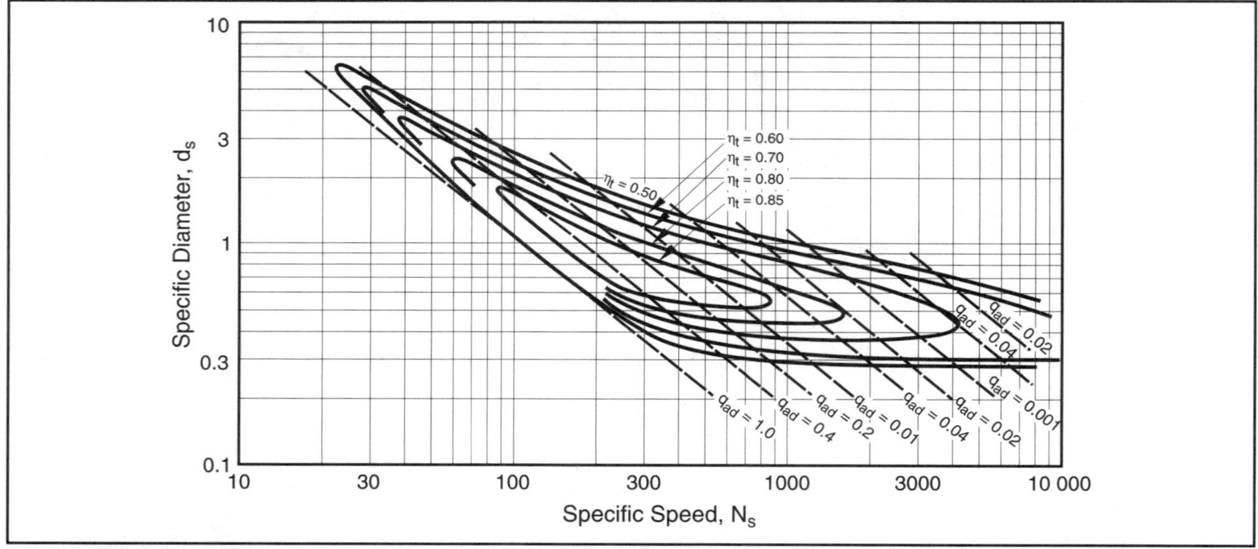

Figure 15.7 Relationship Between Specific Speed and Diameter for Centrifugal Compressors[(15.2)]

Centrifugal Power Calculation

Each of the methods for power calculations shown previously can be used to estimate centrifugal compressor power. Typical polytropic efficiencies vary from 70% to 85% with 78% a good value for planning purposes in natural gas compression. Mechanical losses are normally estimated as follows:

Labyrinth seals and balance piston. The balance piston and labyrinth seals allow high pressure gas to leak back to suction pressure. The additional work required to recompress this internally bypassed gas is usually handled by increasing the design gas flow through the compressor when performing the power calculation. This increased flow usually varies from 2 to 5% of total flow with 3% a useful number for planning calculations.

Bearing and Seals. Mechanical losses at the bearings and seals are a function of compressor size and speed. Figure 15.8 provides a quick approximation of these losses.

The foregoing guidelines are based on standard impeller design and oil seals. For dry seals, mechanical losses may be substantially smaller than indicated by the above.

Gearbox. - Typical efficiencies for speed reducing gears vary from 0.95 to 0.98. For planning purposes, 0.97 is a useful design number.

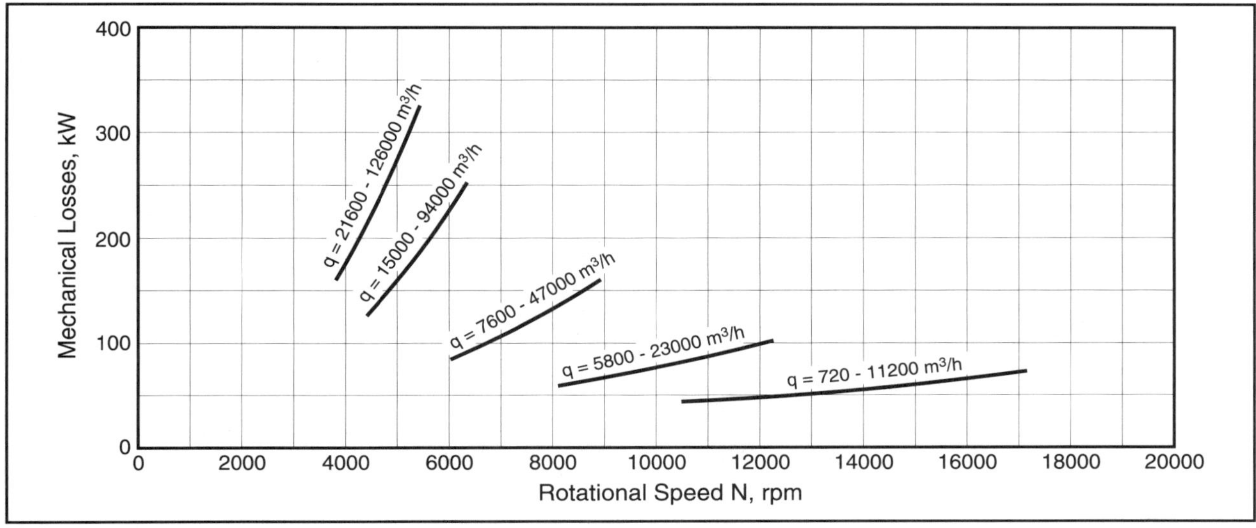

Figure 15.8 Estimated Seal and Bearing Losses in Centrifugal Compressors

Number of Impellers and Speed

The energy imparted to the gas in a centrifugal compressor is termed the "head." The theoretical head may be estimated using an isentropic or polytropic path. The head is the specific enthalpy change of the gas across the compressor. In Imperial units the head is usually expressed in ft (ft-lbf/lbm), and in SI units, kJ/kg. The compressor head is related to the compression ratio and gas properties by Equation 15.5.

The energy imparted to the gas in a centrifugal or axial compressor is related to velocity change through the impeller/diffuser. Like centrifugal pumps, this energy change is a function of impeller tip speed squared and actual flowrate and is essentially independent of the gas compressed. The maximum impeller tip speed is determined by mechanical constraints but a value of 250-300 m/s [820-984 ft/sec] is typical for centrifugal compressors in sweet service. Equation 15.16 relates head per impeller to impeller tip speed.

$$\Delta h = \frac{\mu u^2}{2A} \tag{15.16}$$

Where:

		SI	FPS
Δh	= polytropic head/impeller	kJ/kg	ft
μ	= head coefficient	0.8 to 1.10	
u	= impeller tip speed	m/s	ft/sec
A	= constant	1000 J/kJ	32.18 (lbm-ft)/(lbf-sec^2)

For multistage natural gas compression, a typical design head coefficient is 1.0. Using a fairly conservative tip speed of 275 m/s [900 ft/sec] the maximum head which can be imparted to the gas across one impeller is approximately 38 kJ/kg [12 600 ft]. These values are useful for planning purposes to estimate the number of impellers required to achieve a given pressure rise across a compressor case.

For heavy molecular weight gases (such as refrigerants) the impeller tip speed may be limited by the sonic velocity. Sonic velocity in gases can be estimated from Equation 15.17.

$$u_s = \sqrt{\frac{g_c \, k \, z \, R \, T}{MW}} \tag{15.17}$$

Where:		SI	FPS
u_s	= sonic velocity	m/s	ft/sec
k	= heat capacity ratio (C_p/C_v)	–	–
R	= gas law constant	8314	1545
T	= gas temperature	K	°R
g_c	= mass/force constant	1.0	32.18
MW	= gas molecular weight	–	–
z	= compressibility factor	–	–

Impeller tip speed should not exceed 110% of u_s where u_s is determined at suction conditions.

Equation 15.18 can be used to estimate impeller diameter. It is based on closed backward-bladed impellers typically used in multistage compressors. It is consistent with the results of Equation 15.15 and Figure 15.7.

$$d = \sqrt{\frac{q}{0.050 \, u}} \tag{15.18}$$

Where:		SI	FPS
d	= impeller diameter	m	ft
q	= inlet flowrate	m^3/s	ft^3/sec
u	= impeller tip speed	m/s	ft/sec

Compressor speed can then be determined

$$N = \frac{60 \, u}{d \, \pi} \tag{15.19}$$

Where:		SI	FPS
N	= compressor speed	rpm	rpm
u	= tip speed	m/s	ft/sec
d	= impeller diameter	m	ft

For a centrifugal compressor

1. Flow varies directly with rpm (depending on compression ratio).
2. Head varies with rpm squared.
3. Power varies directly with rpm cubed.

Example 15.7: A centrifugal compressor is required to compress 2.5×10^6 std m^3/d [88 MMscfd] of a 0.65 sp. gr. natural gas from 1500 to 4500 kPa [218-653 psia]. Suction temperature is 35°C [95°F] and the average compressibility factor is 0.95. The compressor efficiency (polytropic) is 78%. Calculate the following:

1. Compressor head
2. Compressor power
3. Compressor discharge temperature
4. Number of impellers required
5. Approximate impeller diameter
6. Approximate shaft speed

SI Solution:

Use isentropic approach –

1. Compressor head (Equation 15.5)

From Equation 15.7, $k = 1.3 - (0.31)(0.65 - 0.55) = 1.27$

$(k - 1)/k = 0.2126$

$$\Delta h = \frac{(308)(0.95)(8.314)}{(0.2126)(0.65)(28.97)}[(3.0)^{0.2126} - 1] = 160 \text{ kJ/kg}$$

2. $E_{isen} = 0.75$ (from Figure 15.5)

$$\underbrace{\text{Actual Gas Rate}}_{\substack{\text{including labyrinth seal} \\ \text{\& balance piston losses}}} = (1.03)(2.5) = 2.575 \times 10^6 \text{std m}^3$$

$$\text{kW} = \left(\frac{11.57}{0.75}\right)\left(\frac{1.27}{0.27}\right)(2.575)\left(\frac{101.3}{288}\right)(308)[(3.0)^{(0.27/1.27)} - 1](0.95) = 5060 \text{ kW}$$

Bearing and Seal Losses = 50 kW (from Figure 15.8)

Total Power = 5060 + 50 = 5110 kW

3. Compressor discharge temperature

$$T_2 = 308\left[1 + \frac{[(3.0)^{(0.27/1.27)} - 1]}{0.75}\right] = 416 \text{ K} = 143°\text{C}$$

4. Number of Impellers (assume 38 kJ/kg)

No. of impellers = 160/38 ≈ 5

5. Approximate impeller diameter (1st stage impeller)

$$q = \left(\frac{2.575 \times 10^6}{86\,400}\right)\left(\frac{101}{1500}\right)\left(\frac{308}{288}\right) = 2.04 \text{ m}^3/\text{s}$$

From Equation 15.18

$$d = \left[\frac{2.04}{(0.0505)(275)}\right]^{0.5} = 0.385 \text{ m}$$

6. Approximate shaft speed, N (assume u = 275 m/s)

$$N = \frac{(275)(60)}{0.385\,\pi} = 13\,600 \text{ rpm}$$

Example 15.7 (Cont'd.):

FPS Solution:

Use isentropic approach –

1. Compressor head (Equation 15.5)

   From Equation 15.7, $k = 1.3 - (0.31)(0.65 - 0.55) = 1.27$

   $(k - 1)/k = 0.2126$

$$\Delta h = \frac{(555)(0.95)(1545)}{(0.2126)(0.65)(28.97)}[(3.0)^{0.2126} - 1] = 53\ 500\ \text{ft-lbf/lbm}$$

2. $E_{isen} = 0.75$ (From Figure 15.5)

$$\underbrace{\text{Actual Gas Rate}}_{\substack{\text{including labyrinth seal} \\ \text{\& balance piston losses}}} = (1.03)(88) = 90.64\ \text{MMscfd}$$

$$hp = \left(\frac{3.03}{0.75}\right)\left(\frac{1.27}{0.27}\right)(90.64)\left(\frac{14.7}{520}\right)(555)[(3.0)^{(0.27/1.27)} - 1](0.95) = 6750\ \text{hp}$$

   Bearing and Seal Losses ≈ 70 hp (From Figure 15.8)

   Total Power = 6750 + 70 = 6820 hp

3. Compressor discharge temperature

$$T_2 = 555\left[1 + \frac{[(3.0)^{(0.27/1.27)} - 1]}{0.75}\right] = 750°R = 290°F$$

4. Number of Impellers (assume 12 600 (ft-lbf)/lbm)

   No. of impellers = 53 500/12 600 ≈ 5

5. Approximate impeller diameter (1st stage impeller)

$$q = \left(\frac{90.64 \times 10^6}{86\ 400}\right)\left(\frac{14.7}{218}\right)\left(\frac{555}{520}\right) = 71.7\ \text{ft}^3/\text{sec}$$

   From Equation 15.18

$$d = \left[\frac{71.7}{(0.0505)(902)}\right]^{0.5} = 1.26\ \text{ft} = 15.1\ \text{in}$$

6. Approximate shaft speed, N (assume u = 902 ft/sec)

$$N = \frac{(902)(60)}{1.26\ \pi} = 13\ 600\ \text{rpm}$$

The temperature will increase in accordance with Equation 15.9. Inter-stage cooling may be necessary, but when used, any condensed liquids must be removed in a properly designed separator. Maximum discharge temperature should not exceed about 150°C [302°F].

In some instances, water may be circulated through the diaphragms within the compressor case between the stages. Cooling is limited by area and metal wall thickness. The most positive approach is external cooling. The gas is withdrawn from the case, externally cooled, and then re-enters.

Surge and Choke

Surge occurs when a compressor impeller can no longer generate sufficient head (energy) to move the gas forward. In other words, the impeller is no longer able to maintain the pressure rise necessary to move gas into the receiving system at the compressor discharge. At surge, the receiving system pressure is temporarily greater than the compressor discharge pressure and a flow reversal occurs in the compressor. This in turn leads to a temporary drop in the receiving system pressure which allows the compressor to again move gas forward until another flow reversal takes place. If the cause of the surge is not eliminated, a series of flow oscillations occur — forwards, backwards, forwards, etc. The frequency of these oscillations depends on the compressor size, piping configuration, placement of non-return valve etc., but can be less than a second. This induces vibration, a rapid temperature rise in the compressor and in severe cases, catastrophic compressor failure.

Surge occurs at low flow conditions. A single compressor impeller may surge at 50-60% of design flow, but in multistage compressors the surge point may be 70-80% of design flow. The most common method of preventing surge is to ensure at all times that the inlet flow to the compressor is greater than the flowrate at which the compressor surges. This is typically done by recycling flow from the compressor discharge back to the compressor section. This recycle system, which is automated using sophisticated instrumentation, is referred to as a surge control or anti-surge system. This system will be discussed further at another point in this chapter.

Choke (the "stonewall" effect) limits compressor capacity. This condition is generally caused by the limiting flowrate of the gas through the "eye" of the first impeller. This flow is always higher than design and usually should not occur below 115-120% of rated capacity. The maximum velocity is limited by the speed of sound of the gas. Theoretically, the choke effect would occur at this value, but usual practice is to limit design to 0.85-0.90 of sonic velocity. At choke, the compressor cannot handle any additional inlet flow regardless of the compressor head. In other words, the performance curve is essentially vertical. Choke is usually not a serious concern for low MW gases like methane. It can be a concern for high MW gases such as refrigerants . Propane, for example, has a sonic velocity of about 220 m/s [720 ft/sec] at -40°C. Methane, by contrast, has a sonic velocity of about 440 m/s [1460 ft/sec] at 20 bar and 30°C [290 psia and 86°F].

Characteristic Curves

Figure 15.9 shows the characteristic curve for a multistage centrifugal compressor.

A performance curve for a centrifugal compressor is similar in shape to a centrifugal pump (Head vs. flow). Figure 15.9 also includes lines of constant efficiency. Note that the maximum efficiency for this compressor is approximately 80%. The compressor must be operated between the two stability limits, surge and choke.

Manufacturers often provide performance curves for a centrifugal compressor which also indicate power consumption and/or discharge pressure. It is important to note that power consumption curves or discharge pressure can only be included in a compressor performance map when the suction conditions are fixed. Suction conditions set the inlet gas density which in turn sets the mass flow through the compressor as well as the head/pressure relationship. A centrifugal compressor is similar to a centrifugal pump in this respect.

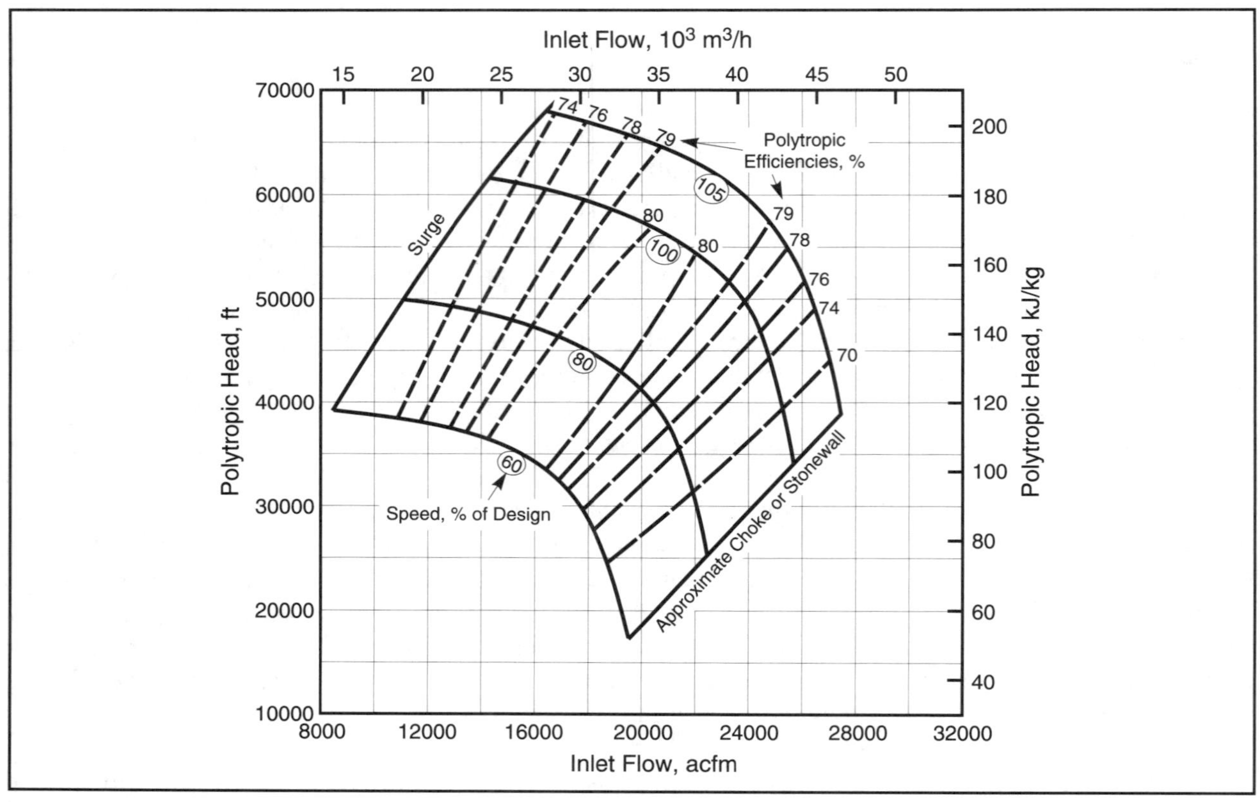

Figure 15.9 Performance Curve for a Centrifugal Compressor

AXIAL COMPRESSORS

An axial compressor is similar in principle to a centrifugal. Rather than using wheels, a series of blades is involved – rotating (rotors) and stationary (stators). About half of the pressure rise is accomplished by the rotors and the other half by the stators.

The rotors are attached to the shaft and add kinetic energy to the gas. Each row of stators acts as a diffuser for the gas flowing off the preceding row of rotors and this converts kinetic energy to pressure. The stators also serve as nozzles to guide the gas into the next row of rotors. Each stage consists of one row of rotors and one row of stators. Often about two stages are needed in an axial compressor for each one stage in a comparable centrifugal.

The axial compressor is used primarily when flowrate is above 102 000 m^3/h [60 000 acfm], at suction conditions, and the discharge pressure is 3500-4000 kPa [500-580 psia] or less.

An axial will tend to have a higher efficiency than a centrifugal, about 8-10% higher. For large flow, medium pressure duties it offers an excellent alternative to the centrifugal compressor. The most common application in the oil and gas industry is air compression in the hot gas generator section of a gas turbine. Axial compressors are seldom used for compression of hydrocarbon gases.

It may be controlled by speed, bypass, suction throttling or by use of variable angle guide vanes. As noted at the top of page 179, it is much less a constant head device than a centrifugal.

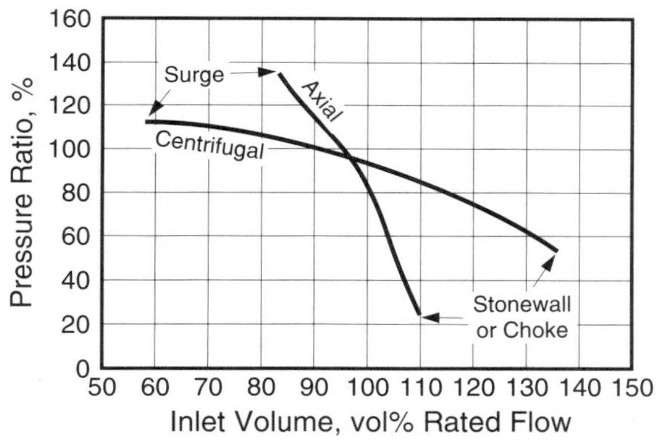

Multistage Axial vs. Centrifugal Compressors

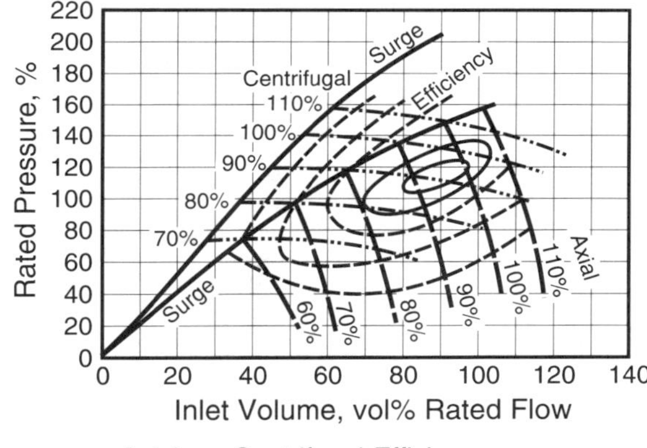

Axial vs. Centrifugal Efficiency

RECIPROCATING COMPRESSORS

Figure 15.10 shows a cutaway of a reciprocating compressor frame and cylinder. The angular motion of the crankshaft is converted into a reciprocating motion at the crosshead. The crosshead is the connection point between the connecting rod and the compressor rod. The distance piece is attached to the crankcase assembly containing the crosshead on one side and to the compressor cylinder on the other. The compressor rod passes through wiper packing on the crank side of the distance piece and through pressure packing on the cylinder side. In enclosed compressor modules the distance piece is often vented outside the enclosure. Double distance pieces may be used in corrosive or toxic gas service.

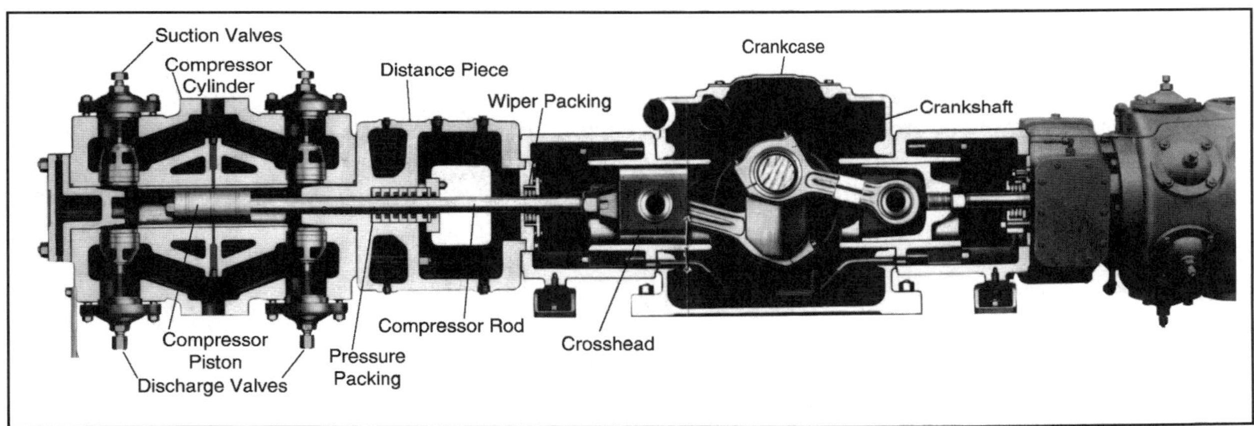

Figure 15.10 Cutaway of a Reciprocating Compressor

Compressor valves are located around the cylinder — one type of compressor valve is shown on page 182. Suction valves are on top and discharge valves on the bottom. The compressor cylinder and rod packing may be cooled by circulation of cooling water. Compressor cylinders are typically lubricated, but non-lubricated cylinders are sometimes used in refrigeration service. The seal between the piston and cylinder wall is provided with rings which are usually metallic in high pressure service and non-metallic or composite material in low pressure service. A set of rider bands supports the piston in the cylinder.

Figure 15.11 shows the basic performance of a reciprocating compressor cylinder. This is explained in greater detail in Figure 15.12.

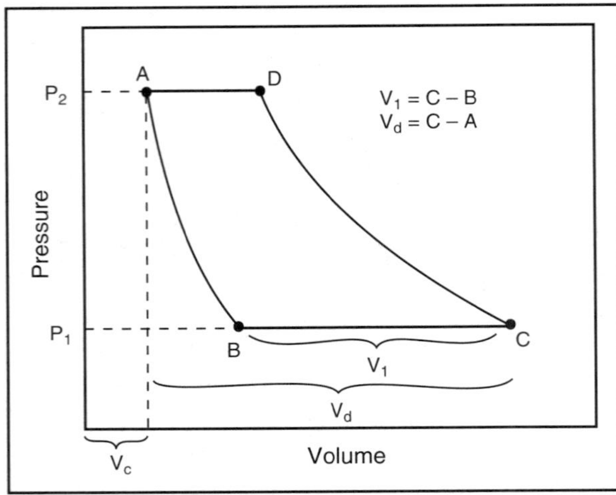

Figure 15.11 Ideal PV Curve for Reciprocating Compressors

Point A represents the end of the compression stroke. Line ABC represents the total suction stroke. Section AB of this line represents expansion of the gas trapped between the piston and the end of the cylinder at pressure P_2. No new gas can enter the cylinder until this gas expands to pressure P_1 (point B). Volume V_1 (Line BC) represents the new gas entering on the suction stroke. The gas handling capacity of the cylinder is fixed by volume V_1. This volume, in turn, depends on the compression ratio (P_2/P_1) and clearance volume, V_c.

The volume represented by line ABC is known as the *piston displacement* – the volume of gas that could be compressed if there was zero clearance. The symbol "V_d" is commonly used to express this volume. It depends on piston size, speed and length of stroke, and whether the piston is single- or double-acting.

The ratio (V_1/V_d) is known as *volumetric efficiency.* It decreases with increasing compression ratio and an increase in clearance volume V_c. Thus, compression ratio is an economic consideration.

Compression Ratio Per Stage

The compression ratio per stage (R) seldom exceeds about 4:1, except when compressing high MW gases. The volumetric efficiency declines, discharge temperature rises, and mechanical stresses imposed on the piston, rod and frame become more pronounced as R increases. In actual practice, R seldom exceeds 3 to 3.5:1 when boosting low MW gases for processing or sale. When the total pressure ratio is greater than this, multiple stages of compression are used.

It may be shown that the total power is minimized when the ratio in each stage is the same. This may be expressed in equation form as

$$R = \left(\frac{P_D}{P_1}\right)^{1/n} \qquad (15.20)$$

Where:

P_D = final discharge pressure
P_1 = suction pressure
n = number of stages required

Where several stages of compression are used, it is necessary to have an interstage cooler and scrubber to cool the gas from the first stage and then collect any liquids formed on cooling. This interstage pressure drop should seldom exceed 35-70 kPa [5-10 psi]. In order to account for this drop, the R found in Equation 15.20 is usually corrected. One common method is to assign half of the pressure drop to each stage. Sometimes the compression ratio is reduced in the higher pressure stages because of rod loading limitations.

The R from Equation 15.20 is theoretical, for it makes no provision for the pressure drop through the interstage piping, cooler, and separator

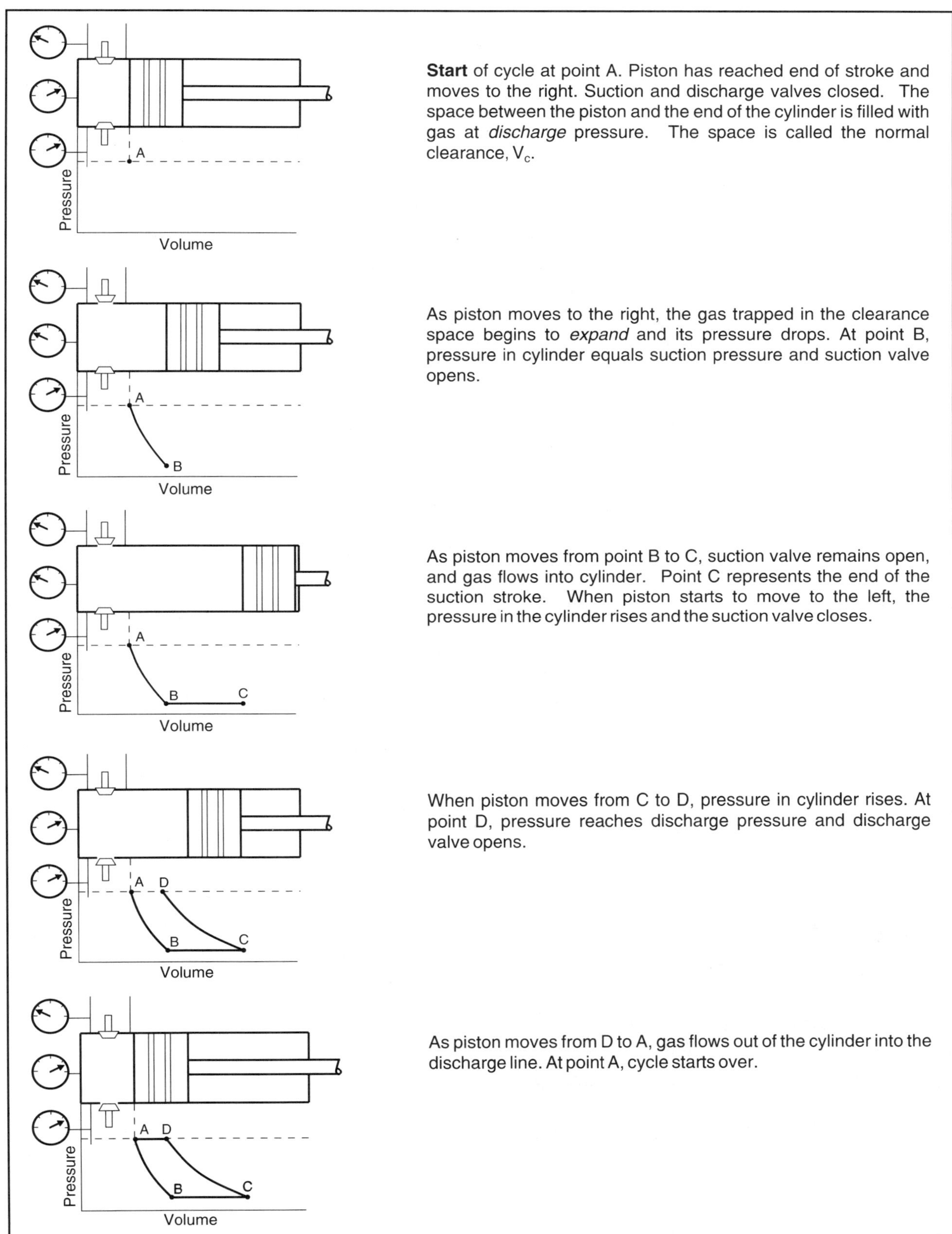

Figure 15.12 Reciprocating Compressor Compression Cycle

The schematic diagram below shows the typical layout for a two-stage compression plant. The general procedure for finding engine horsepower is as follows:

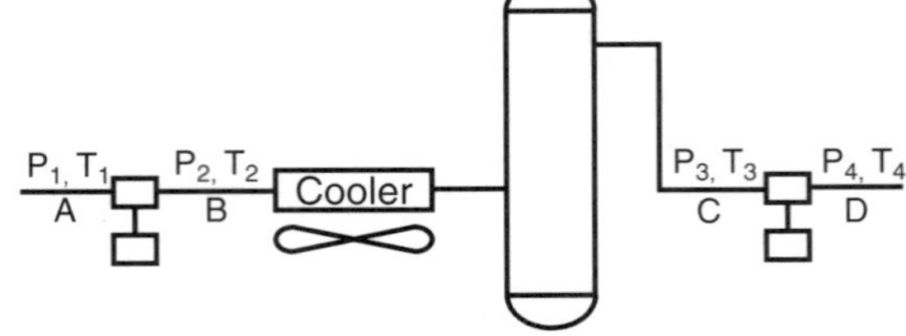

1. Calculate R from Equation 15.20 and multiply by P_1.
2. Estimate pressure drop, $(P_2 - P_3)$.
3. Actual $P_2 = R(P_1) + 1/2$ of pressure drop in Step 2.
4. Actual $P_3 = P_2 - \Delta P$ between stages, using P_2 from Step 3.
5. Estimate T_3, the suction temperature to the second stage. (Ambient dry bulb plus 10-15°C [18-27°F] for air cooling or 5-10°C [9-18°F] for water cooling.)
6. Calculate actual "R" for each stage, calculate power for each stage, and add to get total engine power needed.
7. The total heat load in the cooler is the sum of the gas sensible heat from T_2 to T_3, and the total latent heat of all fluids condensed (water plus hydrocarbons).

Overall Efficiency -"E"

As previously discussed, the overall efficiency of a reciprocating compressor is a function of compressor design details, suction pressure, speed, loading, compression ratio, and general mechanical condition of the unit. For detailed information, the vendor should be contacted. It is, however, possible to estimate efficiency for general planning purposes from general correlations as shown in Figures 15.13 and 15.14.

Figure 15.13 is for air compression. Figure 15.14 is a correction factor correlation for gas relative density that is multiplied by the efficiency of Figure 15.13. Curve A is for large integral or separable units, at low compression ratios, typical of large pipeline operations. Curve B is typical of large integral or separable units operating at speeds of less than 600 rpm. Curve C is for small separables operating generally above 900 rpm. Many units operate between the curves shown. Curves like these are for planning purposes only.

The values shown normally will apply at suction pressures greater than 200 kPa (abs.). For suction pressures lower than this, the efficiency may be 15-20% lower than shown. The compression ratio shown is per stage.

Example 15.8 illustrates the determination of efficiency and power calculation for a two-stage reciprocating compressor.

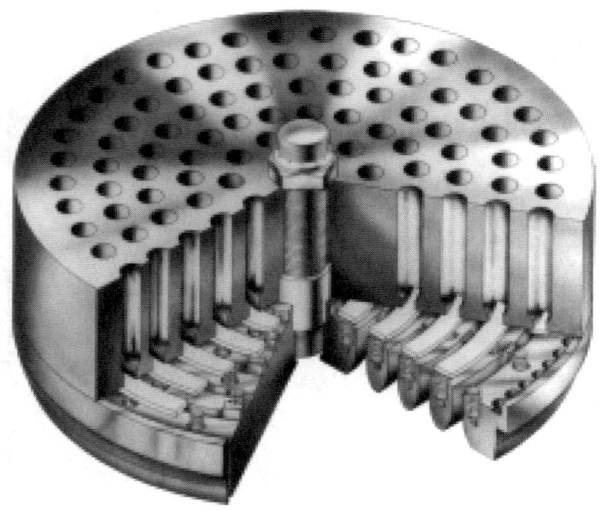

(Courtesy Cooper Industries)

Reciprocating Compressor Valve

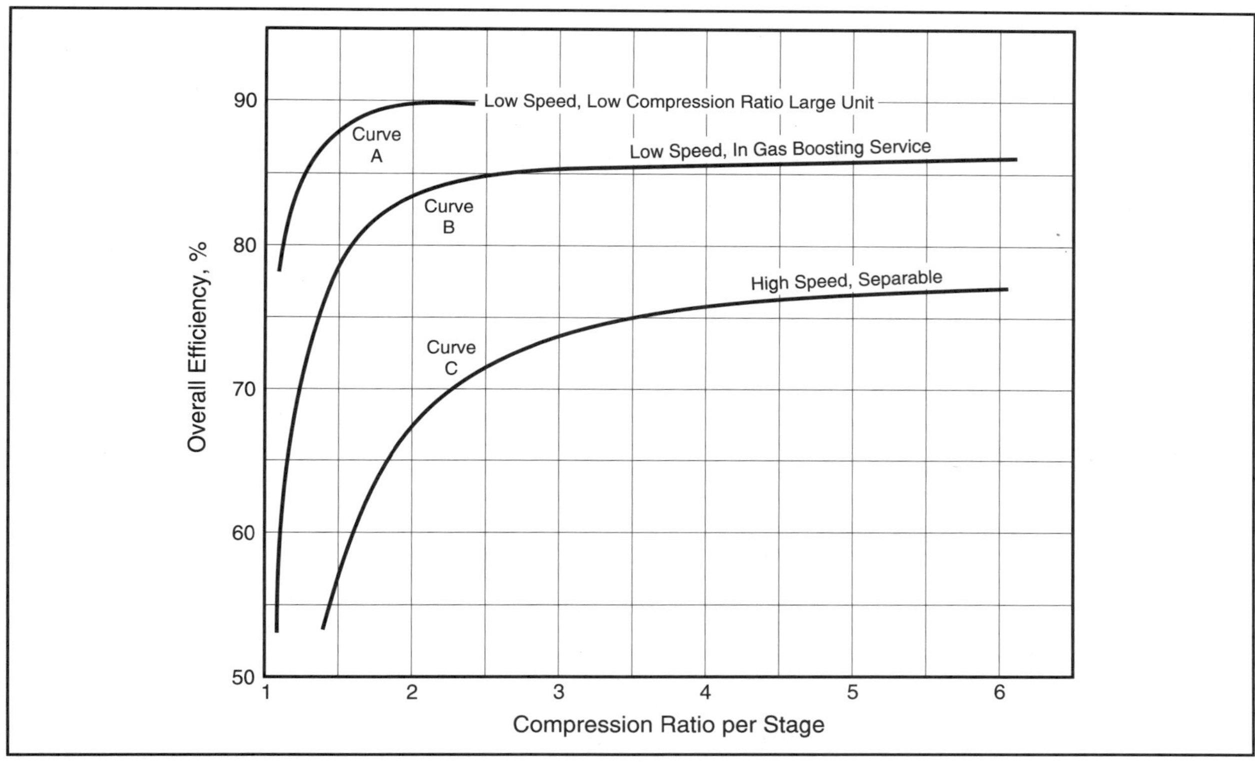

Figure 15.13 Overall Efficiency for Reciprocating Compressors with Air

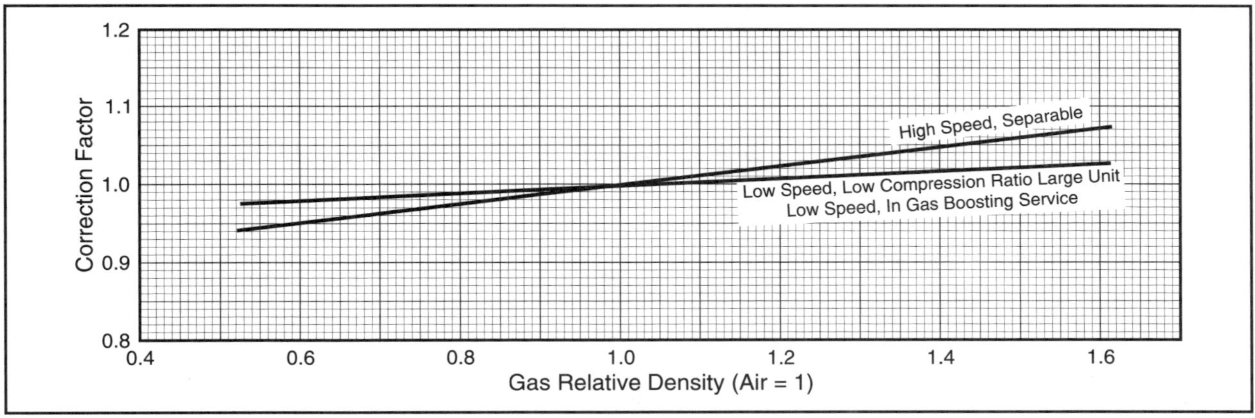

Figure 15.14 Gas Relative Density Correction for Figure 15.13

What do we have at this point? The power that must be transmitted to the compressor to satisfy the gas energy needs, with allowance for efficiency. This must be supplied by a driver (engine) capable of producing this necessary shaft work. Drivers will be discussed in a separate section because different types may be used with a given compressor. The driver power must also be sufficient to handle auxiliary loads such as lubrication, cooling water pumps, fans, etc.

The value of overall efficiency (E) depends on the type of compressor and the model within a given type. It includes both a correction for the isentropic assumption and the actual mechanical efficiency of the equipment.

Example 15.8: 280 000 std m^3/d [9.9 MMscfd] of a 20.3 MW gas is compressed from 700-6000 kPa [101-870 psia] in a low speed integral-type reciprocating compressor. First and second stage suction temperatures are 40°C [104°F] and 45°C [113°F] respectively. Interstage pressure drop is 40 kPa [6 psi]. Estimate the power requirements.

SI Solution:

1. Assume equal compression ratios per stage and calculate the compression ratio including the interstage pressure drop.

   From Equation 15.20, $R = \left(\frac{6000}{700}\right)^{1/2} = 2.93$

   Include interstage pressure drop

   P_1 (1st stage) = (2.93)(700) + (0.5)(40) = 2069 kPa
   P_1 (2nd stage) = 2069 – 40 = 2029 kPa
   R_1 = (2069/700) = 2.96
   R_2 = (6000/2029) = 2.96

2. From Figures 15.13 and 15.14, E = (0.852)(0.982) = 0.837 (use 0.84)

   From Figure 15.3, 1st stage $z_1 = 1.0$, $z_D = 0.98$
   2nd stage $z_1 = 0.955$, $z_D = 0.945$

   Assume k = 1.25

3. Calculate power using Equation 15.8

   1st stage: $$\text{kW} = \left(\frac{11.57}{0.84}\right)\left(\frac{1.25}{0.25}\right)(0.28)\left(\frac{101.3}{288}\right)(313)(2.96^{0.2} - 1)(0.99) = 510 \text{ kW}$$

   2nd stage: $$\text{kW} = \left(\frac{11.57}{0.84}\right)\left(\frac{1.25}{0.25}\right)(0.28)\left(\frac{101.3}{288}\right)(318)(2.96^{0.2} - 1)(0.95) = 500 \text{ kW}$$

FPS Solution:

1. Assume equal compression ratios per stage and calculate the compression ratio including the interstage pressure drop.

   From Equation 15.20, $R = \left(\frac{870}{101}\right)^{1/2} = 2.93$

   Include interstage pressure drop

   P_1 (1st stage) = (2.93)(101) + (0.5)(6) = 299 psia
   P_1 (2nd stage) = 299 – 6 = 293 psia
   R_1 = (299/101) = 2.96
   R_2 = (870/293) = 2.96

2. From Figures 15.13 and 15.14, E = (0.852)(0.982) = 0.837 (use 0.84)

   From Figure 15.3, 1st stage $z_1 = 1.0$, $z_D = 0.98$
   2nd stage $z_1 = 0.955$, $z_D = 0.945$

   Assume k = 1.25

Example 15.8 (Cont'd.)

3. Calculate power using Equation 15.8

1st stage $$hp = \left(\frac{3.03}{0.84}\right)\left(\frac{1.25}{0.25}\right)(9.9)\left(\frac{14.7}{520}\right)(564)(2.96^{0.2} - 1)(0.99) = 680 \text{ hp}$$

2nd stage: $$hp = \left(\frac{3.03}{0.84}\right)\left(\frac{1.25}{0.25}\right)(9.9)\left(\frac{14.7}{520}\right)(573)(2.96^{0.2} - 1)(0.95) = 670 \text{ hp}$$

Volumetric Efficiency

In Figure 15.11 the general behavior of a reciprocating cylinder is shown. The volume of new gas entering the cylinder, V_1 (BC) is less than the piston displacement, V_d (ABC) because of expansion of gas trapped between the piston and the end of the cylinder. The ratio of actual volume entering (V_1) to V_d is called *volumetric efficiency.*

The empirical equation relating the volumetric efficiency to clearance and compression ratio is:

$$E_{vz} = \frac{V_1}{V_d} = 0.96 + (C - L) - (C)(R)^{1/k}\left(\frac{z_1}{z_D}\right) \tag{15.21}$$

Where:

E_{vz} = volumetric efficiency
C = (V_c/V_d) = clearance, expressed as a fraction of displacement volume
R = compression ratio
V_1 = volume of gas per unit time at actual suction conditions
V_d = piston displacement per unit time
L = R/100, where R = compression ratio/stage at speeds of 200-500 rpm
= R/50, where R = compression ratio/stage at speeds above 500 rpm

The value of clearance (C) in Equation 15.21 is expressed as a fraction. As noted in Figure 15.11, it is clearance volume (V_c) divided by piston displacement (V_d).

Equation 15.21 is approximate. The number 0.96 in Equation 15.21 is to account for any loss of volumetric efficiency due to valves, rings, piston rod packing, etc. This number represents a practical average and may be as low as 0.93 for high speed units using lubricated rings and less than 0.9 for non-lubricated cylinders.

Volume "V_1" is the actual flow rate of gas to the suction valves of a given stage at suction P and T.

$$V_1 = \frac{V_s}{1440}\left(\frac{P_s}{P_1}\right)\left(\frac{T_1}{T_s}\right)(z_1) \tag{15.22}$$

Where:

V_1 = volume/min of gas at P_1 and T_1 (suction conditions)
V_s = suction vol. rate per day of flow at standard P_s and T_s
z_1 = compressibility factor at P_1 and T_1
1440 = minutes per day

Displacement volume is dependent on piston area, stroke, speed and whether the cylinder is single or double acting.

$$V_d = (A)(d^2)(\text{Stroke})(\text{rpm})(\text{factor}) \tag{15.23}$$

Where:				SI	FPS
	V_d	=	displacement volume/min	m^3/min	ft^3/min
	A	=	conversion factor	7.85×10^{-7}	0.000 454
	Stroke	=	piston stroke	cm	in
	d	=	cylinder diameter	cm	in
	Factor	=	1.0 for single acting cylinders and slightly less than 2.0 for double acting (to account for piston rod volume). A value of 1.95 is often used for planning purposes		
	rpm	=	revolutions of driving shaft per minute		

Although approximate, Equation 15.21 is a valid model of cylinder performance. If V_1, the gas available for compression changes, the compression ratio, speed, or clearance must change to preserve the equality shown. For example, if V_s decreases, say because a well is taken off production, the compressor will draw down the suction pressure (increasing V_1 and decreasing E_{vz}), until the cylinder capacity equals the amount of gas available for compression. To a large extent the compressor is self controlling. The control of reciprocating compressors will be discussed in a later section.

Equations 15.21-15.23 can be used to size cylinders, estimate the capacity of an existing cylinder or determine the clearance necessary to control the cylinder capacity.

Example 15.9: Using the values from the previous example estimate the cylinder sizes for the following compressor.

Speed = 400 rpm
Stroke = 21.6 cm [8.5 in]

Normal clearance: 1st stage = 7%
2nd stage = 12%

Assume 2 cylinders per stage and all cylinders are double-acting

SI Solution:

1st stage – 1. Estimate E_{vz} from Equation 15.21

$$E_{vz} = 0.96 + (0.07 - 0.0296) - 0.07(2.96)^{0.8}\left(\frac{1.0}{0.98}\right) = 0.83$$

2. Determine V_1 from Equation 15.22

$$V_1 = \left(\frac{280\ 000}{1440}\right)\left(\frac{101.3}{700}\right)\left(\frac{313}{288}\right)(1.0) = 30.6\ m^3/min$$

$$= 15.3\ m^3/min \text{ per cylinder}$$

3. Calculate d from Equation 15.23

$$V_d = \frac{V_1}{E_{vz}} = \frac{15.3}{0.83} = (7.85 \times 10^{-7})(d)^2(21.6)(400)(1.95)$$

Solving for d; d = 37.3 cm

Example 15.9 (Cont'd.):

2nd stage – 1. $E_{vz} = 0.96 + (0.12 - 0.0296) - 0.12(2.96)^{0.8}\left(\frac{0.955}{0.945}\right) = 0.76$

2. $V_1 = \left(\frac{280\ 000}{1440}\right)\left(\frac{101.3}{2029}\right)\left(\frac{318}{288}\right)(0.955) = 10.2\ m^3/min$

$= 5.1\ m^3/min$ per cylinder

3. $V_d = \frac{V_1}{E_{vz}} = \frac{5.1}{0.76} = (7.85 \times 10^{-7})(d)^2(21.6)(400)(1.95)$

$d = 22.5$ cm

FPS Solution:

Speed = 400 rpm
Stroke = 21.6 cm [8.5 in]

Normal clearance: 1st stage = 7%
2nd stage = 12%

Assume 2 cylinders per stage and all cylinders are double-acting

1st stage – 1. Estimate E_{vz} from Equation 15.21

$$E_{vz} = 0.96 + (0.07 - 0.0296) - 0.07(2.96)^{0.8}\left(\frac{1.0}{0.98}\right) = 0.83$$

2. Determine V_1 from Equation 15.22

$$V_1 = \left(\frac{9\ 900\ 000}{1440}\right)\left(\frac{14.7}{101}\right)\left(\frac{564}{520}\right)(1.0) = 1085\ ft^3/min$$

$= 543\ ft^3/min$ per cylinder

3. Calculate d from Equation 15.23

$$V_d = \frac{V_1}{E_{vz}} = \frac{543}{0.83} = (0.000\ 454)(d^2)(8.5)(400)(1.95)$$

Solving for d; $d = 14.7$ in.

2nd stage – 1. $E_{vz} = 0.96 + (0.12 - 0.0296) - 0.12(2.96)^{0.8}\left(\frac{0.955}{0.945}\right) = 0.76$

2. $V_1 = \left(\frac{9\ 900\ 000}{1440}\right)\left(\frac{14.7}{293}\right)\left(\frac{573}{520}\right)(0.955) = 363\ ft^3/min$

$= 182\ ft^3/min$ per cylinder

3. $V_d = \frac{V_1}{E_{vz}} = \frac{182}{0.76} = (0.000\ 454)(d)^2(8.5)(400)(1.95)$

$d = 8.9$ in

The order of these calculations is summarized below:

Size a compressor cylinder, d. Known variables: C, R, k, V_s, P_1, T_1, stroke, rpm, single/double acting.

1. Calculate E_{vz} from Equation 15.21
2. Calculate V_1 from Equation 15.22
3. $V_d = V_1/E_{vz}$
4. Calculate d from Equation 15.23

Estimate cylinder capacity, V_s. Known variables: C, R, k, P_1, T_1, stroke, rpm, d, single/double acting

1. Calculate E_{vz} from Equation 15.21
2. Calculate V_d from Equation 15.23
3. $V_1 = (V_d)(E_{vz})$
4. Calculate V_s from Equation 15.22

Estimate clearance necessary for volume control, C. Known variables: R, k, V_s, P_1, T_2, stroke, rpm, d, single/double acting

1. Calculate V_1 from Equation 15.22
2. Calculate V_d from Equation 15.23
3. $E_{vz} = V_1/V_d$
4. Calculate C from Equation 15.21

Rod, Pin, and Frame Loadings

These are often referred to simply as rod loads, but whatever the name used, this refers to the maximum stress that may be placed on the "weakest link" in the drive mechanism. This will vary with the machine. It may be the rod, the crosshead pin or bushing, or crankshaft. This may be different in compression and tension.

The following equations can be used to determine the rod loads due to differential pressure across the piston:

$$\text{Compression} \qquad L_c = (A_h)(P_2) - (A_c)(P_1)$$
$$\text{Tension} \qquad L_t = (A_c)(P_2) - (A_h)(P_1) \tag{15.24}$$

Where:

L = load (in force units)
A_h = area of piston at head end
A_c = effective area at crank end (area of piston - area of rod)
P_1 = suction pressure
P_2 = discharge pressure

In addition, there are inertial loads associated with the acceleration and deceleration of the running components. In any service, it is critical that the maximum rod load not be exceeded. Allowable rod loads are available from the compressor manufacturer for various standard compressor frames.

General Considerations

Reciprocating compressors have historically been the preferred compressors in low flow – high head applications. They are usually the first choice when the compressor discharge volume is less than about 300 m^3/h [180 acfm] and the discharge pressure exceeds 30-40 barg [435-580 psig], although rotary screw compressors have made inroads into this niche.

Large, heavy-duty reciprocating compressors are usually designed to API 618. This is sometimes not the case for the small packaged compressors units often used in field compressor applications.

The compressor manufacturer builds reciprocating compressors using standard frame sizes. A particular compressor frame will fix the maximum number of cylinders, the maximum power consumption, compressor stroke, rod diameter, maximum rod load, and compressor speed. As an example, one manufacturer offers the compressor frame below:

Number of Cylinders	1-6
Stroke	17.8 cm [7.0 in]
Speed	450-1000 rpm
Rod Diameter	6.4 cm [2.5 in]
Maximum Rod Load	245 kN [55 000 lbf]
Maximum Power	3730 kW [5000 hp] with 6 cylinders
Cylinder Sizes	12.1-67.3 cm [4.75-26.5 in]

Compressor cylinders are manufactured from ductile iron for larger diameter, lower pressure applications and forged steel for smaller diameter, high pressure service. Compressor pistons may be manufactured from aluminum, cast iron, or steel. Aluminum is favored in large diameter pistons and steel in small diameter. Piston speeds should be limited to about 4.5 m/s [15 ft/sec] to ensure a long service life.

Compressor valves are spring loaded, non-return valves at the suction and discharge side of the compressor. There are several types including ring type (see Figure on page 182), channel type and poppet. The valve travel or "lift" is small, usually on the order of 2 mm, and the moving parts – the strips or rings – are subject to fatigue. In a 1200 rpm compressor, compressor valves seat and unseat 20 times a second! Non-metallic materials are increasingly being used in this service. Compressor valves are typically the highest maintenance item on a reciprocating compressor. Their service life is considerably reduced by the presence of liquid droplets and solid particles in the gas, in fact most of the operating problems with reciprocating compressors are a result of inadequate and poorly designed separation equipment upstream of the compressor.

Reciprocating compressors generate flow and pressure pulsations in the piping due to gas velocity variations. This can result in severe vibration if the pipework and structure is in acoustic resonance with these pulsations. These pulsations can be smoothed by providing pulsation bottles at the compressor cylinder suction and discharge nozzles. Sizing of these bottles is beyond the scope of this text and is done based on computer simulation studies, but typically the minimum bottle size is about 10 swept volumes, a swept volume being the piston displacement for a single-acting cylinder. Pulsations can also be reduced by using larger diameter lines. Piping should be sized so that the gas velocity is limited to 6-10 m/s [20-33 ft/sec].

ROTARY SCREW COMPRESSORS

Rotary screw compressors are capturing an increasing share of the compression market in E&P operations. Their primary applications are in air compression, refrigeration service, fuel gas compression, and low to medium pressure gas boosting service. They offer positive displacement characteristics without the vibration and unbalanced forces associated with reciprocating machines. They are manufactured in two configurations — dry (or oil-free) screw and oil flooded screw.

The screw compressor is a rotary, positive displacement machine with continuous internal compression. It consists basically of a housing incorporating two overlapping bores which contain the two spiral-lobed rotors. Most twin screw compressors are built under license from the Swedish company SRM who holds most of the patents referring to screw compressors.[(15.3)] One rotor is the driving rotor and the other the idling rotor. The driving, or male, rotor typically has 3-6 lobes, and the idling rotor, or female rotor, normally has 4-8 lobes. The displacement of the machine is equal to the inter-lobe volume. A designer can control the displacement by changing rotor diameter or length, and/or the speed.

The compression process is illustrated in Figure 15.15. Gas enters the intake nozzle and flows into the helical grooves in the rotors (shaded area). As rotation of the rotors proceeds, the gas is trapped in the interlobe volume. This volume decreases as this *pocket* of gas travels down the rotors. Since the rotors have helical lobes, the decrease of volume is smooth and gradual with the rotation of the rotors. A discharge port is situated in the cylinder side wall and end wall and is located to provide the desired amount of compression.

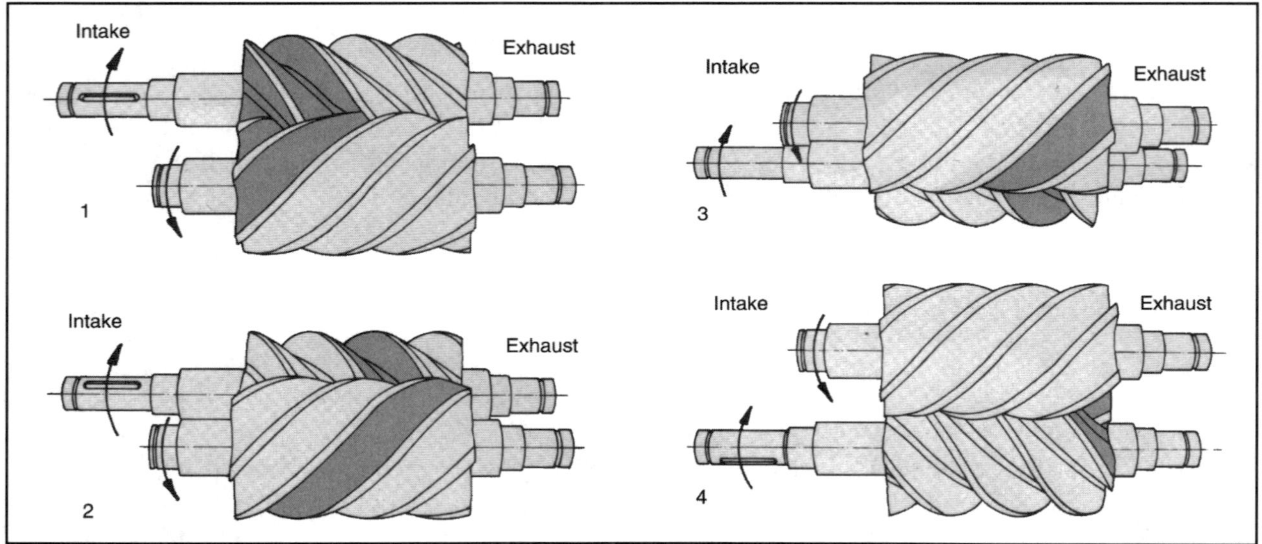

Figure 15.15 Operating Mode of a Twin Screw Compressor

Once the pocket is exposed to the discharge port, the pressure will become the same as that in the discharge line from the machine. Therefore, the gas may be over-compressed or under-compressed in the cylinder prior to delivery to the discharge. Optimum efficiency results when the operating pressure ratio corresponds to the built-in pressure ratio. Figure 15.16 is a P-V diagram illustrating over and under-compression, assuming an adiabatic compression process. Notice that there is no clearance volume expansion causing a decrease in capacity as is characteristic of reciprocating compressors.

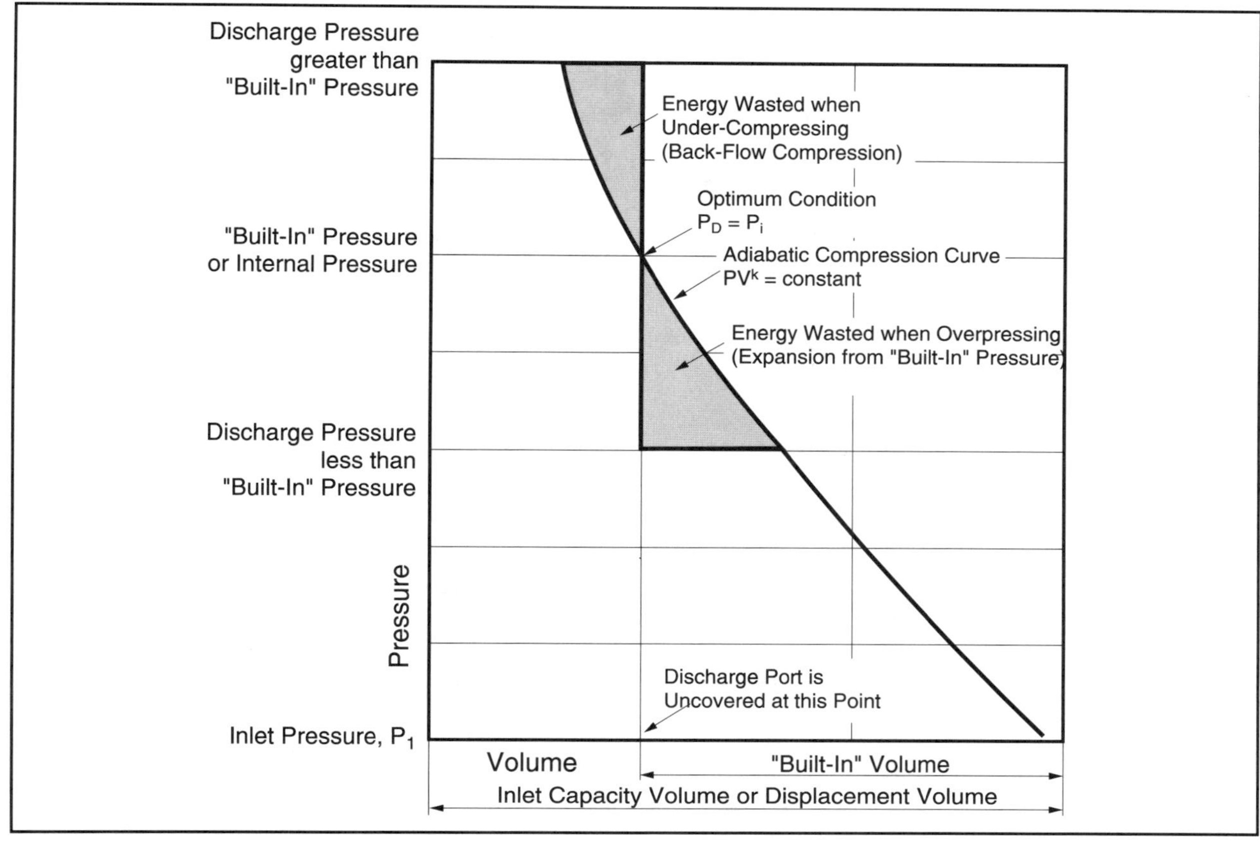

Figure 15.16 Rotary Compressor "P-V" Diagram

The following terms and equations apply for rotary screw compressors.

$$\text{Compression Ratio} = \frac{P_D}{P_1}$$

$$\text{Built-in Compression Ratio} = \frac{P_i}{P_1}$$

$$\text{Built-in Volume Ratio} = \frac{V_d}{V_i} = \left(\frac{P_i}{P_1}\right)^{1/k}$$

Where:

P_D = discharge pressure, absolute
P_1 = suction pressure, absolute
P_i = maximum pressure before rotor lobe passes the discharge port, absolute
V_d = displacement volume
V_i = the compression volume swept before passing the discharge port

There is however a leakage path in a screw compressor. High pressure gas can leak back to the suction side of the compressor between the rotor surfaces and between the rotors and the housing bore. This leakage is called *slip*. The volumetric efficiency of a screw compressor depends on the amount of slip which occurs. Slip is not a function of compressor speed, so the higher the rotating speed of the compressor the higher the volumetric efficiency. Typical volumetric efficiencies vary from about 70% to 95% depending on the compression ratio, built-in volume ratio, and screw type (dry or flooded). Figure 15.17 can be used to estimate the adiabatic and volumetric efficiencies of oil flooded screw compressors.

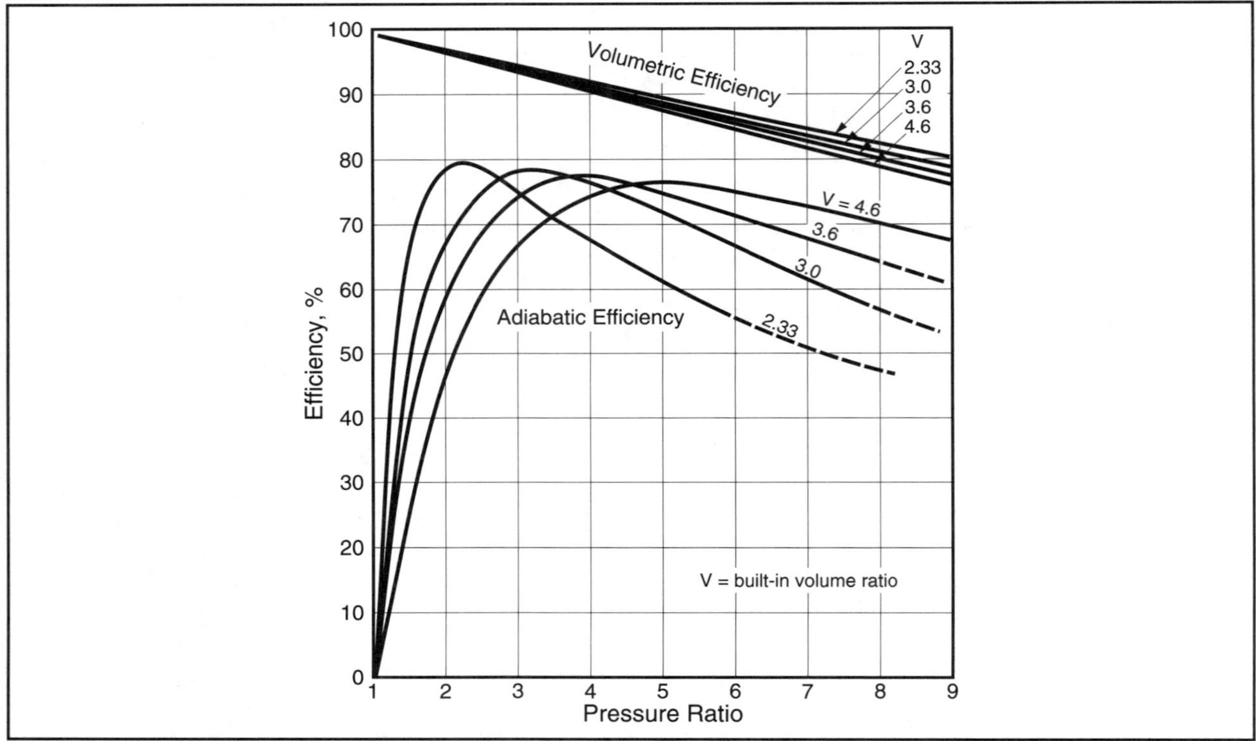

Figure 15.17 Typical Efficiencies for a Oil Flooded, Twin Screw Compressor(15.4)

In oil-free screw compressors the rotors do not contact and the rotation of the rotors is synchronized by timing gears. They have the advantage that no oil is introduced into the process gas and they can handle dirty gases or gases which may contaminate the oil stream. However, the volumetric efficiency of a dry screw will be less (due to higher slip) and the compression ratio is limited by the temperature rise of oil-free gas.

In flooded screw compressors, oil is continually injected along the length of the rotors. This has several advantages over oil-free screws.(15.4)

1. The oil helps to seal the internal leakage clearances, improving volumetric efficiency and overall efficiency. This results in better efficiency at low speeds and high pressure ratios when compared to oil-free operation.
2. The oil removes the heat of compression thereby reducing discharge temperatures. This permits operation at high compression ratios because the mechanical problems caused by high discharge temperatures are significantly reduced.
3. Cool operation permits operation with closer internal clearances since thermal expansion of the rotors and casing are reduced. This also results in improved noise levels.
4. The injected oil dampens pressure pulsations and reduces noise levels.
5. The oil provides inter-rotor lubrication that permits driving the female rotor directly by the male rotor without the use of timing gears. This simplifies the design compared to a non-lubricated compressor.
6. Since the oil is in contact with the gas being compressed, the bearing cavities and lube oil systems are in a gas atmosphere. This eliminates the need for the internal gas-tight seals found in oil-free compressors.

Some disadvantages are that optimum operating speeds are lower than with a dry screw due to oil shearing and pumping losses. At high speeds, pumping losses tend to overcome gains made by improved volumetric efficiencies.

In addition, the oil must be thoroughly separated from the gas in an oil separator. Once the oil has been removed from the gas, the oil is cooled and re-circulated to the compressor by pressure differential. Since the same oil is used to lubricate the bearings as is injected into the compressor, the oil must be filtered to remove any particulates which may cause bearing damage. The oil injection rate is the amount of oil necessary to keep the discharge temperature below about 90°C [194°F].

For planning purpose, the power calculation for screw compressors is similar to reciprocating and centrifugal units. For flooded screw machines the power calculation is complicated by the power necessary to pump the oil and the fact that the compression is no longer adiabatic. The overall efficiency is the product of the volumetric, adiabatic and mechanical efficiency. Mechanical efficiencies are typically 95%.

DRIVERS (ENGINES)

Once the compressor power requirement has been established, it is necessary to choose a driver to supply that amount of power. This driver may be a gas or diesel engine, gas or steam turbine, electric motor or an expansion turbine. The choice depends on the compatibility of compressor and driver, fuel availability, weight and space limitations, etc. If other factors are not critical, compatibility is most important. Comparable speed range, for example, is one consideration. One can use speed increasers and speed reducers satisfactorily but they should be avoided unless their use is better than the other alternatives.

All drivers have a rated capacity. But, this may not be the amount of power that can be produced reliably and continuously under the conditions present. Some adjustment or derating may be necessary. All combustion engines, for example, have an output power which depends to some degree on combustion air density. Both altitude and temperature affect output.

Some of the available engine power may be used to drive accessories such as pumps and coolers. This amount of accessory power, if any, must be provided for in sizing the engine.

On an offshore platform, the accessories may be powered from an independent "power package" used for general purposes. Within a frontier area or an isolated unit compressor, the driver may have to power all accessories. In some cases, lights and other noncompressor auxiliary needs are powered by the driver. In the early planning before such details are known, it is wise to be generous in estimating power requirements.

Reciprocating Gas Engines

This type of engine is the most popular driver for reciprocating compressors in sizes up to about 2.5 MW [3360 hp]. Larger engines are available as compressor drives and recent installations have utilized engines up to 6.0 MW [8000 hp]. Gas engines are also widely used to drive rotary screw compressors.

The most common configuration is a four-cycle gas engine driving a separable compressor either by direct drive or through a gearbox. Direct drive is the preferred choice due to lower installed cost, but in some installations the optimum engine speed may not match the compressor speed and the use of a gearbox provides a more reliable and efficient package.

In North America the use of large, low speed gas engines which are integral with the compressor (see Figure on page 156) was very popular prior to the early to mid 1970's. These engines were typically two cycle and operated at speeds less than 500-600 rpm. While a significant number of these units continue in operation they are rarely specified for new installations.

Reciprocating engines are only available in discreet sizes, so it is necessary for the compressor supplier to "package" the compressor with a compatible driver. Most compressor suppliers have a close relationship with gas engine manufactures and utilize proven standard packages for various compressor applications. Nonetheless, it is helpful to have some familiarity with gas engine specifications to avoid problems associated with a poorly designed compressor package.

Rated Power

For compressor drive applications, the available driver power will be the manufacturer's stated power output for continuous duty service. Most manufacturer's use the rating standard ISO 3046/1 as the basis for establishing the power output for continuous duty. ISO 3046/1 defines the Continuous Power Rating as: "The highest load and speed which can be applied 24 hours/day, 7 days/week, 365 days/year except for normal maintenance." It is permissible to operate the engine at maximum load indicated by the intermittent rating no more than two hours per day. For most engines, the continuous rating will be 10-20% less than the intermittent.

Additional engine de-rating may also be required based on:

1. Experience or severe service
2. Altitude
3. Ambient temperature
4. Fuel gas quality
5. Use of engine power to drive accessories such as water and lube oil pumps, cooler fans, etc.

Derating is an economic compromise. You are buying more potential power which will return the extra capital cost through decreased cost for maintenance and less loss of revenue because of downtime. This is necessarily a judgment call based on experience.

Altitude Correction. Altitude affects the density of the air available for combustion. Any rating must be based to some degree on this density.

The effect of density depends on the engine. There are two basic types:

1. Nonsupercharged or nonturbocharged (naturally aspirated) engines where the air enters the intake manifold at ambient pressure.
2. Supercharged or turbocharged engines where the air is compressed before combustion.

The altitude deration varies with the manufacturer and engine type. The following summary is typical:

Four-cycle gas engines (typical driver on separable compressors)
Nonturbocharged - 3% for each 300 m [1000 ft] above sea level Turbocharged - 2% for each 300 m [1000 ft] above sea level
Two-cycle gas engines (typical driver on integral compressors)
No correction up to 500-800 m [1650-2600 ft] above sea level, then 3% per 300 m [1000 ft] for natural aspirated and 2% per 300 m [1000 ft] for turbocharged (Some large engines are designed for no correction up to 2100 m [6900 ft] above sea level)

The actual altitude deration will be specified by the manufacturer.

Temperature Correction. Temperature also affects air density. Temperature corrections vary with engine type and manufacturer. Deration of 1% per 5.5°C [10°F] above 38°C [100°F] is typical for many engines. Actual temperature factors can be obtained from manufacturer.

Accessories Correction. If the compressor engine drives all of its accessories, not all of the power output is available to compress the gas. The exact amount of accessory power is dependent on the engine and the application.

The following formula is used to correct for altitude and accessories:

$$\text{Rated kW} = \frac{\text{kW}}{(1 - \text{Acc. Corr.} \ - \text{Alt. Corr.} \ - \text{Temp. Corr.})} \tag{15.25}$$

Where:

kW = power required by the compressor
Acc. Corr. = that fraction of the engine power used for accessories
Alt. Corr. = fractional derating for altitude effects
Temp. Corr. = fractional deration for temperature effects

Fuel Gas Correction. Manufacturer's power ratings for natural gas fueled engines are typically based on a fuel with a LHV of 33.4 MJ/std m^3 [900 Btu/scf] which is essentially the heating value of pure methane. For fuel gases containing heavier hydrocarbons, particularly C_4+, some de-rating of the rated power output may be necessary. This is due to the fact that the presence of heavier hydrocarbons can cause engine knocking, pre-ignition and detonation, which can damage cylinder heads and pistons. The problem is more serious in engines with high compression ratios.

It is best to use "processed" gas for fuel, i.e. gas which has undergone some degree of NGL extraction. This is not always possible, particularly in remote locations. In some cases it may be economically viable to install a small fuel gas processing unit especially when the only fuel gas available is low pressure, high heating value rich associated gas.

Other Considerations

1. Brake Mean Effective Pressure (BMEP) – the theoretical average pressure needed in the power cylinder throughout the power stroke to develop the rated power.

$$\text{BMEP} = \frac{(\text{A})(\text{kW})}{(\text{no. cycles})(\text{d})^2(\text{L})(\text{N})} \tag{15.26}$$

Where:			SI	FPS
	BMEP =	in units of	MPa	psia
	L =	length of stroke	cm	in
	d =	power cylinder diameter	cm	in
	A =	conversion factor	77 780	677 200
	N =	rpm for two cycle engines		
	=	(rpm/2) for four-cycle engines		
	kW =	engine power output requirements		

Four-cycle engines must develop a higher BMEP at the same speed, for a given load, due to the additional crankshaft revolution taken per power stroke. Per unit weight, a four-cycle engine must operate at a higher BMEP, rpm, or both, to produce the same amount of power as a two-cycle engine.

Maintenance is dependent to some extent on BMEP. Turbocharged engines can generally operated reliably with a BMEP of 1.10 MPa [160 psia]. BMEP will be less for naturally aspirated engines.

These limits depend to some extent on the heating value of the fuel

2. Cooling – if inter-coolers and/or after-coolers are needed, specify the approach desired.
3. Controls – these should be detailed, even though the control package is fairly standard.
4. Emissions – emissions from gas engines has become a significant environmental issue. Most of the concern is over nitrogen oxides (NO_x) but engines can also emit unburned hydrocarbons and carbon monoxide (CO). Virtually all engine manufacturers now offer "lean burn" engines which consist of a pre-ignition chamber where a fuel-rich mixture is ignited. This, in turn, ignites the primary fuel-air mixture which is leaner than in conventional engines. This process allows for more complete combustion and keeps combustion temperatures low, minimizing NO_x formation.

   Alternatively (or additionally), NO_x may be removed from the compressor exhaust. This is typically done with SCR (selective catalytic reduction) or NSCR (non-selective catalytic reduction) systems. Ammonia or urea based reagent is injected into the compressor exhaust and reacts with the NO_x to form N_2 and H_2O. This method is expensive and often increases the back pressure on the engine, reducing rated power output.

Each of the above factors is open for individual interpretation. Regardless of this, they must always receive some degree of formal consideration.

The choice of supercharged versus naturally aspirated engines is likewise somewhat arbitrary. The former have a lower fuel consumption, are better at high elevations, have a low cost per unit of power, and may require less cooling. Naturally aspirated engines are usually less susceptible to problems associated with high ambient temperatures, are easier to operate unattended, and may operate better on high heating value fuel. They inherently have less overload capacity than supercharged engines.

Gas Turbines

The gas turbine is used where high power output per unit weight is desirable. The capital cost is favorable, although the fuel efficiency is frequently less than a reciprocating engine. In E & P applications, gas turbines are the most popular driver for centrifugal compressors. Gas turbine power outputs vary from less than 1 MW [1340 hp] to over 200 MW [268 000 hp], but in compressor drive applications the power output seldom exceeds about 50 MW [67 000 hp].

The basic mechanical and thermodynamic cycles are shown in Figure 15.18. Part (a) of this figure shows a simple open cycle. Fuel is burned at high pressure in the combustion chamber. A portion of the energy produced is available for net work. The remainder drives the inlet air compressor.

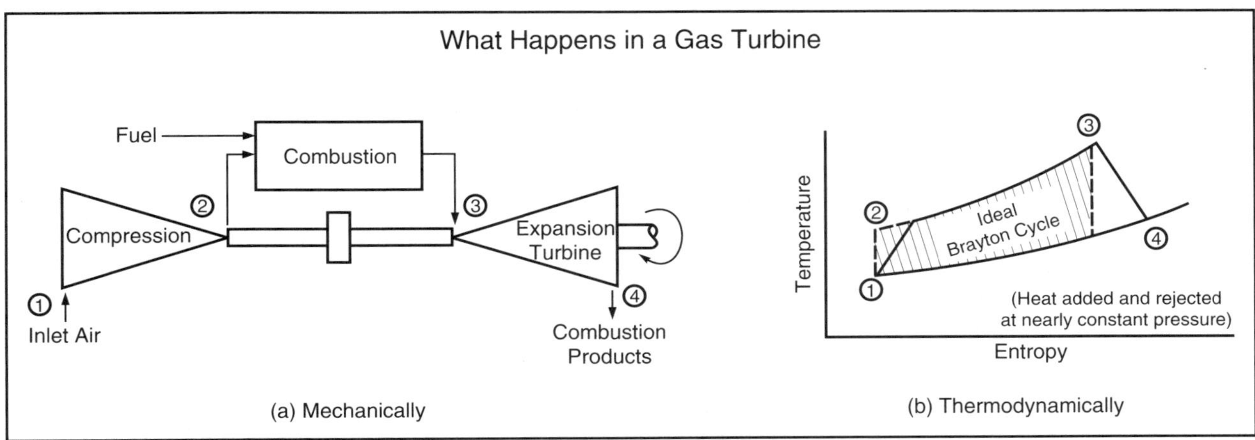

Figure 15.18 Schematic of Simple Turbine Operation

The ideal cycle shown in Figure 15.18 (b) is the Brayton Cycle. Points 1-2-3-4 in that figure show the actual cycle. This ideal cycle may be used to outline pertinent characteristics of a turbine. Figure 15.19 will be used for this purpose. The air enters at Point 1 and is compressed isentropically to temperature T_2 and pressure P_2. Steps 2-3 (combustion) take place at substantially constant pressure. Step 3-4 is an isentropic expansion through the power turbine. In an actual gas turbine, Steps 1-2 and 3-4 will not be isentropic.

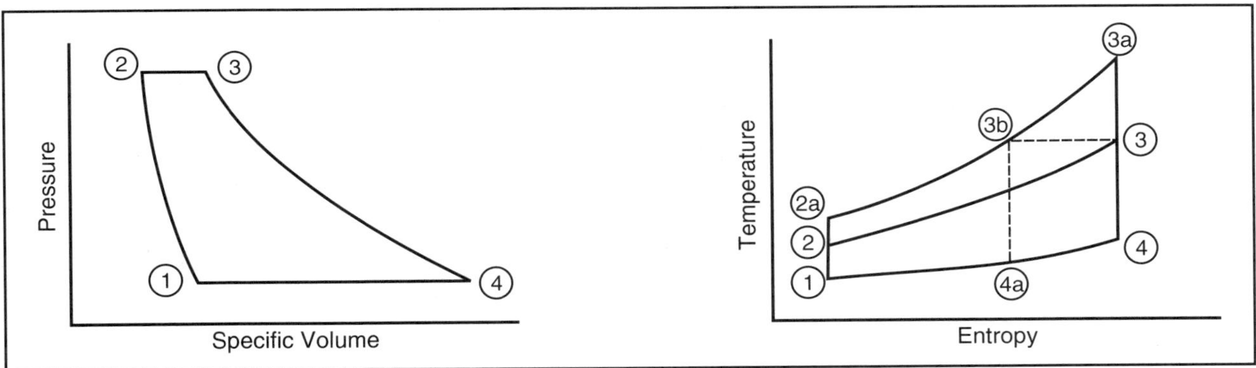

Figure 15.19 Thermodynamic Diagrams for Ideal Brayton Cycle

A gas turbine is a thermally rated device. The thermal efficiency used must be a function of pressure ratio, according to Figure 15.19. For the ideal cycle,

$$E_{theor} = 1 - \left(\frac{T_1}{T_2}\right) = 1 - \left(\frac{1}{(P_2/P_1)^m}\right) = 1 - \left(\frac{T_4}{T_3}\right) \quad (15.27)$$

Where:

E_{theor} = theoretical thermal efficiency
m = (k - 1)/k
T = absolute temperature, K [°R]
P = absolute pressure, kPa [psia]

Note that this thermal efficiency should not be confused with the overall efficiency used previously for compressors.

The thermal efficiency is given by

$$E_{thermal} = \frac{(A)(W)_a}{(m_f)(LHV)} \quad (15.28)$$

Where:			SI	FPS
A	=	constant	1.0	2545
W_a	=	actual shaft work available at power turbine	kW	hp
m_f	=	mass flow rate of fuel	kg/s	lbm/hr
LHV	=	lower heating value of fuel	kJ/kg	Btu/lbm

The thermal efficiency of a gas turbine will decrease about 1.1% for each 10% the power output is reduced below rated output.

If one increases pressure ratio, the cycle will be 1-2a-3a-4 instead of 1-2-3-4. In an actual turbine, the maximum temperature (T_3) is fixed by metallurgical limitations. If T_3 is the allowable maximum temperature, the theoretical cycle will be 1-2a-3b-4a. Thermal efficiency is, therefore, limited by both pressure ratio and temperature. As the pressure ratio increases for Step 1-2, more power is needed for air compression. The most efficient gas turbines available currently employ compression ratios of 20:1 and combustion temperatures in excess of 1200°C [2200°F]. Overall efficiencies vary from 22% for older models to near 40% for state-of-the-art designs.

Metallurgical advances which enable temperature to be raised have dramatic effects on efficiency. Excess air of 200-300% is commonly used to keep this temperature within bounds. The "backwork" necessary to drive the air compressor is therefore large. This backwork typically requires 60-70% of the total power turbine output. The Saturn turbine, for example, produces about 3500 kW [4700 hp] in the hot gas generator turbine to provide about 1185 kW [1590 hp] in net shaft work available for productive use at the power turbine.

Figure 15.20 is a drawing of a turbine driving a pipeline compressor. This is an older model unit but it illustrates the key components in the system. The most critical temperature in the turbine is that entering the high pressure turbine. Current models run at higher temperatures than shown.

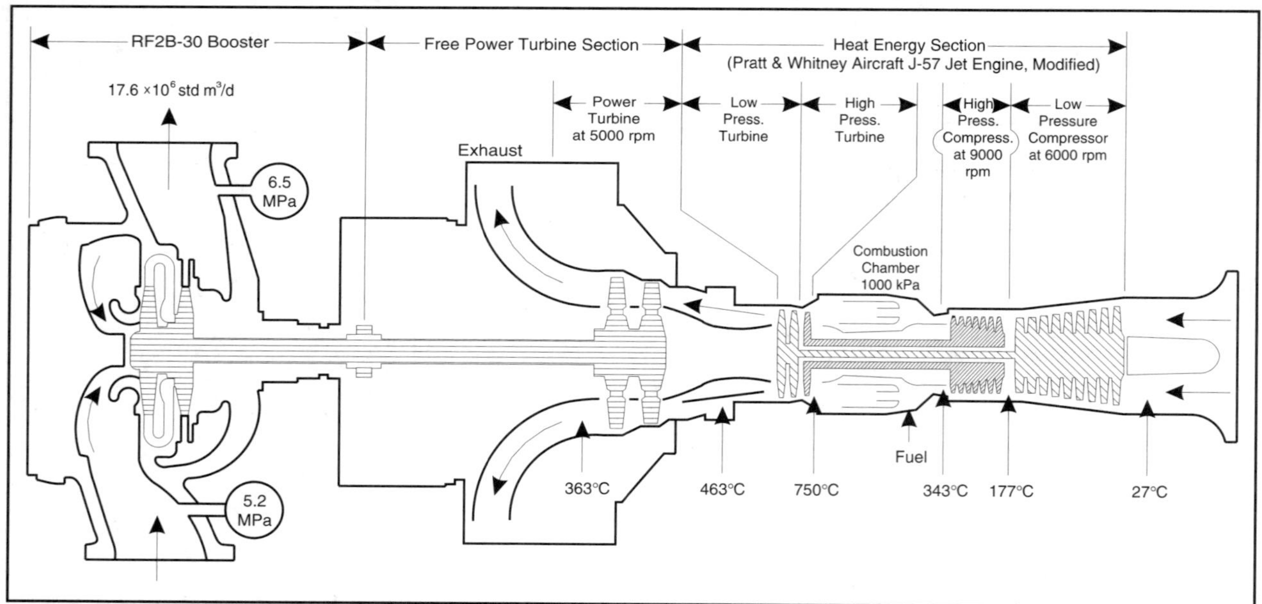

Figure 15.20 Cutaway View of a Gas Turbine *(Courtesy Cooper-Bessemer Corp.)*

Figure 15.20 tends to oversimplify the complexity of the turbine system. The picture below shows a cut-away view of an industrial turbine. Although simple in principle, the turbine involves the proper maintenance of many parts, some operating under severe service conditions.

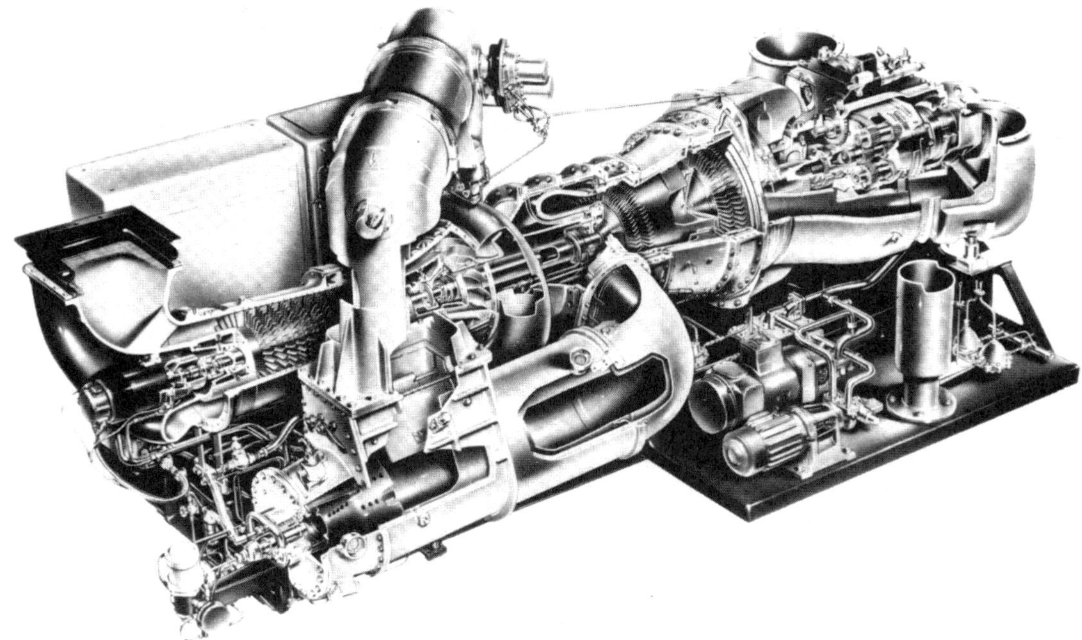

Two types of gas turbines have historically been supplied to the oil and gas industry. These are industrial and aircraft derivative, although the distinction between the two has narrowed in recent years. Industrial turbines typically employ heavy, horizontally split casings designed for long life and long run times between maintenance. They are more durable and tolerate a wider range of fuels. Disadvantages include weight and need for on-site maintenance. Aircraft derivative turbines are lighter, more compact and typically more efficient. They employ modular construction which minimizes downtime for maintenance and repairs.

Aircraft derivative engines consist of an air compressor, combuster and high pressure turbine used to supply power to the compressor. They are often called "hot gas generators" because they supply hot, medium pressure exhaust gas to a power turbine which drives the driven equipment. The power turbine is frequently supplied by a different manufacturer than the hot gas generator. When the power turbine is separate from the high pressure, hot gas generator turbine the unit is called a split shaft or two shaft machine. With the two-shaft turbine it is possible to operate the hot gas generator at a different speed than the power turbine. Several gas turbine configurations are shown in Figure 15.21.

When choosing between single and two-shaft machines, one must compare their operating characteristics with those of the system involved. It must be remembered that flexibility is one requisite. Two shaft machines are preferred for gas compression service because of the ability to vary speed and maintain maximum power output. Single shaft machines are often used in power generation because of the constant speed requirement and the higher inertia of the rotating system.

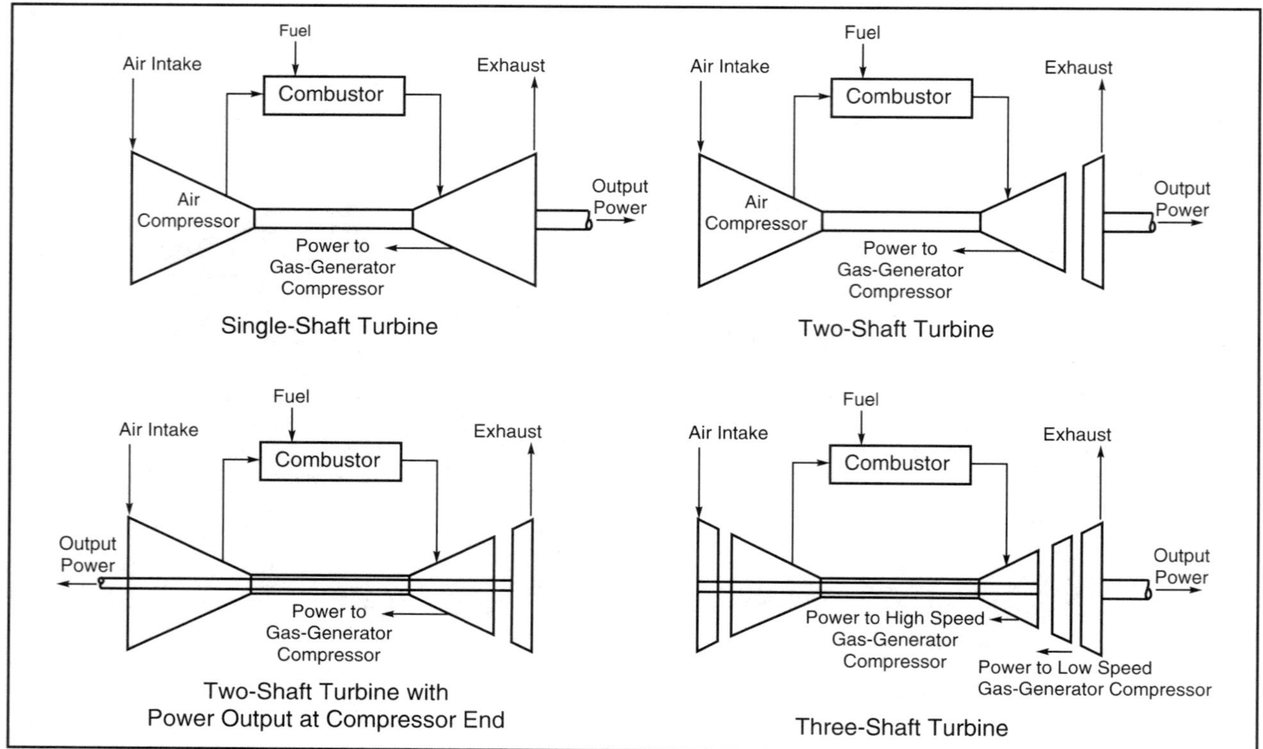

Figure 15.21 Gas Turbine Configurations

Regenerators. The total thermal efficiency of the system is sometimes improved by using a regenerator, as shown in the figure to the right. A regenerator preheats the air to the combustion chamber, improving thermal efficiency. Regenerators are not usually feasible on high efficiency turbines because there is little temperature difference between the turbine exhaust and the air compressor discharge due to the higher compression ratios employed.

Schematic of a Split-Shaft Turbine with Regenerator

Combined Cycles. Gas turbines produce copious amounts of high temperature exhaust gas. For example, a RB-211-24G which has an ISO-rated power output of 31.7 MW [42 600 hp], produces over 90 kg/sec [200 lbm/sec] of exhaust gas at a temperature of 460°C [860°F]. This exhaust gas, which still contains approximately 15% oxygen can be used directly for heat transfer or can be used as combustion air to a supplemental fired waste heat boiler. Utilization of this waste heat can raise the overall thermal efficiency to 50-70%.

When the waste heat is used to generate steam for power generation, the process is referred to as a "combined cycle" or cogeneration system. Combined cycle facilities are the most common method used today in new power plant construction. In many E & P facilities, use of the waste heat in the turbine exhaust is not fully exploited, particularly in offshore installations. This is due to high capital costs, low valuation of fuel gas, and lack of a requirement for high-temperature heat.

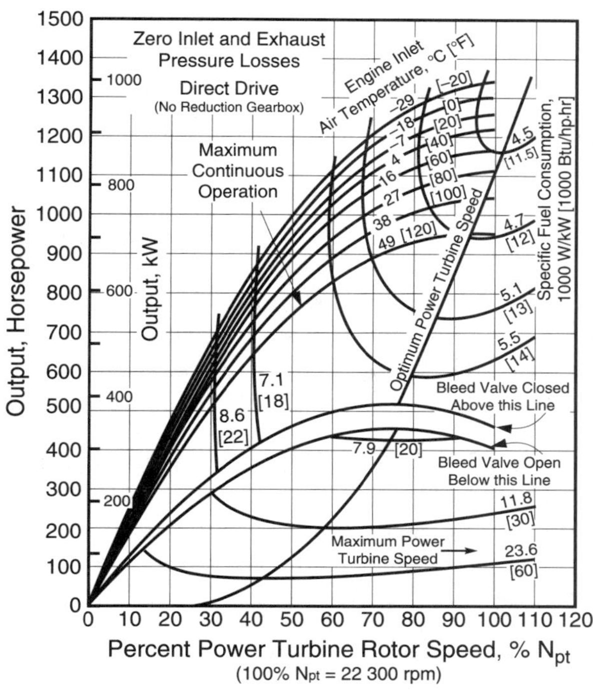

Performance. The curve shown at left is for a single-shaft turbine operating at sea level. Each turbine possesses its own unique characteristics but the curve shown is typical. This turbine provides maximum power output at 22 300 rpm.

The ISO (International Standards Organization) defines rated power output of a gas turbine as the output at design speed, sea level, 15°C ambient temperature and zero inlet and exhaust pressure drops.

The site-rated output of any turbine is governed by altitude and temperature. The ISO has set up standard derating criteria for these factors, at a given speed. If the power rating is established at sea level, deduct about 1.1% for each 100 m [330 ft] the site is above sea level. This altitude deration factor is a straight line and can be applied to any reference altitude.

Temperature of the air entering the air compressor is a critical factor affecting power output. The ISO temperature deration line is also linear. Power output should be derated about 0.9% for each 1°C [0.5% per °F] that the temperature rises above the rating temperature, which commonly is 15°C [59°F]. If the temperature is lower than this, the rating is increased by the same amount. In arctic conditions, where very low temperatures are encountered, the potential power output exceeds rated output and may be limited by mechanical constraints.

Site derating for inlet and exhaust duct losses varies from turbine to turbine but some typical values are 0.2% per mbar for the inlet and 0.1% per mbar for the outlet. Note: 1" $H_2O \approx 2.5$ mbar.

Normal pressure drops for inlet and exhaust ducts range from 5-15 mbar [2-7.5 in H_2O]. Waste heat recovery in the exhaust can significantly increase pressure drop and these losses must be accounted for when site rating a turbine.

The factors affecting turbine speed are the same as those for centrifugal and axial compressors. The front end of a gas turbine is an axial air compressor.

Environmental Impact

Environmental control of a gas-turbine installation in a process plant is governed by local regulations plus company practices at the plantsite. The environmental impact falls into two distinct areas, noise and stack emissions.

To meet the normal noise criterion of 85 to 90 dBa at 1 meter distance, inlet and exhaust silencers together with acoustic enclosures are required. The methods and components used for noise abatement are identical, regardless of the type of gas turbine used, to a point where one can talk about near-standard procedures.

In step with increasing demands for noise control the design and construction of the silencers and acoustic enclosures have undergone a quiet evolution. The noise-abatement components are now of heavy structural design and often of the free-standing type. Outer walls may be fabricated of 3/16 to 1/4-in-thick steel plates, and the materials for the acoustic walls may be heavy-gage galvanized-steel perforated plate for the air intake and enclosures, unless special circumstances dictate otherwise. High-temperature corrosion-resistant materials are used for the exhaust stack. The sound-absorbing materials may be mineral wool or glass fiber.

The operational efforts required for exhaust-emission control for a process-type gas-turbine installation will normally be concerned only with the amount of oxides of nitrogen, NO_x, in the exhaust. This is because the strict turbine fuel specifications, as a side-effect, eliminate most polluting agents at the source, i.e., from the fuel before it is burned. Normal stack pollutants, such as sulfur dioxide, carbon monoxide, and particulates, are controlled by the fuel specifications and the turbine's combustor design, and do not usually pose an operational problem. Unfortunately, this is not the case with NO_x.

Formation of NO_x is complex and not a subject for this writing. However, in simple terms, NO_x production for a gas turbine is influenced by ambient conditions — it decreases with increasing relative humidity and increases with flame temperature and residence time at this temperature. Different fuels produce different levels of NO_x; this again has to do with flame temperature: some low-Btu fuels, e.g., hydrogen, may have high NO_x-emission levels.

The NO_x emission from a gas turbine can be controlled by injecting water or steam of boiler-feed quality into the combustors. In this way, flame temperature can be reduced and NO_x emission held at acceptable levels as dictated by site location and local regulations.

Dry, low-NO_x combustors can also reduce NO_x emissions. These combusters are designed with a premix chamber where most of the fuel gas (75-85%) is mixed with air prior to the main burner. All combustion takes place downstream of the pre-mix chamber in a very lean fuel-air environment. This minimizes flame temperature, hence NO_x emissions. Some gas turbine operators have reported NO_x concentrations below 15-20 ppm with dry, low-NO_x combustors. The combustors are sensitive to fuel gas quality. The presence of liquid droplets in the fuel gas can lead to *"flashback"*, a phenomenon where the flame migrates upstream into the premix chamber. Flashback can damage the combustor.

Another NO_x-control technique is selective catalytic reduction (SCR), in which anhydrous or aqueous ammonia is injected into the fluegas upstream of a catalyst bed. The NO_x and NH_3 combine at the catalyst surface, forming an ammonium salt intermediate that subsequently decomposes to produce elemental nitrogen and water.

In SCR, the NO_x-reduction reactions occur only in a narrow temperature range, generally 260-500°C [500-932°F], depending on the type of catalyst. Operation above the maximum temperature results in oxidation of ammonia to either NO_x or ammonium nitrate and ammonium nitrite; operation below the optimum range does not provide the energy necessary to initiate the reaction. Figure 15.22 shows a schematic of an SCR system.

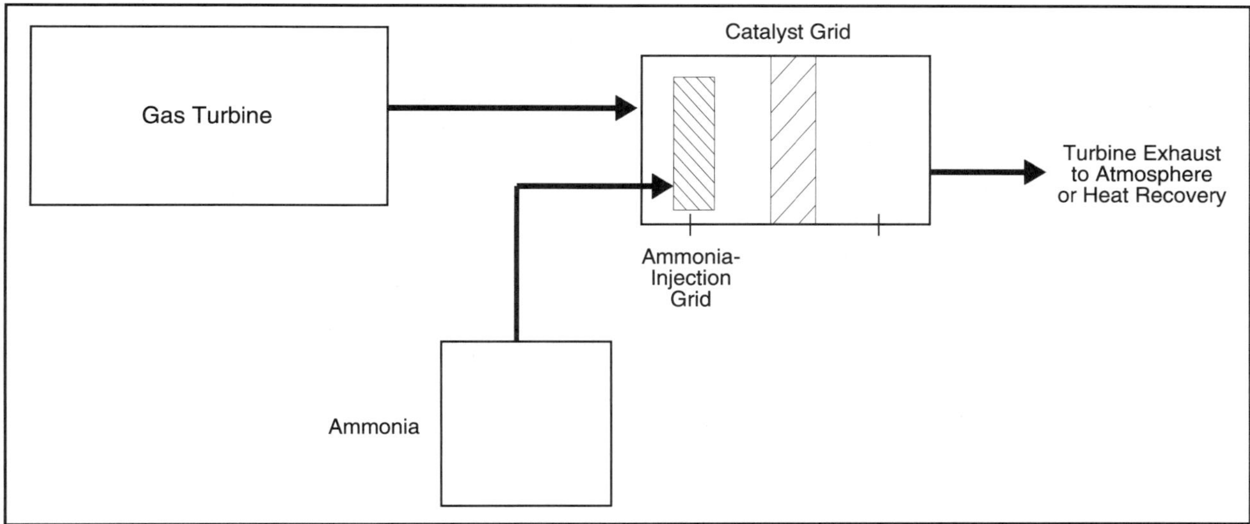

Figure 15.22 Selective Catalytic Reduction for NO_x Emission Control on a Gas Turbine Exhaust

Fuel and Air Quality

Clean fuel and inlet air are absolute necessities if one is to expect trouble-free operation and life. Gas turbines run on a variety of fuels (crude oil, NGL, diesel, natural gas, etc.). In E & P operations, natural gas is the most common fuel. When natural gas is used it is imperative that liquid hydrocarbons not enter the turbine. If available, dry, processed gas should always be used as fuel, if not, the fuel gas should be preheated well above its hydrocarbon dew point. Most gas turbine manufacturers require 15-28°C [27-50°F] of superheat at the fuel gas valve.

A gas turbine ingests massive amounts of air. The air rate to a Solar Mars 100 (11 185 kW [15 000 hp]) turbine is approximately 42 kg/s [92 lbm/sec]. Atmospheric pollutants can cause fouling and corrosion to turbine components. Salt water spray, drilling mud, cement, sandblast are the most common contaminants in offshore service. A properly designed air filter is imperative. It should be designed for easy maintenance and low pressure drop. Rain hoods and de-icing equipment should also be installed if dictated by climate.

Exhaust ducts should be oriented so as not to interfere with other process operations. They should be installed so that the prevailing wind does not carry the hot air plume into fin-fan coolers, air conditioning systems, heliports, other engine intakes, etc.

Expansion Turbines

This form of driver will be discussed in more detail in Chapter 16, Refrigeration.

Electric Motors

Electric motors are probably the simplest driver used for compressors and pumps. They have a high efficiency (however, the electrical power generation plant may not!), are rugged, reliable, and require little maintenance. In larger sizes (> 100-200 kW) the manufacturer will build the motor exactly to specification.

In compressor service, the primary disadvantage of electric motor drivers is their constant speed. However, variable speed motors are becoming increasingly available but are typically economical only in the smaller sizes. When used for centrifugal compressor drives, a gearbox is often be required, increasing cost and weight.

Another disadvantage is starting torque. If, during start-up, the motor torque is less than that required by the driven equipment the motor will "stall," the speed drops rapidly to zero, and the motor will trip due to high temperature or high amperage. The best way to evaluate torque requirements is to superimpose the speed versus torque curve of the motor with that of the driven equipment. These curves are provided by the respective manufacturers. NEMA has established minimum torque requirements for three basic motor designs: Design B, for normal centrifugal loads; Design C, for loads requiring high starting torque (reciprocating compressors); and Design D, for high-slip high-inertia loads such as flywheel drives on machine tools.

Starting current is also a problem. A motor often draws 5-7 times its full load current when starting. This may depress voltage and reduces starting torque in proportion to the voltage drop squared. This is not a problem if the reduced torque is still sufficient to turn the compressor, but the voltage drop may affect other electrical equipment.

Induction motors operating at 1200, 2400, or 3600 rpm (60 Hz) are the most popular driver up to about 1-2 MW [1300-2600 hp]. Synchronous motors are often preferred in the larger sizes. Synchronous motors are often lower in installed cost, more efficient, and have a higher power factor.

In North America, the maximum synchronous speed of a motor will be 3600 rpm [3000 rpm at 50 Hz]. Induction motors will operate slightly below the synchronous speed due to slip. Increasing the number of poles, or pairs of poles, will lower the synchronous speed. Low speed, synchronous motors are frequently used to drive reciprocating compressors.

High speed, induction motors have recently been used in centrifugal compressor drive applications. Speeds up to 18 800 rpm are possible, by increasing the frequency. No gearbox is required, reducing cost and weight.

Motor Enclosures

There are several types of motor enclosures.

1. Open drip-proof – for use indoors in nonhazardous locations
2. Weather Protected II – for use outdoors in nonhazardous locations
3. Totally enclosed, water-air-cooled – for use indoors or outdoors in clean or dirty, nonhazardous locations
4. Totally enclosed, fan-cooled – an alternative for nonhazardous locations
5. Hazardous locations – where flammable vapors are present

A Class I environment contains flammable gas or vapor. Within this class, several groups have been specified. Most hazardous petroleum applications are in Group D, i.e., a Class I, Group D application.

One can use an explosion-proof motor but these are seldom available above 1180 kW [1500 hp], although larger ones have been built. At some point, cost is prohibitive. There are several alternatives.

A common practice is to use a pressurized enclosure. Clean air is drawn in to keep the air free of explosive mixtures and to carry off motor heat. This room is usually isolated from equipment containing hazardous liquid or vapor. As a general rule, minimum air requirements are about 0.3 $m^3/min/m^2$ [1 cfm/ft^2] of floor area. Some value greater than this may be necessary to account for air leakage from the enclosed area.

An alternative to air purging is to use a totally enclosed motor filled with inert gas. These are complicated, an operating nuisance and are seldom used.

Applicable standards differ in the U.S. and Europe because of different philosophies about how to approach safety. One must reconcile these differences in planning.

One must be concerned about many standards, including motor temperature rise. For example, a Class I, Group D explosion-proof motor must maintain a surface temperature below 215°C during any load condition. This and other such standards will be handled by the specialists but we must be generally familiar with the implications of these standards as they affect planning, specification and design.

Flexible couplings are used most commonly with electric motors. They come in many forms: flexible-disc, gear, spring-grid, pin and bushing and rubber-biscuit types. The choice depends on many factors. Do not misinterpret the word "flexible." Precise alignment is critical to achieve satisfactory performance.

CONTROL OF COMPRESSORS

Control systems for compressors are well documented by vendors of the equipment used. Most control problems thus seem to stem primarily from errors of omission. Some vendor-supplied systems are more nearly designed to protect the machinery than to provide an efficient, low-maintenance installation. Design specifications fail to properly denote changes in gas conditions and quantity. Too often, even if the unit runs a satisfactory amount of time, it does so under inefficient conditions. The solution – a pragmatic appraisal of the control system in light of expected process conditions, which vary for each installation. Use of a "standard" control system never yields optimum performance, except by coincidence.

Capacity Control

To a large extent, a compressor is self-controlling with respect to capacity. Any compressor (positive displacement or kinetic) will seek an operating condition such that its throughput matches the gas available to the compressor. If the gas available is reduced, the compressor will draw down the suction pressure until its capacity again matches availability. It is only when the operating conditions fall outside the operating limits of the process or compressor (e.g. surge, choke, driver power, frame loads, discharge temperature, etc.), that explicit capacity control is required.

Constant Speed Centrifugal Compressors

Figure 15.23 shows a typical scheme where a centrifugal compressor takes suction from a production separator with back-pressure control. This discussion is not limited to this scenario, the production separator could be any process unit requiring constant pressure operation. For a given production rate the compressor will seek a suction pressure where its capacity matches the volume of gas available. This is shown in Figure 15.24 using a typical centrifugal compressor performance curve.

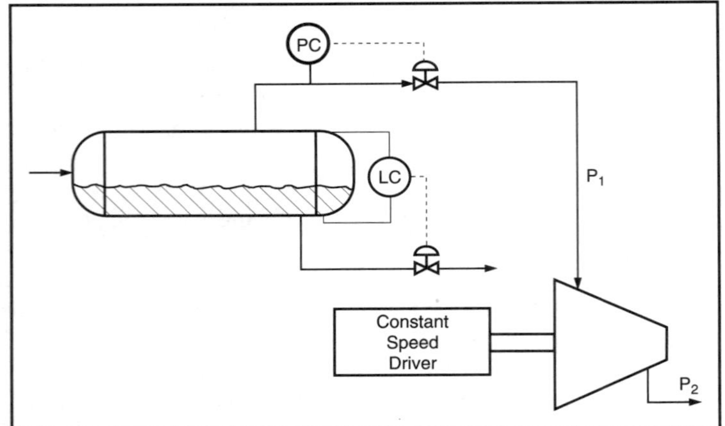

Figure 15.23 Constant Speed Centrifugal Compressor Control Scheme

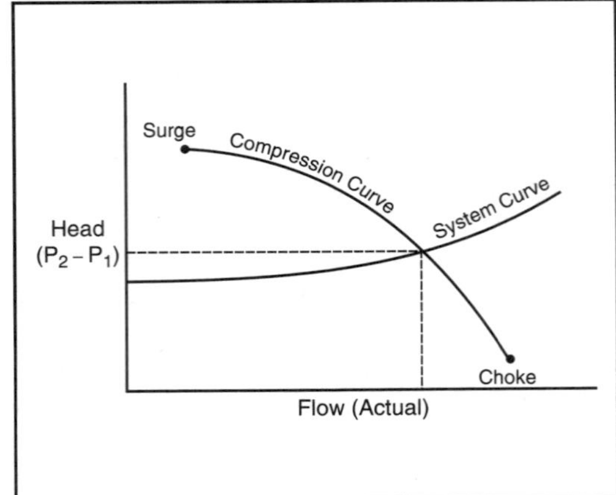

Figure 15.24 Compressor Performance and System Curves for a Constant Speed Centrifugal Compressor

If the production rate is reduced, the compressor will draw down the suction pressure reducing its capacity from q_1 to q_2 as shown in Figure 15.25. Not only has the actual volumetric throughput of the compressor decreased but the mass flowrate has declined even further since the gas density is less at the lower suction pressure. Conversely if the production rate is increased, the suction pressure will rise, increasing compressor throughput.

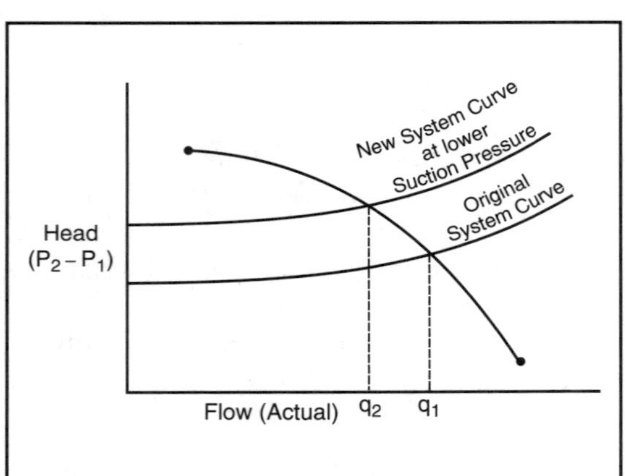

Figure 15.25 Compressor Performance and System Curves for a Constant Speed Centrifugal Compressor

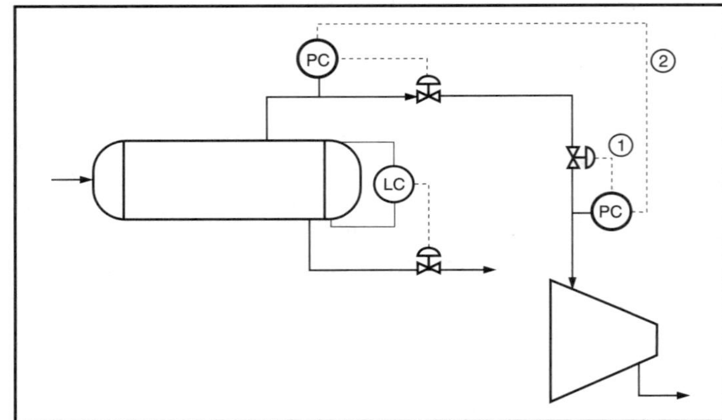

Figure 15.26 Additional Suction Pressure Control Valve to Prevent Driver Overload

Two operating limits are imposed on this system – surge and driver power. Surge is prevented by recycling gas from the discharge to the suction and will be discussed in greater detail later. Driver power limitations may require a second control valve ① in the compressor suction to limit throughput. This valve is manipulated by a suction pressure controller with a setpoint at the suction pressure which coincides with the maximum driver power. Under normal operating conditions this valve is completely open.

Alternatively, a pressure controller at the compressor suction may overide the pressure control valve at the separator ②. This is the preferred option in most E & P applications.

In either case the second pressure controller controls capacity when the compressor throughput is limited by available driver power. When the compressor is power constrained, throughput must be reduced either by reducing production or flaring excess gas.

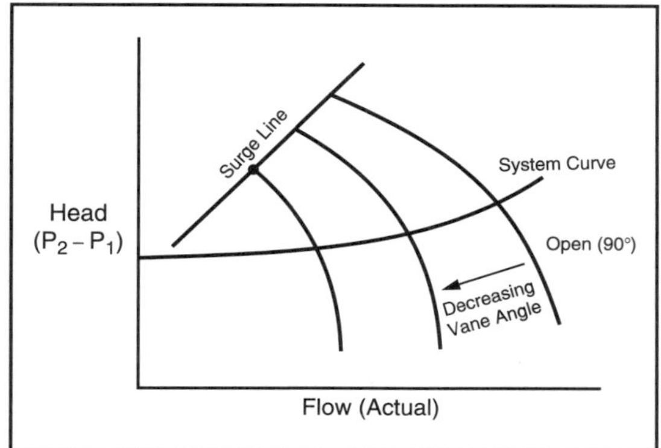

Figure 15.27 Effect of Inlet Guide Vanes on Compressor Performance Curves

Adjustable inlet guide vanes may also be used to control compressor capacity. They are sometimes used on constant speed centrifugal compressors, especially in refrigeration service. Adjustable inlet guide vanes essentially change the aerodynamic characteristics of the first impeller. This is accomplished by *pre-whirling* the gas entering the wheel in the direction of rotation, thereby developing less head than design. The efficiency declines as well but not as much as the head so the net result is lower driver power. These devices are more efficient than suction throttling, but are expensive and can increase maintenance expense. Figure 15.27 shows the effect of adjustable inlet guide vanes on the shape of the compressor performance curves.

Variable Speed Centrifugal Compressors

When using a variable speed driver, such as a gas turbine, the compressor capacity can be controlled by manipulating the driver speed. This has the added advantage of saving one control valve (back pressure valve on the separator) since the separator pressure is now manipulated by the driver speed. The performance curve for such a system is shown in Figure 15.29.

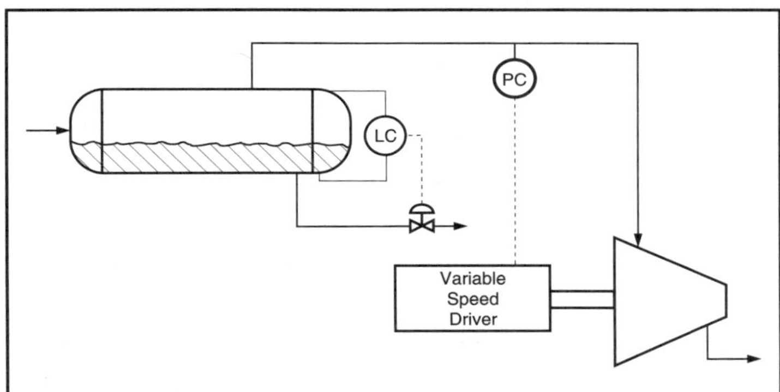

Figure 15.28 Variable Speed Centrifugal Compressor Control Scheme

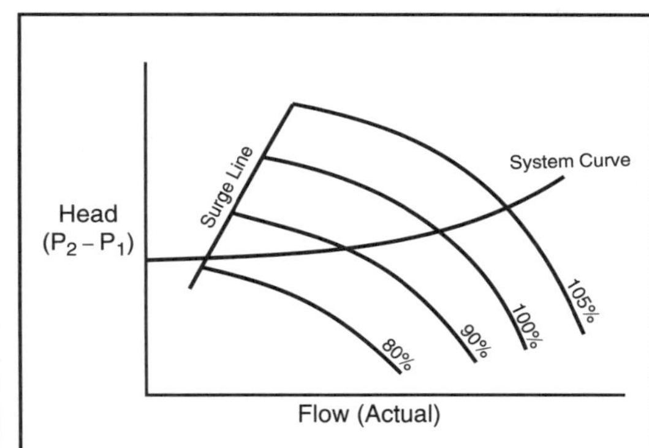

Figure 15.29 Compressor Performance and System Curves for a Variable Speed Centrifugal Compressor

In Figure 15.29, the flowrate through the compressor is represented at any speed by the intersection of the system curve and compressor curve.

As discussed in the constant speed section, some provision for preventing driver overload should be incorporated in the control logic.

Reciprocating Compressors

The following mechanisms may be used to control the capacity of reciprocating compressors.

1. Suction pressure throttling
2. Variation of clearance
3. Speed
4. Valve unloading
5. Recycle

The use of suction pressure throttling to control capacity is virtually identical to that outlined for constant speed centrifugal compressors. A typical compressor capacity curve for a reciprocating compressor operating at constant speed is shown in Figure 15.30. As the suction pressure falls the compressor capacity declines due to the lower P_1 (see Equation 15.22) and the lower volumetric efficiency, E_{vz}.

This compressor is self controlling as long as the operating limits of the compressor or process are not reached. The operating limits for a reciprocating compressor are high rod load or high discharge temperature (this occurs at minimum suction pressure) and high driver load (this typically occurs at maximum suction pressure).

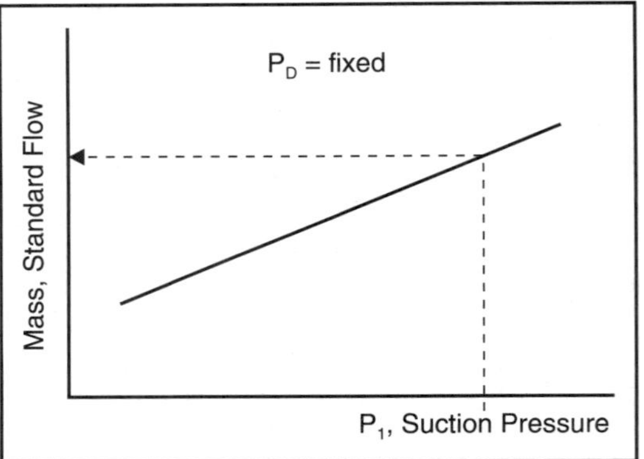

Figure 15.30 Example Compressor Capacity Curve for Constant-Speed Reciprocating Compressor

When the minimum suction pressure is reached, the compressor capacity must be reduced. This can be accomplished by increasing clearance, unloading valves and/or decreasing speed. In any case, the effect is to change the compressor capacity curves as shown in Figure 15.31. Curve A represents the condition where all cylinders are double-acting, all clearance pockets closed and maximum speed. Curves B-E represent a number of conditions where clearance pockets may be open, valves unloaded or speed decreased. Curves B-E are in order of decreasing capacity.

When the maximum suction pressure is reached, valve unloaders and clearance pockets can also be used to reduce compressor capacity as indicated in Figure 15.31. Alternatively, suction flow can be throttled with a conrol valve (similar to that discussed in centrifugal compressors).

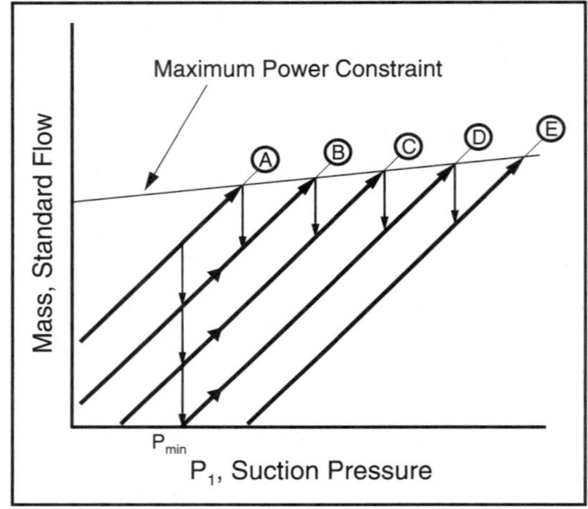

Figure 15.31 Capacity Curves for a Reciprocating Compressor

For some drivers, particularly with electric motors and two-cycle engines, speed control is often not viable. In these cases capacity control is often accomplished with clearance pockets and valve unloaders.

Clearance pockets increase the cylinder clearance, decreasing volumetric efficiency, thus lowering the capacity of a cylinder. They may be fixed volume or variable volume. Fixed volume pockets are more common. They can be actuated pneumatically or manually.

Valve unloaders decrease the capacity by holding the suction valve open through the compression cycle. As a result, no compression takes place since the gas simply flows back and forth across the suction valve. Two types of valve unloaders are shown in Figure 15.32.

A finger unloader uses small fingers to push the valve disks down to the stop plate, thereby holding the valve open. If a pneumatic unloader is used, the standard configuration is air pressure on the actuator unloads the valve.

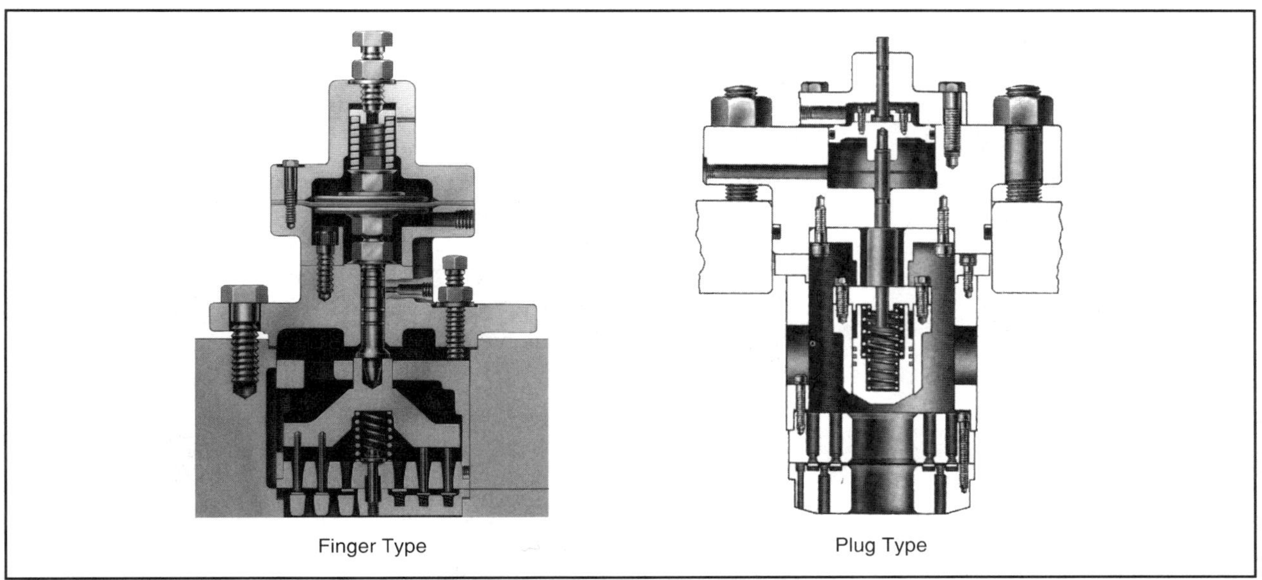

Figure 15.32 Finger and Plug Type Valve Unloaders.[(15.5)]

A plug unloader consists of a port in the center of the valve which can be opened or closed using a plug much like a control valve. Plug unloaders are normally configured so that air pressure on the actuator loads the valve.

Plug unloaders are generally preferred because they are easier to calibrate and maintain; however a compressor valve with a plug unloader will inherently have a smaller loaded capacity than one with a finger unloader.

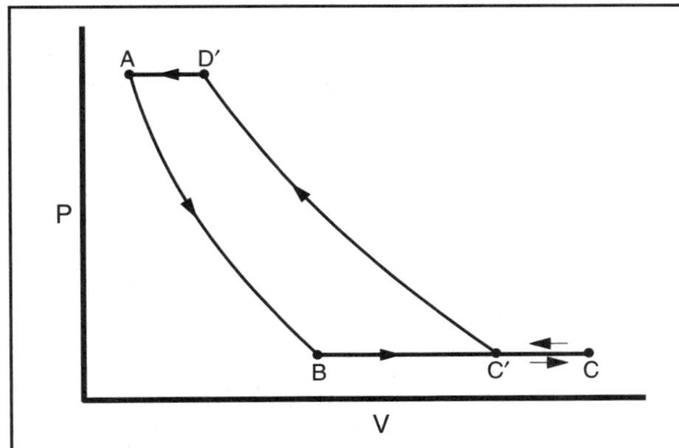

Figure 15.33 P-V Diagram for Stepless Valve Unloaders

Because valve unloading affects the rod loading, valve unloaders should be used only with the manufacturer's approval.

Clearance pockets and valve unloaders are effective capacity control mechanisms, however, they provide a discrete change in capacity. This may lead to unstable control, e.g. cycling. "Stepless" unloaders are available. They hold the suction valve open through only a portion of the compression stroke. When the suction valve closes at point C′, the compression and discharge segments of the cycle begin (BC′ vs. BC). The net effect is lower capacity and power consumption.

Recycle is often used to control reciprocating compressor capacity. In this control scheme, gas is recirculated from the discharge to the suction side of the compressor to maintain constant suction pressure. It must be cooled to avoid increasing the suction temperature. Control is simple, straightforward and continuous, but energy inefficient.

If recycle is used it should be minimized. This can be accomplished by using clearance pockets or valve unloaders in conjunction with recycle. For example, if the recycle valve is more than say 25% open, a clearance pocket will open. This reduces compressor capacity and the recycle valve must close to reduce recycle rate. Control is still continuous and simple, but more efficient.

Screw Compressors

Screw compressor capacity is controlled using the following methods.

1. Suction pressure throttling
2. Speed
3. Slide valve

The principle of suction pressure throttling is similar to centrifugal and reciprocating compressors, i.e. lower inlet pressure reduces the gas density and hence the mass flow for a fixed compressor displacement. Suction pressure throttling changes the compression ratio which can have a significant effect on compressor efficiency for screw compressors.

The most efficient method of capacity control is speed control; however, there is some reduction of volumetric efficiency in screw compressors at lower operating speeds.

Slide valves are unique to screw compressors and can provide a simple, efficient method of capacity control on oil injected machines. A slide valve, shown in Figure 15.34 is an axially moveable segment of the cylinder wall. The valve extends from the discharge port toward the inlet end of the cylinders. When the valve is closed (positioned toward the inlet end) the compressor performs normally. When the valve is moved toward the discharge end, a bypass opens at the inlet end. Gas flows through this opening and returns to the suction side of the machine. The amount of gas bypassed depends on the axial position of the valve. "Stepless" capacity control from 10-100% is possible.

Figure 15.34 Slide Valve for a Twin Screw Compressor

Antisurge Control

Centrifugal and axial compressors have a characteristic combination of maximum head and minimum flow beyond which they will surge. As discussed previously surge is an unstable region where flow reversal occurs inside the compressor. Surge can damage seals and bearings and in some cases can result in a catastrophic compressor failure.

The only way to prevent surge is to recycle or blow-off a portion of the flow to keep the compressor away from its surge limit. Blow-off is not feasible in natural gas compression so recycle is the most common antisurge method.

As discussed earlier, recycle is inefficient and expensive. A well designed antisurge system will protect the compressor from surge but not recycle excessive amounts of flow.

A typical surge line is shown in Figure 15.9. Its location must be determined by test prior to design of antisurge system. An algorithm is then developed to model the relationship between head and flow along the surge line. A surge control line (SCL) is then developed which mirrors the shape of the surge line but is offset to the right (higher flow) by a margin of safety, usually about 10%. The closer the SCL is to the surge line, the less margin of safety and the higher the probability of surge. The further the SCL is from the surge line the greater the protection against surge but the greater the recycle flow resulting in less efficient operation.

The general approach to antisurge control is to measure head (or a surrogate of head, e.g. ΔP, compression ratio, etc.) and reset the setpoint on the antisurge controller as shown in Figure 15.35.

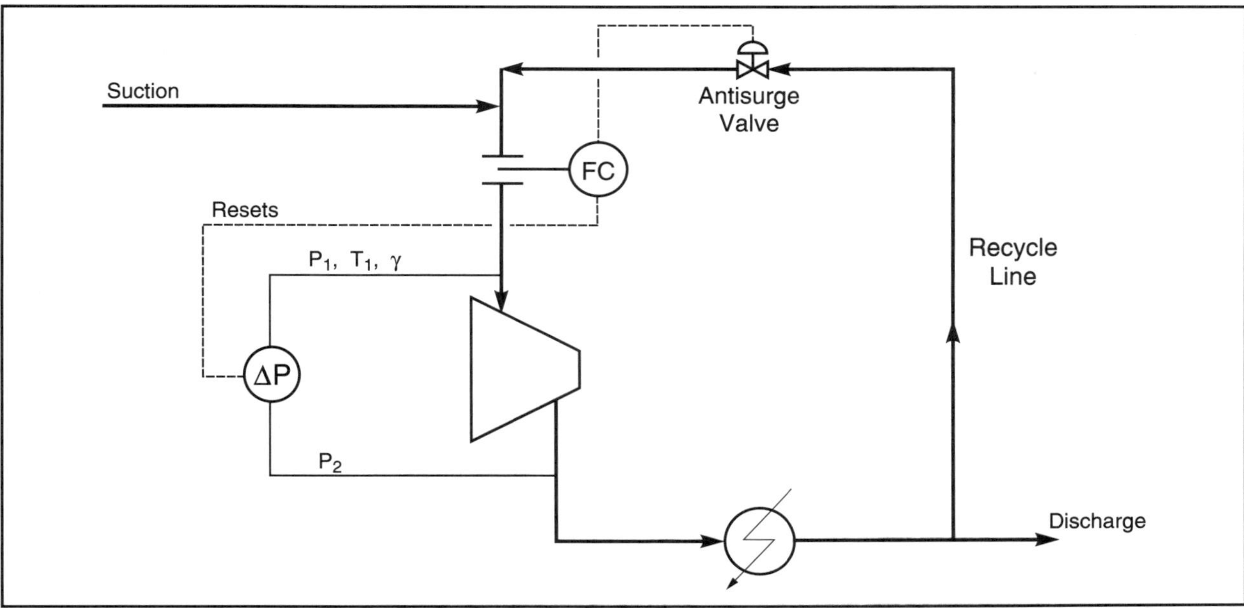

Figure 15.35 Simplified Control Loop for Antisurge System

The actual compressor head can be computed from Equation 15.5. This requires measurement of P_1, P_2, T_1, and γ or MW. Correlations for z and k are included in the software. If variations in gas composition and temperature are minimal, measurement of P_2 and P_1 is often sufficient.

Manufacturers of antisurge controllers provide their own special algorithms and response characteristics to give safe operation with the SCL as close to the surge line as possible. Many integrate surge control with capacity control, particularly on compressors operating in series and/or parallel.

Figure 15.36 shows the preferred antisurge flow scheme for natural gas compression. The recycle line should be located as close to the compressor discharges as possible, but upstream of the non-return valve. The recycle cooler should be installed in the recycle line or on the suction side of the compressor.

It is particularly important to use the flow scheme in Figure 15.36 when compressing gases containing significant amounts of C_5+ hydrocarbons such as flash gas compression from low pressure separators. In these applications, if the recycle line is taken downstream of the aftercooler and discharge scrubber, the MW of the recycle gas will be less than that at the suction because heavy hydrocarbons are removed from the gas in the discharge scrubber. If the compressor operates in recycle for a period of time, the MW of the gas decreases and the compressor head vs. ΔP relationship changes. This can narrow the stable operating window for the compressor and make surge control more difficult.

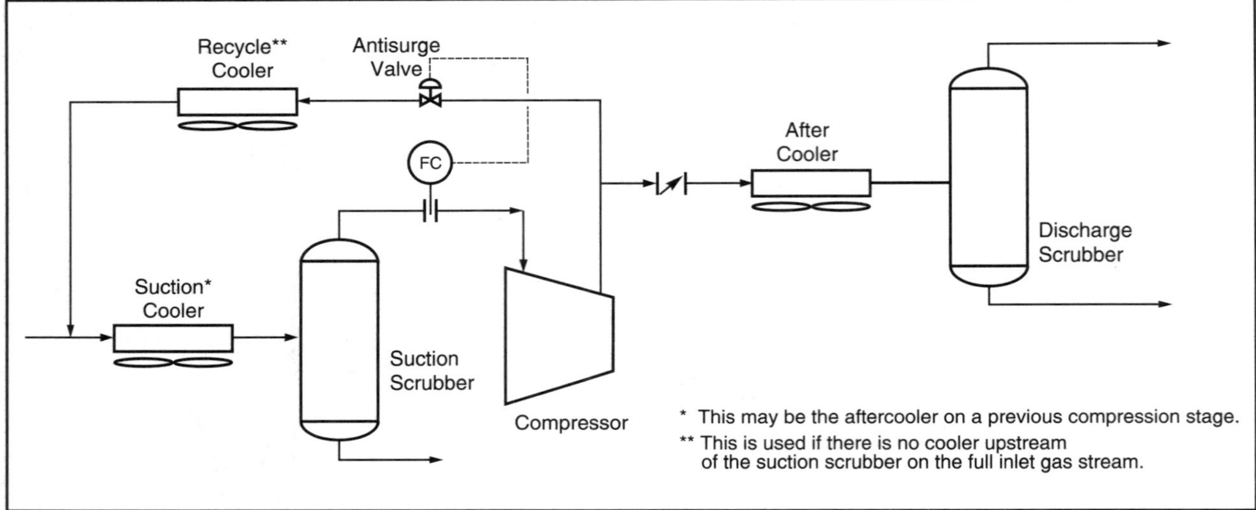

Figure 15.36 An Example Antisurge Piping Scheme

REFERENCES

15.1 Littlefield, R. G., SPE Paper No. 9996, Presented 18-26 March 1982, Beijing, China.

15.2 Balje, O. E., J. *Eng. Power* (Jan. 1962).

15.3 Davidson, J. and O. von Bertele, "Process Fan and Compressor Selection", *Mech. Engr. Pub. LTD, London* (1996)

15.4 Ingran, W. B., "An Introduction to Rotary Screw Compressors," SPE meeting, Tulsa, Oklahoma, USA (Mar 2, 1995).

15.5 Metcalf, J. R., "Unloader Selection for Reciprocating Compressors", *Compressor Tech^{two}* (Mar/Apr 1997), p.40.

16

REFRIGERATION SYSTEMS

A refrigeration system lowers the temperature of a fluid below that possible when using air or water at ambient conditions. A typical building air conditioner cools air to a temperature of 10-15°C. At the other end of the scale is the liquefaction of helium at – 268°C. The temperature produced depends on the process objective. If the objective is to recover marketable liquids (NGLs) from a produced gas stream, basic economics controls the temperature specified. If it is to meet a hydrocarbon dewpoint, that specification and the processing pressure sets the required temperature.

Several basic processes will be discussed herein.

1. Mechanical Refrigeration
    a. compression
    b. absorption
2. Valve Expansion
3. Turbine Expansion

MECHANICAL REFRIGERATION

A refrigeration system is a heat pump. Low temperature heat is removed from the process fluid and is "pumped" to high temperature (ambient) where it is rejected to the environment. Energy is required to pump heat. The amount of energy depends on the quantity of heat to be pumped (chiller duty) and how far the heat has to be pumped (temperature difference between the chiller and the condenser).

Energy used to drive a refrigeration process can be in the form of heat or work. The absorption refrigeration systems use heat to pump heat, compression systems use work.

Ammonia Absorption System

Figure 16.1 shows a flowsheet for a refrigeration system utilizing using two concentrations of ammonia-water solutions. The basic driving force is the heat input to the generator. Ammonia vapor is stripped from the water solutions in the rectifier or stripper. This ammonia vapor is condensed and passes through a receiver, a heat exchanger (optional) and across an expansion valve into the evaporator or chiller. Here it vaporizes while cooling the fluid to be chilled.

The ammonia vapor from the evaporator is absorbed in a weak ammonia-water solution. The result is a strong solution. The ammonia is removed from this strong solution in the generator to begin its cycle all over again.

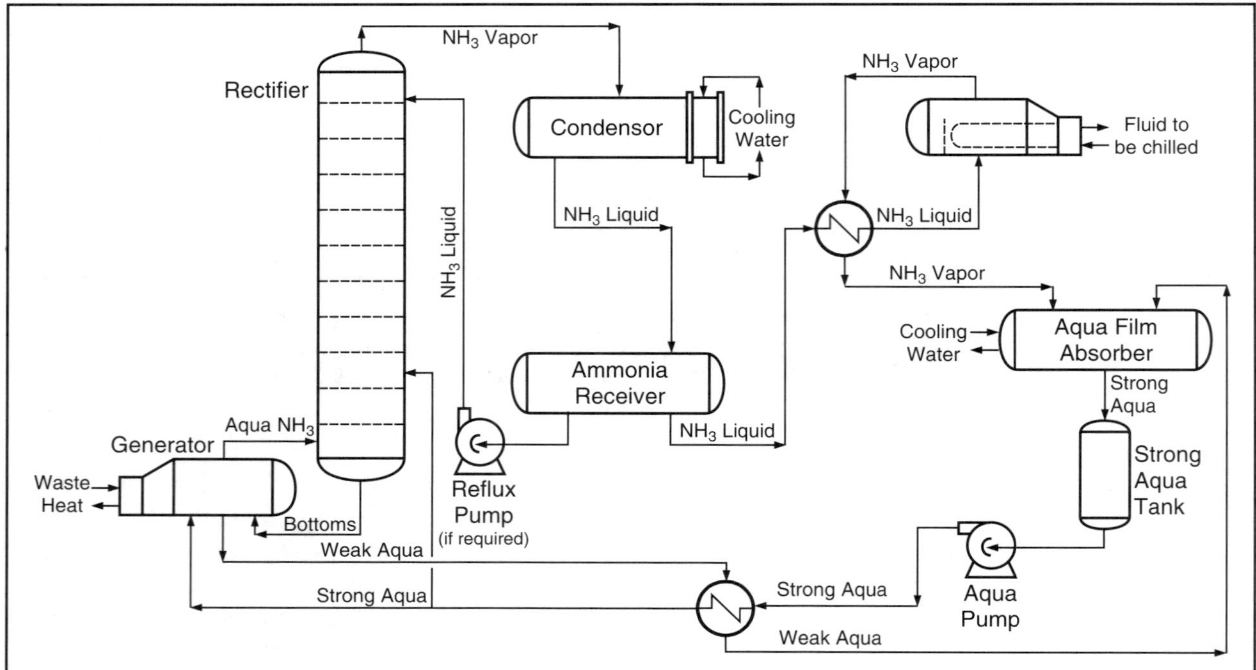

Figure 16.1 Flow Sheet of an Ammonia Absorption System

One can write selected energy balances around this system to determine loading at various points. The starting point is the evaporator. First, determine the total duty required to chill the fluid to its desired temperature. This is the heat load in the evaporator.

The heat absorbed by the ammonia per unit mass is governed by evaporator pressure or temperature. The ΔH is the enthalpy of a saturated ammonia vapor at evaporator conditions minus the enthalpy of the entering ammonia, which should be a saturated or subcooled liquid.

A series of such balances, and an overall balance, enable one to determine sizes and energy loads of each component part. Reference 16.1 summarizes the typical loadings for the ammonia system in Figure 16.1.

The heat for the generator may be obtained from any one of four sources: (1) low pressure steam, (2) fired heater, (3) a hot process stream and (4) waste heat. Since heat loads are large, the availability of (3) or (4) increases the economic attractiveness of this system.

Water cooling is shown for the condenser and absorber but air cooling may be used. As with all refrigeration systems, the higher the cooling temperatures the lower the process efficiency and greater the energy input.

Units have been designed with capacities to about 35 MW [10 000 tons of refrigeration (TR)]. The unit is simple and has few moving parts to maintain. Ammonia solutions are not difficult to handle metallurgically.

A major psychological problem is ammonia smell. In a confined space this can be a nuisance. However, the pungent odor is a safety item that immediately confirms leaks.

Though not widely used in gas processing applications, ammonia systems may be competitive with compression systems, particularly in those situations where significant quantities of waste heat are available. Operating cost comparisons are dependent on the source of heat and the cost of cooling. If the economics are competitive, these systems are a viable alternative to compression systems for producing refrigeration.

COMPRESSION REFRIGERATION

Compression refrigeration is by far and away the most common mechanical refrigeration process. It has a wide range of applications in the gas processing industry.

- Chilling natural gas for NGL extraction
- Chilling natural gas for hydrocarbon dewpoint control
- LPG product storage
- Condensation of reflux in deethanizers/demethanizers
- Natural gas liquefaction (LNG)

Figure 16.2 shows a simple single-stage compression refrigeration system. Saturated liquid refrigerant at Point A expands across a valve (isenthalpically). On expansion some vaporization occurs. The mixture of refrigerant vapor and liquid enters the chiller (sometimes called the evaporator) typically 3-6°C [5-10°F] lower than temperature to which the process stream is to be cooled. The liquid vaporizes. Leaving at Point C is a saturated vapor refrigerant at the P and T of the chiller. This vapor is compressed and then enters the condenser as a superheated vapor.

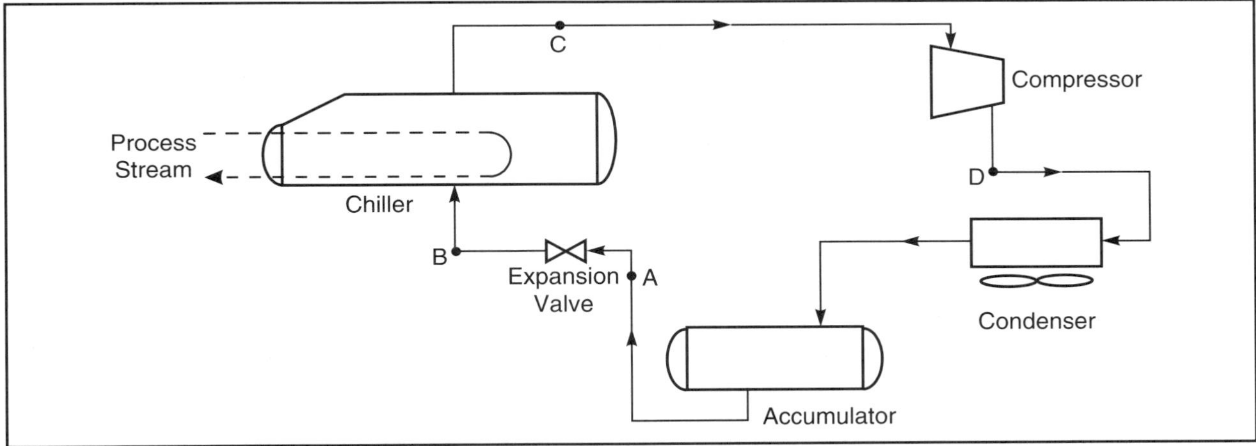

Figure 16.2 Flow Sheet of a Simple Refrigeration System

The refrigerant leaves the condenser as a saturated liquid or slightly subcooled. For air cooling, the condenser temperature will usually be 14-16°C [25-30°F] above the air dry bulb temperature. For water cooling the condensing temperature will be 5-10°C [9-18°F] above the water temperature. The accumulator, sometimes called a surge tank or receiver, merely serves as a reservoir for refrigerant as levels vary in the chiller(s) and condenser.

Calculation of a Simple System

There are several discrete steps in the sizing of the system shown in Figure 16.2. These are summarized below:

1. *Determination of refrigerant circulation rate.*

   Develop an energy balance around the chiller and expansion valve shown at right. At point A the refrigerant is a saturated liquid (or very close to it). At point C it is a saturated vapor. $Q_{chiller}$ is the chiller duty and is set by the process requirements.

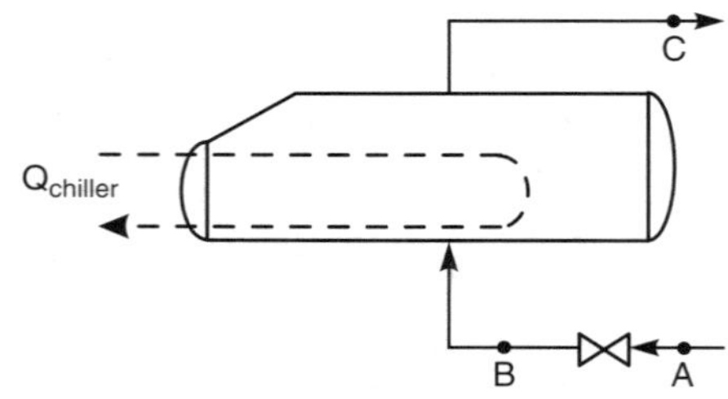

   If one writes an energy balance around the system, $Q_{chiller} + m_A h_A = m_C h_C$. But $m_A = m_C = m$, so

$$m = \frac{Q_{chiller}}{h_C - h_A} \tag{16.1}$$

Where:

		SI	FPS
$Q_{chiller}$	= chiller duty	kW	Btu/hr
h_C	= saturated vapor enthalpy	kJ/kg	Btu/lbm
h_A	= saturated liquid enthalpy	kJ/kg	Btu/lbm
m	= refrigerant circulation rate	kg/s	lbm/hr

2. *Determination of Compressor Power.*

   This is done by any appropriate method as outlined in Chapter 15. Calculate theoretical (isentropic) work and use an efficiency to find actual work. The circulation rate from Step (1) is used.

$$-W = \frac{m(h_D^{isen} - h_C)}{E} \tag{16.2}$$

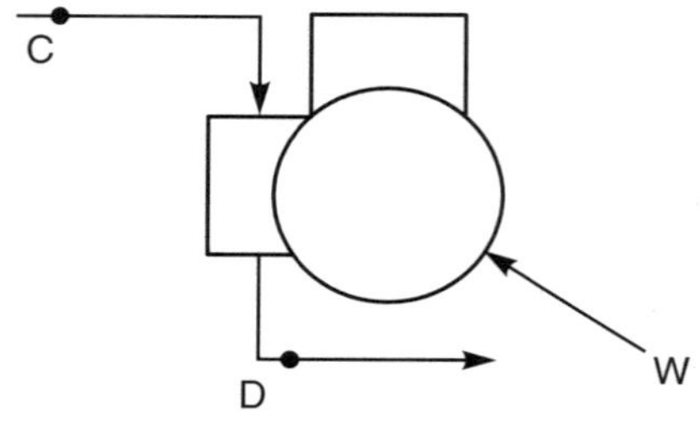

3. *Determination of Condenser Heat Load (Q_{cond}).*

   There are two ways to do this. Knowing $Q_{chiller}$ and W, you can write the overall balance for Figure 16.2 to find Q_{cond}. (Remember, W is negative, so Q_{cond} is the sum of two negative numbers)

$$Q_{cond} = W - Q_{chiller} \tag{16.3}$$

   If you are performing the calculation manually and wish an independent check of the previous work, write the balance shown below.

$$Q_{cond} = m(h_A - h_D) \tag{16.4}$$

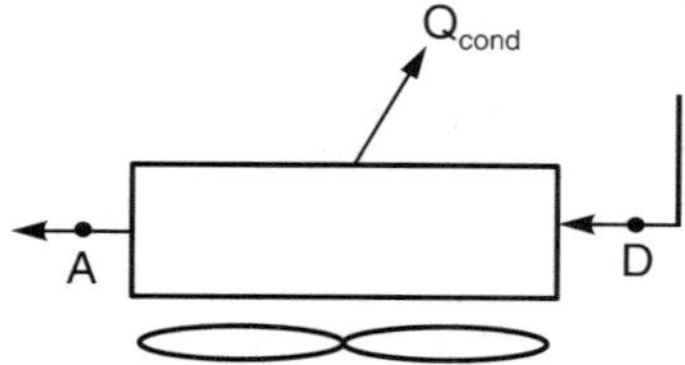

Determination of the Enthalpies

The calculation requires one to find the enthalpy per unit mass at points A, B and C. These can be found from a computer program or from tables and figures published for all of the common commercial refrigerants. Appendix B at the end of this volume contains data on some substances used as refrigerants. Appendix 16A at the end of this chapter contains pressure-enthalpy (P-H) figures for propane and Refrigerant 22 (Freon-22) as well as vapor pressure and physical property information on all common refrigerants.

The P-H diagram is convenient for solving the energy balance for a simple system.

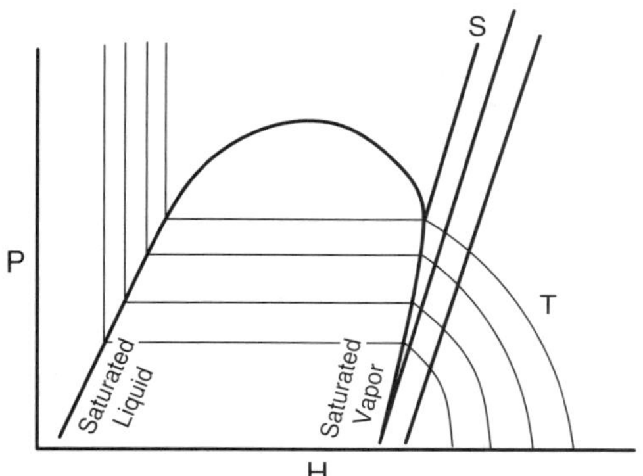

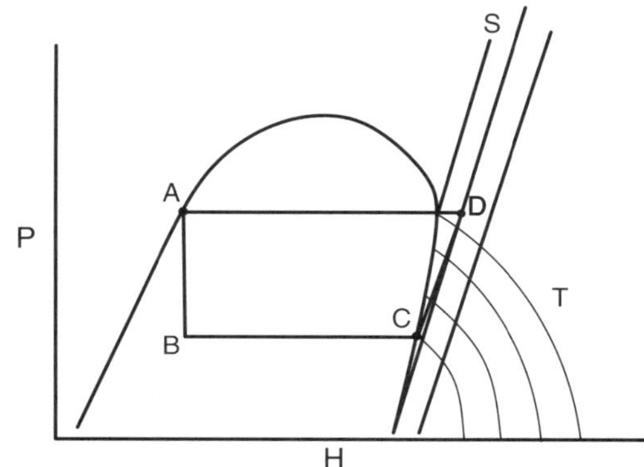

The left-hand figure is a simple representation of the P-H diagrams in Appendix 16A. The refrigerant is liquid to the left of the saturated liquid curve, two-phase inside the envelope and vapor to the right of the saturated vapor curve. The lines of constant temperature (isotherms) are horizontal between the saturated vapor and liquid curves and then rise almost vertically in the liquid region.

The calculation process starts by choosing the temperature of point A. Will water, air or some process fluid be used for condensation of the refrigerant? What temperature can we realistically achieve in the condenser? This temperature sets point A. It is on the liquid saturation curve, since it leaves the condenser as a liquid.

What is the temperature at points B and C? Normally, it will be 3-6°C [5-10°F] less than the minimum desired temperature for the fluid being cooled. This approach fixes the location of point C. It is on the saturated vapor curve, since it is in equilibrium with the liquid in the chiller (evaporator).

The expansion across the valve from point A is an *isenthalpic* process; a vertical line on a P-H diagram. Draw a vertical line from A to B, the pressure of point C, and then go horizontally to C. One can read the Δh required for Equation 16.1.

The theoretical compression is *isentropic*. Starting at point C, follow a constant entropy path until you intersect the pressure line of point A. This is theoretical (isentropic) discharge point D. Compressor power is calculated from Equation 16.2.

The condenser heat load can now be determined from Equation 16.3.

For a commercially pure refrigerant, use of a P-H diagram or a corresponding table is as reliable as any method.

Example 16.1: A refrigeration system is designed to refrigerate a natural gas stream for the purpose of NGL extraction. The following information applies.

Chiller duty	2MW [6.8×10^6 Btu/hr]
Gas temperature exiting chiller	– 15°C [5°F]
Air temperature	35°C [95°F]
Refrig. Compressor efficiency	75%

Calculate the refrigerant circulation rate, compressor power, and condenser duty for this system. Use the P-H diagram for propane on page 258 or 259.

SI Solution: Chiller temperature is 3-6°C below required process temperature. Use – 20°C, note from Figure 16A.1(a) chiller pressure is 240 kPa.

Condenser temperature is 14-16°C above air temperature for aerial cooler. Use 50°C. Note from Figure 16A.1(a) condenser pressure is 1750 kPa.

Point A is a sat. liquid at 50°C, from Figure 16A.1(a) h_A = 33 kJ/kg

Point C is a sat. vapor at – 20°C, from Figure 16A.1(a) h_C = 250 kJ/kg

Refrigerant circulation rate, m:

$$m = \frac{Q_C}{h_C - h_A} = \frac{2000 \text{ kJ/s}}{(250 - 33) \text{ kJ/kg}} = 9.2 \text{ kg/s}$$

Compressor power, W:

$$h_D^{isen} = 341 \text{ kJ/kg @ } s = 5.66 \text{ and } P = 1.75 \text{ MPa}$$

$$-W = \frac{9.2\,(341 - 250)}{0.75} = \underline{\underline{1120 \text{ kW}}}$$

Condenser duty, Q_h:

$$Q_h = W - Q_C = -1120 - 2000 = -3120 \text{ kW}$$

FPS Solution: Chiller temperature is 5-10°F below process temperature. Use –5°F, note from Figure 16A.1(b) chiller pressure is 34 psia..

Condenser temperature is 25-30°F above air temperature. Use 120°F, note from Figure 16A.1(b) condenser pressure is 240 psia..

Point A is a sat. liquid at 120°F, from Figure 16A.1(b) h_A = – 780 Btu/lbm

Point C is a sat. vapor at –5°F, from Figure 16A.1(b) h_C = –686 Btu/lbm

Refrigerant circulation rate, m:

$$m = \frac{Q_C}{h_C - h_A} = \frac{6.8 \times 10^6 \text{ Btu/hr}}{[-780 - (-686)] \text{ Btu/lbm}} = 72\,340 \text{ lbm/hr}$$

Example 16.1 (Cont'd):

Compressor power, W:

$$h_D^{isen} = -646 \text{Btu/lbm} \ @ \ s = \text{slightly less than } 1.36 \text{ and } P = 240 \text{ psia}$$

$$-W = \frac{72\,340[-646-(-686)]}{(2545)(0.75)} = 1515 \text{ hp}$$

Condenser duty, Q_h:

$$Q_h = W - Q_C = (-1515)(2545) - 6\,800\,000 = -10.7 \times 10^6 \text{ Btu/hr}$$

In the above examples, pressure drops were ignored. In an actual system some pressure drop occurs between the compressor discharge and the accumulator as well as between the chiller and compressor suction. These pressure drops increase the required compressor power. Pressure drop on the suction side of the compressor has the greater effect.

Economizer Systems

In the simple refrigeration system presented in Example 16.1, a significant amount of vaporization occurs across the expansion valve. Unlike the multicomponent phase envelopes discussed in Chapter 4, the quality lines inside the two phase region in Figures 16A.1 (a and b) are equidistant. Therefore, it is relatively easy to calculate the quality of a two-phase mixture.

$$x = \text{liquid fraction} = \frac{h_V - h_F}{h_V - h_L} \tag{16.5}$$

$$y = \text{vapor fraction} = \frac{h_F - h_L}{h_V - h_L} \tag{16.6}$$

Where:

h_F = two-phase enthalpy
h_L = saturated liquid enthalpy
h_V = saturated vapor enthalpy

For Example 16.1, $h_F = h_A = h_B$ and $h_V = h_C$.

Example 16.2: Calculate the fraction of liquid entering the chiller in Example 16.1.

SI Solution: $h_F = 33$ kJ/kg, $h_V = 250$ kJ/kg, $h_L = -150$ kJ/kg

$$x = \frac{250-33}{250-(-150)} = 0.54$$

FPS Solution: $h_F = -780$ Btu/lbm , $h_V = -686$ Btu/lbm , $h_L = -860$ Btu/lbm

$$x = \frac{-686-(-780)}{-686-(-860)} = 0.54$$

From Example 16.2, the quantity of vapor entering the chiller is 46%. This vapor provides no refrigeration effect in the chiller, but must be compressed from chiller pressure to condenser pressure, increasing the compressor power requirement.

In refrigeration systems the word "*economizer*" refers generally to any device or process modification which decreases the compressor power requirement for a given chiller duty. Two types of economizers are commonly used — flash tank economizers and heat exchanger economizers.

Figure 16.3 shows a refrigeration system employing one flash tank economizer. In this system, the saturated liquid refrigerant leaving the accumulator is expanded across a valve to an intermediate pressure where vapor and liquid are separated. The separator liquid is expanded across the second valve to chiller pressure while the separator vapor goes to the compressor interstage. The refrigerant entering the chiller now has a higher liquid content. This reduces the refrigerant circulation rate through the chiller as well as the first stage compressor power requirement.

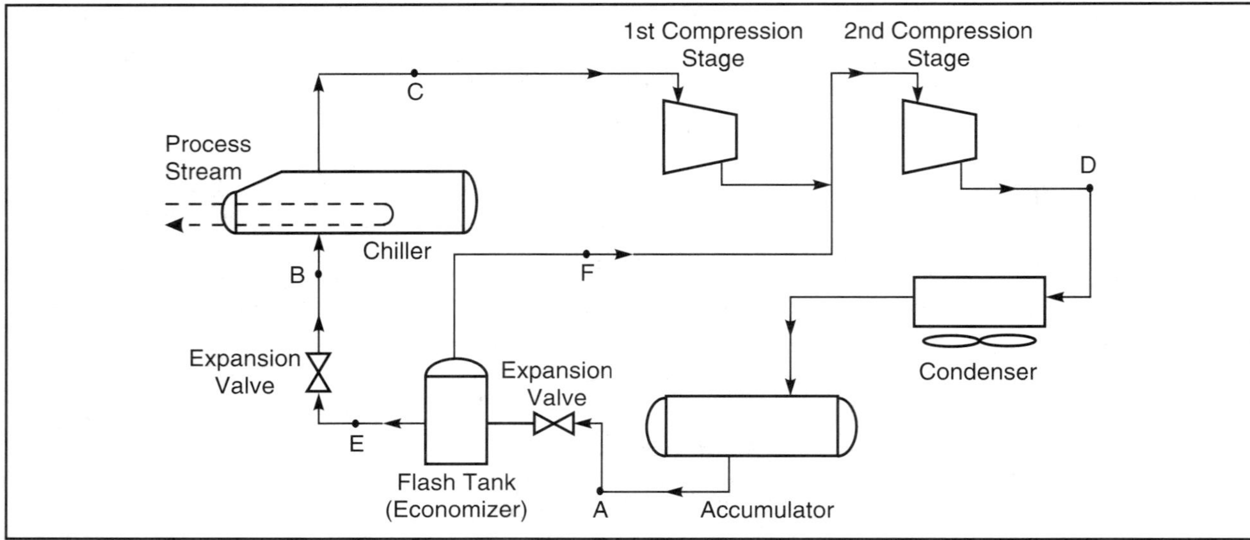

Figure 16.3 Flow Sheet of a Refrigeration System with an Economizer

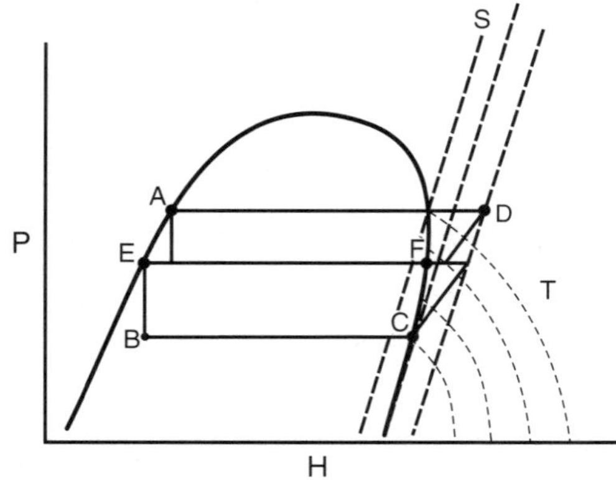

The P-H diagram for the *economizer system* in Figure 16.3 is shown at left. Point E is usually fixed at a pressure that results in equal compression ratios in each stage of compression. Expansion across the economizer expansion valve is isenthalpic. The vapor formed goes to the second stage of compression as a saturated vapor. The saturated liquid, E, leaving the separator is expanded isenthalpically to pressure B. Notice that the Δh available from B to C for this system is larger than for the simple system.

Calculation of Flash Tank Economizer Systems

Determination of circulation rate, work and Q_{cond} for the system in Figure 16.3 follows the same pattern as for the simple system.

The first step is the same in principle. Now, however, the enthalpy of the refrigerant entering the chiller valve is determined by compressor interstage pressure and not by condenser pressure. With this change in h_A, Equation 16.1 may be solved for "m_1," the circulation rate to the chiller (and the amount of vapor to be compressed through Stage 1 of the compressor).

What is the refrigeration circulation rate through the second stage? It is determined from a material balance around the economizer flash tank. The value of m_1 is known from the circulation rate calculation at the chiller.

The second stage circulation rate, m_2, may be calculated from the relationship

$$m_2 = \frac{m_1}{x} \qquad (16.7)$$

where x is the liquid fraction in the economizer vessel and is calculated from Equation 16.5.

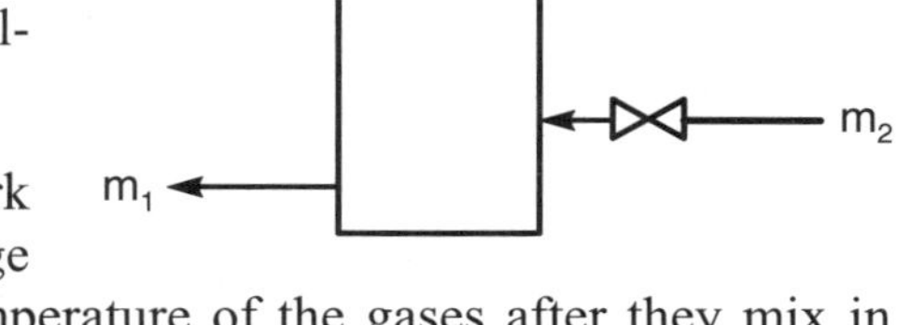

In the compressor power calculation, one finds the work in the first stage (for flowrate "m_1") and adds it to second stage work (for flowrate "m_2") to find total work. The resultant temperature of the gases after they mix in the tee between stages may be found by an energy balance around that tee. In most cases, the temperature effect here is small.

The condenser heat load, Q_{cond}, is found as before.

Flash tank economizer systems are not limited to two stages (one economizer). Many installations employ two economizers (three compression stages). An example of such a system is shown for the propane portion of the cascade refrigeration system in Figure 16.8.

Another advantage of a flash economizer system is the ability to install chillers on the interstages. Figure 16.8 shows this capability. Rather than provide all of the refrigeration at – 40°C [– 40°F], chillers are installed on the low-stage economizer (–19.9°C [– 3.9°F]) and high-stage economizer (6.6°C [43.9°F]). While this scheme is not often used in a straightforward gas chilling plant it is frequently used in more complex processes employing significant heat integration.

Example 16.3: Rework the simple refrigeration system in Example 16.1 assuming a two-stage system with one flash tank economizer.

SI Solution: Estimate economizer pressure assuming equal compression ratios per stage.

$$P_{econ} = P_1\left(\frac{P_2}{P_1}\right)^{0.5} = [(P_1)(P_2)]^{0.5} = [(1750)(240)]^{0.5}$$
$$= 648 \text{ kPa (use 650 kPa)}$$

Note from Figure 16A.1(a) the economizer temperature is approximately 10°C.

1st Stage: Refrigerant circulation rate, m_1

$$m_1 = \frac{Q_C}{h_C - h_B}$$

$h_C = 250$ kJ/kg (sat. vapor @ −20°C)
$h_B = -75$ kJ/kg (sat. liquid @ 650 kPa)

$$m_1 = \frac{2000}{250-(-75)} = 6.15 \text{ kg/s}$$

Compressor power, W_1

refrigerant is compressed in 1st Stage from 240 kPa to 650 kPa

$$-W_1 = \frac{m\,\Delta h^{isen}}{E} = \frac{6.15(295-250)}{0.75} = 369 \text{ kW}$$

2nd Stage: Refrigerant circulation rate, m_T

@ economizer (650 kPa)

$$x = \frac{h_V - h_F}{h_V - h_L} = \frac{282-33}{282-(-75)} = 0.70$$

$$m_2 = \frac{m_1}{x} = \frac{6.15}{0.7} = 8.75 \text{ kg/s}$$

Compressor power, W_2

Note: It is difficult to locate the actual second stage suction point on the P-H diagram. For purposes of this exercise it has been assumed that Δh_2^{isen} is 45 kJ/kg. This results in a total Δh^{isen} of 91 kJ/kg which is identical to the single stage system.

$$-W_2 = \frac{m\,\Delta h^{isen}}{E} = \frac{(8.79)(45)}{0.75} = 539 \text{ kW}$$

Total compressor power = 369 + 537 = 908 kW, a reduction of 18.9% from the simple system.

Condenser duty, Q_h

$$Q_h = W - Q_C = -904 - 2000 = \underline{\underline{-2908 \text{ kW}}}$$

Example 16.3 (Cont'd.):

FPS Solution: Estimate economizer pressure assuming equal compression ratios per stage.

$$P_{econ} = P_1\left(\frac{P_2}{P_1}\right)^{0.5} = [(P_1)(P_2)]^{0.5} = [(240)(34)]^{0.5}$$

$$= 90 \text{ psia}$$

Note from Figure 16A.1(b) the economizer temperature is approximately 50°F.

1st Stage: Refrigerant circulation rate, m_1

$$m_1 = \frac{Q_C}{h_C - h_B} \qquad \begin{array}{l} h_C = -686 \text{ Btu/lbm (sat. vapor @ } -5°\text{F)} \\ h_B = -828 \text{ Btu/lbm (sat. liquid @ 90 psia)} \end{array}$$

$$m_1 = \frac{6.8\times10^6}{-686-(-828)} = 47\,900 \text{ lbm/hr}$$

Compressor power, W_1

refrigerant is compressed in 1st Stage from 34 psia to 90 psia

$$-W_1 = \frac{m\,\Delta h^{isen}}{(2545)(E)} = \frac{47\,900\,[-667-(686)]}{(2545)(0.75)} = 477 \text{ hp}$$

2nd Stage: Refrigerant circulation rate, m_T

@ economizer (90 psia)

$$x = \frac{h_V - h_F}{h_V - h_L} = \frac{(-673)-(-780)}{(-673)-(828)} = 0.69$$

$$m_2 = \frac{m_1}{x} = \frac{47\,900}{0.69} = 69\,420 \text{ lbm/hr}$$

Compressor power, W_2

Note: It is difficult to locate the actual second stage suction point on the P-H diagram. For purposes of this exercise it has been assumed that Δh_2^{isen} is 21 Btu/lbm. This results in a total Δh^{isen} of 40 Btu/lbm which is identical to the single stage system.

$$-W_2 = \frac{m\,\Delta h^{isen}}{E} = \frac{(69\,400)(21)}{(2545)(0.75)} = 764 \text{ hp}$$

Total compressor power = 477 + 764 = 1241 hp, a reduction of 18.1% from the simple system.

Condenser duty, Q_h

$$Q_h = W - Q_C = (-1241)(2545) - 6\,800\,000 = \underline{\underline{-9.96\times10^6 \text{ Btu/hr}}}$$

Heat Exchanger Economizer

A second type of economizer configuration is the heat exchanger economizer shown in Figure 16.4. Cold, low-pressure chiller vapor is used to subcool the saturated liquid refrigerant. This decreases the refrigerant circulation rate, and may reduce compressor power.

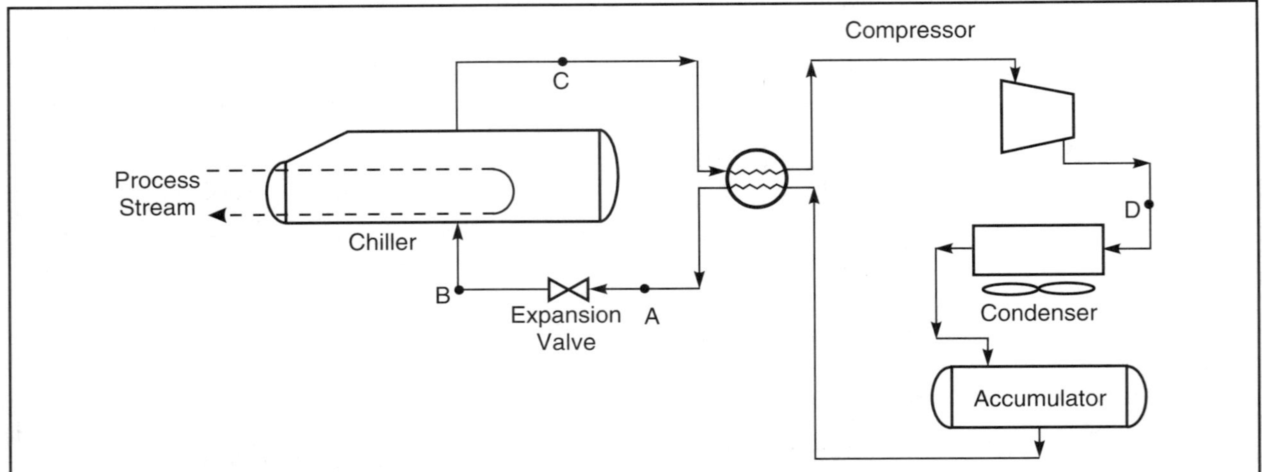

Figure 16.4 Flow Sheet of a Simple Refrigeration System with a Heat Exchanger Economizer

With regard to the compressor power, two factors offset the reduced circulation rate. The first is exchanger pressure drop. The pressure drop on the low pressure side of the exchanger may be 20-50 kPa [3-7 psi]. This can significantly increase compressor power, particularly when the chiller pressure is near atmospheric. Secondly, the refrigerant vapor entering the compressor is now superheated. Although this reduces the likelihood of liquid carryover into the compressor, it results in higher power consumption per unit mass due to the higher suction temperature. This can be seen in Figures 16A.1(a) and 16A.1(b). Note the constant entropy lines are less steep at higher temperatures.

Calculation of Chiller Load ($Q_{chiller}$)

The chiller duty, $Q_{chiller}$, is set by process requirements and is determined by the energy balance calculations presented in Chapters 8 and 13. Figure 13.17 on page 93 of Chapter 13 is a good example of a refrigeration application for natural gas chilling.

Specification of Cold Separator Temperature

The desired temperature of the gas-liquid stream leaving the chiller must be determined by the strategy governing the system. If hydrocarbon dewpoint control is the primary process objective and liquid recovery secondary, this temperature is often 3-5°C [5-9°F] below the temperature required to achieve the specified dewpoint to account for liquid carryover out of the separator. This required temperature may be significantly colder than the specified dewpoint if the gas is processed at high pressure. (See Chapter 4, page 103)

If liquid recovery is the primary function of the unit, what products can be sold? The basic strategy is to condense the least amount of nonsaleable components (usually methane and/or ethane) compatible with the economic criteria. Anything condensed, like methane, that must be revaporized and maybe also recompressed, adds to the operating costs without contributing to liquid revenue.

For a given set of specifications, one should investigate a series of cold separator pressures and temperatures. The pressure for maximum liquid recovery is between 3.0-4.0 MPa [435-580 psia], if C_3+ is the saleable product. As the pressure increases, condensation of methane, ethane, and CO_2 increases. However, 3.0-4.0 MPa [435-580 psia] may not be the optimum pressure economically because of compression requirements. The optimum pressure must minimize total system cost, not merely that of the refrigeration and stabilization system alone. Separation is frequently carried out at the sales gas pressure to eliminate recompression.

For the usual pressures chosen, what is a reasonable temperature? As noted before, this depends on the products desired. If the liquid product must have an RVP less than 1 atm. or is blended with crude oil or condensate sold as crude oil, the common separation temperature may be – 20 to 0°C [– 4 to 32°F]. When propane is the lightest saleable liquid, the temperature may be – 40 to – 20°C [– 40°F to – 4°F].

Choice of temperature (and pressure) is a critical specification. Do not choose arbitrarily? Calculate the economic criteria for a series of conditions and choose the optimum one.

Choice of Refrigerant

The ideal refrigerant is nontoxic, noncorrosive, has PVT and physical properties compatible with the system needs, and has a high latent heat of vaporization. Many substances could be used as a refrigerant. The practical choice reduces to one which has desirable physical properties and will vaporize and condense at reasonable pressures, at the temperature levels desired. In gas processing applications the usual choice is propane, ammonia, or R-22 at chiller temperatures above about – 40°C [–40°F]. At cryogenic conditions, ethylene and methane might be used. In general, the lower practical limit of any refrigerant is its atmospheric boiling point. It is desirable to carry some positive pressure on the chiller to obtain better efficiency in the compressor, reduce equipment size and avoid air induction into the system. Vacuum operation is feasible however, and has been successfully implemented in some systems.

Different refrigerants will exhibit somewhat different power requirements per unit of chiller duty. These differences are typically small, $< 10\%$. The primary factors in refrigerant selection are initial and replacement cost, availability, health and safety considerations and environmental impact.

Propane is by far the most popular refrigerant in the gas processing applications. It is readily available (often manufactured on-site), inexpensive and has a "good" vapor pressure curve. It is flammable but this is not a significant problem if proper consideration is given to the design and operation of the facility.

Freons are widely used as commercial refrigerants. They are non-toxic and nonflammable. CFC (Chlorofluorocarbon) refrigerants like R-11 and R-12 have been phased out due to environmental problems. The replacement for R12 is R-134a. HCFC refrigerants (Chlorofluorocarbons containing at least one hydrogen) are currently considered environmentally acceptable in most locations but the phase-out of some of these refrigerants has been mandated. R-22, an HCFC, is the most commonly used Freon in gas processing. Freons are expensive and system losses can represent a significant operating cost. They are also difficult to ship to remote locations in large quantities. Certain Freons will form hydrates so it is necessary to keep the system dry.

Many replacements for pure CFC or HCFC refrigerants are mixtures. Some of these are aezeotropes (these are designated R-5xx) and some are zeotropes (designated R-4xx). Aezeotropes act as a pure component in the refrigeration system. Zeotropes do not. Zeotropes do not have a constant boiling temperature at a fixed pressure, i.e. the isotherms are not horizontal inside the phase envelope. The temperature difference between the saturated liquid and saturated vapor is referred to as "temperature glide" and may be 3-5°C [5-9°F]. For kettle-type chillers, this effect increases compressor power relative to pure refrigerants. As of this writing the only replacement CFC/HCFC/HFC refrigerants developed for R-22 are zeotropes.

Table 16.1 presents the approximate compressor power requirement and condenser duty for propane, ammonia, and R-22 at various chiller and condenser pressures.

TABLE 16.1

Comparison of Common Refrigerants

Evaporator Temperature, °C ⇒		**-46**	**-40**	**-34**	**-29**	**-23**	**-18**	**-12**	**-6**	**-1**	**4**
Evaporator pressure in kPa	Ammonia	52.9	71.6	95.8	126	163	210	265	331	411	503
	Propylene	111	143	179	221	269	331	400	482	568	662
	Propane	86.8	112	141	176	216	263	320	386	464	551
Condensed Liquid Temperature 35°C; Condenser Pressure in kPa: Ammonia 1357; Propylene 1461; Propane 1220											
kg refrigerant per second per MW of refrigeration	Ammonia	0.967	0.958	0.950	0.941	0.935	0.928	0.922	0.915	0.911	0.907
	Propylene	4.13	4.02	3.93	3.85	3.78	3.70	3.63	3.57	3.50	3.44
	Propane	4.26	4.15	4.04	3.93	3.83	3.74	3.67	3.59	3.50	3.42
m^3/h of refrigerant per MW of refrigeration	Ammonia	7197	5361	4057	3115	2415	1913	1512	1217	985	819
	Propylene	5796	4434	3526	2826	2289	1855	1502	1222	1024	869
	Propane	7487	5796	4526	3521	2797	2304	1869	1507	1251	1029
Kilowatt per MW of refrigeration	Ammonia	793	685	594	511	441	377	318	267	218	177
	Propylene	744	657	570	498	437	369	310	254	212	176
	Propane	727	642	566	492	430	371	316	263	214	170
Condensed Liquid Temperature 52°C; Condenser Pressure in kPa: Ammonia 2088; Propylene 2164; Propane 1792											
kg refrigerant per second per MW of refrigeration	Ammonia	1.047	1.038	1.027	1.019	1.008	1.001	0.995	0.989	0.982	0.976
	Propylene	5.201	5.050	4.900	4.771	4.642	4.534	4.427	4.319	4.234	4.148
	Propane	5.437	5.244	5.072	4.921	4.771	4.642	4.513	4.384	4.277	4.169
m^3/h of refrigerant per MW of refrigeration	Ammonia	7760	5796	4390	3367	2608	2058	1633	1314	1058	879
	Propylene	7438	5555	4540	3536	2826	2280	1840	1488	1256	1053
	Propane	9515	7438	5699	4424	3478	2869	2314	1831	1517	1270
Kilowatt per MW of refrigeration	Ammonia	1052	929	808	706	619	538	464	403	346	293
	Propylene	1128	999	878	776	685	591	511	430	377	329
	Propane	1098	975	861	761	674	596	515	439	375	318

Evaporator Temperature, °F ⇒		**-50**	**-40**	**-30**	**-20**	**-10**	**0**	**10**	**20**	**30**	**40**
Evaporator pressure in psia	Ammonia	7.67	10.4	13.9	18.3	23.7	30.4	38.5	48.0	59.7	73.0
	Propylene	16.1	20.7	26.0	32.1	39.0	48.0	58.0	70.0	82.5	96.0
	Propane	12.6	16.2	20.5	25.5	31.3	38.1	46.4	56.0	67.3	80.0
Condensed Liquid Temperature 95°F; Condenser Pressure in psia: Ammonia 197; Propylene 212; Propane 177											
Pounds of refrigerant per minute per ton of refrigeration	Ammonia	0.450	0.446	0.442	0.438	0.435	0.432	0.429	0.426	0.424	0.422
	Propylene	1.92	1.87	1.83	1.79	1.76	1.72	1.69	1.66	1.63	1.60
	Propane	1.98	1.93	1.88	1.83	1.78	1.74	1.71	1.67	1.63	1.59
CFM of refrigerant per minute per ton of refrigeration	Ammonia	14.90	11.10	8.40	6.45	5.00	3.96	3.13	2.52	2.04	1.70
	Propylene	12.00	9.18	7.30	5.85	4.74	3.84	3.11	2.53	2.12	1.80
	Propane	15.50	12.00	9.37	7.29	5.79	4.77	3.87	3.12	2.59	2.13
Brake horsepower per ton of refrigeration	Ammonia	3.74	3.23	2.80	2.41	2.08	1.78	1.50	1.26	1.03	0.835
	Propylene	3.51	3.10	2.69	2.35	2.06	1.74	1.46	1.20	1.00	0.830
	Propane	3.43	3.03	2.67	2.32	2.03	1.75	1.49	1.24	1.01	0.800
Condensed Liquid Temperature 125°F; Condenser Pressure in psia: Ammonia 303; Propylene 314; Propane 260											
Pounds of refrigerant per minute per ton of refrigeration	Ammonia	0.487	0.483	0.478	0.474	0.469	0.466	0.463	0.460	0.457	0.454
	Propylene	2.42	2.35	2.28	2.22	2.16	2.11	2.06	2.01	1.97	1.93
	Propane	2.53	2.44	2.36	2.29	2.22	2.16	2.10	2.04	1.99	1.94
CFM of refrigerant per ton of refrigeration	Ammonia	16.10	12.00	9.09	6.97	5.40	4.26	3.38	2.72	2.19	1.82
	Propylene	15.40	11.50	9.40	7.32	5.85	4.72	3.81	3.08	2.60	2.18
	Propane	19.70	15.40	11.80	9.16	7.20	5.94	4.79	3.79	3.14	2.63
Brake horsepower per ton of refrigeration	Ammonia	4.96	4.38	3.81	3.33	2.92	2.54	2.19	1.90	1.63	1.38
	Propylene	5.32	4.71	4.14	3.66	3.23	2.79	2.41	2.03	1.78	1.55
	Propane	5.18	4.60	4.06	3.59	3.18	2.81	2.43	2.07	1.77	1.50

Compressor choice is linked to refrigerant choice as well as other considerations. Where weight and size are particularly important, a centrifugal or a screw compressor may be used. A reciprocating compressor is an alternative for accessible land locations. Lube oil selection is particularly important for positive displacement machines. Lube oil carried into the refrigerant can congeal on chiller tubes reducing heat transfer rate.

Effect of Temperature on Cost

Figure 16.5 shows the approximate relative effect of temperature on refrigeration cost. The inset to Figure 16.5 shows the refrigerant often used at various temperature levels. The temperature levels are approximate. Actually, the refrigerant used at a given level depends on economic criteria, which will vary in different circumstances.

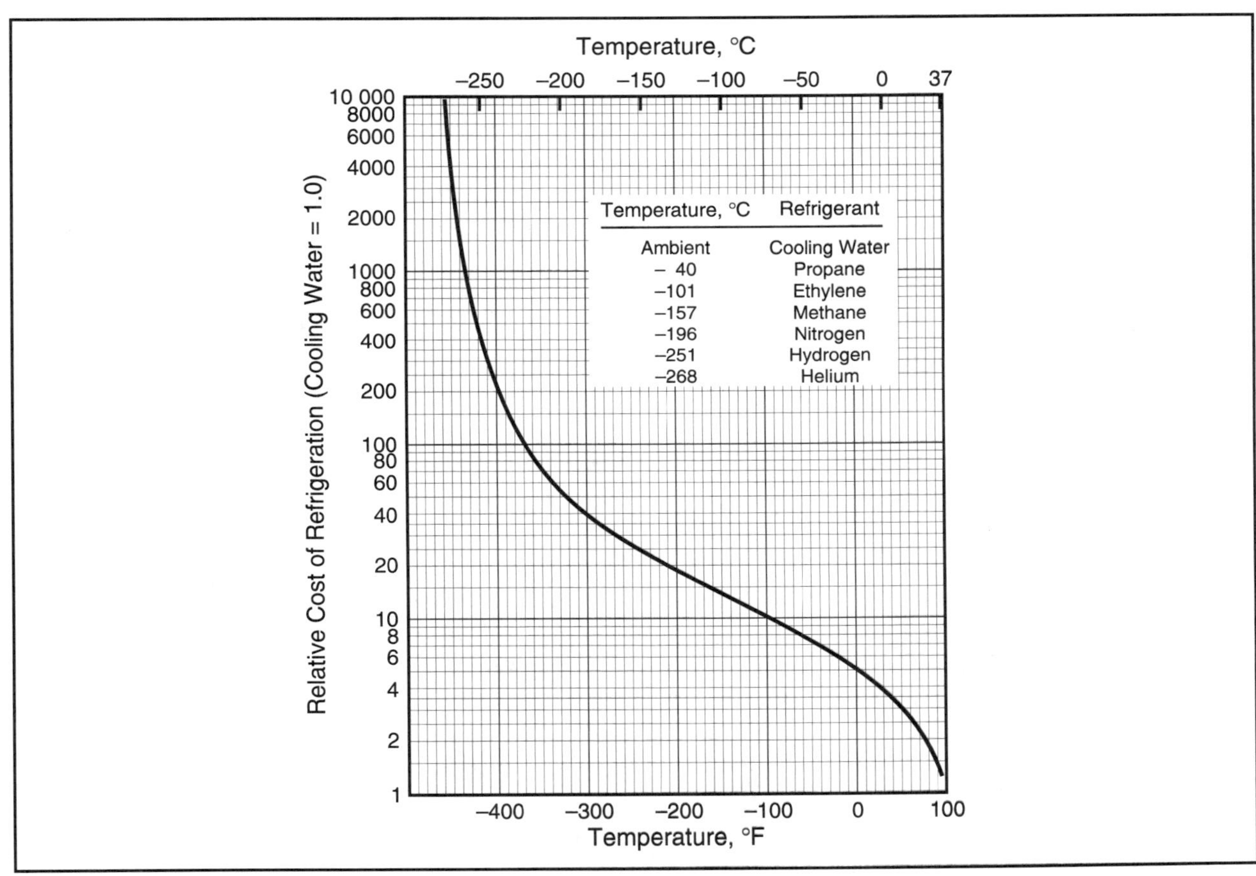

Figure 16.5 Relative Cost of Refrigeration Compared to the Cost of Cooling with Water at Ambient Conditions

Operating Issues

Refrigeration power, hence operating cost, is minimized by minimizing the condensing temperature and operating the chiller as close as possible to the process fluid temperature, i.e. minimum approach. Generally, the condensing medium will be water or air, but in some applications a cold process fluid may provide a lower temperature condensing medium. Some gas processing facilities use the sales gas as the cooling fluid in the condenser, particularly if the inlet gas to the facility is below ambient temperature due to ground or seabed cooling.

Chiller fouling increases the required temperature difference between the process fluid and the refrigerant, and forces the refrigeration system to work harder to deliver the same amount of refrigeration. Fouling can occur on the refrigerant side of the chiller due to contaminants in the refrigerant — usually lube or seal oil from the compressor. Fouling can occur on the process fluid side due to hydrates, wax, or freezing of TEG carried over from a glycol dehydration system.

In climates where significant ambient temperature variations occur between summer and winter, it may be necessary to restrict the condenser cooling on the coldest winter days. There are two reasons for this. First, sufficient pressure differential must exist between the accumulator and the chiller to cause the refrigerant to flow through the piping and valves. This is particularly true in economizer systems. Secondly, if the refrigerant compressor is a centrifugal, a low discharge pressure may cause the compressor to operate in stonewall.

The most common operating problems in refrigeration systems are refrigerant losses, usually due to leaks and refrigerant contamination. Refrigerant losses occur due to leaks at flanges, valve packing, relief valves, etc. If Freons are used, installation of rupture disks below the relief valves can help reduce losses. In addition to lubricating oil, refrigerants can be contaminated by non condensables, usually methane or air. Methane can contaminate the refrigerant due to a leak in the chiller. Air can enter the system if the suction side of the compressor operates at vacuum. Both reduce the efficiency of the refrigeration system because they artificially increase condensing pressure.

The primary objective of refrigeration system control is to cool the process fluid to the desired temperature (in some cases the desired temperature is the coldest temperature attainable given the compressor power and/or metallurgical constraints). The primary load change on the system is typically chiller duty, usually due to process fluid rate changes.

The amount of heat transferred in the chiller is given by the relationship

$$Q = U\,A\,\Delta T_m$$

For control purposes, either area, A or temperature difference, ΔT_m can be manipulated to control Q. Area is manipulated by changing the level of refrigerant in the chiller and temperature difference by changing chiller pressure (hence the boiling temperature of the refrigerant). The latter is preferred in most applications. A typical control scheme is shown in Figure 16.6.

Compressor capacity is controlled using the methods presented in Chapter 15.

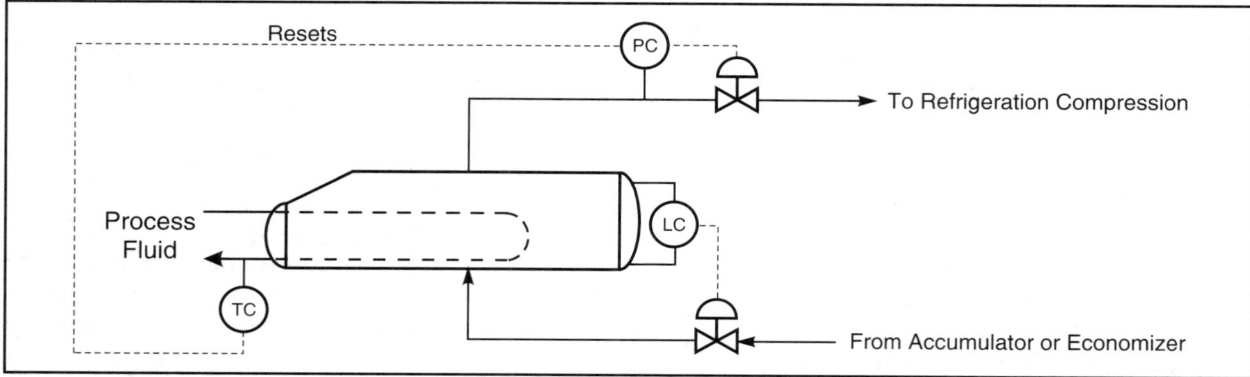

Figure 16.6 Example Control Scheme for a Chiller

Anti-surge control for centrifugal type refrigeration compressors is somewhat different than the natural gas compression due to the phase behavior of the refrigerant. Cooling of the recycle condensation causes condensation of the refrigerant. A common recycle control scheme is shown in Figure 16.7.

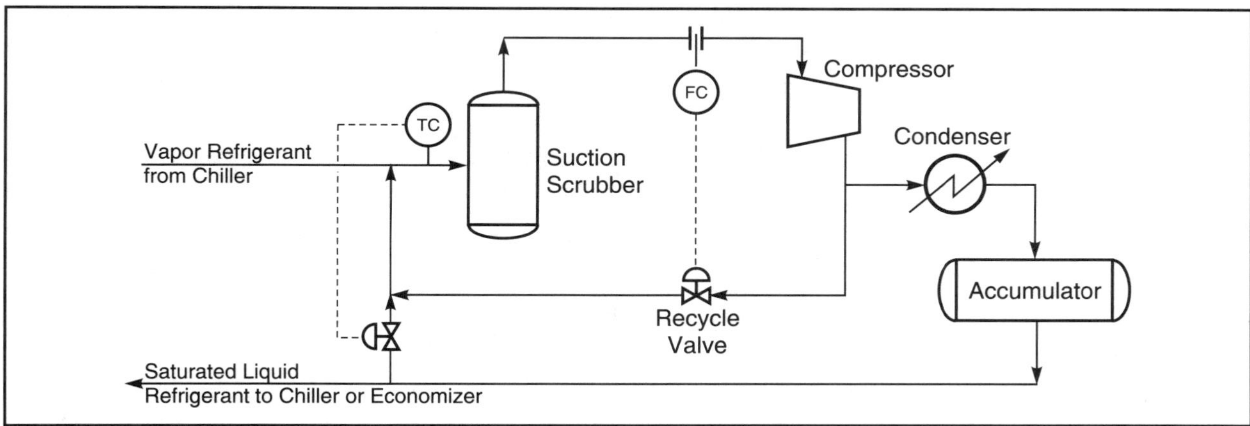

Figure 16.7 Anti-Surge Control on a Refrigeration System

The primary recycle flow comes from the hot discharge gas leaving the compressor. Without cooling this would quickly cause a high discharge temperature trip on the compressor when operating in recycle. Liquid refrigerant is injected into the recycle line to cool the recycle gas. The liquid rate is controlled by the suction temperature.

Cascade Refrigeration

When refrigeration must be provided at very low temperatures, below about < – 40°C [– 40°F], cascade refrigeration systems are sometimes used. Cascade systems employ more than one refrigerant and provide refrigeration at multiple levels. A propane/ethane cascade system is shown in Figure 16.8.

In this system, refrigeration is provided at five levels

°C	°F
7	44
– 20	– 4
– 40	– 40
– 61	– 78
– 84	– 120

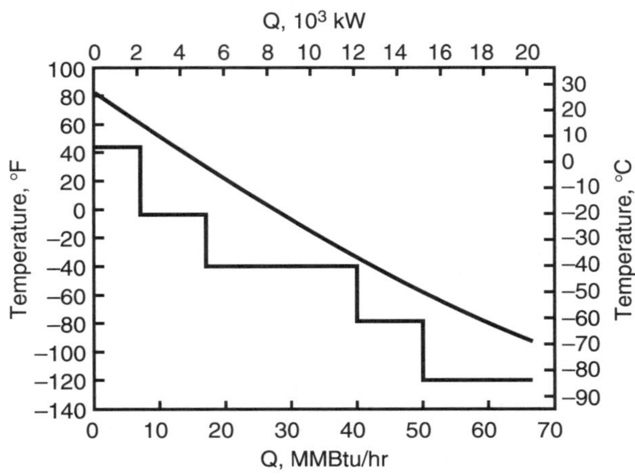

The propane at – 40°C is used to condense the ethane refrigerant. All of the heat picked up in the process is ultimately rejected to the cooling water at the propane condenser. A hypothetical cooling curve for the process fluid has been developed to show the amounts and levels of refrigeration.

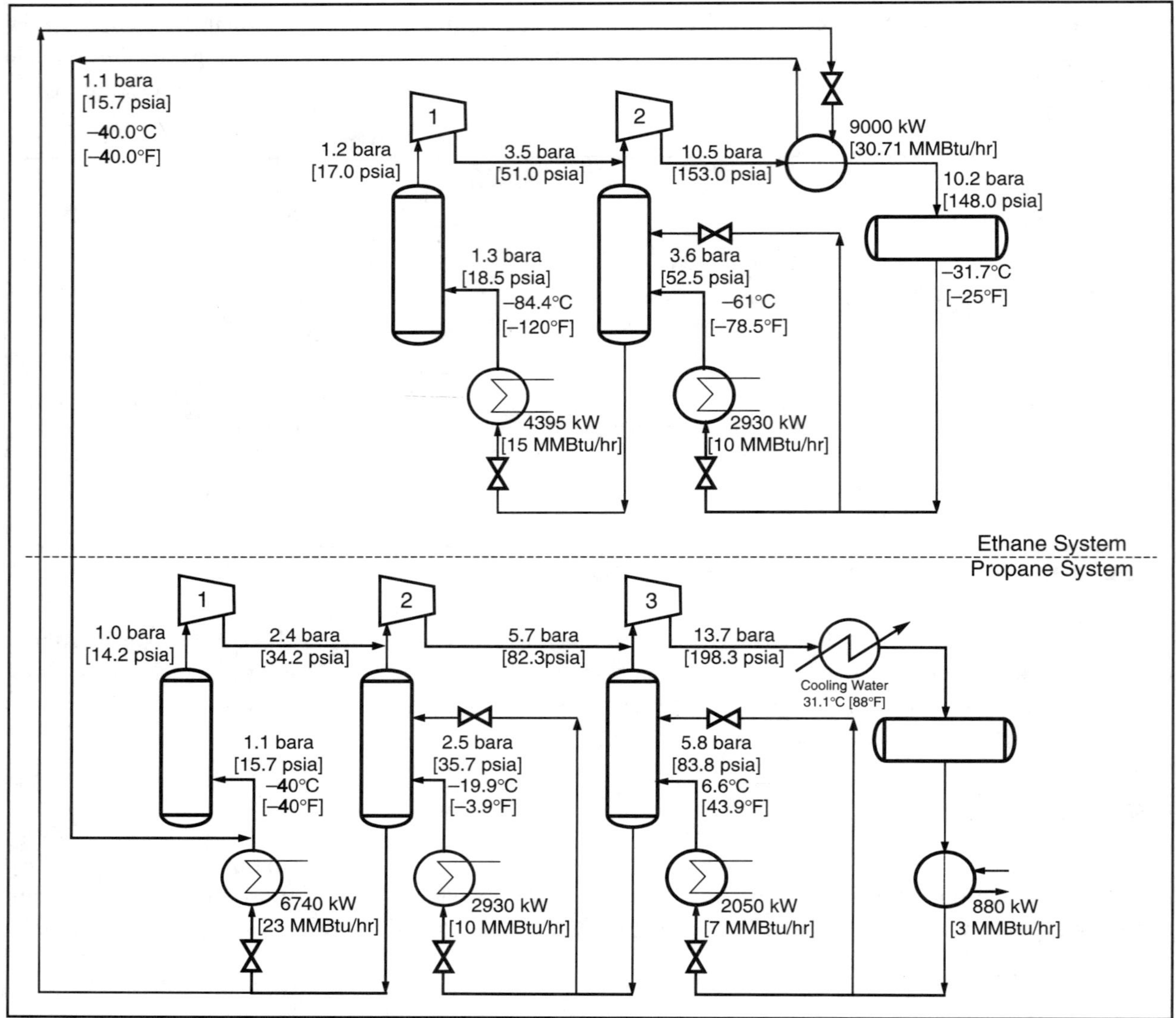

Figure 16.8 Ethane-Propane Cascade Refrigeration System *(Courtesy GPSA Databook)*

Cascade refrigeration systems employing propane, ethylene, and methane are used in ethylene plants and some LNG processes but are generally not common in gas processing. Low level refrigeration is typically provided using expansion or mixed refrigerants.

Mixed Refrigerants

An alternative to cascade refrigeration is to use a mixed refrigerant. Mixed refrigerants are a mixture of two or more components. The light components lower the evaporation temperature and the heavier components allow condensation at ambient temperatures. The evaporation process takes place over a temperature range rather than at a constant temperature as with pure component refrigerants. This is illustrated in the P-H diagram (Figure 16.9) for a mixed refrigerant.

The composition of the mixed refrigerant is sometimes adjusted so that its evaporation curve matches the cooling curve for the process fluid. Heat transfer occurs in a countercurrent exchanger, often an aluminum plate-fin, rather than a kettle-type chiller.

Mixed refrigerants have a somewhat better thermal efficiency than a cascade system because refrigeration is always being provided at the warmest possible temperature.

The amount of equipment is also reduced compared to a cascade system. Disadvantages include a more complex design and the tendency for the heavier components to concentrate in the chiller unless the refrigerant is totally vaporized. The mixed refrigerant process is used in some low-temperature NGL extraction gas processing plants and is the most common process in large LNG plants. LNG processes will be discussed in more detail later in this chapter.

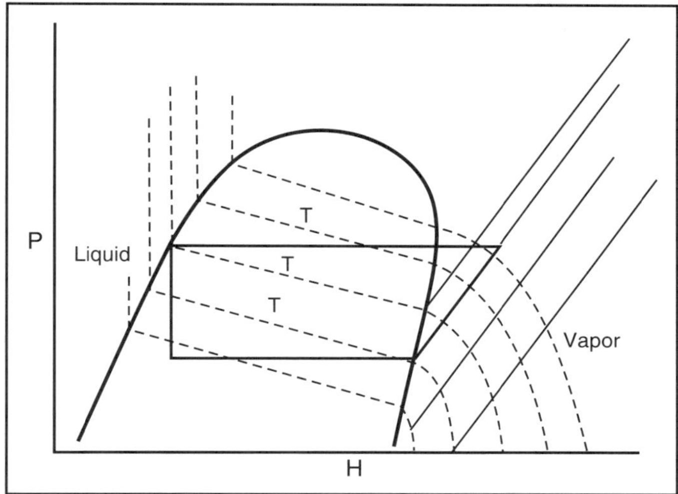

Figure 16.9 Example P-H Diagram for a Mixed Refrigerant

EXPANSION REFRIGERATION

Two types of expansion refrigeration processes are used in gas processing.

1. Valve expansion — often referred to as LTS, LTX or JT process
2. Turbine expansion — turboexpanders

A third process type involves the expansion of gas through a tube where the gas attains sonic or supersonic velocity and liquids are removed from the tube wall. These processes will be briefly discussed in a subsequent section, but are of limited commercial importance at present.

Valve expansion processes have been used in the industry for over 50 years. They are simple, robust and relatively inexpensive, but have a low thermodynamic efficiency and generally require high pressure feed gas (10 000 kPa [1500 psia] or higher) to be economically viable. These processes are most commonly used in processing of gas to meet a hydrocarbon dewpoint.

The turboexpander was first used in gas processing in the late 1960s. It is now arguably the most popular gas processing method. Turboexpanders can be used for very low temperature NGL extraction as well as for hydrocarbon dewpoint control. Turboexpanders (expanders) are very efficient, reliable and relatively inexpensive.

VALVE EXPANSION

Figure 16.10 shows an example LTS process. Feed gas enters a gas-gas exchanger where cooling is recovered from the cold sales gas stream. The gas is often not dehydrated, so hydrate formation in the exchanger and piping must be prevented. This is typically done by injecting an inhibitor (usually MEG) into the gas upstream of the exchanger.

Liquid hydrocarbons and the water-inhibitor mixture are removed from the gas in the Cold Separator. The cold, dry sales gas flows back through the gas-gas exchanger and on to the sales gas system.

The temperature at the cold separator is usually controlled by bypassing a portion of cold sales gas around the gas-gas exchanger.

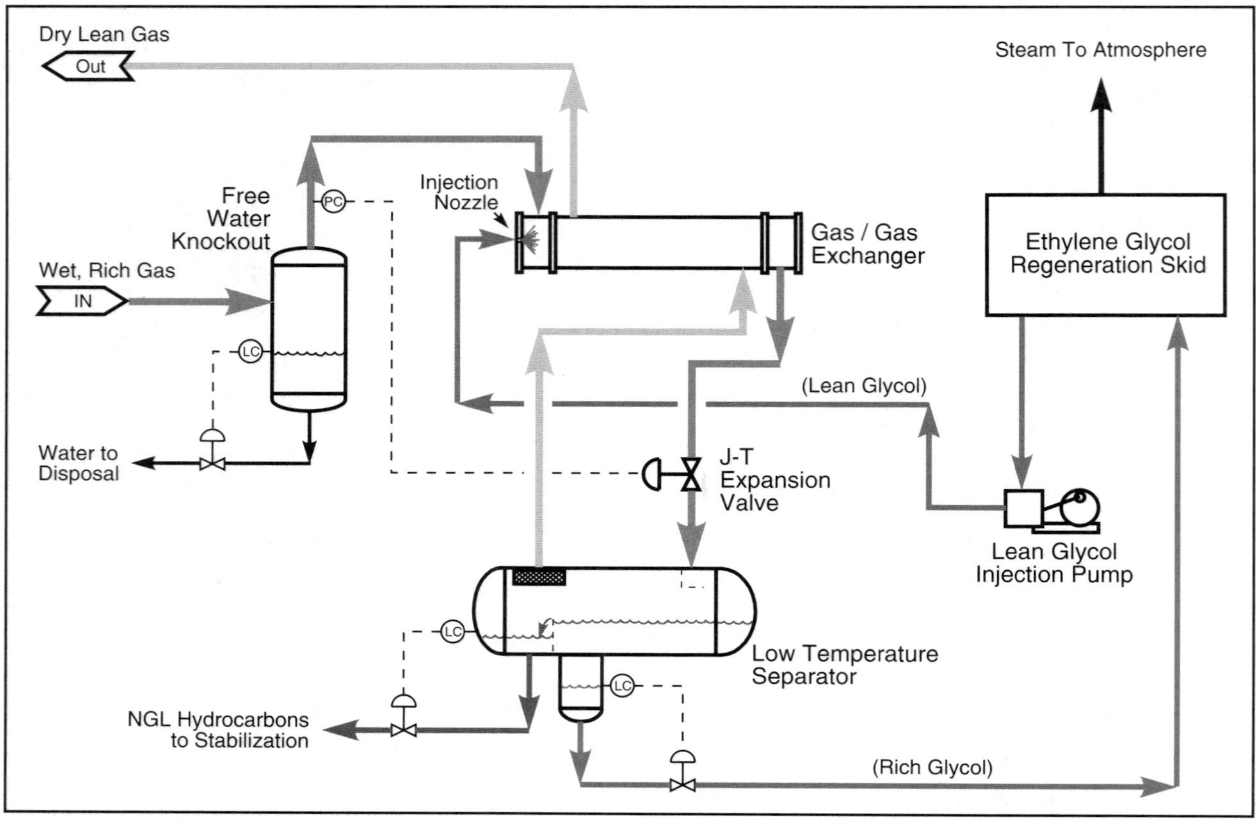

Figure 16.10 Process Flow Diagram for a Typical LTS Process

An LTX process (as opposed to LTS) generally refers to a process where no hydrate inhibitor is injected into the gas. The process is designed so that any hydrate formation occurs downstream of the J-T valve and the cold separator liquid is heated by heat exchange with a warm fluid to melt any hydrates which formed as a result of the valve expansion. The warm fluid is often the inlet gas to the process. LTX processes will not achieve as low a temperature as LTS, but they are simpler and less expensive. They have proven useful for field processing of gas to meet transportation or downstream processing specifications. Figure 16.11 shows an example LTX process.

Pressure drop across a valve is an isenthalpic process, as noted previously. If no liquid forms, the following equation applies:

$$\mu = \left(\frac{\partial T}{\partial P}\right)_H = \frac{\left[T\left(\frac{\partial V}{\partial T}\right)_P - V\right]}{C_P} \tag{16.8}$$

The symbol "μ" is known as the Joule-Thomson coefficient. It is positive or negative, depending on the relative size of the two terms in the numerator.

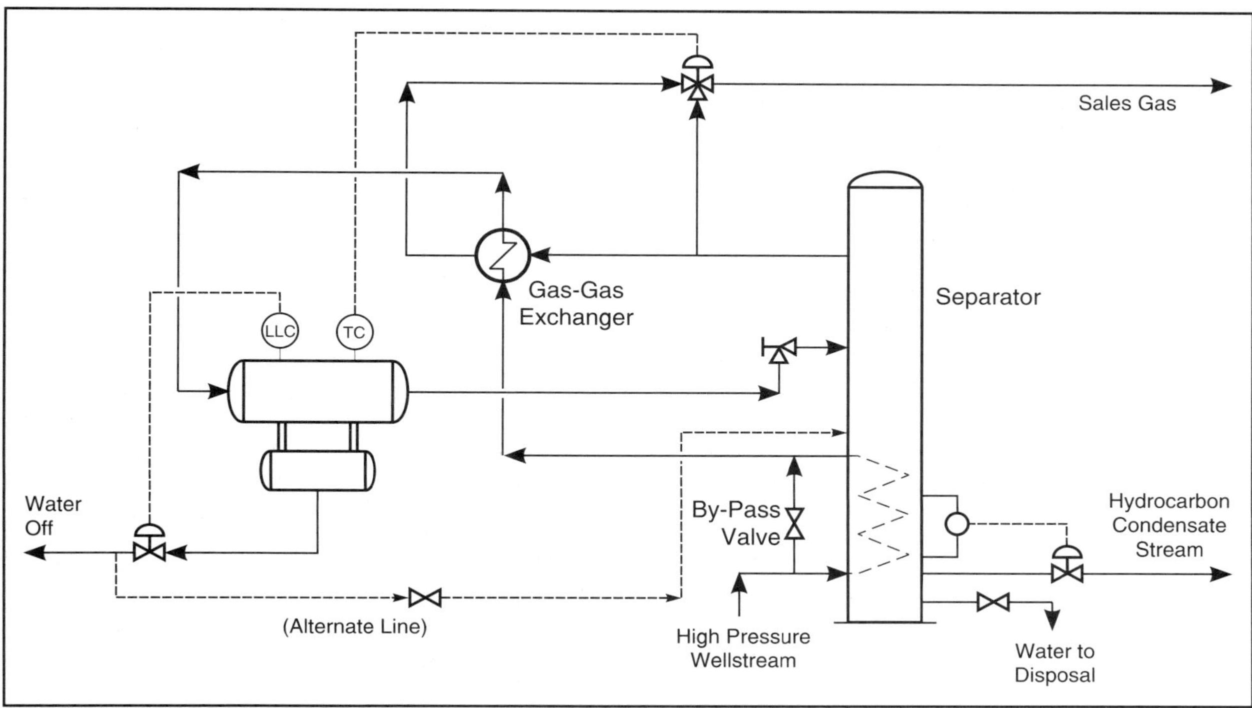

Figure 16.11 Example LTX Process

Curve A below shows a case where the instantaneous slope is greater than the average slope. Therefore, the gas will cool on expansion. The curve C gas is just the opposite and will heat on expansion. Curve B is for an ideal gas, which will not change temperature on expansion.

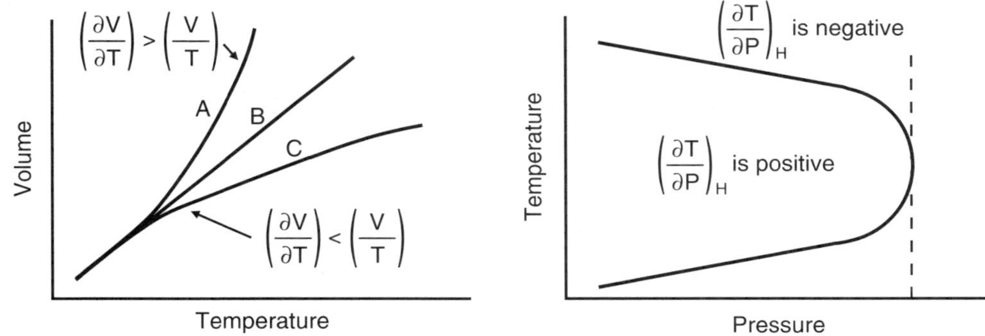

Many gases exhibit a characteristic wherein the slope of the V-T curve changes sign. The temperature at which the slope changes sign ($\mu = 0$) is known as the *inversion temperature.* The right-hand plot above shows inversion temperature versus pressure. The shape shown is general for all actual gases. Outside the curve, the gas represented would heat upon expansion. Inside, it cools on expansion.

Because of the location of the curve, hydrogen heats on expansion at normal pressures, whereas most light hydrocarbons cool. At very high pressures, of the order of 60 MPa [8700 psia], many naturally occurring hydrocarbon gases heat on expansion.

Curves which show the temperature drop expected for a given pressure drop across a valve or choke are only applicable if no liquid forms on such expansion.

Liquid Condensation Across a Valve

When condensation occurs, the calculation of the valve outlet conditions is inherently trial-and-error.

1. Calculate the specific enthalpy of the feed stream at P_1 and T_1. If it is a two-phase stream, the enthalpy is found by the weighted average that of the liquid and vapor phases.
2. Assume the unknown temperature T_2.
3. Make a flash calculation at P_2 and T_2 to find relative amount and analysis of each phase.
4. Find the specific enthalpy at Point Two from the above flash and the assumed T_2.
5. If $h_2 = h_1$, you assumed the right temperature. If not, repeats Steps 2-5 until the h's are equal, within the desired limits of accuracy.

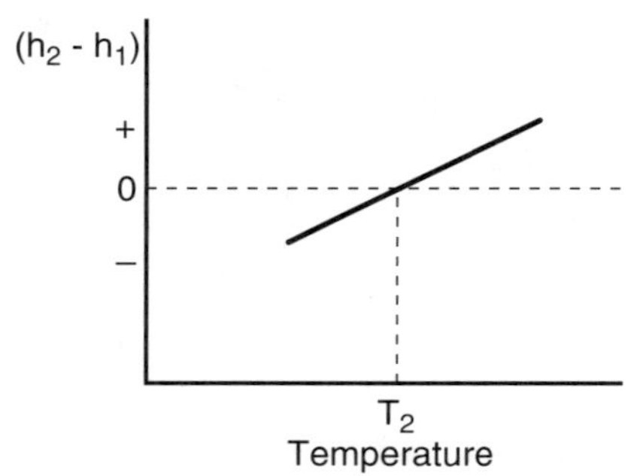

Since the above is quite tedious, the calculation is usually done on a computer using an EOS to calculate thermodynamic properties. For manual calculations, a reasonably good answer can often be obtained by assuming two different temperatures and plotting them on the following type of figure (at left).

A straight line between the $h_2 - h_1$, found for two assumed temperatures, are connected by a straight line. The intersection at $h_2 - h_1 = 0$ gives approximate true T_2.

For the system shown in Figure 16.10, the inlet gas stream is cooled by the sales gas before going to the J-T valve. All one knows are the inlet conditions and composition and the sales gas specifications. These fix the pressure drop across the heat exchanger and valve in series. The drop across the former should not exceed 100 kPa [14.5 psia]. The following type of procedure is needed.

1. Assume temperature of gas downstream of the gas-gas exchanger.
2. Run a flash calculation at this temperature and inlet pressure, minus exchanger pressure drop, say 70 kPa [10 psia].
3. Determine enthalpy of the stream at this point from compositions in Step 2.
4. Use the previous procedure for the valve to find the temperature in the cold separator.
5. Run flash at cold separator conditions.
6. Find the enthalpy of the vapor leaving the separator.
7. Assume a 5-10°C [9-18°F] hot-end approach on the gas-gas exchanger.
8. Perform an energy balance on the inlet and sales gas streams in the gas-gas exchanger.
   a. if $\Delta h_{inlet} > \Delta h_{sales}$, insufficient cooling is available to meet the assumed cold separator temperature. Reduce the exchanger approach or increase the valve pressure drop.
   b. if $\Delta h_{sales} > \Delta h_{inlet}$, excess cooling is available. Bypass sales gas around the gas-gas exchanger or increase the cold separator pressure.
   c. if $\Delta h_{sales} = \Delta h_{inlet}$, a solution has been reached.
9. Once Step 8 is satisfied, the heat exchanger may be sized using heat transfer principles discussed in Chapter 13.

Note: A flash calculation is needed on the inlet gas to the gas-gas exchanger if the inlet is two-phase and composition and relative quantity of each phase is not known.

For applications where only small amounts of liquid condense across the valve, the process may be calculated using a pure component, such as methane. This is useful as a first approximation or reality check. Note: the calculated temperature drop for pure methane will be greater than for a natural gas mixture.

Example 16.4: Using the P-H diagram for methane, Figure 16A.2(a and b), estimate the temperature drop across a valve for a pressure drop to 2000 kPa [290 psia] from 7000 kPa [1015 psia]. Inlet gas temperature is 10°C [50°F].

SI Solution:

h_1 @ 7000 kPa and 10°C = 518 kJ/kg

$h_2 = h_1$, so at P_2 = 2000 kPa and h_2 = 518 kJ/kg,

$T_{2_{isen}} \approx -14°C$

FPS Solution:

h_1 @ 1015 psia and 50°F = –1573 Btu/lbm

$h_2 = h_1$, so at P_2 = 290 psia and h_2 = –1573 Btu/lbm,

$T_{2_{isen}} \approx +7°F$

TURBOEXPANDERS

Turboexpanders are single-stage radial-inflow turbines. A cutaway of a turboexpander/compressor assembly is shown in Figure 16.12. Gas enters the expander impeller radially and leaves axially. Work generated at the expander wheel is absorbed by a single stage centrifugal compressor in most installations. This is because compression is almost always required downstream of the expander to restore the gas to sales gas pressure. In those cases where compression is not required, the expander can drive a generator to produce electricity.

Expander efficiencies are quite high, approaching 85% (isentropic) for new installations operating at design conditions. Since their introduction in the gas processing industry in the late 1960s several thousand expanders have been installed and have generally proven reliable and robust.

Bearings (2-radial and 1-thrust) have historically been oil lubricated. The first magnetic bearings were installed in turboexpanders in 1991. Several expanders currently use magnetic bearings(16.2) and this technology is likely to become standard practice in the future.

Seal gas is required to provide a barrier between the wet, cold process gas and the lube oil or magnetic bearings inside the rotor case. Adequate seal gas flow is necessary to prevent process gas from mixing with the lubricating oil and lubricating oil leaking into the process stream. Early seals were labyrinth as shown in Figure 16.12. Dry gas seals were first used in turboexpanders in 1989 and are commonly used today. They differ somewhat from the design used on centrifugal compressors in that a single seal (rather than tandem seals) is used due to space limitations. Regardless of the seal type, the seal gas must be clean and dry. Several turboexpander failures have been attributed to poor seal gas quality.

The flowrate through the expander is controlled by inlet guide vanes. These are arranged radially around the expander impeller and their movement is synchronized by rotation of a guide vane ring

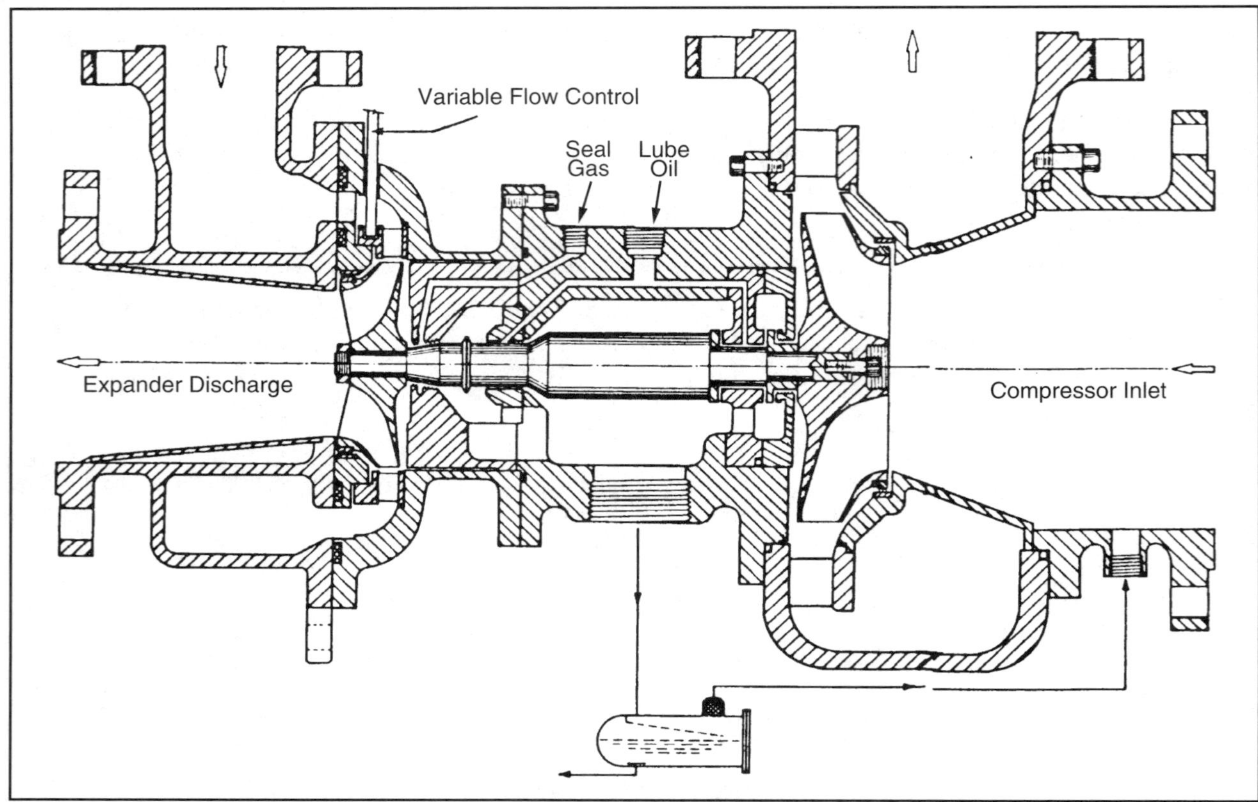

Figure 16.12 Cross-Section of an Expander-Compressor *(Courtesy Atlas Copco)*

which is attached to an actuator. The guide vane position, hence expander flowrate, is usually manipulated by a flow or pressure controller. Figure 16.13 shows expander inlet guide vanes.

Figure 16.13 Expander Inlet Guide Vanes

Turboexpanders are applied both in deep NGL Extraction (C_2+, C_3+) plants and hydrocarbon dewpoint control. The primary difference in these applications is the expansion ratio.

For deep NGL extraction, expansion, ratios are typically 3.0-3.5 and for dewpoint control expansion ratios are usually about 1.3-1.5. An example hydrocarbon dewpoint plant using a turboexpander is shown in Figure 16.14. A simple deep recovery NGL process is shown in Figure 16.15. Modern deep recovery processes are more complex than Figure 16.15, employing significantly more heat integration as well as cooling techniques at the top of the demethanizer/deethanizer to increase NGL recovery. Some of these process schemes will be discussed in more detail in a later section.

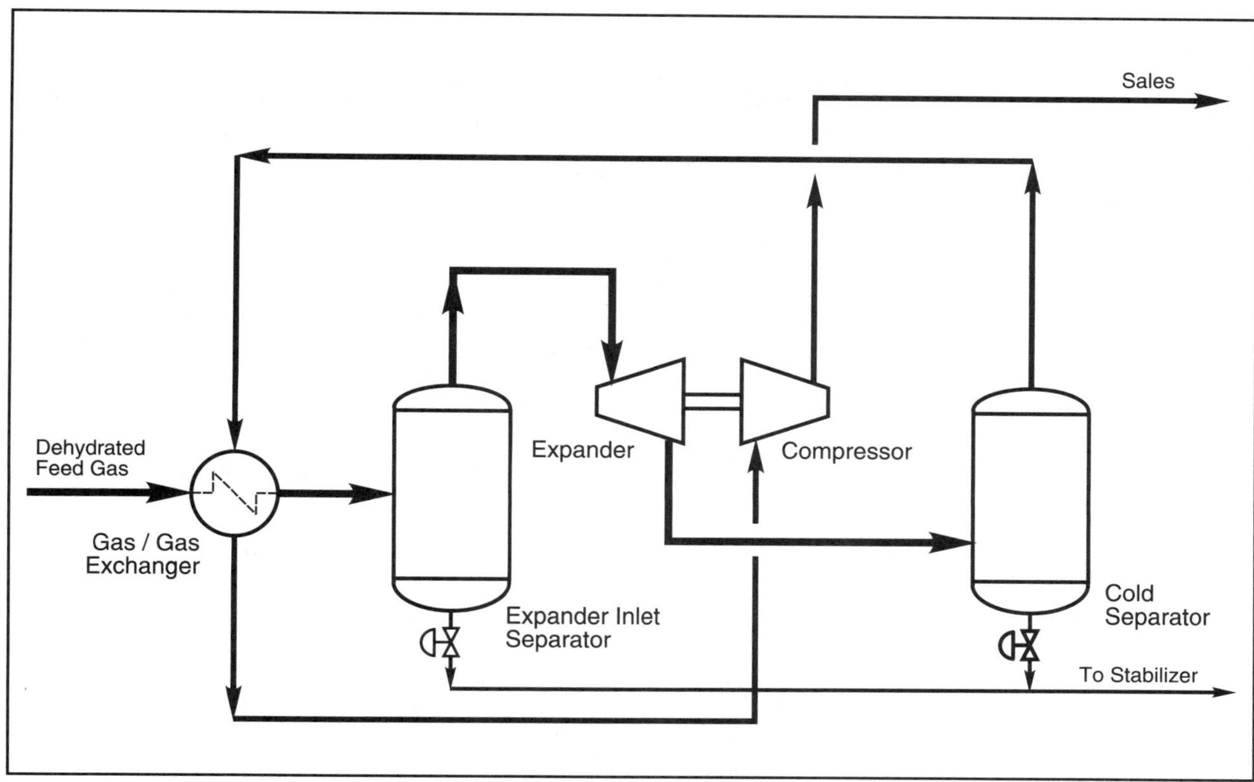

Figure 16.14 Expander Process for Hydrocarbon Dewpoint Control

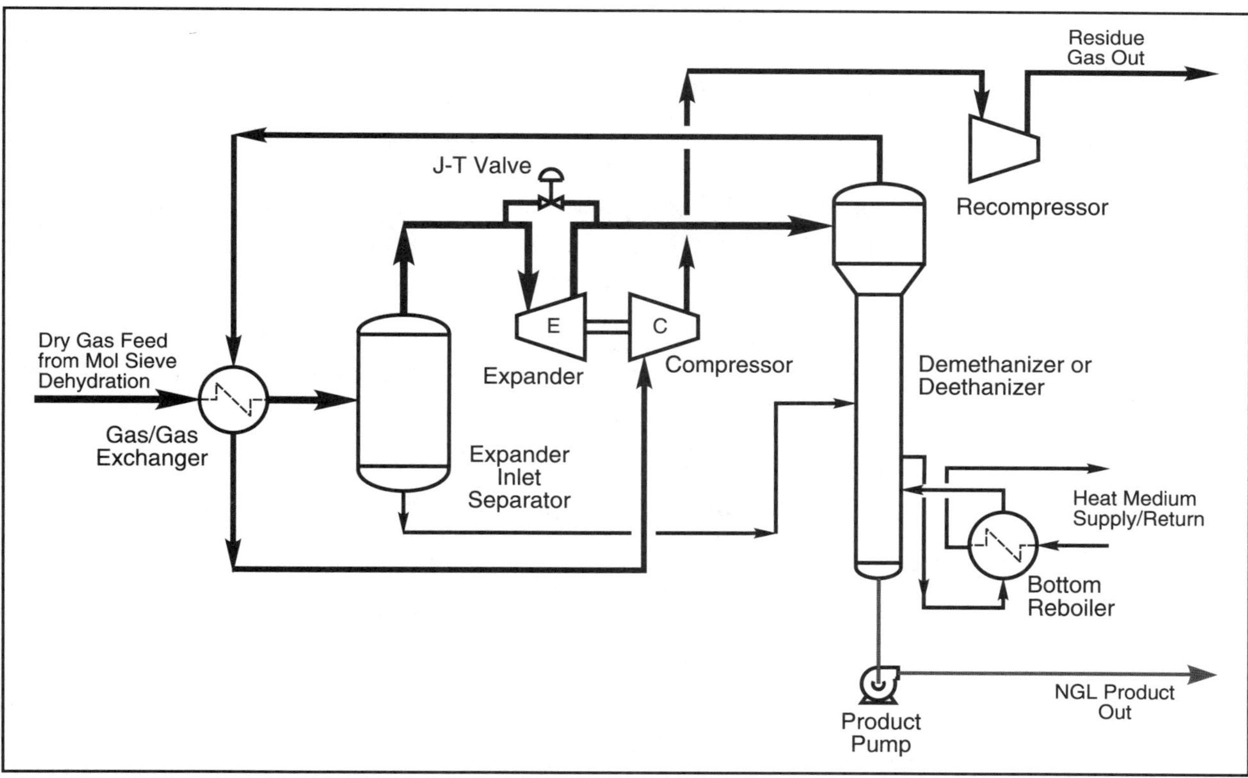

Figure 16.15 Simple Expander Process for Deep NGL Extraction

Expander Calculations

The theoretical thermodynamic path through an expander is isentropic. The theoretical work is calculated from an isentropic path and corrected to the actual work by use of an expander efficiency. For multicomponent streams and condensation across the expander, the calculation is iterative and tedious, and is almost always done on a computer.

The calculations proceed as follows:

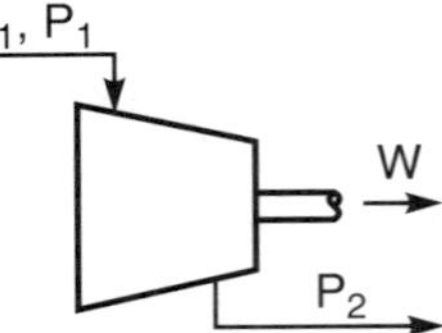

The known variables will typically be feed composition, T_1, P_1, and P_2.

1. From P_1 and T_1, calculate h_1 and s_1.
2. Assume a value of $T_{2_{isen}}$.
3. Run a flash calculation at the assumed T_2 and known P_2 to establish the phase amounts and compositions.
4. Calculate h_2 and s_2. (If the outlet is two-phase, these will be total stream values as discussed in Chapter 8.)
5. If s_2 from (4) equals s_1, you have assumed the right temperature. If not, repeat Steps 2-4 until $s_2 = s_1$.
6. When $s_2 = s_1$, calculated $\Delta h_{isen} = h_{2_{isen}} - h_1$.
7. Calculate expander power, $W = m\Delta h_{isen}\, E_{isen}$.
8. Calculate $h_{2_{actual}} = h_1 + (h_{2_{isen}} - h_1)E_{isen}$.
9. Assume a value of $T_{2_{act}}$ (this will be warmer than $T_{2_{isen}}$).
10. Run a flash calculation at the assumed $T_{2_{act}}$ and P_2.
11. Calculate h_2 ($h_{2_{calc}}$) from the results of the flash.
12. If h_2 from (11) equals $h_{2_{actual}}$ you have assumed the correct temperature. If not, repeat steps 9-11 until $h_{2_{calc}} = h_{2_{actual}}$.

Example 16.5: Using the information in Example 16.4, calculate the outlet temperature and work if the gas was expanded across a turboexpander rather than a valve. $E_{isen} = 0.82$ and the gas flowrate is 1×10^6 std m^3/d [35.4 MMscfd] and MW = 16.

SI Solution:

1. at P_1 and T_1, $h_1 = 518$ kJ/kg and $s_1 = 9.08$ kJ/kg·K
2. $s_2 = s_1$, so at $P_2 = 2000$ kPa and $s_2 = 9.08$,

   $T_{2_{isen}} = -70°C \qquad h_{2_{isen}} = 380$ kJ/kg
3. $\Delta h_{isen} = 380 - 518 = -138$ kJ/kg
4. $m = (1.0)(1762)(16) = 28\ 190$ kg/h $= 7.83$ kg/s

   $-W = m\Delta h_{isen}E_{isen} = (7.83)(380 - 518)(0.82) = -890$ kW

   $W = 890$ kW (work is positive)
5. $\Delta h_{actual} = 518 + (380 - 518)(0.82) = 405$ kJ/kg
6. at $P_2 = 2000$ kPa and $h_2 = 405$ kJ/kg $\qquad T_{2_{actual}} = -60°C$

Example 16.5 (Cont'd.):

FPS Solution:

1. at P_1 and T_1, $h_1 = -1573$ Btu/lbm and $s_1 = 2.17$ Btu/lbm-°R
2. $s_2 = s_1$, so at $P_2 = 290$ psia and $s_2 = 2.17$,

   $T_{2_{isen}} = -94°F \qquad h_{2_{isen}} = -1632$ Btu/lbm
3. $\Delta h_{isen} = [-1632 - (-1573)] = -59$ Btu/lbm
4. $m = (35.4)(110)(16) = 62\ 300$ lbm/hr

   $$-W = \frac{(62\ 300)(-59)(0.82)}{2545} = -1180 \text{ hp}$$

   W = 1180 hp (work is positive)
5. $\Delta h_{actual} = -1573 + [-1632 - (-1573)](0.82) = -1621$ Btu/lbm
6. at $P_2 = 290$ psia and $h_2 = -1621 \qquad T_{2_{actual}} = -76°F$

Estimation of Expander Efficiency

The best way to estimate efficiency is from actual performance data.

It is difficult to correlate efficiency data because many factors affect actual performance. As with compressors, normal manufacturing tolerances can affect performance measurably. Erosion can alter the shape of a wheel and thus efficiency. The presence of liquids likewise may have a dramatic effect. For all of these reasons, there may be a significant error in the estimated efficiency.

In correlating efficiency data, one may use the basic similarity parameters governing all turbomachinery. Similar to centrifugal pumps and compressors, expander efficiency can be correlated against the specific speed of the impeller. Figure 16.16 shows such a correlation.

Specific speed is one of the criteria for determining performance. It was discussed previously in the centrifugal pump section. The basic equation for expanders is:

$$N_s = \frac{N(q_2)^{0.5}}{[(A)(\Delta h)]^{0.75}} \tag{16.9}$$

Where:

			SI	FPS
N	=	shaft speed	rpm	rpm
Δh	=	isentropic Δh	kJ/kg	Btu/lb
q_2	=	turbine exhaust volume	m^3/s	ft^3/sec
A	=	conversion factor	31	778

For a radial inflow turbine, the optimum E is for a specific speed between 70 and 100. However, this is not always possible because of the limitation in available standard expander frame and other parameters which affect performance.

Figure 16.16 is for expanders where the amount of liquid formed is minimal. How much does liquid affect efficiency? This depends on where, and how, it was formed. Data from steam turbines has shown the predicted efficiency to be significantly less than the actual efficiency when too much liquid was formed in an axial turbine.

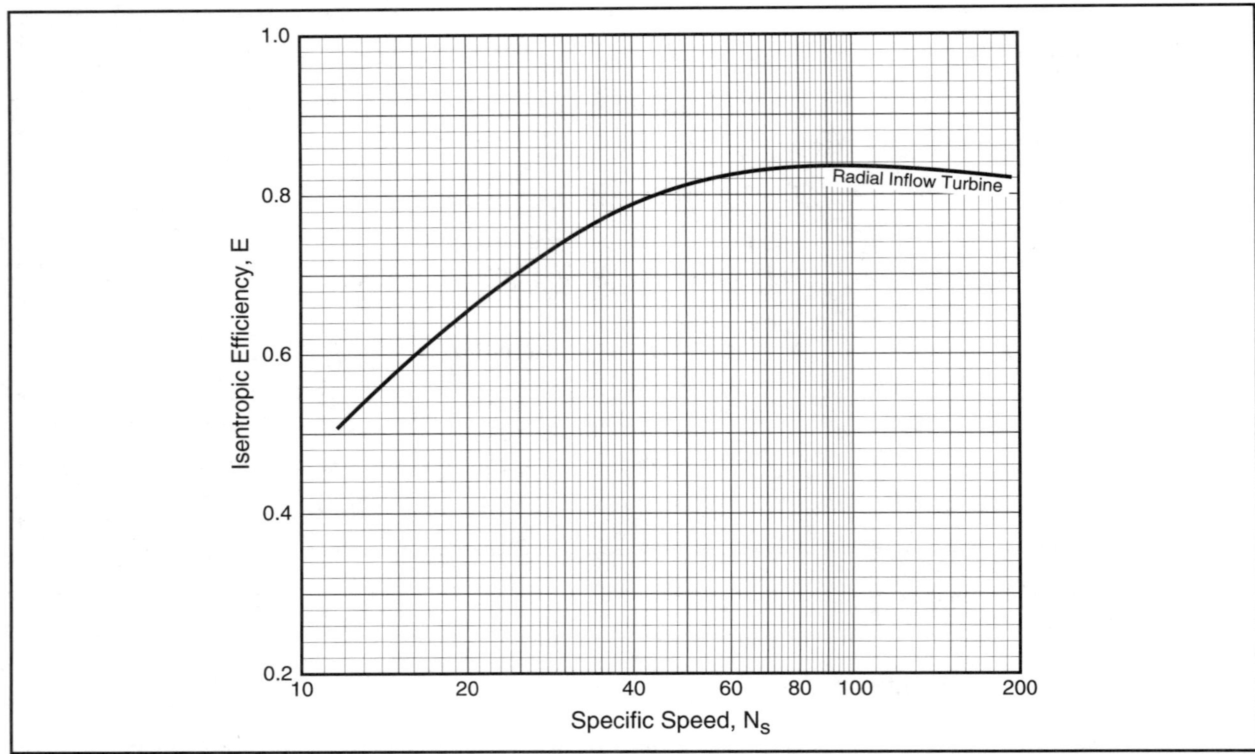

Figure 16.16 Efficiency of Expansion Turbines as a Function of Specific Speed

The question of liquid formation is an important one. There is good reason to believe that liquid droplets do not actually form in the expander due to limited residence time. An equilibrium gas-liquid mixture only occurs at some point and downstream of the impeller. So ... it is likely that most of the liquid forms just downstream of the wheel and does not interact with the wheel proper. Initially, any liquid formed will be of submicron size.

If there was significant liquid within the wheel proper, the denser liquid would be centrifuged outward, which would lead to lower efficiencies than we note because of flow distortions. Also, wheel erosion could be significant.

How much liquid can be handled without significant loss of efficiency? Some units contain above 20 mol% liquid in the exit gas. Certainly 10% is being accommodated and 20% is not unreasonable.

Expander efficiency will deteriorate over time due to normal wear and tear. This deterioration is accelerated by any erosion caused by droplets and particulates in the feed gas. Efficiency can also decline due to variation in flow from the design rates. Figure 16.17 shows how expander efficiency can vary with inlet flowrate.[16.3]

$$\text{Flow Parameter} = \frac{m\sqrt{\dfrac{T}{MW}}}{P}$$

Where:

m = mass flow
T = inlet temperature
MW = gas MW
P = inlet pressure

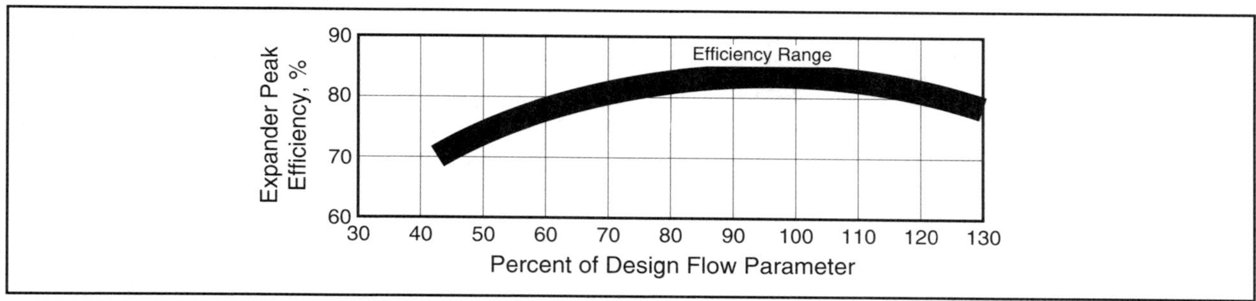

Figure 16.17 Example Peak Efficiency vs. Percent Desgn Inlet Flow Parameter

In the early planning stages, flow rates and gas compositions are rather inexact. Some use a design efficiency about 5-10% lower than that predicted by correlation. Obviously, E affects outlet temperature, liquid condensation and the composition of both phases which, in turn, affects all downstream equipment. Power transmitted to the booster compressor is also affected.

In addition to specific speed there are other parameters affecting expander performance. Specific diameter, presented earlier in Chapters 14 and 15, is one of these, defined by the equation:

$$d_s = \frac{(d)[(A)(\Delta h)]^{0.25}}{(q_2)^{0.5}} \tag{16.10}$$

Where:			SI	FPS
	d =	turbine diameter	m	ft
	Δh =	isentropic Δh	kJ/kg	Btu/lb
	q_2 =	turbine outlet volume	m^3/s	ft^3/sec
	A =	conversion factor	31	778

The specific diameter should be approximately 1.2 to 1.4 to achieve maximum efficiency. This is a useful number for estimating impeller size and shaft speed.

Expander efficiency can be calculated using measured process data. The most convenient (and accurate) method is to use the expander compressor as a dynamometer. The following data is required

- Expander Inlet: gas composition, T and P, and flowrate
- Compressor Suction: gas composition, T, P, and flowrate
- Compressor Discharge: T and P

The calculation procedure proceeds as follows:

1. Accurately measure the pressures and temperature at the suction and discharge of the compressor.
2. Use a computer program to calculate h_{suct} and h_{disch} for the compressor
3. Compressor Work = m ($h_{disch} - h_{suct}$)
4. Add 2-3% to the work in step 3 to account for bearing and seal losses. This represents the actual work generated by the expander
5. Use a computer program to calculate the isentropic work from the expander
6. $E_{isen} = (W_{actual})/(W_{isentropic})$

APPLICATIONS OF REFRIGERATION IN GAS PROCESSING

By far and away, the most common application of refrigeration in gas processing is the cooling of natural gas to meet a sales gas specification and/or the extraction of NGL components for recovery and sale. Selection of a suitable process scheme is based on several parameters including:

- Process objective (hydrocarbon dewpoint control or NGL extraction)
- NGL products to be recovered, sales revenue
- Inlet pressure
- Sales gas pressure
- Capital and operating costs
- Plant location, site conditions

Hydrocarbon Dewpoint Applications

Processing to meet a hydrocarbon dewpoint specification does not typically require low levels of refrigeration. Most specifications can be met by removal of the heavier C_5+ components from the gas. There is little incentive to extract C_2, C_3, and C_4 hydrocarbons from the gas, indeed the extraction of these components may reduce the gas value.

If the process inlet pressure is less than about 8000 kPa [1160 psia] and the sales gas pressure on the order of 6900 kPa [1000 psia] the most common process choice is mechanical refrigeration. There are exceptions to this, particularly for very lean gases where an adsorption system may be a better choice.

If the sales gas pressure is low, e.g. gas turbine fuel (3500-4100 kPa [500-600 psia]) an expansion process is often more economic due to the availability of "free" pressure drop.

Expansion processes are the preferred alternative when the inlet gas pressure is at least 20-30% greater than the sales gas pressure. This is particularly true at pressure above about 8000 kPa [1160 psia] because mechanical refrigeration systems are generally not effective at high pressure due to the retrograde phase behavior of the gas. If the inlet pressure is provided by the reservoir, i.e. no inlet compression is required, the valve expansion process might be more economical when inlet pressure to sales gas pressure ratio exceeds about 1.5.

If the inlet pressure is provided by compression or when the ratio of inlet to sales gas pressure falls between about 1.1 to 1.5 an expander process is often preferred.

Figure 16.18-16.20 shows some process temperatures and pressures for the three refrigeration schemes discussed above. The process path is also plotted on a phase envelope to better visualize the process differences. In each case the sales gas pressure is 6900 kPa [1000 psia], the minimum approach on the gas-gas exchanger is 5°C [9°F], the dewpoint specification is –3°C [26.6°F], and the cold separator carryover is 2%.

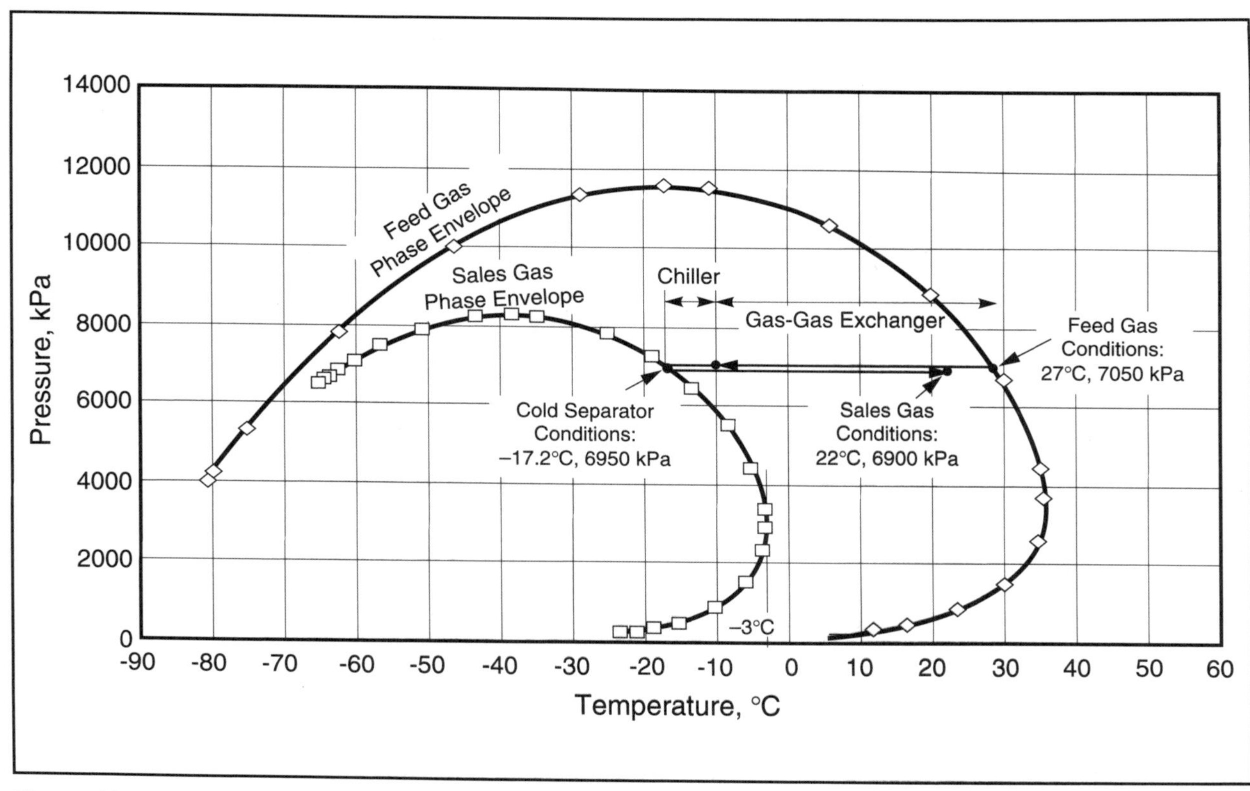

Figure 16.18 P-T Path for a Mechanical Refrigeration Process

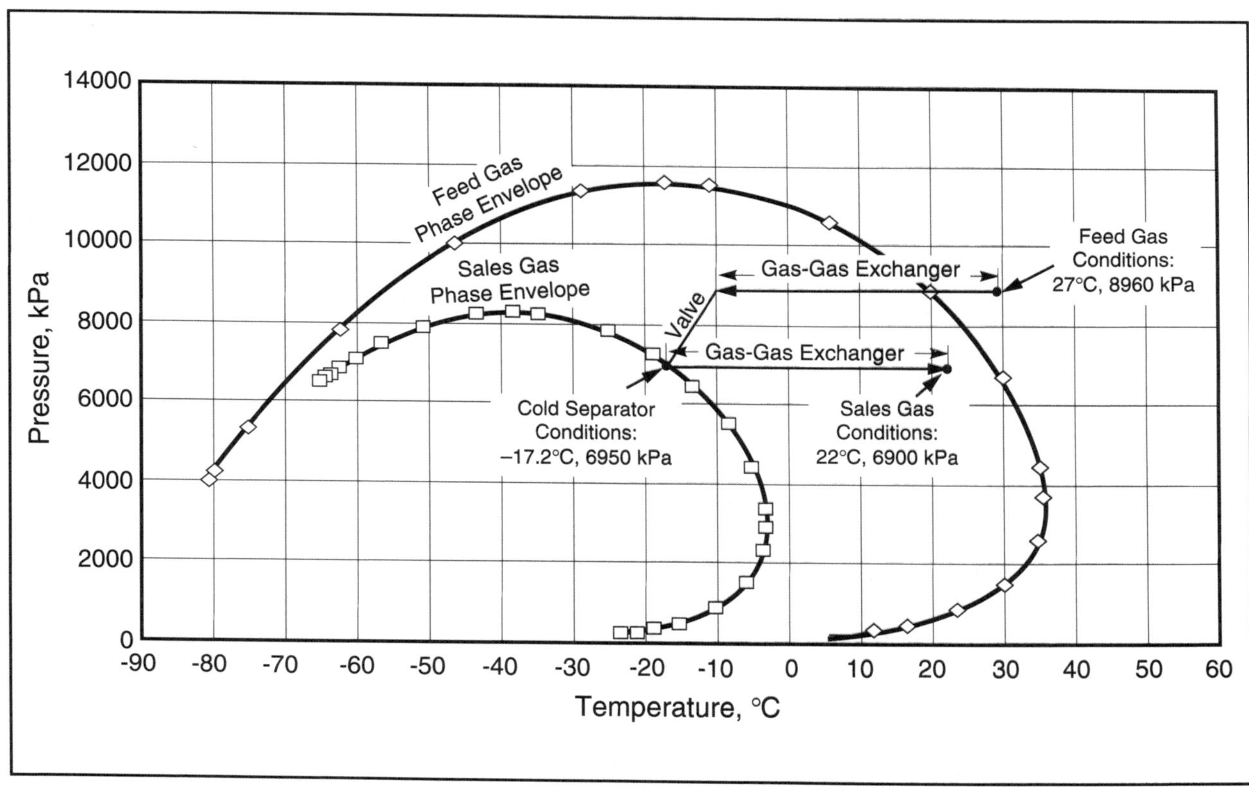

Figure 16.19 P-T Path for a Valve Expansion Process

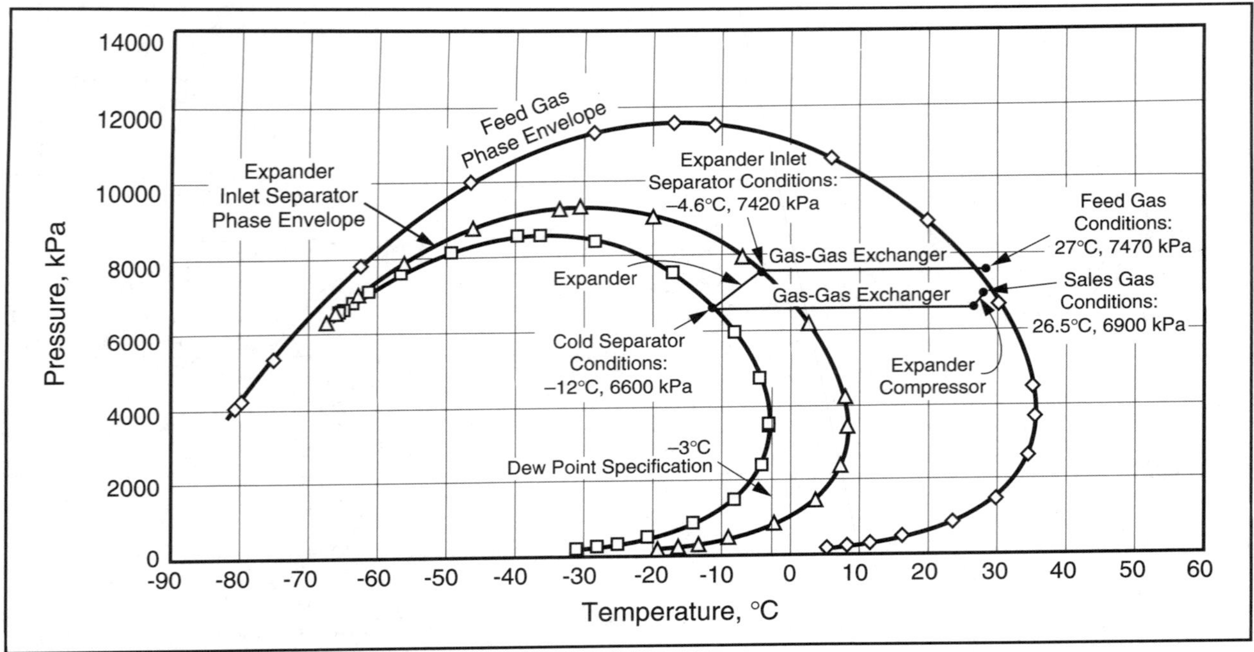

Figure 16.20 P-T Path for a Expander Process

NGL Extraction

The preferred processes for deep NGL extraction from the gas have historically been mechanical refrigeration, refrigerated lean oil absorption and turboexpander. Mechanical refrigeration is generally not effective in achieving high extraction levels unless the gas is relatively rich in C_2+ hydrocarbons and unless cold separator temperatures are below –70 to –80°C [–94 to –112°F]. This requires a cascade or mixed refrigerant type refrigeration system.

Refrigerated lean oil absorption can be very effective at removing NGLs. Ethane recoveries as high as 50% and propane recoveries of 98-99% have been achieved. These processes have high capital costs and are considerably more complex than mechanical refrigeration and expansion type processes. Few plants of this type have been built since the early 1970s, although new process developments have made this process attractive in some applications.[16.4] The lean oil absorption process will be discussed in more detail in Chapter 17.

The Petroflux[16.5, 16.6] process combines mechanical refrigeration and an absorption effect by using a refluxing heat exchanger. A simple flow diagram for the Petroflux process is shown in Figure 16.21. Feed gas is cooled by cross-exchange with cold sales gas and enters the cold separator. Vapor from the cold separator flows to the refluxing heat exchanger sometimes called a dephlegmator. Cooling is provided by external refrigeration. In the refluxing heat exchanger, liquids which condense from the gas flow downwards, countercurrent to the gas flow. The exchanger not only cools the gas but acts like a packed tower. The countercurrent flow through the exchanger also results in a sharper separation between the condensed NGL components (C_2+ or C_3+) and the methane. This mass transfer effect significantly reduces the size and utility demand of the stabilizer. This process can also be combined with a turboexpander.

Expander processes are the preferred process for deep NGL extraction. These processes are relatively simple, compact, can achieve ethane recoveries of 90% (propane 99%+), and have the lowest capital cost. If the sales gas must be returned to the pipeline, at high pressure, the largest plant

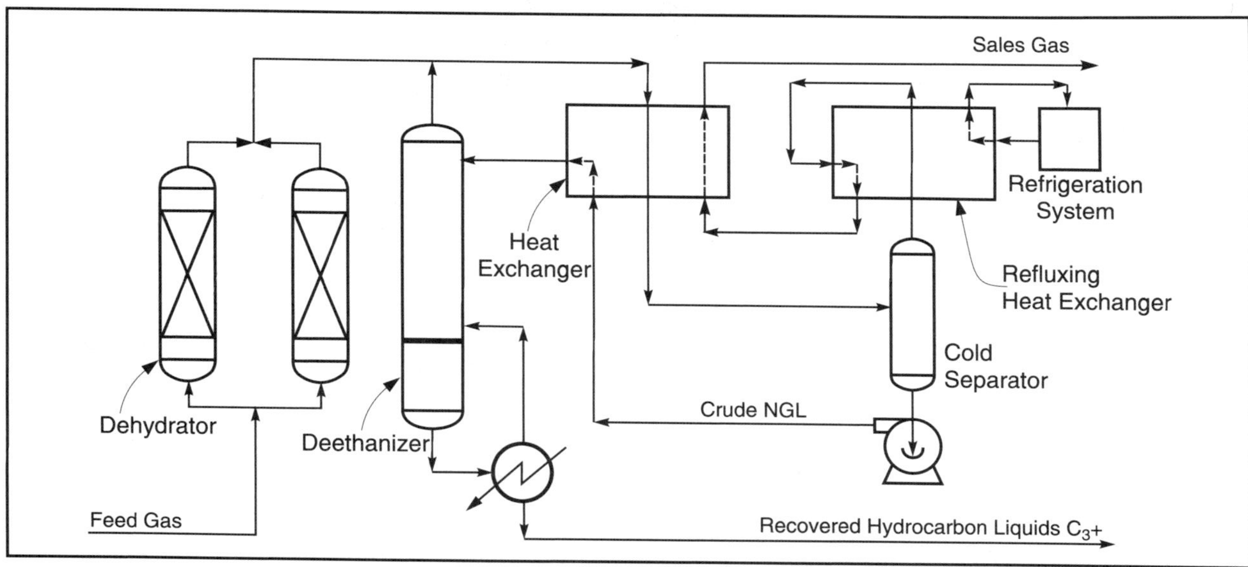

Figure 16.21 Example Petroflux® Process

operating cost is compression. Dehydration upstream of deep NGL recovery plants is almost always adsorption — using molecular sieves, although methanol injection has been used in some designs.

A simple expander plant is shown in Figure 16.22.

Early processes used this scheme. Expansion ratios across the expander were 3 to 3.5 and expander outlet temperatures were typically –90 to –100°C [–130 to –148°F]. While very simple, this process scheme had limited thermodynamic efficiency. Temperature approaches in heat exchangers were large and the potential refrigeration effect of methane revaporization in the demethanizer was not exploited. In addition, because there was effectively no temperature control at the top of the demethanizer, ethane recovery was limited and it was difficult to reject ethane without incurring significant propane losses.

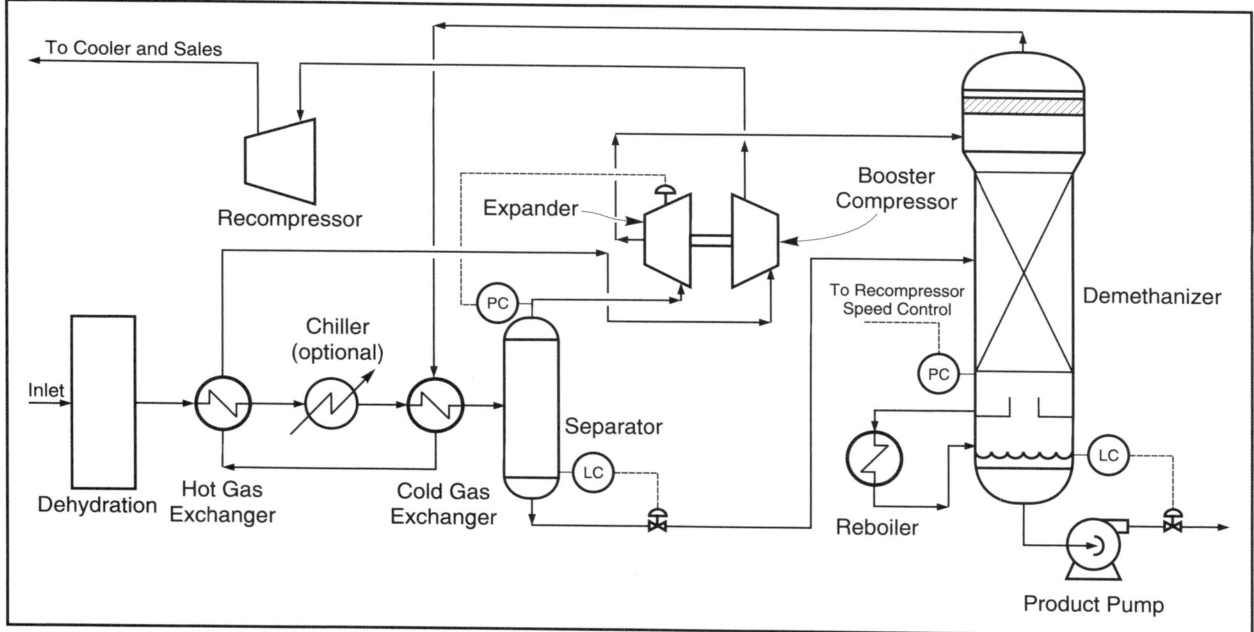

Figure 16.22 Flow Sheet for Simple Expander Plant

The use of side reboilers on the demethanizer improved heat recovery. A portion of the feed gas was used as the heating medium, resulting in lower expander outlet temperatures (hence higher NGL recovery levels) for the same expansion ratio. In addition, the use of BAHXs decreased temperature approaches in heat exchangers to 2-3°C [4-6°F].

Figure 16.22 shows an optional chiller. External refrigeration is sometimes used to assist in feed gas cooling. This is often the case when the feed gas is quite rich in C_2+ components. The reason for this is that less sales gas is available for back exchange in the gas-gas exchanger.

Basic turboexpander plants, like the one shown in Figure 16.22 are limited to about 70% ethane recovery and 95% propane recovery. In addition, the basic plant is subject to CO_2 freezing and cannot maintain high levels of propane recovery during those periods of operation when ethane is rejected.

The design of turboexpander plants has evolved further for four reasons.

1. increase ethane recovery
2. increase CO_2 tolerance of the process
3. increase propane recovery during periods of ethane rejection
4. improved energy efficiency

Several process designs have emerged, but they all share one common theme — provide a "reflux" stream at the top of the demethanizer.

One process which is now widely used in the industry is Ortloff's Gas Subcooled Process (GSP).(16.7) A simple process flow diagram is shown in Figure 16.23. In this process, a portion of the expander feed, say 20-30%, is bypassed around the expander and subcooled by heat exchange with the

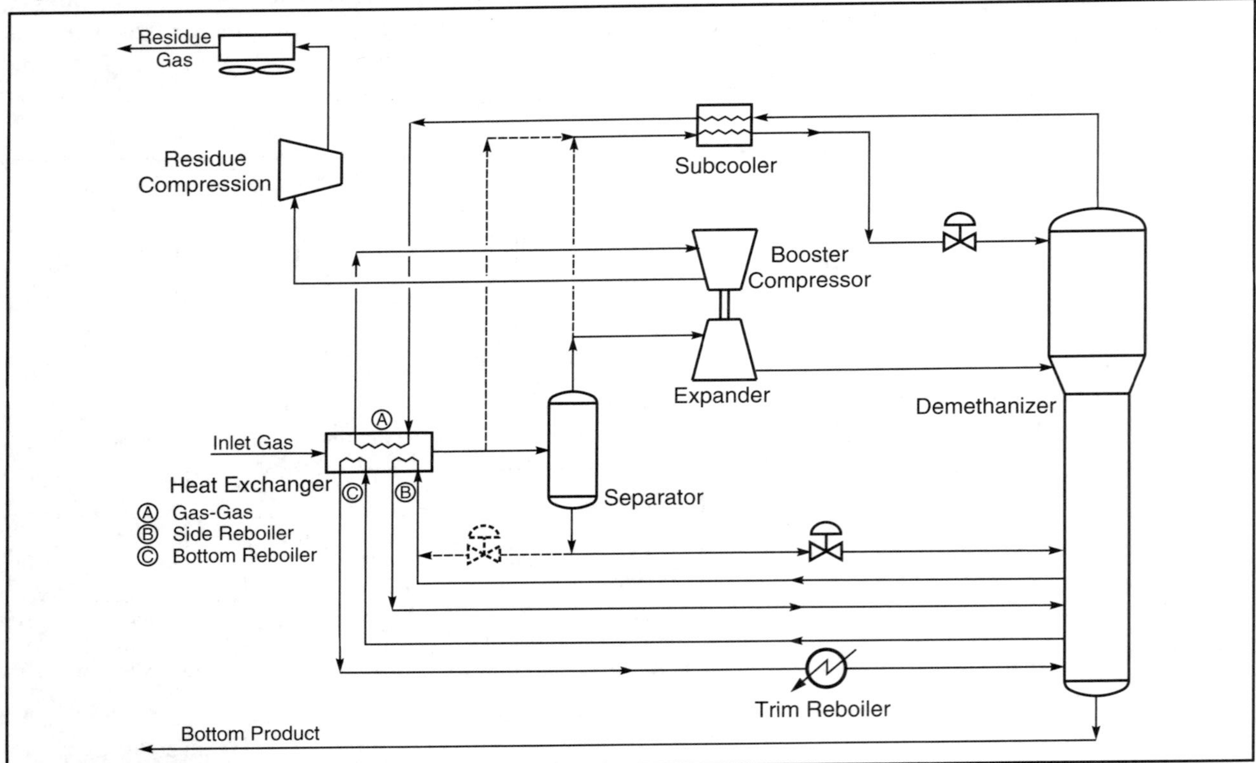

Figure 16.23 Gas Subcooled Process (GSP) Plant Design(16.7)

demethanizer overhead. The high pressure subcooled dense fluid is then expanded across a valve into the top of the demethanizer. The temperature of this stream can be very cold (< –100°C [–148°F]) and increases NGL recovery by condensing and absorbing NGL from the boil-up vapors in the tower. This design offers higher NGL recovery levels for the same recompression power when compared to the simple design. It is also more CO_2 tolerant. The C_2+ liquids in the top of the demethanizer help suppress the CO_2 freezing point. The bottoms product also contains less CO_2 than in the simple design.

One of the limitations of the GSP process is that the sub-cooled feed contains C_2, C_3 and C_4 components. This limits NGL recovery due to equilibrium compositions at the top of the tower. The primary advantage of the CRR process, shown in Figure 16.24, is that the subcooled split-flow feed is used to condense reflux. In ethane recovery mode, the condensed reflux stream is nearly pure methane. This allows higher ethane and propane recovery.

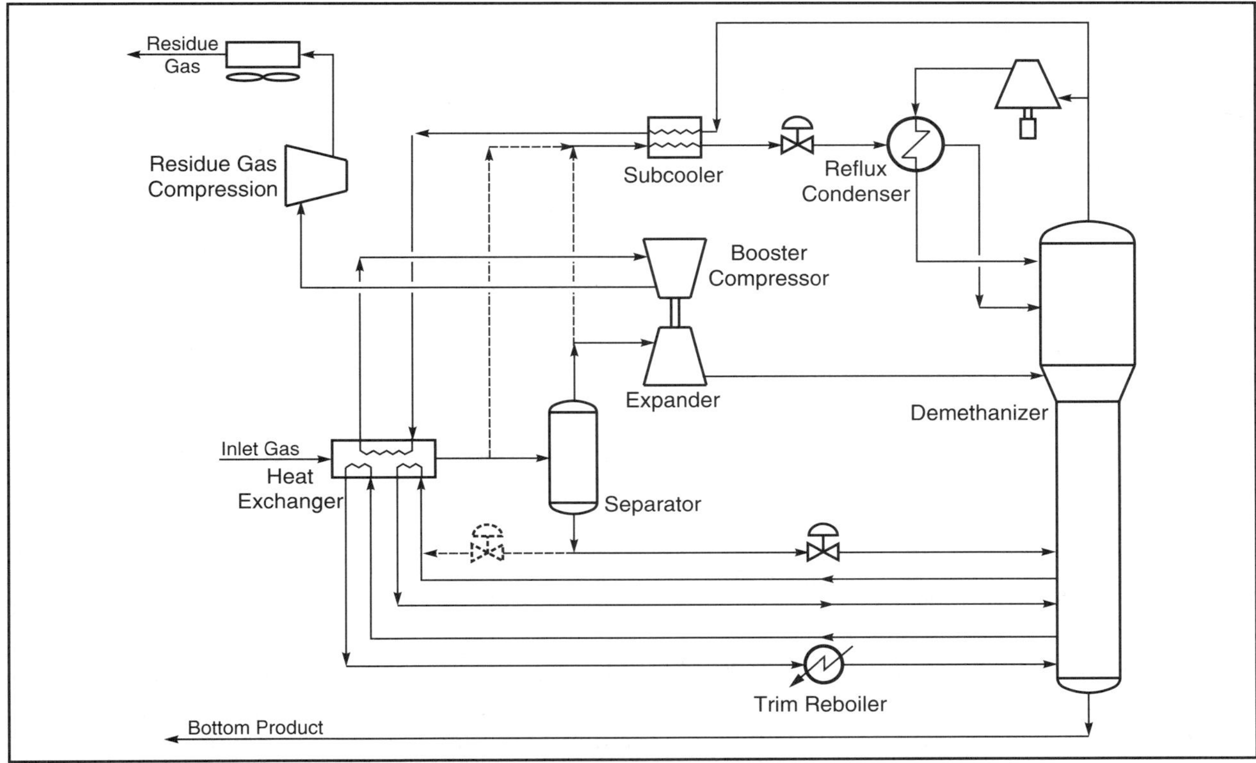

Figure 16.24 Cold-Residue Reflux (CRR) Plant Design

The split-vapor stream is not quite cold enough to liquefy a pure methane stream at demethanizer pressure; however, the inclusion of a small compressor can boost the demethanizer overhead to a slightly higher pressure so that methane can be condensed. The condensed methane "reflux" stream is fed to the top of the tower with the slightly warmer split-vapor stream fed below. Ethane recoveries in excess of 90% are claimed for the CRR process and propane recoveries above 99% can be achieved while rejecting almost all of the ethane.

A simpler and sometimes less expensive version of the CRR process is the Recycle Split Vapor (RSV) scheme. In this process, shown in Figure 16.25 a portion of the high-pressure recompressed sales gas stream is subcooled, flashed to tower pressure and fed to the demethanizer top as a "reflux" stream. This process is also used in some plants without the split flow feed. This process is effective in achieving high ethane recoveries, but requires more power than the CRR process for a given ethane recovery.

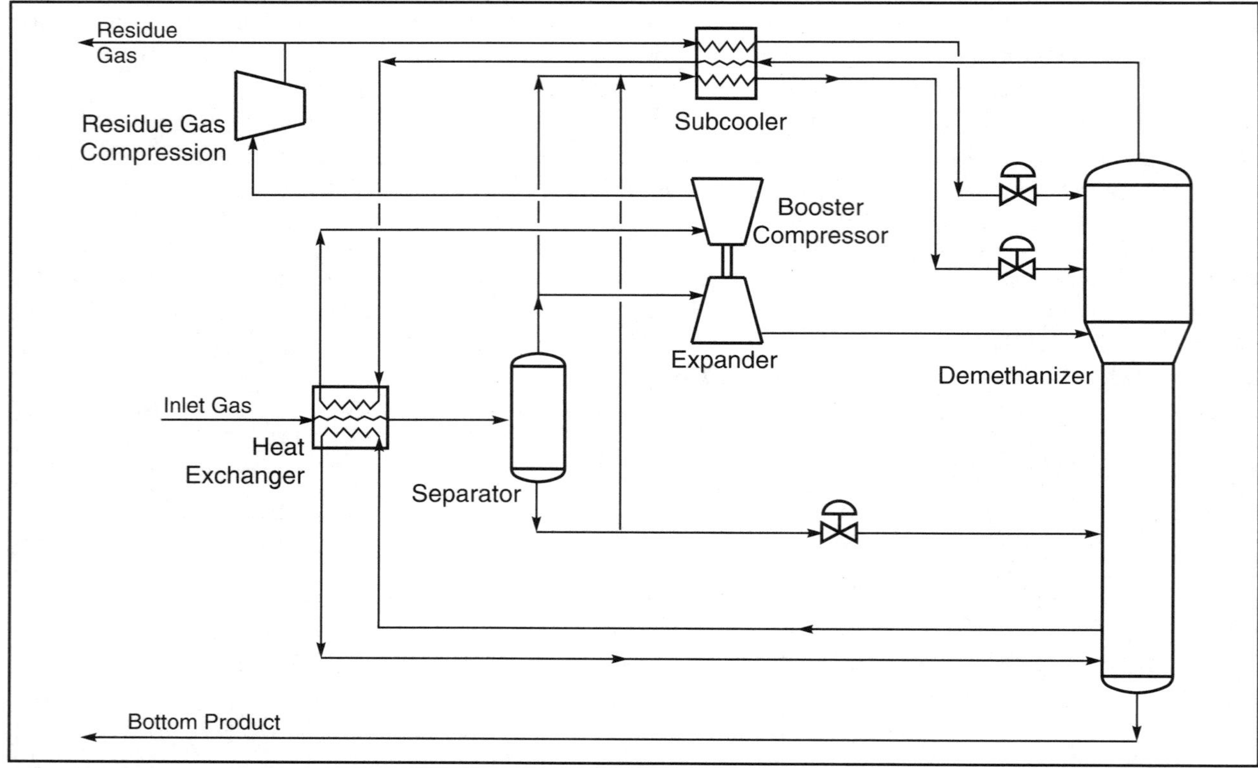

Figure 16.25 Recycle Split Vapor (RSV) Process

There are several other variations of the split-flow concept. An excellent summary of these processes is presented in Reference 16.8.

Sonic/Supersonic Expansion Processes

Another type of expansion process has been developed for gas cooling. This process employs expansion in a tube where velocities can exceed Mach 1. Temperatures and pressures at the highest velocity point in the tube can be quite low. Liquid is directed to the tube wall by centrifugal force and removed from the tube. The gas exits the tube and flows through a diffuser where a portion of the pressure drop is recovered. The overall temperature drop (inlet to outlet) is comparable to a J-T valve, but the instantaneous temperatures inside the tube can be much colder and actually approach an isentropic process.

Overall process efficiency depends on the effectiveness of liquid separation from the gas inside the tube at the minimum temperature. Separation effectiveness is limited by the very short retention times within the tube.

The primary advantages of this process are:

- compact, small weight and footprint
- high internal velocities and short retention times often preclude dehydration or hydrate inibition
- no emissions

Disadvantages include:

- separation efficiency at minimum temperature is limited by short retention time and droplet aggloneration and coalescence
- outlet hydrocarbon and water dewpoints limited (currently around 0°C)
- turndown is poor per tube – many tubes in parallel required for high turndown

Further development of these processes continues and it is expected that recovery efficiencies will improve. The process is a non-equilibrium process and cannot be simulated on a steady-state process simulator.

LIQUEFIED NATURAL GAS (LNG)

Beyond a certain distance it is more economical to transport natural gas by ship than pipeline. LNG tankers carry cargoes of LNG at –162°C [– 250°F] from remote gas fields to areas of demand such as Europe, Japan, and Korea. Roughly 4% of the world's total gas consumption and 25% of the internationally traded gas moves as LNG. Typically 1/4% of the cargo vaporizes each day and is used as boiler fuel for the ship's engines.

Overall Flow Scheme

Figure 16.26 shows a simplified overall LNG facility utilizing the mixed refrigerant process. The cold section is based on the propane pre-cooled multicomponent refrigerant (MCR) technology developed by Air Products and Chemicals (APCI). This is used in most of the world's base-load LNG plants.

The entering gas is first treated in a sweetening unit to remove CO_2 and any H_2S to low levels. H_2S is removed to meet sales gas specifications, CO_2 to prevent freezing in the process. Three levels of propane refrigeration reduce the sweet gas temperature to around –32°C [–26°F]. During these cooling steps water is removed to < 1 ppmv, CO_2 to below 50-100 ppmv and H_2S to typically < 4 ppmv. Benzene and other heavy hydrocarbons are removed to very low levels in the Heavies Removal Column or "Scrub Column" to prevent freezing in the LNG exchanger.

The purified and pre-cooled gas then enters the large APCI vertical spiral-wound exchanger where the gas is condensed and subcooled to LNG. The LNG exits at the top of the Cryogenic Heat Exchanger (sometimes called Main Heat Exchanger or MHE) under high pressure. It is then depressured in a variety of ways such as staged end-flashes, liquid expanders, etc. before entering the large LNG storage tanks at the shipping terminal.

Example LNG Composition	
	mol%
N_2	1.0
C_1	92.0
C_2	5.0
C_3	1.5
C_4+	0.5
CO_2	< 50 ppm
H_2S	< 4 ppm
H_2O	< 1 ppm

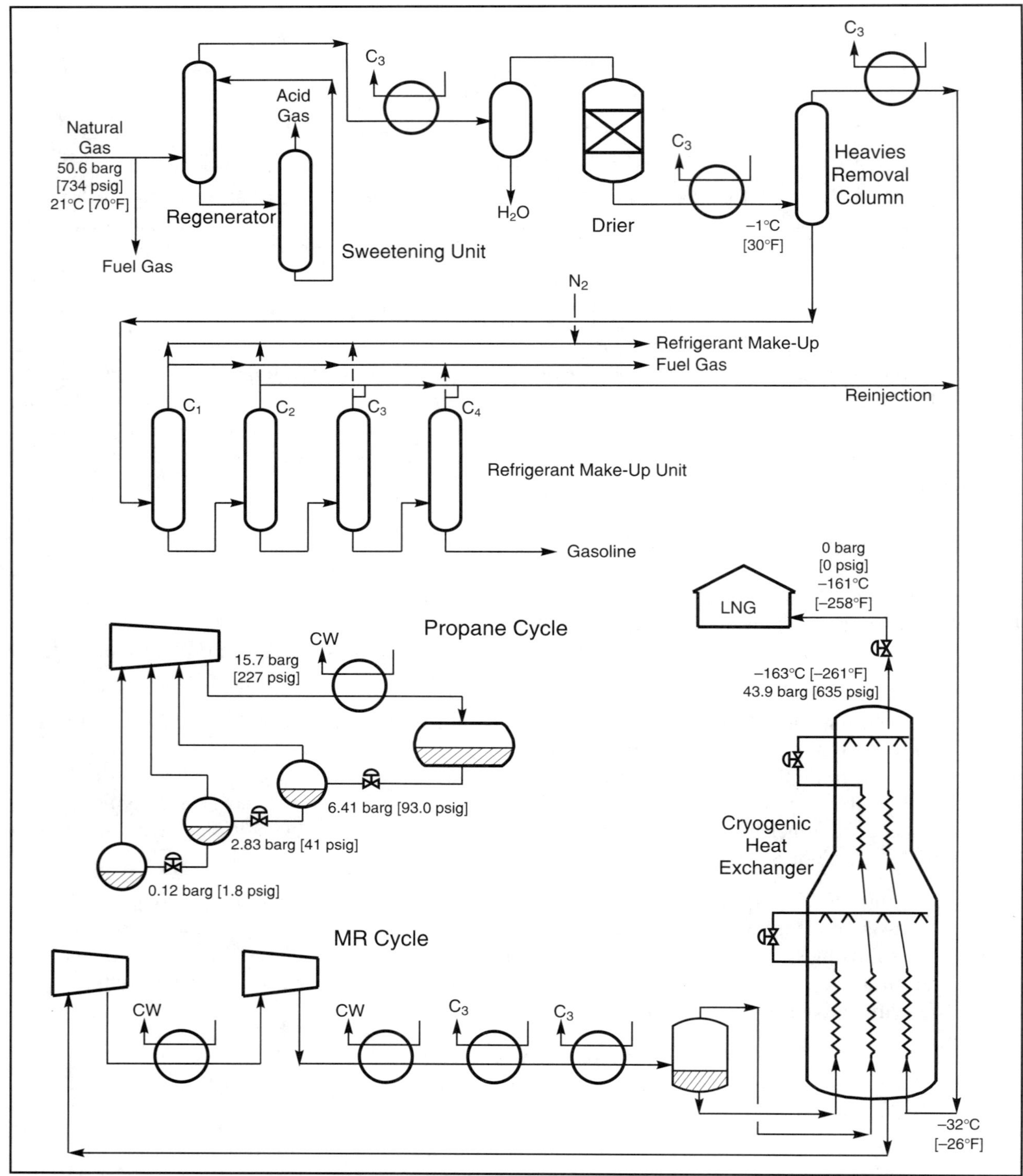

Figure 16.26 Basic LNG Process *(Courtesy Air Products and Chemicals, Inc.)*

Mixed Refrigerants

Mixed refrigerants were discussed briefly in a previous section.

An example mixed refrigerant might be a 50:50 mixture of ethane and propane. Such a mixed refrigerant will vaporize (i.e. boil) over a temperature range and not at a single temperature as shown in the P-H diagram, Figure 16.27. The temperature range is set by both pressure *and* composition. Table 16.2 summarizes the atmospheric boiling temperatures of the common components in LNG mixed refrigerants. Figure 16.28 shows the P-H diagram for the following mixture: 5% N_2, 50% C_1, 30% C_2, 15% C_3.

This means the refrigeration effect will be distributed over a range of temperatures as opposed to a constant temperature, e.g. kettle-type chillers. This produces a smaller overall temperature difference between the natural gas (being cooled) and the mixed refrigerant (being warmed up). These smaller temperature differences lead to a more efficient liquefaction process. Contrast such close temperature approaches with the "stair-step" temperature differences in the cascade refrigeration (Figure 16.8).

As can be seen by comparing Figure 16.27 and 16.28 adding methane and nitrogen to the simple 50:50 C_2-C_3 mixture (mixed refrigerant or MR) extends the boiling temperature range to lower temperatures. Most gas industry mixed refrigerants (MR) are a mixture of C_3, C_2, and C_1, often with an additional small amount of N_2.

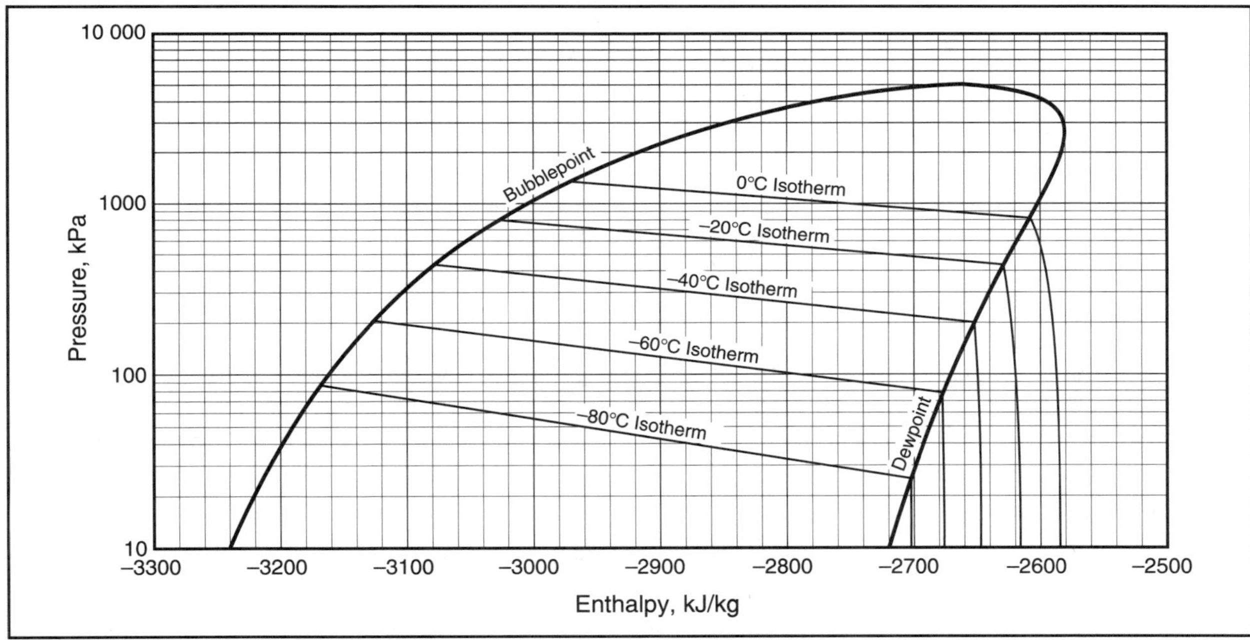

Figure 16.27 P-H Diagram for Simple Mixed Refrigerant, 50% C_2, 50% C_3

TABLE 16.2

Refrigerants – Boiling Points (Single Component)

Component	Boiling Point	Temperature
C_3	@ 1 atm	–42°C [–44°F]
C_2	@ 1 atm	–89°C [–127°F]
C_1	@ 1 atm	–162°C [–259°F]
N_2	@ 1 atm	–196°C [–320°F]
C_1	@ 13.1 bara [190 psia]	≈ –120°C [–180°F]

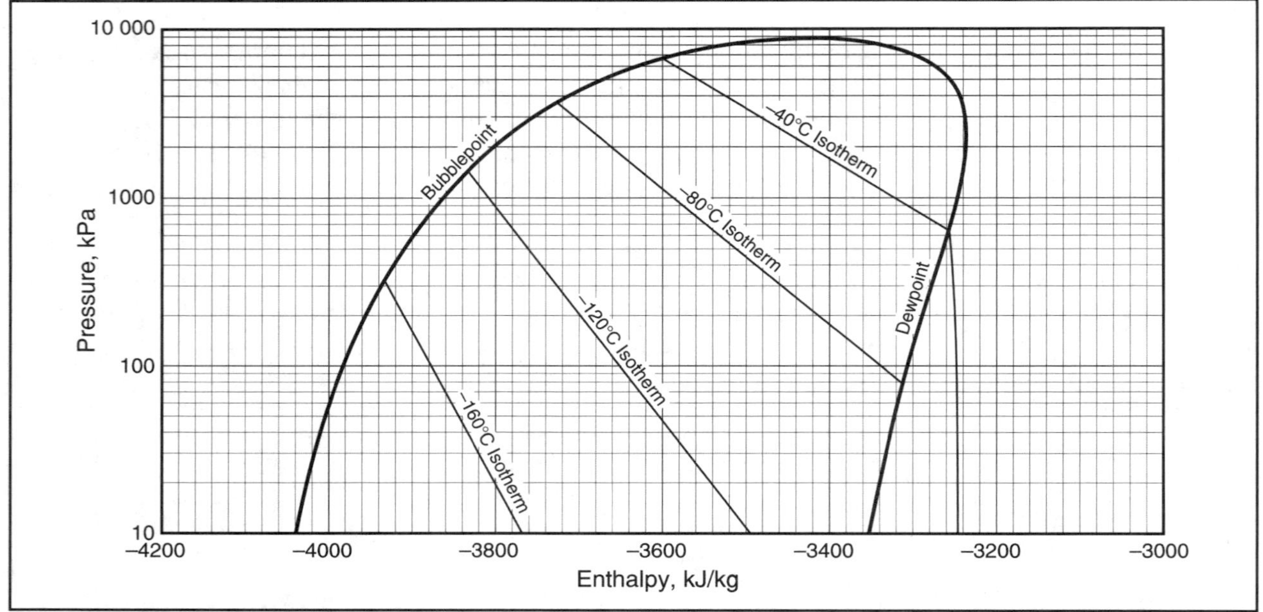

Figure 16.28 P-H Diagram for Example Mixed Refrigerant, N_2, C_1, C_2, C_3

Cryogenic Heat Exchanger

The main heat exchanger (MHE) is the cryogenic exchanger that efficiently condenses the pre-cooled natural gas to LNG, Figure 16.29. In its simplest form it contains two heat exchanger bundles with extremely large surface area. Standing vertically it is fed two mixed refrigerant (MR) streams: "Light MR" to the cold bundle on top and "Heavy MR" to the warmer bottom bundle. Some MHEs contain three bundles.

The high-pressure mixed refrigerant supply (T) is already partially condensed in discharge coolers downstream of the MR compressor (see Figure 16.26). The saturated vapor (U) is the light MR containing predominantly the lighter components C_1 and C_2 plus any nitrogen, if present. The heavy MR is saturated liquid (V), predominantly C_2 and C_3, which is subcooled in the warm or bottom bundle before being depressurized at point X and distributed across the warm bundle cross-section. The heavy MR then progressively vaporizes in downflow to the bottom outlet (DD) which in turn leads to the MR compressor suction (see Figure 16.26). The refrigerating effect of the heavy MR is used on three streams:

1. subcooling itself, points V to W
2. first stage of cryogenic natural gas cooling and condensing, M to N
3. condensing the light MR stream, U to Y

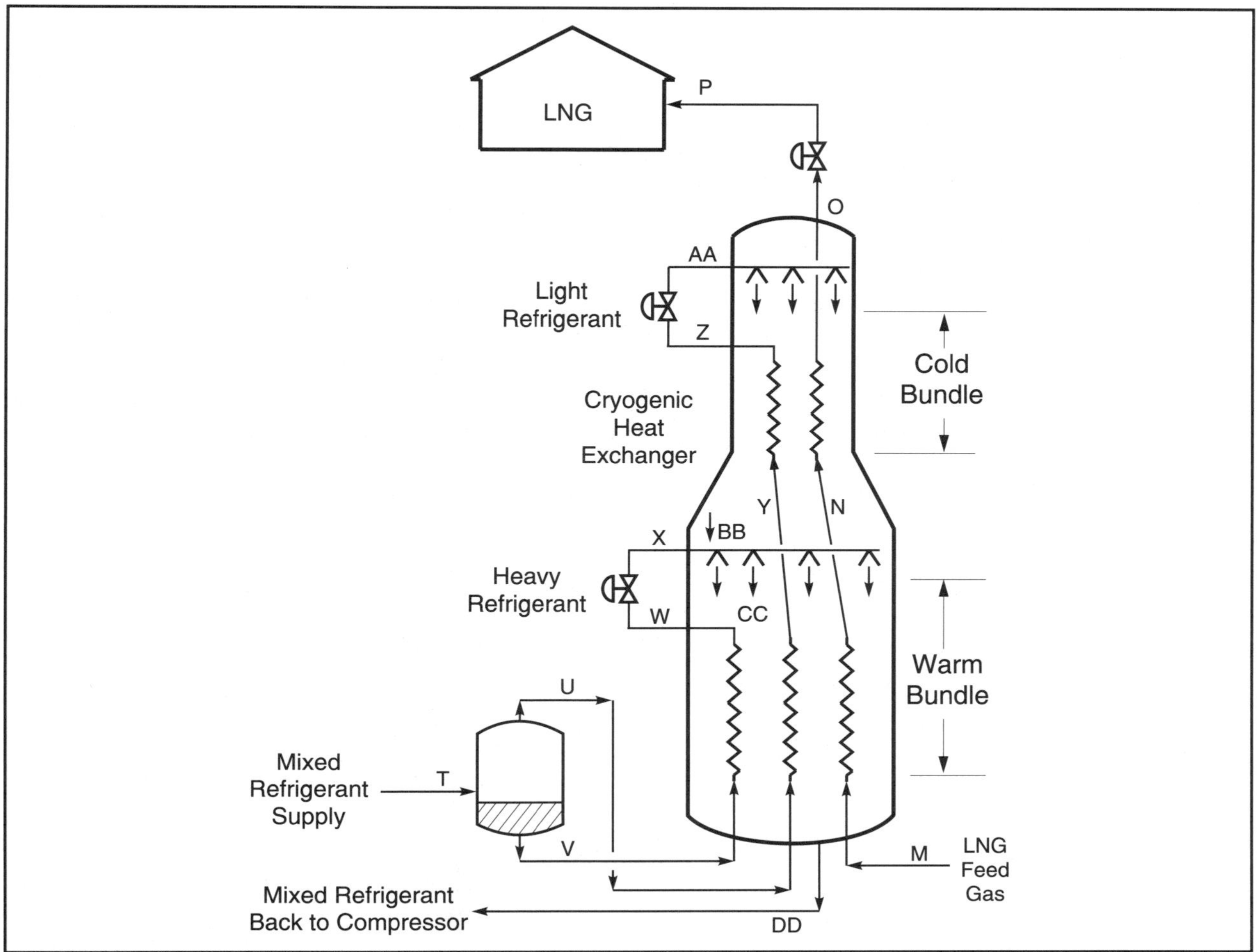

Figure 16.29 Main Heat Exchanger

The light MR, predominantly C_1, having been condensed at point Y is further subcooled in the top bundle to point Z. It is then depressured and distributed across the cold top bundle at point AA. It then downflows through the top bundle and progressively vaporizes to point BB. The refrigerating effect of the light MR is used on two streams:

4. subcooling itself, Y to Z
5. final stage of cryogenic natural gas cooling and condensing, N to O

Note that the refrigeration effect generated in the warm (bottom) bundle is actually due to the combined cooling effect of the heavy refrigerant (X) and the vaporized light refrigerant leaving the cold bundle (BB). This combined refrigeration effect in the warm bundle broadly comes from:

a. boiling (vaporizing) of the heavy MR
b. plus a superheating of the already vaporized light MR

The vaporized light MR and the liquid heavy MR that combined at CC eventually exit the bottom as vapor stream, DD. In base-load LNG plants this vapor line leading to the MR compressor suction can be more than 1 m [3.3 ft] diameter.

Cascade Refrigeration for LNG

A second type of LNG refrigeration system is the Phillips Optimized Cascade LNG Process. As the name suggests, it is a cascade system and employs three refrigerants: propane, ethylene, and methane.

A simplified process flow diagram for the Phillips Process is shown in Figure 16.30. Following the typical front-end treatment (acid gas removal, propane chilling/liquids removed, dehydration and mercury removal) the gas enters the propane refrigeration cycle. Here the gas is cooled to about –40°C [–40°F]. The propane refrigeration system also condenses the ethylene refrigerant and cools the methane refrigerant. The gas then enters the ethylene refrigeration system where it is further cooled to about –90°C [–130°F]. The ethylene refrigerant also condenses the methane refrigerant. The feed gas is finally cooled by the methane refrigerant to produce LNG.

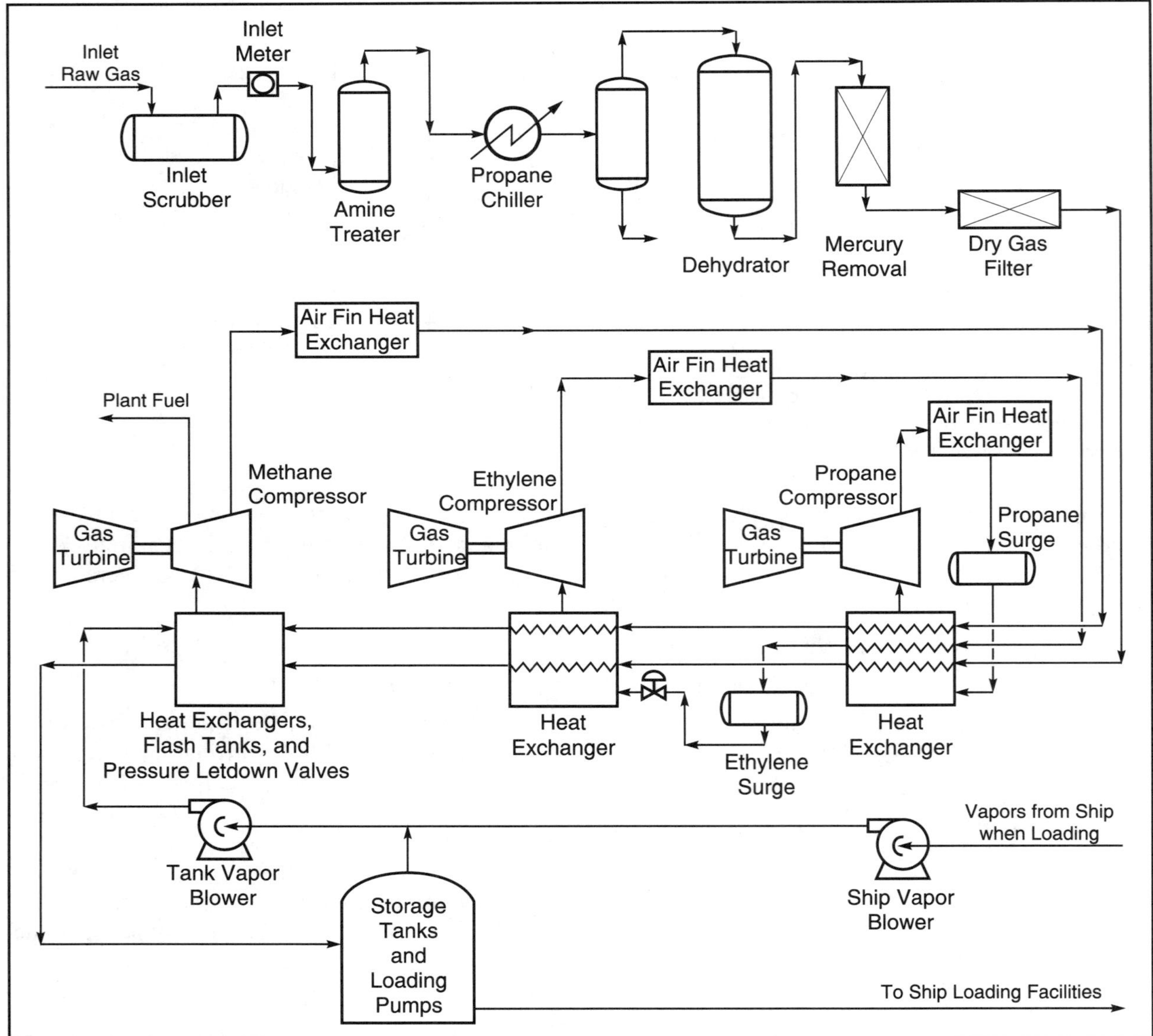

Figure 16.30 Phillips Optimized Cascade LNG Process *(Courtesy Phillips Petroleum Co.)*

The methane refrigeration system is open-cycle, i.e. the methane refrigerant stream is taken from the gas being liquefied. This allows the rejection of nitrogen (if required) and allows boil-off gases to be reintroduced back into the liquefaction process without the necessity for a large boil-off gas compressor.

If necessary, NGLs can be removed from the gas downstream of the propane refrigeration system. Fuel gas is taken off the methane refrigeration compressor.

Rather than one large proprietary spiral wound exchanger, the Phillip's process uses conventional BAHXs. This allows more flexibility in design capacity and may reduce major equipment delivery time.

As in the case of the mixed refrigerant process the primary energy input is refrigeration compression. Modern LNG plants typically have an overall efficiency (LNG out vs. feed gas in) of about 90-92% depending on feed composition and pretreatment requirements.

REFERENCES

16.1 Briley, G. C., *Hydr. Proc.* (May 1976), p. 173.

16.2 Agahi, R. A., "Turboexpanders with Dry Gas Seals and Active Magnetic Bearings in Hydrocarbon Processing," 78th Annual GPA Convention, Nashville, Tennessee (Mar. 1999).

16.3 McIntire, R., "Increased Productivity of Turboexpanders," 61st Annual GPA Convention, Dallas, Texas (Mar. 1982).

16.4 Mehra, Y. R., "Mehra Process Flexibility Improves Gas Processing Margins," 66th Annual GPA Convention, Denver, Colorado (Mar. 16-18, 1987).

16.5 Limb, D. I. and B. A. Czarnecki, "The Petroflux Process in NGL Recovery - Experience to Date and New Developments," 66th Annual GPA Convention, Denver, Colorado (Mar. 16-18, 1987).

16.6 Finn, A.J. *et.al,* "Modern Process Designs for Very High NGL Recovery," 78th Annual GPA Convention, Nashville, Tennessee (Mar. 1-3, 1999).

16.7 Wilkinson, J. D. and Hank M. Hudson, "Improved NGL Recovery Designs Maximized Operating Flexibility and Product Recoveries," 71st Annual GPA Convention, Anaheim, California (Mar. 16-18, 1992).

16.8 Pitman, R.N. *et.al,* "Next Generation Processes for NGL/LPG Recovery." 77th Annual GPA Convention, Dallas, Texas (Mar. 16-18, 1998).

NOTES:

APPENDIX 16A

PROPERTIES OF COMMON REFRIGERANTS

(See also Appendix B at the back of this book.)

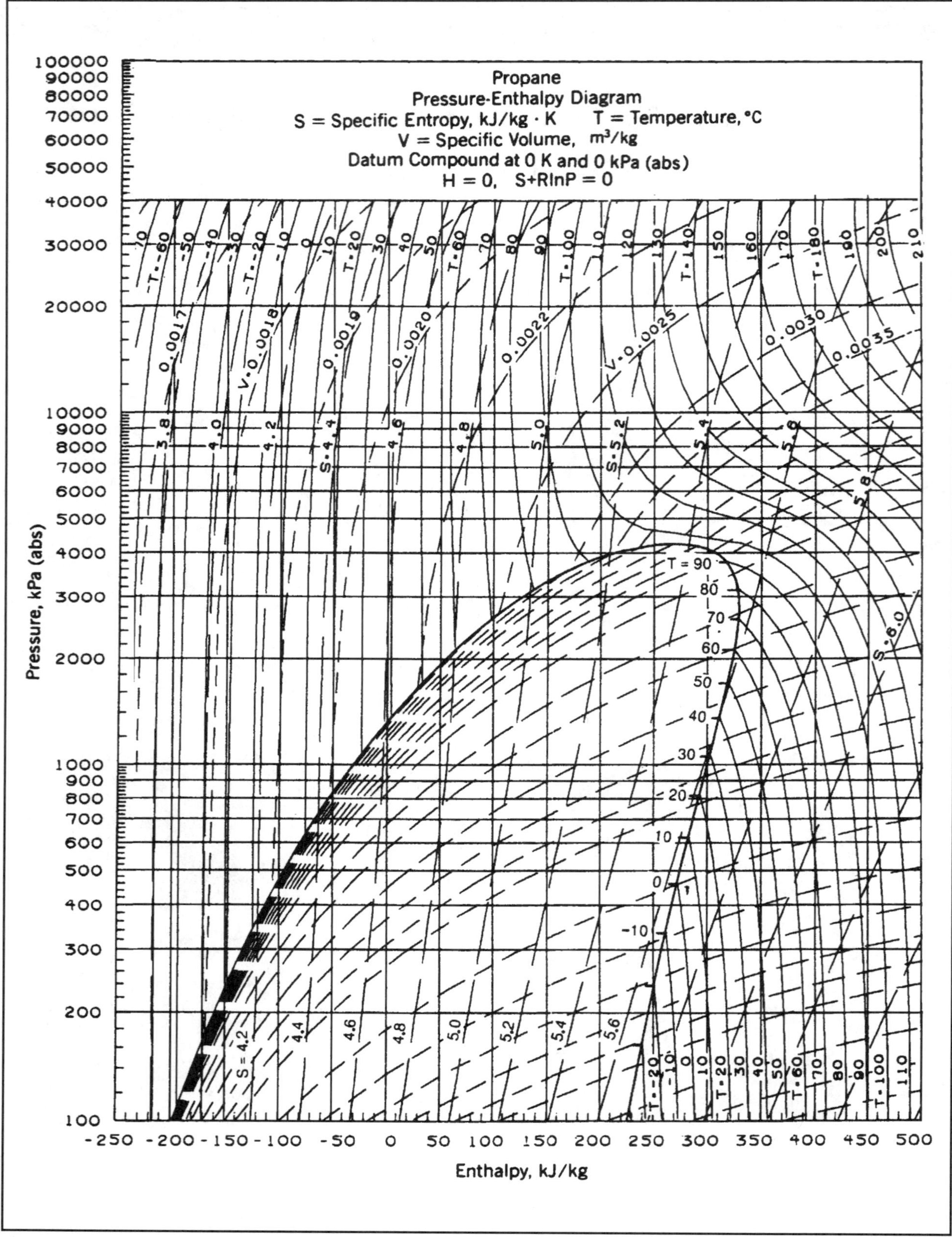

Figure 16A.1(a) Pressure-Enthalpy Diagram for Propane (SI Units)

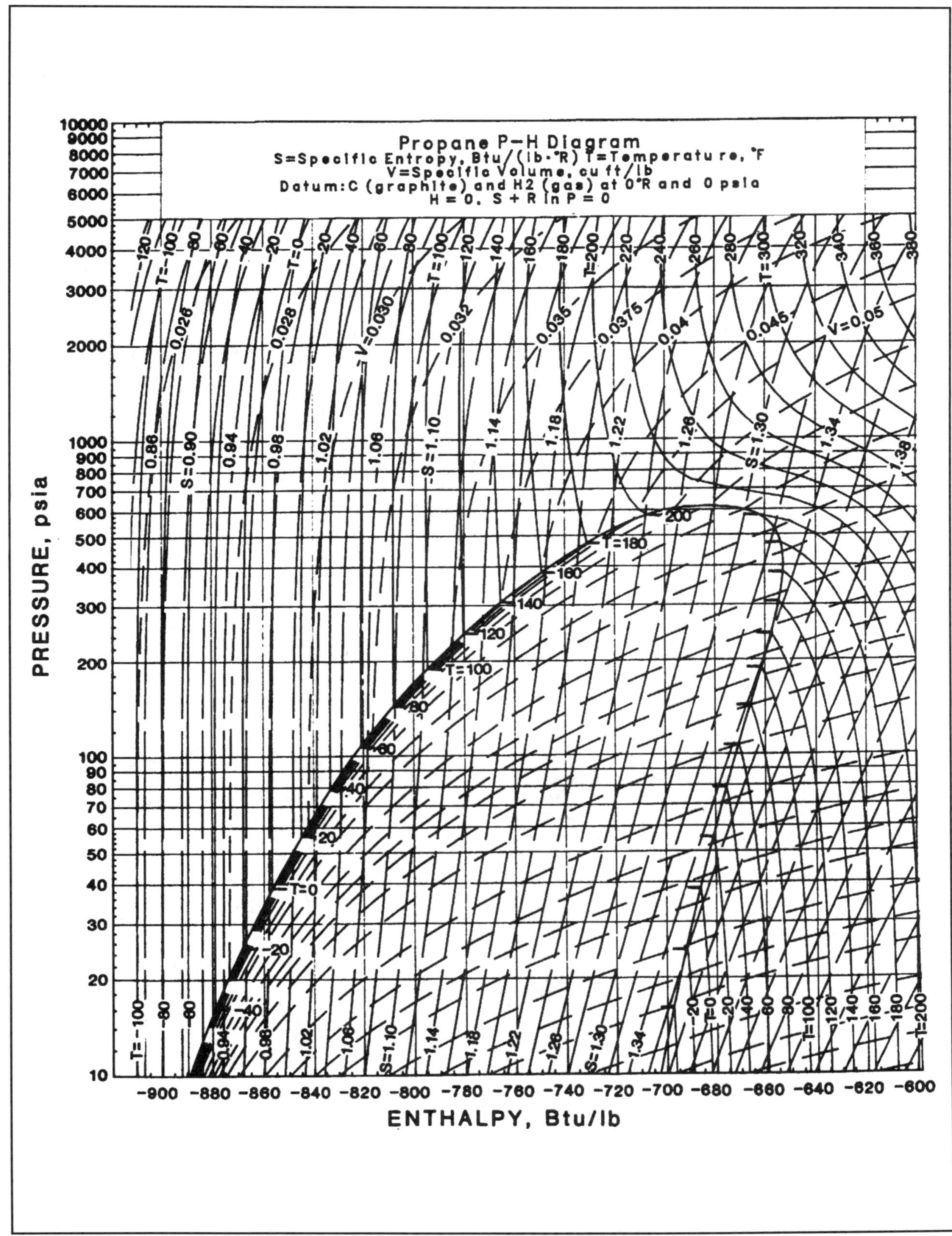

Figure 16A.1(b) Pressure-Enthalpy Diagram for Propane (FPS Units)

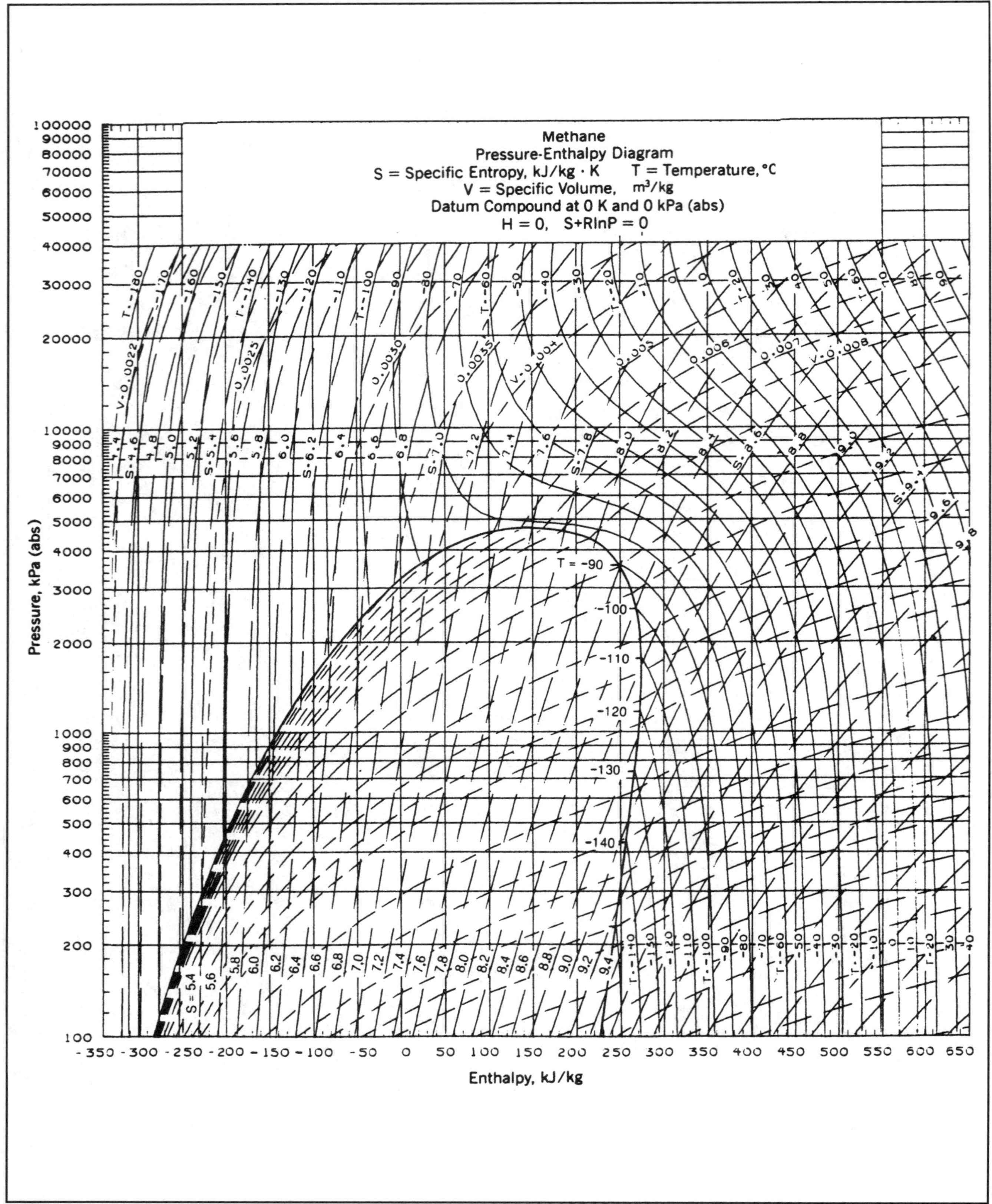

Figure 16A.2(a) Pressure-Enthalpy Diagram for Methane (SI Units)

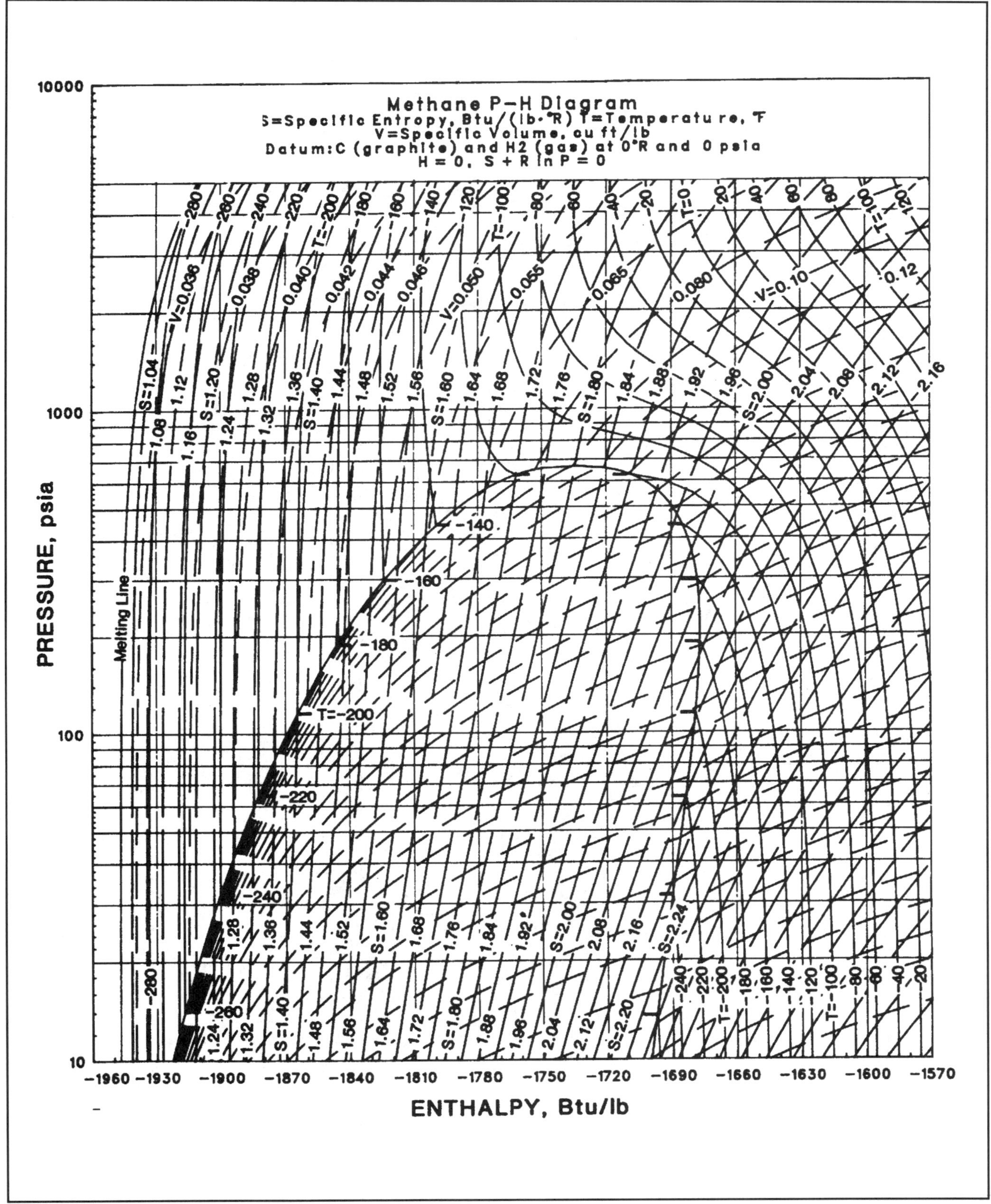

Figure 16A.2(b) Pressure-Enthalpy Diagram for Methane (FPS Units)

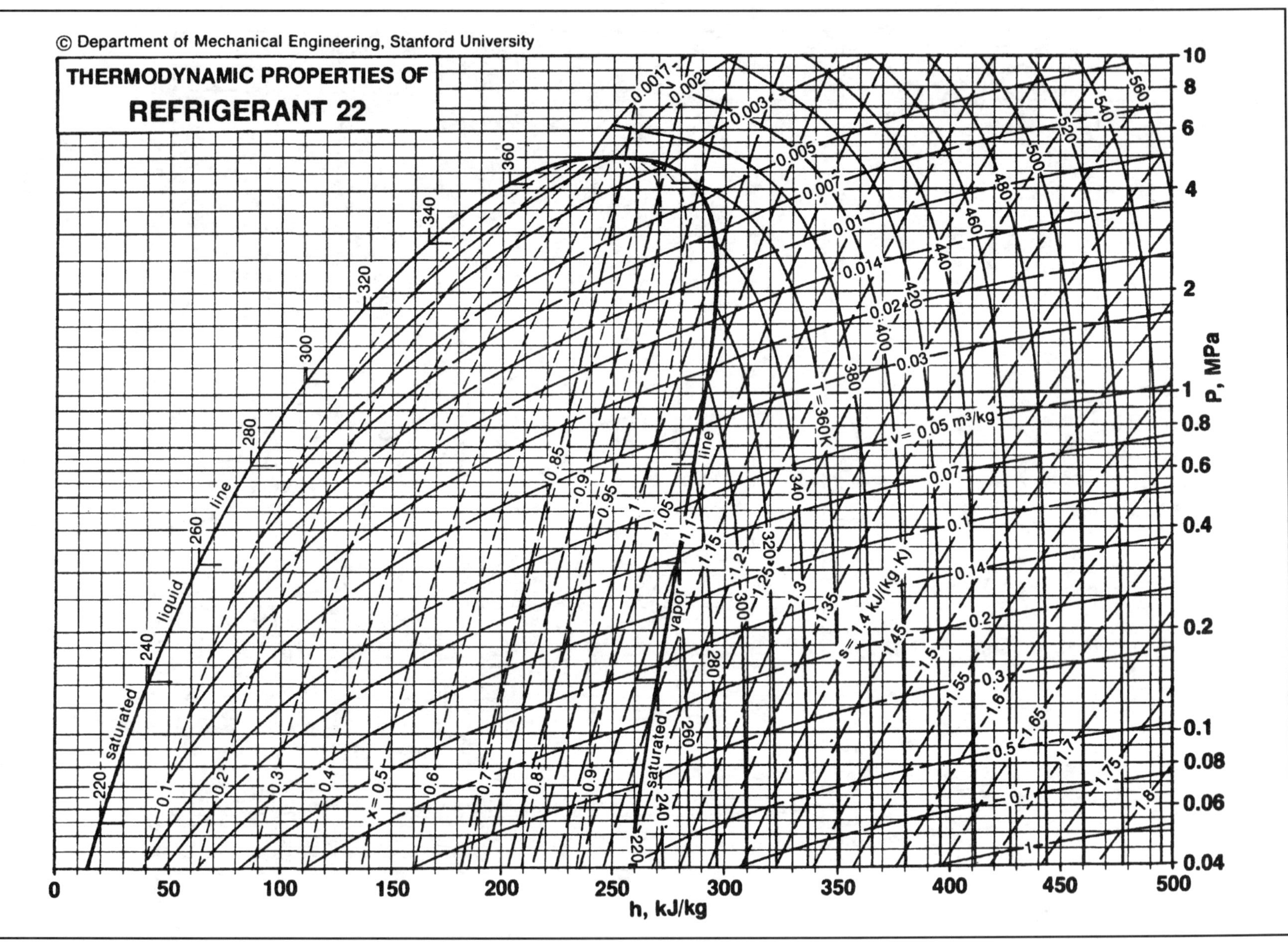

Figure 16A.3(a) Pressure-Enthalpy Diagram for Refrigerant 22 (SI Units)

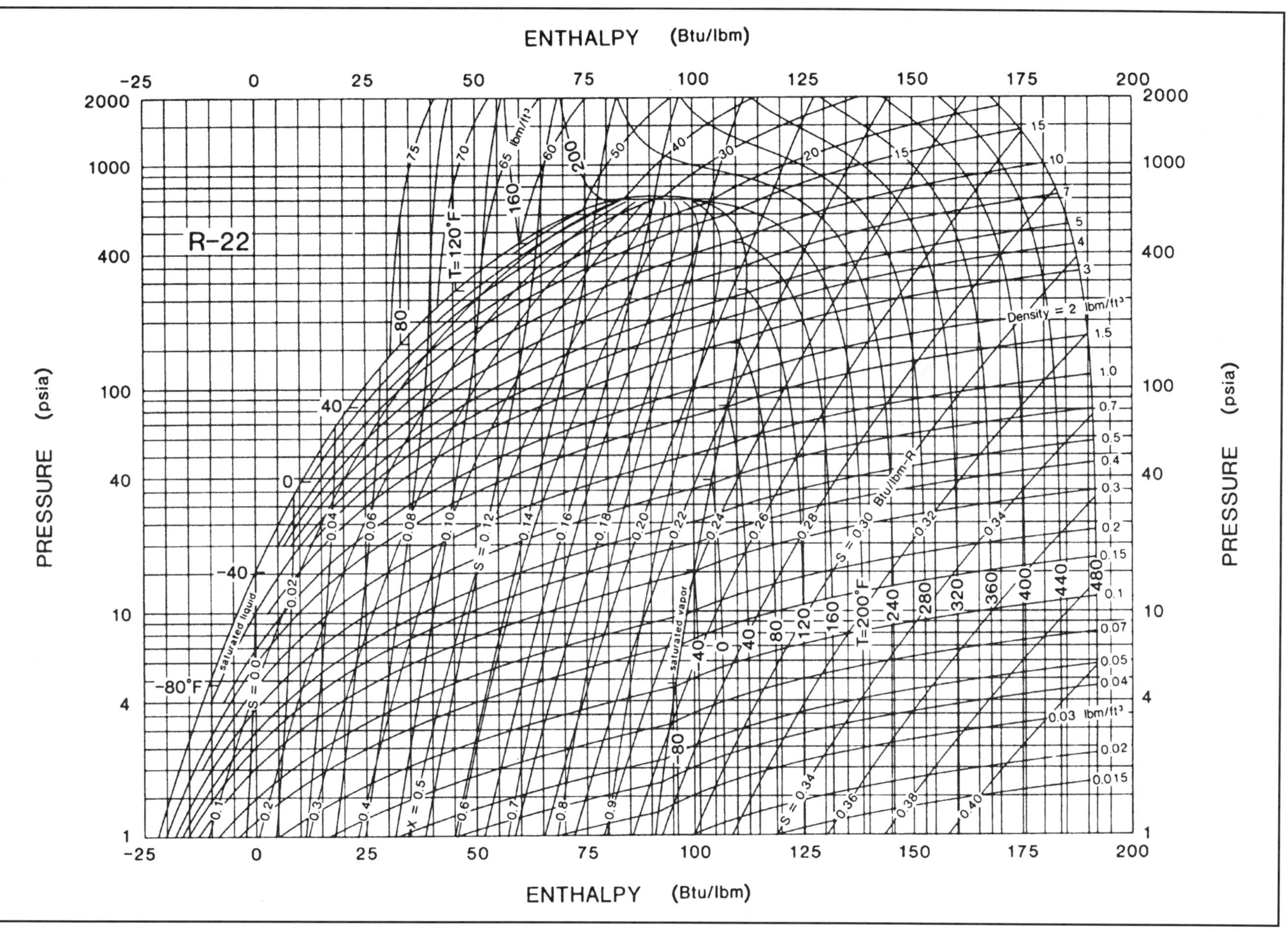

Figure 16A.3(b) Thermodynamic Properties of Refrigerant 22 (FPS Units)

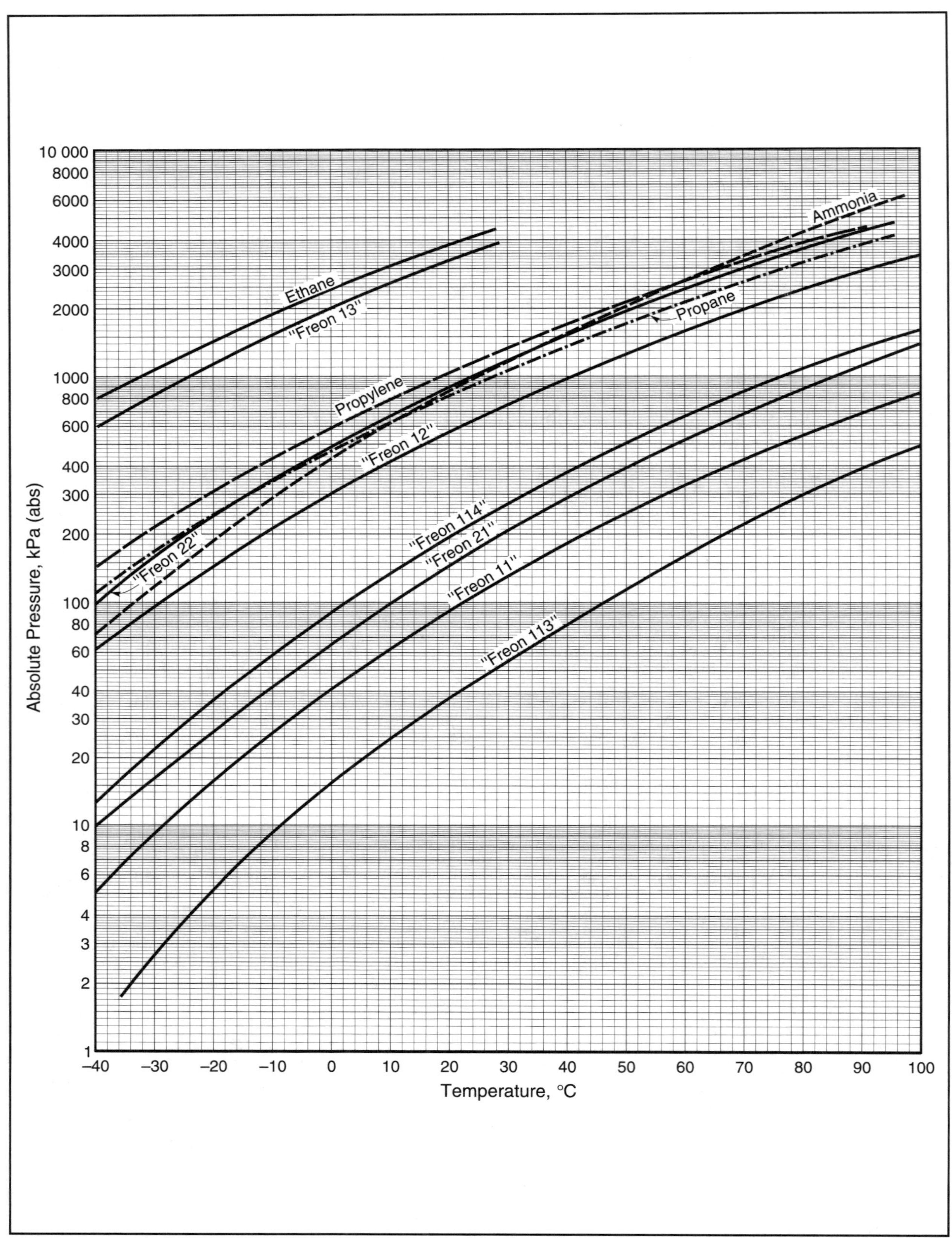

Figure 16A.4(a) Pressure-Temperature Relationships of Refrigerants (SI Units)

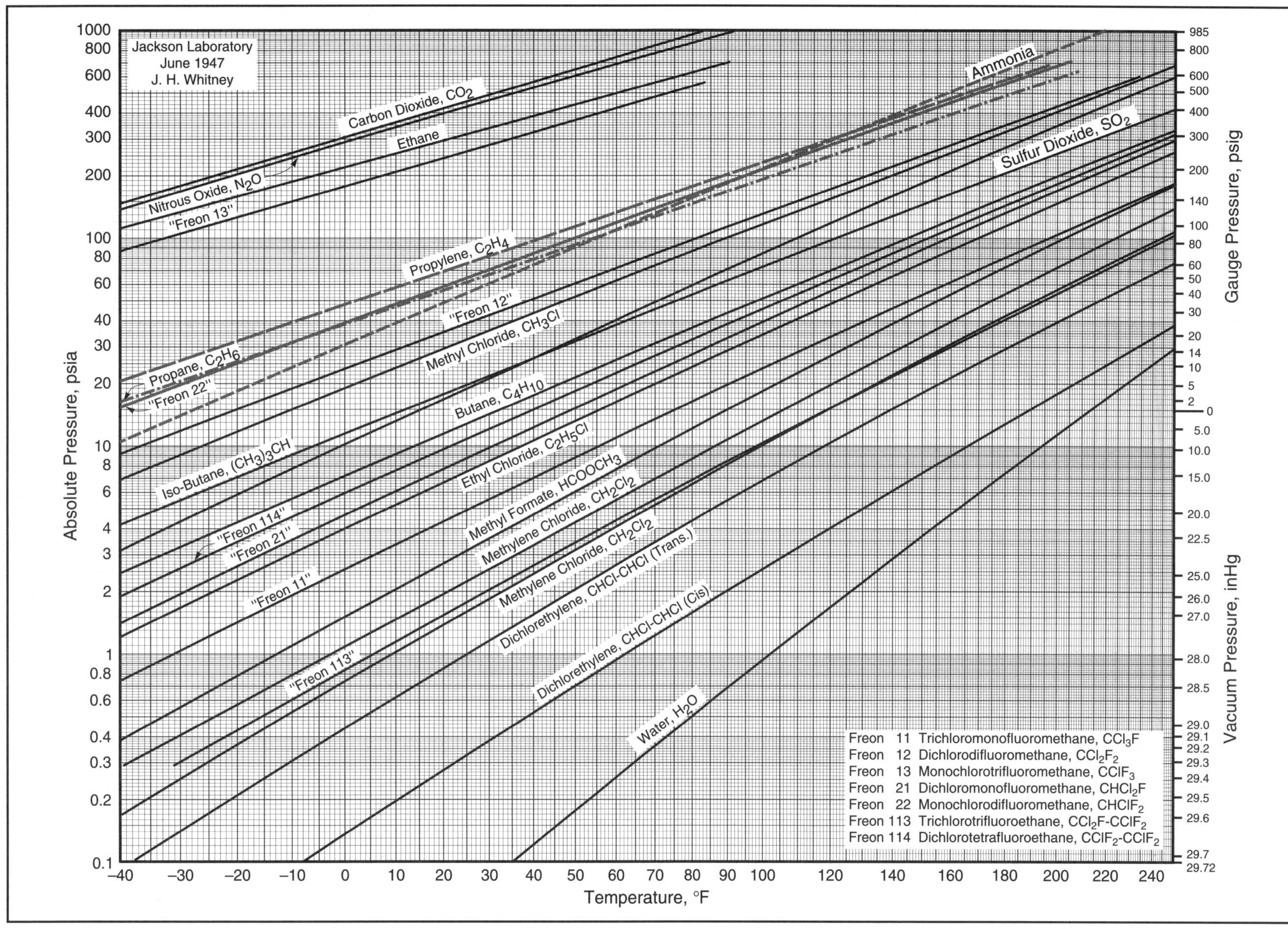

Figure 16A.4(b) Pressure-Temperature Relationships of Refrigerants (FPS Units)

TABLE 16A.3

Comparative Data of Refrigerants

Refrigerant Number ASHRAE Designation	R-134a	R-407C	R-410A	R-22	R-23
Replaces	R-12	R-22	R-22	N/A	R-13, R-503
Chemical Formula/Composition	CH_2FCF_3	R32/R125/R134a 23/25/52 wt%	R32/R125 50/50 wt%	$CHClF_2$	CHF_3
Molecular Weight	102.03	86.2	72.58	86.47	70.01
Boiling Point at 1 atm, °C	-26.5	-43.56	-51.53	-40.8	-82.03
Liquid Density at 25°C, kg/m^3	1210.	1134	1062	1195	669.91
Vapor Pressure of Sat'd. Liquid at 25°C, kPa	661.9	1174	1653	1043	4728
Heat Capacity of Liquid at 25°C, kJ/kg·K	1.42	1.54	1.84	1.24	1.55
Heat Capacity of Vapor at 1 atm at 25°C, kJ/kg·K	0.854	0.829	0.833	0.657	0.737
Thermal Conductivity of Liquid at 25°C, W/m·K	0.0824	0.0819	0.0886	0.0849	0.0774
Thermal Conductivity of Vapor at 101.3 kPa, W/m·K	0.0145	0.01314	0.01339	0.01074	0.0131
Critical Temperature, °C	101.1	86.74	72.13	96.24	25.83
Critical Pressure, kPa	4060	4619	4926.1	4981	4836
Boiling Point at 1 atm, °F	-15.7	-46.4	-60.76	-41.4	-115.66
Liquid Density at 77°F, lb/ft^3	75.02	70.8	66.32	74.53	41.82
Vapor Pressure of Sat'd. Liquid at 77°F, psia	96	170.3	239.7	151.4	685.7
Heat Capacity of Liquid at 77°F, Btu/lb-°F	0.339	0.367	0.44	0.296	0.37
Heat Capacity of Vapor at 1 atm at 77°F, Btu/lb-°F	0.204	0.198	0.199	0.157	0.176
Thermal Conductivity of Liquid at 77°F, Btu/hr-ft-°F	0.0478	0.0455	0.0511	0.0458	0.045
Thermal Conductivity of Vapor at 1 atm, Btu/hr-ft-°F	0.00836	0.00758	0.00772	0.00621	0.0076
Critical Temperature, °F	213.9	188.13	161.83	205.24	78.5
Critical Pressure, psia	588.9	669.95	714.5	722.39	701.4
Ozone Depletion Potential (ODP), CFC-12=1	0	0	0	0.05	0
Greenhouse Warning Potential (GWP), CO2=1	1300	1526	1725	1500	11700

17

FRACTIONATION AND ABSORPTION FUNDAMENTALS

The comprehensive design of fractionators and absorbers is a specialty area that involves details outside the scope of this book. Many references discuss the particulars of fractionation design. In addition, tray and packing vendors can provide detailed information regarding tower internals. This chapter is an overview of the basic factors governing performance for those who specify, buy and operate such units.

An absorber is any device where a liquid (absorbent) flows counter-current to a gas stream for the purpose of removing one or more constituents (absorbate) from that gas. Most absorbers are vertical with the liquid entering at the top and the gas at the bottom. The amount of contact that must be provided in this absorber tower depends on the system, the relative flowrate of liquid and gas, and the concentrations involved.

A stripper is any device where a gas flows counter-current to a liquid stream for the purpose of removing one or more components from the liquid.

A fractionator is a device for separating a mixture into two or more parts, at least one of which will have a controlled composition or vapor pressure. In crude oil or condensate systems such a fractionator is often called a *stabilizer* and is an alternative to stage separation. The fractionator is essentially a constant pressure device that uses heat, absorption and stripping to separate components based on the difference in their boiling points.

Each of the above are mass transfer devices that must operate according to the rate criteria outlined in Chapter 12.

$$Q_m = k_G A (c_1 - c_2) \qquad (17.1)$$

Where:

Q_m = mass transfer rate between phases
k_G = mass transfer coefficient (empirical)
A = area of contact between vapor and liquid
c_1 = concentration (high)
c_2 = concentration (low)

(In using Equation 17.1 one never truly knows a value of "A." Consequently, the mass transfer coefficient often is written k_{GA}, found by dividing measured mass rate by the concentration difference.)

Mass always flows from high concentration to low concentration. In an absorber the concentration of the absorbate must be higher in the vapor than in the liquid for absorption to occur. In *stripping*, the opposite of absorption, the liquid concentration must be the higher of the two. Concentration is the *driving force*. The term "concentration" as used in this paragraph means concentration relative to equilibrium.

The resistance to transfer is the reciprocal of area. The empirical mass transfer coefficient includes the effect of all variables except concentration. For a given system operating at specified conditions with required characteristics of the effluent streams, the efficiency of mass transfer depends on area of contact between vapor and liquid, and also on those factors (e.g. viscosity, surface tension, relative volatility, etc.) which affect diffusion of molecules across the vapor-liquid interface. Mass transfer area, temperature gradient, and vapor-liquid ratios, are the primary variables available to the designer to meet the processing objective economically.

For a fractionator with a given feed composition and phase condition, at a fixed pressure, the basic variables available to the designer are control of the temperatures at top and bottom, feed location and the amount of vapor-liquid contact. In an absorber the basic variables are absorbent rate and composition, and the amount of vapor-liquid contact.

There are three basic ways to maximize the amount (area) of vapor-liquid contact per unit volume of tower:

1. Bubble the gas through liquid.
2. Spray the liquid as droplets into the gas stream.
3. Flow the liquid as a thin film over a surface with a high area.

Many different mechanical configurations have been developed to accomplish these. Suppliers continually search for new ways to obtain advantage by increasing contact efficiency. In fractionation and absorption, (1) and (3) are the primary methods employed.

TRAY (PLATE) TYPE TOWERS

This is a vertical vessel containing a series of trays. Liquid on the top tray is provided by the absorbent or from partial or total condensing of the vapor leaving the tower. The liquid flows across the tray and then by gravity to the next tray below, using what is called a *downcomer*. The liquid flows across the second tray and repeats the same path until it emerges from the bottom of the tower. On each tray a mechanism is provided for the upward flowing gas to bubble through the liquid crossing the tray.

The left portion of Figure 17.1 shows the basic pattern, using two different types of downcomers. Weirs are used to help maintain the proper level of liquid on the trays.

As the tower diameter increases and/or high liquid rates are involved, it may be impractical to flow the liquid across the entire tower diameter because of the liquid gradient involved. In these cases the tray is split so that more than one path is used, as shown in the right part of Figure 17.1. This shortens the flow path and reduces liquid gradient per tray. The liquid gradient is the difference in liquid height from the side of the tray where the liquid enters to the side where liquid leaves and flows into the downcomer. Large liquid gradients cause maldistribution of gas flow and increase tray pressure drop. As a general rule, the liquid gradient should not exceed 40-50 mm [1 1/2-2 in].

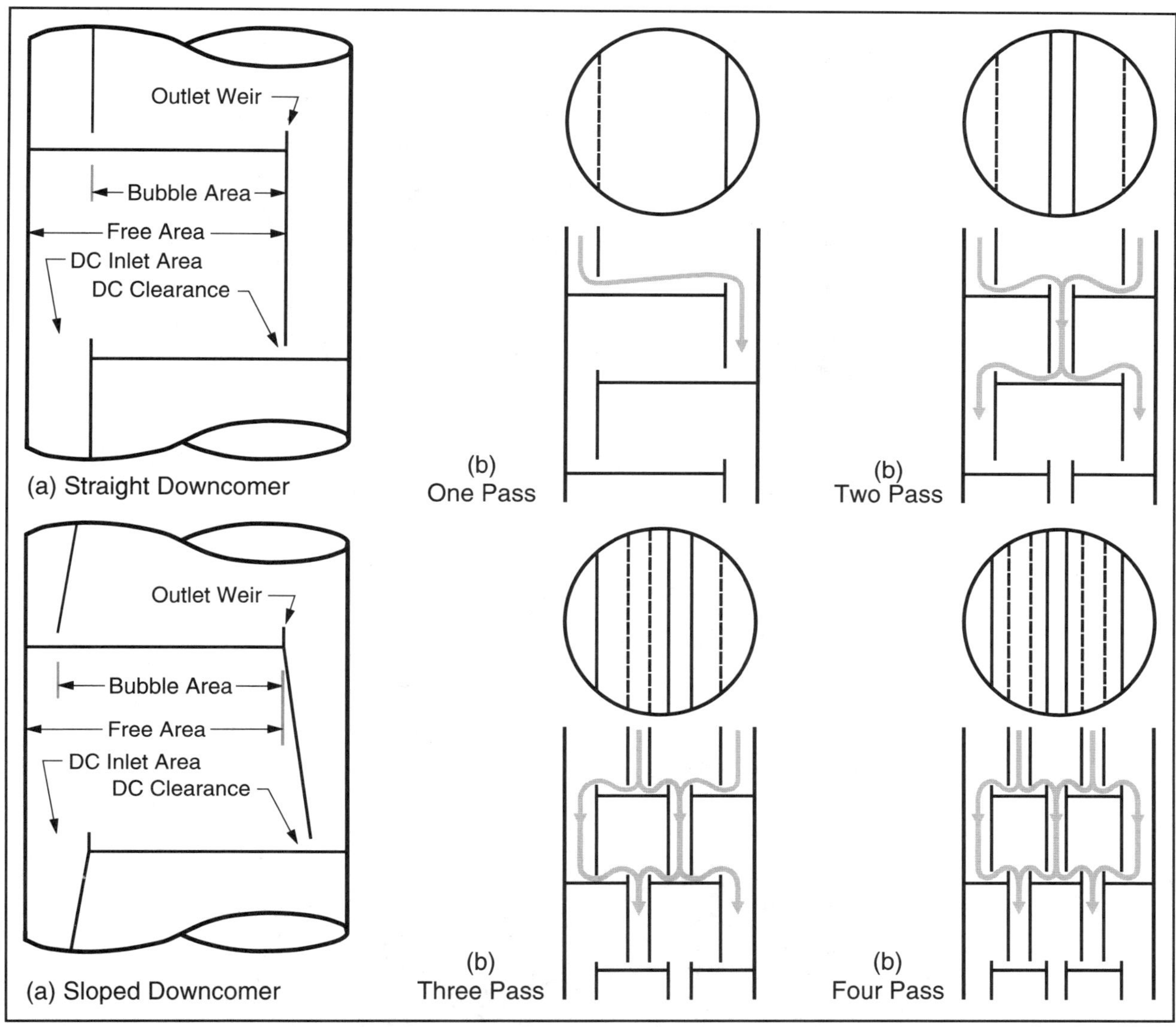

Figure 17.1 Basic Layout of Tray Type Towers

Figure 17.2 shows three common types of trays used. *Bubble caps* have a riser in the middle of the cap through which the gas passes. This gas hits the top and flows out through the slots. The riser keeps the liquid from flowing through the gas opening, which is called *weeping*. As the name implies, *sieve trays* are nothing but a series of holes drilled in flat metal. Gas bubbles through the holes. The size, number and spacing of holes must be proper to obtain good gas-liquid contact and prevent weeping.

The *valve tray* is a generic description of a tray containing devices where the opening for gas flow device varies with gas rate. There are many types and configuration which are proprietary to a given manufacturer. The "valve" rests on the tray deck in the closed position until the rising gas flow is sufficient to cause it to open. It is, in effect, a variable orifice whose opening depends on gas rate up to the limit imposed by the mechanism which holds it in place.

Both sieve and valve trays are less expensive than and can be more efficient than bubble caps, but weeping can present a problem at low gas or liquid rates. Factors governing choice and design will be discussed in later sections of this chapter.

Figure 17.2 Three Common Types of Contact Trays *(Courtesy Glitch, Inc.)*

PACKED TOWERS

These are basic alternatives to tray-type units. The packing may be *structured* or *random*. Whatever the type, said packing must be capable of producing a thin liquid film and a long, tortuous flow path even at low flowrates. It also should exhibit a low pressure drop and liquid holdup. Structurally, it should not degrade thermally, it should be chemically impervious to the fluids involved and not crush or powder mechanically. There has been a substantial evolution in packing in recent years.

The right-hand portion of Figure 17.3 shows some common types of random packing. Random packing can be made from plastics like fiberglass reinforced polypropylene and corrosion-resistant metals.

Random packing consists of individual pieces, varying in size from 10-15 mm up to 75-80 mm [½-¾ in up to 3-3 ½ in]. These are typically "dumped" into the vessel and come to rest in a random manner. Structured packing comes in sections made for placement in the unit. It is made from knitted mesh and all sorts of metal plate and tubing which is crimped, twisted, wound, stamped or somehow arranged to present a large area per unit volume. Each is described by a name which is a registered trademark of its owner. Names like Mellapak, Gempack, Flexipac, Montz, and Flexigrid are just a few of the structured packings available.

The left side of Figure 17.3 shows all of the elements that may exist in a packed tower.[17.1] Shown are various types of liquid distributors and support plates. Liquid distribution can be a problem. The type of distributor chosen and the number used may be critical to obtain satisfactory tower performance.

In some service there is a tendency to retrofit some tray contactors with modern packing. This may expand tower capacity and efficiency. However, packed towers have some inherent limitations. It is more difficult to predict the performance of packed towers. In some applications their flexibility is limited. A later section discusses further packing details.

The table below summarizes some of the advantages and disadvantages of trays and packing.

Packing		Trays	
Advantages	**Disadvantages**	**Advantages**	**Disadvantages**
Lower pressure drop	Turndown ratio limited	Higher turndown	Higher pressure drop
Good for low pressure and vacuum operation	Fouling and plugging in dirty fluid service	Less subject to fouling on dirty fluids	Greater tendency for foaming
Can be manufactured from a number of materials – metallic, ceramic, plastic – more adaptable to corrosive environments	More difficult to handle side draws and side feed streams	Easier to handle side draws and side feeds	Limited materials in corrosive service
Less subject to foaming	Liquid redistribution required in tall towers		
Less liquid holdup in tower			

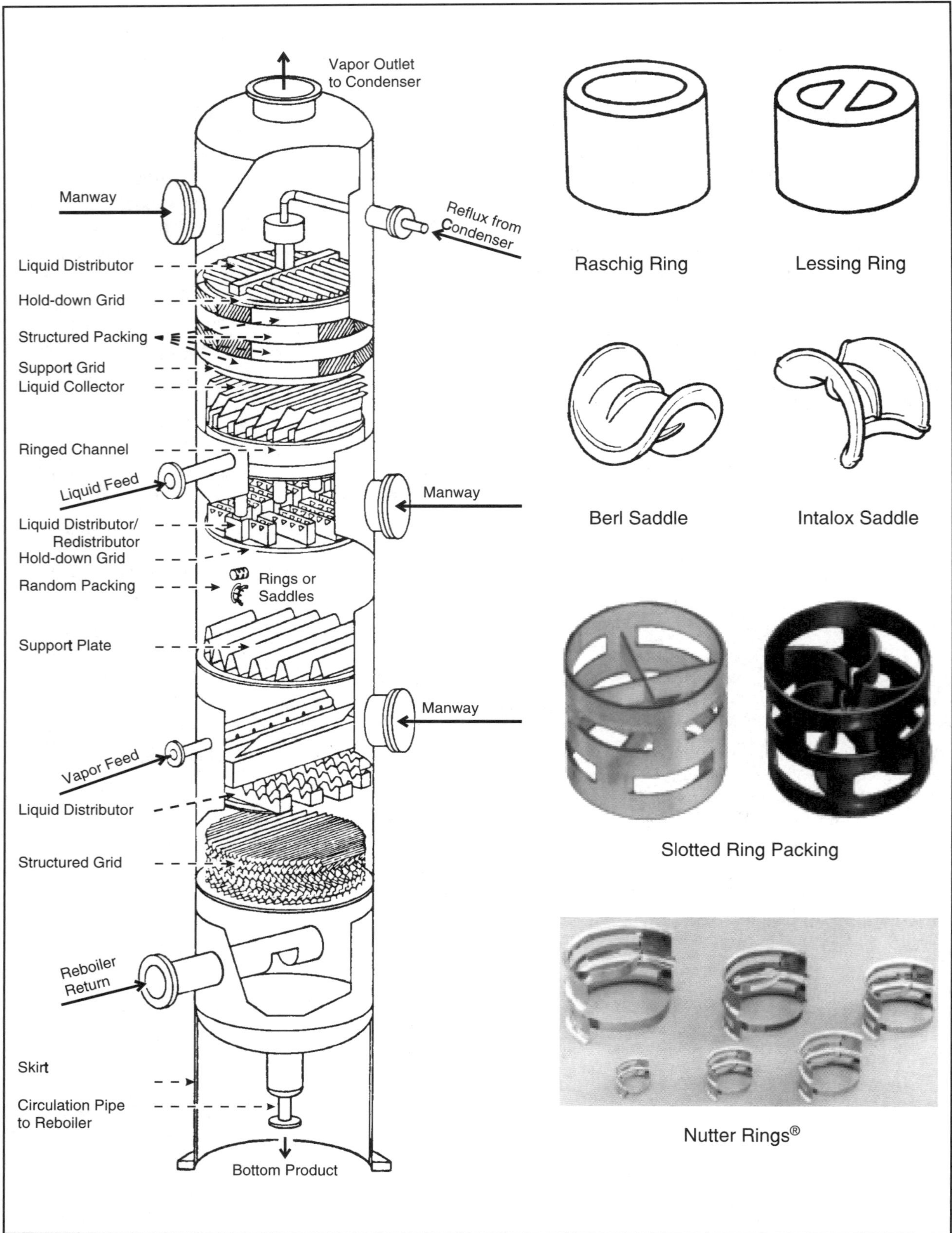

Figure 17.3 Towers Using Structured and/or Random Packing

FRACTIONATION

Regardless of how fluids are extracted from natural gas, fractionation is often necessary to produce saleable products meeting market specifications. Fractionation produces a sharper separation than a series of separator flashes (see Figure 5.6 in Volume 1). The number of fractionating columns required depends on the number of products to be sold and the character of the feed. The single tower system shown in Figure 17.4(a) ordinarily produces one specification product (the bottom stream), with all other components in the feed passing overhead to the distillate product. This type of fractionator (stabilizer) is ordinarily used where a C_5+ condensate stream with an atmospheric vapor pressure is being produced for sale or storage.

The two-tower system shown in Figure 17.4(b) is most commonly used to produce an LPG (C_3-C_4 mixture) distillate product and a C_5+ natural gasoline bottoms product. In this system, the deethanizer must remove all methane, ethane, and other constituents not saleable in the two product streams from the second tower. Any component that enters the second tower must necessarily leave in one of the product streams.

The three-tower system shown in Figure 17.4(c) commonly produces commercial propane (C_3), commercial butane (iC_4 and nC_4) and natural gasoline (C_5+) as products. In this system also, the deethanizer must remove the volatile constituents that cannot be sold in one of the three products. The sequence of fractionation following the deethanizer may be varied. In the second tower, an LPG mixture could be produced overhead with natural gasoline produced as bottoms. The third tower would then split the LPG into commercial propane overhead and commercial butane as bottoms. This

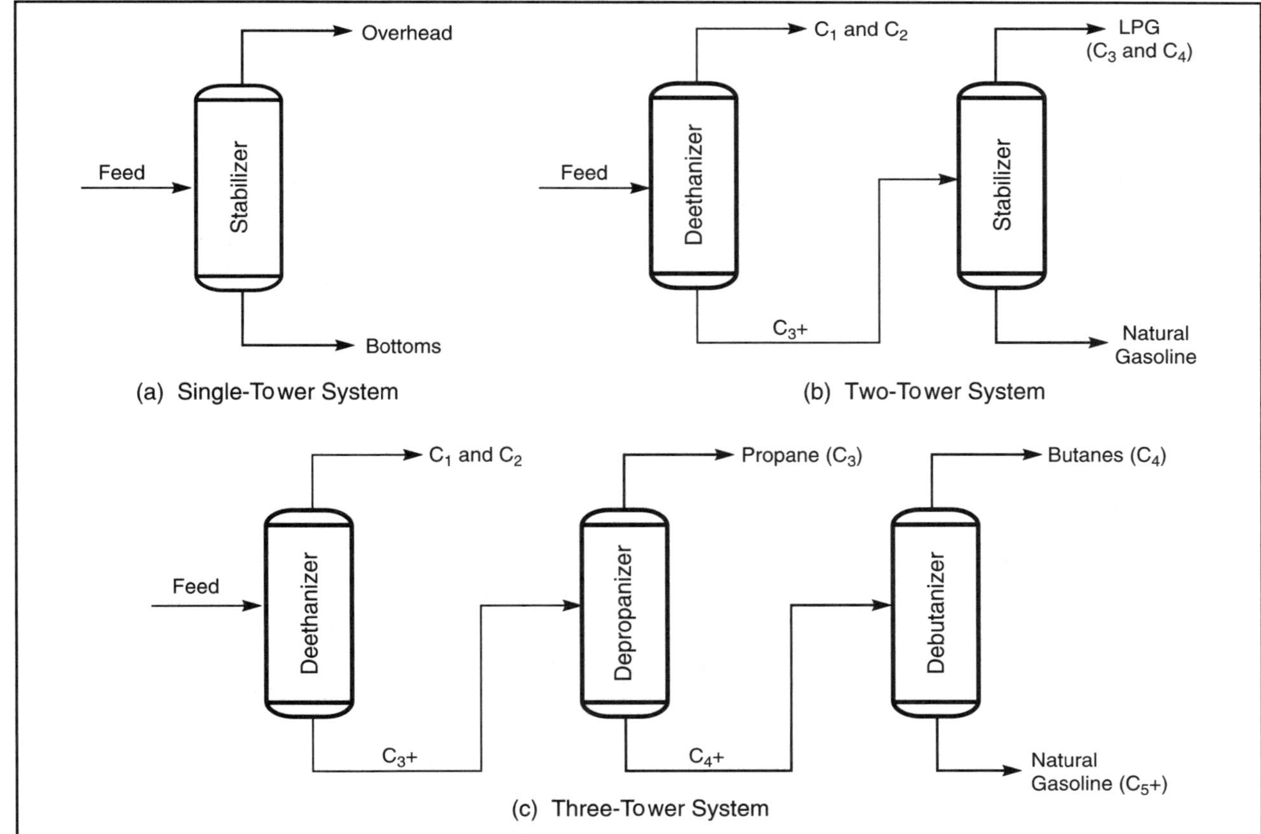

Figure 17.4 Typical Fractionation Systems

sequence is favored sometimes where the market situation is variable and a market for LPG only exists during a portion of the year. During this period, the third tower would be shut down and not operated.

If ethane product was saleable, the deethanizer would be proceeded by a demethanizer which would remove methane and lighter components from the NGL. Due to the difficulty in condensing reflux for a demethanizer, this column is often a stabilizer with a top tray feed (Figure 17.4) rather than a conventional fractionator.

Figure 17.5 shows an example distribution of hydrocarbon components for four separation schemes – single stage flash, three stage flash, stabilizer, and three tower fractionator systems. Note the much sharper separation between components in the stabilizer/fractionation system compared to stage separation.

Equilibrium Concept

Fractionation or distillation refers generally to those processes used to separate components based on differences in their boiling points. Binary distillation involves separation where only two components are present. Multicomponent distillation, on the other hand, refers to systems where many components are present in the feed, even though the separation point is between two components. These components are often referred to as key components – the more volatile component is called the "light key" and the less volatile, the "heavy key." So, even in multicomponent distillations, the primary separation parameters are set by the key components.

The design of fractionators (and absorbers and strippers) is based on the assumption of equilibrium stages. In the final design this assumption is corrected by the use of tray efficiencies and packing heights equivalent to an equilibrium stage, but the primary factor affecting distillation design is the equilibrium behavior of the components being separated.

The phase behavior of binary systems is very important to the understanding of fractionation. Even though most of our separations involve multicomponent systems, the actual separation occurs between two pure components. For example, in a depropanizer, there are several components present in the feed ... C_2, C_3, iC_4, nC_4 and C_5+; but the actual separation occurs between C_3 and iC_4. So, if one has a clear understanding of the phase behavior of the binary C_3-iC_4 system, one can predict (with reasonable accuracy) the fractionation system required to make this separation.

The T-x diagram is very useful in visualizing what happens in the distillation process. It is the plot of the phase behavior of a binary system at constant pressure. It shows the bubblepoint and dewpoint temperature for all possible binary compositions.

To gain a better appreciation of how a T-x diagram is developed, refer to Figure 4.5 (Volume 1) which shows the T-P-x diagram for the C_2-nC_7 system.

Now, imagine that we enter Figure 4.5 at a pressure of say 2068 kPa [300 psia] and "slice off" the tops of the phase envelopes. The remaining three-dimensional diagram would appear as a large mesa or plateau. When we view this "mesa" from above, we see the "football" shaped envelope shown in Figure 17.6. This is the T-x diagram at 2068 kPa [300 psia] for the C_2-nC_7 binary system.

The top of the envelope is the locus of all the dewpoints at 2068 kPa [300 psia] for all possible combinations of C_2 and nC_7. Likewise, the bottom of the envelope is the locus of all the bubblepoints. Any point inside the envelope will be in the two phase region. Any point above will be all vapor; any point below, all liquid.

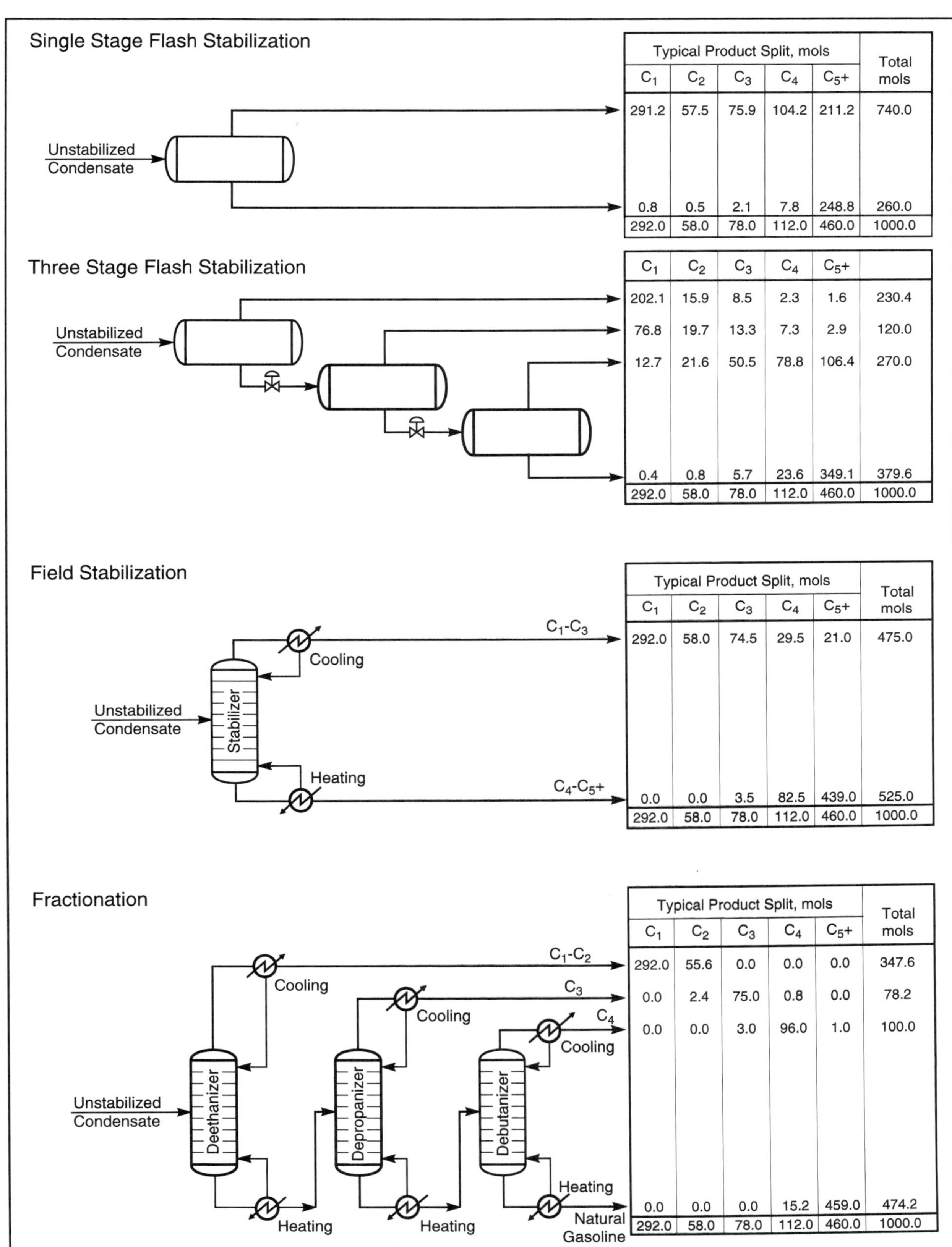

Single Stage Flash Stabilization

Typical Product Split, mols					Total mols
C_1	C_2	C_3	C_4	C_5+	
291.2	57.5	75.9	104.2	211.2	740.0
0.8	0.5	2.1	7.8	248.8	260.0
292.0	58.0	78.0	112.0	460.0	1000.0

Three Stage Flash Stabilization

C_1	C_2	C_3	C_4	C_5+	
202.1	15.9	8.5	2.3	1.6	230.4
76.8	19.7	13.3	7.3	2.9	120.0
12.7	21.6	50.5	78.8	106.4	270.0
0.4	0.8	5.7	23.6	349.1	379.6
292.0	58.0	78.0	112.0	460.0	1000.0

Field Stabilization

Typical Product Split, mols					Total mols
C_1	C_2	C_3	C_4	C_5+	
292.0	58.0	74.5	29.5	21.0	475.0
0.0	0.0	3.5	82.5	439.0	525.0
292.0	58.0	78.0	112.0	460.0	1000.0

Fractionation

Typical Product Split, mols					Total mols
C_1	C_2	C_3	C_4	C_5+	
292.0	55.6	0.0	0.0	0.0	347.6
0.0	2.4	75.0	0.8	0.0	78.2
0.0	0.0	3.0	96.0	1.0	100.0
0.0	0.0	0.0	15.2	459.0	474.2
292.0	58.0	78.0	112.0	460.0	1000.0

Figure 17.5 Comparison of Stage Separation to Fractionation-Type Separation

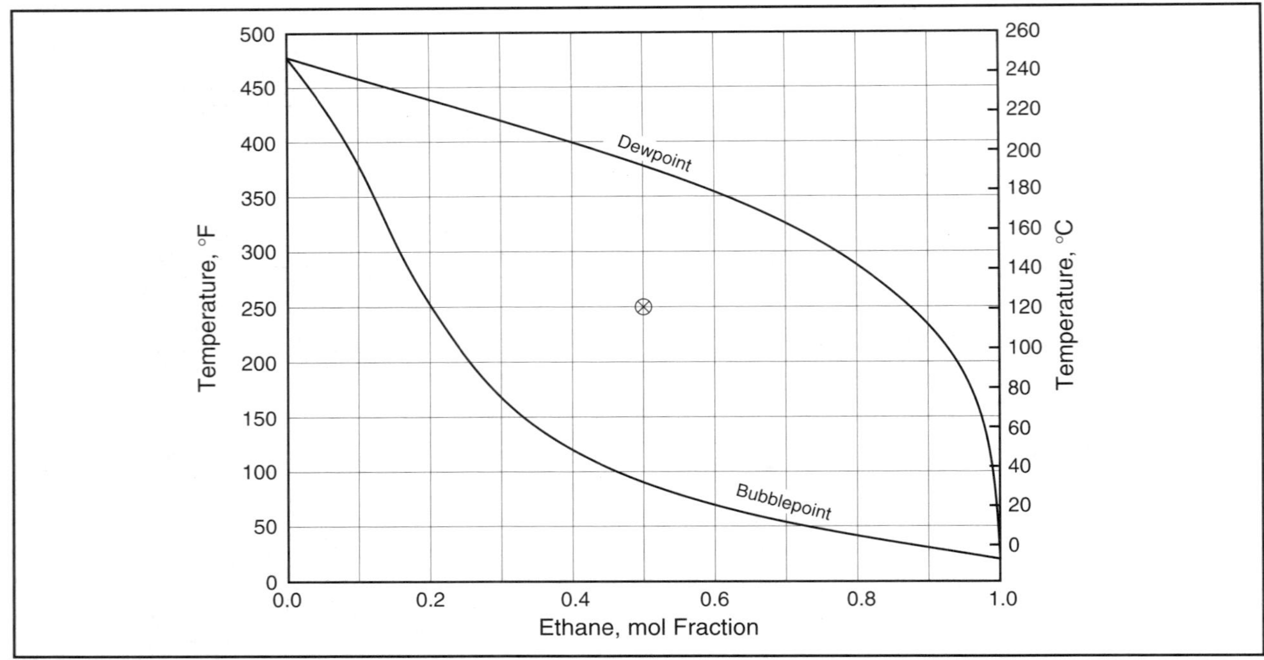

Figure 17.6 T-x Diagram for Ethane, n-Heptane at 2068 kPa [300 psia]

Imagine an equimolar mixture of C_2 and nC_7 (50% C_2, 50% nC_7) at 2068 kPa [300 psia] and 10°C [50°F]. The phase condition for this mixture is a compressed liquid. Say we heat this mixture to 121°C [250°F], and feed it to the separator in Figure 17.7. As you can see from Figure 17.6, the phase condition of the feed is two phase. The composition of the vapor leaving the separator is about 87% C_2 and 13% nC_7; the composition of the liquid is 20% C_2 and 80% nC_7. The vapor is at its dewpoint and the liquid at its bubblepoint. This is an example of equilibrium separation and is identical to what happens in one equilibrium stage in a fractionation column.

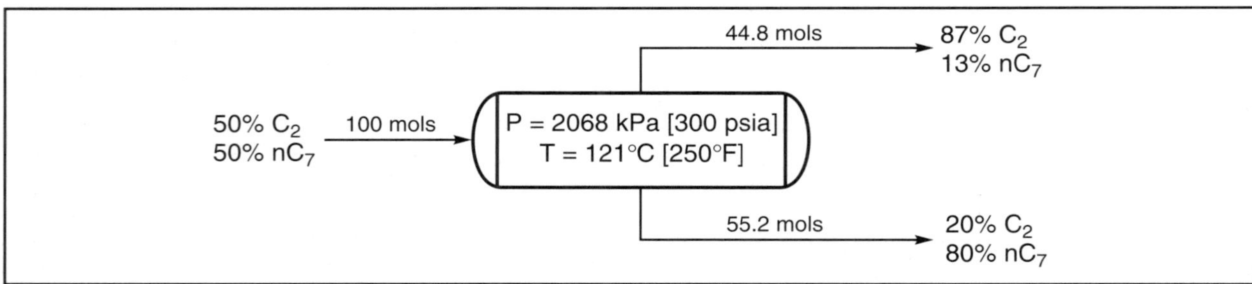

Figure 17.7 Single Stage Separation System

If we perform a material balance around this equilibrium separator we can determine the flowrate of the liquid and vapor streams.

Let: V = mols of vapor L = mols of liquid

C_2 balance: $50 = 0.87\ V + 0.2\ L$

nC_7 balance: $50 = 0.13\ V + 0.8\ L$

Solving simultaneously, V = 44.8 mols

L = 55.2 mols

Relative Volatility

A significant amount of separation between C_2 and nC_7 occurred in the single-stage equilibrium separator shown in Figure 17.7. This is because C_2 is significantly more volatile than nC_7. The volatility ratio of two components is called the relative volatility and is calculated from Equation 17.3 below.

$$\alpha_{ij} = \frac{K_i}{K_j} \tag{17.3}$$

Where:

α_{ij} = relative volatility of component "i" to component "j"
K_i = K value for component "i" at a specified T and P
K_j = K value for component "j" at a specified T and P

Relative volatility is perhaps the single most important parameter in fractionator design.

Example 17.1: Calculate the relative volatility of C_2 to nC_7 at the conditions in Figure 17.7. The K values for C_2 and nC_7 in our separator example are:

$$K_{C_2} = \frac{y}{x} = \frac{0.87}{0.20} = 4.35$$

$$K_{nC_7} = \frac{y}{x} = \frac{0.13}{0.8} = 0.162$$

$$\alpha_{ij} = \frac{4.35}{0.162} = 26.9$$

In other words, at 121°C [250°F] and 2068 kPa [300 psia] we can say that C_2 is about 27 times more volatile than C_7.

The greater the relative volatility the easier the separation.

The relative volatility of two components varies with pressure, temperature, and composition. For example, at 177°C [350°F] and 2068 kPa [300 psia] the relative volatility of C_2 to nC_7 is only 10.5. Significant changes in the relative volatility within a fractionation tower can have a profound effect on the difficulty of the separation and column design and operation.

The degree of separation which takes place in any equilibrium stage depends on the width of the envelope in the T-x diagram. Wide envelopes indicate larger relative volatilities, narrow envelopes indicate smaller relative volatilities.

The smaller the relative volatility, the smaller the composition change that will occur in any equilibrium separator. In most of the hydrocarbon fractionation systems, the relative volatility is less than 3; in some cases it is less than 1.5.

Figure 17.8 shows the T-x diagram for the C_3-iC_4 system which represents the separation made in a depropanizer, and the iC_4-nC_4 system which represents the separation made in a deisobutanizer, or as it is sometimes called, a butane splitter. At normal operating conditions the average relative volatility of C_3 to iC_4 is about 1.8. The average relative volatility of iC_4 to nC_4 is about 1.3.

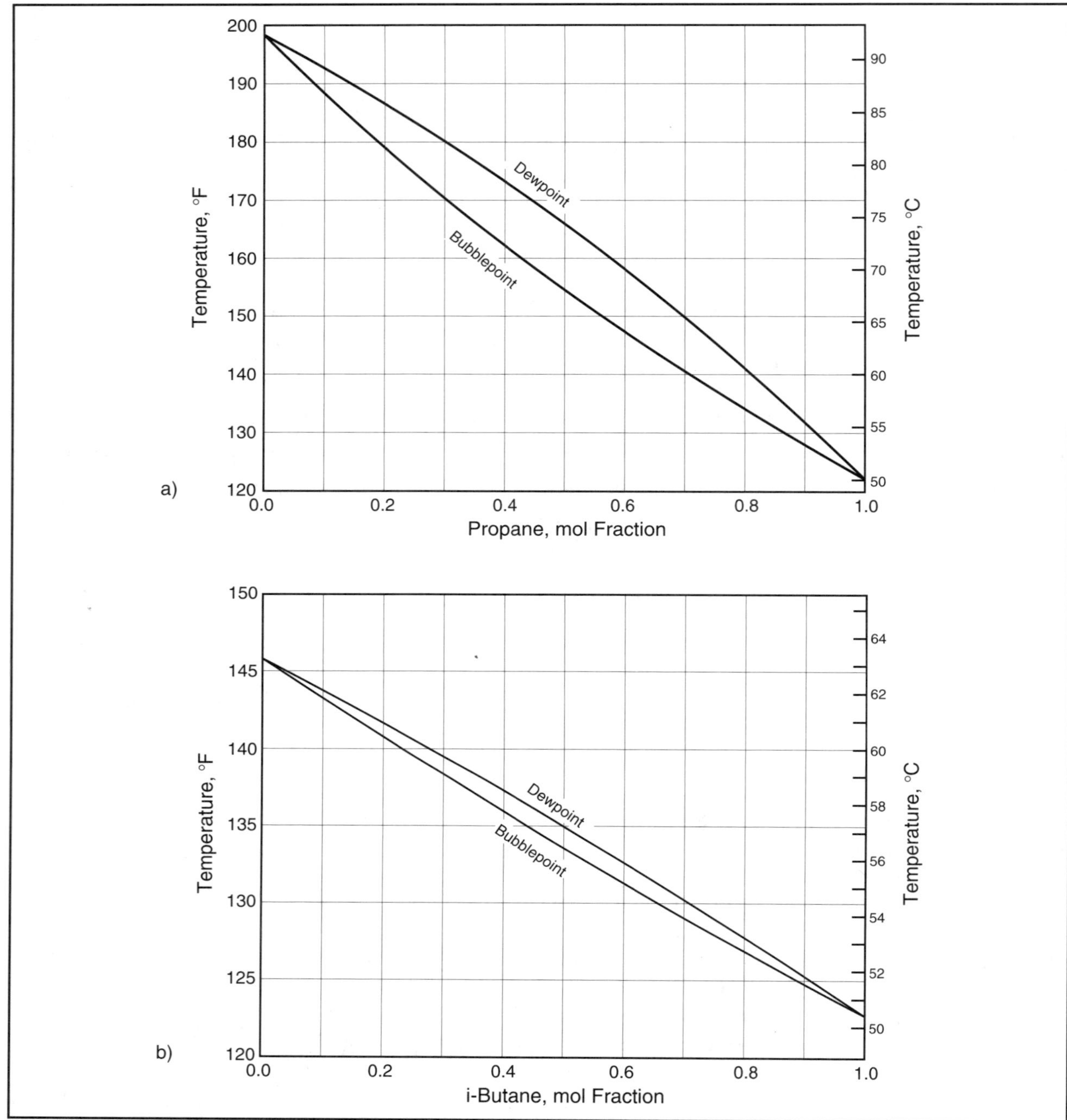

Figure 17.8 T-x diagrams for (a) C_3-iC_4 at 1500 kPa [218 psia] and (b) iC_4-nC_4 at 700 kPa [102 psia] Systems

Some binary mixtures exhibit a relative volatility of 1.0 at certain compositions. These systems are called aezeotropes. An aezeotrope boils like a pure component, i.e. the vapor phase composition is identical to the liquid phase composition. Aezeotropes are not common in the gas processing industry, but they do occur. A T-x diagram for the CO_2-C_2 system at three different pressures is shown in Figure 17.9. Note that at the aezeotrope point (approximately 65% CO_2, 35% C_2), the vapor and liquid compositions are the same.

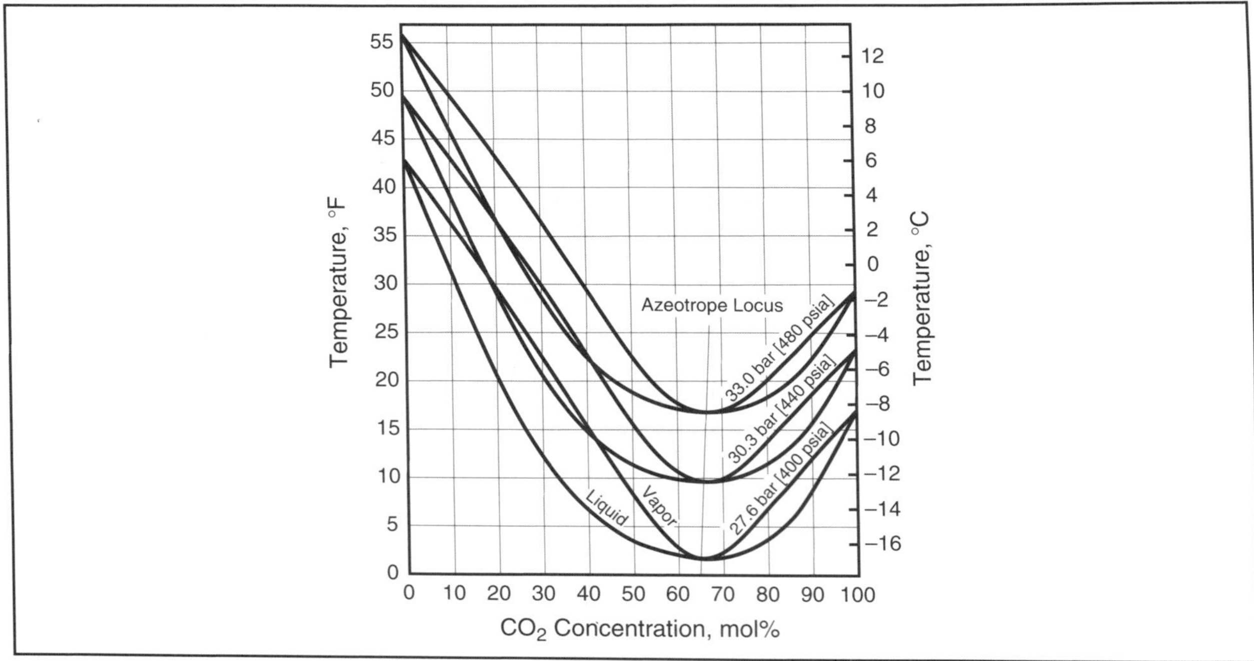

Figure 17.9 T-x Diagram for CO_2-C_2 System at 400, 440, and 480 psia[(17.2)]

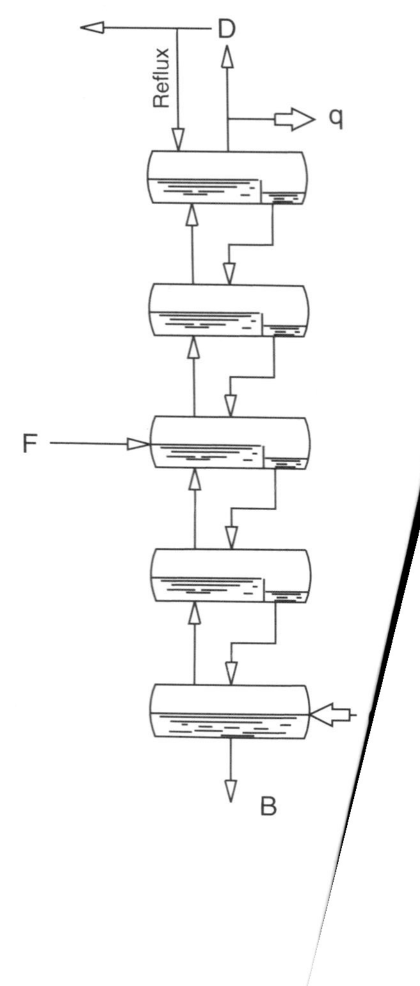

A fractionator consists of a series of equilibrium separators in series. Vapor and liquid flow counter-current to each other, theoretically reaching equilibrium in each separator. The compositions achieved in any separator can be estimated from a T-x diagram. One fractionator design method for binary systems which uses this approach is the McCabe-Thiele method summarized in Appendix 17A.

Figure 17.10 is a schematic view of a typical fractionator. The bottom product (bottoms) leaves the tower, or in some cases the reboiler, at its bubblepoint. The overhead product (distillate) leaves the reflux accumulator as a bubblepoint liquid in the case of a total condenser and a dewpoint vapor in the case of a partial condenser. Although not common, the distillate product can also be a mixed phase stream. With a total condenser, the reflux and distillate product will have the same composition. In the case of a partial condenser the reflux will be a bubblepoint liquid in equilibrium with the distillate vapor product. The partial condenser acts as an equilibrium separation stage. The total condenser does not.

Heat input at the reboiler is usually provided by a high temperature heating medium such as hot oil, steam, or by direct heat transfer in a fired heater. A process fluid, e.g. hot compressor discharge vapor may also be used. The reboiler provides hot stripping vapor in the bottom portion of the tower, and the section of the tower between the reboiler and feed tray is referred to as the "stripping section."

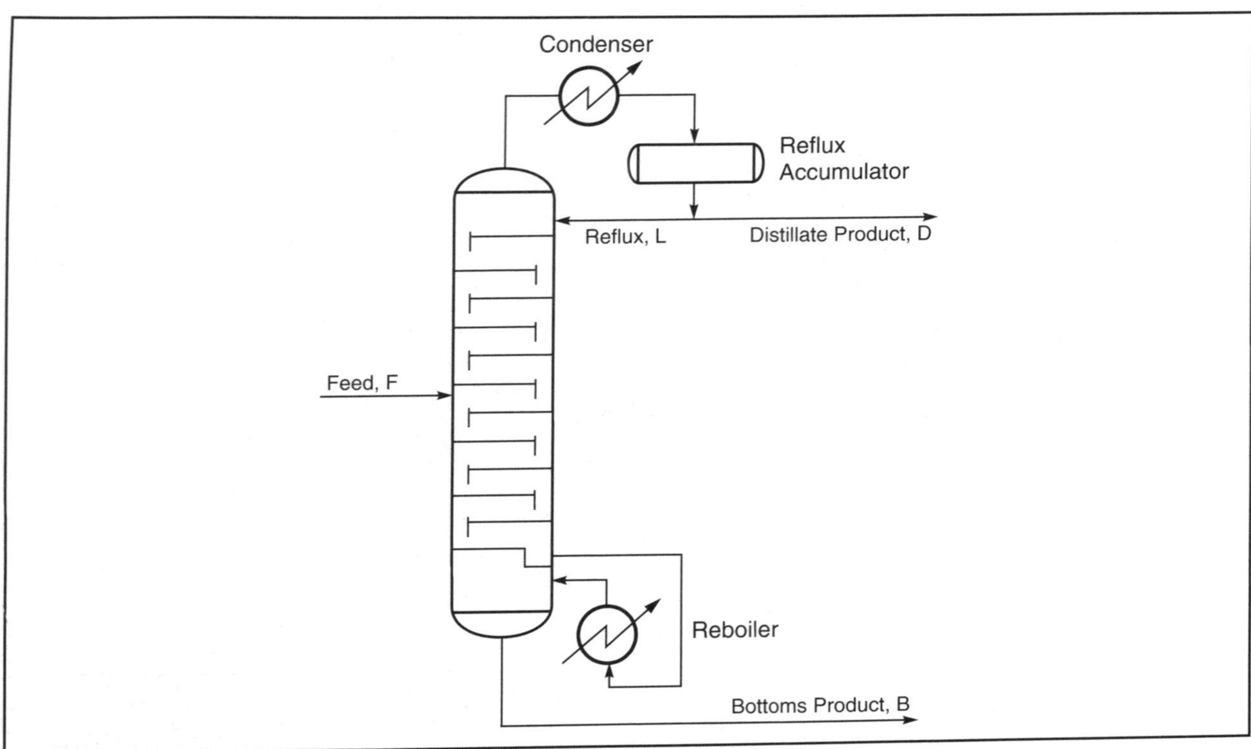

Figure 17.10 Schematic of a Fractionation Column

Heat is removed from the tower at the condenser. A portion of the condensed liquid is returned to the tower as reflux. Reflux provides cooling in the top portion of the tower. This section is often called the "rectifying" or "enriching" section. The ratio of the reflux to the distillate rate, L/D, is referred to as the reflux ratio.

The operating pressure of a fractionation tower is ordinarily fixed by the condensing tempera- of the distillate product. In turn, the condensing temperature is controlled by the cooling medium. condenser temperature is fixed by the designer by allowing for practical temperature difference be- the cooling medium and the overhead product. In the case of a liquid distillate, the bubblepoint is then calculated; for a vapor distillate product the dewpoint pressure would be calculated. sure is the minimum pressure at which the tower can operate at the chosen condenser temper- still meet product specifications.

choice between a total and partial condenser for a tower is a practical matter involving temperature and disposition of the distillate product. As a practical matter, if the distil- subsequently processed, stored, or transported as a liquid, the condenser will usually be . If not, a partial condenser is used.

FRACTIONATOR DESIGN CONSIDERATIONS

The design of a fractionation column involves the following steps. These steps can also reinforce fractionator operating and control principles. An example exercise, a depropanizer, is included to illustrate concepts and calculations.

1. Establish the Design Basis

- Feed rate, composition, condition (T&P)
- Product specifications: recovery, purity, vapor pressure, etc.
- Available utilities: heating medium – steam, hot oil, process fluid, fired heater; cooling medium – air, water, refrigerant, process fluid

Example 17.2 Consider the following depropanizer feed stream.

Component	Mol %
C_2	1.0
C_3	46.0
iC_4	10.0
nC_4	21.0
C_5+	22.0

The feed rate is 1000 kmol/h [2200 lbmol/hr]. The feed condition is 1724 kPa [250 psia] and the feed is at its bubblepoint.

The product specifications require 99% of the propane to be recovered in the distillate product and the composition of propane in the distillate must be a minimum of 95%.

Hot oil is available at a temperature of 204°C [400°F] for reboiler heat. The condensing medium is air (aerial cooler) at 35°C [95°F]. The design condensing temperature is 49°C [120°F] and is based on an approach of 14°C [25°F].

2. Perform a Material Balance Around the Column to Establish the Distillate and Bottom Product Rates and Compositions

This requires the assumption that only the key components distribute, i.e. appear in both the distillate and bottom product streams. This means that any component lighter than the light key will appear only in the distillate product. Likewise any component heavier than the heavy key will appear only in the bottoms product.

Selection of the light and heavy key components is not usually a difficult task; they will be defined by the product specifications, i.e., deethanizing, depropanizing, etc. The light and heavy key components are usually adjacent components, e.g., ethane/propane, propane/isobutane, etc. However, in unusual circumstances, e.g. when the composition of one of the "normal" keys is significantly smaller than the two adjacent components or an azeotropic mixture is encountered, split keys may be used. An example of split keys would be propane (light key) and n-butane (heavy key).

The total fractionation system must be considered when establishing the product specifications. It is common to design as close to specification limits as practical. For example, one way to reduce column cost may be to reduce the percentage recovery of the light key component in the distillate. However, if there is a second column downstream of the first, too much of the light key component may remain in the bottoms of the first column making it impossible (or impractical) to meet the distillate purity specification in the second column. Consequently, the entire fractionation system must be considered.

Example 17.3: Using the feed composition and product specifications from Example 17.2, determine the distillate and bottom product rates and compositions.

Product Specifications: Percentage recovery of C_3 in the distillate is 99%

Purity specification for C_3 in the distillate product is 95%

Assume that non-key components do not distribute, e.g. all of the C_2 goes to the distillate, all of the nC_4+ goes to the bottoms.

		Distillate		Bottoms	
Component	**Feed mols**	**mols**	**mol %**	**mols**	**mol %**
C_2	10	10.0	2.09	0.0	0.00
C_3	460	455.4	95.00	4.6	0.88
iC_4	100	14.0	2.91	86	16.52
nC_4	210	0.0	0.00	210	40.34
C_5+	220	0.0	0.00	220	42.26
	1000	479.4	100.00	520.6	100.00

mols of C_3 in distillate = (460)(0.99) = 455.4

mols iC_4 in distillate is calculated by difference as follows,

total distillate mols = (455.4/0.95) = 479.4

mols iC_4 = 479.4 – 455.4 – 10.0 = 14.0

Other product specifications yield a different calculation approach but the principle is the same. Once the feed composition, feed rate, and product specifications are known, the column material balance is fixed.

The equations below are useful for performing material balance calculations when both specifications are purity specifications. The equations are valid for a fractionator operating at steady state.

$$\frac{D}{F} = \frac{x_F - x_B}{x_D - x_B} \quad \text{and} \quad \frac{B}{F} = \frac{x_F - x_D}{x_B - x_D} \tag{17.4}$$

Where:

D = distillate rate, mols
B = bottoms rate, mols
F = feed rate, mols
x_D = mol fraction of a component in the distillate
x_B = mol fraction of a component in the bottoms
x_F = mol fraction of a component in the feed

The product specifications are usually set by contractual requirements, e.g. 95% C_3 in the distillate, 1% C_3 in the bottoms, C_1/C_2 ratio = 0.02, etc. It is obvious from Equation 17.4 that at a given feed composition, the distillate and bottoms product specification set the overall material balance for the column. In other words, in order to exactly meet the distillate and bottoms specifications there is one (and only one) distillate or bottoms rate which can be drawn from the tower. From a control standpoint, this is very difficult (if not impossible) to achieve, on a continuous basis so some flexibility is often allowed in the composition of one of the products. This will be discussed in more detail in a later section.

For both design and operation of a fractionator two independent specifications are required. For non-refluxed distillation only one specification is required.

Vapor pressure is sometimes used as a specification in place of composition. If this is the case, the material balance can be made using the procedure outlined in Chapter 5.

In some cases, particularly in existing towers, one of the specifications or constraints can be reflux rate, reboiler or condenser duty. In other words, an equipment limitation constrains the operation of the tower. When this happens, one (or both) of the product specifications is no longer a controlling parameter.

3. Determine Tower Pressure

The tower pressure is determined from a bubblepoint or dewpoint calculation on the distillate product at the condensing temperature. This will fix the pressure at the reflux accumulator.

For a total condenser a bubblepoint calculation is performed.

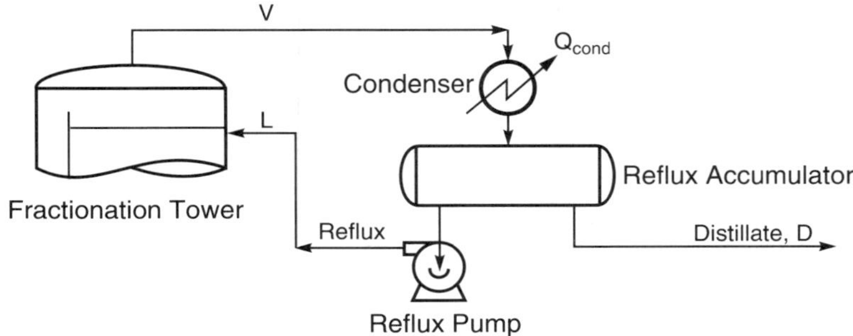

For a partial condenser a dewpoint calculation is performed.

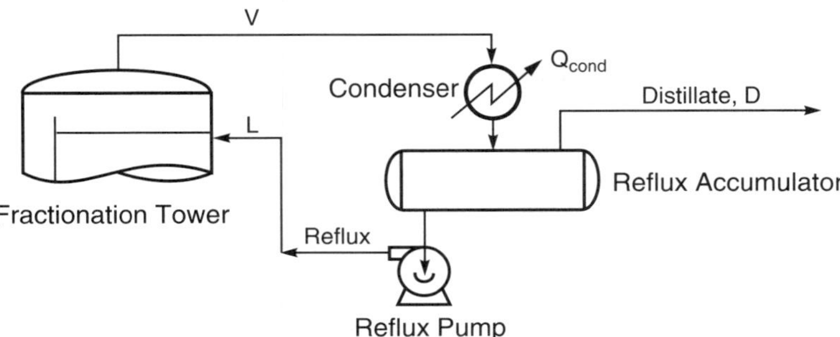

The actual tower pressure will typically be 35-100 kPa [5-15 psi] higher than the calculated accumulator pressure to account for pressure drop through the condenser and piping. The condensing temperature will depend upon the cooling medium. Typical approaches are shown below.

	Approach	
	°C	°F
Refrigeration	3-6	5-10
Cooling Water	5-10	9-18
Pressurized Fluid	5-10	9-18
Air	10-15	18-27

Example 17.4: Estimate the tower pressure using distillate product composition from Example 17.3. From step (1) the condenser is a total condenser using air. The condensing temperature is 49°C [120°F].

Perform bubblepoint calculation at 49°C [120°F] to determine the accumulator pressure – from SRK EOS.

Component	x_i, mol%	K_i @ 49°C [120°F] & 1737 kPa [252 psia]	$K_i x_i$, mol%
C_2	2.09	2.33	4.88
C_3	95.00	0.985	93.61
iC_4	2.91	0.520	1.51
	100.00		100.00

Accumulator pressure is 1737 kPa [252 psia], assume a tower pressure of 1793 kPa [260 psia].

It is useful at this point to also determine the top tray temperature. This will be needed to estimate average relative volatilities as well as to calculate condenser duty. This cannot be done for a partial condenser until the reflux rate has been established. For a total condenser, however, the top tray vapor and the distillate product composition are the same. Since the tower operating pressure is known, one can estimate the top tray temperature by performing a dewpoint calculation on the distillate product at the tower pressure.

Example 17.5: Estimate the top tray temperature for Example 17.4 using a tower pressure of 1793 kPa [260 psia]

Perform dewpoint calculation at 1793 kPa [260 psia] to determine top tray temperature.

Component	x_i, mol%	K_i @ 52°C [126°F] & 1793 kPa [260 psia]	y_i/K_i, mol%
C_2	2.09	2.34	0.90
C_3	95.00	1.013	93.76
iC_4	2.91	0.545	5.34
	100.00		100.00

Top tray temperature is 52°C [126°F].

4. Calculate the Reboiler Temperature

The reboiler temperature can be estimated once the tower pressure has been established. Since pressure drop exists across the tower, it is common practice to assume that the reboiler pressure is 20-35 kPa [3-5 psi] higher than the top tray pressure. The reboiler temperature is calculated by performing a bubblepoint calculation on the bottoms product at the reboiler pressure.

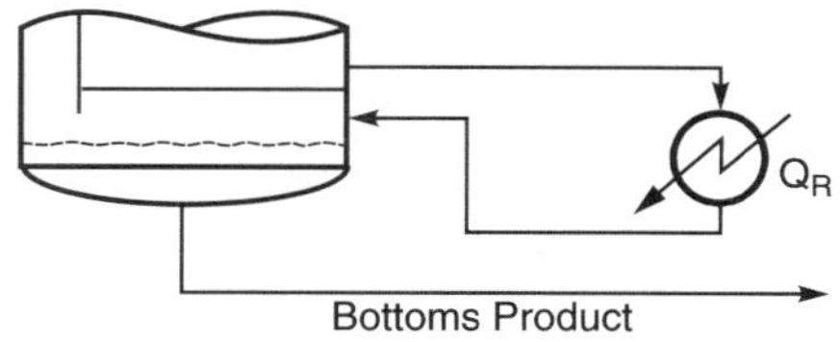

Example 17.6: Estimate the reboiler temperature for the conditions in Example 17.5. Assume a 20 kPa [3 psi] pressure drop across the tower.

Perform a bubblepoint calculation at 1813 kPa [263 psia] on the bottoms product to determine the reboiler temperature.

Component	x_i, mol%	K_i @ 134°C [273°F] & 1813 kPa [263 psia]	K_ix_i, mol%
C_3	0.88	2.31	2.03
iC_4	16.52	1.52	25.19
nC_4	40.34	1.32	53.33
iC_5+	42.26	0.460	19.44
(assume nC_6)	100.00		100.00

Reboiler temperature is 134°C [273°F].

5. Determine the Reflux Ratio and Number of Theoretical Stages Required to Make the Separation

This step involves stage-to-stage calculations inside the column. These calculations are most frequently performed by computer using a rigorous computer tray-to-tray simulation. Manual shortcut calculations may be used as well. There are infinite combinations of these two variables that will give the desired separation, as illustrated in Figure 17.11.

The curve is essentially a hyperbola. At one asymptote, it approaches the minimum reflux rate and at the other it approaches the minimum number of trays. The minimum reflux rate occurs at that value of reflux with an infinite number of trays in the column to produce the desired separation. The minimum number of trays occurs at an infinite reflux ratio, often referred to as total reflux, when the separation per stage in the tower is a maximum.

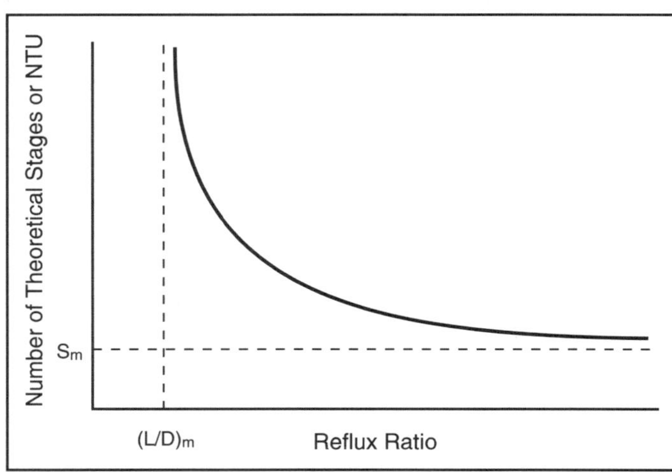

Figure 17.11 Relationship between Theoretical Stages (S) and Reflux Ratio (L/D)

Neither of these two limits of operation in the column has practical value in itself. However, they do serve as useful correlation tools for predicting the relationship between the operating reflux ratio and actual number of theoretical plates. Most hydrocarbon fractionators operate at an actual reflux from 1.05 to 1.25 times the minimum. This is based on economic criteria. Operation is typically near the minimum reflux rate to minimize heating and cooling costs. Condenser heat load is a direct function of reflux rate; as this increases, the reboiler load also has to increase to maintain the energy balance around the tower.

Short-cut calculations

Short-cut calculations are often used to establish the variables in Figure 17.11. Short-cut methods are suitable for the evaluation of most hydrocarbon fractionation systems. Those which follow have been found useful and reliable for the types of fractionation systems discussed in this book. They calculate minimum reflux (L_m), minimum theoretical stages (S_m), the relationship between reflux ratio and theoretical stages (Figure 17.11). A multicomponent system is treated as a pseudo-binary consisting only of the light key and heavy key components.

Appendix 17A reviews the McCabe-Thiele method for binary mixtures, which can be used also for multicomponent mixtures. It is not widely used commercially, but it contributes visually to an understanding of fractionator performance.

Shortcut calculations are useful for:

1. Scoping studies suitable for preliminary costs
2. Parametric evaluation of operating variables
3. Separations having coarse purity requirements (i.e., contaminants > 0.5 mol%)
4. Detailed designs for ideal and close-to-ideal systems
5. Designs for systems for which equilibrium data are unavailable
6. Evaluation or reality check of computer simulations

On the other hand, *rigorous* design procedures should be applied for the following:

1. The separation is for a multicomponent system, requiring high product purity.
2. The system is highly nonideal, but good equilibrium data are available.
3. The relative volatility between key components is less than 1.3.
4. One or more of the components is near the critical pressure.

Fenske's Method for Minimum Theoretical Plates

The Fenske Equation is a convenient and very useful technique for calculating the minimum number of theoretical plates required for most multicomponent separations. When properly used, it can be applied between the top and bottom of the distillation column and give an accurate estimate of the minimum number of theoretical plates. With a total condenser, it is applied between the top stage of the tower and the reboiler; for a partial condenser, it is applied between the reboiler and the distillate product. The Fenske Equation can be written in several forms. A convenient form is:

$$S_m = \frac{\log\left[\left(\frac{x_{LK}}{x_{HK}}\right)_D\left(\frac{x_{HK}}{x_{LK}}\right)_B\right]}{\log \alpha_{avg}} \qquad (17.5)$$

Where:

S_m = minimum number of theoretical plates
x_{LK} = mol fraction of light key component
x_{HK} = mol fraction of heavy key component
α_{avg} = relative volatility at average tower temperature (top tray temperature + bottom tray temperature)/2 (see Equation 17.3)

Subscripts:
D = distillate product
B = bottom product

Example 17.7: Applying Fenske's Equation for the example depropanizer. Calculate S_m, minimum theoretical stages.

$T_1 = 52°C$ (top tray temperature)
$T_2 = 134°C$ (reboiler temperature)
$T_{avg} = (T_1 + T_2)/2 = (52 + 134)/2 = 93°C$

At average tower conditions

$K_{C_3} = 1.683$
$K_{iC_4} = 0.993$
$\alpha_{avg} = 1.683/0.993 = 1.695$

$$S_m = \frac{\log\left[\left(\frac{0.950}{0.0291}\right)\left(\frac{0.165}{0.0088}\right)\right]}{\log 1.695} = 12.2$$

The minimum number of theoretical trays for the separation is 12.2. This represents the total number of equilibrium contacts including reboiler and partial condenser (if any), at total reflux.

Distribution of Non-Key Components

One of the assumptions made in the material balance is that non-key components do not distribute. This is seldom the case, although in many cases the distribution of the non-keys is minimal.

The Fenske equation can be adapted to predict the non-key component distribution.

$$\left(\frac{b}{d}\right)_i = \frac{\left(\frac{b}{d}\right)_{HK}}{\alpha_i^{S_m}} \tag{17.6}$$

Where:

α_i = relative volatility of a non-key component (relative to the heavy key)
S_m = minimum theoretical stages
b = mols of non-key component in bottoms product
d = mols of non-key component in distillate product

The distillate rate for each component can be determined by a rearranged form of the overall material balance.

$$d_i = \frac{f_i}{1+\left(\frac{b}{d}\right)_i} \tag{17.7}$$

and the bottoms rate for each component by:

$$b_i = f_i - d_i \tag{17.8}$$

The primary use of this calculation is the ability to quickly check the validity or feasibility of the original material balance for a complete fractionation train before proceeding further. While not exactly perfect, the agreement between this short cut calculation and predicted product distributions from rigorous tray-to-tray is usually very good. Thus, the Fenske equation is a powerful tool for getting preliminary estimates of product distribution as well as the minimum number of trays.

Underwood Methods for Minimum Reflux

An infinite number of theoretical stages is required at minimum reflux. This means that a large number of these stages are performing negligible separation of components between vapor and liquid. Characteristics of these stages are such that there is no significant change of composition from tray to tray and, as a result, no change of temperature from tray to tray. Computer tray-to-tray calculations will calculate the required reflux ratio at the desired number of theoretical stages. It is useful, however, to estimate the minimum reflux using shortcut techniques. Once the minimum reflux has been determined, the actual reflux ratio can be established based on economic criteria, i.e. $(L/D)_a$ = 1.05 to 1.25 $(L/D)_m$.

The Underwood method involves use of two equations.[(17.3)]

$$\sum_{i=1}^{i=n} \frac{\alpha_i f_i}{\alpha_i - \varphi} = F(1-q) \tag{17.9}$$

$$\sum_{i=1}^{i=n} \frac{\alpha_i d_i}{\alpha_i - \varphi} = L_m + D \tag{17.10}$$

Where:

α_i = relative volatility of component at average tower temperature (relative to heavy key)

φ = constant

f_i = mols of component in feed

d_i = mols of component in distillate

q = total heat needed to convert one mole of feed into a saturated vapor divided by the molal latent heat of the feed

for bubblepoint feed, $q = 1.0$
for dewpoint feed, $q = 0$
for 2 phase feed, $0 < q < 1.0$

L_m = minimum reflux rate, mols

D = distillate rate, mols

Use of Equations 17.9 and 17.10 to calculate the minimum reflux rate involves determination of the value of φ that will satisfy Equation 17.9. This is a trial-and-error solution, as shown in

Example 17.8. The procedure is simplified considerably by the limitation that the value of φ falls between 1.0 and the relative volatility of the light key component. The value of φ that satisfies Equation 17.9 is used in Equation 17.10 to calculate the minimum reflux. In this calculation, the distillate composition is obtained from the material balance performed in Example 17.3.

For a bubblepoint feed, Equation 17.9 becomes

$$\sum_{i=1}^{i=n} \frac{\alpha_i f_i}{\alpha_i - \varphi} = 0 \qquad (17.11)$$

This calculation of minimum reflux is a practical, convenient way to begin the analysis of tower performance. It provides one with a reference point for the adjustment of reflux rate. In operational troubleshooting, it provides an indication if reflux rate is a potential problem.

Example 17.8: Calculation of φ. Assume φ = 1.130

Comp.	Feed mols	K @ 93°C 1803 kPa	α_i	$\alpha_i f_i$	$\alpha_i - \varphi$	$\frac{\alpha_i f_i}{\alpha_i - \varphi}$
C_2	10	3.437	3.461	34.61	2.331	14.85
C_3	460	1.683	1.695	779.7	0.565	1380.72
iC_4	100	0.993	1.000	100.0	-0.130	-767.63
nC_4	210	0.823	0.829	174.09	-0.302	-576.93
iC_5+	220	0.211	0.213	46.86	-0.918	-51.01
	1000					0.00

Calculation of L_m

Comp.	Distillate mols	α_i	$\alpha_i d_i$	$\alpha_i - \varphi$	$\frac{\alpha_i d_i}{\alpha_i - \varphi}$
C_2	10.0	3.461	34.61	2.331	14.8
C_3	455.4	1.695	771.89	0.565	1366.9
iC_4	14.0	1.000	14.00	-0.130	-107.5
	479.4				1264.2

$1264.2 = L_m + D$ D = 479.4 from material balance in Example 17.3

$L_m = 1264.2 - 479.4 = 784.8$

Minimum reflux ratio = $(L_m/D) = 784.8/479.4 = 1.64$

Actual Reflux Rate and Theoretical Stages

Two useful correlations are available for relating actual reflux rate and number of theoretical trays from the corresponding minimum values.[17.4, 17.5]

The Gilliland correlation is widely used. It was originally presented graphically, as shown in Figure 17.12. Once the minimum theoretical stages and minimum reflux have been calculated, the Gilliland correlation can be used to establish an actual reflux rate for a fixed number of stages or conversely, the number of stages for a fixed reflux rate.

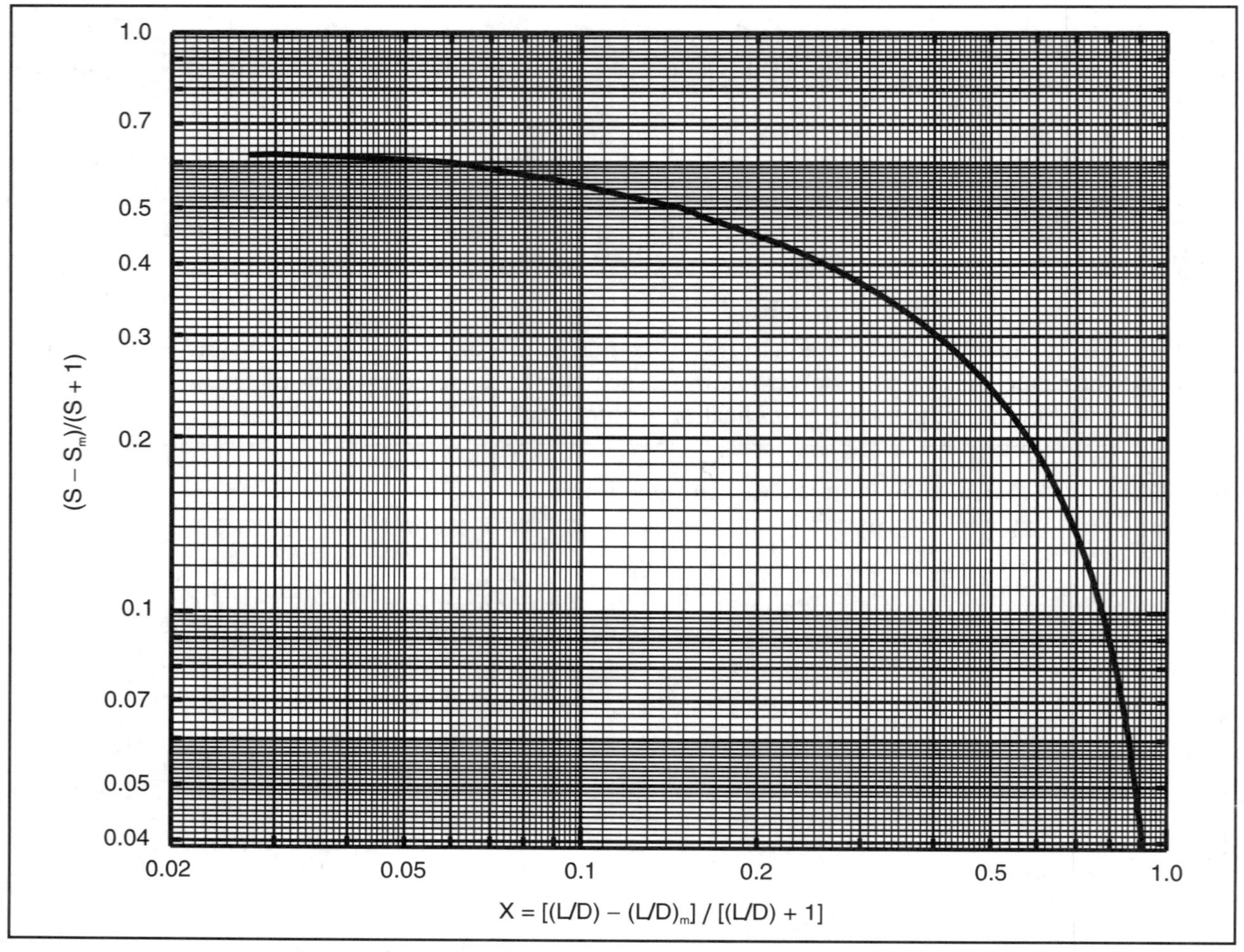

Figure 17.12 Gilliland Correlation for Relationship Between Reflux Rate and Number of Theoretical Plates

An approximate correlation of Gilliland for tray-type towers may be represented by the equation(17.6)

$$Y = 0.75(1 - X^{0.5668}) \tag{17.12}$$

Where:

Y = $(S - S_m)/(S + 1)$
S = actual theoretical trays
S_m = minimum theoretical trays
X = $[(L/D) - (L/D)_m] / [(L/D) + 1]$

The calculation for packed towers is somewhat different. The following equation may be used(17.7, 17.8)

$$Y = 0.763(1 - X^{0.5806}) \tag{17.13}$$

Where:

Y = $(NTU - NTU_m)/(NTU + 2A)$
A = $\ln \alpha/(\alpha - 1)$
α = relative volatility
X = $(R - R_m)/(R + 1)$
NTU = number of transfer units

Example 17.9: The depropanizer considered in the previous examples is designed to operate at an actual reflux rate 10% above the minimum reflux. How many equilibrium stages will be required to make the separation.

$$(L/D) = 1.1\ (L/D)_m = (1.1)(1.64) = 1.8$$

$$\frac{L/D-(L/D)_m}{(L/D)+1} = \frac{1.8-1.64}{1.8+1} = 0.06$$

From Figure 17.12

$$(S - S_m)/(S + 1) = 0.6$$

From Fenske equation (Example 17.7) $S_m = 12.2$, so

$$S = 32 \text{ equilibrium stages}$$

The reboiler is often equivalent to one equilibrium stage, so for this tower 31 equilibrium stages must be installed in the column proper.

A rigorous computer tray-to-tray simulation for the above example depropanizer gives a reflux ratio of 1.91 for 32 equilibrium stages. This compares favorably with the 1.80 value calculated from shortcut calculations.

6. Convert Theoretical Stages to Actual Trays Using Tray Efficiencies

The calculations outlined above provide an estimate of the numbers of theoretical stages required to achieve the desired separation. Since equilibrium is not achieved on an actual tray, the number of actual trays must be greater than the number of theoretical stages.

Tray efficiencies can be defined several ways; Murphree plate efficiency, overall efficiency, etc. For planning and feasibility calculations, the overall tray efficiency is typically used.

$$E = \frac{\text{number of theoretical stages}}{\text{number of actual trays}} \qquad (17.14)$$

TABLE 17.1

Typical Fractionator Parameters

* Reflux ratio relative to distillate product, mol/mol	Operating Pressure, psig	Number of Actual Trays	Reflux* Ratio	Tray Efficiency, %
Demethanizer	200-400	18-26	Top Feed	45-60
Deethanizer	375-450	25-35	0.9-2.0	50-70
Depropanizer	240-270	30-40	1.8-3.5	80-90
Debutanizer	70-90	25-35	1.2-1.5	85-95
Butane Splitter	80-100	60-80	6.0-14.0	90-110
Rich Oil Fractionator (Still)	130-160	20-30	1.75-2.0	Top 67 Bottom 50
Rich Oil Deethanizer or Demethanizer	200-400	40	–	Top 25-40 Bottom 40-60
Condensate Stabilizer	100-400	16-24	Top Feed	40-60

Many factors affect tray efficiency:

1. Relative vapor and liquid loading of the tray
2. Physical characteristics of liquid (foaming, viscosity, surface tension, etc.)
3. Characteristics of the tray
4. Mechanical design and installation of the tray
5. Thermodynamic properties used to determine the number of theoretical trays.

As a result of these effects, there is no good prior method of estimating tray efficiencies for unusual or different separations. For known separations (depropanizers, debutanizers, etc.), FRI data and/or operating experience is probably the best way to estimate efficiency. Table 17.1 provides estimates of tray efficiencies for NGL Fractionators. The next best alternative is the O'Connell correlation[(17.9)] (Figure 17.13). While fairly old and based on a relatively small sample of data, this correlation gives reasonable results.

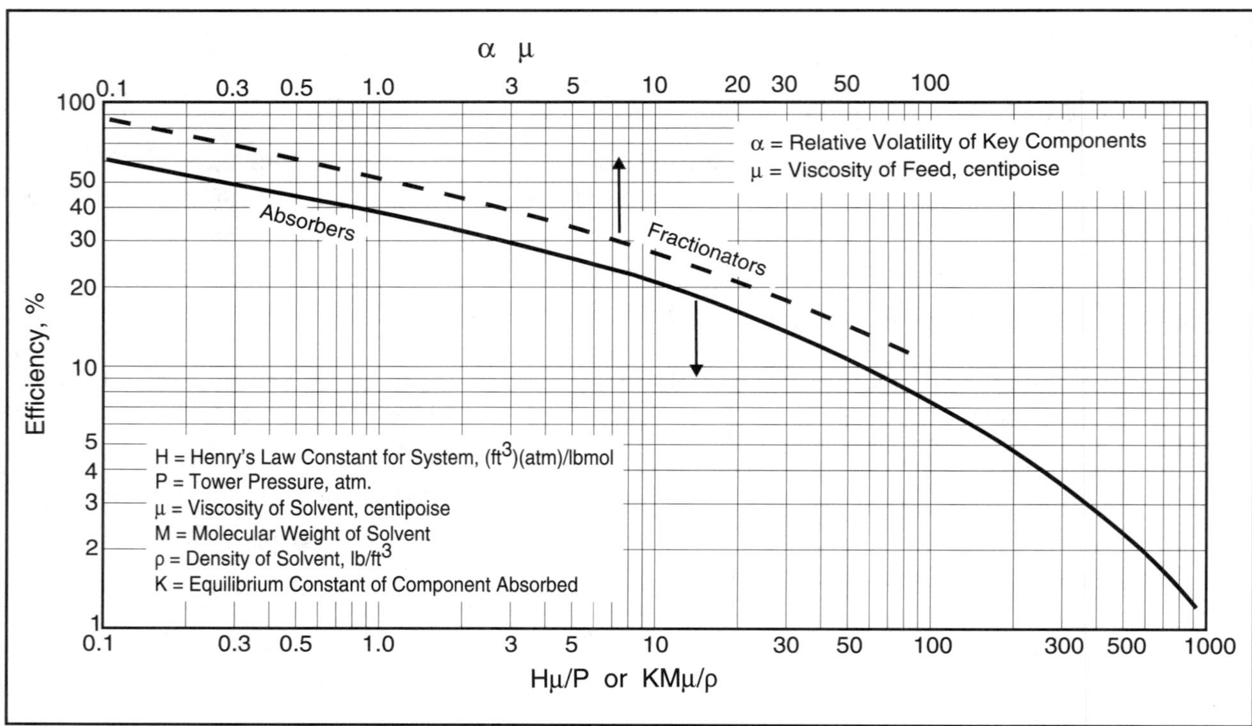

Figure 17.13 Correlation for Overall Efficiency of Absorbers and Fractionators

Example 17.10: Estimate the tray efficiency of the example depropanizer and the number of actual trays required.

$\mu = 0.088$ cp $\qquad \alpha = 1.695$

$\alpha\mu = (1.695)(0.088) = 0.15$

From Figure 17.13, E = 80%

$$\text{Number of actual trays} = \frac{32\text{ theor. stages} - 1\text{ reboiler}}{0.8}$$

$$= 39\text{ actual trays}$$

Be cautious when estimating tray efficiencies. Consider the advice of supplier and operating company experts. Don't ignore foaming problems, etc., that may make your system unusual, and if in doubt, guess low.

If packing is used rather than trays, the conversion from theoretical stages to packing height involves the determination of an HETP (Height Equivalent to a Theoretical Plate) or an HTU (Height of a Transfer Unit).

7. Determine Optimum Feed Tray Location

Selection of the proper feed tray location is extremely important in order to optimize the operation of the fractionator. Placing the feed tray too high in the tower can result in excessive reflux to meet distillate product specification. Too low and excessive reboiler heat may be required to meet bottom product specification.

Improper feed plate location is usually manifested by a sharp discontinuity in the temperature profile. Multiple feed nozzles and or a feed preheater are typically used to provide flexibility to adjust to changing feed conditions.

Unfortunately, predicting the optimum feed plate location in the design phase is not easy, particularly if a shortcut calculation is used. Virtually all of the shortcut calculation methods of estimating feed plate location are based on the assumption of total reflux. A convenient empirical correlation is[17.7] Equation 17.15.

$$\log\left(\frac{N}{M}\right) = 0.206\log\left[\left(\frac{B\,x_{HK_F}}{D\,x_{LK_F}}\right)\left(\frac{x_{LK_B}}{x_{HK_D}}\right)^2\right] \qquad (17.15)$$

Where:

N = no. theoretical stages in rectifying section
M = no. theoretical stages in stripping section
B = bottoms rate, mols
D = distillate rate, mols
x_{HK_F} = composition of heavy key in the feed
x_{LK_F} = composition of light key in the feed
x_{LK_B} = composition of light key in the bottoms
x_{HK_D} = composition of heavy key in the distillate

Example 17.11: Estimate the optimum feed tray location for the example depropanizer.

For Case I of the example depropanizer:

$$\log\left(\frac{N}{M}\right) = 0.206\log\left[\left(\frac{(479.4)(0.46)}{(520.6)(0.10)}\right)\left(\frac{0.0088}{0.0291}\right)^2\right]$$

$$\log\left(\frac{N}{M}\right) = -0.0848$$

$$N/M = 0.822$$

$$N + M = 32 \qquad N = 14,\ M = 18$$

Optimum feed tray is about tray 14.

From Example 17.11, approximately 18 equilibrium stages will be required below the feed tray (including reboiler) and 14 equilibrium stages above.

Feed Stage	(L/D)
12	1.924
13	1.910
14	1.911
15	1.928

A tray-to-tray computer simulation may also be used to locate the optimum feed tray. Several feed tray locations are simulated and the one yielding the lowest reflux ratio (reboiler duty) is the optimum location. For the example depropanizer, the optimum feed tray is tray 13 or 14. The results of the simulation are tabulated at left.

8. Calculate the Condenser and Reboiler Duties

Energy is required to effect the separation of components in a distillation column. High temperature heat enters the distillation at the reboiler and low temperature heat is withdrawn at the condenser. The amount of energy required depends on several factors: difficulty of the separation (relative volatility, product specifications), feed condition and feed rate.

The overall energy balance around the fractionator is

$$Q_B + Q_c = h_D D + h_B B - h_F F \tag{17.16}$$

Where:
Q_B = reboiler heat load
Q_c = condenser heat load
h_D = enthalpy of distillate product
h_B = enthalpy of bottoms
h_F = enthalpy of feed
D, B, F = rate of flow of distillate, bottom and feed streams, respectively

This equation has two unknowns, Q_c and Q_B. All values of "h" are in energy/mass or mol; "Q" is in energy/time. D, B and F are in mols or mass/time.

The energy balance for the condenser is:

$$Q_c = L\,(h_L - h_1) + D(h_D - h_1) \tag{17.17}$$

Where:
h_L = enthalpy of reflux stream
h_1 = enthalpy of vapor from top tray
L = reflux rate

For a total condenser, Equation 17.17 simplifies to

$$Q_c = V_1\,(h_D - h_1) \tag{17.18}$$

Where:
V_1 = top tray vapor rate = L + D

Reflux ratio is often used as a dimensionless measure of the energy required to make the separation. An increase in the reflux ratio increases V_1, which in turn, increases Q_c. Increases in condenser duty, Q_c increase the reboiler heat input, Q_B, and this heat directly affects operating costs since it is usually supplied by heaters or boilers.

In some facilities, reboiler heat is supplied by high temperature gas, *e.g.* compressor discharge, waste heat from gas turbine exhaust, by condensing steam from steam turbine exhaust. In these cases the cost of reboiler energy may be low or zero. This fact should be incorporated in the economic analysis used in the trays vs. reflux decision in step 5.

In many fractionation applications the feed stream is preheated by heat exchange with the bottoms product. This reduces the reboiler duty requirement.

There are many implications of the energy balance equations. If the condenser duty changes, the reboiler duty must change accordingly if the bottom product is to remain on specification. Further, there is a unique condenser and reboiler duty for a given feed rate and product specification.

Example 17.12: Calculate the condenser and reboiler duties for the example depropanizer. From the column material balance and SRK EOS, the following numbers apply.

F = 1000 kmol/h [2205 lbmol/hr]
D = 479.4 kmol/h [1057 lbmol/hr]
B = 520.6 kmol/h [1148 lbmol/hr]
L = (1.83) (479.4) = 862.9 kmol/h [1902.7 lbmol/hr]
h_F = 117.93 kJ/kg [50.70 Btu/lbm] MW_F = 57.56
h_D = 41.52 kJ/kg [17.85 Btu/lbm] MW_D = 44.21 (also MW of reflux L and top tray vapor V_1)
h_B = 256.56 kJ/kg [110.30 Btu/lbm] MW_B = 69.86
h_1 = 331.59 kJ/kg [142.56 Btu/lbm]

$$\begin{aligned}Q_c &= V_1(h_D - h_1)\\ &= (479.4\ +\ 862.9)(44.21)(41.52-331.59)\\ &= -17.2\times10^6\ \text{kJ/h} = -4782\ \text{kW}\quad[-16.3\times10^6\ \text{Btu/hr}]\end{aligned}$$

$$\begin{aligned}Q_B &= h_D D + h_B B - h_F F - Q_c\\ &= (41.52)(479.4)(44.21) + (256.56)(520.6)(69.86)\\ &\quad -(117.93)(1000)(57.56) - (-17.2\times10^6)\\ &= 20.62\times10^6\ \text{kJ/h} = 5728\ \text{kW}\quad[19.5\times10^6\ \text{Btu/hr}]\end{aligned}$$

9. Estimate the Tower Diameter

This topic will be covered in a later section "Tower Mechanical Design."

FRACTIONATOR CONTROL

The fractionator operates by using a controlled temperature gradient from top to bottom. The composition of the distillate product is fixed by its bubblepoint or dewpoint. The bottom product composition is controlled by its bubblepoint.

The fractionator always must operate so that the material and energy balances around it are satisfied on a steady-state basis. Any momentary upsets will be reflected by internal unstable operation which causes "upsets." Furthermore, it is a "sluggish" device. Liquid "hold-up time" is fairly large since flowrates are relatively low compared to its fluid inventory. Therefore, an inherent time lag occurs when controlling at the tower extremities.

The fractionator must process the particular feed inlet rate, temperature and composition that comes to it at a given moment. Feed rate is set by the upstream extraction process and/or product demand. Feed rate is controlled on a short-term basis to avoid temporary fluctuations in flow. This is usually accomplished by use of a fractionator feed surge tank. Feed composition is set by the upstream process and is not normally a controlled variable. Feed temperature may be controlled if a feed heat exchanger is installed.

A demethanizer and/or deethanizer normally is used to remove noncondensables that are prohibited in the saleable bottom product, e.g. NGL. The problem is to keep these noncondensables from passing out the bottom (fairly easy) with only minimum loss of saleble products out the top (more difficult). The distillate composition can only be controlled if the tower is refluxed.

Usually, a depropanizer and/or debutanizer produces a commercial distillate and bottoms product that must meet certain specifications. At this point no noncondensables should be in the system. In the usual situation the propane, butane or LPG mix are less valuable per unit volume than the heaviest product (natural gasoline, condensate, etc.).

Two rules are generally followed in fractionator control:

1. The lesser of the two streams should be manipulated (control wise) to obtain the greatest sensitivity in product quality.
2. Separation parameters should be manipulated to control the purity of the purest product; the material balance should be manipulated to control the quality of the less pure product.

Older control systems attempt to accomplish these functions by the use of pressure, temperature (compositions), level and flow controls on each stream independently. However, these streams are not really independent and the control system must address the interaction between them. A simple analog system may be used to accomplish this, but modern systems "marry" all of these to a computer which has been properly programmed. All streams being sensed feed their information into this computer or programmable logic controller which runs through a dynamic simulation and then tells the controls what to do.

A computer does not solve the control problem. It can only react within the limits imposed on it by its creator. It also represents an expensive control system. Many installations simply cannot justify the cost. In others, this degree of sophistication simply is not needed. The difficult decision is finding that proper system that minimizes operating costs and maximizes revenue while meeting product specifications.

As a guide in this endeavor, a series of control systems will be shown. These systems should be viewed as examples to illustrate the principles involved, not as firm recommendations.

Feed Surge Control

Regardless of the upstream process used to extract liquids, both flowrate and composition will vary to the first fractionator. A combination surge drum and vessel to flash-off a portion of the methane and ethane might be used ahead of this fractionator. The level must be allowed to fluctuate in this vessel to maintain a relatively stable flow to the fractionator. One simple approach is to use a liquid level control with a long displacement type float. By using proportional control with a wide proportional band (100-200%), large level fluctuations will dampen out rate changes. It is imperative that the feed surge drum have a minimum retention time of 10-15 minutes.

Diagrams (1) to (3) show several possible arrangements. The pump could be eliminated in the three arrangements if the tank is at a high enough pressure. If the tank is large, (1) could be used. A level indicator with level alarms would be required on the tank to guard against low and high levels. Method (2) shown would use the wide proportional band (and maybe a long displacer) with or without the pump. Method (3) is frequently used when feed fluctuations are frequent and/or the pressure on the accumulator is not constant. The level controller (LC) resets the flow recording controller (FRC). This is a cascade control system.

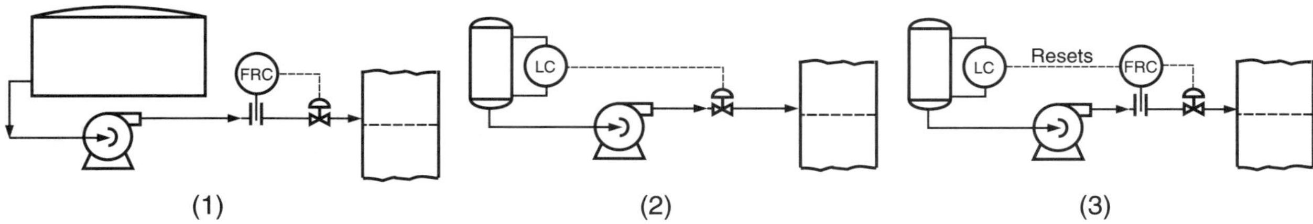

Column Pressure Control

Regardless of the column control system, it must contain some provision for pressure control. Column pressure can be controlled by manipulating the material balance (rate of distillate product) or by manipulating the condensing temperature (bubble/dewpoint pressure of distillate). Diagrams 4-7 show several arrangements.

Diagram (4) shows a simple back pressure control on the vapor from the partial condenser. In this case, only enough liquid is condensed to provide reflux. The pressure tap could be on the tower, as shown, or on the reflux accumulator. A proportional plus integral (P+I) controller might be used, although a proportional controller with a narrow proportional might be adequate since the offset would be small.

Diagram (5) shows a system for a total condenser that has proven suitable for a narrow boiling range product. The disadvantage is that a large control valve must be placed in the overhead line and

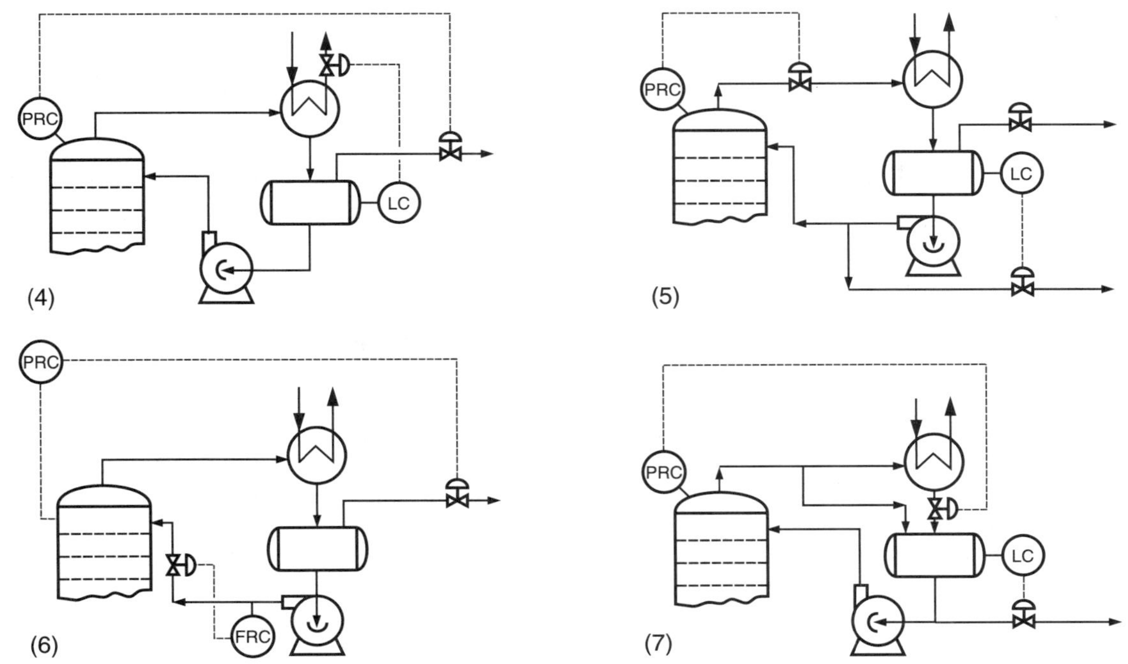

the tower pressure is higher than the condenser pressure making separation more difficult. It also increases the required head (ΔP) of the reflux pump.

Diagram (6) shows a flooded condenser system for a total condenser. In this system the accumulator runs completely full of liquid and pressure is controlled by manipulating the level (hence the heat transfer area) in the condenser. This method is often used in depropanizers and debutanizers with water-cooled condensers. Flooded condensers will almost always result in subcooled reflux.

Diagram (7) is an example of one type of hot vapor bypass to control tower pressure. The condenser is partially flooded and tower pressure is controlled by adjusting the level in the condenser. The vapor bypass equalizes the pressure between the accumulator and the tower.

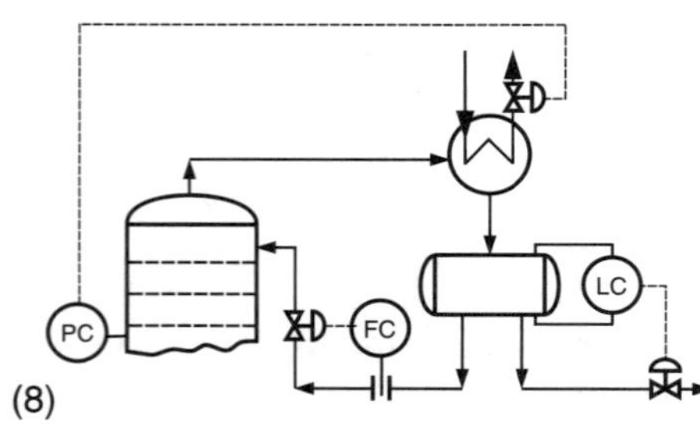

The temperature of the condensed product in the accumulator can also be controlled by controlling the cooling medium. This is shown schematically in Diagram (8). This method is not recommended if the cooling medium is water as it can cause fouling and scaling in the condenser due to low water flowrates. If the cooling medium is air, louvres or variable pitch for blades can be used to control air flow. Induced draft coolers are preferred in this service because the tube bundle is not exposed to the weather, e.g. rain, snow, etc.

Most pressure control systems are based on manipulating the cooling rate at the condenser. If the condenser is allowed to operate without restriction, the column pressure will be as low as possible given the cooling medium and operating conditions. This is called "floating pressure control" and has the benefit of reducing the difficulty of separation, as relative volatilities tend to increase with decreasing pressure for most hydrocarbon separations. Reference 17.8 quotes energy savings of 1% for every 1°F reduction in condenser temperature for a separation of iC_5/nC_5. Reference 17.10 presents one scheme for a floating pressure control system.

The rate of reflux is manipulated by a flow controller, cascaded with a temperature or composition controller on the distillate product. It may also be set manually by the operator. In feed-forward systems it is sometimes ratioed to the feed rate.

Column Control

There are a number of ways to instrument a fractionation tower, but the ultimate objectives are

1. Meet the product specifications for the distillate and bottoms products
2. Optimize profitability by minimizing operating costs and maximizing the yield of the most profitable product

In addition the above objectives must be satisfied with a viable, cost effective and safe control system.

To the casual observer, fractionation control appears quite complex but if we ignore feed rate and feed condition, there are only five "knobs to twist" on a fractionation column.

1. Distillate rate
2. Bottoms rate
3. Heat input (reboiler)
4. Heat removal (condenser)
5. Reflux rate

With these five "knobs" we must control:

1. Bottoms composition (temperature)
2. Distillate composition (temperature)
3. Tower pressure
4. Liquid level in tower
5. Liquid level in accumulator

Theoretically, there are 120 ways of pairing controlled and manipulated variables to achieve column control. Obviously many of these are not feasible, but several are viable control alternatives.

Historically, the control system has been set up as follows.

Controlled Variable	Manipulated Variable
Bottoms composition (temperature)	Heat input (reboiler)
Distillate composition (temperature)	Reflux rate
Tower Pressure	Heat removal (condenser)
Liquid level in tower (or reboiler)	Bottoms rate
Liquid level in accumulator	Distillate rate

If a flooded condenser is used, the liquid level in the accumulator is not controlled and the distillate rate is manipulated to control heat removal (condenser surface area), hence tower pressure. This is illustrated in diagram (9).[17.11]

Ideally one would like to operate the fractionator so that both the distillate and bottoms product specification are met exactly. We know from our steady-state calculations that this condition allows no degrees of freedom in the fractionator. In an operating fractionator, the distillate rate, bottoms rate, reboiler duty, reflux rate and condenser duty cannot all be fixed. How does one cope with the upsets – changes in feed rate, composition, ambient temperature, etc. Frequently, the operator will open the control loop on either/both the distillate or bottom composition and make better than specification product. This allows the operating flexibility to handle upsets as well as a cushion of "better than specification" product in the storage tank. This option is certainly viable from a control standpoint but can be expensive in terms of energy utilization.

What is preferred is a control system that minimizes energy utilization while meeting product/yield specifications. The system in diagram (9) works well at low reflux rates, but it is highly interactive. The material balance and heat balance interact because distillate flow is used to control the heat removal rate. However, at low reflux rates (L/D < 1.0) this system has proven satisfactory.

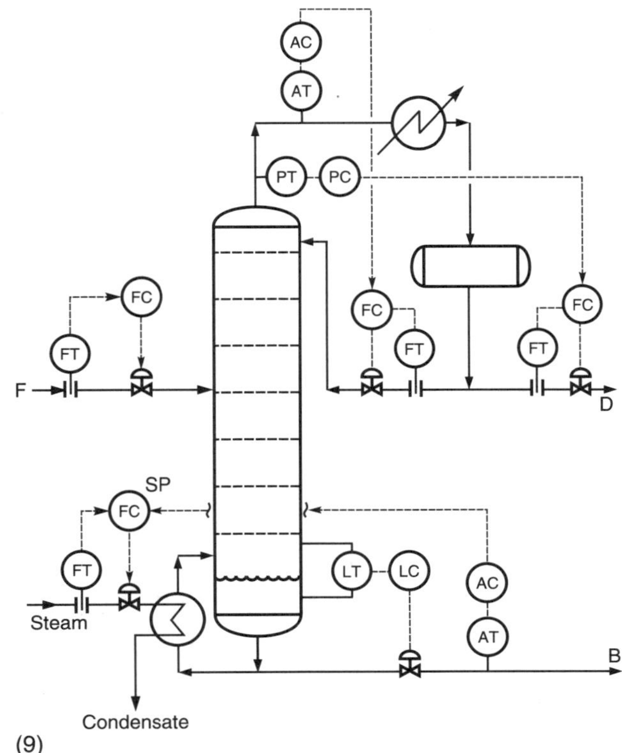

(9)

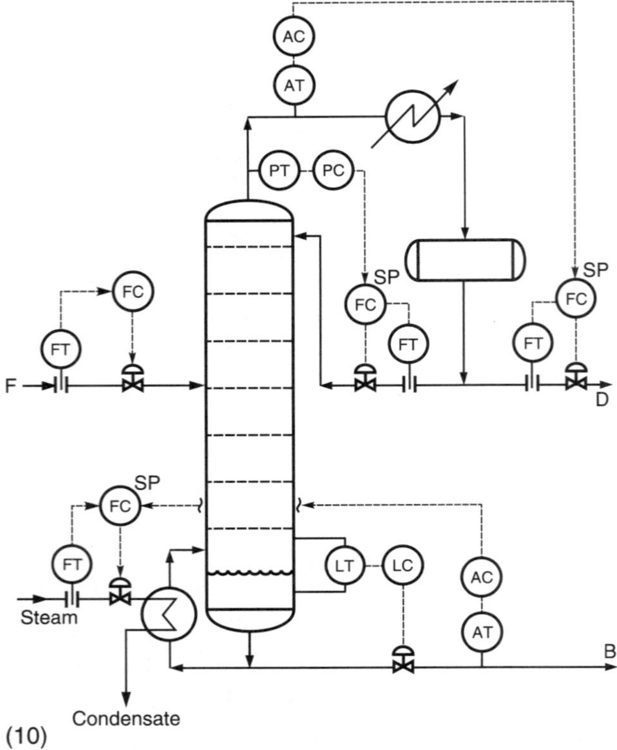

(10)

At higher reflux ratios (L/D > 5.0), the conventional system may be inadequate. A material balance system has been used to reduce interaction, as shown in Diagram (10).

Notice that the distillate composition is controlled by manipulating distillate rate. If the distillate product contains too much heavy key, too high a distillate rate is indicated. As the distillate rate is reduced the liquid level rises in the condenser increasing tower pressure. The pressure controller responds by increasing the reflux rate.

A material balance control scheme can also be set up to control bottoms composition by manipulation of the bottoms rate. In these schemes the heat input at the reboiler is used to control either the liquid level in the tower bottoms or the tower pressure. When material balance control is used it is common to control the composition of the most important product by manipulating the heat balance and control the composition of the other product by manipulating the material balance.

Feed-forward control is used to minimize the upsets caused by changes in the feed rate, composition, temperature. An effective feed-forward control system allows the column to operate as closely as possible to product specifications with a minimum of "over fractionation" (exceeding product specifications).

One example of a feed-forward system is shown in diagram (11). The feed rate signal is fed forward to both the reboiler (heat input) and distillate rate. Both variables are adjusted based on feed rate changes. Feed-forward control can improve the response of the control system to upsets saving energy and minimizing off-specification fractionation.

Feed composition is generally not monitored in a feed-forward control system as composition changes tend to be of lower magnitude and frequency than rate changes. Also, analyzers are more expensive and less reliable than flowmeters.

The open-loop response of a fractionator to feed changes is not instantaneous. Liquid flow across the trays must change and in a column with

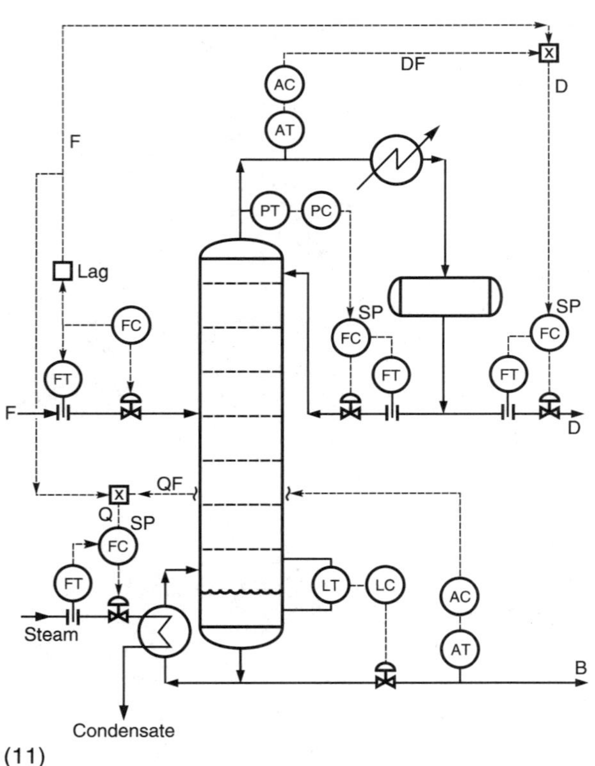

(11)

many trays this may induce considerable lag. Temperature changes are subject to lags due to contact efficiency, resistances to heat transfer, etc. When feed-forward control is used the feed-forward signal is often "dynamically lagged" so that it arrives at the manipulated variable at the correct time and provides the right amount of correction.

Most DCS control systems are often used to provide this dynamic lagging. This is easily done today with the wide availability of programmable controllers and microcomputers. The fractionation process is essentially simulated by the computer and the magnitude and timing of the feed-forward signals is computed. The use of computers in fractionation control is not universal by any means. However, the increased cost and complexity of the control system must be economically justified by lower operating costs or more profitable product yields. This justification is sometimes more imagined than real.

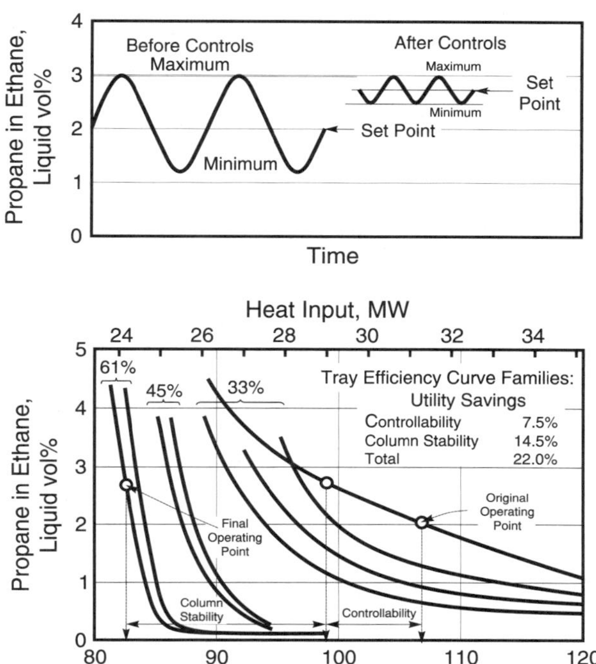

The figures at right show an example for a deethanizer.(17.12) The top figure shows the deviation in the volume percentage of ethane in the bottoms product before and after installing more sophisticated control. Narrowing the deviation obviously improves controllability and results in greater savings (and ease of operation).

The bottom figure shows a summary of performance for three different overall tray efficiencies 61, 45 and 33%. By moving (and holding) closer to the set point, a 22% savings in utility consumption is achieved. This alone is significant. It also results in greater tray efficiency since the column is more stable.

Condensers and Reboilers

The exact control of condensers and reboilers depends on the type. Condensers may be either the open type which are located in the bottom of a cooling tower and do not contain a shell, or they may be of the closed type, i.e. shell and tube exchanger, generally located near the bottom of the fractionator. In condensing a vapor with water, the greatest resistance to heat transfer will be the water film. For this reason, efforts to increase the overall coefficient should center around increasing the heat transfer coefficient on the water side. A water velocity of about 1.5 m/s [5 ft/sec] is considered to be about optimum. An average overall coefficient for condensers used on light hydrocarbon fractionators would be 570 $W/m^2 \cdot °C$ [100 $Btu/hr\text{-}ft^2\text{-}°F$].

Reflux accumulators are often sized for a liquid retention time of 5-10 minutes. Large accumulators offer the disadvantage of increasing the time lag in the control system. Smaller accumulators are indicated from a process standpoint, particularly when a "flooded" condenser is used; however, if the accumulator is too small it can be difficult to operate the column, particularly during start-up because the accumulator level changes too rapidly.

The feed preheater may use steam for the heating medium or it may make use of a hot hydrocarbon stream such as the column bottoms. If steam is used, a value of about 3400 $W/m^2 \cdot °C$ [600

Btu/hr-ft^2-°F] for the overall coefficient would represent a fair average. When using another hydrocarbon liquid stream to heat the feed, the value of U would be about 570 W/m^2·°C [100 Btu/hr-ft^2-°F]. This value would also represent a fair average for product coolers. When using a liquid hydrocarbon stream, the optimum velocity of the hydrocarbon through the heat exchanger will be between 2-3 m/s [7-10 ft/sec].

Reboilers on light hydrocarbon fractionators can be divided into three categories or types: internal reboilers, external "kettle type" reboilers, and external "heat exchanger type" reboilers. Heat exchanger type may be "pump-through" or thermosyphon. In some applications, the "heat exchanger type" reboilers are showing themselves to be superior to the other two types for hydrocarbon fractionation service. All four are shown in the following figure.

Internal reboilers (a) are generally used only on small diameter columns. They are simple, inexpensive and minimize footprint; but require entry into the tower to perform maintenance.

Kettle reboilers (b) are quite common in NGL distillation. Liquid from the bottom tray is gravity fed to the reboiler from a draw-off at the bottom of the column. A weir maintains the liquid level in the reboiler above the tube bundle. Vapor disengaging space is provided in the reboiler and the vapor is routed back to the column. The bottom product is drawn from the reboiler. The kettle

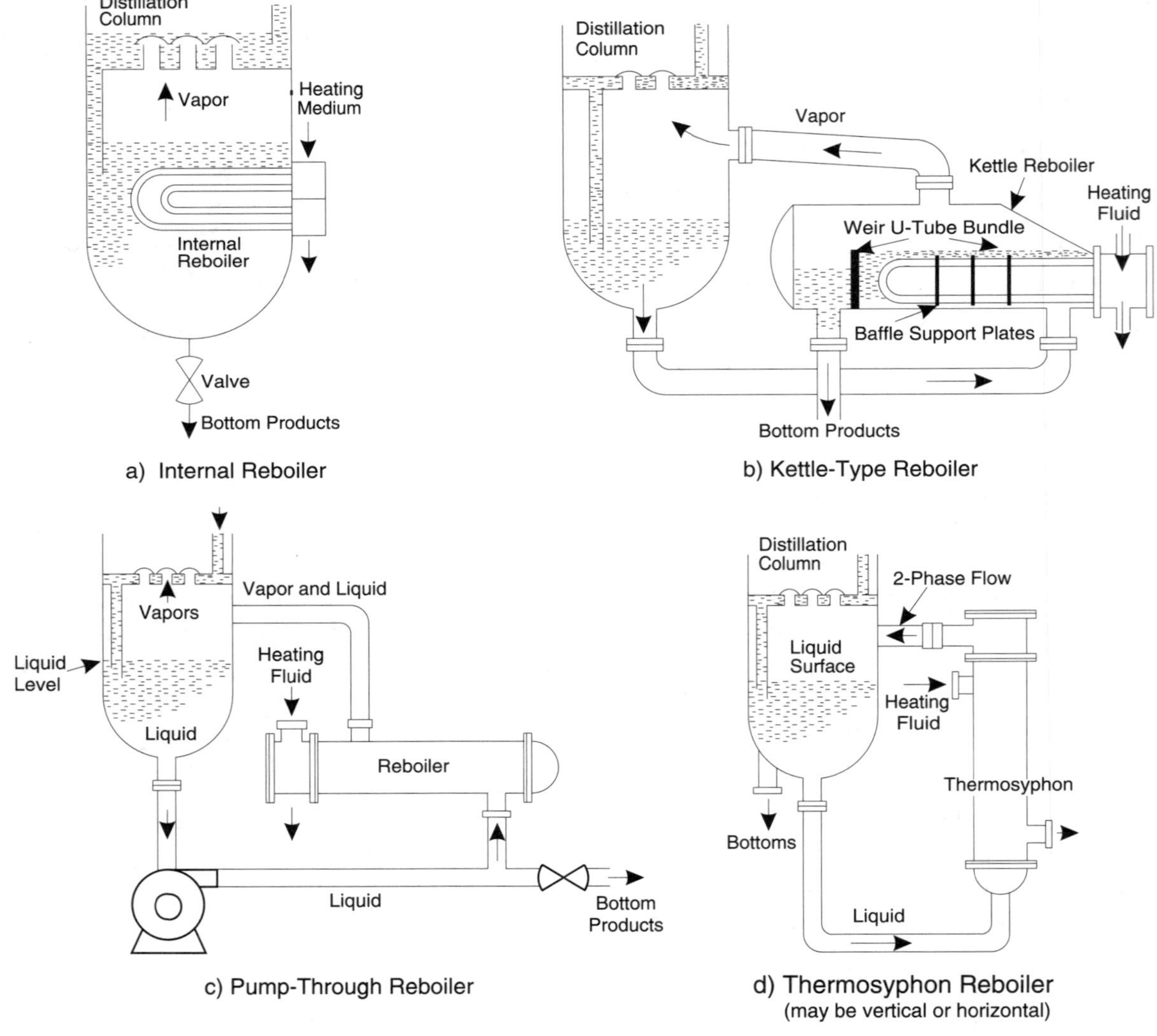

reboiler is equivalent to one theoretical stage. It is also simpler to design and operate than pump-through reboilers since there is no two-phase flow.

Pump-through heat exchanger reboilers (c) utilize a pump to pump the product through the reboiler. These are sometimes referred to as forced circulation reboilers. The pump rate will typically be much higher than the bottom product rate. The bottom product may be withdrawn from the pump discharge or from the column itself. This reboiler scheme is generally used where the reboiler heat is provided by a remote heat exchanger, such as a fired heater, and the pressure drop through the reboiler loop is too great to support natural circulation.

Natural circulation or thermosyphon reboilers (d) utilize the density difference between the liquid entering the reboiler and the two-phase stream exiting the reboiler to provide the pressure differential necessary to create flow through the reboiler. The reboiler can be installed vertically or horizontally. Horizontal reboilers provide easier maintenance access. In vertical reboilers the heating medium is typically on the shell side and in horizontal reboilers on the tube side. TEMA type G and H reboilers are often used in thermosyphon service because of the low shell-side pressure drop. The flow through the reboiler can either be once-through or recirculation. In once-through reboilers the liquid flow exits a tray in the tower and returns to a tray (or distributor) or the tower bottom. This type is typically used in side-reboiler applications in demethanizer and deethanizers. Recirculating reboilers are generally used in bottom reboiler applications. The circulation rate is whatever density difference and pressure drop allow.

It is critical that any reboiler have sufficient vapor disengaging space, adequate nozzle sizes, and return piping. If not, vapor binding may limit capacity instead of area.

For thermosyphon type reboilers, size of the vapor-liquid return line also is critical. Detailed design of thermosyphon reboilers is beyond the scope of this text. The normal liquid velocity in the inlet line is 0.7-2.0 m/s [2.3-6.5 ft/sec]. The velocity in the return line may be roughly approximated by the equation

$$v = \left(\frac{A}{\rho_m} \right)^{0.5} \tag{17.18}$$

Where:		SI	FPS
A	= constant	6000	4000
ρ_m	= mixture density	kg/m^3	lbm/ft^3
v	= velocity	m/s	ft/sec

The heat transfer medium can be a circulating fluid such as hot oil, Dowtherm®, Therminol®, etc. or steam. Steam provides higher heat transfer coefficients but steam systems are more expensive to build and operate. Steam is popular in facilities where low pressure steam is available from another part of the process such as steam turbine exhaust, or Claus sulfur recovery unit. Hot oil systems are the most popular in light hydrocarbon distillation and crude oil/condensate stabilization.

Reboiler heat flux must be limited to avoid film-wise vaporization discussed in Chapter 13. For standard tube designs Q/A should be limited to about 32 000-39 000 W/m^2 [10 000-12 000 $Btu/hr\text{-}ft^2$] and temperature differences to below 40°C [72°F]. Many reboilers are installed with proprietary tube designs e.g. twisted, finned, grooved, etc. to increase heat transfer rate and to encourage nucleate boiling.

CRUDE OIL AND CONDENSATE STABILIZERS

Fractionation-type stabilization of crude oil and condensate from production separators can be economically attractive relative to traditional flash stabilization. This is particularly true in the case of high API gravity crudes or low-temperature separation processes. However, stabilization is not limited to these systems and is not nearly so widely utilized as it should be. The primary reason for this lack of use seems to be the apparent complexity of this approach compared to stage separation and, in many cases, mere habit. However, more use is now being made on large oil projects because of higher efficiency than stage separation and the ability to control vapor pressure for tanker shipment. Stabilization of heavier crudes can produce extensive fouling in the tower.

Figure 17.14 is a view of an example non-refluxed stabilizer. In most production system applications the column will operate as a non-refluxed tower. This type of operation is simpler, but sometimes less efficient, than the refluxed tower operation shown in the inset to Figure 17.14. Refluxed operation is preferred when the feed temperature-pressure relationship exceeds that shown in Figure 17.15. The non-refluxed tower requires no external cooling source, so it is particularly applicable to remote locations. When a condenser is used in a stabilizer separation, it will nearly always be a partial condenser because of the quantities of methane and ethane that must be removed from the tower feed.

A refluxed stabilizer is designed using normal fractionation methods. In general, the specification product will be the bottoms product, and in most cases the specification on this will be a vapor pressure limitation. A non-refluxed tower cannot be designed using the shortcut calculations described for fractionators. There is no external reflux for the tower and so one degree of control over the tower has been lost.

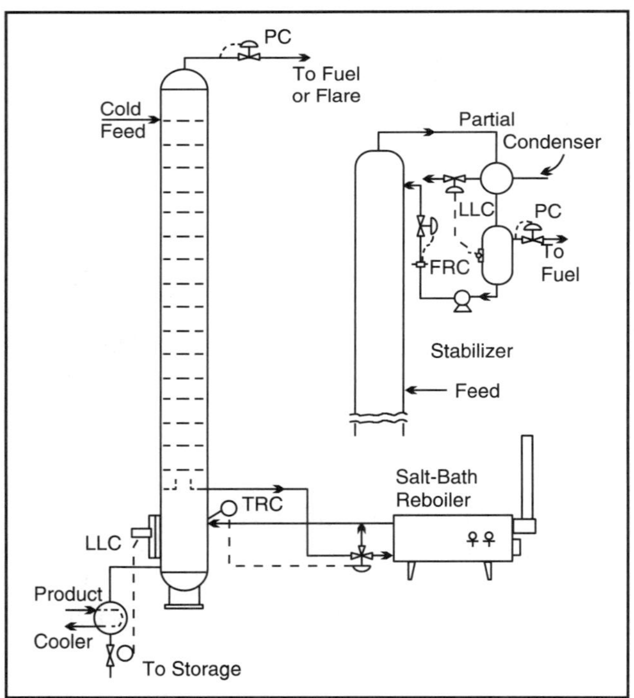

Figure 17.14 Schematic View of a Stabilizer

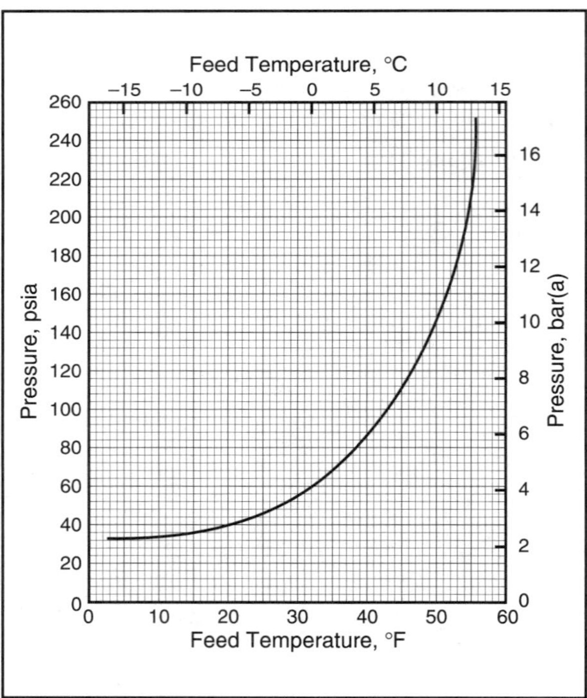

Figure 17.15 Maximum Recommended Feed Temperature to Cold-Feed Stabilizer

Figures 17.15-17.17 show convenient guides for estimating proper operating range of a non-refluxed stabilizer. Figure 17.15 shows the maximum recommended feed temperature to a stabilizer as a function of operating pressure. In some stabilizers, the feed is split so that a portion of the feed is fed to the top tray for cooling, and the remainder of the feed is preheated with the bottoms product to minimize reboiler duty.

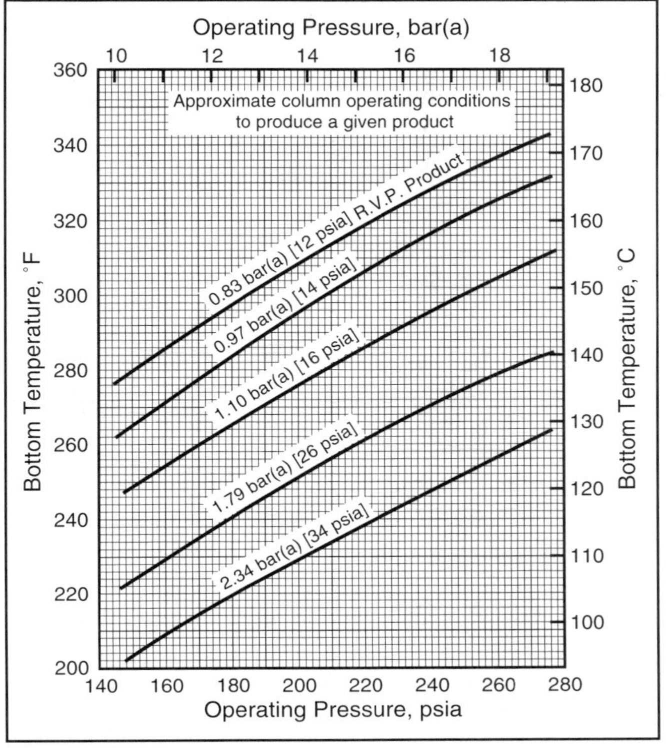

Figure 17.16 Estimation of Stabilizer Bottom Temperature

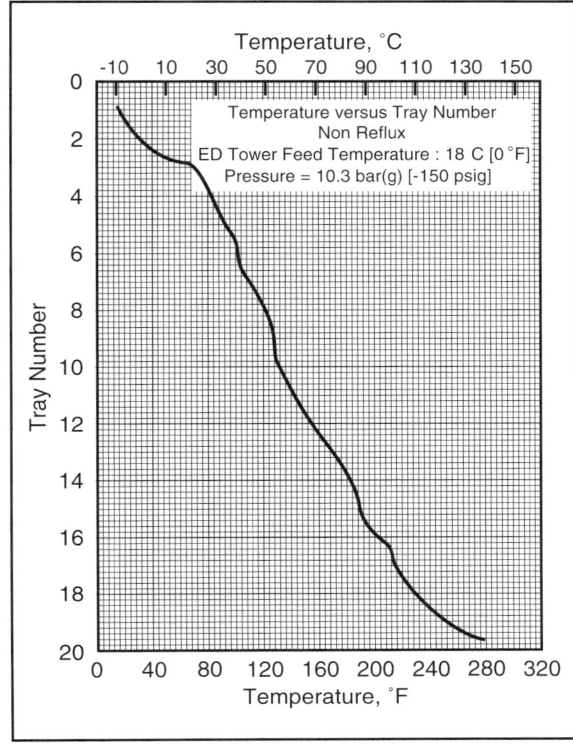

Figure 17.17 Example Temperature Profile of Cold-Feed Stabilizer

In some cases the reboiler for the stabilizer will be an indirect salt bath heater or a steam heated exchanger. Figure 17.16 shows suggested bottom (reboiler) temperatures for producing a specified Reid vapor pressure product. The temperature for a stable crude oil or a stable stock tank condensate will be in the vicinity of the 12 psi curve.

After the operating temperatures and the pressure have been established through use of Figures 17.9-17.11, the preliminary split in the tower can be predicted. Computer simulation is routine today, however, there are shortcut methods which give reasonable results. One of the most convenient and accurate involves utilization of pseudo-K values for each component between the top and the bottom of the tower. Using this concept, the separation that can be achieved across a non-refluxed stabilizer can be estimated by use of the pseudo-K values and a simple flash calculation.

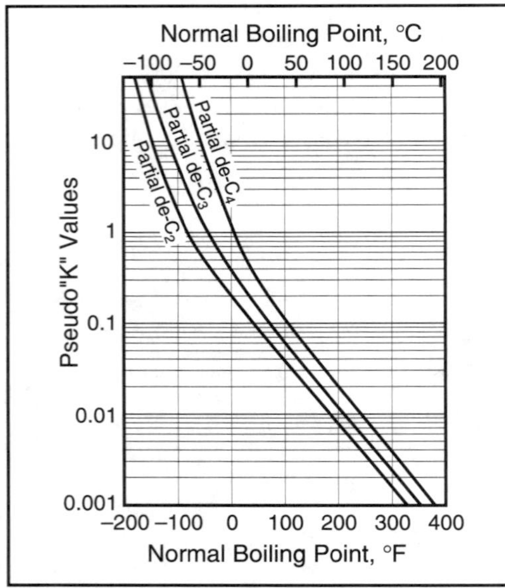

Figure 17.18 Pseudo K Values for Cold-Feed Stabilizers

When using this approach, the following procedure is recommended:

1. Using Figures 17.15 and 17.16, estimate the operating pressure of the stabilizer.
2. Obtain pseudo-K values from Figure 17.18 for each of the components in the feed mixture.
3. Perform a flash calculation to estimate the product split in the stabilizer tower. The vapor from the flash calculation will be the composition of the distillate product and the liquid from the flash will be the composition of the bottom product.
4. Determine the top tower temperature as the dewpoint of the calculated vapor and the bottom tower temperature as the bubblepoint of the calculated bottom product.
5. Make an energy balance around tower to find reboiler heat duty.

Use of stabilization will ordinarily enable one to increase the liquid recovery and/or crude API gravity, as well as reduce weathering in storage and during shipment and transfer. Stabilization should always be considered an economically attractive alternative to stage separation. In heavy crude areas where "spiking" with light components is common, stabilization of the mix is frequently desirable. Obtaining good mixing when "spiking" is usually much more difficult than one would first anticipate.

Fractionation-type stabilization is being used more extensively offshore in lieu of separator-type stabilization discussed in Chapter 5. Fractionation stabilization is often a more effective method of retaining the C_3-C_5 hydrocarbons in the crude or condensate product while meeting the RVP specification. Other advantages cited include reduced compression costs and better water/oil separation.

Based on computer simulation, Reference 17.13 suggests fractionation-type stabilization be evaluated when

1. If the crude must have an export quality of 50-60 ppm H_2S, maximum.
2. If tanker-quality crude is required (70-84 kPa [10-12 RVP]) and compressor-interstage condensate cannot be directly exported.
3. If gas cannot be exported via sales line and must be re-injected.
4. If upstream oil-gas separation temperatures are low.

The use of a stabilizer is usually not recommended if:

1. A pipeline export crude at an RVP < 12 is not specified.
2. Small amounts of vapor are flowing as may occur in low GOR crudes.

One common operating problem in fractionation-type stabilization systems is the presence of water (and salt) in the system. Depending on the top tray temperature water can accumulate in the tower. It builds up in the tower and should be removed. Many stabilizers have water-draw trays near the top of the tower to collect water. It is periodically drained from the tray. The source of the water is carryover with the oil or condensate. If the water has a high salt content, salt can precipitate out at the reboiler causing heat transfer problems. A water wash upstream of the stabilizer can reduce this problem.

ABSORPTION

Physical absorption is used primarily in gas conditioning for the removal of water, hydrogen sulfide and carbon dioxide. Chapter 18 covers glycol dehydration; Volume 4 of this series addresses sweetening.(17.14)

At one point in time, oil absorption plants were the primary source of natural gas liquids. They have been supplanted, generally, by one of the refrigeration processes. Although the discussion that follows is based on lean oil examples, the word "oil" could be replaced with glycol, amine, Selexol, etc.

Rich gas (sometimes called wet gas) enters the bottom of the absorber and flows upward through the absorber counter-current to the lean oil. The absorbable components are transferred from the gas to the oil. The dry gas (sales gas) leaves the top of the absorber. Lean oil enters on the top tray of the absorber and flows downward counter-current to the rich gas, picking up absorbable materials. Rich oil leaves the bottom of the absorber and flows from there to the top of the stripper. In the stripper, heat is added and sometimes a stripping agent such as steam or dry gas. The absorbed material is removed overhead from the stripper, and the lean oil from the bottom of the stripper is recycled to the top tray of the absorber. The figure is a block diagram only, and there are heat exchangers, pumps and perhaps even fractionating towers (a demethanizer or deethanizer) that process the rich oil between the absorber and the stripper.

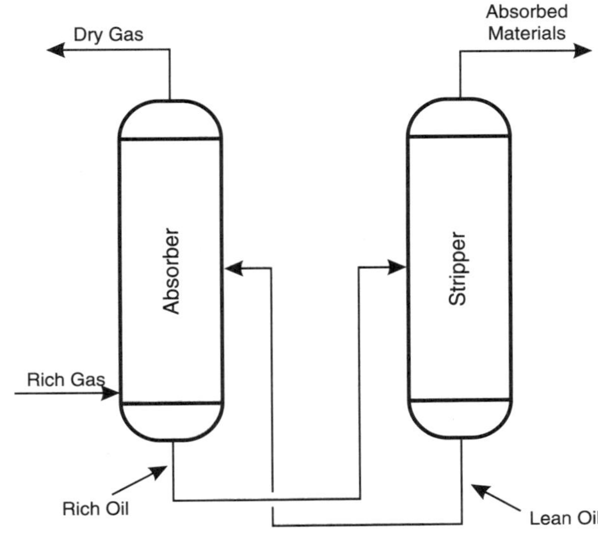

The absorber-stripper combination are "twins." They go together. The absorber recovers components, and then the stripper removes them so the oil can return to the absorber in proper condition.

A glycol dehydration plant is an absorber-stripper combination. The principles outlined herein will be re-applied in Chapter 18.

In the counter-current contact of the gas and oil, a component is absorbed when its "partial pressure" in the gas phase exceeds the "partial pressure" in the liquid phase. (Partial pressure is used here for discussion purposes, the actual driving force is chemical potential.) The oil acts as a means of lowering the vapor pressure of the components by dilution. The leaner the oil is in the components to be absorbed, the more efficient the absorption process. For a given gas rate and composition, the oil circulation rate, oil composition and number of equilibrium contacts are the prime variables. Oil rate and number of contacts are interdependent because as oil rate increases it remains leaner for a given absorption and less contacts will be needed.

As the components change phase from vapor to liquid, an amount of energy known as the *heat of absorption* is released. In magnitude, it is slightly greater than the latent heat of condensation. The energy released must be absorbed by the oil and gas as they flow through the tower. Consequently, both the dry gas and the rich oil leave at temperatures higher than the rich gas and lean oil. The total heat release is almost proportional to the amount of gases absorbed because the heat of absorption of light hydrocarbons does not vary significantly from component to component. In some cases, the oil must be cooled externally at some point in the absorber to maintain the temperature at the desired level.

The lean oil used in absorbers will usually have a molecular weight in the range of 100-200, depending on the average tower temperature. Below –18°C [0°F], a 120-140 molecular weight oil is commonly used. At absorber temperatures near 40°C, the molecular weight will probably be about 180-200. A properly stripped lean oil will contain few components lighter than pentane in measurable amounts.

A number of absorber calculation methods have been proposed and are summarized in various texts, articles, and handbooks. Only one short-cut method and a tray-by-tray method will be considered here. These will suffice for most applications for those charged with specification and operation of absorption facilities.

Nomenclature

In order to simplify the algebra, a slightly different nomenclature is often employed in absorption than that used in standard equilibrium calculations. This is shown in Figure 17.19.

In an absorber the plates are numbered from top to bottom. The bottom plate is plate (N). The subscript on any value is the plate number involved. The absorbent (lean oil) enters from above plate (1) so it carries the subscript (0). The gas enters from below plate (N) so it carries the subscript (N+1). In some equations the subscript (n) is used. This refers to any particular plate (n) from the range (1) through (N).

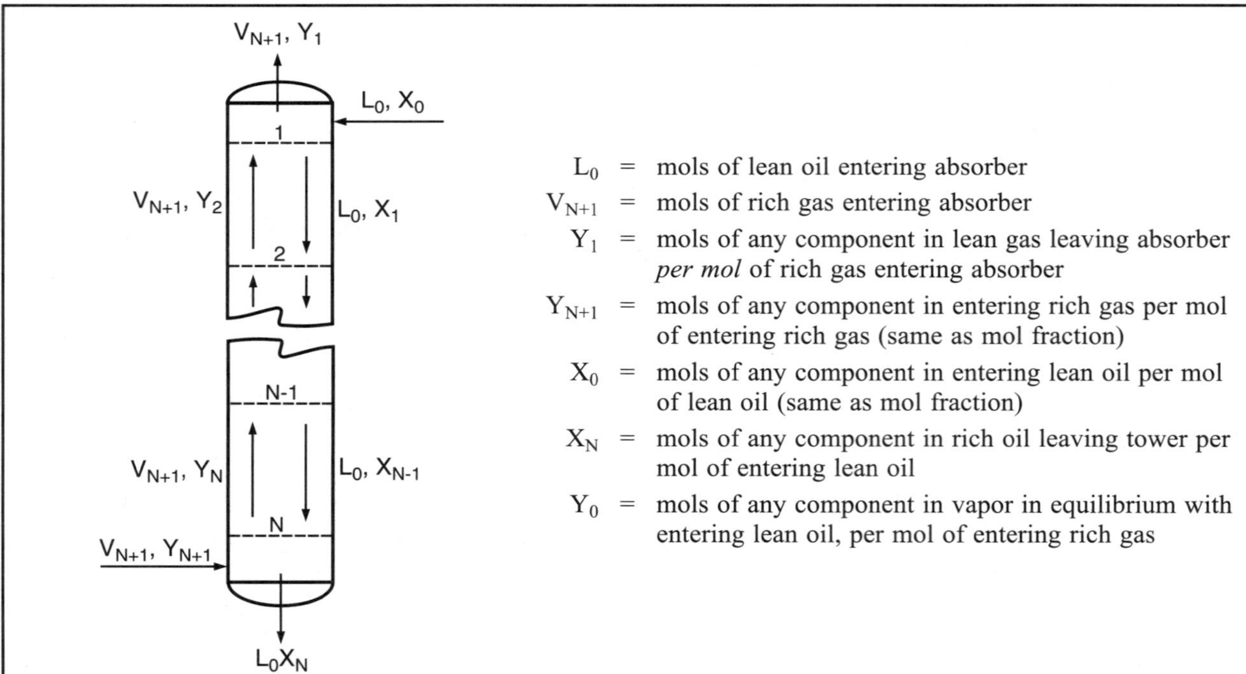

Figure 17.19 Schematic View and Nomenclature for an Oil Absorber

Notice that Y is the relative amount of a component in the gas leaving any plate using the inlet gas rate as a basis. The value of X is the relative amount of a component in the liquid leaving any tray using the entering amount of lean oil (L_0) as a basis. Y and X are not mol fractions; they are relative amounts, or ratios, where the denominator is V_{N+1} and L_0, respectively.

The value of Y_0 is the relative amount of any component in the lean gas leaving if this gas is in equilibrium with the lean oil entering. It is not the case, of course. So, it is merely a boundary condition number necessary for the calculation. It would be found by an equilibrium calculation. If the lean oil contains a negligible amount of the component involved, Y_0 is negligible and may be taken as zero for approximate calculations.

Basic Equations

Using the above nomenclature, the equilibrium relationship on any theoretical plate may be written as

$$K_n = \frac{y_n}{x_n} = \left[\frac{V_{N+1} Y_n}{V_n}\right]\left[\frac{L_n}{L_0 X_n}\right] \tag{17.19}$$

A material balance around any plate (n) and the top or bottom of the tower ties together the vapor and liquid streams passing each other between plates. If plates (n–1) and (n+1) are the plates above and below (n) respectively, then for any component

$$\frac{L_0}{V_{n+1}} = \frac{Y_{n+1} - Y_1}{X_n - X_0} = \frac{Y_n - Y_{n+1}}{X_{n-1} - X_n} \tag{17.20}$$

Equations 17.19 and 17.20, together with the corresponding enthalpy balance, can be used as a basis for rigorous tray-to-tray calculations. This becomes a multiple trial-and-error process involving the following steps:

1. Assume a top tray temperature. The entering lean oil temperature plus 3-6°C [5-10°F] is a good guess.
2. Make a dewpoint calculation on the estimated exit gas composition. (The short-cut procedure which follows will provide a good estimate for dry gas composition.) Add enough oil to the gas to make the dewpoint calculation. (This also serves to estimate oil losses.)
3. Find composition of liquid leaving top tray using K values at assumed top tray pressure and temperature.
4. Find composition of vapor leaving Plate 2 from liquid leaving Plate 1 by means of Equation 17.20.
5. Run enthalpy balance around top plate using an assumed vapor temperature from Plate 2. If balance checks, continue a similar calculation for Plate 2. If not, repeat steps 1-5.
6. Proceed down the absorber in this manner until reaching the bottom plate. Then, an overall heat balance must be run. If it checks, solution is complete. If not, the whole procedure must be repeated until the overall heat balance checks.

An aid to estimating liquid and vapor rates for the calculation procedure is to consider that the total absorption is constant on each plate. This leads to the equation

$$\frac{V_n}{V_{n+1}} = \left[\frac{V_1}{V_{N+1}}\right]^{1/N} \tag{17.21}$$

A material balance around Plate (n) and the top of the tower yields

$$L_n = L_0 + V_{N+1} - V_1 \tag{17.22}$$

The temperature on Plate (n) may be estimated by assuming that the temperature change is proportional to the volume of gas absorbed.

$$\frac{T_N - T_n}{T_N - T_0} = \frac{V_{N+1} - V_{n+1}}{V_{N+1} - V_1} \tag{17.23}$$

The above procedure seldom can be justified for routine applications, except on a computer. Certainly, it is too lengthy and time consuming to be used frequently on a hand calculator. Consequently, much shorter and more convenient solutions are desired.

Kremser-Brown Approach

If one assumes that (L/V) does not change very much from plate to plate and that temperature changes are nominal, one can simplify the tower balances. This is a reasonable assumption for most natural gas and glycol absorbers where gas shrinkage is small.

If Equations 17.19 and 17.20 are applied to each plate of an absorber containing a total of (N) plates, Equation 17.24 results.

$$\frac{Y_{N+1} - Y_1}{Y_{N+1} - Y_0} = \frac{A_1 A_2 \ldots A_N + A_2 \ldots A_N + \ldots + A_N}{A_1 A_2 \ldots A_N + A_2 \ldots A_N + \ldots + A_N + 1} \tag{17.24}$$

The absorption factor "A" is defined as A = L/KV. For a specified plate, n, $A_n = L_n/KV_n$. Utilizing a mathematical identity, Kremser and Brown simplified Equation 17.24 to:

$$\frac{Y_{N+1} - Y_1}{Y_{N+1} - Y_0} = \frac{A_e^{N+1} - A_e}{A_e^{N+1} - 1} = E_a \tag{17.25}$$

where A is the average of "effective" absorption factor.

Kremser-Brown recommend defining the effective absorption factor (A_e) as:

$$A_e = \frac{L_0}{K V_{N+1}} \tag{17.26}$$

The ratio (L_0/V_{N+1}) is the mols of lean oil per mol of entering gas – a constant for a given calculation. "K" would be the value for the key component at average column conditions – average column temperature and pressure. The key component will be:

- The component whose recovery is specified or the component whose "A" is closest to unity where no recovery of a key component is specified.

If the lean oil is essentially free of a given component, $Y_0 = 0$ for that component. In this case, the left-hand side of Equation 17.25 represents the fraction of that component absorbed (E_a).

For analyzing an existing tower, the top and bottom temperatures may be averaged to find the average "A." For approximate design, the bottom temperature may be assumed to be about 3-8°C [5-15°F] above that of the entering rich gas. The top tray temperature should be estimated to be 3-6°C [5-10°F] above the entering oil temperature.

One can plot Equation 17.25 in graphical form. The result is a figure possessing the general characteristics shown at left.

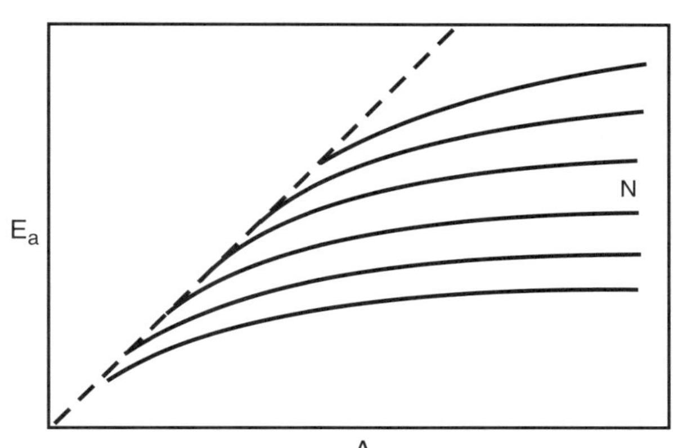

All of the lines representing a constant value of N become coincident with the 45° dashed line representing an infinite number of theoretical plates. For a given value of E_a, the combination of A and N used must be to the right of the dashed line for an actual unit.

The lines for N become rather flat as A increases. The most economical unit would be designed to operate a reasonable distance to the right of the dashed line in the curved portion of these lines.

There are three variables that one can control in absorption, assuming the quantity, composition and pressure of the inlet gas are fixed – oil circulation rate, number of theoretical plates and oil temperature. Figure 17.20 is a plot of Equation 17.25. From Figure 17.20, the relationship between absorption factor, number of theoretical plates and fraction absorbed can be obtained for each component. The average "K" should be found by taking the arithmetic average of the lean oil and entering gas temperature and adding 5°C [9°F]. The pressure drop in most absorbers is small and tower pressure may be assumed constant.

If one wishes to find the oil rate for 85% recovery of propane in an absorber having 8 theoretical trays, one would locate 0.85 on the ordinate of Figure 17.20, read horizontally to 8 trays and drop vertically to read the value of A. Knowing K, V_{N+1}, and A, calculate L_0, the oil rate in mols of lean oil per mol entering rich gas. Now, in similar fashion, if the oil rate is known, the value of A can be determined; at the intersection of the horizontal line from 0.85 and the vertical line from the value of A, the number of theoretical trays necessary for that oil rate can be found.

Note that the lines of constant N in Figure 17.20 are not smooth curves. This is because the scale for A on the abscissa is not uniform. The scale range has been varied to cover wider ranges of A.

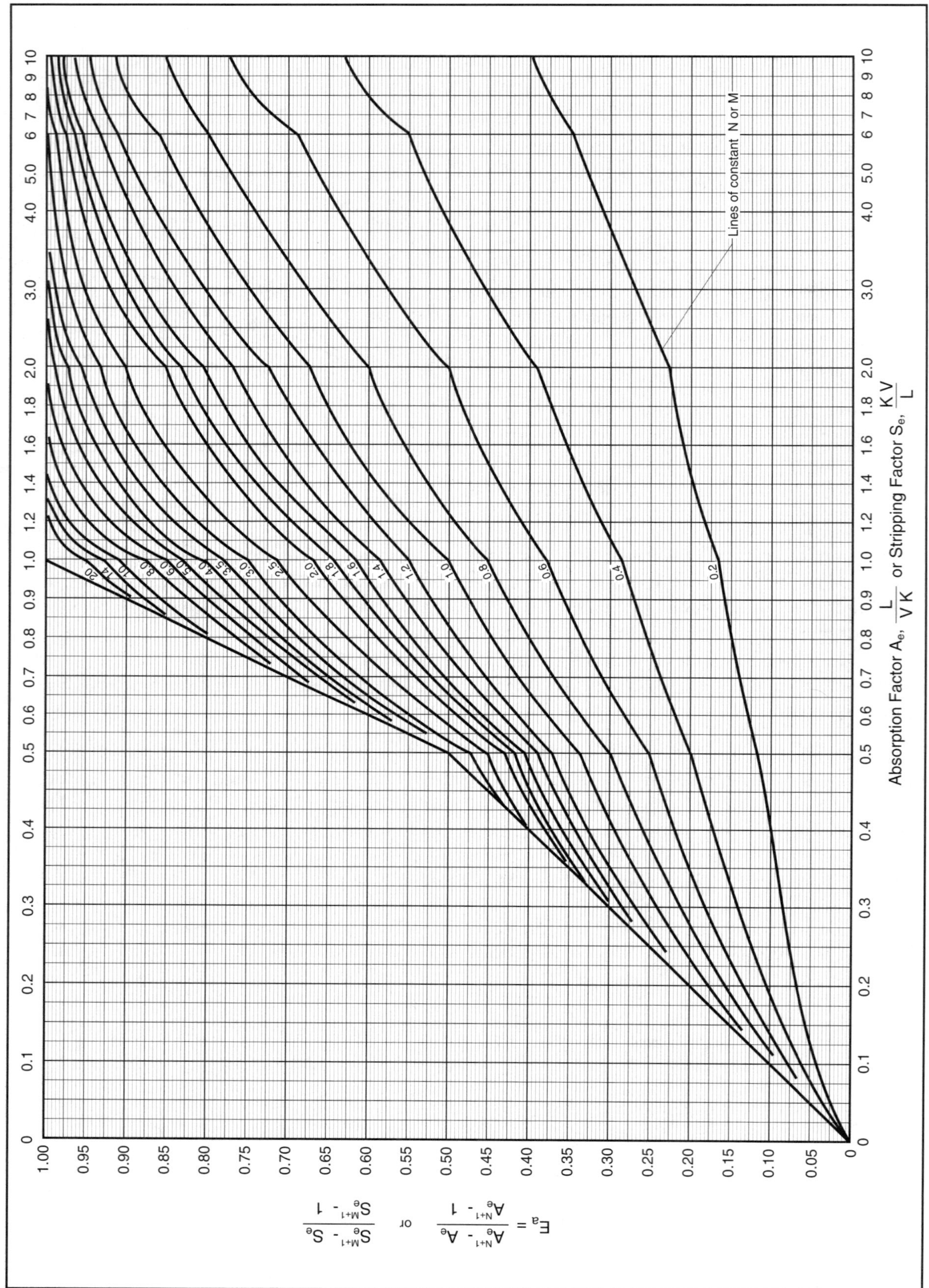

Figure 17.20 Absorption Factor Chart

Units

The L and V for this correlation are molar rates. Actual units of measurement or specification normally will be in other units and conversion to mols is necessary. The following conversions are rather common. Those for gas are shown for the standard condition specified.

Calculation of V in Mols

SI: $P_s = 101.3$ kPa and $T_s = 15°C$

V in kmol/hr = 1762 (10^6 std m^3/d)

FPS: $P_s = 14.7$ psia and $T_s = 60°F$

V in lb mol/hr = 110 (MMscfd)

Calculation of L in Mols

SI: L in kmol/h = 1000 (m^3/h)(γ)/(mol wt)

FPS: L in lb mol/hr = 350 (API bbl/hr)(γ)/(mol wt) = 500 (US gpm)(γ)/(mol wt)

The results of this calculation may be used as a first guess for a trial-and-error solution (such as plate-to-plate) or as an approximate picture of the expected performance.

Example 17.13: An absorber containing 6 theoretical plates uses a lean oil with a relative density of 0.825 flowing at the rate of 1136 m^3/d [263 US gpm]. The molecular weight is 161. The tower pressure is 500 kPa and the average temperature (for calculation) is 30°C. The gas flowrate is 510 700 std m^3/d [18.1 MMscfd]. Estimate the recovery of each component.

For this case: $L_0 = (1136/24)(825)/161 = 242.5$ kmol/h

$V_{N+1} = (1762)(0.5107) = 899.8$ kmol/h

So, $(L_0/V_{N+1}) = 242.5/899.8 = 0.27$

A table may now be prepared using these values. The analysis of the gas in question is shown in Column 2.

(1)	(2)	(3)	(4)	(5)	(6)	(7)	(8)	(9)	(10)
Comp.	Y_{N+1}	K	A	E_a	kmol/h In	kmol/h Absorbed	X_N	kmol/h Overhead	Y_1
Methane	0.83	34.66	0.0078	0.0078	746.83	5.83	0.024	741.0	0.824
Ethane	0.084	6.78	0.040	0.040	75.58	3.02	0.013	72.56	0.081
Propane	0.048	2.04	0.132	0.132	43.19	5.70	0.024	37.49	0.042
i-Butane	0.009	0.81	0.333	0.333	8.10	2.70	0.011	5.40	0.006
n-Butane	0.017	0.59	0.458	0.458	15.30	7.01	0.029	8.29	0.009
i-Pentane	0.004	0.25	1.08	0.885	3.60	3.19	0.013	0.41	–
n-Pentane	0.008	0.19	1.42	0.96	7.20	6.91	0.029	0.29	–
					899.80	34.36	0.1430	865.44	0.962

Example 17.13 (Cont'd):

Column 8 says that 0.1430 mols are absorbed in the tower per mol of entering lean oil. So, the mols of rich oil leaving per mol entering is 1.1430. At the top of the tower $(L_0/V_1) = 242.5/865.44 = 0.28$. At the bottom $(L_0/V_0) = (1.1430)(242.5)/899.8 = 0.31$. This shows that L/V is rather constant. If a second calculation were made, a new L/V of about 0.295 could be used (the average for the top and bottom).

Isopentane is the key component since its value of A is closest to one. It would be used as the key component for any further calculations necessary.

The gas shrinkage was 3.8%, (1 – 0.962). For this low a value, Kremser-Brown is a good approximation. In fact, it tends to predict conservative values, i.e., the tower will perform slightly better than predicted.

The recipe for the above example follows:

1. Write analysis of entering rich gas in Column (2).
2. Tabulate K for each component at average column conditions (pressure is substantially constant throughout) in Column (3).
3. Calculate A in Column (4) using Equation 17.26. For this example, A = 0.27/K.
4. Calculate E_a from Equation 17.25 or Figure 17.20 and tabulate in Column (5). (Assume $Y_0 = 0$).
5. Column (6) is merely Column (2) times the mols per hour of rich gas entering (899.8 kmol/h).
6. Column (7) = Column (5) times Column (6).
7. Each entry in Column (8) = corresponding entry in Column (7) divided by L_0 (242.5 kmol/h).
8. Column (9) = Column (6) minus Column (7).
9. Column (10) = each entry in Column (9) divided by total mols of rich gas entering.

For the case where recovery of a key component is specified, for a given gas rate and number of plates, the unknown is L_0. It is found by entering the absorption factor at the known value of E_a, move horizontally to N and then drop vertically to the A axis. Look up K for the key component. Knowing A, K, and V_{N+1}, you can solve for L_0. This can then be used to complete the table for all other components.

If oil and gas rate is specified, as well as recovery of a key component, one can solve for N, the number of theoretical plates needed. These types of calculations may be used to analyze an existing column to pinpoint any problems or to adapt it for different service conditions.

Most natural gas absorbers contain 7-10 theoretical plates (20-30 actual plates). In planning, the assumption of 8 theoretical plates is a good first guess if a high percentage of propane recovery is desired.

Figure 17.20 shows that as oil rate declines the number of plates tends to infinity. Also, it indicates that more than about 8 theoretical plates gives a minimum decrease in oil rate. The left-hand line in Figure 17.20 represents infinite theoretical plates. Along this 45° line, $E_a = A$. All curves for a finite number of plates coincide with this 45° line at some value of "A," and follow it at lower values.

As a matter of practical economics one must balance absorber cost with oil rate. The former is a one-time capital investment. The latter involves continuing fuel, pumping and cooling costs. The optimum is usually one which uses the minimum possible oil rate, with an absorber of feasible size, for the desired recovery. Cost is sensitive to oil rate and rough estimates may be based on it.

The lowest molecular weight oil should be used that conditions permit. This minimizes the mass to be pumped. The limit is fixed by oil vapor pressure and the fact the lean oil should be essentially denuded of the key component. If the molecular weight of the oil is too low, vaporization losses of the oil can be excessive. Some plants employ a heavier MW "sponge oil" to absorb the light lean oil on the top 2 or 3 trays of the absorber.

Since pressure is usually fixed by contract or other practical factors, temperature is the primary process variable. K decreases with decreasing temperature, which in turn increases "A" and more recovery is obtained per unit of oil circulated. The use of subambient temperatures therefore decreases the cost of the absorption-stripping portion of the recovery plant, but cooling (refrigeration) costs increase. The optimum temperature may be found by plotting refrigeration costs, absorption-stripping costs, and product revenue as a function of average absorption temperature. For a given recovery, fractionation costs will be essentially constant. The cost of the absorption section then becomes the cost of the absorber-stripper, oil cooling, pumping and accessories. Most refrigerated absorption plants operate the absorber between –40° and –7°C [–40° and 20°F]. At temperatures in this range, stripping can be obtained by fractionation rather than by the use of steam or a stripping gas. This is an added advantage of subambient absorption temperatures. For high recovery of ethane (40-60%), temperatures down to –50°C [–58°F] are sometimes employed.

In refrigerated lean oil plants, both the lean oil and gas are chilled. The lean oil is often presaturated with methane by contacting it with a portion of the sales gas upstream of the oil chiller. This minimizes the absorption of methane in the absorber and allows the absorber to operate cooler, particularly at the top, thereby minimizing lean oil losses.

Other than mechanical problems, most problems in absorber operation center around oil quality and circulation rate. Proper stripping of the oil is mandatory to minimize losses in the exit gas and insure proper absorption. In most circumstances, the lightest possible molecular weight oil should be used. Vaporization losses limit the oil molecular weight. Also, the oil should contain a minimum quantity of the components for which recovery is desired. In most cases, it should be composed primarily of pentanes and heavier components. Reference 17.15 summarizes desirable absorption oil characteristics.

The procedures outlined determine only the number of theoretical plates. There are many methods available for conversion of theoretical plates to actual plates. As a rule of thumb, an efficiency of 25-40% is a reasonable estimate for hydrocarbon absorbers with standard tray designs and in the absence of foaming.

STRIPPER CALCULATIONS

The purpose of the stripper is to remove the absorbed components from the lean oil. Stripping of the absorbed components from the absorption oil is sometimes accomplished by a material immiscible in hydrocarbons because this increases the vaporization tendency of the absorbed components. Steam was commonly used. Steam stripping, however, saturates the lean oil with water, requiring that the products or the oil be dehydrated. For this reason, dry gas stripping is sometimes used. This utilizes the leanest gas available (frequently the sales gas). With dry gas stripping efficiency is somewhat less, and condensation of the overhead product is more difficult.

In refrigerated lean oil plants stripping is almost always done by fractionation. This vessel is often called the lean oil still and the reboiler heat is provided by fired heaters. Stripping may be done by fractionation. In this case, the design procedure will follow that outlined in the fractionation section of this chapter.

Fundamentally, a stripper is simply an upside-down absorber. Trays are numbered from the bottom up with the top plate designated as "M." For convenience, the previous absorption equations will be repeated in proper form for stripping calculations.

$$S = \frac{KV}{L} \tag{17.27}$$

$$\frac{X_{M+1} - X_1}{X_{m+1} - X_0} = \frac{S^{M+1} - S}{S^{M+1} - 1} \tag{17.28}$$

Where:

S = stripping factor

X_1 = mols of component in stripped lean oil leaving bottom of the stripper per mol of rich oil entering the stripper

X_{M+1} = mols of component in rich oil entering the stripper per mol of rich oil entering the stripper

L_{M+1} = mols of rich oil entering the stripper

Y_0 = mols of component in stripping medium per mol of stripping medium entering

V_0 = mols of stripping medium entering

X_0 = mols of component in liquid in equilibrium with stripping medium per mol of entering rich oil

L_1 = mols of lean oil leaving stripper

Equations 17.27 and 17.28 and Figure 17.20 may be used to characterize gas strippers with sufficient accuracy for most operational and planning needs. Glycol and amine type strippers utilize reboilers and calculating methods similar to fractionating columns. The current use of purely gas strippers is very limited.

TOWER MECHANICAL DESIGN

A column does not know if it is an absorber or a fractionator. It merely reacts to the vapor and liquid loads imposed on it. We will not discuss detailed design herein since that is a specialty area of the vendors and manufacturers. However, the customer must know enough about the design in order to properly prepare specifications and operate the resultant unit.

There are five basic choices of vapor-liquid contact devices: bubble cap, sieve or valve trays, and structured or random packing. There are some general guidelines for the first choice between trayed and packed towers. These are not absolute because in some services they might both be suitable, the choice being economic. Generally, though, a packed tower is suitable when: (1) the tower diameter is small, (2) corrosive fluids require special materials, (3) a low pressure drop is needed, (4) the liquid rate is high enough to minimize distribution problems, (5) the depth of packing required does not exceed about 8 m [26 ft]. (If the column is taller than this, multiple packing sections are used).

The exact impact of these is different between structured and random packing. For example, it is not a good practice to use random packing with glycol contactors because of the low liquid circulation rate. The problem is liquid distribution. However, structured packing is widely used in this service.

It is difficult to justify a trayed tower when the diameter is less than 600 mm [2 ft]. By the same token, random packing is not often used in large diameter towers.

The pressure drop per theoretical stage is much higher in a trayed tower than a packed tower, but this is a factor only in low pressure or vacuum towers. Since it is expensive to manufacture trays from alloy metals, packing sometimes offers an advantage with corrosive fluids.

The depth of random packing is limited by the crushing or deformation characteristics of the packing. Some plastics are limited to bed depths of 3-4 meters. Liquid maldistribution may also limit bed depth.

When foam is present, it tends to fill the interstices of the packing and enhance flooding; however, foaming is more likely to occur in trayed towers due to the shearing nature of vapor-liquid contact. The presence of solids is a cause for concern because of the potential for packing plugging. In production operations where salt water may be present, salt plugging is a problem if the temperature is high enough to vaporize the water.

Sieve or valve trays are used in most columns. They are more efficient than bubble caps when operated at, or near, design capacity. At this capacity they possess about the same efficiency, but the sieve tray is less expensive. However, at low vapor flowrates the sieve tray efficiency declines faster than either a valve or bubble cap tray because of weeping. So, when high turndown is anticipated, the valve tray is the typical choice, even though it is somewhat more expensive.

Bubble caps are now used only in those cases where liquid rates are low and high turndown ratios are required. One example is the glycol absorber.

Flooding

The terms "flood capacity, flooding point, etc." refer to a condition wherein excess *liquid holdup* occurs in a tower. It is a design limit. When it occurs in operation, excess loss of liquid occurs out of the top of the tower. In addition, process efficiency decreases rapidly.

Flooding can result from several causes. *Jet or entrainment flooding* results from too high a vapor velocity through the tray or bed. The gas space becomes full of aerated liquid or foam; not all of the liquid can flow downward by gravity.

Downcomer flooding occurs when the downcomer can no longer accommodate the liquid leaving the tray. There are two type of downcomer flooding – back-up and choking. Both occur at high liquid rates.

Downcomer back-up occurs when liquid fills the downcomer due to high ΔP across the trays and/or inadequate flow area under the downcomer. Downcomer choking occurs when the area at the top of the downcomer is inadequate to handle the liquid entering. Figure 17.21 shows the normal operating region for a tray in a distillation column.

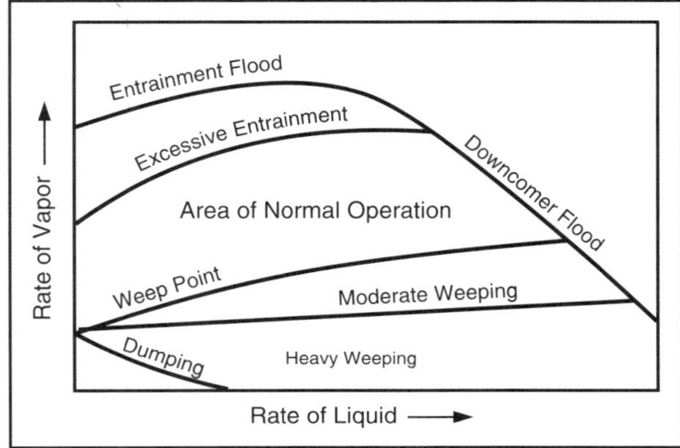

Figure 17.21 Performance Diagram for Sieve Tray

Most will specify a tower size that does not exceed 75-80% of flood. In some cases, it may be judicious to use not over 50% where flexibility is required or foaming fluids are present. The calculation of flooding capacity is made by the vendor.

Most are participants in Fractionation Research, Inc. (FRI), a cooperative of many firms. FRI performs tests and issues reports which often are a part of design specifications. These are considered acceptable standards, but some company standards exceed those of FRI.

Tower diameter is affected by both gas and liquid rates, choice of contact device and mechanical features of that device. Tower length is a function of the amount of contact required, the effectiveness of each contact stage and mechanical considerations. The discussion which follows provides summary guidelines only and is not offered for use in detailed designs.

TRAY TYPE TOWERS

Diameter

The diameter of a distillation column depends on both vapor and liquid rates. In general, the vapor rate sets the active tray area – the area where the vapor and liquid contact. The downcomer area is set by the liquid rate. The total cross sectional area of the column is the sum of the active tray area and the downcomer area, as shown at right for a single pass tray.

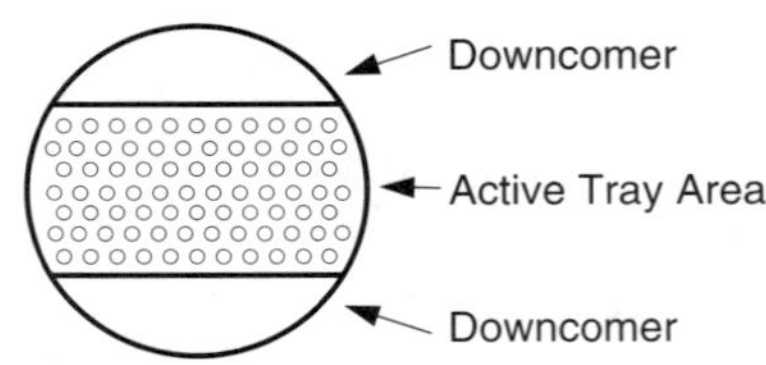

Sizing of the active tray area is similar to separator sizing. The allowable K_s factor depends on tray spacing, liquid to gas ratios, gas density, liquid density and surface tension. Most vendor correlations establish a K_s value for the flood point of the tray. The design K_s value is some percentage of the flood K_s value. This ratio (K_s design/K_s flood) is called the *system factor* and varies from about 0.85 to 0.9 for high-pressure light hydrocarbon distillation to 0.5 to 0.7 for amine and glycol contactors. For light hydrocarbon distillation using 610 mm [24 in] tray spacing and operating at about 85% of flood, a K_s value of 0.05 to 0.07 m/s [0.16 to 0.23 ft/sec] is typical. Glycol and amine contactors use somewhat lower values due to foaming.

Equation 17.29 and 17.30 can be used to calculate the active tray area.

$$v = K_s \left(\frac{\rho_L - \rho_g}{\rho_g} \right)^{0.5} \tag{17.29}$$

Where:		SI	FPS
ρ_L	= liquid density	kg/m^3	lb/ft^3
ρ_g	= gas density	kg/m^3	lb/ft^3
v	= allowable gas velocity	m/s	ft/sec
K_s	= an empirical constant	m/s	ft/sec

$$A_T = \frac{q_a}{v} \tag{17.30}$$

Where:		SI	FPS
A_T	= active tray area	m^2	ft^2
q_a	= vapor rate	m^3/s	ft^3/sec
v	= allowable velocity from Equation 17.29	m/s	ft/sec

The portion of the tower cross-sectional area required for the liquid involves tray design. This is beyond our scope herein, but the following are typical tray characteristics.

Sieve – Hole size varies from 0.64-2.5 cm [0.25-1.0 in] with about 1.27 cm [0.5 in] being the most common. The distance between holes will usually be about 2-3 times the hole diameter. In most cases, the outlet weir will be about 5 cm [2 in] high. The inlet weir and downcomer seal must be of a height and area compatible with flowrate.

Valve – The same basic guidelines apply as for sieve trays. The valves vary in size but the most commonly used types are circular (about 5 cm [2 in] in diameter) or 2.5 cm × 12.5 cm [2 in × 5 in] rectangles. About 130 circular valves or about 75 rectangular valves will be used per square meter of active plate area.

Bubble Cap – The slot (gas) velocity is about 3-6 m/s [10-20 ft/sec], depending on gas density. Since the riser area is slightly larger than slot area, riser velocities will be slightly lower. Most plates are designed so that the liquid pressure drop across the plate does not exceed 0.35-1.00 kPa [1.4-4.0 in H_2O].

Downcomer Size

Downcomers are sized using velocity, residence time and height of liquid as factors. From these, downcomer area can be found which, in turn, can be used for approximate tower sizing.

Figure 17.22 shows the basic pressure balance around a tray. There must be sufficient head of liquid in the downcomer to flow the liquid across the tray at a proper rate. This height must be less than tray spacing.

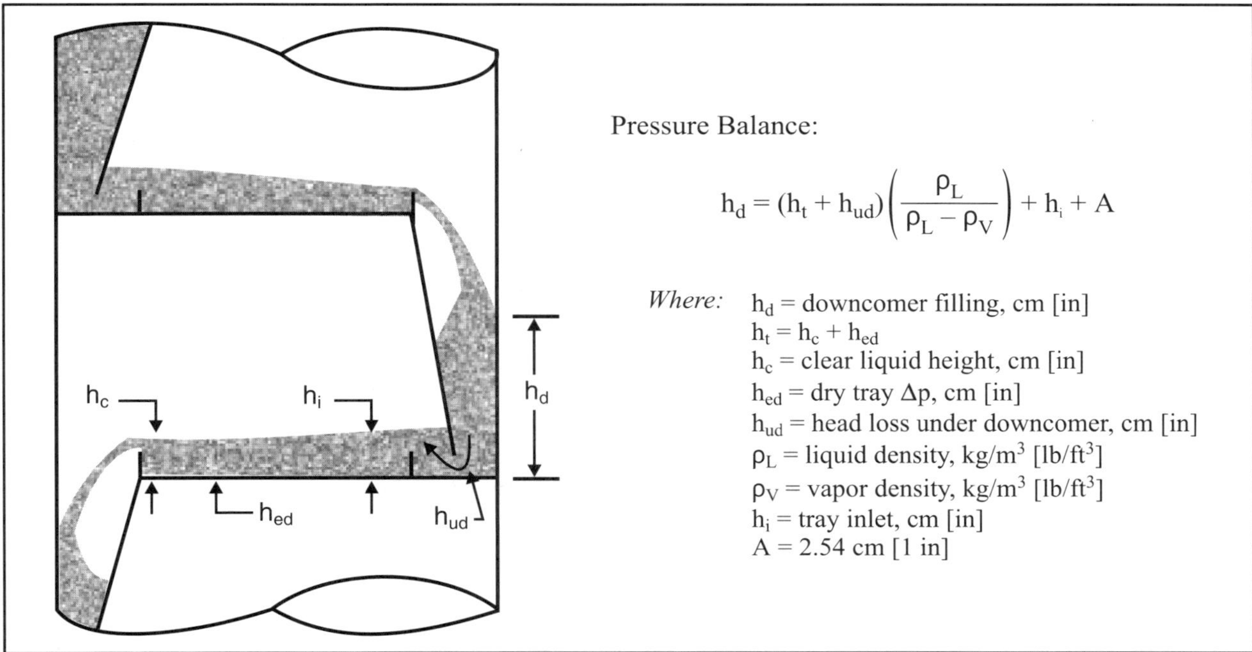

Figure 17.22 Pressure Balance Around a Tray

There are many methods for estimating allowable velocity in a downcomer. As a general rule, the velocity should not exceed 0.15 m/s [0.5 ft/sec], regardless of the correlation used. As an approximation, use the lesser of this value and that found from Equation 17.31.

$$v_D = [A(\rho_L - \rho_V)^{0.5} + B]\,FF \qquad (17.31)$$

Where:		SI	FPS
v_D	= downcomer velocity	m/s	ft/sec
ρ_L	= liquid density	kg/m³	lbm/ft³
ρ_V	= vapor density	kg/m³	lbm/ft³
A	=	0.0064	0.083
B	=	0.0026	0.0086
FF	= liquid foaming factor	–	–

The values of FF can be taken as 1.0 for normal light hydrocarbons and 0.70-0.85 for amines and glycols, depending on the amount of foaming.

An alternative approach uses residence time. This will vary from 3-10 seconds, depending on the liquid and its foaming tendencies. The following are representative values.

Amine Type Absorbers	6-8 sec	Glycol Still	5-6 sec
Glycol Absorbers	6-8 sec	Sour Water Stripper	6-8 sec
Oil Absorbers	4-5 sec	Oxygen Stripper	3-4 sec
Amine Still	5-6 sec		

A residence time of around 4 sec is suitable for most hydrocarbon fractionators.

The allowable velocity may be found from residence time by the equation

$$v_d = \frac{h}{\text{(Residence Time)}} \qquad (17.32)$$

Where: h = height of liquid in downcomer

A given residence time and a given velocity fix downcomer height, which in turn fixes plate spacing. What about the percentage of this height filled with clear liquid? As a general rule, the following guidelines are realistic.

Pressure, MPa	Max. Liquid Level as % of Tray Spacing
Greater than 2.5	35
0.7-2.5	40
Less than 0.7	50

Once again, foaming is a consideration.

The total downcomer area normally will be at least 10-12% of total column area unless the liquid rate is unusually low.

The total area will be the active tray area plus the downcomer area. The table below summarizes some effective K_s values which include both the active tray and downcomer areas. The tower size can be calculated using the relationships in Chapter 11.

Service	Tray Spacing					
	SI			FPS		
	46 cm	**61 cm**	**76 cm**	**18 in**	**24 in**	**30 in**
Absorbers - oil	0.0594	0.0678	0.0719	0.194	0.222	0.236
Absorbers - glycol*	–	0.0425	0.0467	–	0.139	0.153
Absorbers - amine	–	0.0297	0.0333	–	0.0972	0.110
Fractionators	0.0372	0.0458	0.0508	0.122	0.150	0.167
* assumes bubble cap trays						

Tray spacing of at least two feet is recommended for glycol and amine absorbers. This spacing also depends on downcomer design.

In some columns there is a significant change in the vapor or liquid rate at a particular point in the column – usually at the feed tray. This often results in a tower with two different diameters. Demethanizers and deethanizers frequently exhibit this characteristic.

High Capacity Trays

High capacity trays can be a viable way to increase the capacity of a fractionator which is capacity constrained. Capacity increases of 10-25% compared to a well designed standard tray have been reported. Unfortunately much of the information needed to design and evaluate high capacity trays is proprietary. Reference 17.16 provides a good summary of high capacity tray designs.

Most high capacity tray designs increase the active tray area by reducing the area taken up by the downcomer where the liquid exits the downcomer. This area can be minimized by using sloped or stepped downcomers, but high capacity trays will often use hanging downcomers.

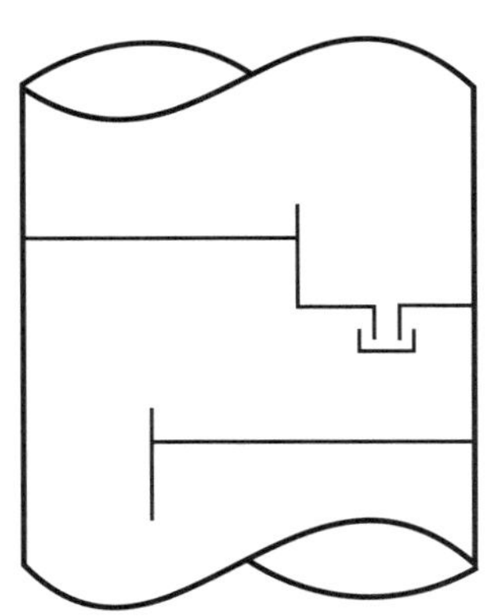

With a hanging downcomer, the downcomer does not actually extend to the tray below as shown in the figure to the right.

This provides additional active tray area where a standard downcomer would normally occupy tray space. Some type of seal pan is necessary at the bottom of the downcomer to prevent vapor from entering.

Other high capacity trays use special valve designs to limit entrainment flooding, thus allowing increased vapor velocity. Still others use multiple downcomers with long weir lengths. This reduces the liquid level on the tray allowing decreased tray spacing and/or higher vapor rates.

Several success stories have been published in the literature. However, high capacity trays can reduce the operating flexibility of the column, particularly turndown ratio. In some applications it has been necessary to recycle liquids back into the feed to maintain vapor and liquid traffic in the design region.

PACKED TOWERS

Packed tower calculations are identical to trayed tower calculations up to the point of tower sizing. The determination of both packed tower diameter and height require special correlations.

Packed Tower Diameter

The diameter of packed towers has historically been calculated by estimating the gas velocity at the *flood point* then sizing the tower so that the actual velocity is 50-80% of the flooding velocity. The flood point can be defined several ways, but a widely accepted definition is the point where measured liquid hold-up increases abruptly. It is an unstable condition which will result in poor tower performance and/or liquid carryover. The flood point is a function of liquid rate, packing characteristics, gas and liquid densities and liquid viscosity.

Sherwood[17.17] was the first to develop a correlation by which flooding velocity could be determined. This correlation was extended to include several curves for various pressure drops through the tower. This model is called the Generalized Pressure Drop Correlation (GPDC) and is widely used today for sizing packed towers. Several versions of the GPDC have been developed. One version is shown in Figure 17.16.[17.18] Some correlations show an additional curve above the 1.5 in H_2O/ft curve to represent flooding.

$$Y = \frac{A\, G^2\, F\, \nu^{0.1}}{\rho_g(\rho_L - \rho_g)} \qquad X = \left(\frac{L}{G}\right)\left(\frac{\rho_V}{\rho_L}\right)^{0.5}$$

				SI	FPS
Where:	L	=	liquid mass velocity	$kg/m^2 \cdot s$	lbm/ft^2-sec
	G	=	gas mass velocity	$kg/m^2 \cdot s$	lbm/ft^2-sec
	ρ_L	=	liquid density	kg/m^3	lbm/ft^3
	ρ_V	=	gas density	kg/m^3	lbm/ft^3
	ν	=	liquid viscosity	centistokes	
	F	=	packing factor	From Table 17.2	
	A	=	units conversion factor	10.76	1.0

TABLE 17.2

Norton Company – Generalized Pressure Drop Correlation, Jan. 1986 (Packing Factors, F_P)

		Nominal Packing Size					
Packing Type	**Material**	**5/8" or #15**	**1" or # or #25**	**1.5" or #1.5 or #40**	**2" or #2 or #50**	**3" or #70**	**3.5" or #3**
IMTP®	Metal	51	41	24	18	12	
Hy-Pak®	Metal		45	29	26		16
Pall Rings	Metal	70	56	40	27		18
Pall Rings	Plastic	75	55	40	26		17
Super Intalox® Saddles	Plastic		40		28		18
Super Intalox® Saddles	Ceramic		60		30		
Intalox® Saddles	Ceramic		92	52	40	22	
Raschig Rings	1/16" Metal	300	144	93	62	43	

The packing factor is determined empirically based on experimental data. Packing factors can be obtained from the packing manufacturer. Table 17.2 shows packing factors published by one packing manufacturer (Norton) for their products. It is important to recognize that packing factors also depend on the GPDC which was used to correlate results. It is always best to check with the a manufacturer to find the GPDC which is consistent with their published factors.

Historically the packing factor was estimated numerically by dividing the specific area of the packing (m^2/m^3, ft^2/ft^3) by the cube of the fractional void space. These values are reasonable for preliminary sizing and/or loading calculations. Reference 17.19 provides a nomograph for estimating packing factors for various packings.

Figure 17.23 can be used in several ways. The design pressure drop depends on the service. The following values may serve as a guide.

Service	ΔP, in H_2O/ft packing
Absorbers/Regenerators Liquids with foaming tendency Light hydrocarbon distillation	0.25-0.50
Atmospheric and H.P. distillation non-foaming fluids	0.50-1.0
Minimum ΔP	0.05
Maximum ΔP	1.0

For a given service, the vendor should be consulted for a recommended ΔP value.

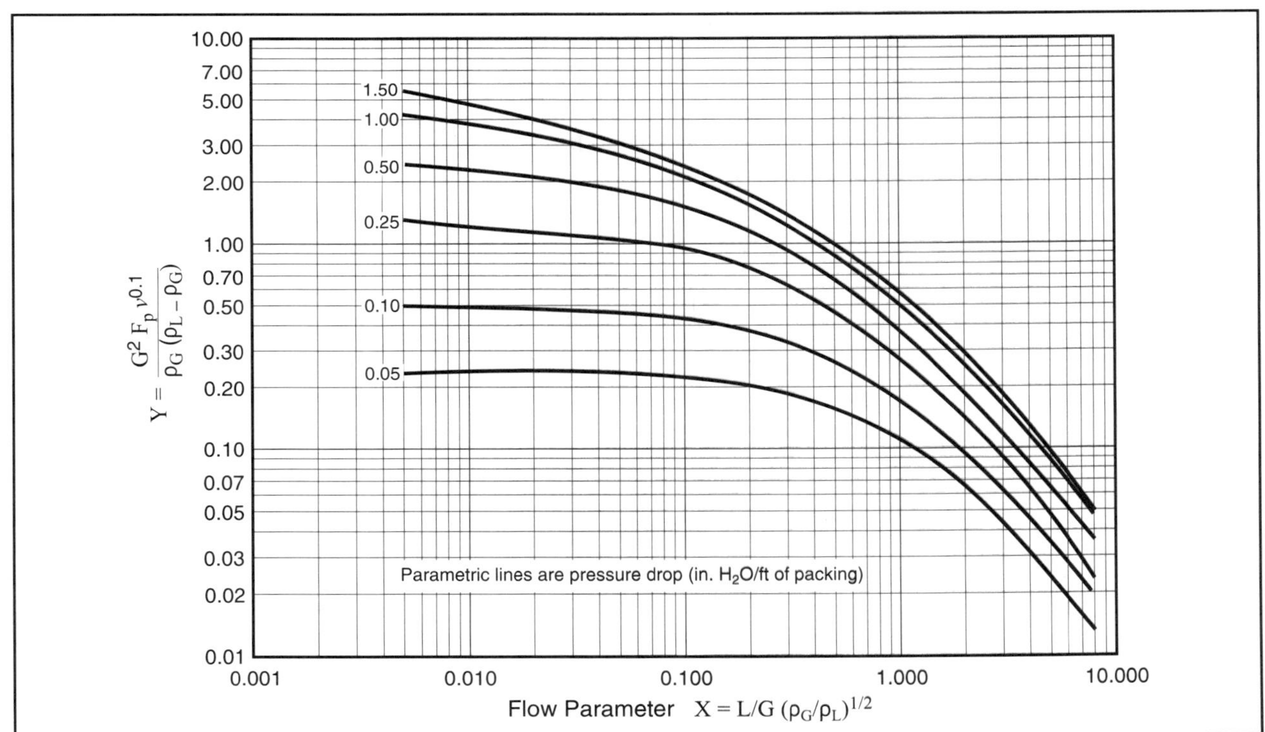

Figure 17.23 Generalized Pressure Drop Correlation

Once the design ΔP has been determined, sizing the tower proceeds as follows:

1. Calculate a value of the abscissa parameter "X" (mass rates for L & G can be used since the area term cancels out)
2. Proceed vertically to the design ΔP curve then horizontally to read a value of ordinate parameter "Y"
3. From "Y" calculate the gas mass velocity G
4. From G calculate the tower diameter, d, remembering that

$$d = \left(\frac{4\,m}{\pi G}\right)^{0.5}$$

Where: m = mass flowrate, kg/s [lbm/sec]

Figure 17.23 can also be used to estimate reasonable vapor or liquid loading conditions for an existing tower.

Example 17.14: Calculate the diameter of a glycol contactor, packed with 2 in. Hy-Pak® rings. Gas rate is 0.28×10^6 std m^3/d [10 MMscfd]. The glycol rate is 0.45 m^3/h [2 gpm]. The following data apply:

ρ_V = 60 kg/m^3 [3.7 lbm/ft^3] ν = 16 centistokes

ρ_L = 1120 kg/m^3 [69.9 lbm/ft^3 γ_V = 0.62

SI Solution: Design the contactor to operate at a pressure drop of 2.0 mbar/m packing.

Calculate "X"

$$m_L = \left|\frac{0.45\ m^3}{h}\right|\frac{1120\ kg}{m^3}\left|\frac{1\ h}{3600\ s}\right| = 0.14\ kg/s$$

$$m_V = \frac{(0.28)(51060)(0.62)}{3600} = 2.5\ kg/s$$

$$X = \left(\frac{m_L}{m_V}\right)\left(\frac{\rho_V}{\rho_L}\right)^{0.5} = \left(\frac{0.14}{2.5}\right)\left(\frac{60}{1120}\right)^{0.5} = 0.013$$

For ΔP = 2.0 mbar/m , Y = 1.2

From Table 17.2, F = 26

$$Y = 1.2 = \frac{10.76\ G^2 (26)(16)^{0.1}}{60\ (1120 - 60)} \qquad G = 14.38\ kg/m^2 \cdot s$$

$$d = \left(\frac{4\ m}{\pi\ G}\right)^{0.5} = \left(\frac{(4)(2.5)}{(3.14)(14.38)}\right)^{0.5} = 0.47\ m$$

Example 17.14 (Cont'd):

FPS Solution: Design the contactor to operate at a pressure drop of 0.25 in H_2O/ft packing.

Calculate "X"

$$m_L = \left|\frac{2 \text{ US gal}}{\text{min}}\right|\frac{9.33 \text{ lbm}}{\text{US gal}}\left|\frac{1 \text{ min}}{60 \text{ sec}}\right| = 0.31 \text{ lbm/sec}$$

$$m_V = \frac{(10)(3180)(0.62)}{3600} = 5.5 \text{ lbm/sec}$$

$$X = \left(\frac{m_L}{m_V}\right)\left(\frac{\rho_V}{\rho_L}\right)^{0.5} = \left(\frac{0.31}{5.5}\right)\left(\frac{3.7}{69.9}\right)^{0.5} = 0.013$$

For $\Delta P = 0.25$ in H_2O/ft , $Y = 1.2$

From Table 17.2, $F = 26$

$$Y = 1.2 = \frac{G^2(26)(16)^{0.1}}{3.7\,(69.9 - 3.7)} \qquad G = 2.93 \text{ lbm/ft}^2\text{-sec}$$

$$d = \left(\frac{4\,m}{\pi\,G}\right)^{0.5} = \left(\frac{(4)(5.5)}{(3.14)(2.93)}\right)^{0.5} = 1.55 \text{ ft} \quad \text{(use 18 in diameter tower)}$$

Tower Height

The height of a packed tower must be sufficient to provide enough contact between the vapor and liquid to give the desired result. In a trayed contact this requires determination of the number of actual trays to be installed. In a packed tower this calculation requires determination of an HETP (Height Equivalent to a Theoretical Plate) or an HTU (Height of a Transfer Unit). These two concepts are related but based on different treatment of the driving force – concentration difference. Some feel the HTU approach is more rigorous, but the HETP approach is frequently used by process engineers because it relates to the equilibrium stage calculations discussed earlier in this chapter.

The actual packing height, h, is then calculated as follows:

$$h = (HTU)(NTU) \qquad h = (HETP)(N)$$

Where:

N = number of theoretical stages

NTU = number of transfer units

The HETP is determined experimentally in laboratory or pilot plant tests. It is a function of packing type, vapor and liquid densities, liquid viscosity and surface tension diffusivity, vapor and liquid loading. References 17.20 and 17.21 provide a review of a few correlations available to calculate HETP.

Few generalized methods for calculating HETP are available in published literature. The packing manufacturer can provide reasonable estimates of HETP for a particular service. For preliminary planning and sizing calculations the following guidelines have proved useful for high efficiency random packing such as slotted rings in hydrocarbon distillation service.

Packing Size	HETP
2.54 cm [1 in]	0.46 m [18 in]
3.81 cm [1.5 in]	0.66 m [26 in]
5.08 cm [2 in]	0.89 m [35 in]

For glycol dehydration an HETP of 1.5-2.0 m [5-6.5 ft] can be used to estimate contactor height for both random and structured packing.

Packing performance is often limited by the liquid distribution. At low liquid rates the packing may not be fully irrigated which can have a disastrous effect on performance. When the liquid rate is less than 5 $(m^3/h)/m^2$ [(2 US gpm/ft^2)], a special distributor design may be required.

REFERENCES

17.1 Chen, G. K., *Chem. Eng.* (Mar. 5, 1984), p. 40.

17.2 Schendel, Ronald L. (Fluor Engineers, Inc.), "Processing Gases Associated with CO_2 Miscible Flood", Gas Conditioning Conference (Mar. 1983), Norman, Oklahoma, USA.

17.3 Erbar, R. C. and Maddox, R. N., *Chem. Eng. Prog.*, 44, No. 8 (1948), p. 603.

17.4 Gilliland, E. R., *Ind. Eng. Chem.*, 32 (1940), p. 1220.

17.5 Erbar, J. H. and R. N. Maddox, *Pet. Ref.*, 40 (1961), p. 183.

17.6 Eduljee, H. E., *Hydr. Proc.* (Sept. 1975), p. 120.

17.7 Kirkbride, C. G., *Petr. Ref.*, 23 (1944), p. 321.

17.8 Shinskey, F. G., *Chem. Eng. Prog.* (May 1976).

17.9 O'Connell, *Trans.* AIChE, 42 (1946), p. 741.

17.10 Shinskey, F. G., *Process Control Systems,* McGraw Hill (1979), Chapter 11.

17.11 Ryskamp, C. J., *Hydr. Proc.* (June 1980), p. 51.

17.12 Kemp, D. W. and D. G. Ellis, *Oil and Gas J.* (Aug. 11, 1975), p. 60.

17.13 Morris, J. K. and R. S. Smith, *Oil and Gas J.* (May 7, 1984), p. 112.

17.14 Maddox, R. N. and D. J. Morgan, *Gas Treating and Sulfur Recovery*, Vol. 4 of "Gas Conditioning and Processing" series, John M. Campbell and Company, Norman, Oklahoma (1998).

17.15 *Oil and Gas J.* (July 19, 1965), p. 99.

17.16 Sloley, A. W., "Should You Switch to High Capacity Trays," *CEP* (Jan. 1999), p. 23.

17.17 Sherwood, T. K., *et. al.*, *Ind. Eng. Chem.*, 30 (1938), p. 765.

17.18 Hausch, G. W., "Tower Packings in the Gas Processing Industry," 36th Gas Cond. Conf. (Mar. 3, 1986), Univ. Okla., Norman, Oklahoma.

17.19 Zanker, A., *Chem. Eng.* (Sept. 3, 1973), p. 126.

17.20 Diab, S. and R. N. Maddox, *Chem. Eng.* (Dec. 27, 1982), p. 38.

17.21 Vital, T. J., *et. al., Hydr. Proc.* (Dec. 1984), p. 75.

17.22 Johnson, J. E. and D. J. Morgan, "Graphical Techniques for Process Engineering," *Chem. Eng.*, v. 92, no. 14 (July 8, 1985), p. 72-83.

17.23 Underwood, A. J. V., *Trans. Inst. Chem. Eng.* (London), 10 (1932), p. 112.

APPENDIX 17A

McCABE-THIELE METHOD

This method is a graphical representation of the equilibrium and material balance relationships, and how reflux rate relates to theoretical stages. It was developed for binary mixtures but can be used for multi-component mixtures by the use of key components.

One can write the equilibrium relationship as

$$y = \frac{\alpha x}{1 + (\alpha - 1) x} \tag{17A.1}$$

Where:

α = relative volatility, K_L/K_H

K_L = K for light key component

K_H = K for heavy key component

This assumes that relative volatility is relatively constant in most fractionation towers. Figure 17A.1 is a plot of Equation 17A.1.

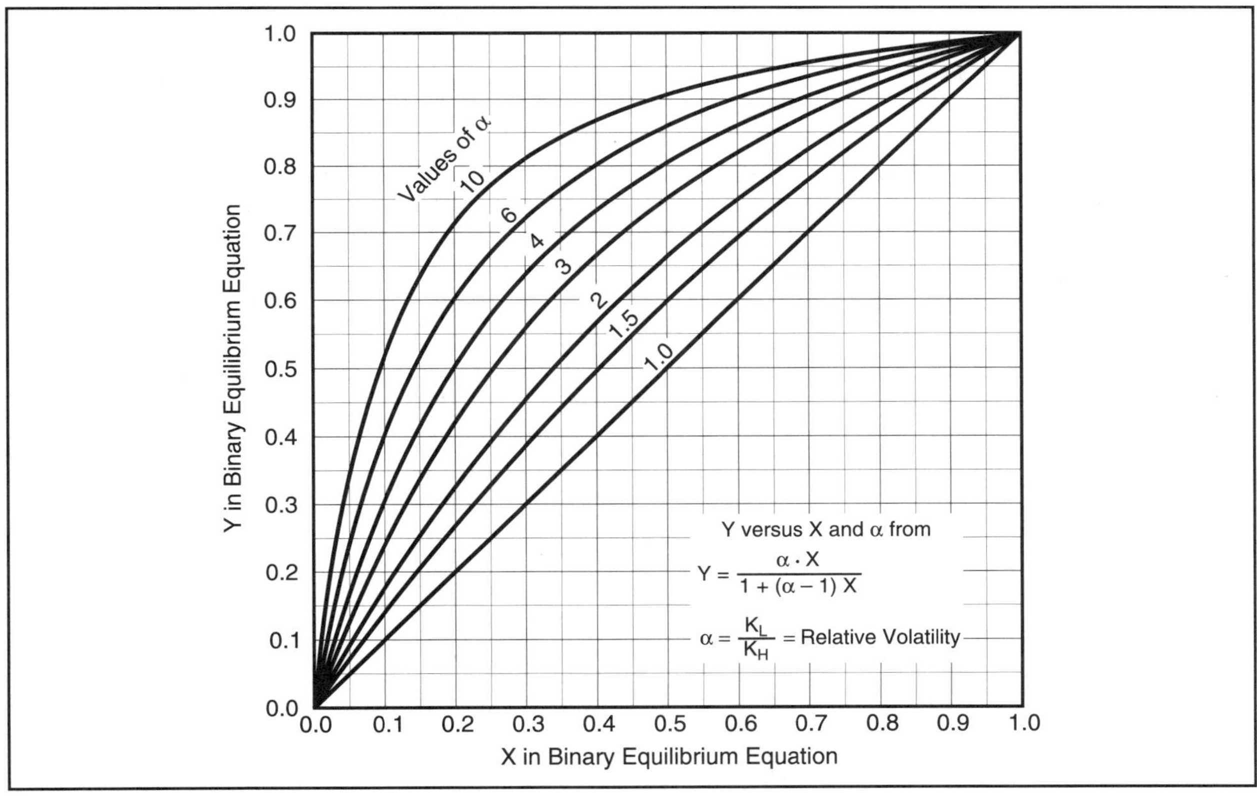

Figure 17A.1 Plot of Equation 17A.1

Material Balance Around Top

A material balance below plate "n" and around the top of the column relates molar flowrates of vapor and liquid streams below plate "n."

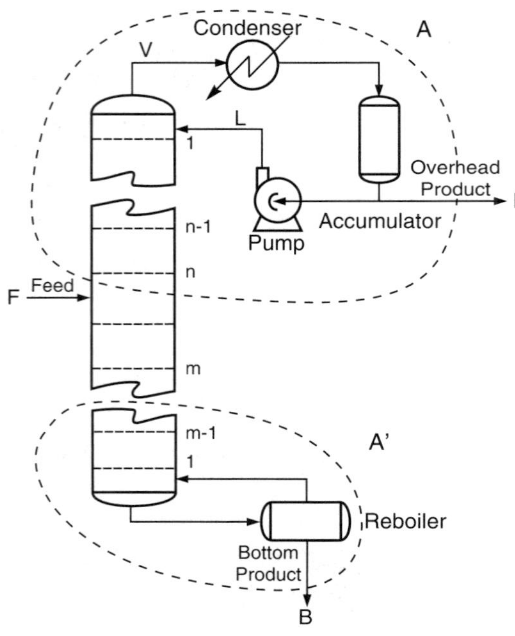

$$V_{n+1} = L_n + D \qquad (17A.2)$$

A material balance (using the same envelope) for one component is

$$(V_{n+1})(y_{n+1}) = L_n x_n + D x_D \qquad (17A.3)$$

Rearranging,

$$y_{n+1} = \left(\frac{L_n}{V_{n+1}}\right)(x_n) + \left(\frac{D}{V_{n+1}}\right)(x_D) \qquad (17A.4)$$

Where:
- V_{n+1} = mols vapor entering plate "n"
- L_n = mols liquid leaving plate "n"
- D = mols distillate overhead
- y_{n+1} = mol fr component in vapor entering plate "n"
- x_n = mol fr component in liquid leaving plate "n"
- x_D = mol fr component in overhead product
- L = mols reflux

The basic assumption of the McCabe-Thiele procedure is that the liquid overflow from plate to plate is constant. If we look at Equation 17A.2 and assume that L has the same magnitude for all values of n, we see that V_{n+1} will also be the same for all values of n. Equation 17A.3 then takes the form y = mx + b. If plotted on coordinates of y and x, this will be the equation of a straight line with a slope of "m" (L/V) and a y-value at x = 0 of Dx_D/V. When plotted on Figure 17A.2, Equation 17A.4 is a straight line located as shown.

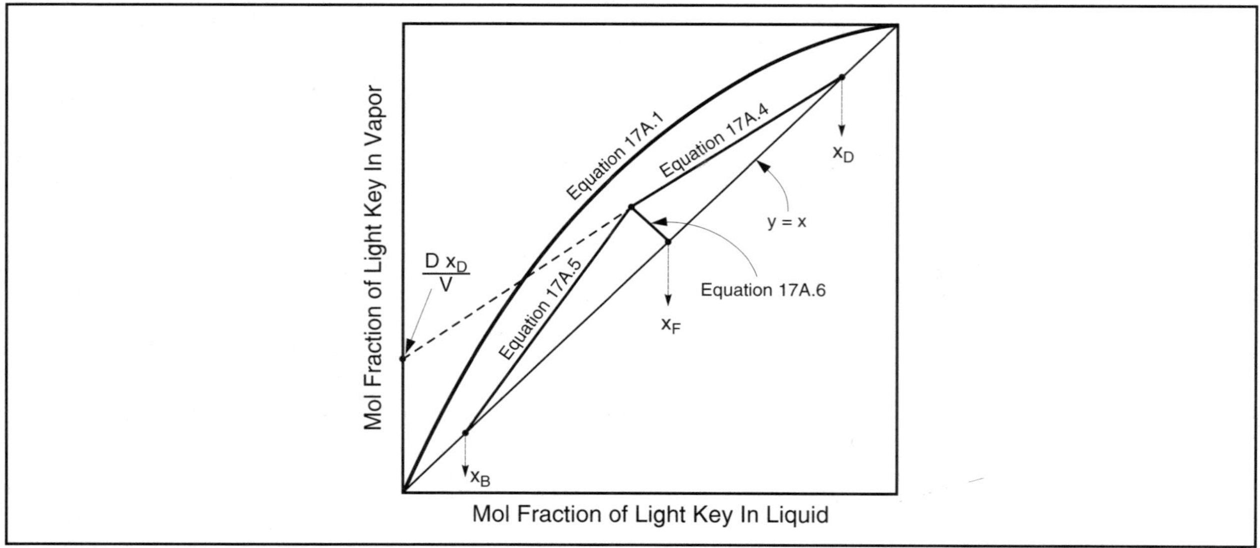

Figure 17A.2 Construction of McCabe-Thiele Diagram

Material Balance Around Bottom

In similar fashion, a material balance around the stripping section of the column will give the material balance line for the stripping section.

$$x_m = \left(\frac{L_m}{V_{m-1}}\right)(x_m) - \left(\frac{B}{V_{m-1}}\right)(x_B) \qquad (17A.5)$$

Where:

L_m = mols liquid leaving plate "m"
V_{m-1} = mols vapor leaving plate "m-1"
B = mols bottom product
x_m = mol fr of component in liquid from plate "m"
x_B = mol fr of component in bottoms
y_{m-1} = mol fr of component in vapor from plate "m-1"

Assuming constant molal overflow in the stripping section leads to the same line of reasoning for the rectifying section. The resulting straight line on y-x coordinates is shown in Figure 17A.2.

The two straight lines represented by Equations 17A.4 and 17A.5 are termed the *"operating" lines* for the rectifying and stripping sections of the column, respectively. They are given this title because their location (slope) is governed by the amount of reflux (L) pumped back to the column – a function of the manner in which the column is operated.

Effect of Feed Condition

A combined heat and material balance around the feed plate may be written in the form

$$x_f = \left(\frac{q}{q-1}\right)(x_f) - \left(\frac{x_F}{q-1}\right) \qquad (17A.6)$$

Where:

q = total heat needed to convert one mole of feed into saturated vapor divided by the molal latent heat
x_F = mol fr of component in feed
x_f = mol fr of component in liquid leaving feed plate
y_f = mol fr of component in vapor leaving feed plate

Consideration of Equation 17A.6 will make clear that it also is a straight line when plotted on y-x coordinates. It has a slope of (q/q–1). Also, when x is equal to the feed composition (x_F), y also has the value of x_F ($y = x_F$). The behavior of the so-called "q" line around the point $x = x_F$, $y = x_F$, is shown in Figure 17A.3.

The sequence of steps in constructing a McCabe-Thiele diagram is outlined below.

1. With knowledge of feed composition and condition and the products desired, make product splits for the tower and complete the material balance.
2. Using Equation 17A.1, calculate several points on the equilibrium curve. For an assumed value of x, the corresponding value of y is calculated.
3. Draw the equilibrium curve (for the more volatile component) and the 45° line as shown in Figure 17A.4.
4. Mark x_B, x_D, and x_F on the x-axis and the 45° line.

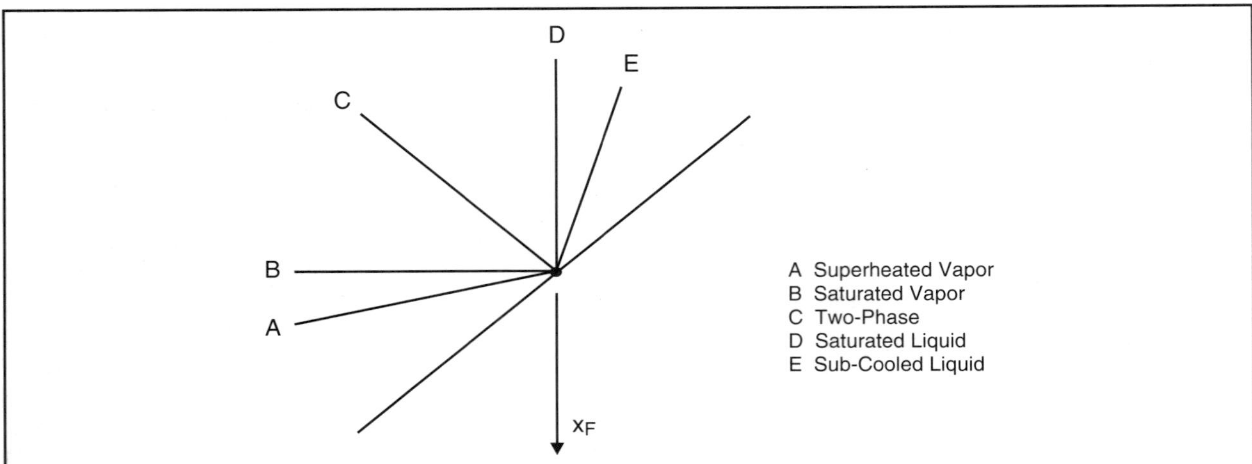

Figure 17A.3 Effect of Feed Condition on "Q" Line

5. Draw the rectifying section operating line (Equation 17A.4) as a straight line between the points ($x = x_D$, $y = y_D$) and ($x = 0$, and $y = Dx_D/V$).
6. Draw the "q" line as a straight line from the point ($x = x_F$, $y = x_F$) with a slope equal to q divided by $q - 1$.
7. Draw the stripping section operating line as a straight line from the intersection of the rectifying section of operating line and the "q" line to the point $x = x_B$ and $y = x_B$.
8. Step off the number of theoretical plates required to make the specified separation. This is done as shown in Figure 17A.4 by starting at the point representing the distillate composition (x_D) and alternating horizontal and vertical straight lines until the value of the bottoms composition (x_B) is reached.

This stepwise process is equivalent to calculations carried out tray-by-tray. The numbered points (1, 2, etc.) on the equilibrium curve represent the step of determining the liquid in equilibrium with a given vapor. Point 1, for example, represents (by horizontal line to the y-axis) the vapor leaving tray 1 and (by a vertical line drawn to the x-axis) the liquid in equilibrium with that vapor. The

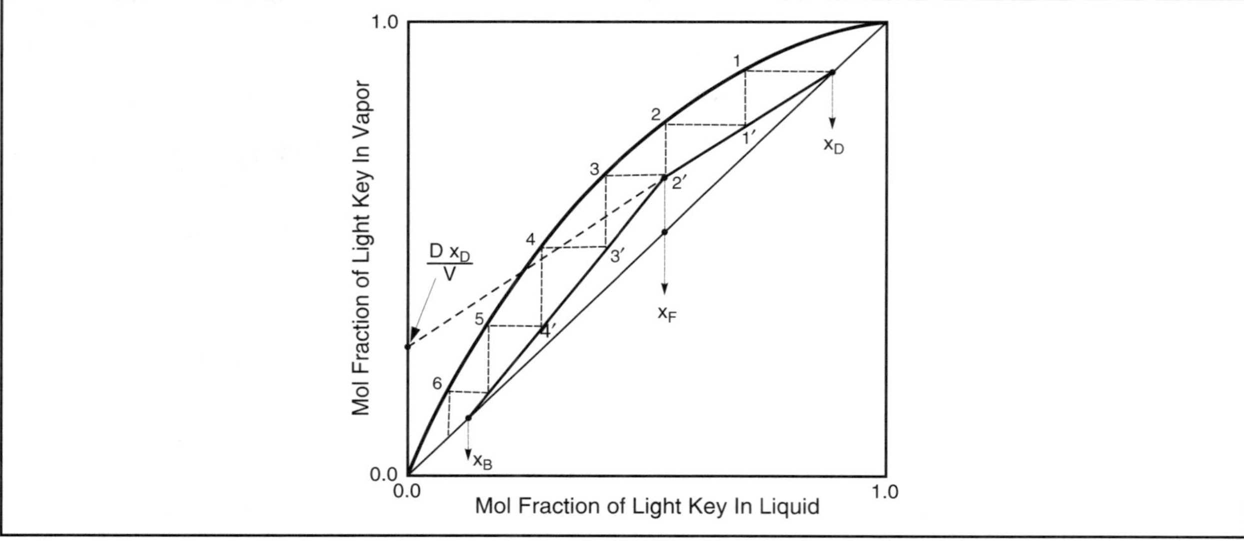

Figure 17A.4 McCabe-Thiele Diagram at a Finite Reflux Ratio

numbered points on the operating lines (1′, 2′ etc.) represent (by a horizontal line to the y-axis) the composition of the vapor leaving Plate 2 and (by a vertical line to the x-axis) the liquid leaving Plate 1. The operating line represents a material balance around the rectifying (stripping) section and must be used to determine the composition of passing streams.

The limits of column operation, minimum plates and minimum reflux, can be illustrated very clearly through use of the McCabe-Thiele diagram.

Consider the overall material balance for the rectifying section (Equation 17A.2). As L_n increases with respect to D, V_{n+1} approaches the value of L_n. When D = 0, $V_{n+1} = L_n$, but L_n can never be greater than V_{n+1}. This means that the rectifying section operating line (Equation 17A.4) can never be greater than unity. The same line of reasoning applies to the stripping section operation line (Equation 17A.5). This means that when D= 0 (and B = 0) the operating line for both sections of the column coincides with the 45° line. This condition is known as "total reflux" and under this condition each plate in the column makes the maximum possible separation. Consequently, the minimum number of plates for the specified separation will occur at total reflux. This situation and the method of stepping off trays at total reflux is illustrated in Figure 17A.5.

Consider what happens when one of the operating lines intersects the equilibrium curve. As we are stepping off theoretical trays and draw closer to the point of intersection, the separation per tray becomes smaller and an infinite number of theoretical trays is required to actually reach the point of intersection. This will be true any time an operating line intersects the equilibrium curve. The one of particular interest to us is the time at which the "q line," the rectifying section operating line, the stripping section operating line and the equilibrium curve intersect at a common point. Under this condition, only with an infinite number of theoretical plates in the column would we be able to make the specified separation of the feed into the distillate and bottom products. The construction of the McCabe-Thiele diagram under conditions of minimum reflux is shown in Figure 17A.6. The minimum reflux rate may be determined from the slope (L/V) of the rectifying section operating line or from its intercept (Dx_D/V) with the y-axis.

Although this method is not used for regular calculations, it is an excellent instructional tool to show the effect of changing operating variables on tower performance and to graphically evaluate computer simulations.[17.22]

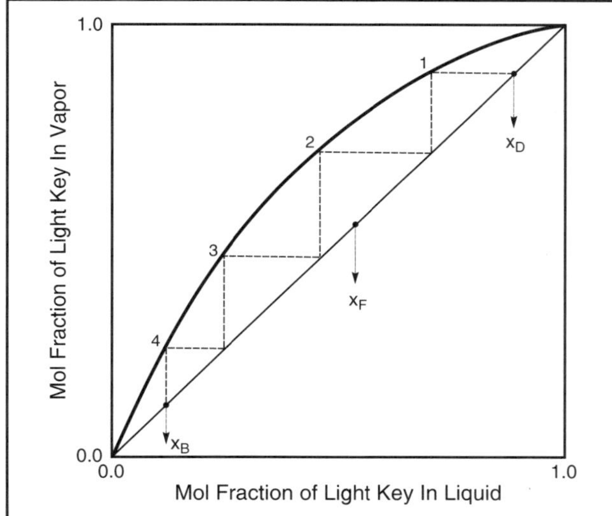

Figure 17A.5 McCabe-Thiele Diagram at Total Reflux

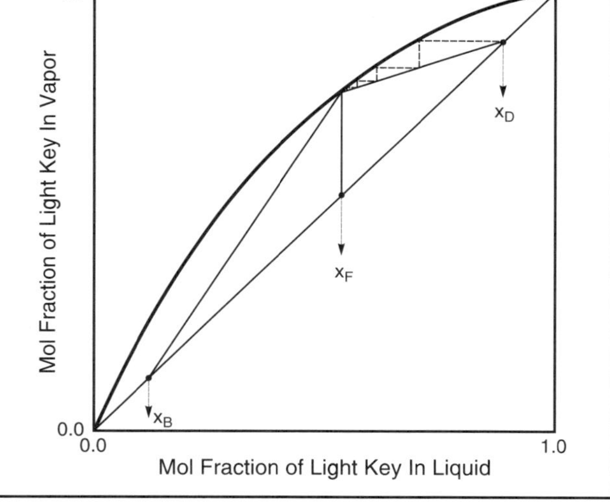

Figure 17A.6 McCabe-Thiele Diagram at Minimum Reflux

APPENDIX 17B

SHORT-CUT UNDERWOOD METHOD(17.23)

This method uses the equation

$$\alpha_f = \left[\frac{(L+D)x_{cf} + (q-1)D x_{cD}}{(L+D)x_{df} + (q-1)D x_{dD}}\right] = \frac{L x_{cf} + qD x_{cD}}{L x_{df} + qD x_{dD}} \quad (17B.1)$$

Where:

α_f = ratio of Ks for the key components at the feed plate temperature (temperature of feed at its bubblepoint would be a fair approximation of the feed plate temperature)

L = mols of liquid leaving any plate above the feed at minimum reflux

D = mols of overhead product

x_{cf} = composition, in mol fraction or mol percent, of the *light key* component in *feed*

x_{df} = composition, in mol fraction or mol percent, of the *heavy key* component in *feed*

x_{cD} = composition, in mol fraction or mol percent, of the light key component in overhead product

x_{dD} = composition, in mol fraction or mol percent, of the heavy key component in overhead product

The above equation tends to give high values of minimum reflux. In some cases, the value found may be considered as a minimum in the actual range of reflux rates considered.

If the feed is at its bubblepoint, q = 1, this equation is simplified since some terms fall out. This is the case in our continuing example.

For Case I, Equation 17B.1 may be solved for L, where $\alpha_f = 1.81$

$$(1.81)\left(\frac{8.9}{63.3}\right) = \left(\frac{8.9L + (7.8)(92.3)}{63.3L + (7.8)(5.1)}\right)$$

$$16.11\ L + 10.12 = 8.9\ L + 720\ , \quad L = 709.9/7.21 = 98.5 \text{ mols/100 mols feed}$$

If we write a material balance around the top of the tower, V = L + D or V = 98.5 + 7.8 = 106.3 mols/100 mols feed. Thus,

$$(L/V)_m = 98.5/106.3 = 0.927, \qquad (L/D)_m = 98.5/7.8 = 12.63$$

By the other Underwood method, L = 79.8 mols/100 mols feed, which is usually rather close to what one would obtain from a more complex model. So, Equation 17B.1 in this example gives results which are high by 23%.

18

GLYCOL DEHYDRATION

Dehydration is the process of removing water from a gas and/or liquid so that no condensed water is present in the system. Inhibition is the process of adding some chemical to the condensed water so hydrates cannot form.

Dehydration is generally preferred, if economically and mechanically feasible, because it prevents water from condensing in the system. This water is the source of both hydrates and corrosion/erosion problems. In some instances, inhibition may be the preferred process, particularly in sweet systems employing moderate refrigeration (-40°C [-40°F] and above).

Natural gas is commercially dehydrated in one of three ways.

1. Absorption — Glycol Dehydration
2. Adsorption — Mol Sieve, Silica Gel, or Activated Alumina
3. Condensation — Refrigeration with Glycol or Methanol Injection

Glycol dehydration (absorption) is the most common dehydration process used to meet pipeline sales specifications and field requirements (gas lift, fuel, etc.). Adsorption processes are used to obtain very low water contents (0.1 ppm or less) required in low temperature processing such as deep NGL extraction and LNG plants. Condensation is commonly used as a dehydration process when moderate levels of refrigeration are employed or in pipeline transportation. An inhibitor such as ethylene glycol (EG) or methanol is used to prevent hydrate formation, but it should be noted that the actual water extraction mechanism is condensation.

Four glycols are used for dehydration and/or inhibition:

1. Monoethylene glycol (MEG) – often referred to as ethylene glycol (EG)
2. Diethylene glycol (DEG)
3. Triethylene glycol (TEG)
4. Tetraethylene glycol (TREG)

Triethylene glycol (TEG) is the most common glycol used in absorption systems. Monoethylene glycol (MEG) is the most common glycol used in glycol injection systems. All glycol are hygroscopic, which means they have an affinity for water.

The basic properties of these glycols are shown in Appendix 18A. The major properties governing the choice of glycol for a given application are:

1. Viscosity
2. Vapor pressure
3. Solubility in hydrocarbons

In absorption dehydration systems the solvent (glycol) should be hygroscopic, noncorrosive, non-volatile, easily regenerated to high concentrations, insoluble in liquid hydrocarbons, and unreactive with hydrocarbons, CO_2 and sulfur compounds.

Several of the glycols come close to meeting all of these criteria. Diethylene (DEG), triethylene (TEG) and tetraethylene (TREG) glycols all possess suitable traits. However, almost 100% of the glycol dehydrators use TEG.

DEG is somewhat cheaper to buy and sometimes is used for this reason. It is also sometimes used as an inhibitor in addition to an absorbent. However, by the time it is handled, stored, and added to the units there is often no real savings. Compared to TEG, DEG has a larger carry-over loss, offers lower dewpoint depression, and regeneration to high concentrations is more difficult. For these reasons, it is difficult to justify a DEG unit, although a few are built each year.

TREG is more viscous and more expensive than the other processes. The only real advantage is its lower vapor pressure which reduces absorber carry-over loss. It may be used in those relatively rare cases where glycol dehydration will be employed on a gas whose temperature exceeds about 50°C [122°F].

In recent years some units have employed propylene glycol (PG). PG is the least toxic glycol and has a lower affinity for aromatics, but PG has a much higher vapor pressure than TEG, and a much lower flash point.

This chapter will concentrate on TEG, even though property data are shown in Appendix 18A for several glycols. Some of the system characteristics apply also for all glycols.

THE BASIC GLYCOL DEHYDRATION UNIT

Figure 18.1 shows the basic glycol unit, regardless of the glycol used. Not shown is any inlet gas cooling equipment that may be a part of the dehydration unit. When it is possible to cool the entering wet gas with air, water or another process stream ahead of the absorber, do so. Such cooling is the least expensive form of dehydration.

The entering wet gas, free of liquid water, enters the bottom of the absorber (contactor) and flows countercurrent to the glycol. Glycol-gas contact occurs on trays or packing. Bubble cap trays have been used historically but structured packing is more common today. The dried gas leaves the top of the absorber.

The lean glycol enters on the top tray or at the top of the packing and flows downward, absorbing water as it goes. It leaves rich in water.

It is convenient to use the word "rich" to describe the bottom of the absorber and the word "lean" for the top. At the bottom, both the entering gas and glycol leaving are rich in water; at the top end they both are lean in water.

The rich glycol leaves the bottom of the absorber and flows to a reflux condenser at the top of the still column. The rich glycol then enters a flash tank where most of the volatile components (entrained and soluble) are vaporized. Flash tank pressures are typically 300-700 kPa [44-102 psia]. Leaving the flash tank the rich glycol flows through the glycol filters and the lean-rich exchanger where it exchanges heat with the hot lean glycol. The rich glycol then enters the still column where the water is removed by distillation.

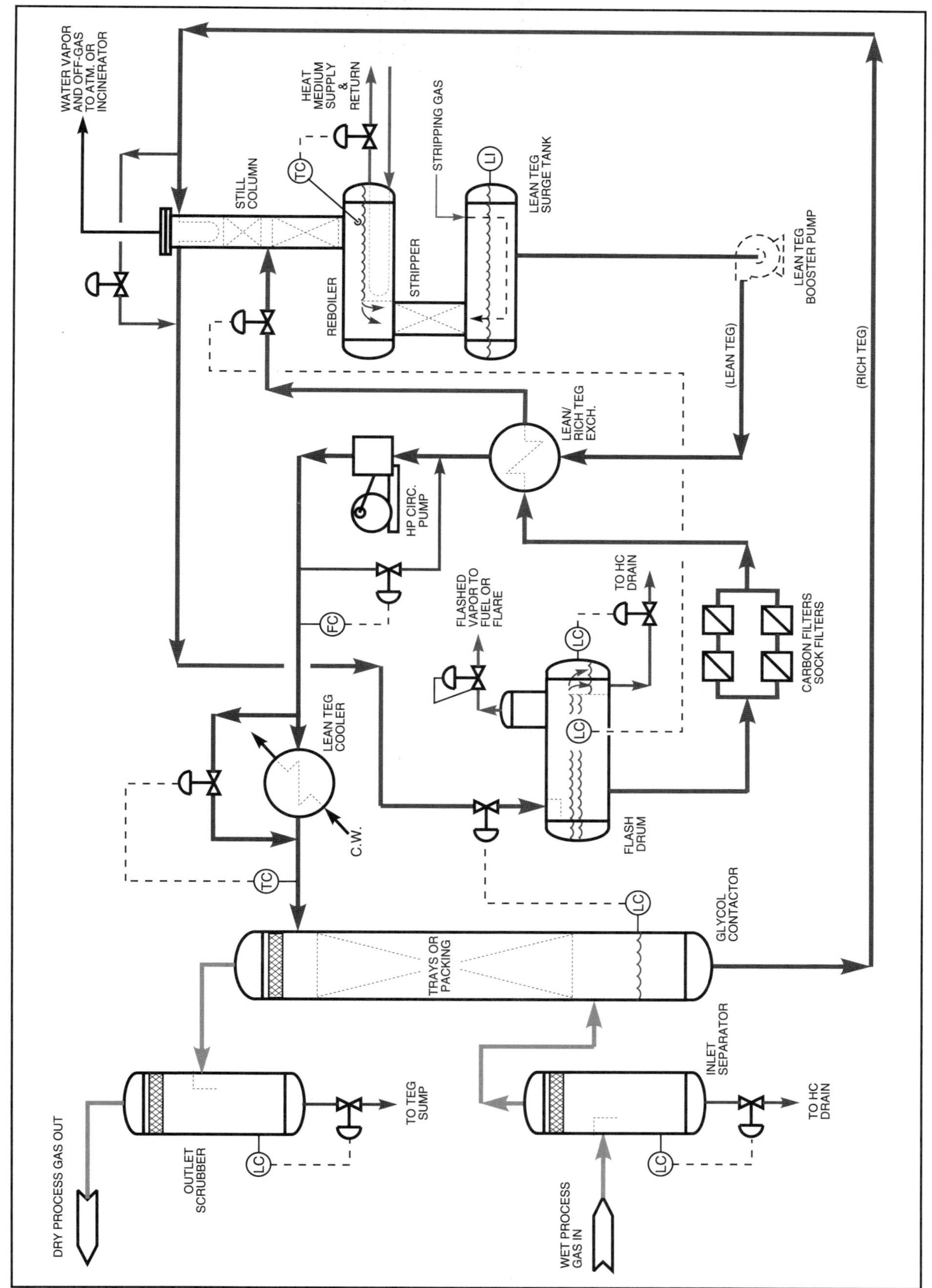

Figure 18.1 Basic Glycol Dehydration Unit

The still column and reboiler are often called the regenerator or reconcentrator. This is where the glycol concentration is increased to the lean glycol requirement.

The regeneration unit shown is designed to operate at prevailing atmospheric pressure. The initial thermal decomposition temperatures of the glycols are shown in the table below:

Glycol	Decomposition Temperature	Lean Glycol* Concentration, wt%	Equilibrium** Water Dewpoint @ 38°C [100°F]
EG	165°C [329°F]	96.0	3°C [37°F]
DEG	164°C [328°F]	97.1	3°C [37°F]
TEG	206°C [404°F]	98.7	-8°C [18°F]
TREG	238°C [460°F]	> 99	-18°C [0°F]

* Equilibrium concentration at decomposition temperature and 1 atm

** At concentration in column at left

These are the temperatures at which measurable decomposition begins to occur in the presence of air. DEG is no more stable than EG because it pyrolyzes in contact with carbon steel.

In the normal unit containing no air (oxygen), it has been found that one can operate the reboiler very close to the above temperatures without noticeable decomposition. The composition of the lean glycol is set by its bubblepoint composition at the regenerator pressure. Maximum concentrations achievable in an atmospheric regenerator operating at decomposition temperature are also shown in the above table.

If the lean glycol concentration required at the absorber (to meet the dewpoint specification) is higher than the maximum concentrations above, then some method of further increasing the glycol concentration at the regenerator must be incorporated in the unit. Virtually all of these methods involve lowering the partial pressure of the glycol solution either by pulling a vacuum on the regenerator or by introducing stripping gas into the regenerator.

A typical stripping gas system is shown in Figure 18.2(a). Any inert gas is suitable. A part of the gas being dehydrated, or exhaust from a gas-powered glycol pump (if used), is suitable. The quantity required is small. The stripping gas may be introduced directly into the reboiler or into a packed "stripping column" between the reboiler and surge tank. In theory, adding gas to a packed unit between the reboiler and surge tank is superior and will result in lower stripping gas rates. If introduced directly to the reboiler, it is common to use a distributor pipe along the bottom of the reboiler.

A second stripping gas alternative is to close the stripping gas loop and use a material like iso-octane, as shown in Figure 18.2(b). It vaporizes at reboiler temperature but can be condensed and separated from the water in a three-phase separator. The stripping solvent is then pumped back to the regenerator to complete the stripping loop. Sold under the trade name DRIZO®, this unit has the advantage of providing very high stripping gas rates with little or no venting of hydrocarbons. Glycol concentrations in excess of 99.99 wt% have been achieved with the DRIZO® process. It has an added advantage of condensing and recovering aromatic hydrocarbons from the still column overhead. In practice, these units often operate with a stripping solvent which is not iso-octane but a mixture of aromatic, naphthenic and paraffin hydrocarbons in the C_5-C_{10} range.

Figure 18.2(c) shows a third regenerator alternative called a COLDFINGER®. The COLDFINGER® process achieves glycol enrichment by passing a cooling medium (often the rich glycol) through a cool "finger" inserted in the surge tank vapor space. This condenses a water-TEG

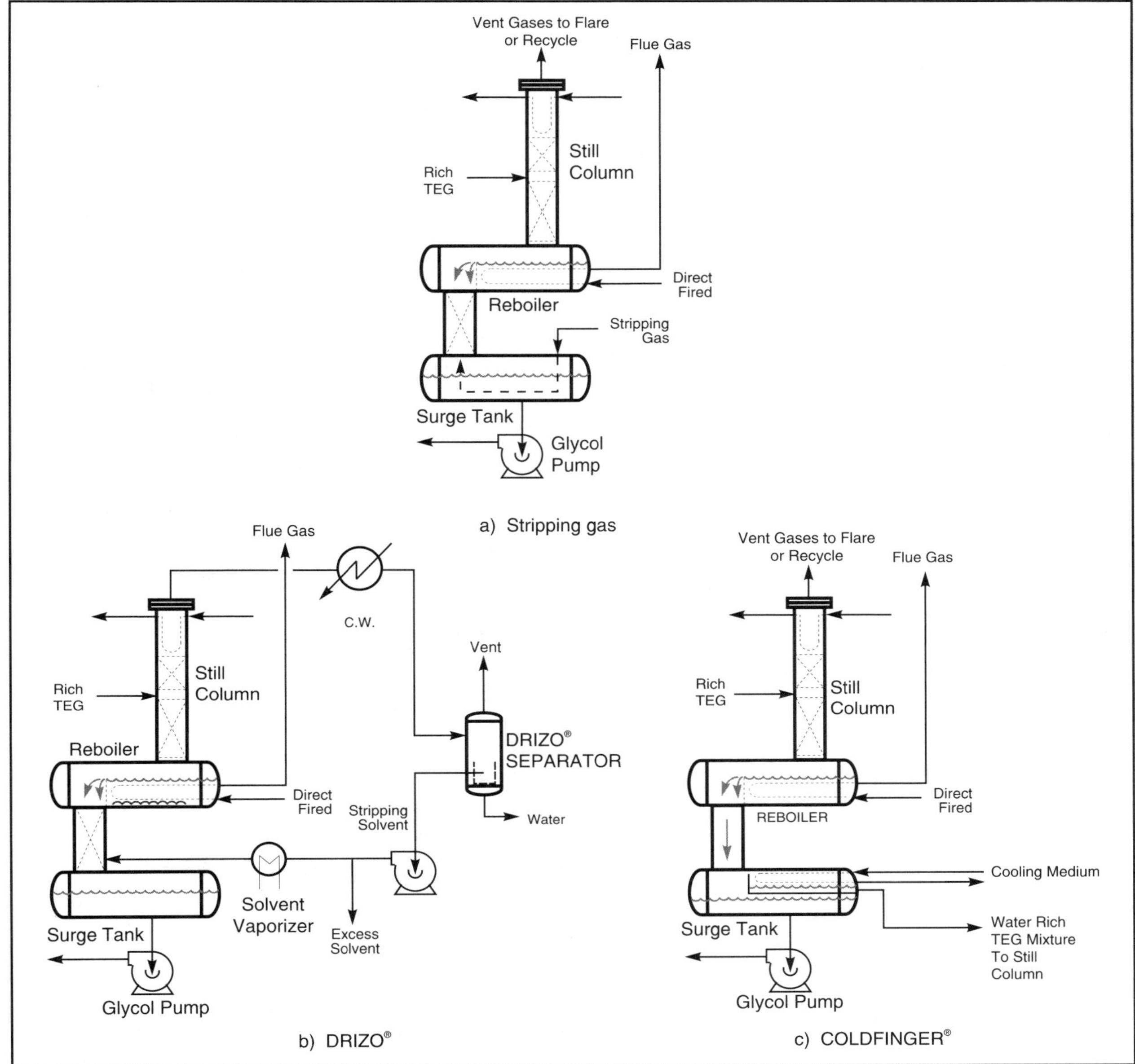

Figure 18.2 TEG Regeneration Alternatives

mixture which is rich in water. This mixture is drawn out of the surge tank by means of a trough below the "coldfinger" and is recycled back to the regenerator. The H_2O partial pressure in the surge tank vapor space is thus lowered and the lean glycol concentration increased. It is claimed that lean TEG concentrations of 99.5-99.9 wt% have been achieved in COLDFINGER® units without the use of stripping gas, although a small amount of gas is introduced into the surge tank for pressure balancing.

The unit shown in Figure 18.1 is typical. Figure 18.3 shows an example of an offshore unit. The inlet scrubber is in the bottom of the absorber. Three-phase separation is required. The gas rises through a "chimney tray" to the absorber. The hydrocarbon and water are separated as shown. Three-phase separation saves on deck space and is less expensive, but many of the existing units are unsatisfactory because they provide inadequate separation. This is particularly true when the absorber utilizes structured packing.

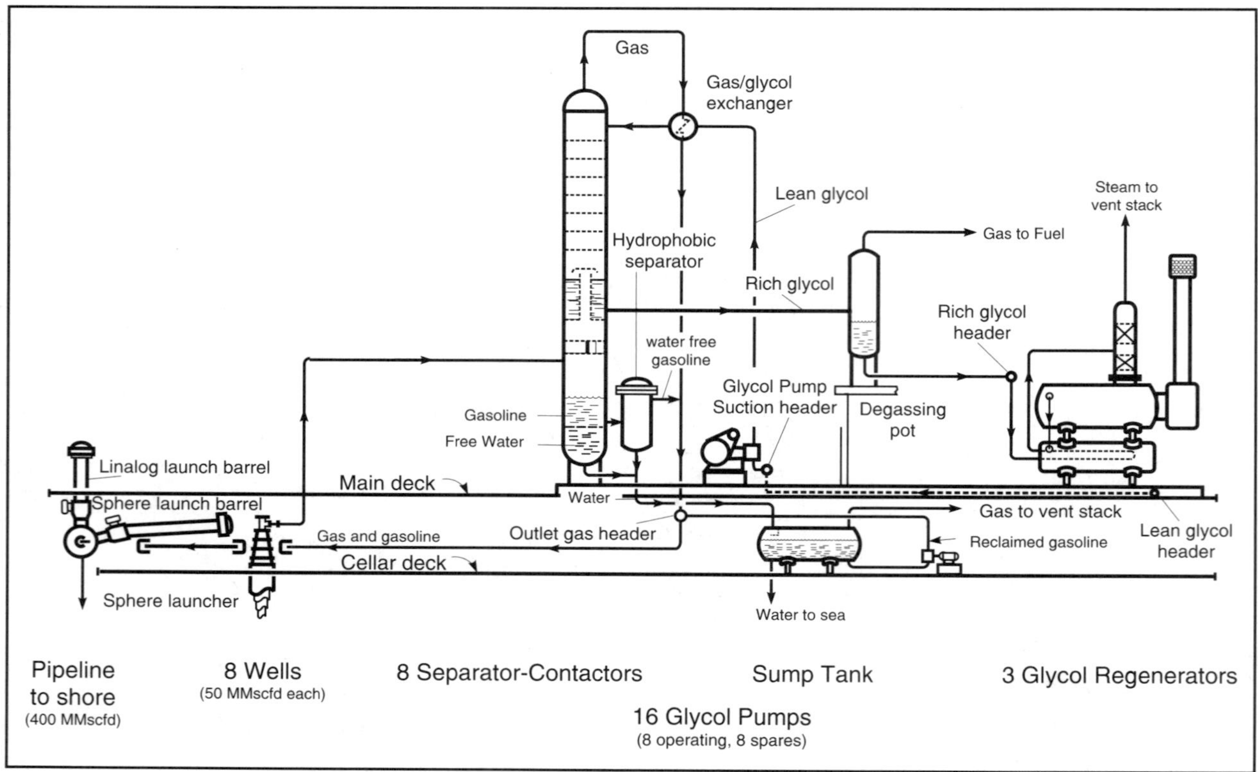

Figure 18.3 Example Offshore Glycol Dehydration Installation

The rich TEG from the chimney tray goes to a degassing pot (flash tank) which is operated at a high enough pressure to send the gas to fuel or recompression. In some systems the pressure is sufficient merely to enter the main flare system. The purpose of this is several fold: (1) use or dispose safely any volatile components picked up by the TEG in the absorber, and (2) minimize the presence of corrosive sulfur compounds and carbon dioxide in the high temperature reboiler.

The true solubility of paraffin hydrocarbons is very low in the glycols. But, separator carry-over and entrainment does introduce hydrocarbons into the rich glycol. Many of these are "heavier" than air and can be a safety problem unless disposed of properly. In addition, aromatic components are very soluble in TEG. These can also be a safety and environmental concern when discharged to atmosphere at the top of the regenerator.

Both sulfur compounds and carbon dioxide are soluble in water and react to some degree with the glycols. The degassing in the flash tank prior to the stripping column reduces their concentration in the glycol somewhat. This degassing is more efficient if the rich glycol is first preheated, usually in the rich-lean glycol exchanger.

The glycol cooler is a gas-TEG exchanger in Figure 18.3, as opposed to cooling medium in Figure 18.1. This is a common method of cooling the lean glycol. It is simple and inexpensive and ensures the glycol enters the contactor warmer than the gas. The gas-glycol exchanger can result in excessively high lean TEG temperatures at low gas flowrates.

The rich-lean exchanger in Figure 18.3 is a coil in the surge tank. This is inexpensive to build but often results in poor heat transfer and higher reboiler duties. Pipe-in-pipe exchangers are widely used in this service.

Figure 18.3 shows a gas fired reboiler. The use of hot oil, steam, waste heat or electrical resistance coils are all suitable if they are readily available at the site. The use of electrical resistance heating on offshore installations is cost-effective and more convenient if area classification does not allow a fired heater.

No filter is shown in Figure 18.3. I prefer locating the particulate filter at some point ahead of the reboiler to minimize the "gunk" accumulating therein. For effective operation it is imperative that full-flow, glycol filters be installed in the system.

BASIC PROCESS DESIGN FACTORS

All factors controlling the behavior of absorption systems also apply for TEG dehydration. In fact, from a process viewpoint, TEG is one of the simpler absorption processes being employed in the petroleum industry.

To properly design a unit one needs to know maximum and minimum gas flow rate, maximum and minimum temperature and pressure, gas composition and required water dewpoint or water content of the outlet gas. From these one can calculate:

1. The *minimum concentration* of TEG in the lean solution entering the top of the absorber required to meet outlet gas water specification.
2. The lean TEG circulation rate required to pick up the necessary amount of water from the gas to meet the outlet gas water content specification.
3. The amount of absorber contact required to produce the necessary approach to equilibrium required in (1) above at the chosen circulation rate.

To obtain these answers it is necessary to have a vapor-liquid equilibrium correlation for a TEG-water system. From this basic input, one can evaluate or size equipment and develop mechanical specifications.

The procedure that follows is straightforward and can be performed manually. In all but a few exceptional applications, it will give results as reliable as more rigorous methods. Following an outline of the basic calculation procedure, each major equipment component will be reviewed.

MINIMUM LEAN TEG CONCENTRATION

If water-saturated gas is placed in a static cell with a given concentration of TEG-water solution at a fixed P and T, equilibrium would be attained in time. Assuming the liquid had a sufficiently low water concentration, water would transfer to this liquid from the gas. At equilibrium, the mol fraction water in the gas divided by its mol fraction in the liquid equals the K value for this system.

Figures 18.4(a & b) are based on equilibrium data published by Parrish, *et al.*[(18.1)] Several equilibrium correlations[(18.1-18.6)] have been presented since 1950. Previous editions of the Campbell texts and the GPSA engineering Data Book presented an equilibrium correlation based on the work of Worley[(18.2)]. In general, the correlations of Worley[(18.2)], Rosman[(18.5)] and Parrish[(18.1)] agree reasonably well and are adequate for most TEG system designs. All are limited by the ability to accurately measure the equilibrium concentration of water in the vapor phase above TEG solutions. The Parrish correlation has been included in this edition because equilibrium water concentrations in the vapor phase were determined at infinite dilution (essentially 100% TEG). The other correlations use extrapolations of data at lower concentrations to estimate equilibrium in the infinite dilution region. The effect of pressure on TEG-water equilibrium is small up to about 13 800 kPa [2000 psia][(18.4)].

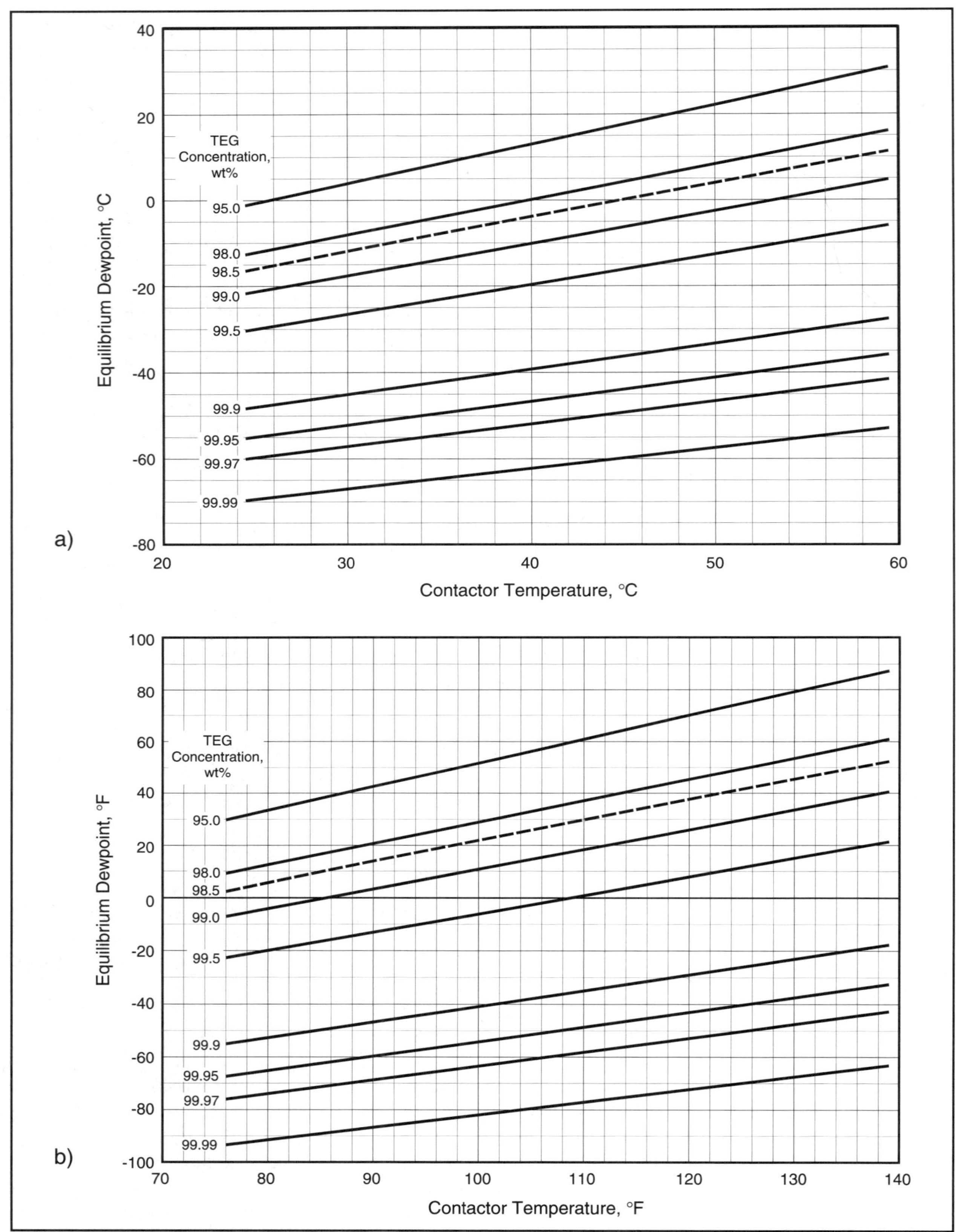

Figure 18.4 Equilibrium H_2O Dewpoint vs. Temperature at Various TEG Concentrations

A TEG absorber is essentially isothermal. The heat of solution is about 211 kJ/kg [91 Btu/lbm] of water absorbed in addition to the latent heat. But, the mass of water absorbed plus the mass of TEG circulated is small relative to the mass of gas. So, the inlet gas temperature controls. The temperature rise due to heat of absorption and heat transfer between the glycol and the gas seldom exceeds 1-2°C [2-4°F] except when dehydrating at pressures below about 1000 kPa [145 psia]. In low pressure service some temperature adjustment may be desirable.

Example 18.1: What equilibrium water dewpoint could be obtained at 40°C [104°F] with a lean glycol solution containing 99.5 wt% TEG?

In Figure 18.4, locate 40°C [104°F] on the abscissa, go vertically to the 99.5 wt% line and then horizontally to the ordinate. Read -19°C [-2°F].

This water dewpoint could be attained in a test cell but not in a real absorber. The gas and TEG are not in contact for a long enough time to reach equilibrium. In addition, the gas theoretically leaves the top tray of the absorber in equilibrium with the TEG *leaving* the tray, not entering. Numerous tests show that a well designed, properly operated unit will have an actual water dewpoint 5-10°C [9-18°F] higher than the equilibrium dewpoint. This "approach" to equilibrium depends on the glycol circulation rate and number of contacts in the absorber and is used to specify minimum lean glycol concentration. The procedure is as follows.

1. Establish the desired outlet water dewpoint needed from sales contract specifications or from minimum system temperature.
2. Subtract the approach from (1) to find the corresponding equilibrium water dewpoint.
3. Enter the value in (2) on the ordinate of Figure 18.4 and draw a horizontal line.
4. Draw a vertical line from the inlet gas temperature on the abscissa.
5. The intersection of the lines in Steps (3) and (4) establishes minimum lean TEG concentration required to obtain the water dewpoint in Step (1).

If water content is specified or calculated in mass per unit gas volume, a water content, pressure, dewpoint temperature correlation is required (see Figure 6.1 in Volume 1 or Figure 18C.1 in Appendix 18C). Note that the equilibrium water dewpoints on the ordinate of Figure 18.4 are based on the assumption the condensed water phase is a metastable liquid. At low dewpoints the true condensed phase will be a hydrate. The equilibrium dewpoint temperature above a hydrate is higher than that above a metastable liquid. Therefore, Figure 18.4 may predict dewpoints which are lower than can actually be achieved. The difference is a function of temperature, pressure and gas composition but can be as much as 8-12°C [15-20°F]. Figure 6.5 shows the relationship between the metastable and hydrate dewpoints for one gas. When dehydrating to very low dewpoints, such as those required upstream of a refrigeration process, the TEG concentration should be sufficient to dry the gas to the hydrate dewpoint and not just the metastable dewpoint.

The dashed line in Figure 18.4 at about 98.5 wt% represents the concentration of lean TEG that can be produced routinely in a regenerator operating at standard atmospheric pressure and 204°C [400°F]. This is a safe value for design and specification purposes. Concentrations of 98.7-98.8 wt% are common; some to 99.1 wt% have been reported but represent a special case where incoming hydrocarbons provided natural stripping and/or the atmospheric pressure was lower than 101 kPa [14.7 psia].

Since the initial capital cost of ordinary gas stripping accessories is small, they always should be included. Conditions can change to where they may be required.

Example 18.2: The gas sales contract specifies an outlet water content of 100 kg/10^6 std m^3 [6 lbm/MMscf] at a pressure of 69 bar [1000 psia]. The inlet gas temperature is 40°C. What minimum lean TEG concentration is required?

SI Solution: For 100 kg/10^6 std m^3 and 69 bar, the equivalent dewpoint from a water content correlation (Figure 18C.1(a)) is -2°C. If we use an 8°C approach the equilibrium dewpoint is -10°C. From Figure 18.4(a) at -10°C and 40°C contact temperature, lean TEG concentration = 99.0 wt%

FPS Solution: For 6 lb/MMscf and 1000 psia the equivalent dewpoint from a water content correlation (Figure 18C.1(b)) is 28°F. An approach of 14°F gives an equilibrium dewpoint of 14°F. From Figure 18.4(b), lean TEG concentration equals 99.0 wt%.

It is necessary to fix a lean TEG concentration for subsequent calculations. For the first consideration, use the results from Figure 18.4. If the concentration obtained is less than 98.5 wt%, use 98.5 wt% for the calculation unless you plan to reduce the reboiler temperature below 204°C [400°F].

The minimum lean TEG concentration may not be the one eventually used. A higher concentration than this may be specified to minimize circulation rate and optimize cost.

Absorber Design

The absorber (contactor) is where the water is removed from the gas by the process of physical absorption. The amount of water removed depends on three factors:

- glycol concentration
- glycol circulation rate
- number of contacts in absorber

The effect of glycol concentration on water removal is illustrated in Figure 18.4. As the TEG concentration increases, the equilibrium water dewpoint decreases. In an absorber this would mean a lower water concentration in the gas at the absorber outlet, hence higher water removal.

In a real absorber the gas leaving the absorber does not reach equilibrium with the incoming lean TEG. This is the reason for the "approach" discussed in the previous section and applied in Example 18.2. Approach depends on two parameters — circulation rate and number of contacts. The theoretical outlet water dewpoint (water content) is set by the lean glycol purity. The actual water dewpoint depends on the circulation rate and number of trays or height of packing. This is illustrated in Figure 18.5.

As you can see in Figure 18.5, adding contacts or increasing the circulation rate increases water removal. However, the curves in Figure 18.5 reach an asymtotic value of about 0.94. This water removal represents the equilibrium dewpoint condition at the top of the contactor. If we have an infinite number of contacts or an infinite circulation rate, the water removal percentage would still be 94% and the approach would be zero. At this point, the water removed depends exclusively on the lean TEG concentration. Our goal is to design a glycol system which is economically viable. We have several options. We can select a contactor with several contacts and a low circulation rate or one with few contacts and a high circulation rate. In fact, there exist an infinite number of choices to meet our outlet dewpoint specification.

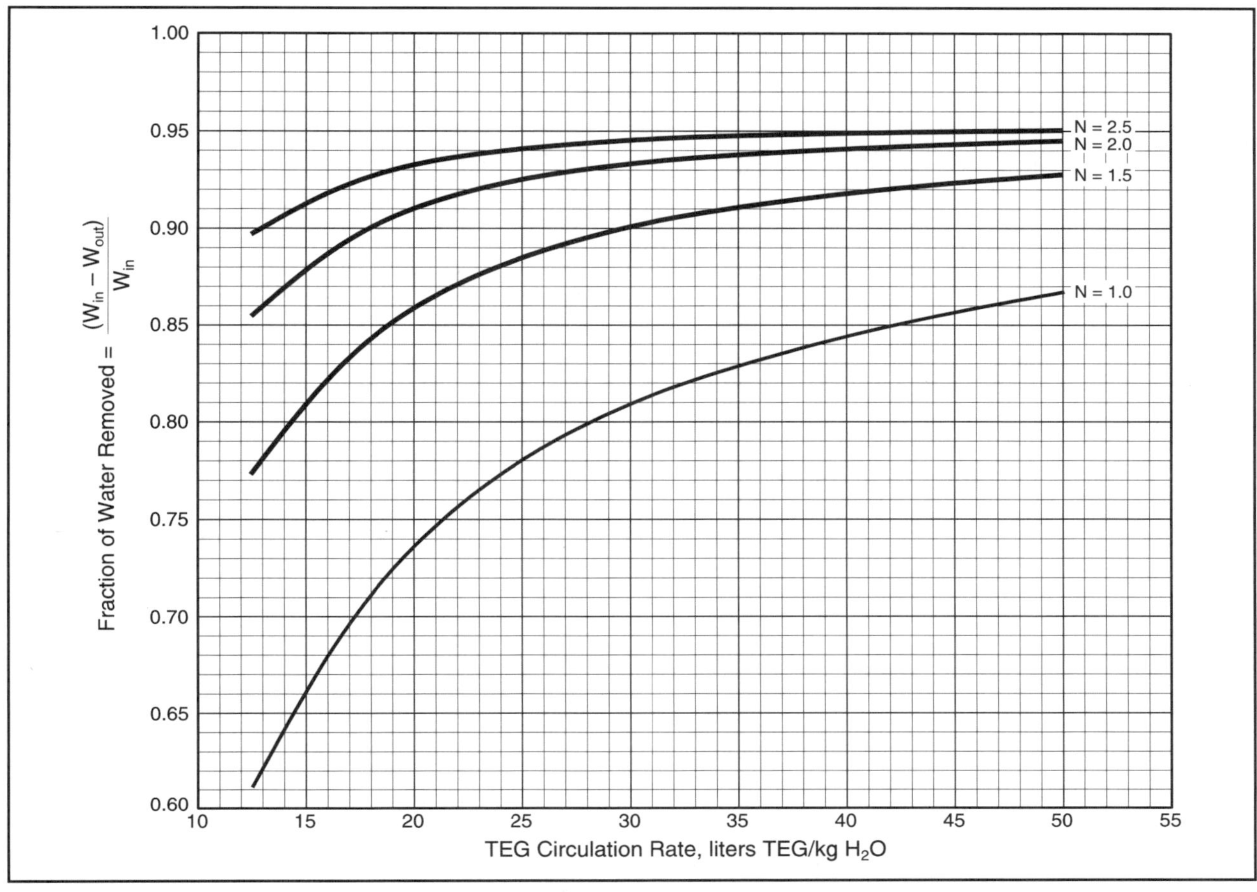

Figure 18.5 Water Removal vs. TEG Circulation Rate for Several Contactor Theoretical Stages (99.0 wt% TEG)

Most designs use a circulation rate of 15-40 liters TEG/kg H_2O absorbed [2-5 US gal/lb H_2O absorbed]. This is near the economic optimum. Higher circulation rates result in a larger regeneration system, higher energy consumption, and higher coabsorption of aromatic hydrocarbons. Lower rates require a taller contactor and may result in poor tray/packing hydraulics.

The relationship between circulation rate and the number of contacts can be quantified by use of the Kremser-Brown equation (see Chapter 17), a shortcut absorber calculation.

$$E_a = \frac{Y_{N+1} - Y_1}{Y_{N+1} - Y_0} = \frac{A^{N+1} - A}{A^{N+1} - 1} \tag{18.1}$$

Where:

E_a = efficiency of absorption
Y_{N+1} = mols of water in entering (wet) gas per mol of gas entering
Y_1 = mols of water in leaving (dry) gas per mol of gas entering
Y_0 = mols of water in equilibrium with the incoming lean glycol per mol of gas entering
A = absorption factor, A = L/VK
L = glycol circulation rate, mols/unit time
V = gas flow rate, mols/unit time
K = vapor liquid equilibrium ratio (y/x) for water in a water-gas-TEG system
N = number of theoretical contacts in absorber

In TEG systems, the mol fraction, y, may be used instead of the regular absorption parameter, Y, due to the concentrations involved.

This means that the left hand side of Equation 18.1 becomes

$$\frac{y_{N+1} - y_1}{y_{N+1} - y_0}$$

where "y" refers to the mol fraction of water rather than a rate per mol of incoming gas.

Further, since the mol fraction of water can be converted to mass of water per standard gas volume by a fixed conversion factor,

$$W = B\,y_{H_2O}$$

Where:			SI	FPS
	W =		$kg/10^6$ std m^3	lbm/MMscf
	B =	conversion factor	761 000	47 400
	y =	mol fraction H_2O		

Equation 18.1 may be rewritten in terms of W.

$$E_a = \frac{W_{N+1} - W_1}{W_{N+1} - W_0} = \frac{A^{N+1} - A}{A^{N+1} - 1} \tag{18.2}$$

Where:			SI	FPS
	W_{N+1} =	water content of entering (wet) gas	$kg/10^6$ std m^3	lbm/MMscf
	W_1 =	water content of leaving (dry) gas	$kg/10^6$ std m^3	lbm/MMscf
	W_0 =	water content of gas in equilibrium with incoming lean TEG	$kg/10^6$ std m^3	lbm/MMscf

W_0 is a function of the lean TEG concentration and contactor T and P.

Figure 18.6 is a plot of Equation 18.2 and is convenient for manual calculations. This uses what could be called an *overall absorption factor.* L_0 is the rate of lean TEG entering the top tray and V_{N+1} is the gas rate entering the bottom tray. Even though the (L/V) changes slightly throughout the absorber the effect of these changes are effectively canceled by changes in the "K" value, so use of $(L_0/V_{N+1}K)$ as the average absorption factor has little effect on the accuracy of the method.

The left-hand ordinate of Figure 18.6 is called *absorption efficiency of absorption*, E_a. It is the actual amount of water removed, divided by the maximum amount theoretically removable. The values of N encompass the range of theoretical stages usually employed in TEG contactors.

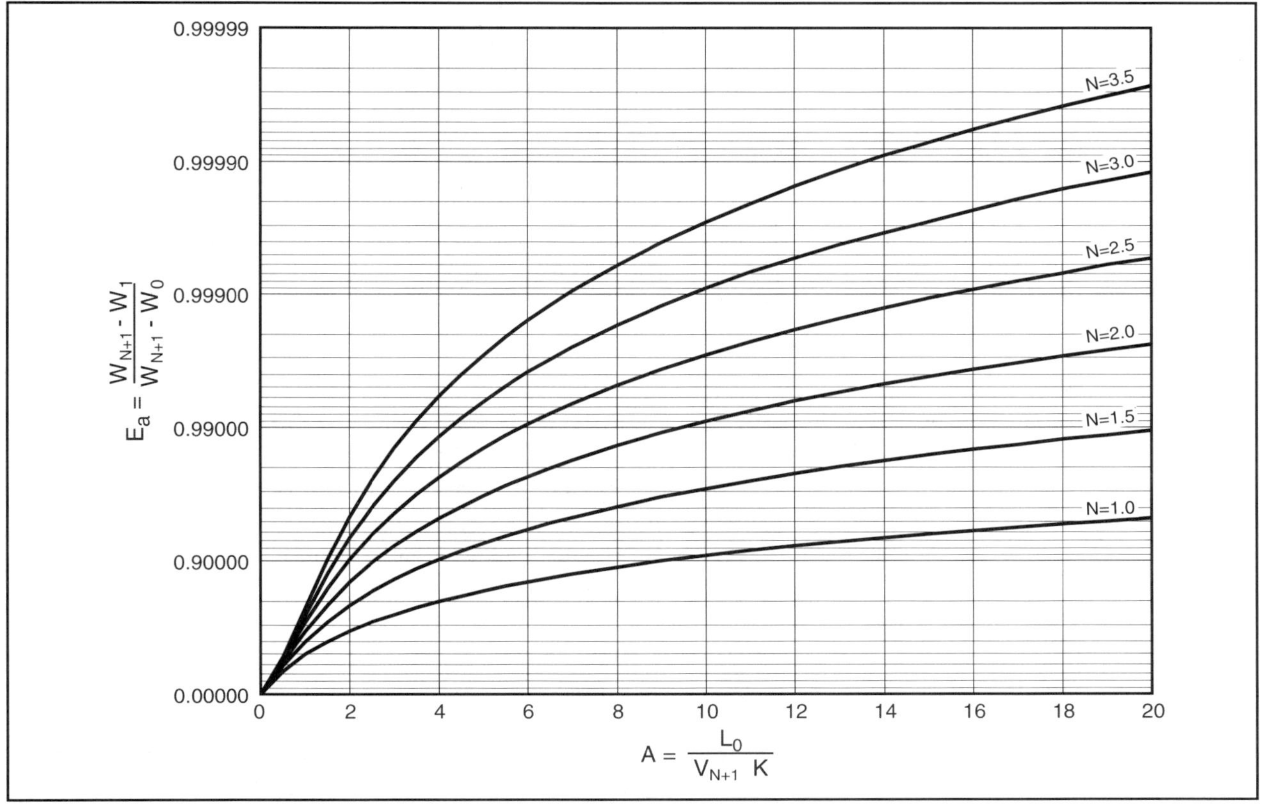

Figure 18.6 Efficiency of Absorption, E_a vs. Absorption Factor, A

Calculation of Lean TEG Rate for a Given Absorption Efficiency and N —

1. Calculate y_0 (or W_0)
2. Determine absorption efficiency
3. Use Equation 18.2 or Figure 18.6 to find absorption factor A for a given value of N
4. Knowing V_{N+1} and K, solve A for L_0, the lean TEG circulation rate

Calculation of N for a Given Lean TEG Rate and Absorption Efficiency —

1. Calculate y_0 (or W_0)
2. Determine absorption efficiency
3. Calculate absorption factor A
4. Determine N from Equation 18.2 or Figure 18.6

It is usual to repeat the calculation to obtain three lean glycols rate/absorber contact values that satisfy the required absorption efficiency. The final choice is based on economic considerations. This usually involves selection of standard designs.

This calculation should be made at the lowest pressure and highest temperature anticipated for the entering wet gas, to obtain the maximum water loading. Unfortunately, the tendency is to choose a design temperature lower than that actually incurred.

The overall tray efficiency in a well-designed TEG unit will vary from 25-40%. It is recommended that 25% be used for most applications. This provides an affordable safety factor to help compensate for the inherent errors in the design specifications.

Equilibrium Relationships

Various studies have been made of the equilibrium behavior of water in the TEG-water system.(18.1, 18.3-18.6) All provide rather consistent data. The use of an activity coefficient (γ) is a convenient and reliable method for calculating K. Using this relationship

$$K = (y_w)(\gamma) = \frac{(W)(\gamma)}{B} \quad (18.3)$$

Where:

K = equilibrium constant for water in a TEG-water system

y_w = mol fr. water in the gas at saturation over 100% liquid water (from regular water content correlation)

γ = activity coefficient for water in the TEG-water system as found from Figure 18.7

W = water content on a mass per volume basis, at saturation, as found from a regular water content correlation

B = 761 000 when W = kg/10^6 std m^3 47 400 when W = lbm/MMscf

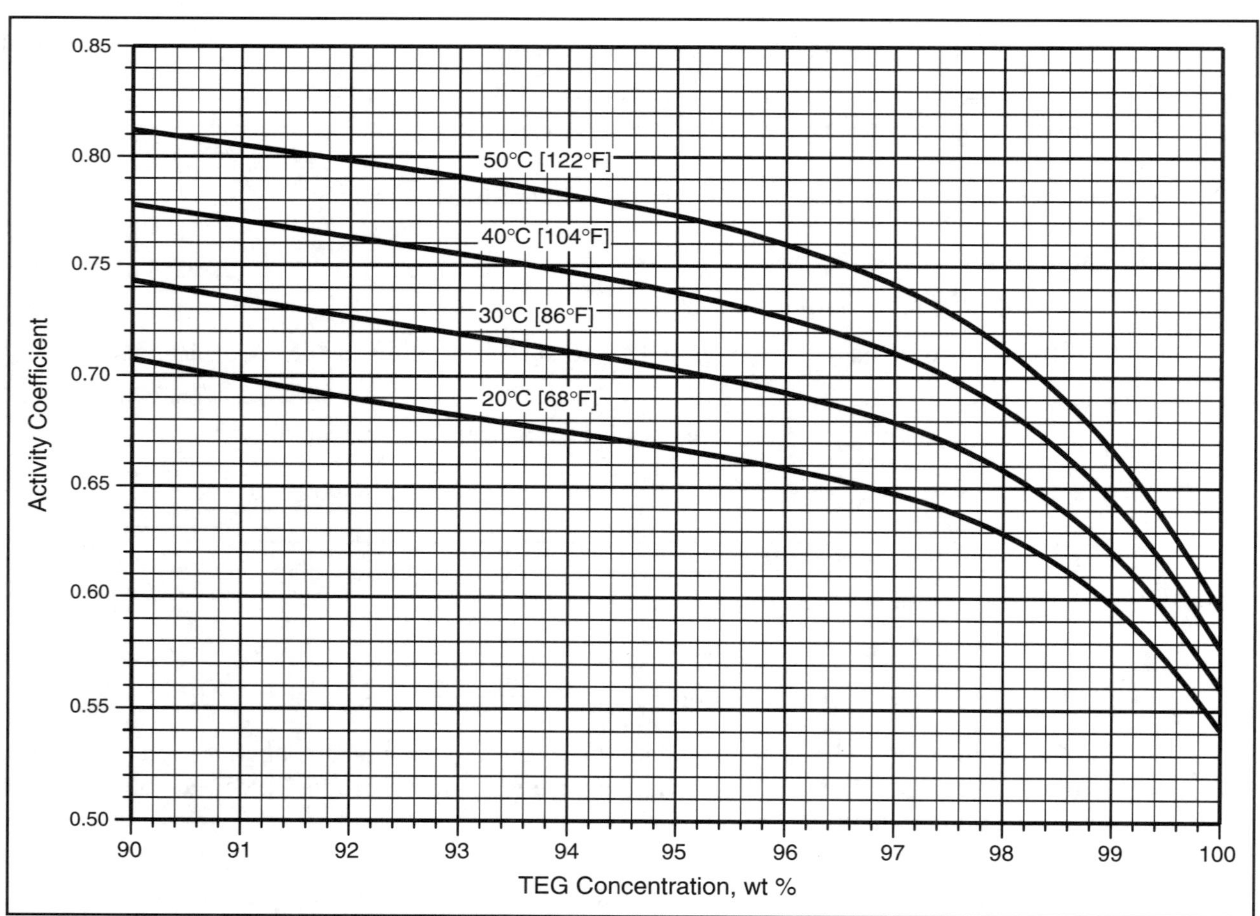

Figure 18.7 Activity Coefficient for H_2O Concentration at Various Temperatures(18.1)

Notice that γ (and thus K) varies with TEG concentration, which in turn changes across the absorber. So, a mean or average K cannot be determined until the circulation rate is fixed. This involves a trial and error calculation. In most cases both L and K are determined at top tray conditions, i.e. use the lean TEG concentration and lean TEG circulation rate. As stated previously, the increase in K from the lean to rich TEG is roughly proportional to the increase in L/V, so the absorption factor (A = L/VK) remains relatively unchanged throughout the absorber.

In the absorption efficiency term,

$$y_0 = K x_0 \quad \text{and} \quad W_0 = (W)(\gamma)(x_0) \tag{18.4}$$

Where: x_0 = mol fr. water in the lean TEG entering the absorber

This may be calculated from X_{gl}, the weight percent TEG in the lean solution entering the absorber. This must be not less than the minimum value required from Figure 18.4.

$$x_0 = \frac{\dfrac{100 - X_{gl}}{18}}{\dfrac{100 - X_{gl}}{18} + \dfrac{X_{gl}}{150}} \tag{18.5}$$

Where:
x_0 = mol fr. of H_2O
X_{gl} = wt% TEG in lean glycol

Equation 18.5 is shown graphically in Figure 18.8.

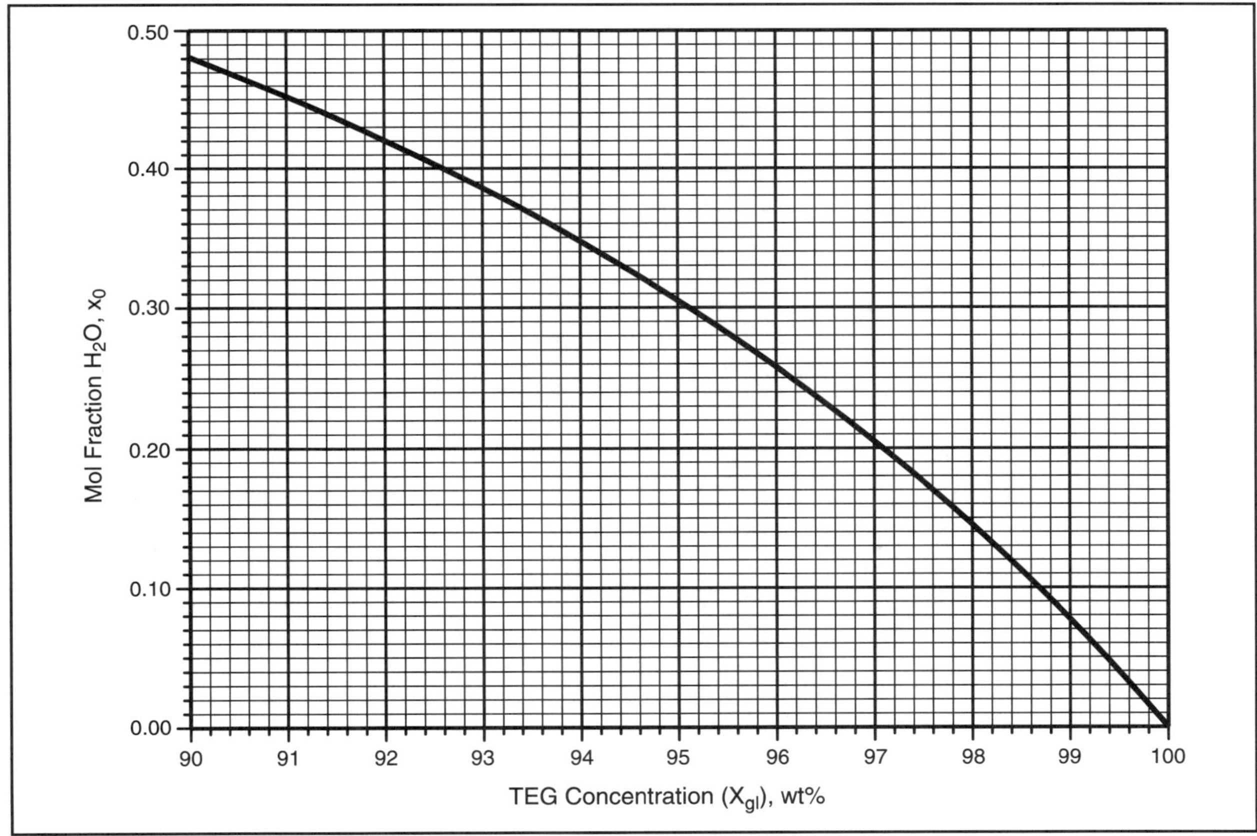

Figure 18.8 Mol Fraction H_2O, x_0, vs. TEG Concentration, X_{gl}

The above procedure using the Kremser-Brown approach gives results comparable to computer simulation.

The use of other equilibrium K values will have some effect on contactor design. The required lean glycol concentrations may differ but the difference is normally less than the random error in process specifications and is statistically insignificant.

Example 18.3: Calculate the circulation rate of 98.7 wt% lean TEG needed to dry 10^6 std m^3/d [35.4 MMscfd] of gas at 7.0 MPa [1000 psia] and 40°C [104°F] in a six tray absorber (1.5 theor. stages) to achieve an exit gas water content of 117 kg/10^6 std m^3 [7 lbm/MMscf]. The inlet water content is 1100 kg/10^6 std m^3 (saturated gas) [68.5 lbm/MMscf].

1. From Equation 18.5, $x_0 = 0.099$
2. From Figure 18.7, $\gamma = 0.66$
3. W is the water content of saturated gas at 7.0 MPa and 40°C or 1100 kg/10^6 std m^3 in this case.
4. From Equation 18.4, $W_0 = (1100)(0.66)(0.099) = 71.9$ kg/10^6 std m^3
5. The left-hand side of Figure 18.6 is:

   $$(1100 - 117)/(1100 - 71.9) = 983/1028 = 0.956$$

6. From Figure 18.6, for N = 1.5, A = 7.3
7. $L_0 = (A)(K)(V_{N+1})$

   From Equation 18.3, $K = \dfrac{(1100)(0.66)}{761\ 000} = 0.000\ 954$

   V = 1762 kmol/h

   So, $L_0 = (7.3)(9.54 \times 10^{-4})(1762) = 12.3$ kmol/h
8. MW lean glycol = (0.099)(18) + (0.901)(150) = 137 kg/kmol
9. kg TEG/h = (12.3)(137) = 1685
10. Density of TEG = 1.12 kg/liter

    Circulation rate is 1685/1.12 = 1504 liter/h

    In one hour (1100 – 117)/24 or 41.0 kg H_2O is absorbed

    Circulation ratio is 1504/41 = 36.7 liter TEG/kg H_2O absorbed

In FPS units the calculation follows the same format.

1 & 2. The same

3. W = 68.5 lbm/MMscf
4. $W_0 = (0.66)(68.5)(0.099) = 4.48$ lbm/MMscf
5. (68.5 – 7)(68.5 – 4.48) = 0.960
6. A = 7.5

Example 18.3 (Cont'd):

7. $K = \frac{(68.5)(0.66)}{47\ 400} = 0.000\ 954$

   V = (35.4)(110) = 3894 lbmol/hr

   $L_0 = AKV_{N+1} = (7.5)(9.54 \times 10^{-4})(3894) = 27.9$ lbmol/hr

8. MW = 137 lbm/lbmol
9. lbm TEG/hr = (27.9)(137) = 3817
10. Density of TEG is 9.33 lbm/US gal

    Circulation rate is 3817/9.33 = 409 US gal/hr

    In one hour a total of 90.7 lbm of water is absorbed

    Circulation ratio is 409/90.7 = 4.5 US gal TEG/lbm water absorbed

When using the Kremser-Brown method, the terms V and L must be expressed in molar units. This requires calculation of the MW of the TEG solution. The molecular weight of a TEG-water solution may be calculated as follows:

$$MW = 18x_0 + 150(1 - x_0) \tag{18.6}$$

The actual circulation rate used in the unit may be different from this because operating conditions always differ to some degree from those specified in the design. In addition, TEG circulation pumps come in discrete capacities. It is sound practice to choose other components based on the capacity of the circulation pump.

For the previous example, would we buy a 1.5 theoretical tray absorber? Probably not! The circulation rate calculated is toward the high end of the economic range. An absorber with 1.75-2.0 theoretical trays might well be specified to provide valuable flexibility and inexpensive "insurance."

The shortcut calculation method presented in this section is still somewhat tedious to do by hand. Appendix 18B presents graphical solutions to this shortcut method taking into account the water removal, circulation ratio, lean TEG concentration and number of theoretical stages in the contactor.

Example 18.4: Rework the previous example using the absorber performance curves in Appendix 18B.

SI Solution:

$$\text{Water removal} = \frac{W_{in} - W_{out}}{W_{in}} = \frac{1100 - 117}{1100} = 0.894$$

From Figure 18B.2 (N = 1.5), at 98.7 wt% TEG

The circulation ratio is 34-35 liters TEG/kg H_2O which is very close to the 36-37 value calculated in Example 18.3.

FPS Solution:

$$\text{Water removal} = \frac{W_{in} - W_{out}}{W_{in}} = \frac{68.5 - 7}{68.5} = 0.898$$

From Figure 18B.2 (N = 1.5), at 98.7 wt% TEG

The circulation ratio is 4.3-4.4 US gal/lb H_2O which is very close to the 4.5 value calculated in Example 18.3.

Absorber Design

The design of the absorber (contactor) is based on two parameters:

1. gas rate which determines the contactor diameter, and
2. number of contacts, which determine the contactor height.

In some cases the contactor may also contain an integral scrubber designed to remove entrained droplets and solids from the gas prior to entering the absorption section. In addition, an internal gas-glycol exchanger is included at the top of the contactor in some designs.

The contactor diameter depends almost exclusively on the gas rate and is virtually independent of the glycol rate. This is due to the low liquid loadings employed in glycol contactors. Two types of contactor internals are used in glycol systems — trays (usually bubble cap) and packing (usually structured).

Bubblecap trays have been historically used in TEG systems. This was certainly the case until the mid-1980s. Since then, many contactors have been installed with structured packing. This is primarily due to the superior gas handling capacity of structured packing relative to trays. Swirl Tube Trays, a proprietary Shell technology, have been recently introduced.[18.7] Shell claims significant capacity increases over structured packing.

When trayed contactors are used, bubblecaps are preferred over other types of trays (valve and sieve) due to their higher turndown ratio and generally better efficiency at low liquid rates. Likewise, when a packed contactor is employed, structured packing is preferred over random packing because of its higher capacity, better turndown and superior performance at low liquid rates. Random packing is sometimes used in small diameter contactors (less than 0.6 m [24 in]) for convenience.

Figure 18.9 shows a bubblecap tray, section of structured packing and random packings.

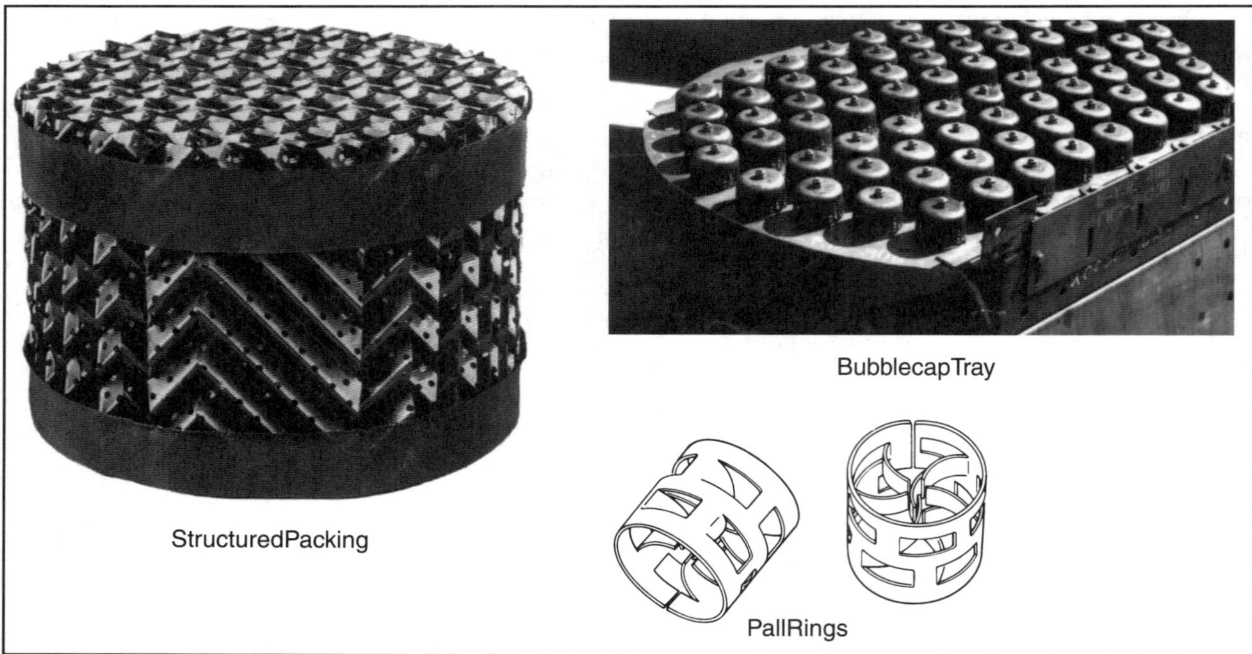

Figure 18.9 Bubblecap Tray, Random, and Structured Packing

Contactor Diameter

As stated earlier, the diameter of a glycol contactor is determined, almost exclusively by the gas rate. If a contactor is operating near flood, changes in the glycol rate can have a noticeable effect on the glycol carryover, but for design purposes the liquid rate is typically not a factor in absorber sizing.

The calculation of diameter can proceed two ways. The first method employs the Souders Brown equation (Equation 11.11), frequently used to size separators.

$$v = K_s \left(\frac{\rho_L - \rho_g}{\rho_g} \right)^{0.5} \qquad (18.7)$$

Where:			SI	FPS
	v =	allowable gas velocity	m/s	ft/sec
	K_s =	sizing parameter: bubblecaps	0.055 m/s	0.18 ft/sec
		structured packing	0.09-0.105 m/s	0.30-0.34 ft/sec
	ρ_g =	gas density	kg/m³	lbm/ft³
	ρ_L =	liquid density	kg/m³	lbm/ft³
	ρ_L =	for TEG systems	1120 kg/m³	69.9 lbm/ft³

The calculation of the diameter follows

$$d = \sqrt{\frac{4 q_a}{\pi v}} \qquad (18.8)$$

Where:			SI	FPS
	d =	contactor diameter	m	ft
	q_a =	actual gas flowrate	m³/s	ft³/sec

The actual gas flowrate can be calculated from either the mass flowrate or the standard volumetric rate.

$$q_a = \frac{m}{\rho_g} \qquad (18.9)$$

Where:			SI	FPS
	m =	mass flowrate	kg/s	lbm/sec
	ρ_g =	gas density	kg/m³	lbm/ft³

$$q_a = \frac{q_{std}}{86\,400} \left(\frac{P_{std}}{P_a} \right) \left(\frac{T_a}{T_{std}} \right) z \qquad (18.10)$$

Where:			SI	FPS
	q_{std} =	gas flowrate in standard volumes	std m³/d	scf/day
	P_{std} =	standard pressure	kPa	psia
	P_a =	actual flowing pressure	kPa	psia
	T_a =	actual flowing temperature	K	°R
	T_{std} =	standard temperature	K	°R
	z =	gas compressibility factor at flowing conditions		

A second sizing equation uses an F_s value which is related to the kinetic energy of the gas (ρv^2). This is the more popular method for sizing packed towers.

$$v = \frac{F_s}{\sqrt{\rho_g}} \tag{18.11}$$

Where:		SI	FPS
v	= allowable gas superficial velocity	m/s	ft/sec
F_s	= sizing parameter	$Pa^{0.5}$	$lbm^{0.5}/sec\text{-}ft^{0.5}$

The sizing parameter, F_s, depends on the type of packing and the packing density, but for most structured packings, $F_s = 3.0$ in SI units [2.5 in FPS units].

It should be noted that for contactors containing structured packing the gas handling capacity may be limited by the mist extractor, not the packing. This is particularly true when the glycol viscosity exceeds 15-20 cp. This is the approximate viscosity range of TEG at 32-38°C [90-100°F].

In an actual design it is sound engineering practice to size the contactor for a gas rate 20-30% higher than the expected rate. This contingency provides contactor capacity for changes in flow rate and pressure and for pessimistic reservoir engineers.

Example 18.5: A glycol contactor is to be designed to handle 1×10^6 std m^3 [35.4 MMscfd] of gas at 40°C and 70 bar [1015 psia]. The gas compressibility factor is 0.85 and the MW = 19.0. Size the contactor for both bubblecaps and structured packing.

SI Solution:

1) Calculate the gas density

$$\rho_g = \frac{(P)(MW)}{ZRT} = \frac{(7000)(19)}{(0.85)(8.314)(313)} = 60 \ \text{kg/m}^3$$

2) Calculate allowable velocity, v

Bubblecaps,
$$v = K_s\left(\frac{\rho_L - \rho_g}{\rho_g}\right)^{0.5} = 0.055\left(\frac{1120-60}{60}\right)^{0.5} = 0.23 \text{ m/s}$$

3) Calculate the actual volumetric rate, q_a

$$q_a = \frac{q_{std}}{86\,400}\left(\frac{P_{std}}{P_a}\right)\left(\frac{T_a}{T_{std}}\right)(z) = \frac{1\,000\,000}{86\,400}\left(\frac{101}{7000}\right)\left(\frac{313}{288}\right)(0.85)$$
$$= 0.154 \ \text{m}^3/\text{s}$$

4) Calculate contactor diameter, d

$$d = \sqrt{\frac{4\,q_a}{\pi v}} = \sqrt{\frac{(4)(0.154)}{(0.23)(3.14)}} = 0.92 \text{ m}$$

5) For structured packing

$$v = \frac{F_s}{\sqrt{\rho_g}} = \frac{3.0}{\sqrt{60}} = 0.39 \ \text{m/s}$$

Example 18.5 (Cont'd):

$$d = \sqrt{\frac{4\, q_a}{\pi\, v}} = \sqrt{\frac{(4)(0.154)}{(0.39)(3.14)}} = 0.71 \text{ m}$$

FPS Solution: 1) Calculate the gas density

$$\rho_g = \frac{(P)(MW)}{Z\,R\,T} = \frac{(1015)(19)}{(0.85)(10.73)(564)} = 3.75 \text{ lbm/ft}^3$$

2) Calculate allowable velocity, v

Bubblecaps, $$v = K_s\left(\frac{\rho_L - \rho_g}{\rho_g}\right)^{0.5} = 0.18\left(\frac{69.9 - 3.75}{3.75}\right)^{0.5} = 0.756 \text{ ft/sec}$$

3) Calculate the actual volumetric rate, q_a

$$q_a = \frac{q_{std}}{86\,400}\left(\frac{P_{std}}{P_a}\right)\left(\frac{T_a}{T_{std}}\right)(z) = \frac{35.4 \times 10^6}{86\,400}\left(\frac{14.7}{1015}\right)\left(\frac{564}{520}\right)(0.85)$$

$$= 5.47 \text{ ft}^3/\text{sec}$$

4) Calculate contactor diameter, d

$$d = \sqrt{\frac{4\, q_a}{\pi\, v}} = \sqrt{\frac{(4)(5.47)}{(0.756)(3.14)}} = 3.0 \text{ ft}$$

5) For structured packing

$$v = \frac{F_s}{\sqrt{\rho_g}} = \frac{2.5}{\sqrt{3.75}} = 1.3 \text{ ft/sec}$$

$$d = \sqrt{\frac{4\, q_a}{\pi\, v}} = \sqrt{\frac{(4)(5.47)}{(\pi)(1.3)}} = 2.3 \text{ ft}$$

Contactor Height

The contactor height is determined by the number of equilibrium contacts required and efficiency of the mass transfer. For trayed contactors the conversion from equilibrium stages to actual trays is accomplished by using a tray efficiency. The tray efficiency is measure of the approach to equilibrium and can be calculated from either vapor phase or liquid phase compositions.

The overall tray efficiency is defined as follows:

$$E_{overall} = \frac{\text{No. of Equilibrium Stages}}{\text{No. of Actual Trays}} \qquad (18.12)$$

For glycol contactors using bubblecap trays $E_{overall}$ typically ranges from 25-30%. (This is equivalent to a Murphree plate efficiency of approximately 45-50%). For most engineering calculations an overall tray efficiency of 25% will yield satisfactory results.

A minimum spacing of 610 mm [24 in] is recommended. It is essential that a stable foam not fill the gas space between trays to prevent excessive glycol loss. This spacing also allows a suitable liquid level in the downcomers.

Tray hydraulics design is critical because of the low circulation rate. Liquid can bypass caps or valves in some areas of the tray, ineffective gas-liquid contact can occur with low gas rates, and tray liquid levels can be unstable. In situations like this, absorber performance can vary markedly with gas and liquid rate. A higher than calculated liquid rate may be necessary to provide the tray efficiency required. A minimum weir length of 50% of the tower diameter is required to ensure adequate liquid distribution across the tray.

For packed towers, equilibrium stages are converted to packing heights using an HTU (Height of a Transfer Unit) or HETP (Height Equivalent to a Theoretical Plate (Stage)).

HETP and HTU are related concepts and depend primarily on the gas and liquid properties, gas and liquid rates as well as the surface characteristics and density of the packing.

The mass transfer in a packed tower is continuous, and does not take place in discrete steps as in trayed towers. It is for this reason that many packing manufacturers prefer to work in HTUs. The HTU is multiplied by the Number of Transfer Units (NTUs) to arrive at the packing height. For overall mass transfer controlled by resistance on the gas side, the number of transfer units may be calculated from the following equation.

$$NTU = \int_{y_t}^{y_b} \frac{dy}{y - y^*} \qquad (18.13)$$

Where:

NTU = number of transfer units
y_b = concentration of water in the bottom of the contactor
y_t = concentration of water at the top of the contactor
y^* = equilibrium water concentration

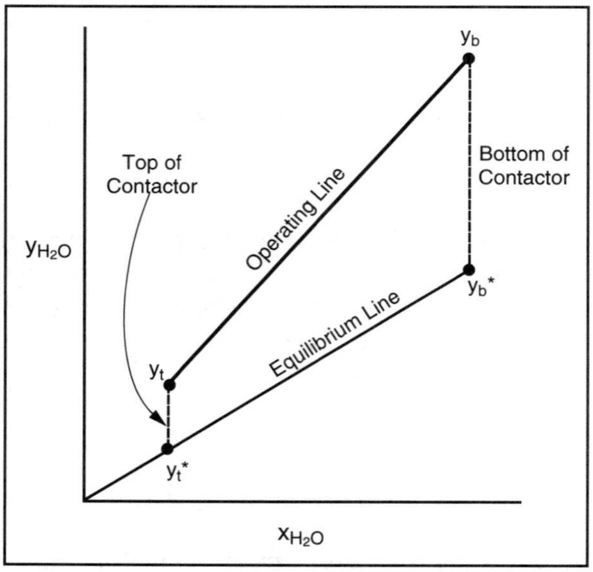

Figure 18.10 Operating and Equilibrium Lines for a Typical Glycol Contactor

A similar concept is employed in heat transfer calculations. If we assume the operating and equilibrium lines to be straight, as shown in Figure 18.10, then

$$NTU = \frac{y_b - y_t}{\Delta y_b - \Delta y_t} \ln\left(\frac{\Delta y_b}{\Delta y_t}\right) \qquad (18.14)$$

Where:

$\Delta y_b = y_b - y_b^*$
$\Delta y_t = y_t - y_t^*$

A typical HTU for structured packed towers depends on gas density and packing type but for packing with a specific area of 250 m^2/m^3 [76 ft^2/ft^3] an HTU of 0.8 m [2.6 ft] gives good results for preliminary calculations.

Despite the technical superiority of the HTU/NTU approach many companies still use the HETP/NTS method.

NTS stands for the Number of Theoretical Stages and is identical to the "N" values in Equations 18.1 and 18.2. Even though mass transfer in packed towers is continuous the HETP/NTS method is more common.

The relationship between NTS and NTU can be estimated from Equation 18.15.

$$\frac{NTS}{NTU} = \frac{HTU}{HETP} = \frac{(1 - 1/A)}{\ln A} \tag{18.15}$$

Where: A = absorption factor ($L_o/V_{N+1}K$)

For TEG contactors containing structured packing with a specific area of 250 m^2/m^3 [76 ft^2/ft^3], typical HETP values range from about 1.6-2.0 m [5.3-6.5 ft]. The HETP of structured packing varies with contactor operating parameters as follows:

Increasing Pressure	Increases HETP
Increasing Gas Flowrate	Increases HETP
Increasing Liquid Flowrate	Decreases HETP
Increasing Specific Area of Packing	Decreases HETP
Note: Increases in HETP mean poorer mass transfer Decreases in HETP mean better mass transfer	

Example 18.5: Calculate the height of packing required to provide 2 equilibrium stages in a TEG contactor.

Assume HETP = 1.75 m [5.74 ft]

Packing height = (2)(1.75) = 3.5 m

Packing height = (2)(5.74) = 11.5 ft

It is customary to add about 10% additional packing (usually 1-2 layers) to account for operating contingencies and allow for distribution of the gas and glycol at the top and bottom of the packings.

Liquid Distributor Design

In packed contactors, the design of the liquid distributor is critical. The liquid distributor ensures that the incoming lean glycol is uniformly distributed across the packing. Several different distributor designs are used but in general all consist of a main header box and a series of rectangular lateral flow channels, as shown below.

1. Distributor must be level
2. Weir openings or drip tubes must be resistant to plugging
3. Drip point density should be 80-100/m^2 [7-9/ft^2]
4. Minimum glycol circulation rate should be about 1 (m^3/h)/m^2] [0.4 US gpm/ft^2]
5. Area available for gas flow should be a minimum of 40-50% of cross sectional area
6. Drip points should be less than 10 mm [1/2 in] from the top of the packing to avoid splashing and droplet re-entrainment

Lateral Distributor
Main Header Box

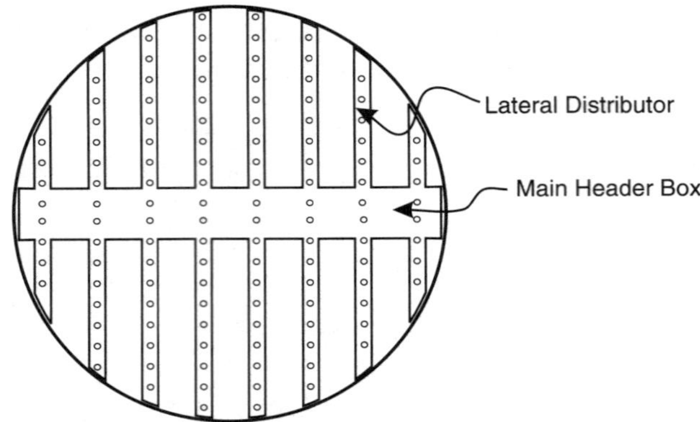

The distance between the top of the distributor and the mist eliminator should be a minimum of 0.6 m [2 ft]. An example of a structured packed tower is shown in Figure 18.11.

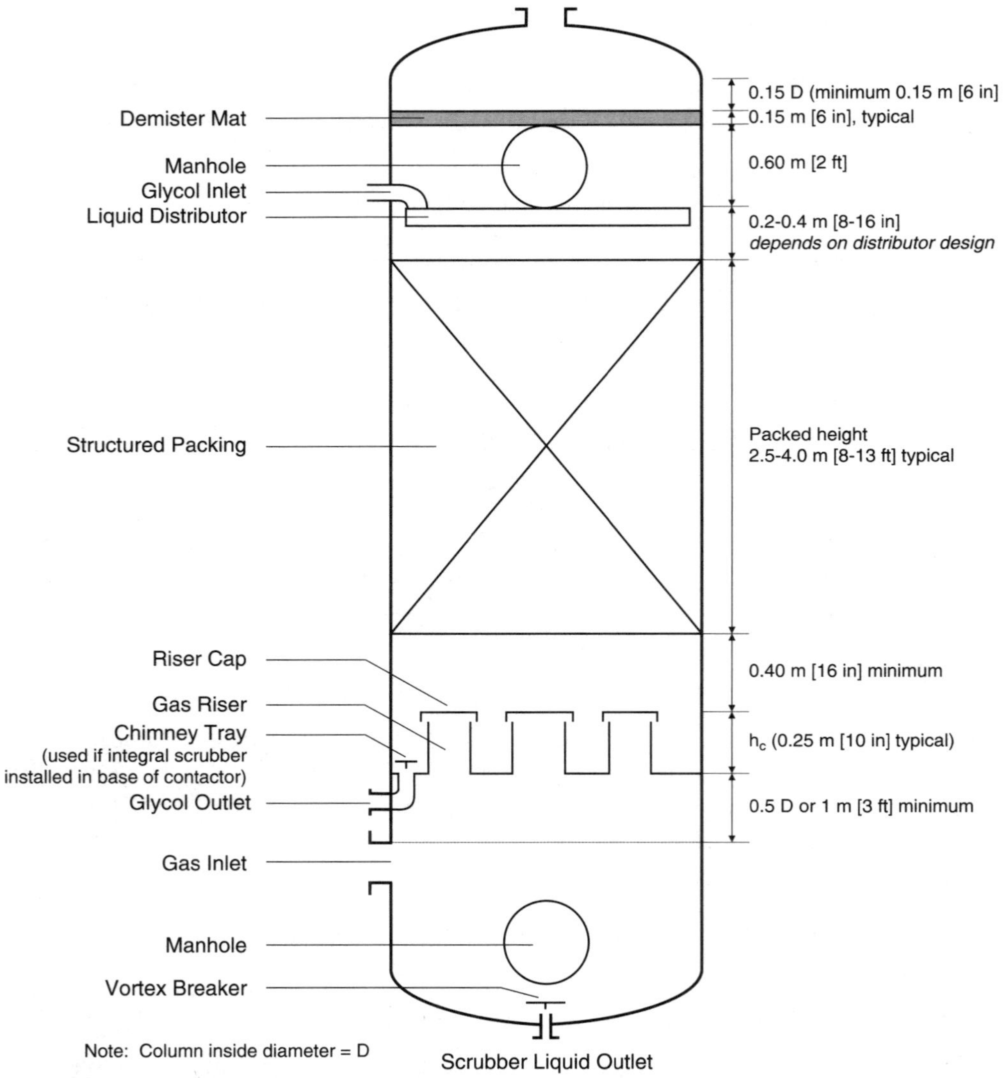

Gas Outlet
Demister Mat
Manhole
Glycol Inlet
Liquid Distributor
Structured Packing
Riser Cap
Gas Riser
Chimney Tray
(used if integral scrubber installed in base of contactor)
Glycol Outlet
Gas Inlet
Manhole
Vortex Breaker
Note: Column inside diameter = D
Scrubber Liquid Outlet
0.15 D (minimum 0.15 m [6 in]
0.15 m [6 in], typical
0.60 m [2 ft]
0.2-0.4 m [8-16 in]
depends on distributor design
Packed height
2.5-4.0 m [8-13 ft] typical
0.40 m [16 in] minimum
h_c (0.25 m [10 in] typical)
0.5 D or 1 m [3 ft] minimum

Figure 18.11 Example Structured Packed Absorber

Some vendors offer an absorption system other than a conventional vertical unit. One design uses two or three cocurrent contactors in series separated by scrubbers. The cocurrent contractors are static mixers. This design may have a lower installed weight when compared to a conventional contactor, and may offer a cost advantage when a high dewpoint depression is not required.

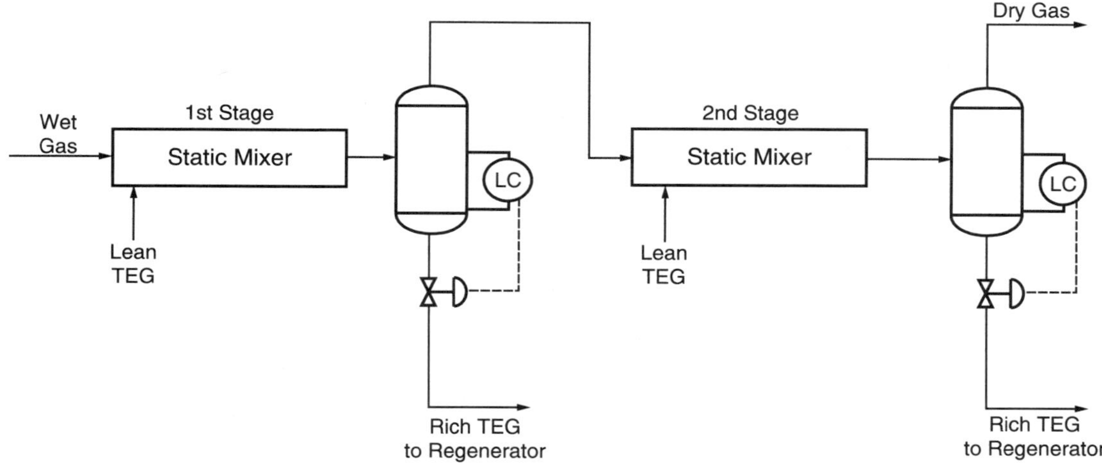

Static mixers have also been used upstream of conventional contactors to add some portion of a theoretical stage to the unit.

TEG REGENERATION

The required lean TEG concentration is produced in the regenerator. The regenerator consists of a reboiler, still column and in some cases a gas stripping column. The lean TEG concentration is controlled by adjustment of reboiler temperature, pressure and the possible use of a stripping gas. So long as no stripping gas is used, the concentration of the lean TEG leaving the reboiler is independent of the rich TEG entering.

The concentration of rich TEG leaving the absorber is found by a water material balance around that absorber. By definition

$$\text{wt\% Rich TEG} = \frac{\text{mass lean TEG}}{\text{mass lean TEG} + \text{mass water absorbed} + \text{mass water in lean TEG}} (100)$$

The mass quantities in this equation may be found per unit of time or per unit of gas or glycol flow. In any case, the values used depend on circulation rate. As we have seen in previous sections, TEG rate depends on dewpoint requirements, lean TEG concentration, number of absorber contacts and economics.

The rich TEG concentration may be calculated from Equation 18.16.

$$\text{Rich TEG} = \frac{(\rho)(\text{lean TEG})}{\rho + \left(1/\text{CR}\right)} \quad (18.16)$$

Where:

			SI	FPS
ρ	=	lean TEG density	1.12 kg/liter	9.3 lb/US gal
CR	=	circulation ratio	liters TEG/kg H_2O	US gal TEG/lb H_2O
rich TEG	=	wt% TEG in rich TEG solution		
lean TEG	=	wt% TEG in lean TEG solution		

Regeneration Parameters

Because regeneration takes place near atmospheric pressure, under essentially ideal gas conditions, the calculation is routine. Figure 18.12 has been calculated to predict regenerator performance.(18.8)

The minimum wt% lean TEG on the top abscissa is found from Figure 18.4. The wt% of rich TEG on the bottom abscissa is found from Equation 18.16. The diagonal lines in the lower left portion of Figure 18.12 represent various amounts of stripping gas for units where the stripping is sparged into the reboiler.

Three temperature lines are shown. Where high concentrations are desired, the specification of 204°C [400°F] is normal unless the gas being dehydrated contains oxygen. This is close to the thermal decomposition temperature (in air). In the usual case where the natural gas oxygen free, the use of 204°C [400°F] has proven satisfactory.

The diagonal lines at upper right in Figure 18.12 represent the effect of regeneration pressure in mmHg.

Unless a vacuum is being used, it is customary to use the 760 mmHg line for design calculations.

Notice that at 760 mmHg pressure and a reboiler temperature of 204°C, Figure 18.12 shows a lean TEG concentration of 98.7 wt%. If in using Figure 18.4 you obtain a concentration less than this, use 98.7 wt% as the desired concentration when utilizing Figure 18.12.

The general procedure for using Figure 18.12 is as follows:

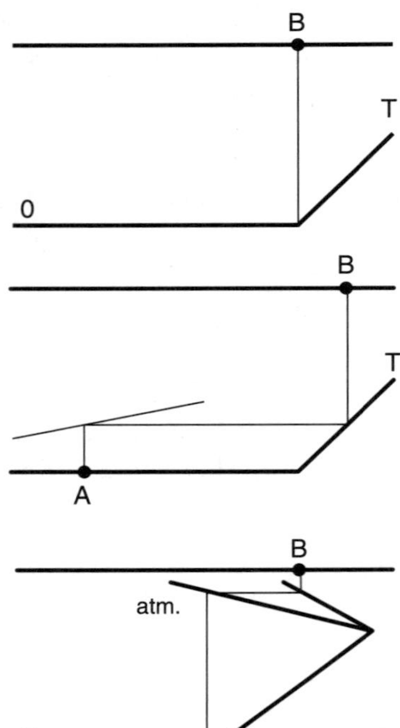

1. Atmospheric Pressure, No Stripping Gas.

   wt% rich glycol is not a variable. Proceed vertically from 0 stripping gas and temperature line intersection. You will read 98.7 wt% TEG at 204°C; 98.4 wt% at 193°C.

2. Atmospheric pressure, Stripping Gas.

   a. Proceed vertically from B to temperature line and then horizontally.

   b. Proceed vertically from A.

   c. Intersection of two lines from A and B fixes amount of stripping gas.

3. Vacuum, No Stripping Gas.

   a. Proceed vertically from intersection of 0 gas line and temperature line to atmospheric line (760 mmHg).

   b. Proceed horizontally from point in (a) to pressure line necessary to fix value of point B.

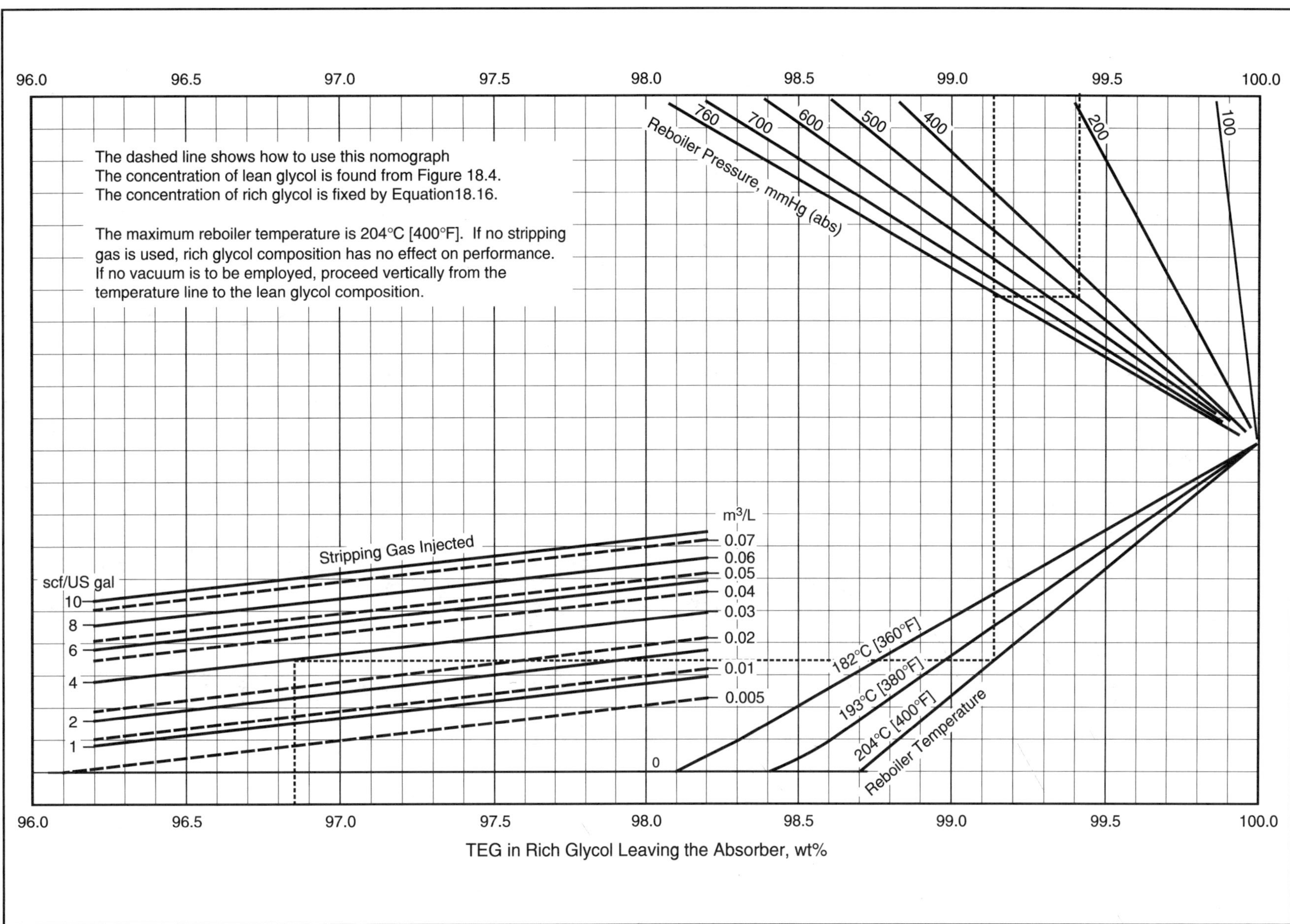

Figure 18.12 Nomograph for Estimating Regenerator Performance as a Function of Pressure, Reboiler Temperature, and Stripping Gas

In that rare case where both stripping gas and vacuum are used, procedures (2) and (3) are combined.

Example 18.4:	An example is shown in Figure 18.12 for use of stripping gas and vacuum. A 96.85 wt% rich glycol enters a regenerator using 0.03 m^3 of stripping gas per liter of glycol solution [4 scf/US gal]. Proceeding to 204°C and then vertically, one reads 99.14 wt% if atmospheric pressure is used. If a vacuum is employed and the absolute pressure is 500 mmHg, the lean glycol concentration is 99.41 wt%.

As a general rule, vacuum is avoided unless necessary to simplify unit operation. Vacuum pumps can be a nuisance and air (oxygen) ingress accelerates glycol degradation. An ejector can be used to produce necessary vacuum in the right circumstances.

Figure 18.12 is based on 1 equilibrium stage in the regenerator. Most regenerators will contain more than 1 equilibrium stage, particularly if a stripping column is installed between the reboiler and surge tank. Figure 18.13 shows the lean TEG concentration vs stripping gas rate for units employing a stripping column. In Figure 18.13, "N" refers to the number of theoretical stages in the stripping column. The approximate HETP in the stripping column is 0.8 m [2.7 ft] for 5/8 in pall rings and 0.4 m [1.3 ft] for structured packing.(18.9)

Stripping gas rates seldom exceed 75 m^3 (std)/m^3 TEG [10 scf/gal] unless lean TEG concentrations in excess of 99.99 wt% are required. If these concentrations are required, an alternate design such as DRIZO® or an adsorption system should also be considered.

Reboiler

Heat input to the regenerator is provided in the reboiler. The heat source is usually direct fired with the fire-tubes immersed in a glycol bath. Other heat sources include hot oil (or other heat transfer fluid) steam, or electric resistance heating.

In any event, the temperature of the glycol in the reboiler should not exceed 204°C [400°F] due to the degradation of TEG at higher temperatures. However, in order to maintain the glycol bath at a temperature at or slightly below 204°C [400°F] it is necessary to maintain the heat transfer surface at a temperature above this value. This can lead to some degradation of the glycol in contact with the heat transfer surface. For this reason the following maximum flux rates should not be exceeded. These flux rates, in turn, will set the heat transfer area.

Direct Fired	19 kW/m^2 [6000 Btu/hr-ft^2]
Steam	24 kW/m^2 [7600 Btu/hr-ft^2]
Hot Oil	24 kW/m^2 [7600 Btu/hr-ft^2]
Electric	12.5 kW/m^2 [4000 Btu/hr-ft^2]

Most glycol reboilers maintain the bath temperature near 204°C [400°F]. Lower temperatures may reduce degradation but result in lower lean TEG concentrations which, in turn, necessitate higher circulation rates or higher stripping gas rates.

Pressure effects on regenerator operation are not often fully appreciated. Lean TEG concentrations are reduced by backpressure on the regenerator. Typical backpressure for venting to a LP flare system or condenser system may be as much as 3.5-7 kPa [0.5 to 1 psi]. For a regenerator operating at

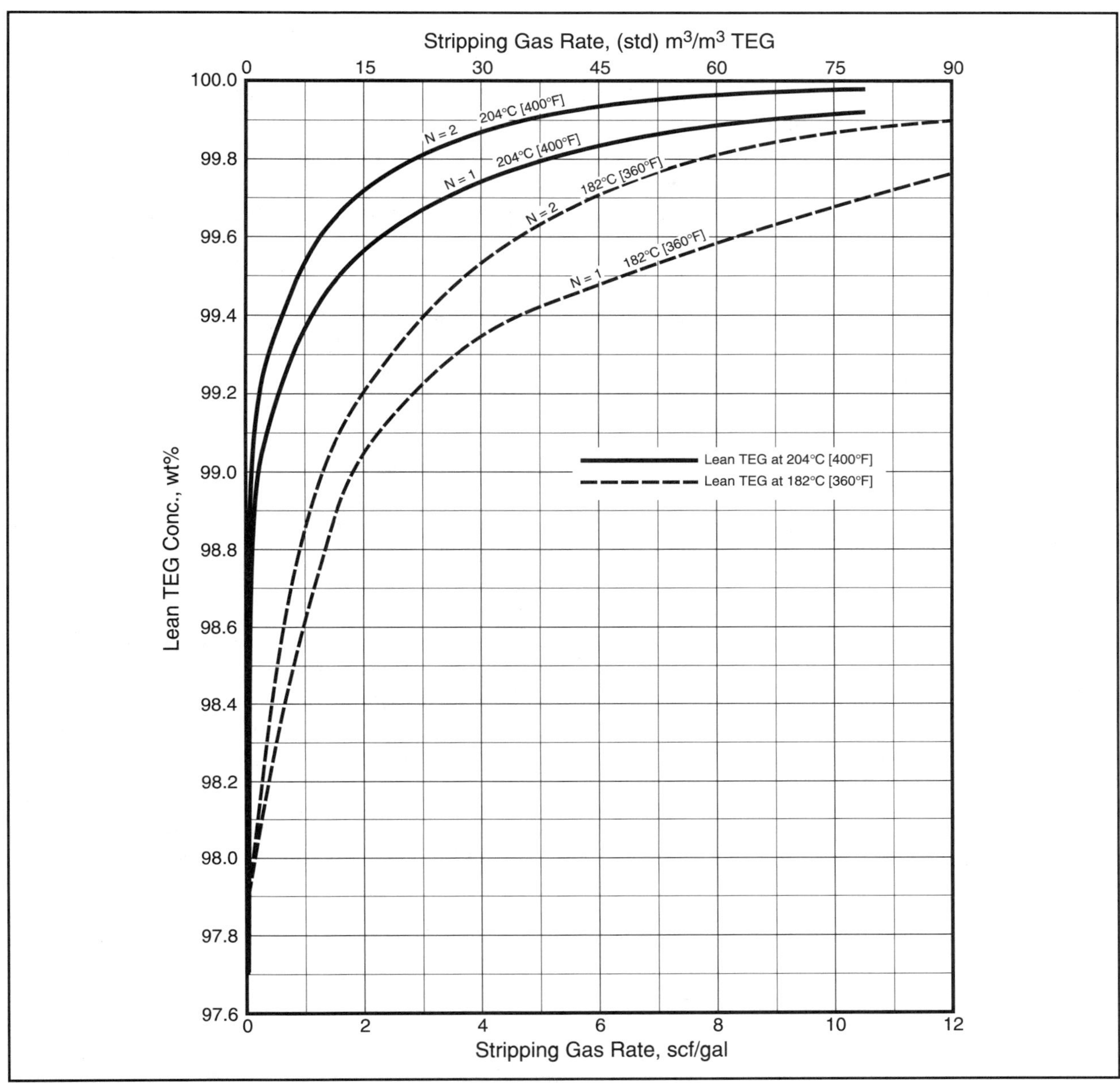

Figure 18.13 Effect of Stripping Gas on Lean TEG Concentrations for Regenerators with Stripping Columns

204°C [400°F] with no stripping gas, the effect of pressure in lean TEG concentration is about 0.014 wt %/kPa [0.1 wt %/psi].

The reboiler duty depends on the TEG circulation ratio (liters TEG/H_2O [gal TEG/lbm H_2O]), the efficiency of the rich-lean TEG exchanger, the reflux ratio, stripping gas rates and effectiveness of the insulation. Heat balances indicate a required reboiler duty of 250-300 kJ/liter [900-1075 Btu/US gal]. The reboiler duty should actually be sized to deliver 130-140% of the expected duty to provide for start-up, insulation losses, etc. A design value of 400 kJ/liter [1430 Btu/US gal] will typically provide sufficient heat input flexibility to meet any expected operating condition. The reboiler duty should always be sized based on circulation pump capacity, not the expected circulation rates.

Still Column

The still column is the "fractionator" portion of the regenerator. The column may be packed or trayed. Packed columns are more common and the packing is typically a random packing such as stainless steel slotted rings. Packing sizes range from 16 mm [5/8 in] to 51 mm [2 in] depending on the still column size with the larger packing used in the larger diameter columns. Structured packing has also been used, and in very large units (diameters > 1 m [3 ft]), trays have been installed.

The still column is sized based on standard packed tower sizing correlations. Since the vapor loading is often tied to the glycol circulation rate many correlations have been developed which estimate the still column diameter as a function of TEG circulation rate. One such correlation is shown below and is based on 25 mm [1 in] slotted ring packing.

$$d = (A)(q_{TEG})^{0.5} \tag{18.17}$$

Where:			SI	FPS
d	=	diameter of packed tower	mm	in
q_{TEG}	=	glycol circulation rate	m^3/h	US gal/min
A	=	empirical constant based on (1 in. pall rings)	210	4.0

The reflux ratio employed in TEG systems is very small. L/D values typically range from 0.1 to 0.2 mols/mol. This is equivalent to condensing 10-20% of the total overhead vapor stream. The reflux rate should be the minimum required to maintain the still overhead temperature at the boiling point of water for the partial pressure of water at the top of the regenerator. When stripping gas is used, the water partial pressure will be less than 1 atm. In fact, this is the principle which results in lower water concentrations in the lean TEG. Consequently, when stripping gas is used, the partial pressure of H_2O will be less than 1 atm, hence the boiling point of water will be lower as well.

Figure 18.14 shows the recommended still column overhead temperature as a function of the TEG circulation ratio and stripping gas rate for a still column operating at 1 atm.

Reflux is normally supplied by using the rich glycol stream circulating through a condensing coil inserted in the top of the still column. Cooling water or a fin-fan cooler could also be used. However, using the rich glycol stream is the preferred method for cost, simplicity and energy efficiency. Heat transfer coefficients in the reflux coil typically range from 110-230 $W/m^2 \cdot K$ [20-40 $Btu/hr\text{-}ft^2\text{-}°F$].

Glycol-Glycol (Lean-Rich) Heat Exchanger

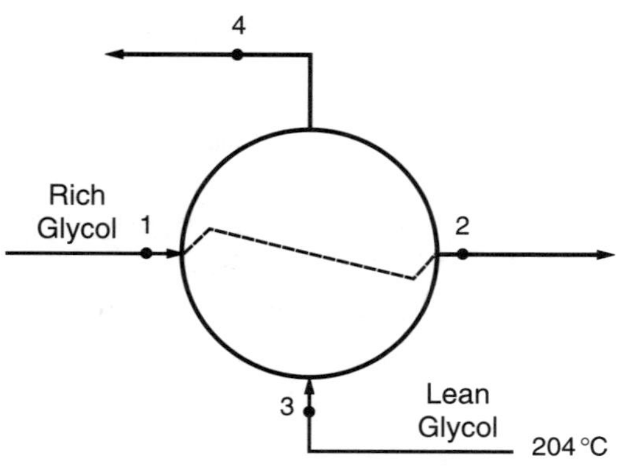

This is a basic heat exchanger. Its efficiency has a direct effect on reboiler heat load. The rich glycol from the absorber (Pt.1) enters at a temperature 5-13°C [9-23°F] warmer than the inlet gas due to the reflux condenser duty. The lean glycol from the regenerator (Pt. 3) enters usually at about 204°C [400°F]. The exchanger is designed so the temperature of the lean glycol at Pt. 4 should not be greater than 60-65°C [140-149°F].

In most cases a 15-20°C [27-36°F] approach in the heat exchanger is desirable. If it is too high, the reboiler and glycol cooler duties will increase.

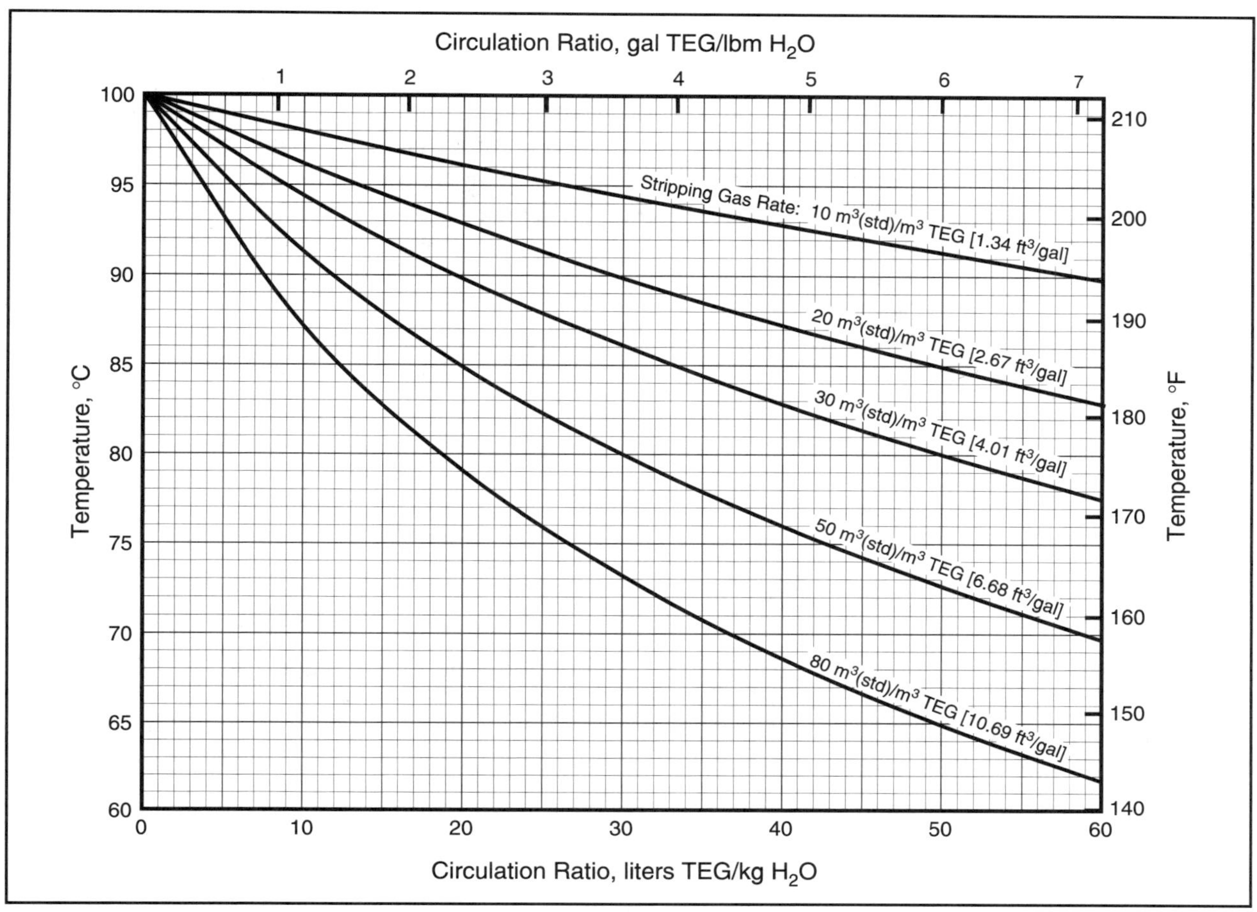

Figure 18.14 Recommended Still Column Top Temperatures

In the lean-rich exchanger

$$Q_{3-4} = Q_{1-2} = m_1(h_2 - h_1) = m_3(h_3 - h_4) \qquad (18.18)$$

The easiest way to perform this calculation is to look up the average heat capacity of the glycol in Appendix 18A and multiply it by ΔT across the heat exchanger to find Δh for the lean mixture. Remember $\Delta h = C_p \Delta T$.

Double-pipe type, with bare or finned tubes, or plate-type heat exchangers are frequently used. Plate exchangers are preferred, especially for offshore or large units, because they are more compact, lighter and cheaper. They are, however, susceptible to fouling and plugging and it is imperative the glycol is clean and filtered.

Some small dehydration units use a coil in the surge drum to exchange heat between the rich and lean glycol stream. Not only are these coils limited in surface area but the overall 'U' values tend to be low. As a result they may not be adequate for applications where the available reboiler heat is limited. They also may not cool the lean glycol adequately to meet the pump temperature limit. They should only be considered for small packaged units having a reboiler duty less than 150 kW [500 MBtu/hr].

Lean Glycol Cooler

A final glycol cooler is required so that the lean glycol entering the top of the contactor is cooled to within 5-10°C [9-18°F] of the gas temperature entering the top tray. The lean glycol is often cooled with a lean glycol-gas heat exchanger. Gas-glycol exchangers are inexpensive and compact but the lean TEG temperature can increase to unacceptable levels at low gas rates. Although air or water-cooled glycol coolers risk over-cooling the lean glycol, they are often used since control of the glycol temperature is then independent of the gas flow.

If a gas-glycol exchanger is used we do not recommend the type which employs an integral coil in the top of the absorber column due to less efficient heat transfer and problems of inspection and maintenance.

FILTERS

Good filtration is critical. The full-flow type is preferred. I recommend two filters in parallel, with no by-pass lines, so that full filtration is assured.

A cloth fabric element that is capable of reducing solids to about 100 ppm by weight is preferred. Paper and fiberglass elements generally have proven unsatisfactory. Filter size in a properly operated glycol system should be 5-10 μm. Larger sizes (25-50 μm) may be required during startup and in dirty service.

It may be impossible to judge the effectiveness of filtration by color alone. Even well filtered glycol will often be black. But, removal of most of the solids will reduce corrosion, plugging and deposits in the reboiler, and may reduce foaming losses. Good filtration is critical for satisfactory performance. It is desirable to measure the pressure differential across the filter and change the elements when it reaches the filter supplier's recommended maximum ΔP which is often 170 kPa [25 psi].

The use of a carbon purifier downstream from the filter often is recommended. This can produce essentially water-white glycol. Maintenance of this color has proven desirable because it tends to increase dehydration efficiency and minimize foaming, a major source of glycol loss.

Aromatic hydrocarbons are often present in the rich glycol entering the regenerator and will be adsorbed on the carbon filter. These components will quickly reach equilibrium loading on the carbon filter although it is likely they are eventually displaced (to some extent) by heavier hydrocarbons. Because of this high aromatic content, changeout of carbon filters requires special precautions to avoid unnecessary exposure of workers to BTEX components.

In some units, the carbon filter is installed in the lean TEG stream, upstream of the contactor, to avoid significant aromatic loading.

Coal-based activated carbon should be used because wood-based charcoal tends to break up in use. This carbon can be placed in a metal canister or installed as fill into a vessel. In either case, good screens are needed to prevent carbon loss into the system. Carbon particles, much like iron sulfide, tend to promote a stable foam.

Glycol filters are only effective when used. In especially dirty glycol systems, filters are often bypassed to avoid frequent filter change-out. The problems with this should be obvious. If filter plugging is excessive, try larger filter size and look for source of problem (e.g., poor inlet separation, degradation, corrosion products, etc.).

SURGE DRUM

The surge drum should be sized to provide the following:

- a retention time not less than 20 minutes between low and normal levels, based on design circulation rate;
- hold-up capacity between normal level and high level;
- a reasonable length of time between glycol additions;
- sufficient volume to accept the glycol drained from the reboiler to allow repair or inspection of the firetube or heating coil.

It is often vented to the reboiler but a small amount of N_2 or dry fuel purge gas is sometimes needed to prevent water vapor from the reboiler flowing through the vent line and being absorbed by the glycol in the surge drum. If it is not located directly below the reboiler, it should be provided with a separate purge gas supply.

Provisions should also be made to facilitate:

- make-up of the glycol inventory from glycol storage by means of a simple hand or air-driven pump with a check valve and filter;
- the batch addition of chemicals to the glycol surge vessel, e.g. for pH control, corrosion inhibition, etc.

GLYCOL FLASH VESSEL

The glycol flash vessel is used to remove light hydrocarbons, CO_2 and/or H_2S, that have been absorbed or entrained with the glycol as well as recover the spent gas from gas-glycol powered pumps. It also serves to separate any liquid hydrocarbons from the glycol to prevent them from entering the reboiler and causing fouling and foaming.

Both the sulfur compounds and carbon dioxide are very soluble in water and react to some degree with the glycols. Degassing in the flash vessel upstream of the regenerator reduces their concentration somewhat and helps mitigate corrosion in the regenerator. Degassing is more efficient if the glycol is preheated first. Preheating, to about 60-70°C [140-158°F] is often done to decrease viscosity and facilitate degassing.

When gas-glycol powered pumps are used, the operating pressure of the flash vessel must be compatible with the maximum pump exhaust pressure which is 15 percent of the contactor operating pressure. Thus, with 70 bar [1015 psia] contactor pressure the flash vessel must operate at a pressure lower than 10 bar [145 psia].

The flash vessel should not contain liquid hydrocarbons. Unfortunately this is often not the case due to poor inlet gas separation upstream of the contactor. Therefore, it is prudent to install a hydrocarbon skim nozzle, bucket or trough and weir to collect and separate the liquid condensate from the rich glycol.

GLYCOL CIRCULATION PUMPS

Glycol circulation pumps may be electric driven, gas, or gas-glycol powered. The pumps should be sized to provide a minimum of 25 percent excess capacity, and in critical service two glycol pumps shall be provided, each designed for 100 percent duty.

Pumps utilizing conventional electric motor drives are normally reciprocating multiplex type. A conservative, slow piston speed (0.6 m/s [120 ft/min]) pump should be used since the lubricating properties of glycol are poor. Due to the expected turn-down of glycol systems, variable speed drives are often used on larger units. They can provide the flexibility to increase the glycol circulation rate if needed to meet dewpoint requirements and reduce the glycol rate to operate more economically. Low speed centrifugal booster pumps are sometimes used when NPSHA to the reciprocating pumps is marginal.

The gas-glycol powered pump (Kimray pump) utilizes the rich glycol under pressure plus supplemental gas from the contactor to furnish the driving energy. Since the pumping rate is proportional to the volume of the return glycol and gas, the pumping rate is controlled by adjusting this flow.

A cutaway of a Kimray pump is shown in Figure 18.15.

INLET SEPARATION

Without a doubt, most operating problems with glycol dehydration systems are a direct result of inadequate inlet gas treatment upstream of the contactor. The upstream separator should remove liquid hydrocarbons, liquid water, solids, corrosion inibitors, etc. prior to the glycol unit. John Campbell Sr. says "You cannot afford the dehydration if you cannot afford to place an effective separator on the gas inlet." A glycol unit is a closed system. Non-volatile contaminants remain in the glycol and must be removed by filtration or blowdown.

Some of the more deleterious contaminants are salt and high boiling point hydrocarbons. These remain in the glycol. Salt precipitates in the reboiler, still column and rich-lean exchangers. This can cause plugging increasing pressure drop and decreasing flowrates. In the reboiler, salt can coat the firetube causing hot spots and eventual firetube failure.

Heavy hydrocarbons can "coke" in the reboiler causing hot spots on the firetube and plugging in the still and stripping columns. In addition, these burnt hydrocarbons cause the glycol to become black and promote foaming in the contactor, and cause frequent filter change outs. If sulfur compounds are present in the gas, these can combine with the heavy hydrocarbons to form a corrosive "sludge" which accumulates in the system.

Certainly, a properly sized impingent separator is the first step in preventing unwanted carryover. Many units also utilize coalescing filter separators and/or superheaters downstream of the primary separator. Frequently, a heat exchanger designed to increase the gas temperature by 5°C [9°F] is installed immediately upstream of the contactor. This serves to vaporize any entrained liquid particles in the gas.

Many designs utilize an integral scrubber installed in the base of the contactor. While this may provide some removal of entrained particles, these "scrubbers" are typically under-sized, particularly when the contactor is packed with structured packing. These integral scrubbers should never be used as a primary separator — only for secondary scrubbing.

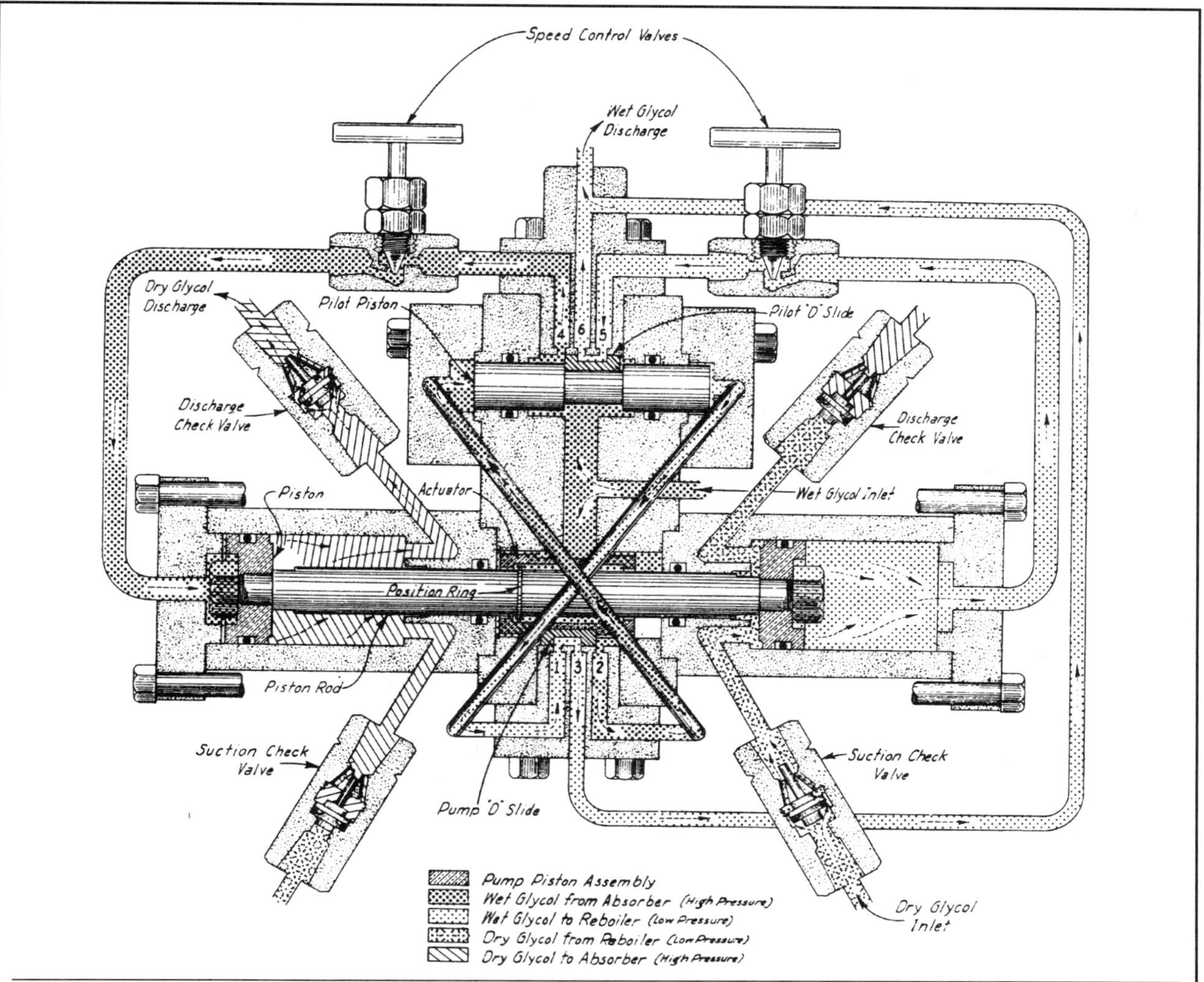

OPERATION: The Kimray glycol pump is double acting, powered by Wet Glycol and a small quantity of gas at absorber pressure.

Wet Glycol from the absorber flows through port #4 and is throttled through the SPEED CONTROL VALVE to the left end of the Pump Piston Assembly, moving this assembly from left to right. Dry Glycol is being pumped from the left cylinder to the absorber while the right cylinder is being filled with Dry Glycol from the reboiler. At the same time Wet Glycol is discharging from the right end of the Pump Piston Assembly to a low pressure or atmospheric system.

As the Pump Piston Assembly nears the end of its stroke, the POSITION RING on the PISTON ROD contacts the right end of the ACTUATOR. Further movement to the right moves the ACTUATOR and PUMP "D" SLIDE to uncover port #1 and communicate ports #2 and #3. This exhausts Wet Glycol from the left end of the PILOT PISTON through ports #2 and #3 to the low pressure Wet Glycol system. At the same time port #1 (which was communicated with port #3) admits Wet Glycol to the right end of the PILOT PISTON. This causes the PILOT PISTON and PILOT "D" SLIDE to be driven from right to left.

In its new position the PILOT "D" SLIDE uncovers port #5 and communicates ports #4 and #6. This exhausts Wet Glycol from the left end of the Pump Piston Assembly through ports #4 and #6 to the low pressure Wet Glycol system. Port #5 (which was communicated with port #6) now admits Wet Glycol through the right hand SPEED CONTROL VALVE to the right end of the Pump Piston Assembly.

The Pump Piston Assembly now starts the stroke from right to left. It follows the above procedure with reversed directions of flow.

Figure 18.15 Cutaway and Operation of a Kimray Glycol Pump

OPERATING PROBLEMS

The glycol unit operation should be essentially trouble-free. It seldom is. Many of the problems stem from inadequate design and/or operational faults. The basic simplicity of the unit and the availability of "standard" units tends to obscure the need for attention to mechanical design details. The glycols are chemically reactive and need to be protected from contamination.

One common symptom of many problems is excess glycol loss. This loss is due to one, or a combination, of the following:

1. Foaming
2. Degradation
3. Salt plugging the regenerator still column
4. Inadequate mist extraction
5. Inadequate absorber design for flow conditions
6. Loss of glycol from pinholes in a gas-glycol coil in the top of the absorber or in a chimney tray above a separator section in the bottom of the absorber
7. Spillage of glycol or pump leakage
8. Lean glycol to absorber is too hot
9. Inadequate reflux (temperature too high at top of still column)

Glycol likes to foam. It will foam whenever allowed to. Ordinary foaming may not be critical if the unit is carefully designed. Any foam tends to be more stable when aromatics and/or sulfur compounds are present. Metallic sulfides and sulfites, and degradation products, all contribute to the problem.

Foams are only broken using surface and time, or chemicals. Tray spacing must be large enough so that foam cannot fill up the space between trays and form a continuous liquid phase. A mist extractor does not break foam effectively. Once foam fills the absorber, there is a continuous liquid phase for glycol to go out overhead.

Use of an antifoam agent can reduce the problem. There are many antifoam agents available. One that works in one unit may not work equally well in another. Some trial-and-error testing of an antifoam agent, and concentration of that agent, is often necessary.

Avoid adding too much antifoam agent. If too much is added it may accelerate foaming. Set up a careful control policy so operators keep unit concentration within the limits specified.

Some degradation is natural, but it is accelerated in the presence of oxygen and sulfur compounds. The answer is effective filtration. Degradation products contribute to foaming but they also are major sources of corrosion problems.

Salt is a continuing problem. Good separation ahead of the absorber is mandatory. Any salt arriving at the regenerator deposits either in the still column, reboiler, or rich-lean glycol exchanger. It is common for packed still columns to plug up to the point that glycol is lost overhead. Even if this does not occur, salt can precipitate on the heat transfer surface in the reboiler and cause failure. Not providing good separation is inexcusable.

The water vapor in gas is relatively fresh but can be slightly saline. NaCl is soluble in TEG to some degree. At 50°C about 3.3 kg will dissolve in 100 kg of TEG. So, some salt is always present. A portion of the soluble salt may hydrolyze to HCl and lower the pH of the glycol.

Corrosion-Erosion

Glycols can react with sulfur compounds. The resultant materials tend to polymerize and form "gunk" which is very corrosive. Also, the glycol pH becomes lower. Corrosion inhibitors alone cannot solve the problem satisfactorily. The real solution is good mechanical design and good filtration supplemented by a corrosion inhibitor.

Good design involves factors like control of fluid velocities, long radius ells and a host of little details that too seldom are done properly. In many cases, good mechanical design will eliminate the need for expensive alloy steels.

If feasible to do so, the glycol pH should be maintained above 6.0. Some become so preoccupied keeping it at 7.0 or above that they add copious amounts of caustic, sodium carbonate and the like to the unit. The result is seldom satisfactory. Adjustment of pH is proper but the cure can be worse than the disease if it is overdone.

Corrosion inhibitors which plate out on metal surfaces and form a film can be effective in minimizing corrosion. Judicious use of stainless steel can also be effective in minimizing corrosion. This is particularly true for dehydration of high CO_2 gases. When the CO_2 partial pressure is high, the rich TEG pH can be less than 5. It is often more effective to use stainless steel on the "rich" side of the unit than to try to combat corrosion with inhibitors.

In a corrosive environment, the total elimination of corrosion is an unrealistic goal. The proper goal is reducing it to economically tolerable levels.

Figure 18.16 shows the solubility of H_2S in TEG.(18.10) This is true absorption that takes place in the absorber. Figure 18.17 shows solubility of CO_2 in a 96.5 wt% TEG solution.(18.11) Solubility of

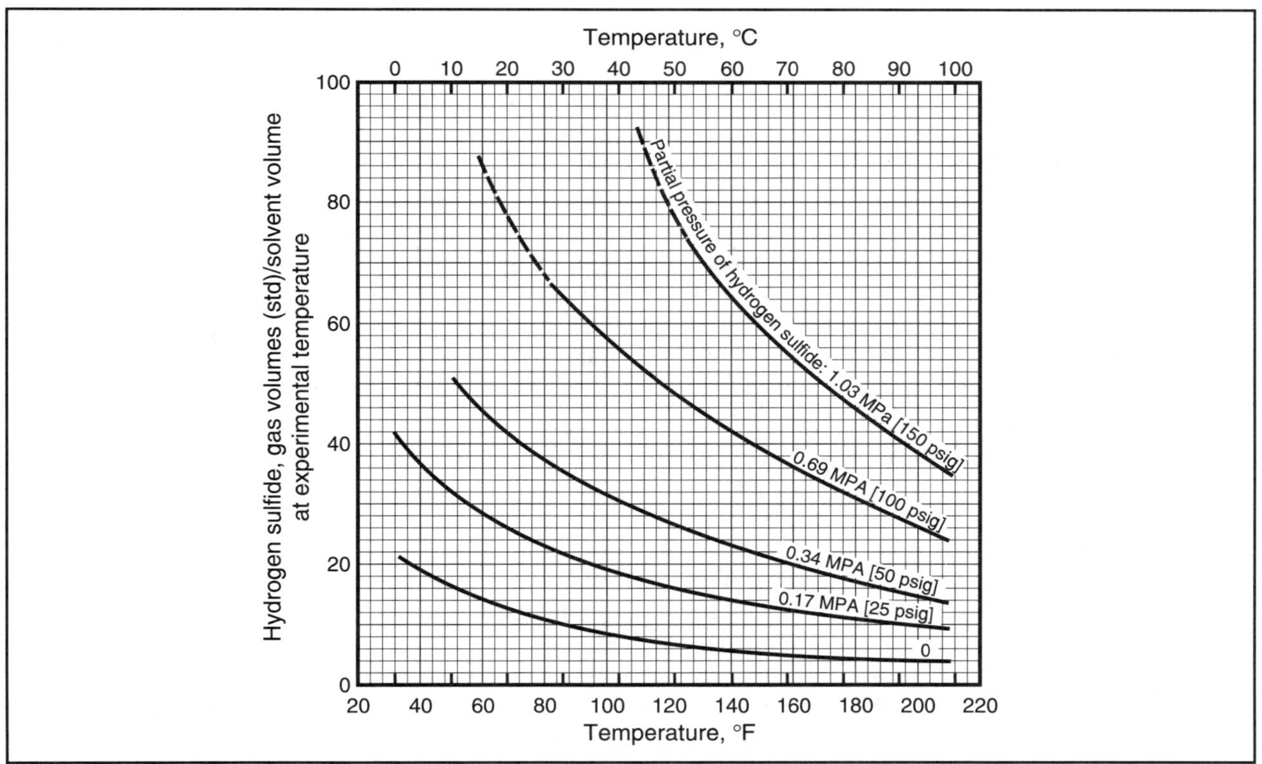

Figure 18.16 Solubility of Hydrogen Sulfide in TEG at Various Partial Pressures

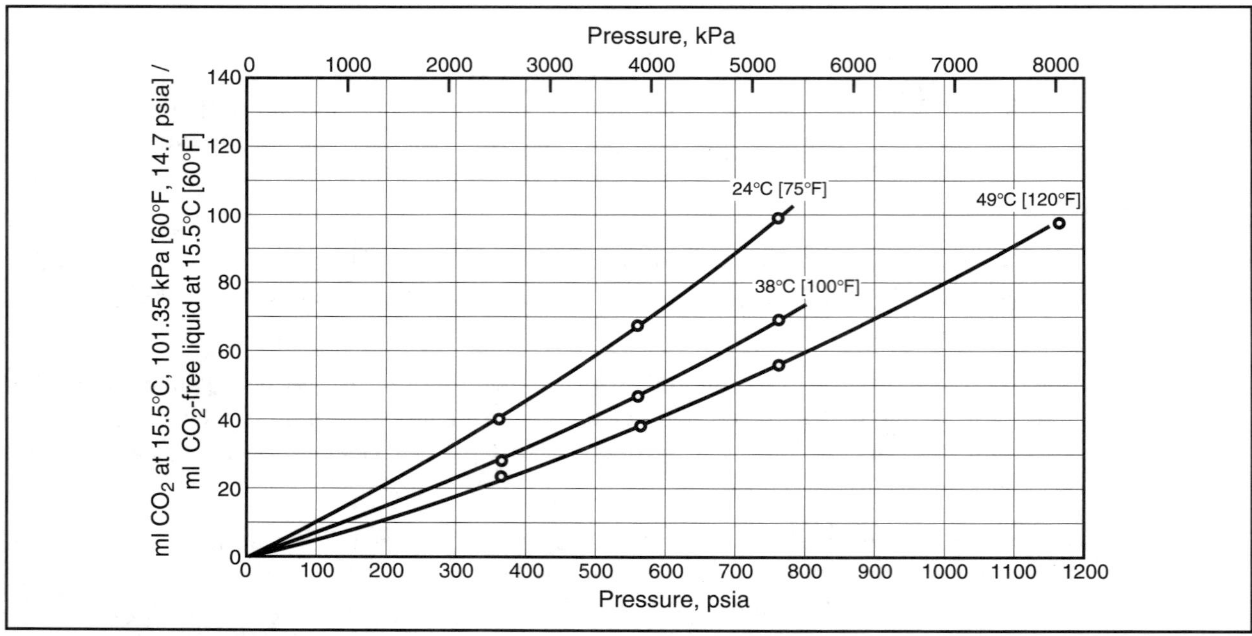

Figure 18.17 CO_2 Solubility in TEG 3.5 wt% H_2O

CO_2 in pure TEG is approximately 20% higher. Reference 18.12 provides additional data on the solubilities of H_2S and CO_2, as well as C_1, C_2 and C_3 in TEG.

AROMATIC ABSORPTION

The affinity for aromatics by TEG has long been recognized. The UDEX process was used for many years in refineries and chemical plants to extract aromatic hydrocarbons from paraffins with TEG.

In gas dehydration service, TEG will absorb limited quantities of aromatic hydrocarbons (benzene, toluene, ethylbenzene and xylene) from the gas. These components are often abbreviated as BTEX. Quantifying the absorption levels has become more important in recent years due to increased restrictions on aromatic emissions from glycol units. Published equilibrium data is available.[18.13,18.14]

Based on data from Reference 18.14, predicted absorption levels for BTEX components vary from 5-10% for benzene to 20-30% for ethylbenzene and xylene. Figure 18.18 shows approximate absorption percentages for BTEX components vs. TEG circulation rate and contactor temperature at 6900 kPa [1000 psia]. Figure 18.18 was generated using the process simulator PROSIM®. Absorption is favored at lower temperatures, increased TEG concentration and circulation rate. Pressure does not have a strong effect on aromatic absorption. Absorption can actually decrease at higher pressures due to retrograde behavior.

The bulk of absorbed aromatics will be vented with the water vapor at the top of the regenerator. In some cases these aromatics can be condensed and recovered, however the effectiveness of condensation declines rapidly with increasing stripping gas rates.

In these cases, many operators use a partial condenser to condense the water and pipe the volatile hydrocarbons to an incinerator or to the reboiler as fuel. Back-pressure on the regenerator may be minimized by the use of an eductor (which is often the fuel gas valve on the reboiler) or by a downward-sloping water-cooled condenser.

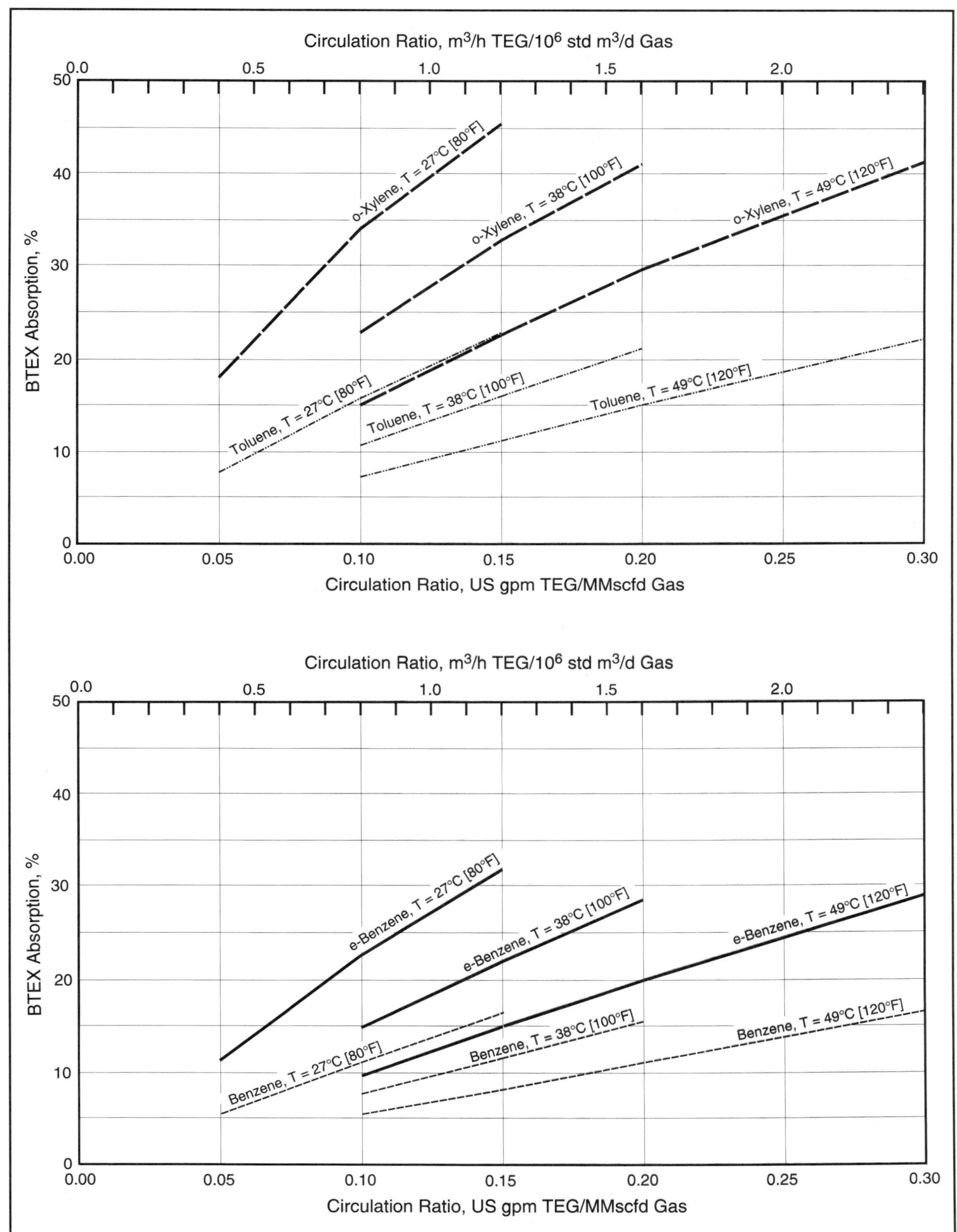

Figure 18.18 Approximate BTEX Absorption in TEG

Another mitigation measure which looks promising is to strip the aromatics from the TEG with a small stripping gas stream at the flash tank. The stripping vapors are then at a sufficiently high pressure to allow recycling to a compressor or use in a remote fuel gas system.

Others have proposed using membranes or activated carbon but these methods have not proven commercial.

This problem is one which requires careful attention in the design phase. Environmental considerations are increasingly driving the selection and operation of process equipment. In some cases the use of a dry desiccant unit (albeit at a higher capital and operating cost) may be a lower cost alternative to glycol given the environmental impact of these BTEX emissions.

REFERENCES

18.1 Parrish, W. B., et al., Proceedings GPA 65th Annual Meeting, San Antonio, Texas (1986).

18.2 Worley, M. S., Gas Cond. Conf., Univ. of Oklahoma (Apr. 1967).

18.3 Townsend, F. M., Ph.D. Thesis, Univ. of Oklahoma, Norman, Oklahoma (1955).

18.4 Scauzillo, F. R., *Jour. Petr. Tech.* (July 1961), p. 697.

18.5 Rosman, A., Ibid. (Oct. 1973), p. 297.

18.6 Worley, M. S., *Cdn. Petr.* (June 1967), p. 34.

18.7 Bravo, J. Wallace, C., and M. Curole, "Mission to Mars: Double the Capacity of a Glycol Contactor Beyond that Available with Structured Packings," Proceedings Laurance Reid Gas Conditioning Conference, Norman, Oklahoma (1998).

18.8 Perry, C. R., personal communication.

18.9 Hubbard, R. A., "An Independent Appraisal of Gas Dehydration Processes," 1997 European GPA, Antwerp, Belgium.

18.10 Blake, R. J., *Oil and Gas J.* (Jan. 9, 1967), p. 105.

18.11 Takahashi, S., et al., GPA Technical Publication TP-9 (1982).

18.12 Jou, F.Y., et al., Fluid Phase Equilibrium, 36 (1982), p. 121.

18.13 Fitz, C. W. and R. A. Hubbard, "Quick Manual Calculation Estimates Amount of Benzene Absorbed in Glycol Dehydrator," *Oil and Gas J.* (Nov. 23, 1987), p.72.

18.14 Robinson, D.B., Chen C.-J., and H-J Ng, RR-131, "The Solubility of Selected Aromatic Hydrocarbons in TEG", Gas Processors Association (May 1991).

APPENDIX 18A

TABLE 18A.1

Physical Properties† of Glycols

	Ethylene Glycol	Diethylene Glycol	Triethylene Glycol	Tetraethylene Glycol	Propylene Glycol[2]
Formula	$C_2H_6O_2$	$C_4H_{10}O_3$	$C_6H_{14}O_4$	$C_8H_{18}O_5$	$C_3H_8O_2$
Molecular weight	62.1	106.1	150.2	194.2	76.1
Boiling Point[1] at 760 mm Hg °F	387.3	473.8	550.0	618.1	369.3
Boiling Point[1] at 760 mm Hg °C	197.4	245.5	287.8	325.6	187.4
Vapor Pressure[1] at 77°F [25°C]	<0.1	<0.01	<0.01	<0.01	0.13
Density (g/cc) at 77°F [25°C]	1.110	1.111	1.120	1.123	1.032
(g/cc) at 140°F [60°C]	1.085	1.087	1.093	1.096	1.006
lbm/gal at 77°F [25°C]	9.26	9.27	9.35	9.37	8.62
Freezing Point, °F	7.9	16.4	19.0	15.1	Supercools
Freezing Point, °C	–13.4	–8.7	–7.2	–9.4	Supercools
Viscosity in centipoise at 77°F [25°C]	16.9	25.3	39.4	43.0	48.6
at 140°F [60°C]	5.2	7.3	10.3	11.4	8.42
Surface Tension at 77°F [25°C] dynes/cm	48	44	45	–	36
Refractive Index at 77°F [25°C]	1.430	1.446	1.454	1.457	1.430
Specific Heat at 77°F [25°C] Btu/lb/°F	0.58	0.55	0.52	0.52	0.60
Flash Point, °F (PMCC)[3]	247	281	325	400	220
Fire Point, °F (C.O.C.)[4]	245	290	330	375	220

†Note: These properties are laboratory results on pure compounds or typical of the product, but should not be confused with, or regarded as, specifications.

[1]Pure compound

[2]Available in industrial and U.S.P. grades

[3]Penskey-Martins Closed Cup

[4]Cleveland Open Cup

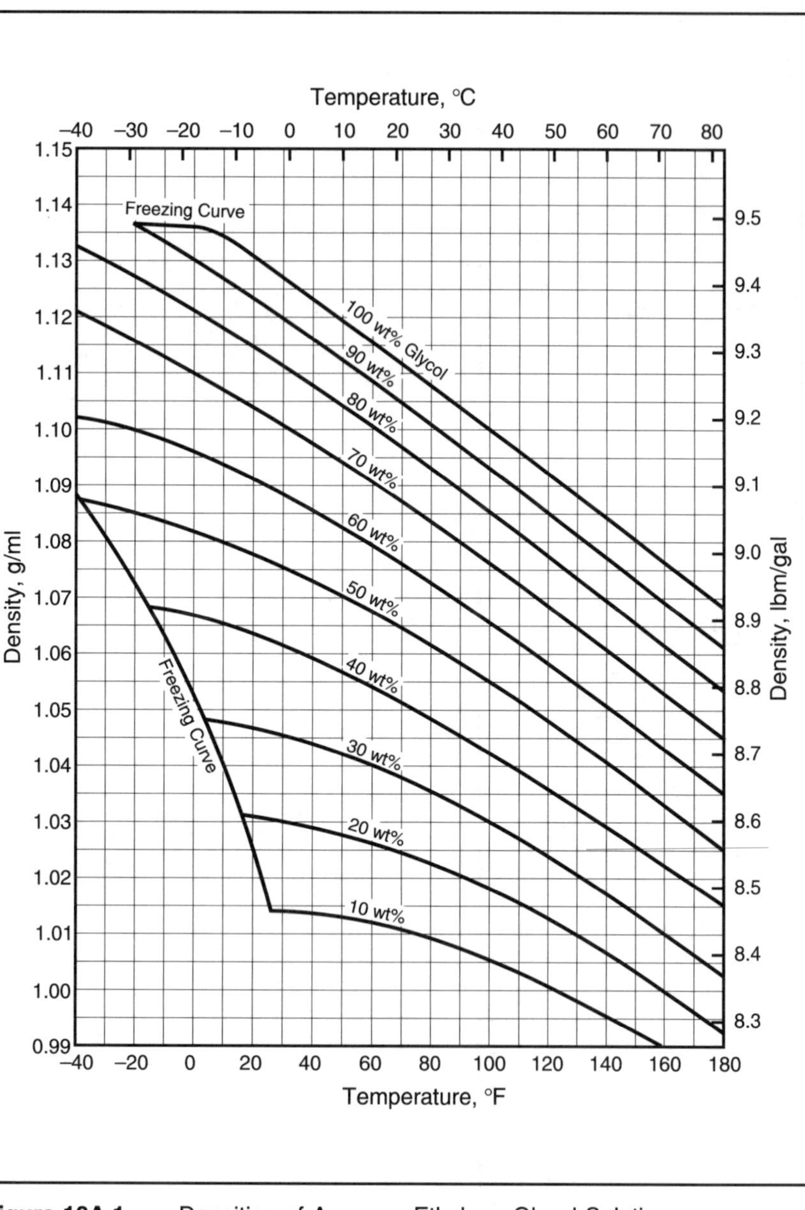

Figure 18A.1 Densities of Aqueous Ethylene Glycol Solutions (percent by weight)

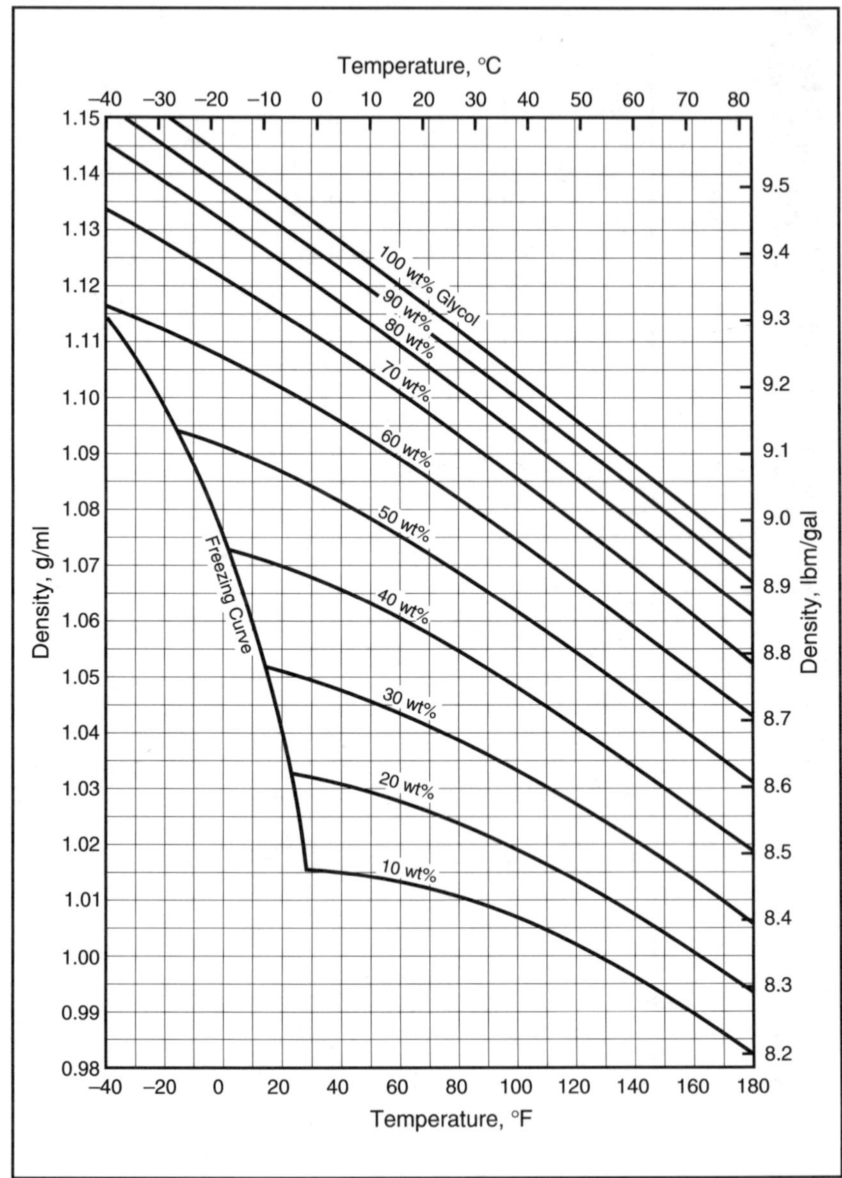

Figure 18A.2 Densities of Aqueous Diethylene Glycol Solutions (percent by weight)

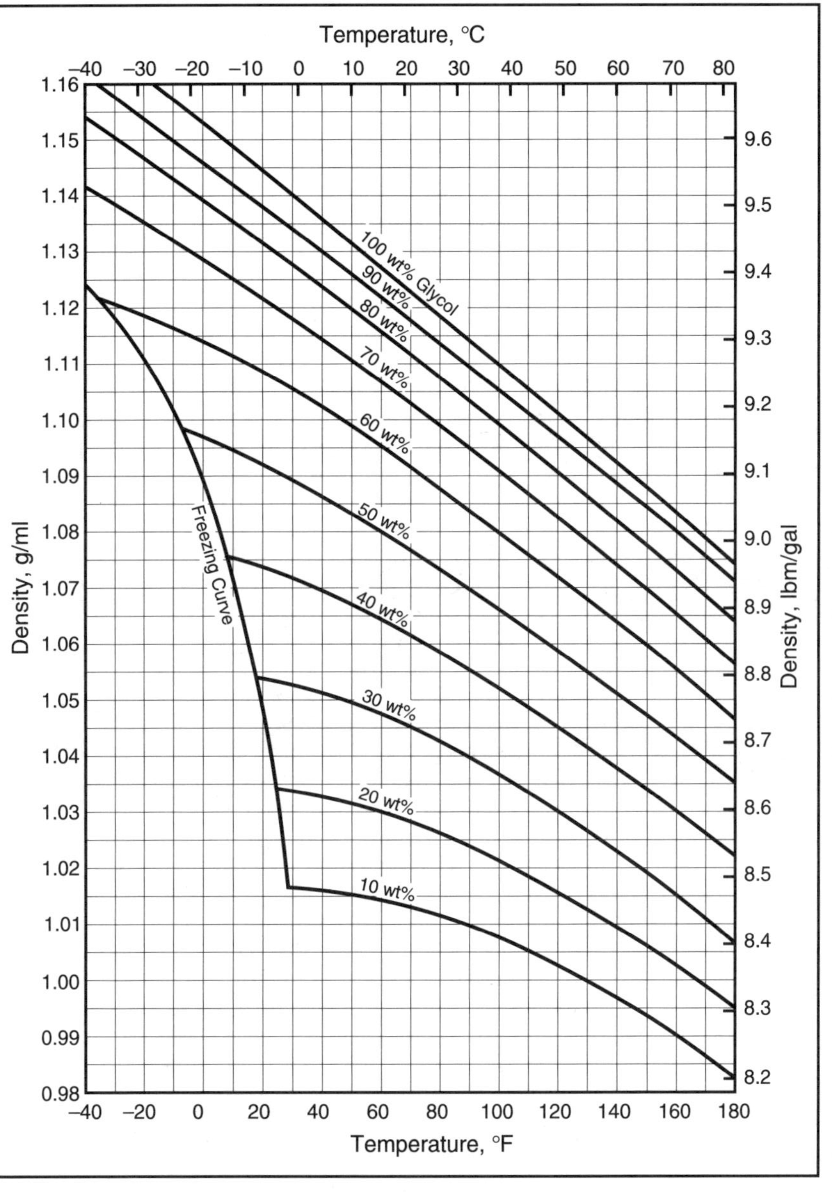

Figure 18A.3 Densities of Aqueous Triethylene Glycol Solutions (percent by weight)

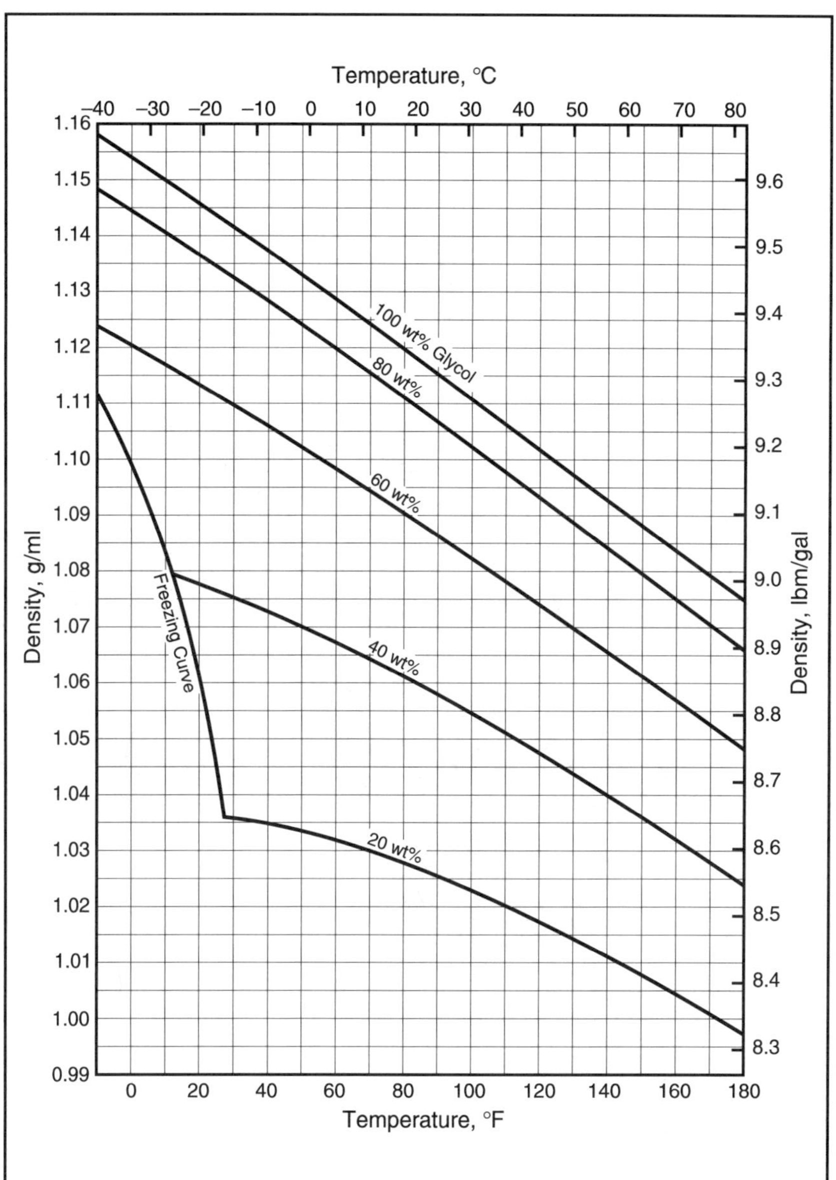

Figure 18A.4 Densities of Aqueous Tetraethylene Glycol Solutions (percent by weight)

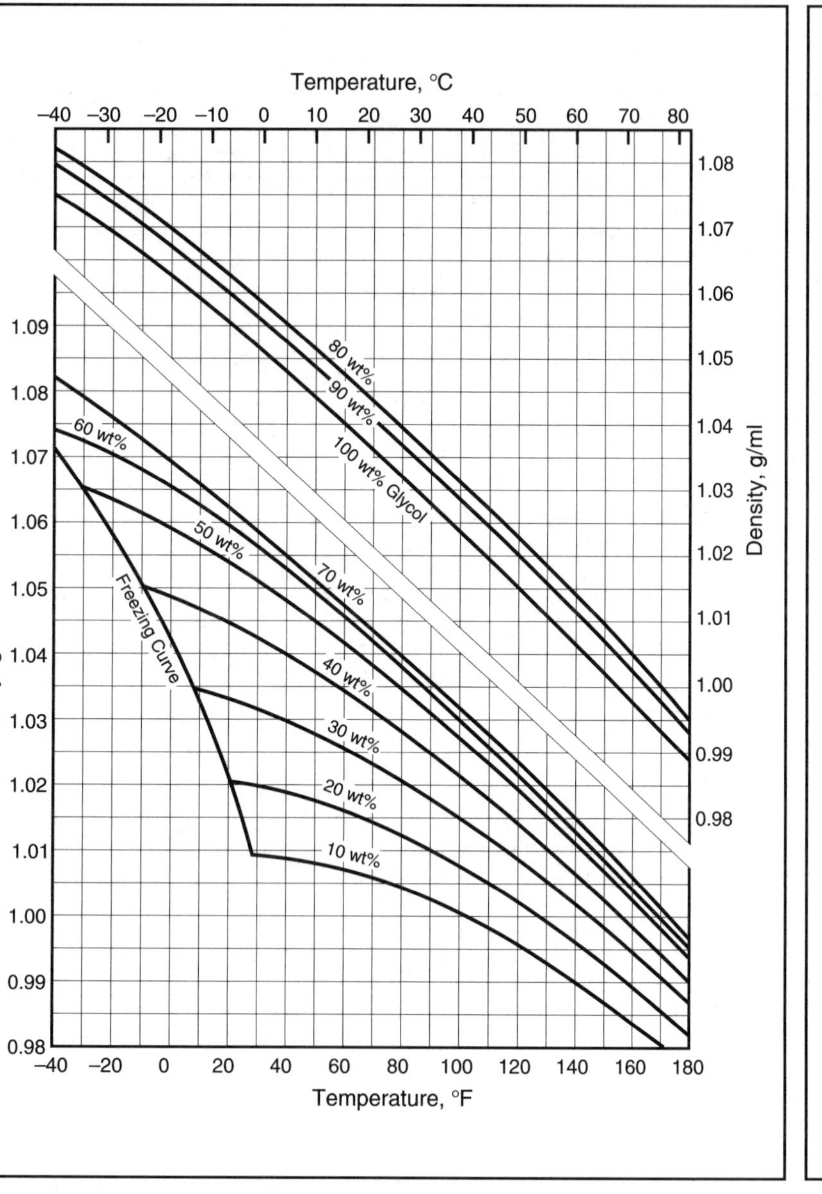

Figure 18A.5 Densities of Aqueous Propylene Glycol Solutions (percent by weight)

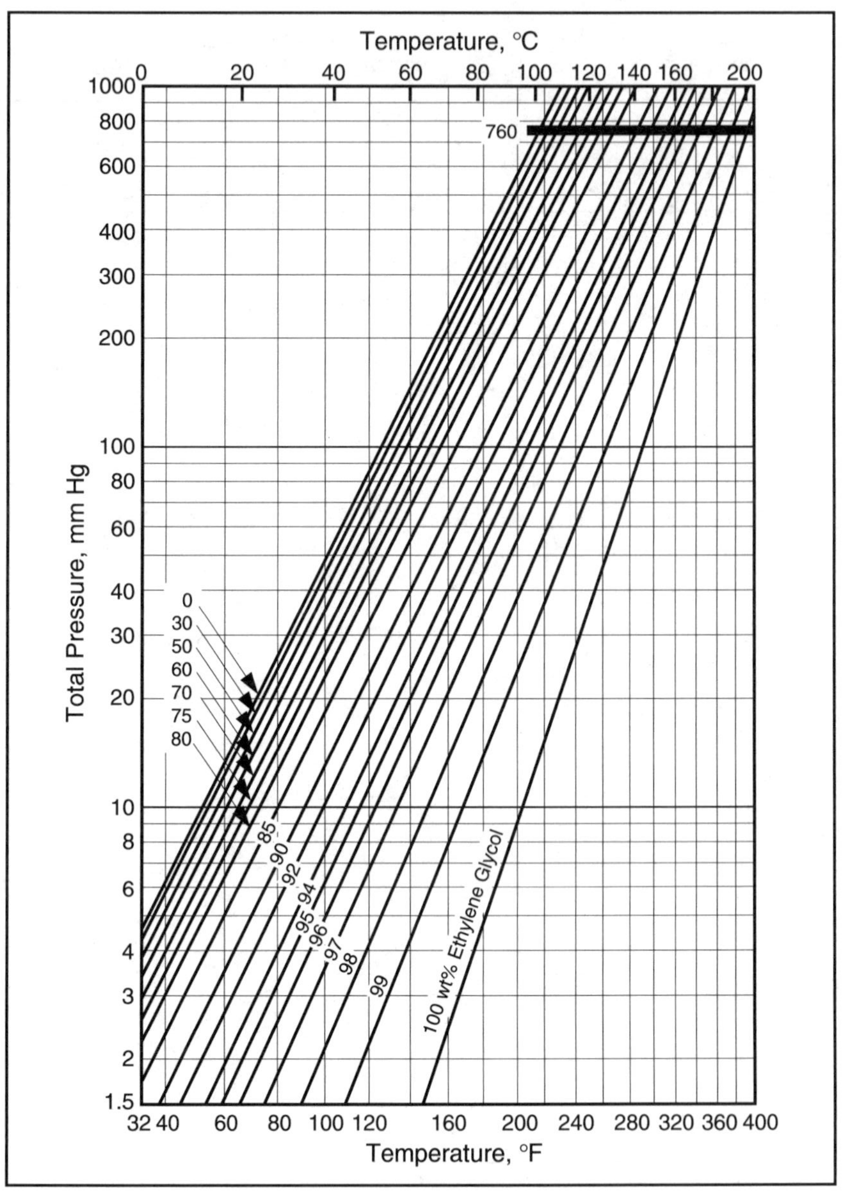

Figure 18A.6 Total Pressure over Aqueous Ethylene Glycol Solutions vs. Temperature

Figure 18A.7 Total Pressure over Aqueous Diethylene Glycol Solutions vs. Temperature

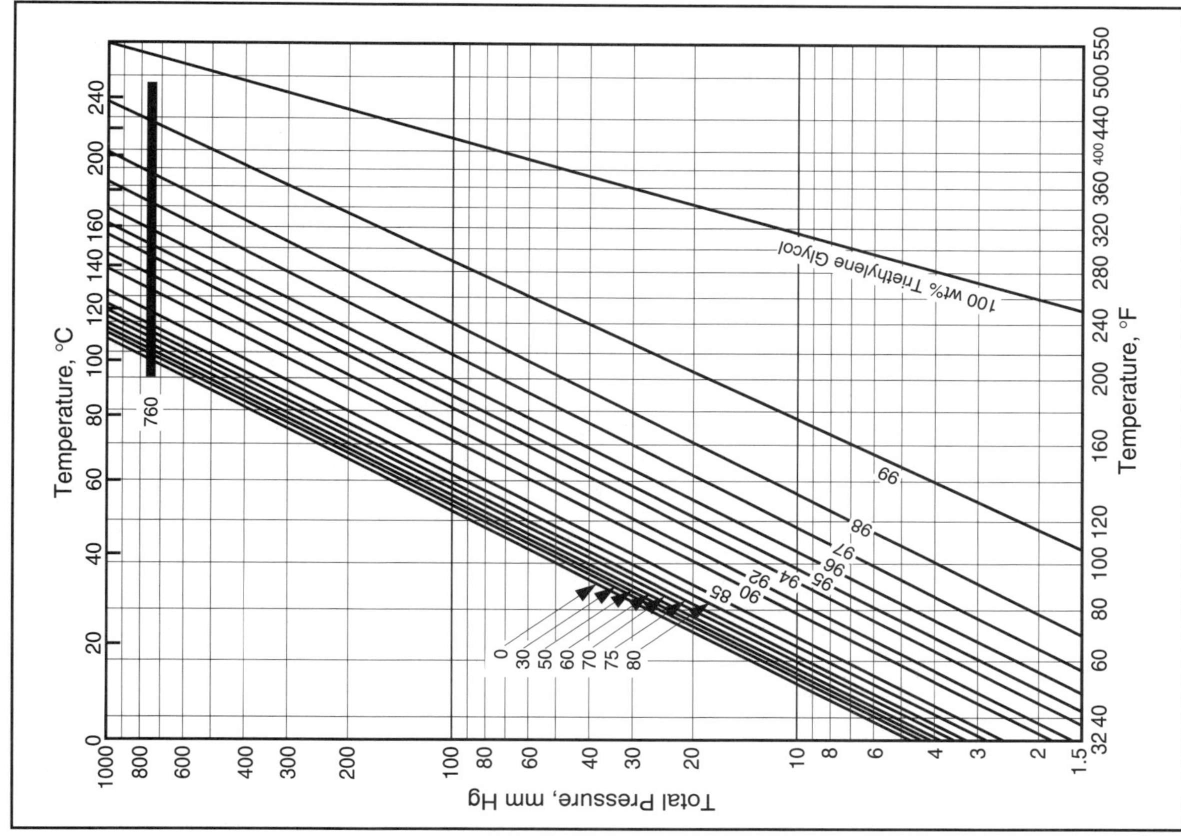

Figure 18A.8 Total Pressure over Aqueous Triethylene Glycol Solutions vs. Temperature

Figure 18A.9 Total Pressure over Aqueous Propylene Glycol Solutions vs. Temperature

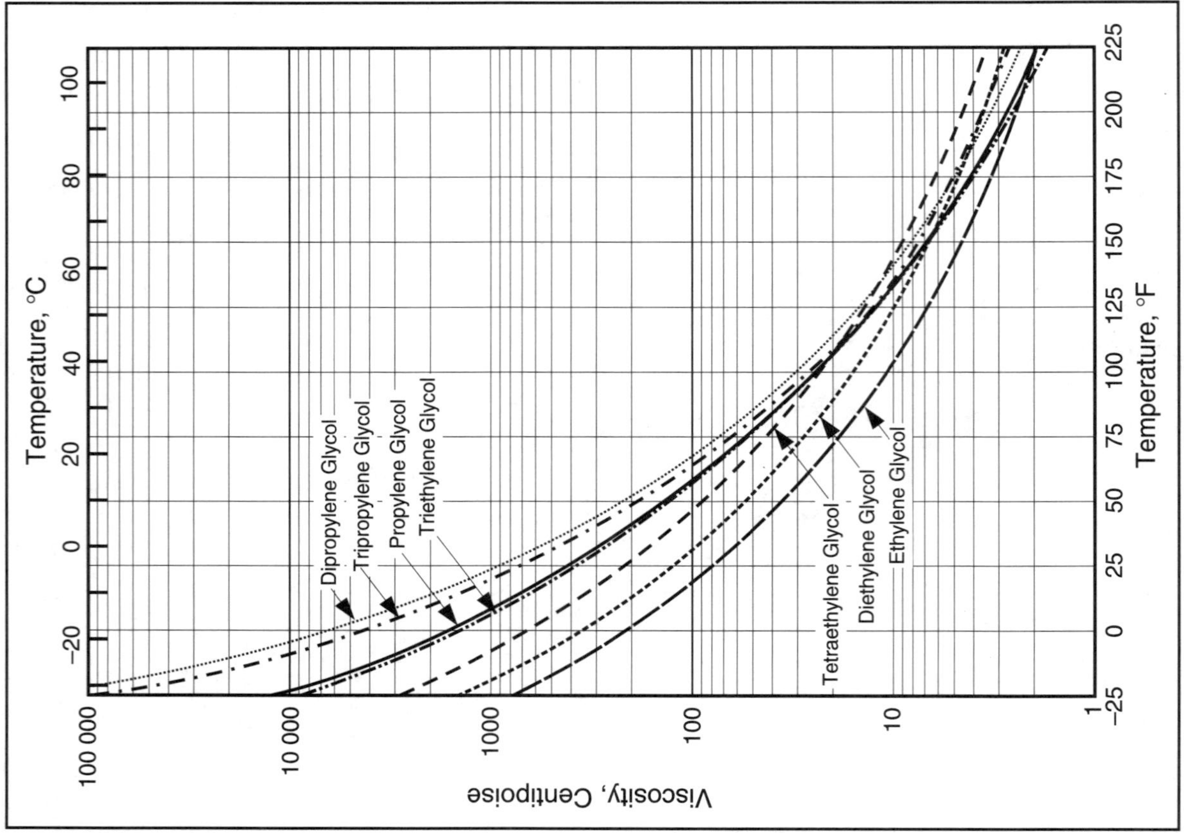

Figure 18A.10 Viscosities of Anhydrous Glycols

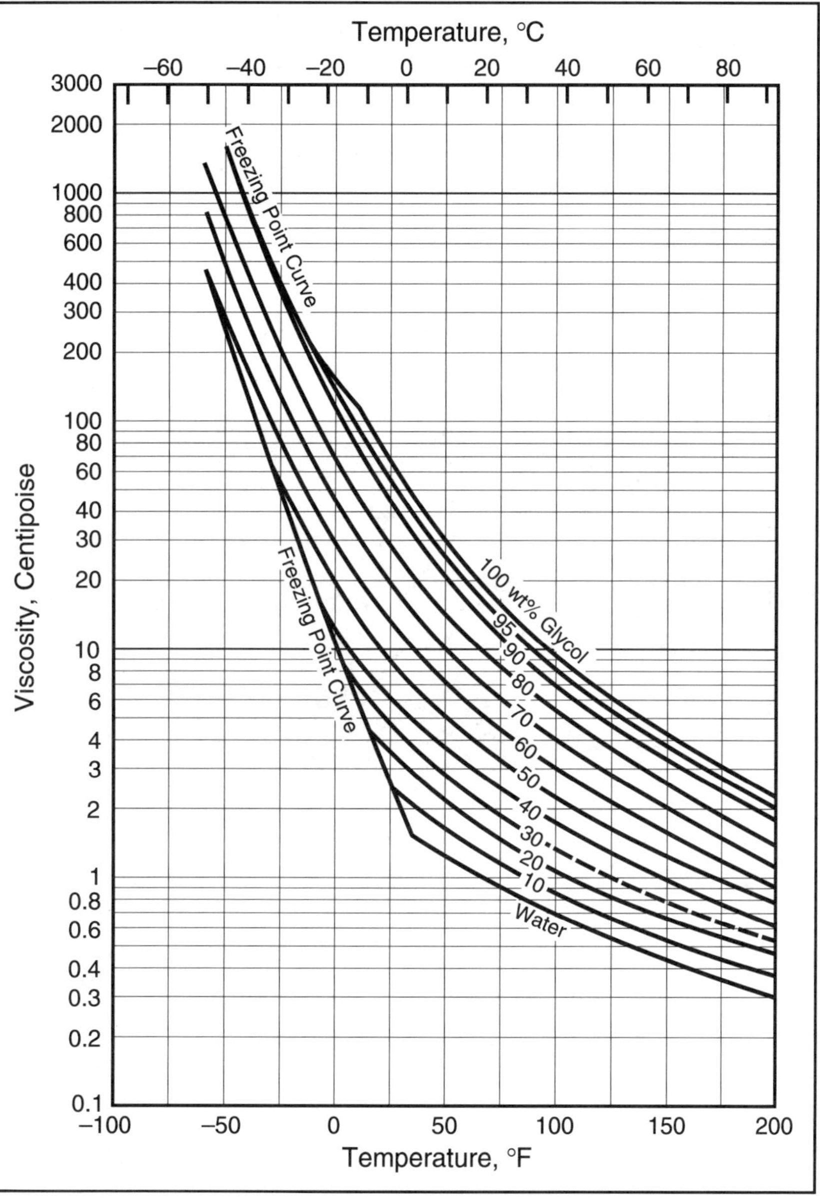

CHAPTER 18

Figure 18A.11 Viscosities of Aqueous Ethylene Glycol Solutions

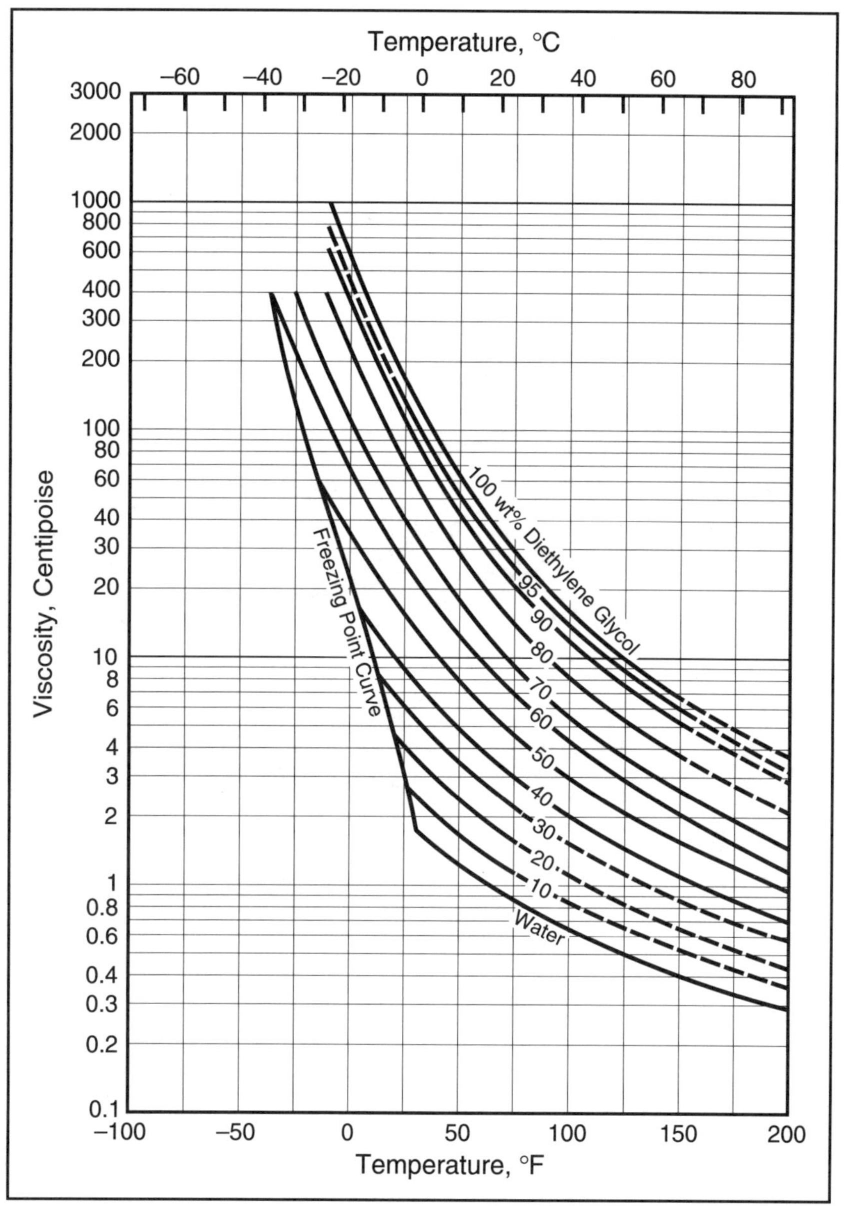

Figure 18A.12 Viscosities of Aqueous Diethylene Glycol Solutions

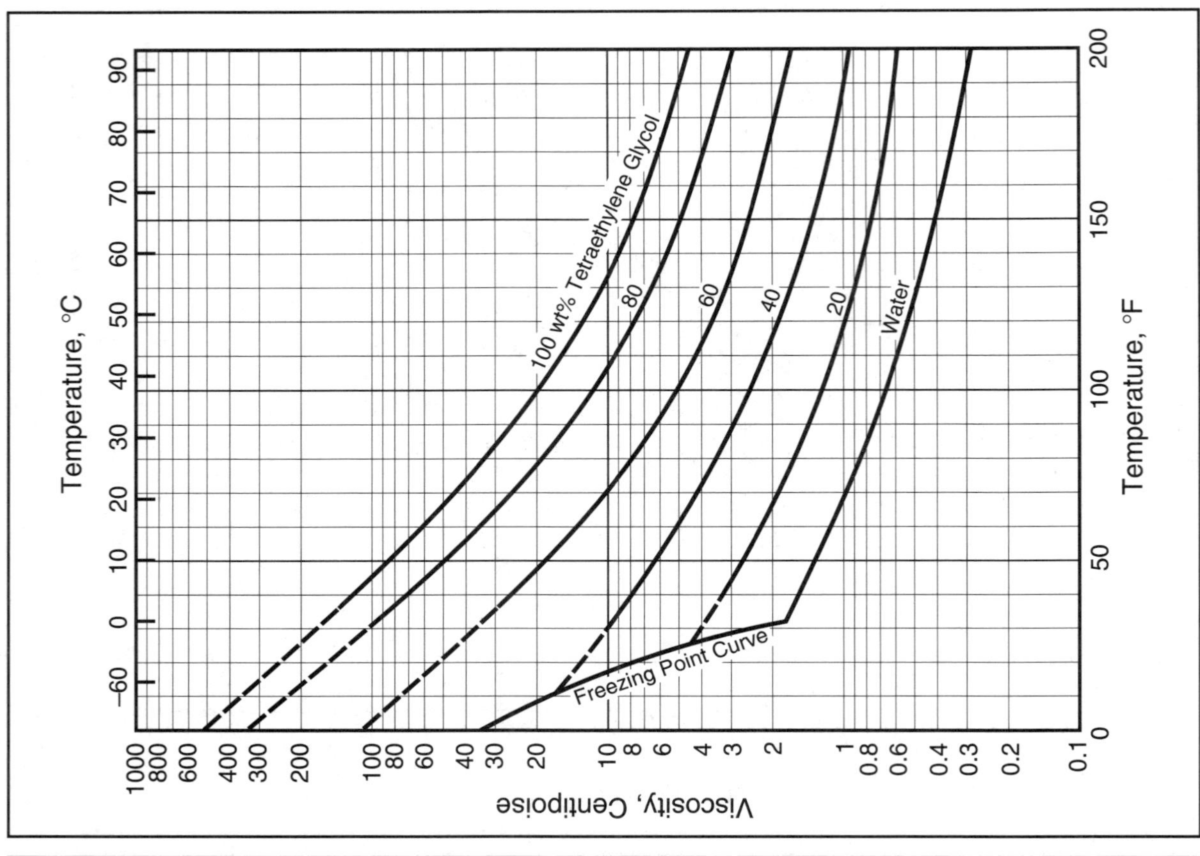

Figure 18A.14 Viscosities of Aqueous Tetraethylene Glycol Solutions

Figure 18A.13 Viscosities of Aqueous Triethylene Glycol Solutions

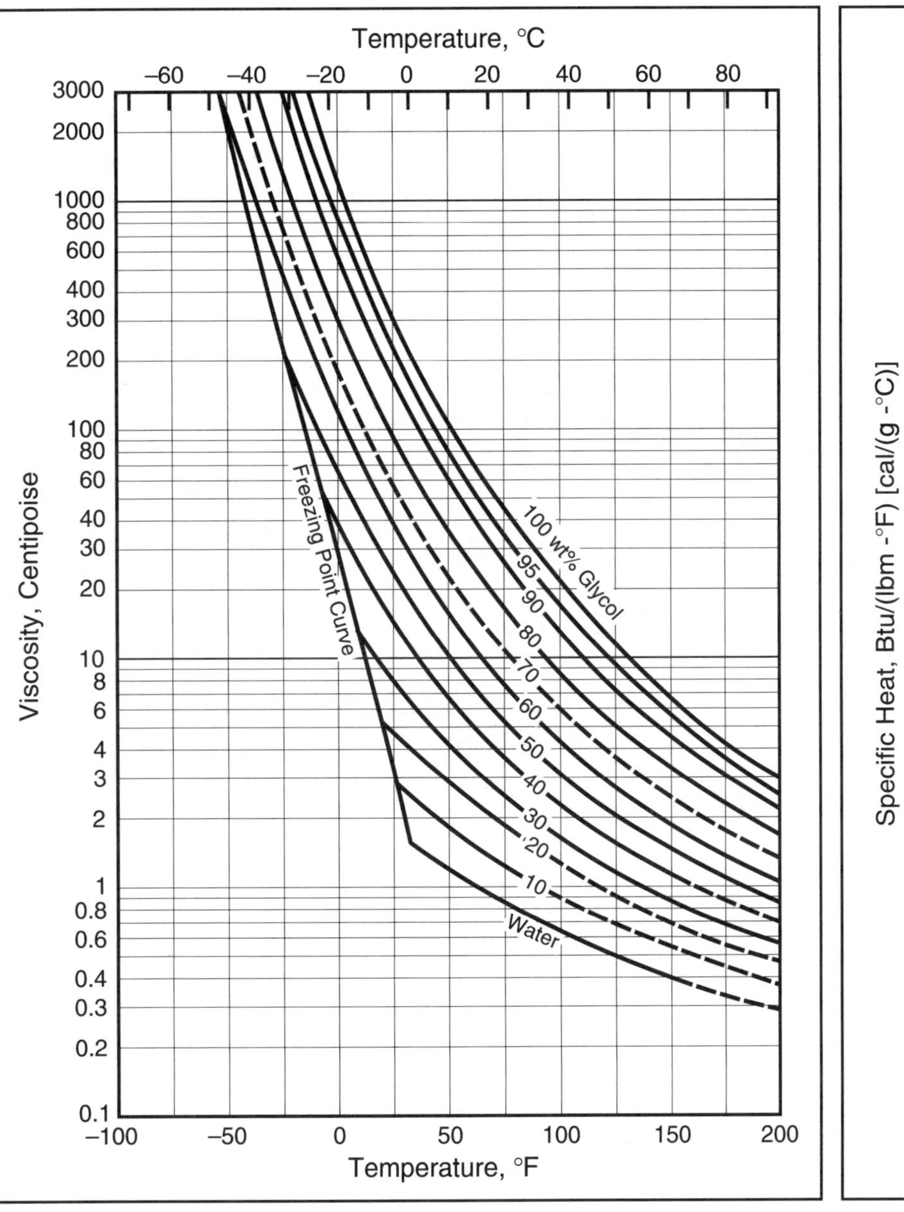

Figure 18A.15 Viscosities of Aqueous Propylene Glycol Solutions

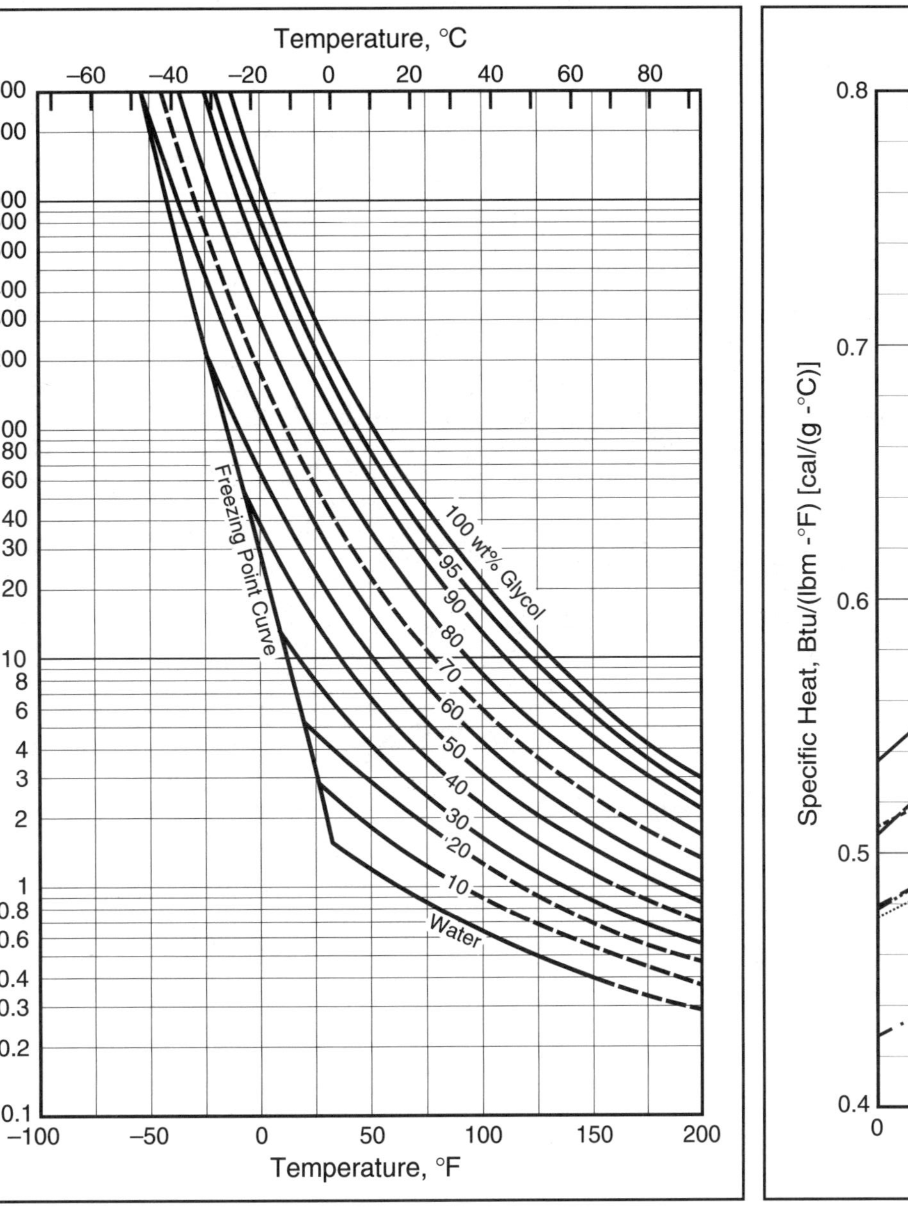

Figure 18A.16 Heat Capacity of Anhydrous Glycols

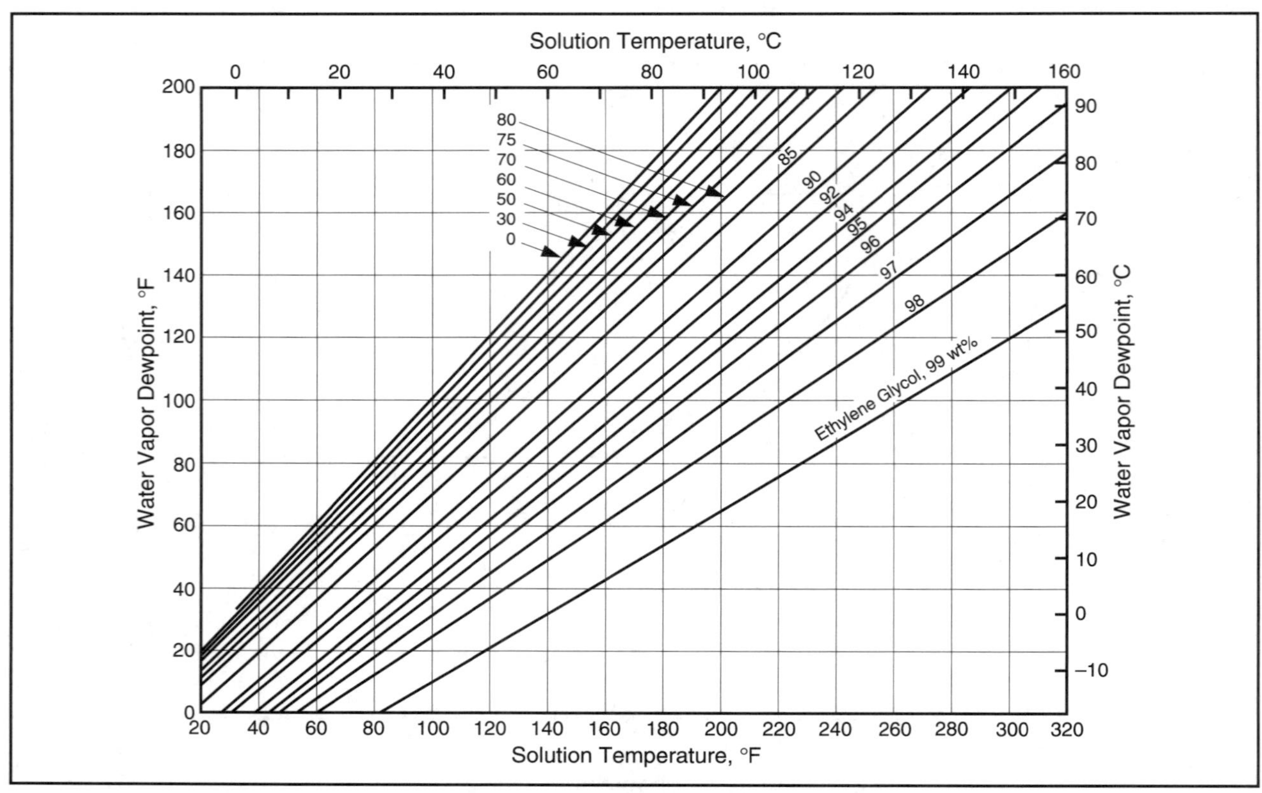

Figure 18A.17 Water Vapor Dewpoints over Aqueous Ethylene Glycol Solutions

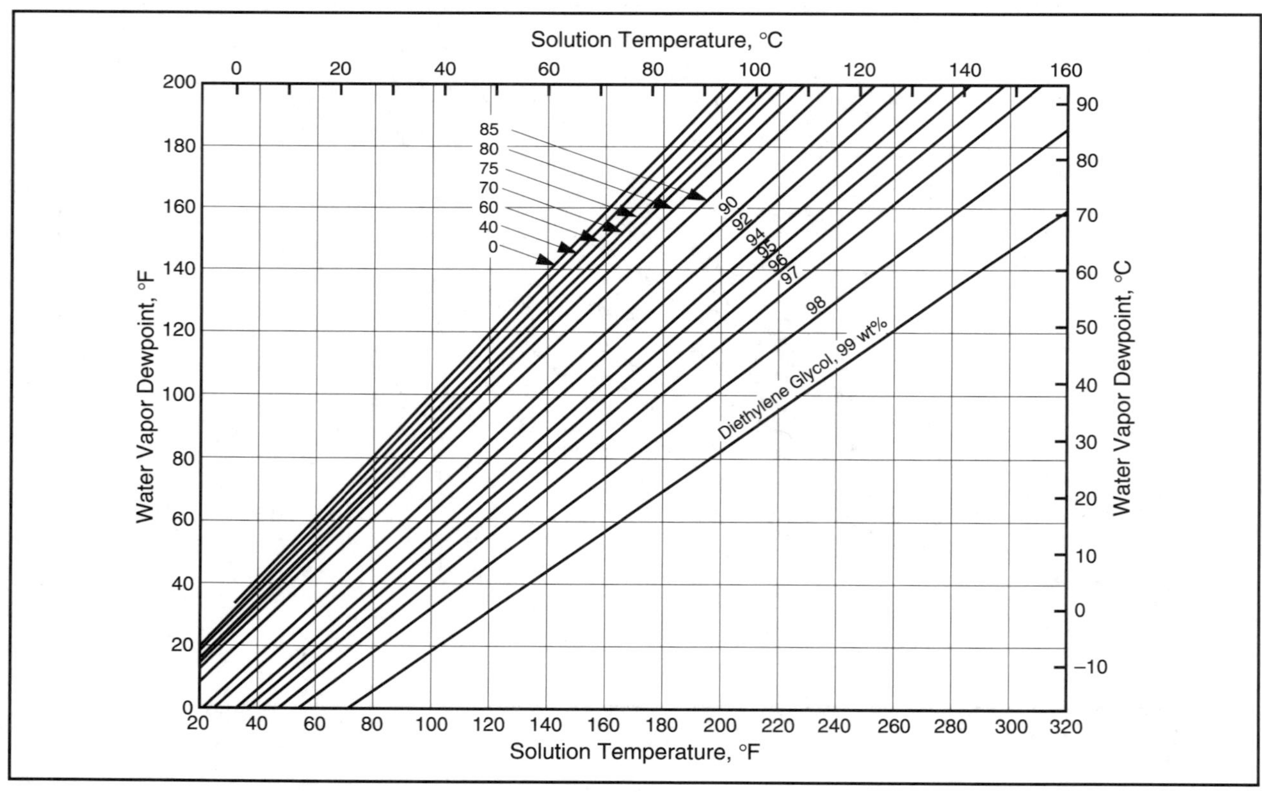

Figure 18A.18 Water Vapor Dewpoints over Aqueous Diethylene Glycol Solutions

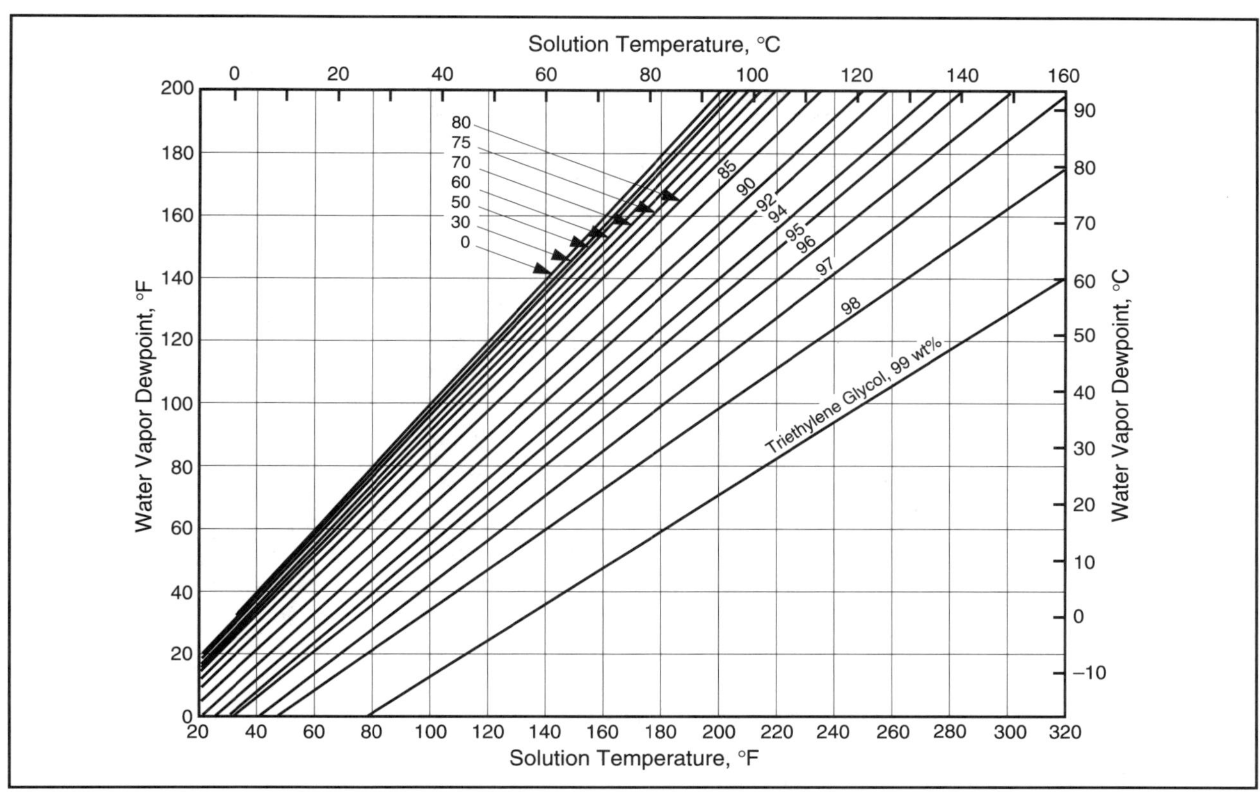

Figure 18A.19 Water Vapor Dewpoints over Aqueous Triethylene Glycol Solutions

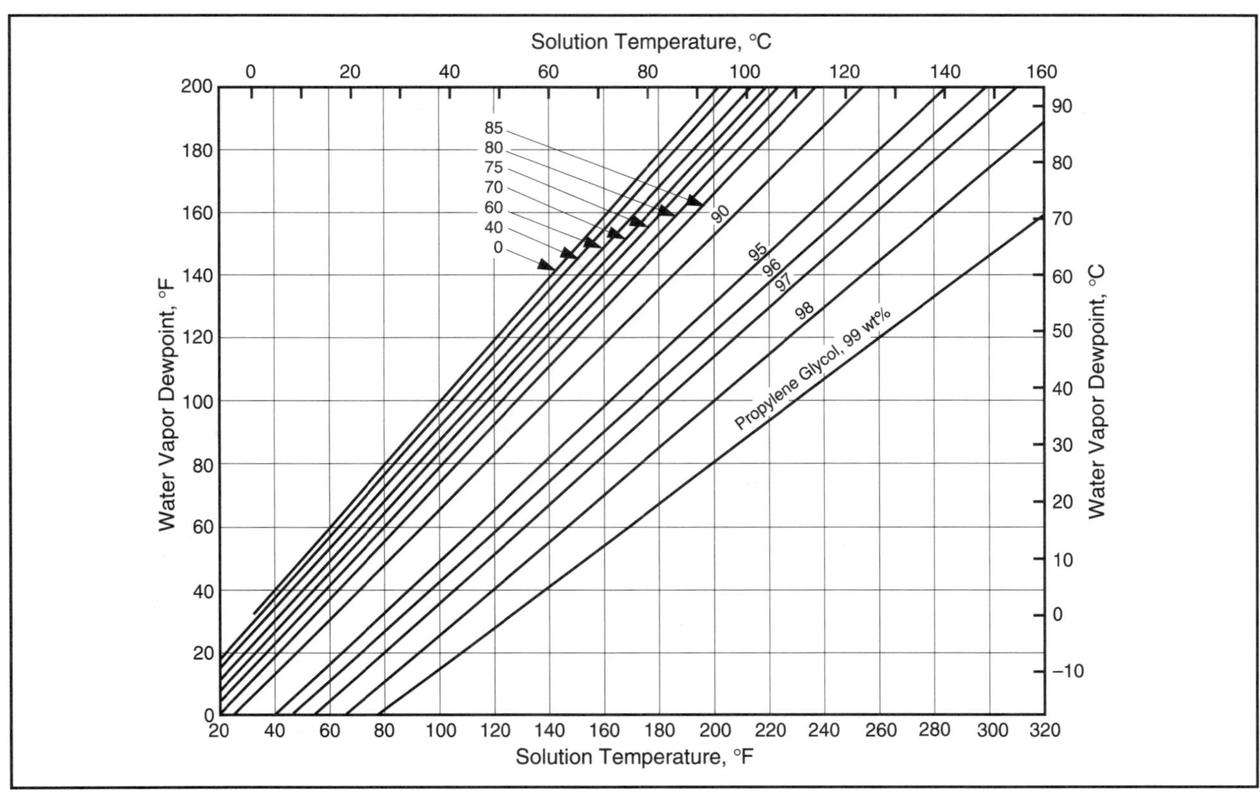

Figure 18A.20 Water Vapor Dewpoints over Aqueous Propylene Glycol Solutions

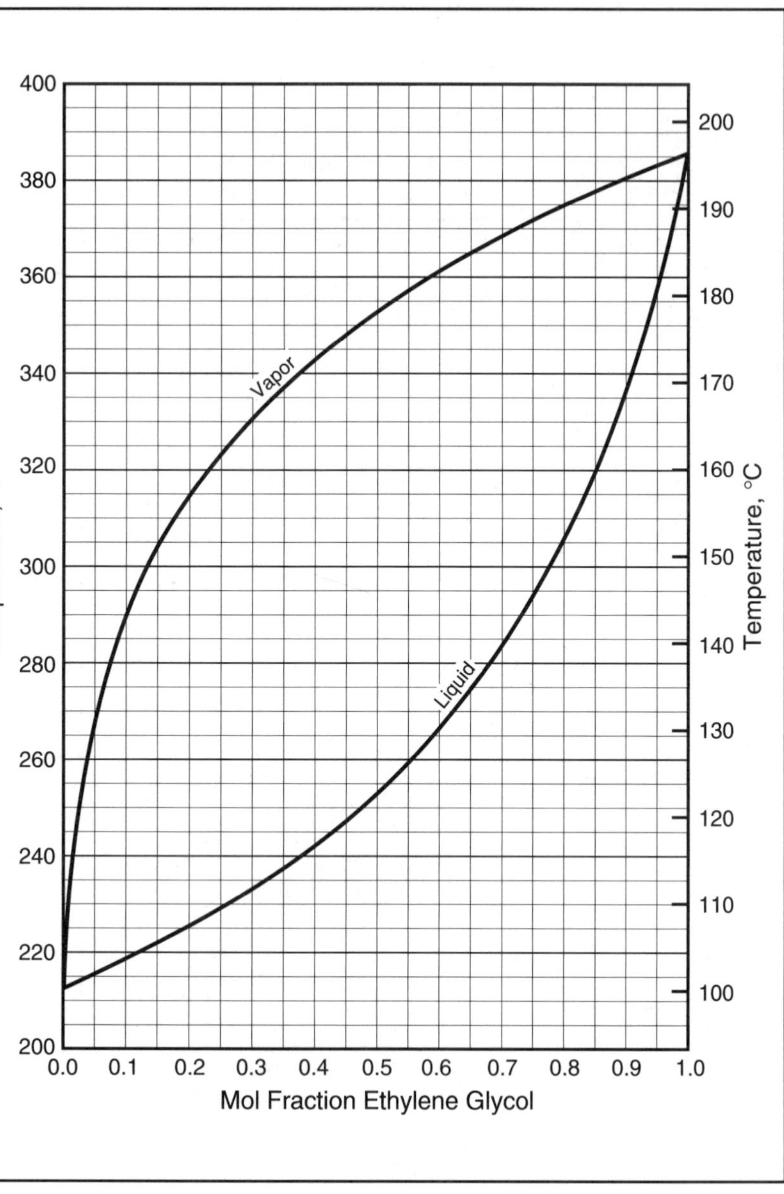

Figure 18A.21 Vapor-Liquid Composition Curves for Aqueous Ethylene Glycol Solutions (760 mmHg Pressure)

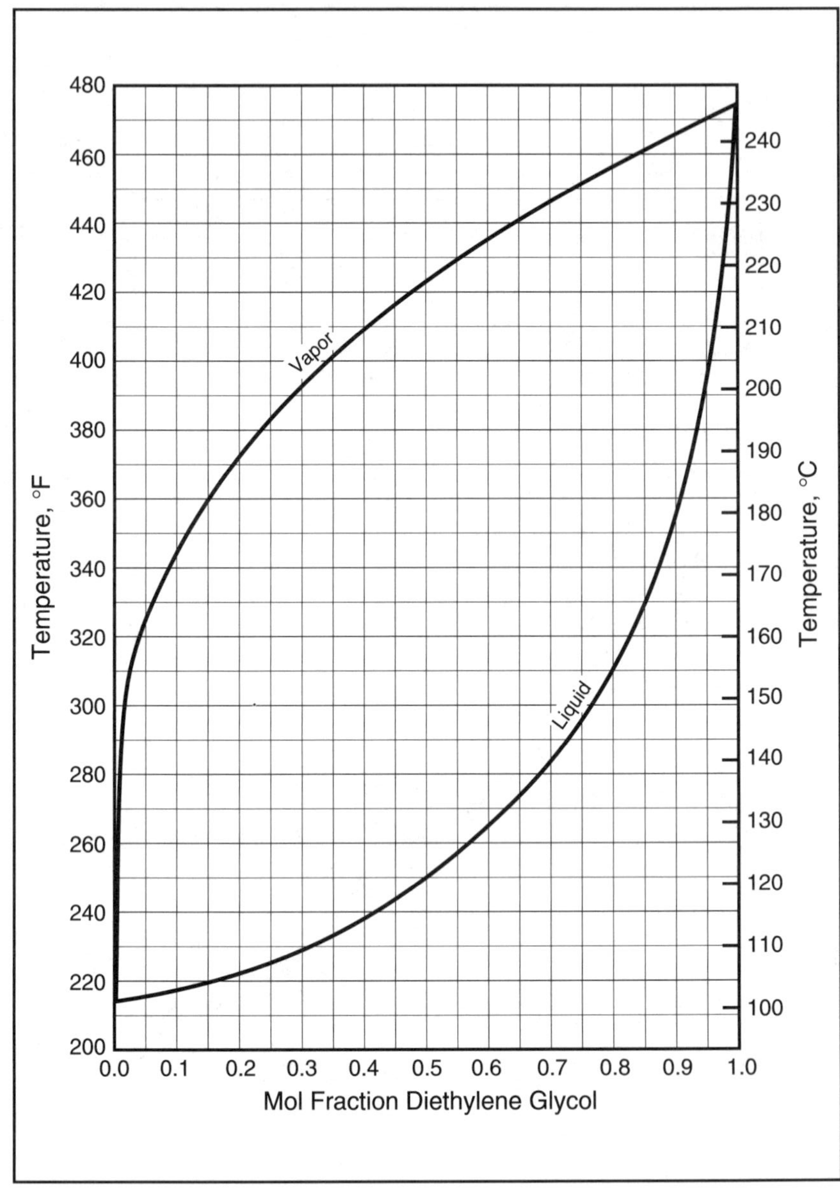

Figure 18A.22 Vapor-Liquid Composition Curves for Aqueous Diethylene Glycol Solutions (760 mmHg Pressure)

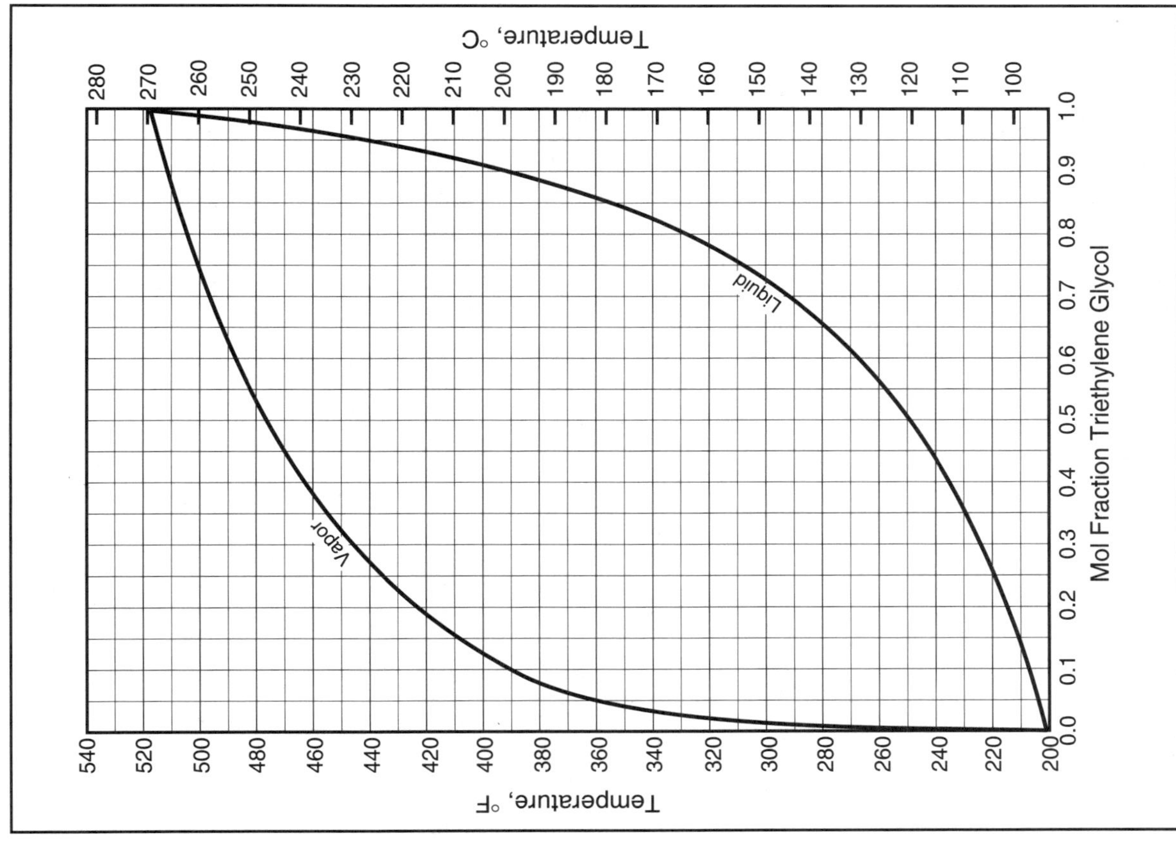

Figure 18A.24 Vapor-Liquid Composition Curves for Aqueous Triethylene Glycol Solutions (600 mmHg Pressure)

Figure 18A.23 Vapor-Liquid Composition Curves for Aqueous Triethylene Glycol Solutions (300 mmHg Pressure)

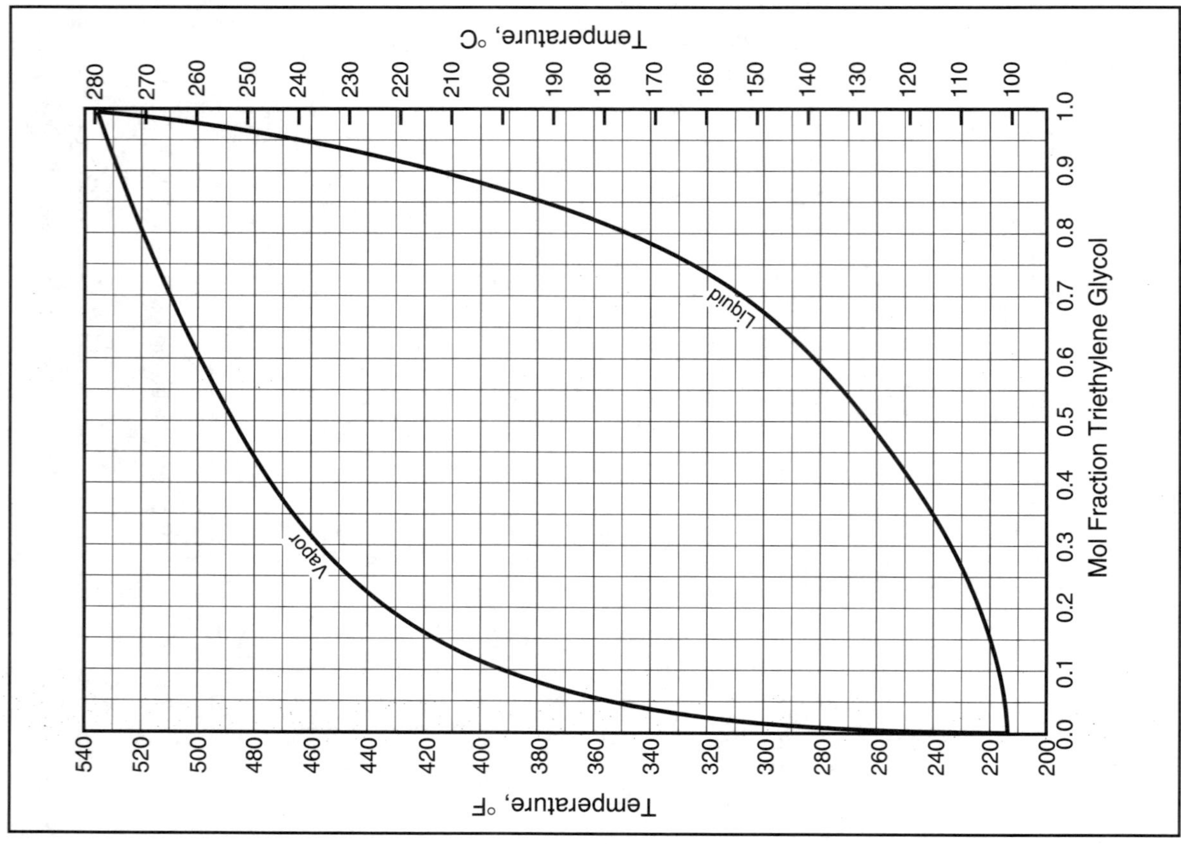

Figure 18A.25 Vapor-Liquid Composition Curves for Aqueous Triethylene Glycol Solutions (760 mmHg Pressure)

NOTES:

APPENDIX 18B

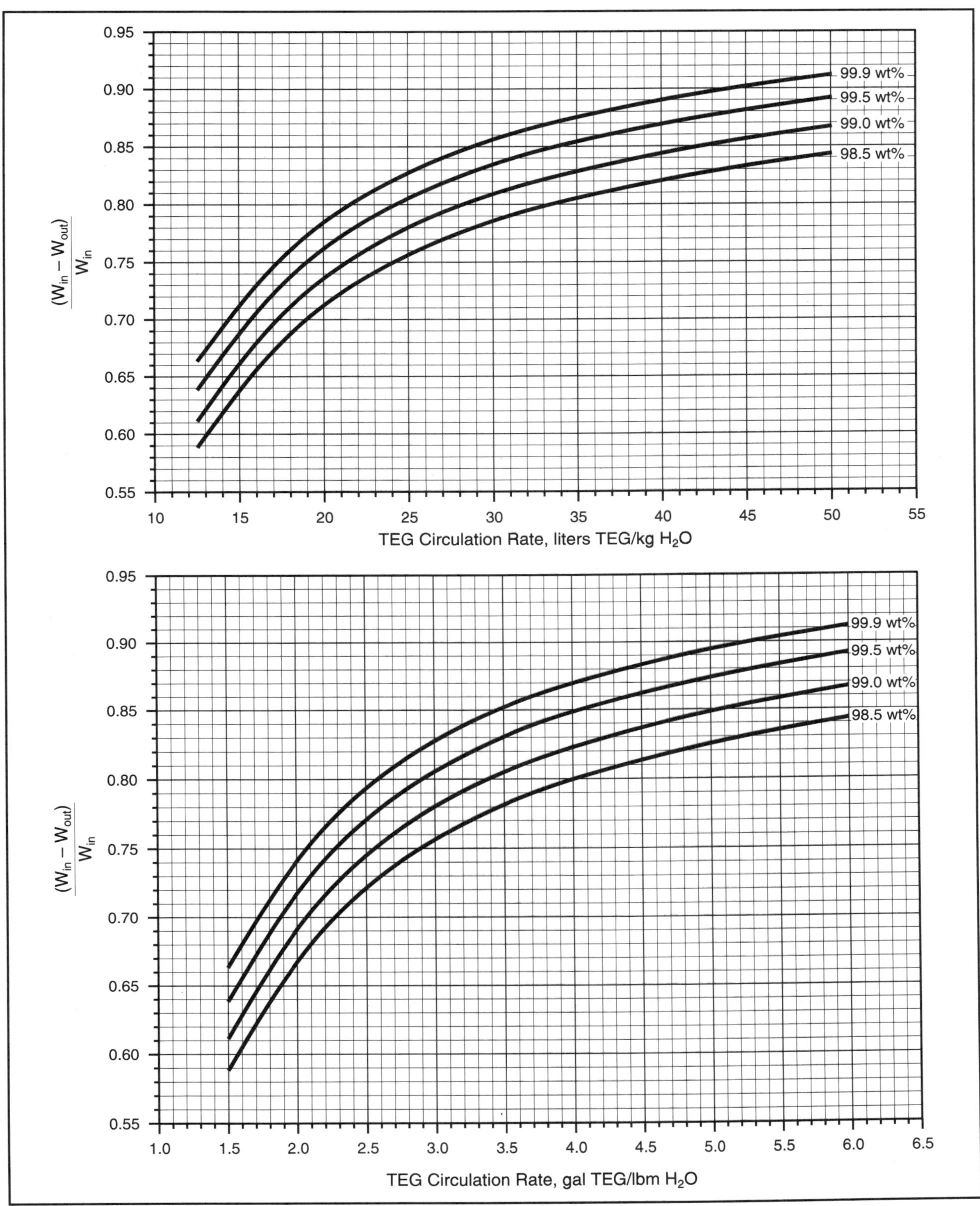

Figure 18B.1 Water Removal vs. TEG Circulation Rate at Various TEG Concentrations (N = 1.0)

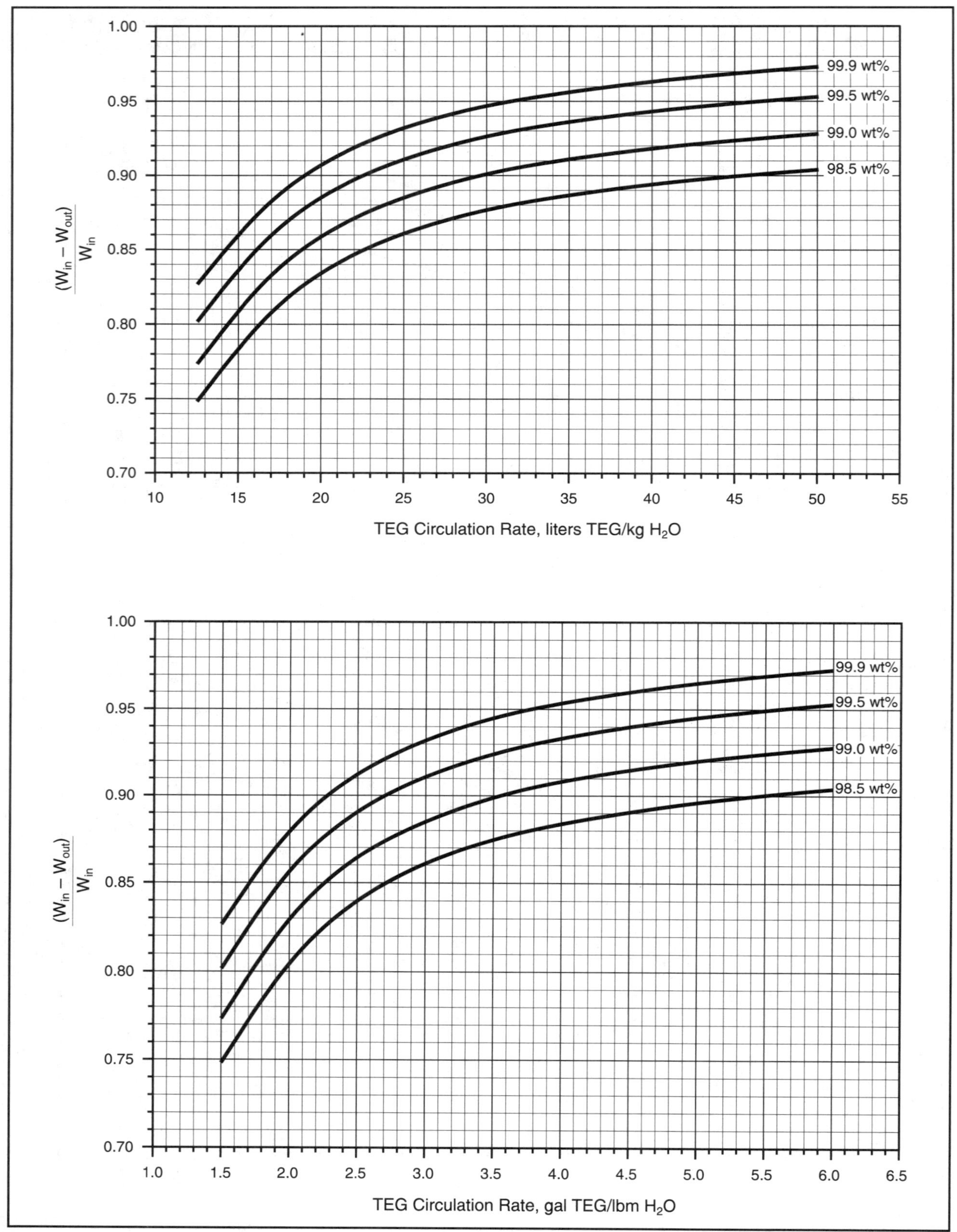

Figure 18B.2 Water Removal vs. TEG Circulation Rate at Various TEG Concentrations (N = 1.5)

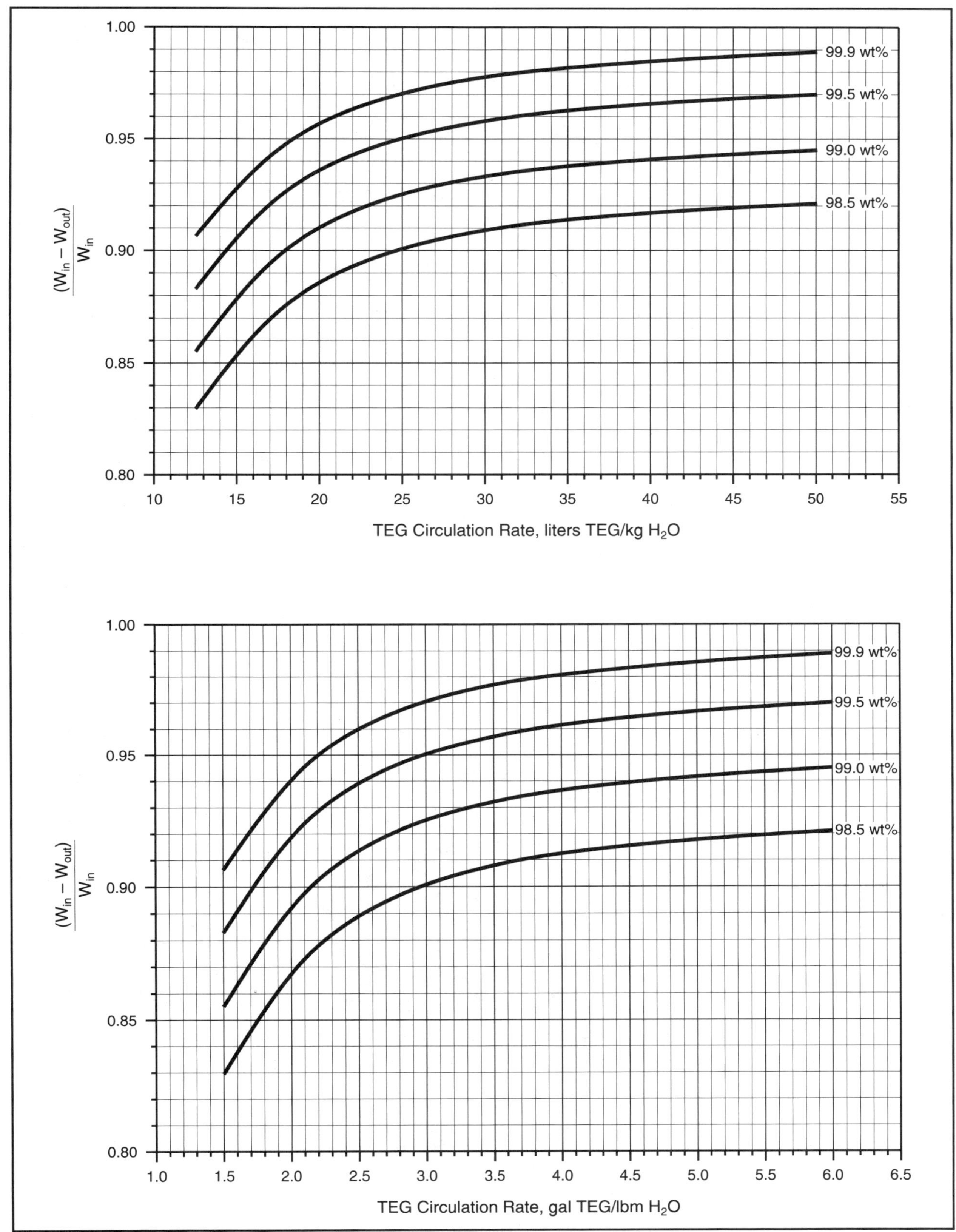

Figure 18B.3 Water Removal vs. TEG Circulation Rate at Various TEG Concentrations (N = 2.0)

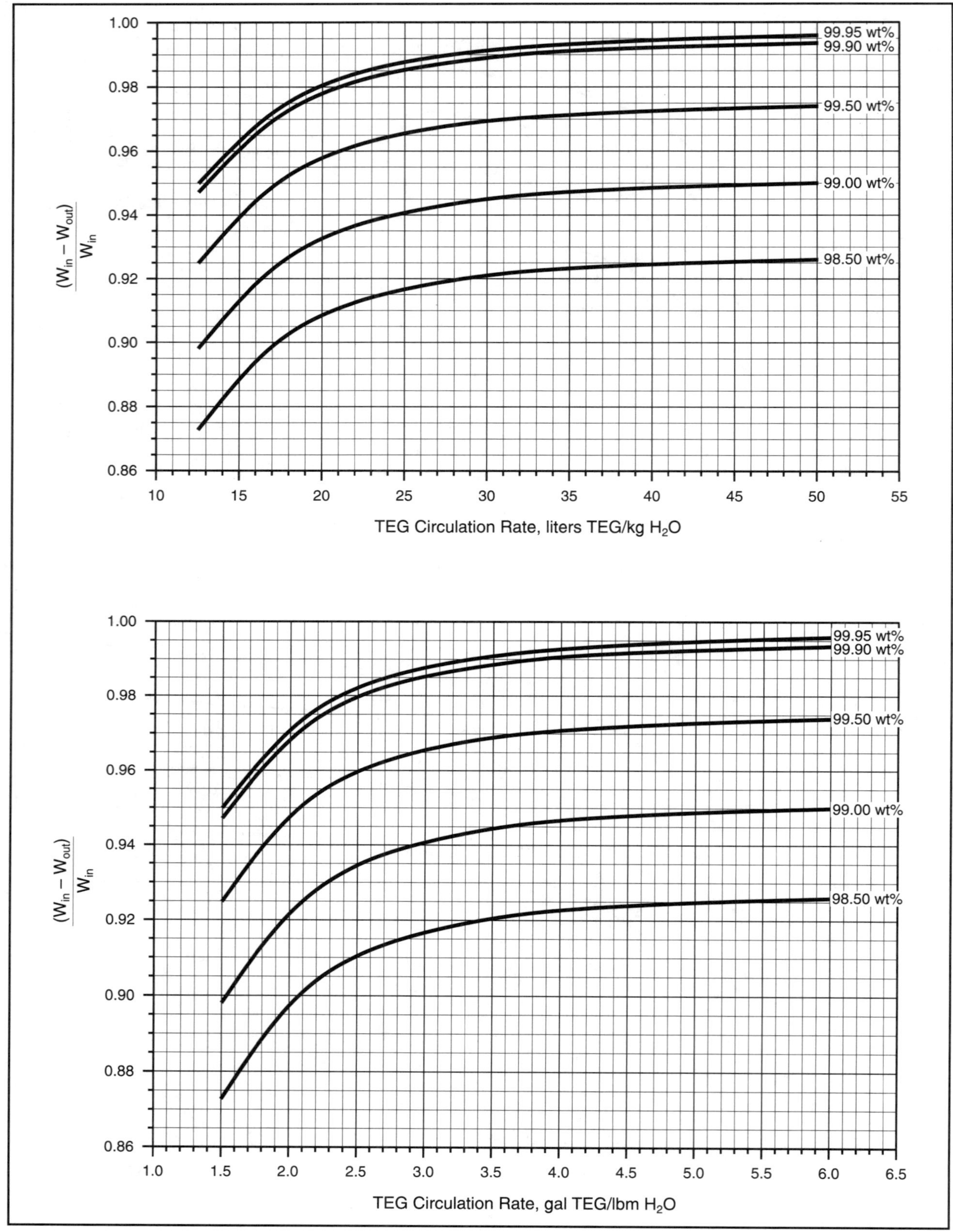

Figure 18B.4 Water Removal vs. TEG Circulation Rate at Various TEG Concentrations (N = 2.5)

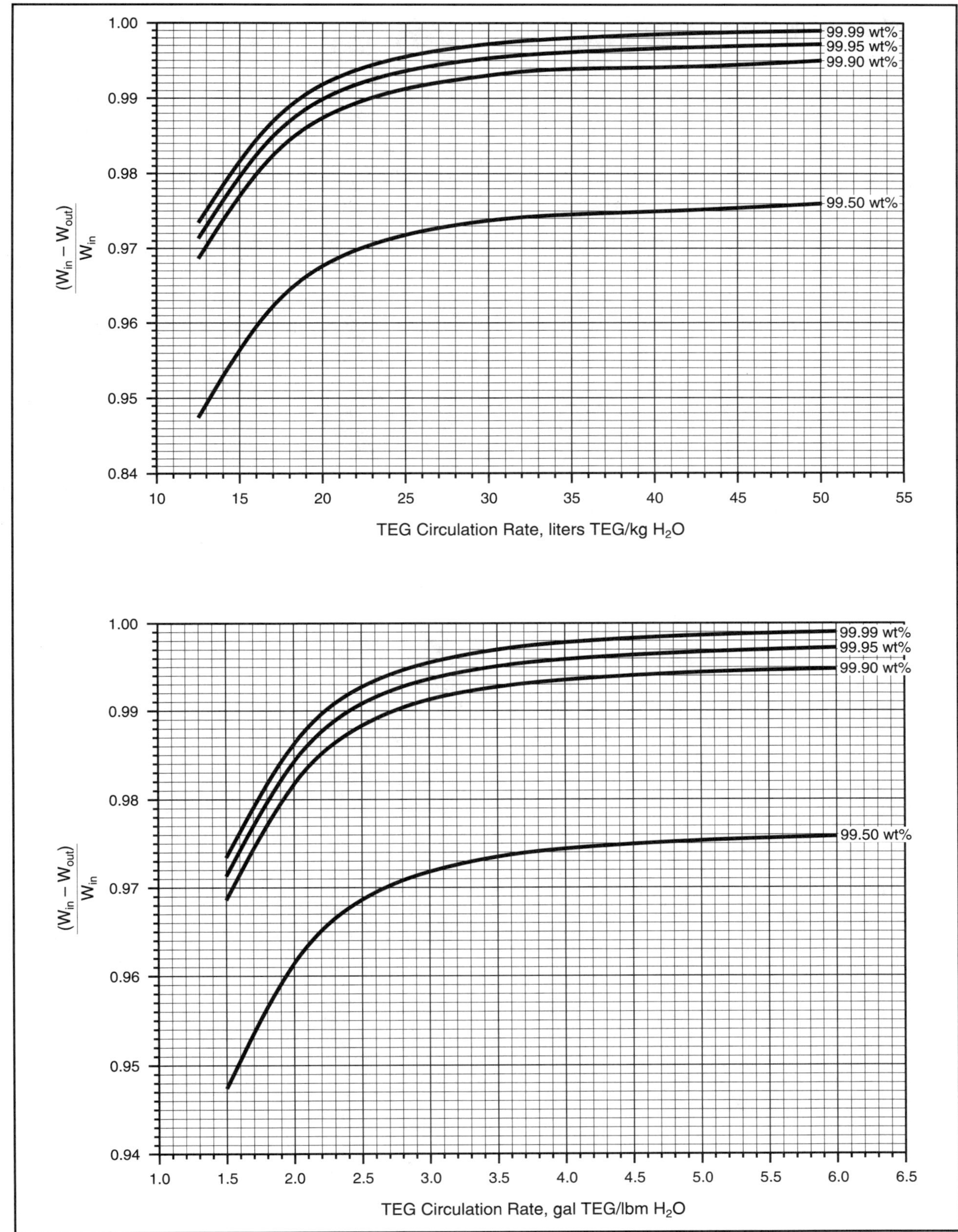

Figure 18B.5 Water Removal vs. TEG Circulation Rate at Various TEG Concentrations (N = 3.0)

APPENDIX 18C

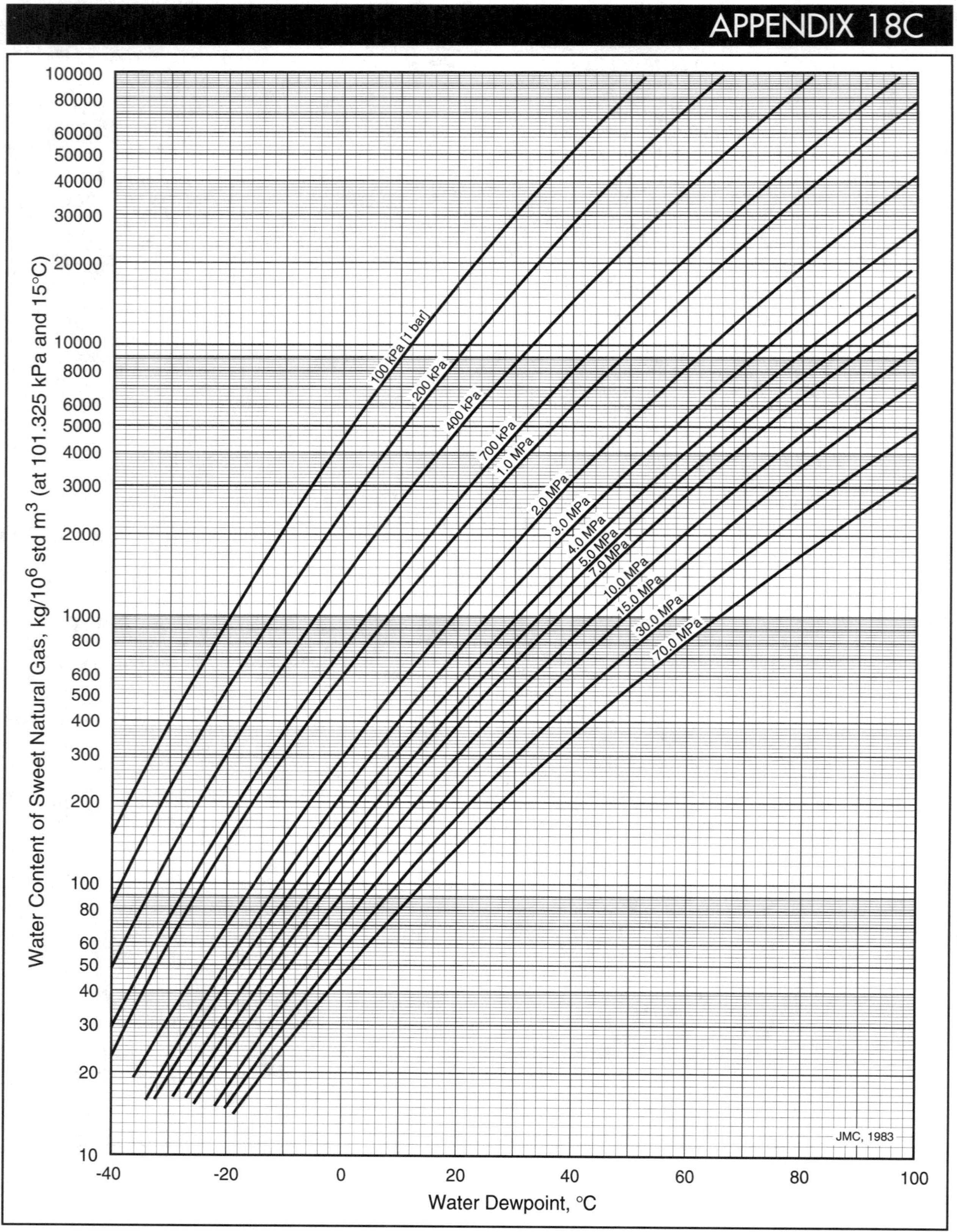

Figure 18C.1(a) Water Content of Sweet, Lean Natural Gas

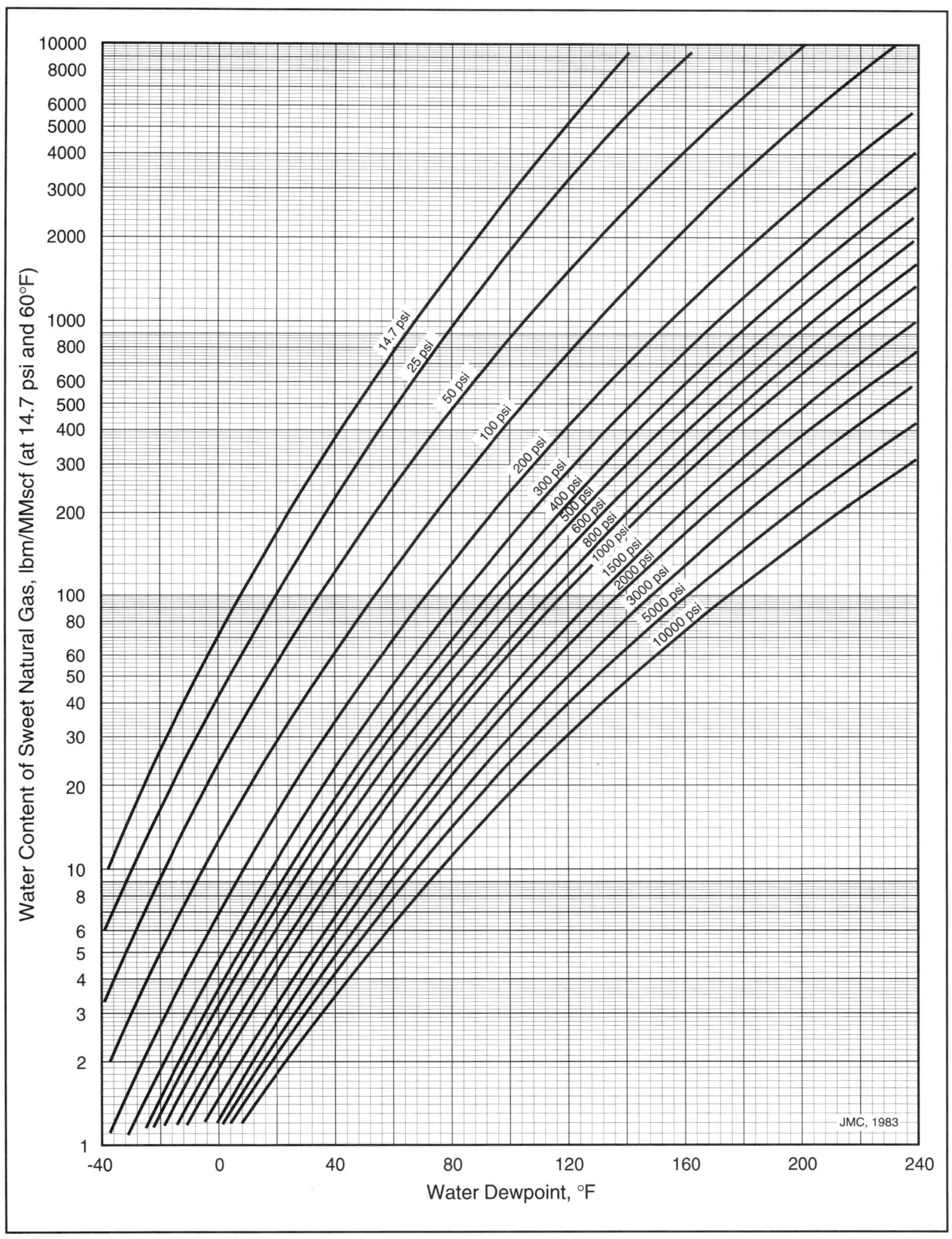

Figure 18C.1(b) Water Content of Sweet, Lean Natural Gas

NOTES:

19

ADSORPTION DEHYDRATION

Glycol dehydration or refrigeration with hydrate inhibition is used for most applications where dehydration of natural gas to pipeline specification is required. Solid bed dehydration (also called dry desiccant or adsorption dehydration) is often the superior alternative in applications such as:

1. Dehydration to water dewpoints less than -40 to -50°C [-40 to -58°F], such as those required upstream of NGL extraction plants utilizing expanders and LNG plants.
2. Hydrocarbon dewpoint control units where simultaneous extraction of water and hydrocarbon is required to meet both of the respective sales specifications — well suited for hydrocarbon dewpoint control on lean, high pressure gas streams.
3. Simultaneous dehydration and sweetening of natural gas.
4. Dehydration of gases containing H_2S where H_2S solubility in glycol can cause emission problems at the regenerator.
5. Dehydration and trace sulfur compound (H_2S, COS, CS_2, mercaptan) removal for LPG and NGL streams.

Adsorption describes any process wherein molecules from the gas are held on the surface of a solid by surface forces. Adsorbents may be divided into two classes – those which owe their "activity" to surface adsorption and capillary condensation, and those which react chemically. The latter group finds limited application in natural gas processing and will not be discussed herein. They are discussed in Volume 4, "Gas Treating and Sulfur Recovery," of this series. The former (physical adsorption) requires use of an adsorbent material which probably has the following characteristics:

1. Large surface area for high capacity.
2. Possesses "activity" for the components to be removed.
3. Mass transfer rate is high.
4. Easily and economically regenerated.
5. Good activity retention with time.
6. Small resistance to gas flow.
7. High mechanical strength to resist crushing and dust formation.
8. Inexpensive, non-corrosive, non-toxic, chemically inert and possesses a high bulk density.
9. No appreciable change in volume during adsorption and desorption, and should retain strength when "wet."

Most commercial adsorbents will have a total surface area of 500 to 800 m^2/g [2 400 000 to 3 900 000 ft^2/lbm]. One pound can easily be held in your cupped hands. This exceptionally large area is only achieved by producing a material with large interior surface resulting from capillaries or a crystalline-type lattice. The exterior surface of the particles is almost negligible.

The materials which meet the above requirements may be divided into several general categories:

- Bauxite – naturally occurring mineral composed primarily of Al_2O_3.
- Alumina – a purer, manufactured version of bauxite.
- Gels – composed largely of SiO_2 or alumina gel; manufactured by chemical reaction.
- Molecular Sieves – a calcium-sodium alumino-silicate (zeolite).
- Carbon (charcoal) – a carbon product treated and activated to have adsorptive capacity.

(Only listed are those materials commonly used for bulk treating.)

All but carbon are used for dehydration. Carbon has desirable properties for hydrocarbon removal and adsorption of certain impurities but possesses negligible water capacity.

Desiccant Properties

Table 19.1 presents properties of common desiccants. The SI values in Table 19.1 can be converted to FPS units using the following conversion factors:

$$1\ m^2/g = 4885\ ft^2/lbm$$
$$1\ cm^3/g = 27.7\ in^3/lbm$$
$$1\ kg/m^3 = 0.0624\ lbm/ft^3$$
$$1\ kJ/kg{\cdot}°C = 0.239\ Btu/lbm\text{-}°F$$

The potential capacity per unit volume is a product of bulk density times the available area for adsorption. In essence, monolayer adsorption occurs. Gels have a higher effective capacity than the aluminas because of their larger surface area.

The pore opening at the surface of the desiccant must be large enough to admit the molecules being adsorbed to the interior of the particle where most of the surface area exists. In the internal pores of the gels exist capillaries of the diameter range shown. With molecular sieves, the internal pores are crystalline cavities larger than the openings on the surface.

TABLE 19.1

Summary of Typical Desiccant Properties

Property	Grade 03 Silica Gel	Mobilbead R	Mobilbead H	Alcoa F-200 Alumina	Alcoa H-156 Alumina	4A-5A Molecular Sieves
Surface Area, m^2/g	750-830	550-650	740-770	340-360	340-360	650-800
Pore Volume, cm^3/g	0.40-0.45	0.31-0.34	0.50-0.54	0.50	0.50	0.27
Pore Diameter, Å	21-23	21-23	27-28	43	26	(see note below)
Bulk Density, kg/m^3	721	785	721	769	769	689-721
Heat Capacity, kJ/(kg·°C)	0.92	1.05	1.05	1.0	1.0	1.0

Note: Types 4A and 5A contain cavities 11.4 Å in diameter with circular openings 4.2 Å in diameter (opening size for adsorption). 10^8 angstroms (Å) = 1 cm.

Table 19.2 shows the *nominal diameter* of common molecules involved in hydrocarbon system adsorption. This is called the nominal diameter because the molecules are not spheres and their ability to enter a given size opening depends on their direction of approach. Also, they are flexible and can "squeeze" through an opening to some degree.

TABLE 19.2

Molecular Diameters

Molecule	Diameter (A)	Molecule	Diameter (A)
Helium	2.0	Propylene	5.0
Hydrogen	2.4	Ethyl Mercaptan	5.1
Acetylene	2.4	Butene-1	5.1
Carbon Monoxide	2.8	Butene-2 Trans	5.1
Carbon Dioxide	2.8	1,3-Butadiene	5.2
Nitrogen	3.0	Chlorodiflouromethane (R-22)	5.3
Water	3.2	Thiophene	5.3
Ammonia	3.6	i-Butane to $iC_{22}H_{46}$	5.6
Hydrogen Sulfide	3.6	Dichlorodiflouromethane (R-12)	5.7
Argon	3.8	Cyclohexane	6.1
Methane	4.0	Benzene	6.7
Ethylene	4.2	Toluene	6.7
Ethylene Oxide	4.2	p-Xylene	6.7
Ethane	4.4	Carbon Tetrachloride	6.9
Methanol	4.4	Chloroform	6.9
Ethanol	4.4	Neopentane	6.9
Methyl Mercaptan	4.5	m-Xylene	7.1
Propane	4.9	o-Xylene	7.4
n-Butane to $nC_{22}H_{46}$	4.9	Triethylamine	8.4

The various commercial desiccants can be divided into three broad categories: alumina, gel and molecular sieves. Within each are a series of trade names.

Alumina is a hydrated form of aluminum oxide (Al_2O_3). When manufactured it is essentially iron free. In its natural state (bauxite) it contains varying amounts of iron. It is *activated* by driving off part of the hydrated water adsorbed on the surface.

Gel is a granular, amorphous solid. Silica gel is the generic name for a gel manufactured from sulfuric acid and sodium silicate. It is essentially 100% silicon dioxide (SiO_2). Other gels like *alumina gel* may be largely a form of Al_2O_3. Other gel type desiccants are some combination of these two.

Molecular sieves are alkali metal crystalline aluminosilicates very similar to natural clays. 4A molecular sieves are composed of Na_2O_3, Al_2O_3 and SiO_2. Types 3A and 5A are produced by ion exchange of about 75% of the Na ions by potassium and calcium ions, respectively. Type 10X is produced from 13X by ion exchange of about 75% of the Na ions by Ca ions. All types have a pH of about 10 and are stable in the pH range of 5-12.

The affinity for water is based on the previous environment. However, polarity of the water molecule also plays an important part. Molecular sieves have electric charges on the inner surfaces of

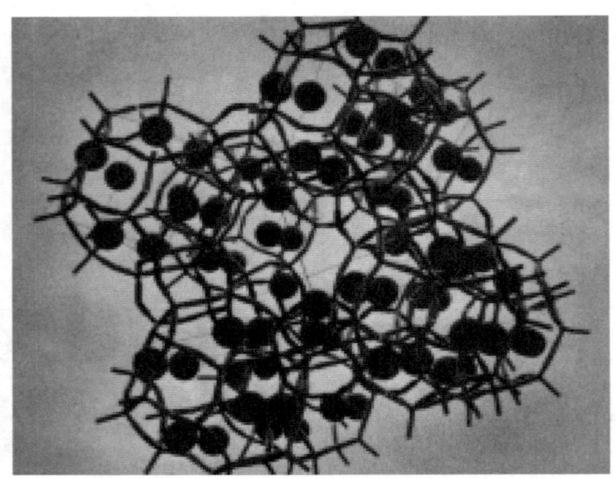

the crystal cavities, which are attracted to similar charges on polar molecules. Such molecules, including hydrogen sulfide, ammonia, carbon monoxide, methylamine, and the alcohols, are adsorbed in preference to non-polar molecules. Similarly, molecular sieves show a preference for "unsaturated" hydrocarbons, in which some of the carbon atoms are joined together by double or triple chemical bonds. This is because these compounds contain loosely bound electrons which give them polar characteristics resembling those of water molecules. As an example, if a mixed stream of ethane (a saturated hydrocarbon) and ethylene (an unsaturated hydrocarbon) is passed through a molecular sieve bed, eighty percent of the molecules adsorbed will be ethylene.

The A type sieves have a crystalline zeolite structure consisting of intracrystalline voids as shown above. All adsorption takes place in these voids. The voids are 11.4 Å in diameter and are connected by openings 4.2 Å in theoretical diameter (pore diameter). The effective pore diameter is determined by the cation and its position in the structure. The maximum diameter of molecules that can enter the crystalline structure and be adsorbed are as follows:

Type	Molecule Diameter - Å
3A – potassium zeolite	3
4A – sodium zeolite	4
5A – calcium zeolite	5
10X – calcium zeolite	8
13X – sodium zeolite	10

The X type sieves vary from the A type in the internal character of the crystalline structure. Their adsorption characteristics are the same. The X type can adsorb all molecules adsorbed by the A type with somewhat higher capacity. 13X can adsorb large molecules such as heavy mercaptans and aromatics.

The selective capacity of molecular sieves for different sizes of molecules is important. To a degree, one can exclude those sizes too large to enter the crystal. This is why a 3A or 4A sieve might be used for drying. Sieves are likewise used at high temperatures because their capacity does not decrease as much as gel or alumina above 38°C [100°F]. Table 19.3 summarizes the characteristics of common molecular sieves.

Most molecular sieve suppliers also offer special types of sieve for unique applications. Some of these are listed below.

1. 5A sieves, commonly used for removal of trace sulfur compounds from natural gas, are manufactured to minimize COS formation.
2. 4A sieves, commonly used for natural gas dehydration, are manufactured to increase the CO_2 removal capacity of the desiccant for LNG applications.
3. Acid resistant sieves, for dehydration of natural gas containing high concentrations of acid gas ($H_2S + CO_2$).

TABLE 19.3

Basic Characteristics of Molecular Sieves

Basic Type	Nominal Pore Diameter (Angstroms)	Available Form	Equilibrium H_2O Capacity (% wt)	Molecules Adsorbed	Molecules Excluded	Applications
3A	3	Powder 1/16 in Pellets 1/8 in Pellets 8-12 Beads 4-8 Beads	23 20 20 20 20	Molecules with an effective diameter <3 angstroms, including H_2O and NH_3	Molecules with an effective diameter >3 angstroms, e.g. ethane	Dry olefins, methanol, ethanol, and natural gas.
4A	4	Powder 1/16 in Pellets 1/8 in Pellets 8-12 Beads 4-8 Beads 14 × 30 Mesh	28.5 22 22 22 22 22	Molecules with an effective diameter <4 angstroms, including ethanol, H_2S, CO_2, SO_2, C_2H_4, C_2H_6, and C_3H_6	Molecules with an effective diameter >4 angstroms, e.g. propane	Dry natural gas, remove H_2S.
5A	5	Powder 1/16 in Pellets 1/8 in Pellets 8-12 Beads 4-8 Beads	28 21.5 21.5 21.5 21.5	Molecules with an effective diameter <5 angstroms, including n-C_4H_9OH, n-C_4H_{10}, C_3H_8 to $C_{22}H_{46}$, R-12	Molecules with an effective diameter >5 angstroms, e.g. iso compounds and all 4 carbon rings	Separates normal paraffins from branched-chain and cyclic hydrocarbons through a selective adsorption process, remove H_2S, and light mercaptans.
10X	8	Powder 1/16 in Pellets 1/8 in Pellets 8-12 Beads 4-8 Beads	36 28 28 28 28	Iso paraffins and Olefins, C_6H_6, Molecules with an effective diameter <8 angstroms	Di-n-butylamine and larger	Aromatic hydrocarbon separation
13X	10	Powder 1/16 in Pellets 1/8 in Pellets 8-12 Beads 4-8 Beads	36 28.5 28.5 28.5 28.5	Molecules with an effective diameter <10 angstroms	Molecules with an effective diameter >10 angstroms, e.g.	Remove mercaptans and H_2S from hydrocarbon liquids, remove H_2O and CO_2 from air plant feed.

Note: 8-12 and 4-8 refers to the Tyler screen size
4-8 beads is equivalent to a nominal diameter of 3 mm [1/8 in]
8-12 beads is equivalent to a nominal diameter of 1.5 mm [1/16 in]

Desiccant Selection

The selection of a desiccant for a particular application depends on several factors – water dewpoint specification, presence of contaminants (especially sulfur compounds), coadsorption of heavy hydrocarbons and cost.

All commercial desiccants are capable of producing water dewpoints below – 60°C [–76°F].

In a well designed and properly operated unit, the following dewpoints are achievable:

Desiccant	Outlet Dewpoint
Alumina	–73°C [–100°F]
Silica Gel	–60°C [–76°F]
Molecular Sieves	–100°C [–150°F]

For most gas drying applications upstream of low temperature NGL extraction plants and LNG plants, molecular sieve will be the first choice due to the very low outlet water dewpoints and higher effective capacity. Molecular sieves are also used in applications requiring removal of sulfur compounds. Molecular sieves are more expensive than gels or alumina and require higher regeneration heat loads.

In the presence of a gas stream saturated with water, aluminas have a higher equilibrium capacity for water than sieves, but the water loading declines rapidly as the relative saturation of the gas stream decreases. Aluminas also have a lower heat of regeneration than sieves. However, the limited outlet water dewpoints achievable with alumina preclude their use in very low temperature gas processing applications. Aluminas are sometimes used in conjunction with sieves in a compound bed application – alumina on top and sieves on the bottom. This scheme takes advantage of alumina's higher equilibrium water loading, but also uses the sieve to achieve lower outlet water dewpoints.

Silica gel is sometimes used when both a water and hydrocarbon dewpoint must be met. Some silica gels (Sorbead H) have an appreciable capacity for C_5+ hydrocarbons as well as for water. This allows both dewpoints to be achieved in a single unit. The equilibrium capacity of gel for hydrocarbons is lower than for water, consequently the bed saturates with hydrocarbons much more quickly than for water. This results in short adsorption cycle times – sometimes less than 1 hour – hence the name Short Cycle Units is often applied to these installations.

THE BASIC SYSTEM

Figure 19.1 shows the simplest dry desiccant system. It consists of two towers containing desiccant. One is drying while the other is regenerating.[19.1] During regeneration all adsorbed materials are desorbed by heat to prepare the tower for its next cycle on-stream.

At the time shown, Tower 2 is drying. The main gas stream flows into the top of the tower and out the bottom. The filter shown is not used in all systems. As later discussions will detail, the regeneration cycle consists of two parts – heating and cooling. During the heating portion the regeneration gas is heated to 200-315°C [400-600°F]. The temperature depends on the desiccant being used and the character of the material to be desorbed.

The regeneration gas by-passes the heater to cool down the bed once the desiccant bed has been heated to the desired level. This cooling normally ceases when the bed is 10-15°C [18-27°F] higher than the inlet gas temperature.

Some units are designed to use "pulse" regeneration. In these units, the heating cycle is shortened and the first portion of the cooling gas is used to displace the hot regeneration gas from the top of the bed. This strategy results in shorter regeneration cycles and is often used in hydrocarbon recovery applications.

The regeneration gas leaving the tower is cooled to condense the materials desorbed. After separation the regeneration gas can return to the main inlet gas stream or if specification allows, can be routed to sales. In dehydration applications, this regeneration gas rate will normally be 5-15% of the total throughput, with 10% being a good average. In gas dehydration, adsorption flow is almost always downward because of the higher allowable velocity in this direction. Upward regeneration is preferred even though it requires more valves and piping. Most bed contamination occurs at the top. By regenerating upward, the "steam" produced from the lower part of the bed helps remove the contamination without spreading it throughout the bed. The flow direction for cooling is optional. Upflow cooling saves two switching valves per tower (since unheated regeneration gas may be used), but requires dry gas. Downflow cooling (same direction as adsorption) is preferred if the cooling gas contains water.

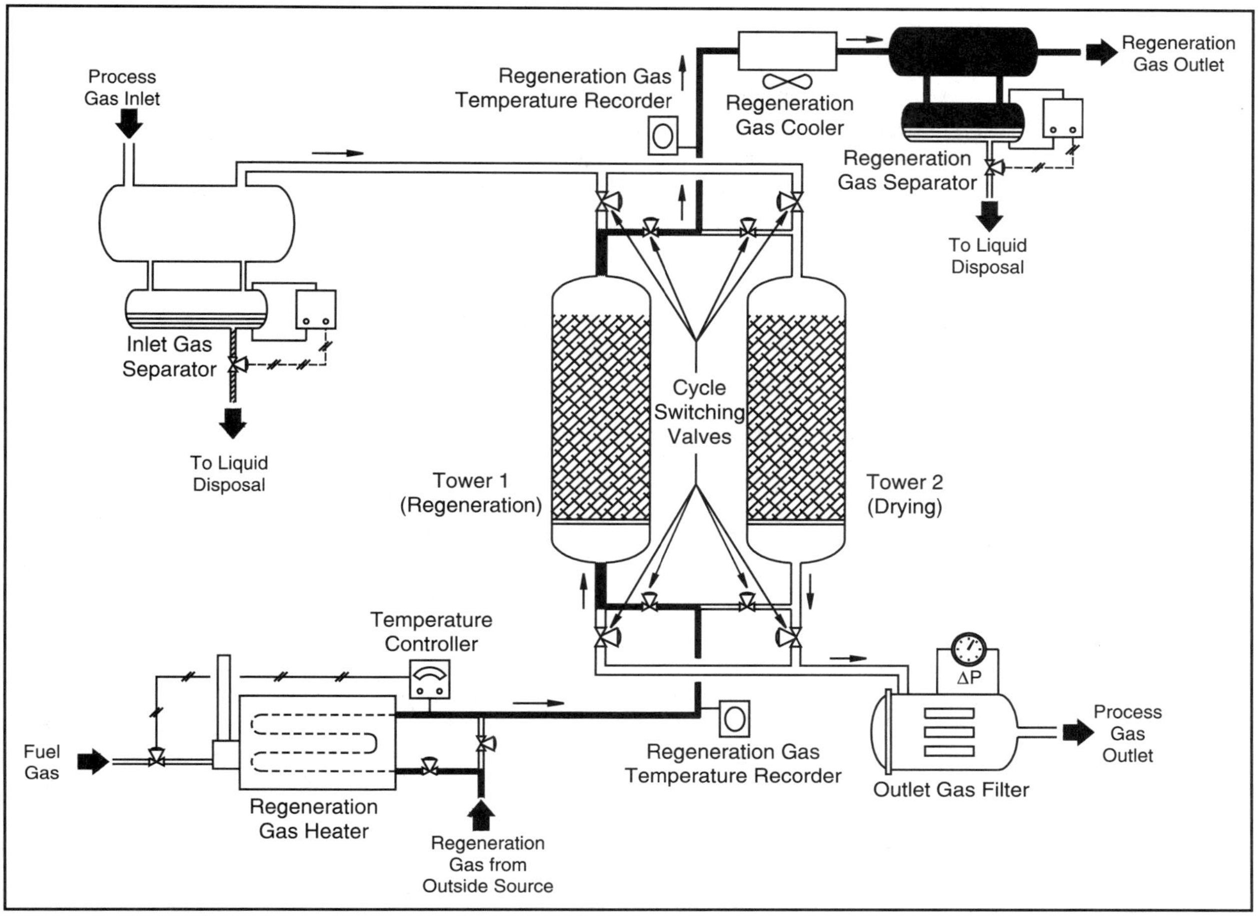

Figure 19.1 Flow Sheet of a Basic Two-Tower Dry Desiccant Unit

There are three basic sources of regeneration gas in gas dehydration:

1. Inlet gas.
2. A closed cycle separate from the stream being dehydrated.
3. Dry effluent (tail) gas from the unit.

(1) involves some degree of re-saturation of the bed during cooling which limits the useful bed capacity and if upflow cooling is used the minimum water dewpoint achievable; (2) requires a separate piping system and a high speed in-line centrifugal compressor; (3) is the most efficient and is the norm in cryogenic drying service.

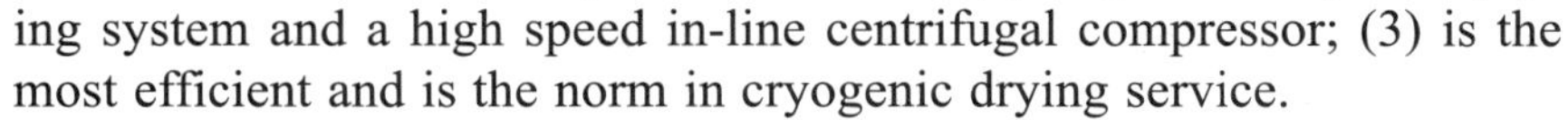

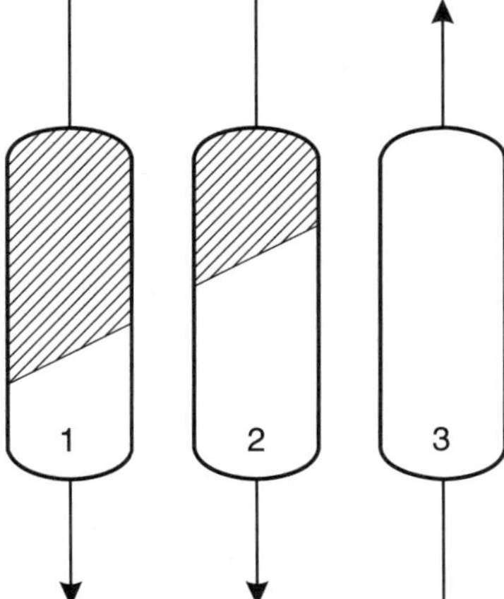

One design factor is the number of towers. Most large dry desiccant units for natural gas drying contain more than two towers to optimize the economics.

There are several ways to use multiple towers. As shown in the illustration at left using three towers, two are operating in adsorption in parallel with the third being regenerated and cooled.

In this illustration, the shaded area inside towers 1 and 2 shows the progress of water adsorption in the bed or the portion of the bed which is essentially saturated with water. Below this area, the desiccant is capable of adsorbing more water. The bottom of this area represents the position of the adsorption front as it moves down through the bed with time.

The adsorption front in bed 1 is lower than in bed 2 because it has been on-stream longer. When the leading edge of this front reaches the outlet, bed 1 will be switched to regeneration, and beds 2 and 3 will be on adsorption. Thus, at any one time, the two dehydrating towers possess different degrees of saturation. By the time bed 2 is ready for regeneration, bed 1 must be ready to go back on-stream.

The operating sequence of the towers on stream is:

1 and 2, 2 and 3, 1 and 3, 1 and 2, ad infinitum.

A similar arrangement could be used with four towers, with three on stream at a time. Obviously, the flow arrangement affects the cycle time chosen.

The adsorption beds could also be arranged in series, with the fresher bed (lag bed) downstream of the partially saturated bed (lead bed). This allows longer adsorption cycles and fully utilizes the equilibrium capacity of the desiccant; however, at a given flowrate, larger diameter towers would be required and the system pressure drop would be greater.

Another variation on the 3-tower system is shown in Figure 19.2. In this system one bed is in adsorption, one in heating, and one in cooling. This scheme is often used in Short Cycle or Hydrocarbon Recovery Units when simultaneous adsorption of C_5+ hydrocarbons and water is required. The

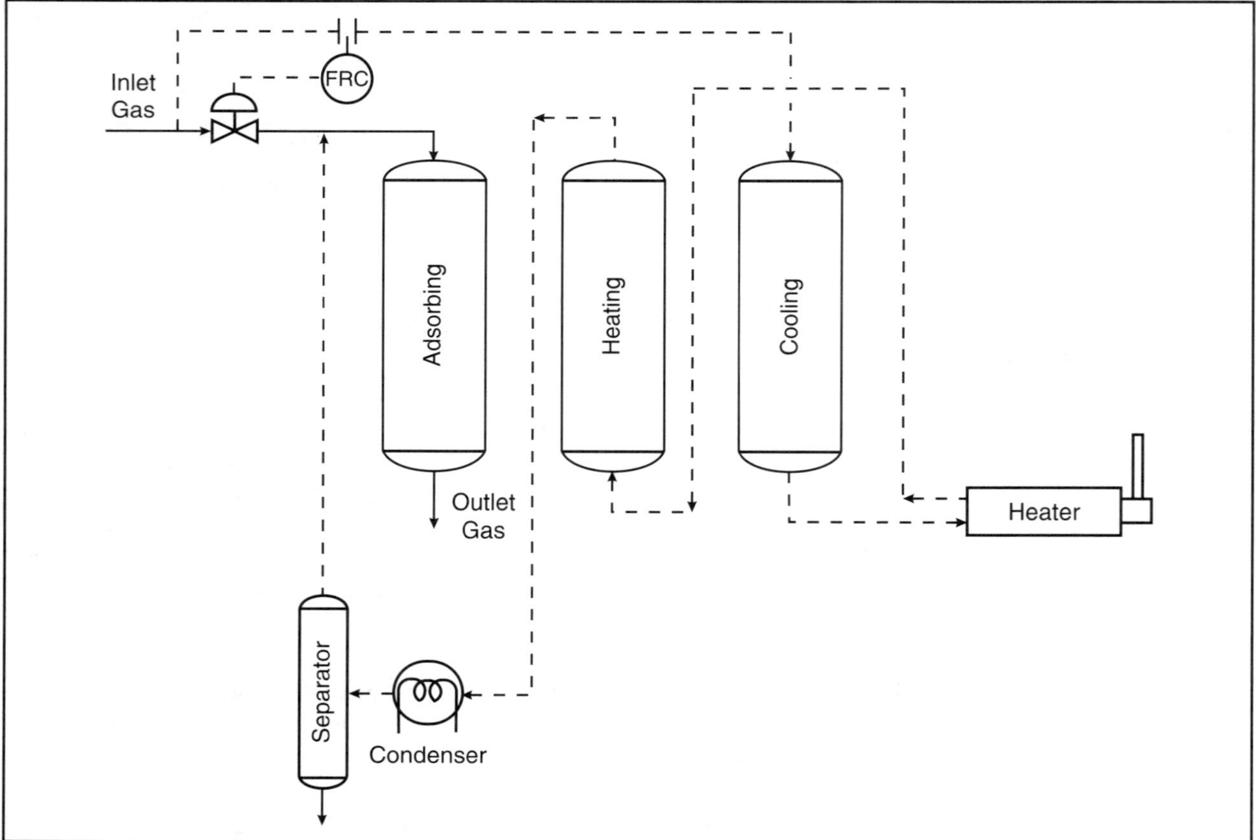

Figure 19.2 Schematic View of a Typical Three-Tower Plant Using Cooling and Heating in that Order

adsorption cycle in these facilities is sometimes less than one hour. This does not leave sufficient time to completely heat and cool one bed during the regeneration cycle, hence the use of separate heating and cooling beds.

In Figure 19.2 the regeneration gas flow is created by holding a small amount of back pressure on the inlet side of the unit and using the control valve ΔP to force the regeneration gas through the loop. Alternatively, a small in-line high speed centrifugal compressor can be installed in the regeneration loop. The latter requires higher maintenance but is more energy efficient.

Figure 19.3 shows the piping manifold for a three tower plant using a general configuration like that shown in Figure 19.2. Two-way switching valves are shown. The most common switching valves employed are Orbit® ball valves, although others are used.

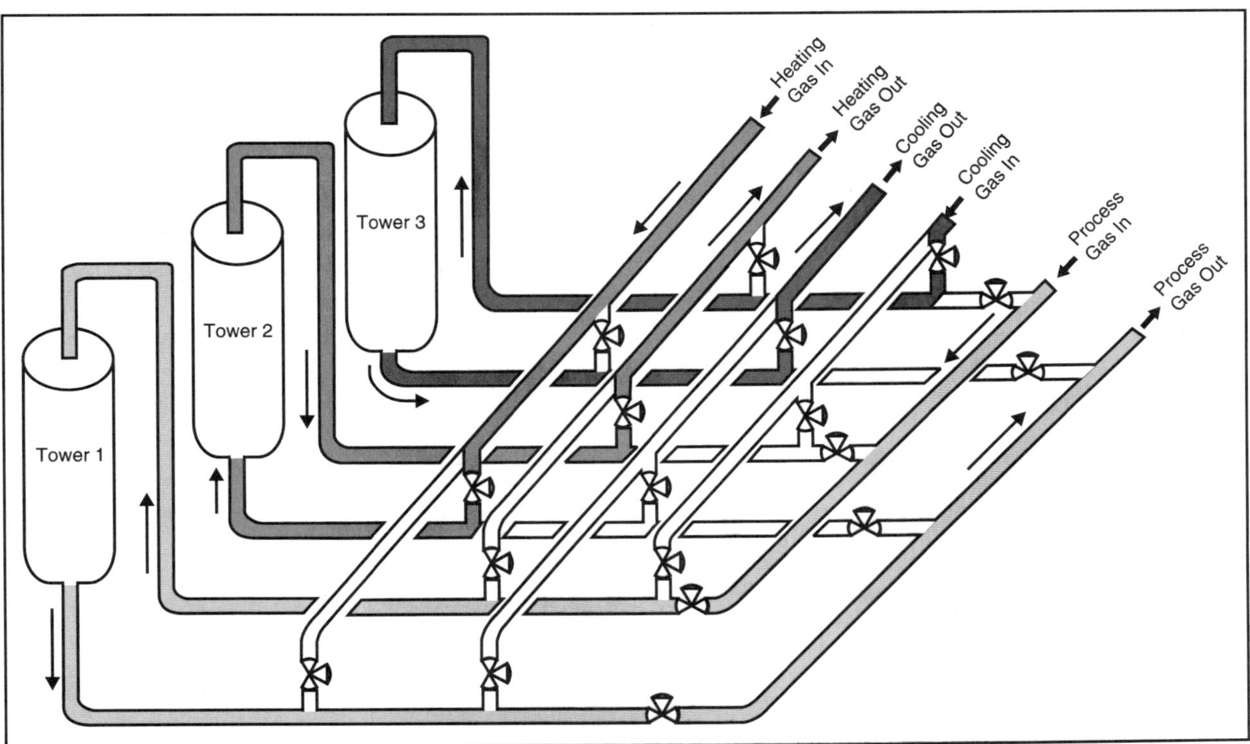

Figure 19.3 Pipe Manifold for Three-Tower Adsorber Plant (Tower 1 is Adsorbing, Tower 2 is Heating, and Tower 3 is Cooling)

Regeneration may occur at full adsorption pressure or at a reduced pressure. If the beds are regenerated at full adsorption pressure, the wet regeneration gas can sometimes be directly routed to the sales gas line without dehydrating it. Also, since the regeneration and adsorption pressures are nearly the same it is not as critical to equalize the system pressure with the bed pressure before the switching valves are opened.

Some operators regenerate at a lower pressure than the adsorption pressure. In a turboexpander plant a common source of regeneration gas is the recompressor suction (expander booster compressor discharge). The primary advantages are the higher water carrying capacity of the low pressure gas and the higher regeneration velocity through the bed for the same mass flowrate. This results in a lower regeneration gas rate compared to high pressure regeneration. When low pressure regeneration gas is used, a pressure equalization system to equalize the bed pressure with the system pressure is necessary to avoid damage to the bed from a gas flow surge when the switching valves are opened. In addition, separate dehydration of the regeneration stream is frequently required (usually with a small glycol unit) and an in-line centrifugal compressor is often used to force the regeneration gas through the regeneration loop. If the feed gas contains H_2S, special handling of the regeneration gas may be required due to the H_2S "spike" which occurs during the heating cycle.

THE NATURE OF ADSORPTION

Figure 19.4(a) illustrates the basic behavior of an adsorbent bed in gas dehydration service. During normal operation in the drying (adsorbing) cycle, three separate zones exist in the bed: 1) equilibrium zone, 2) mass transfer zone (MTZ), and 3) active zone.

In the equilibrium zone the desiccant is saturated with water. It has reached its equilibrium water capacity based on inlet gas conditions and has no further capacity to adsorb water.

Virtually all of the mass transfer takes place in the MTZ. A concentration gradient exists across the MTZ. This is illustrated in Figure 19.4(b) for various times throughout the cycle. Curves 1-3 show the formation of the MTZ; curve 4 reflects the concentration gradient for the MTZ position in Figure 19.4(a). Curve 6 shows the concentration gradient at breakthrough. Notice the adsorbate (water) bed saturation is 0% at the leading edge of the MTZ and 100% at the trailing edge.

The third zone is the active zone. In the active zone the desiccant has its full capacity for water and contains only that amount of residual water left from the regeneration cycle.

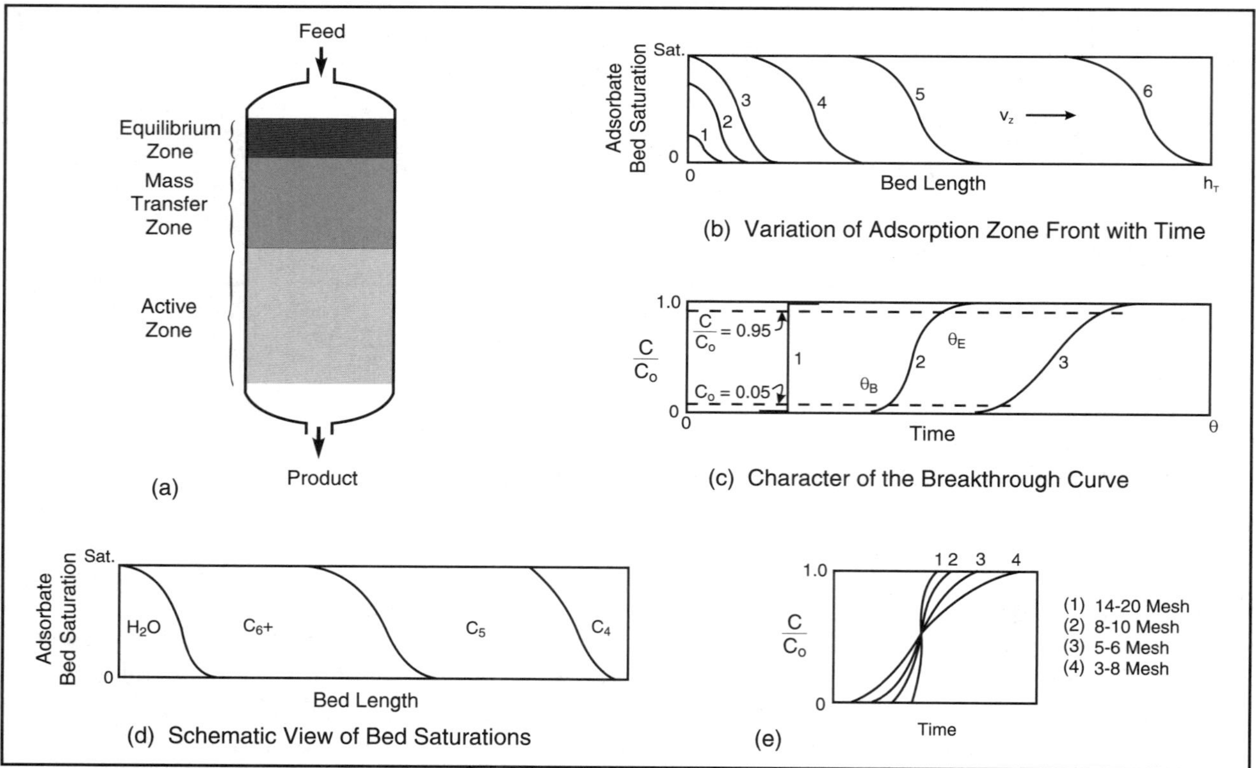

Figure 19.4 Schematic Portrayal of Adsorption Process

When the leading edge of the MTZ reaches the end of the bed, breakthrough occurs. If the adsorption process is allowed to continue, the water content of the outlet gas will increase following the traditional "S" curve. Breakthrough curves are illustrated in Figure 19.4(c) for three MTZ lengths.

Figure 19.4(d) shows the location of MTZ's in multicomponent adsorption typical of hydrocarbon and water adsorption on silica gel. As the gas enters a dry desiccant bed, all of the adsorbable components are adsorbed at different rates. After the process has proceeded for a very short period of time, a series of adsorption zones will appear. These zones represent that portion of the tower involved in the adsorption of any component. Behind the zone all of that component entering has been adsorbed on the bed. Ahead of the zone, the concentration of that compound is zero (unless some is left from a previous adsorption or regeneration). These zones form and move down through the desiccant bed. Water would be the last zone formed. On all materials except carbon it will displace the hydrocarbons if enough time is allowed to do so. If 3A or 4A molecular sieve is used, adsorption of the C_4-C_6+ fractions will not occur because these molecules cannot fit in the desiccant structure.

100% of any component is adsorbed on the desiccant until the front of its zone reaches the outlet of the bed. When the back of its zone reaches the outlet of the bed, no more adsorption of that component will occur. It will furthermore be displaced almost entirely by the component in the zone following it down the bed if the cycle is continued. If the process continues long enough, no effective amount will remain on the bed.

With silica gel, at typical flowrates and tower configurations, pentane may have a breakthrough time of 15-45 minutes. Methane and ethane break out almost instantaneously. In most installations, if the process cycle proceeds beyond 1-2 hours, only the C_7-C_8+ components will remain on the bed. From this time on, primarily dehydration is taking place. Thus, the performance of a given unit is dependent on the cycle length used. For very short cycles, both hydrocarbon adsorption and dehydration occur.

Part (e) shows the effect of desiccant size on the length of the zone. The steeper the zone, the sharper the separation, the higher the effective desiccant capacity. Therefore, the desiccant used should always be the smallest compatible with the drop limitations. The most common sizes are 4-8 mesh (Tyler screen scale) – nominally 3 mm [1/8 in] and 8-12 mesh – nominally 1.5 mm [1/16 in].

Other factors which affect the length of the MTZ include gas velocity (increasing velocity increases length), contaminants, water content and relative saturation of the inlet gas. Contaminants are particularly insidious because they can slow the mass transfer process (lengthen MTZ) by providing additional resistance to adsorption.

The length of the MTZ has a significant effect on the useful capacity of the desiccant since the MTZ is left in the bed at the end of the adsorption cycle. Remember, the desiccant in the MTZ is only partially saturated with water!

Desiccant Capacity

The capacity of a desiccant for water is expressed normally as mass of water adsorbed per mass of desiccant. There are three capacity terms used.

Static Equilibrium Capacity – the water capacity of new, virgin desiccant as determined in an equilibrium cell with no fluid flow. The static equilibrium capacity of several commercial desiccants is shown in Figure 19.5.

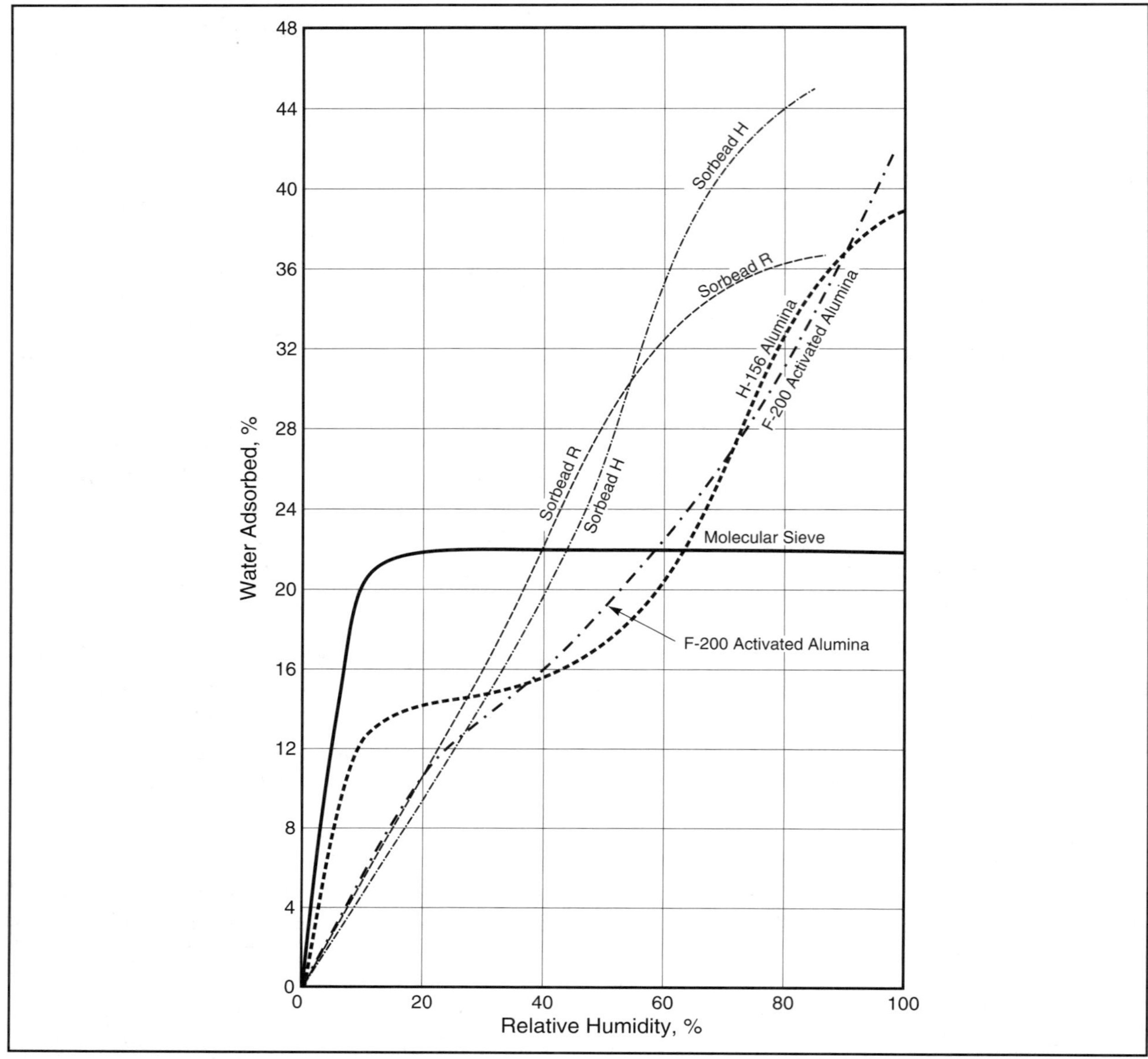

Figure 19.5 Static Equilibrium Curves for Various Commercial Desiccants

Dynamic Equilibrium Capacity – the water capacity of new, unused desiccant where the fluid is flowing through the desiccant at a commercial rate.

Useful Capacity – the design capacity that recognizes loss of desiccant capacity with time as determined by experience and economic considerations and the fact that the desiccant bed never can be fully utilized (the MTZ is left in the bed).

The static equilibrium capacity has no direct use in design although it shows the effect of P, T and water saturation on capacity. As later calculations will illustrate, dynamic and useful capacity are used directly in calculations. Dynamic capacity is typically 50-70% of static equilibrium capacity.

All desiccants degrade in service. Figure 19.6 is a typical curve for silica gel. Other desiccants will have the same shaped curves in normal service, although the values will vary. 4A and 5A molecular sieves tend to degrade more slowly because their pore size is such that heavy

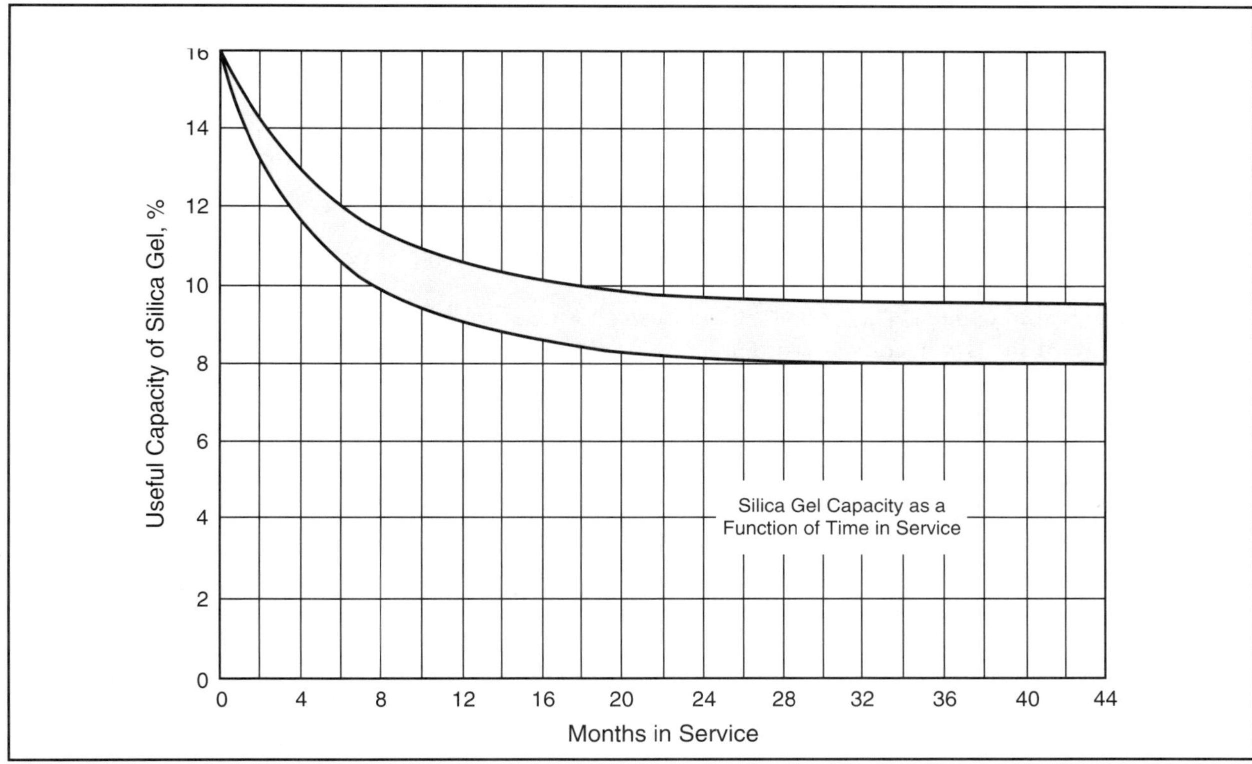

Figure 19.6 Silica Gel Capacity as a Function of Time in Service

hydrocarbon molecules are excluded from the adsorption. However, a heavy external coating of hydrocarbons in the binder macropores will still compromise their performance.

Normal degradation occurs through loss of effective surface area on repeated regeneration. This loss is rapid at first and then becomes more gradual as the desiccant "matures." Abnormal degradation occurs primarily through blockage of the small capillary or lattice openings which control access to the interior surface area. Heavy oils, amines, glycols, corrosion inhibitors and the like, which cannot be removed by regeneration, can reduce the capacity to uneconomic levels in short periods of time. There is no room for wishful thinking. If these contaminants are present ahead of the unit, provision to handle them must be made just ahead of the unit. In addition to a normal impingement separator, a coalescing filter separator or other high efficiency separation device is required.

If the gas is being processed at a pressure above the cricondentherm pressure, retrograde condensation can occur across the switching valves, piping manifold and the bed itself. In these cases, the gas should be heated slightly (5-10°C) to avoid condensation of hydrocarbons.

Free water is always a problem. Salt water entering will evaporate and fill the bed with salt. With the gels this water will cause bead breakage unless a guard section is provided at the inlet. The only permanent solution is to let no liquid water enter the bed.

Another form of degradation can occur if liquid water enters the bed. Some desiccants explode in the presence of liquid water. The *fines* thus produced increase pressure drop, reduce effective capacity, and may leave the bed and damage or plug equipment downstream. A layer of water resistant desiccant may thus be placed on top of the bed to minimize this problem. The most positive solution is effective inlet scrubbing. In addition, dust filters are frequently installed downstream of the unit to remove these fines.

The useful capacity of a desiccant may be estimated from Equation 19.1.

$$(x)(h_B) = (x_s)(h_B) - (0.45)(h_Z)(x_s) \qquad (19.1)$$

Where:

x = maximum desiccant useful capacity, wt%
x_s = dynamic capacity at saturation, wt%
h_Z = MTZ length
h_B = bed length (or length of bed to front of adsorption zone)

The dynamic capacity "x_s" must reflect desiccant condition and other such factors. It is the effective capacity of the desiccant – for water – *behind* the adsorption zone. Since desiccant degrades in service, the value used must reflect a capacity at some future time to optimize desiccant replacement cost. A value of 50-70% of the static equilibrium capacity (from Figure 19.5) is a reasonable estimate.

The value of x_s is a function of the relative water saturation of the gas. Usually the gas being dehydrated will be saturated with water (100% RH). This will be true for gas leaving a separator upstream of the dehydration system where water is drained from the separator. Exceptions include those instances where the gas may have been preheated (to prevent hydrocarbon condensation) or where the gas was dehydrated with glycol upstream of the adsorption system.

For gels and aluminas the value of x_s is also a function of temperature. No temperature correction is required for molecular sieves for temperatures less than 70°C [158°F]. Figure 19.7 shows the capacity multiplier for silica gel and activated alumina. This is applied to the dynamic (not static) equilibrium capacity.

As Equation 19.1 shows, the useful capacity of the desiccant will be less than the dynamic equilibrium capacity because, at breakthrough, the MTZ remains in the bed. The longer the MTZ, relative to the bed height, the lower the useful capacity of the desiccant. Tall, slender beds have a higher useful capacity than short, fat beds.

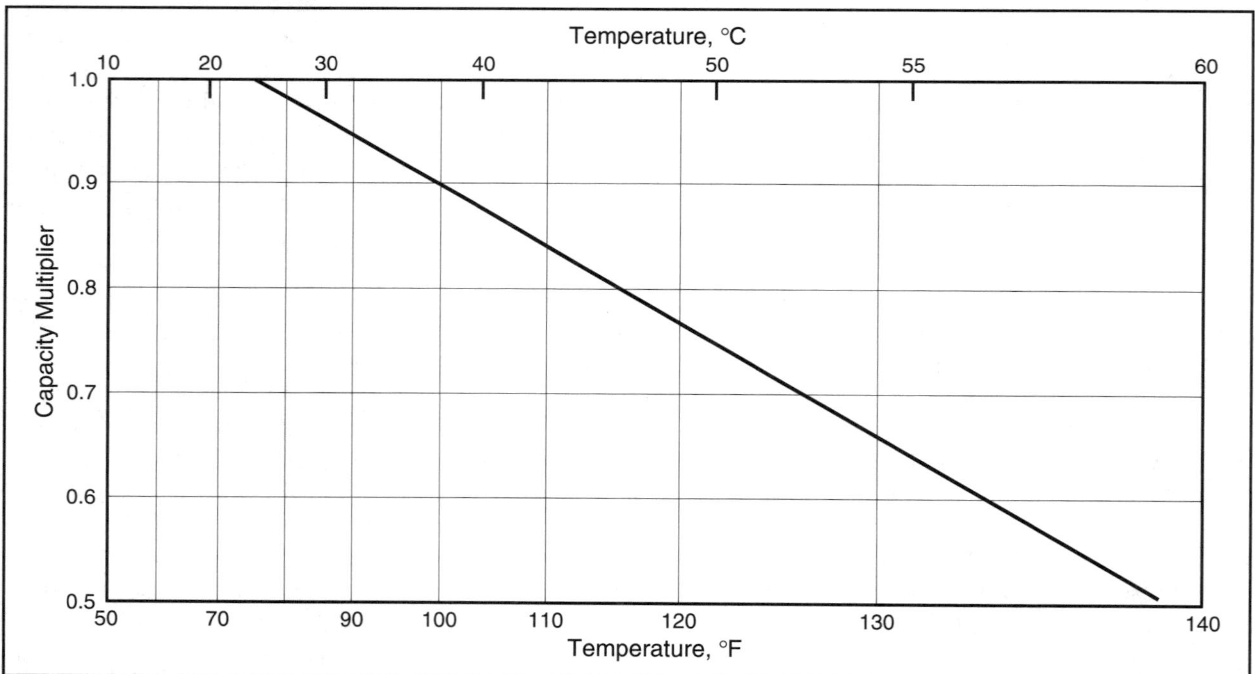

Figure 19.7 Effect of Temperature on the Dynamic Capacity of Silica Gel and Activated Alumina

Incorporating the effect of both the MTZ and the loss of desiccant capacity over time gives useful capacity values shown below:

	x, wt%
Activated Alumina	5-12
Silica Gel	5-8
Molecular Sieve (4A)	7-14

MTZ Length

The MTZ length depends on gas analysis, gas velocity, size of desiccant, coabsorption of other components, relative water saturation and bed contamination. It can vary from 0.1 to 0.2 meters up to 1.5 to 2.0 meters. The precise determination of MTZ is beyond the scope of this book and is left to the desiccant suppliers.

Equation 19.2 is a simple relationship and is based on MTZ values for air drying(19.2). It can be used to estimate MTZ lengths for natural gas dehydration. Reference 19.3 presents a slightly more complex approach.

For silica gel, the MTZ length may be estimated from Equation 19.2.

$$h_Z = A\left[\frac{(\text{flowrate})^{0.2389}(W)^{0.7895}(RH)^{0.5249}}{d^{0.4778}}\right]\left[\frac{P}{Tz}\right]^{0.5506} \qquad (19.2)$$

Where:

		SI	FPS
h_Z =	MTZ length	mm	in
A =	constant	0.155	0.165
flowrate =	gas flowrate	10^6 std m^3/d	MMscfd
W =	saturated water content of gas at dehydrated T & P	kg/10^6 std m^3	lbm/MMscf
RH =	relative humidity of gas	%	%
d =	dehydrator vessel diameter	m	ft
P =	pressure	kPa	psia
T =	temperature	K	°R
z =	compressibility factor	—	—

In natural gas service, when using Equation 19.2, the following multipliers are suggested for alumina and molecular sieves:

Alumina – 0.8 times h_Z for gel

Molecular Sieve – 0.6 times h_Z for gel

Shorter zones are obtained with these materials because they have less capacity for hydrocarbons. When using h_Z in Equation 19.1, the numerical value of 0.45 is an average number based on test. It is a function of MTZ but only varies from 0.40-0.52 in a wide range of applications. The value used is the mode of the distribution curve for most services.

ADSORBER DESIGN

The design of fixed bed adsorption systems is complex. The desiccant manufacturers will typically perform these calculation with their own proprietary software. Unlike many other processes, these calculations cannot be done with a process simulator. The manual calculations presented here can be used for preliminary designs and feasibility studies.

1. The first step in planning and specifying a dry desiccant unit is to establish the number of adsorbers and the adsorption cycle time. For small units (< 1.5×10^6 m^3/d [50 MMscfd]) a two tower system is generally a good assumption. For larger systems a 3 or 4 tower system may be more economic. Increasing the number of adsorbers allows for better bed geometry (tall thin beds) and increases operating flexibility. The drawback is higher capital costs and reduced time for regeneration since the regeneration time, t_r, is equal to the adsorption time, t_a divided by n – 1 where n = number of beds.

$$t_r = \frac{t_a}{n-1} \tag{19.3}$$

   Shorter regeneration times increases the regeneration gas rate and the size of the regeneration equipment.

   Once the number of beds has been decided, the cycle time will determine the mass of desiccant per tower. If the feed gas is saturated with water a good assumption for cycle time is 8-16 hours. If the feed gas has been dehydrated upstream with a glycol unit, a cycle time of 24-30 hours may be feasible. There are a limitless number of beds and cycle times which result in a technically feasible design, the goal is to find the one which minimizes capital and operating costs.

2. For the unit configuration established in step 1, the second step is to size the adsorbers. The diameter of the adsorbers is set by gas velocity and allowable pressure drop. Several methods of determining the bed diameter are presented in the literature. Equation 19.4, below is based on the Ergun equation.[19.4]

$$\frac{\Delta P}{L} = B\mu v_g + C\rho_g v_g^2 \tag{19.4}$$

Where:			SI	FPS
	$\Delta P/L$ =	pressure drop/length	kPa/m	psi/ft
	μ =	gas viscosity	cp	cp
	ρ_g =	gas density	kg/m^3	lbm/ft^3
	v_g =	superficial gas velocity	m/min	ft/min

Constants B and C for Equation 19.4 are:

	SI		FPS	
Particle Type	B	C	B	C
1/8″ bead	4.16	0.001 35	0.0560	0.000 088 9
1/8″ extrudate	5.36	0.001 89	0.0722	0.000 124
1/16″ bead	11.3	0.002 07	0.152	0.000 136
1/16″ extrudate	17.7	0.003 19	0.238	0.000 210

Most designs are based on a $\Delta P/L$ of about 7-10 kPa/m [0.3-0.44 psi/ft].

The particle diameter is found from the mesh size of the desiccant used. Alumina and bauxite are granular materials; the gels are somewhat spherical. Their size is determined by screening through a series of screens bearing a *mesh size*. Different scales are used. The most common is the Tyler Screen Scale. The table below shows the common mesh sizes used for most desiccants.

Tyler Mesh	Screen Opening	
	mm	inches
3	6.680	0.263
4	4.699	0.185
5	3.962	0.156
6	3.327	0.131
7	2.794	0.110
8	2.362	0.093
9	1.981	0.078
10	1.651	0.065
12	1.397	0.055
14	1.168	0.046

A typical sieve or gel will have a size like 3-8 mesh or 4-8 mesh. The first number is the size of screen all particles pass through; the second number is the size opening all particles are retained on. The size distribution is never uniform but for calculation purposes an average size for the range may be used for D_p.

Molecular sieves are supplied in spherical and pellet (extrudate) form, as well as a powder. Pellets 1.59 mm and 3.18 mm [1/16 and 1/8 in] are available in most grades. Comparable sphere sizes are offered by some suppliers.

When Equation 19.4 is written in terms of velocity it becomes

$$v = \frac{-B\mu + \left[(B\mu)^2 + 4C\rho_g\left(\frac{\Delta P}{L}\right)\right]^{0.5}}{2C\rho_g} \tag{19.5}$$

For preliminary calculations, Equation 19.6 gives a useful approximation of gas velocity.

$$v_g = \frac{A}{\sqrt{\rho_g}} \tag{19.6}$$

Where:			SI	FPS
v_g =	superficial gas velocity		m/min	ft/min
ρ_g =	gas density		kg/m^3	lbm/ft^3
A =	constant	1/8″ spheres	67	55
		1/16″ spheres	48	39

For most moderate to high pressure designs using 4-8 Tyler screen mesh (3 mm [1/8 in] diameter) desiccant the superficial gas velocity will be 9-12 m/min [30-40 ft/min].

Calculation of tower diameter follows:

$$d = \sqrt{\frac{4 q_a}{\pi v_g}} \tag{19.7}$$

Where:			SI	FPS
	d =	tower diameter	m	ft
	q_a =	actual gas flowrate	m^3/min	ft^3/min
	v_g =	superficial gas velocity	m/min	ft/min

The actual flowrate, q_a, can be calculated from Equation 19.8.

$$q_a = \frac{q_s}{1440}\left(\frac{P_s}{P}\right)\left(\frac{T}{T_s}\right) z \tag{19.8}$$

Where:			SI	FPS
	q_s =	standard gas flow	std m^3/d	scf/day
	P_s =	standard pressure	kPa	psia
	P =	actual pressure	kPa	psia
	T =	actual temperature	K	°R
	T_s =	standard temperature	K	°R
	z =	gas compressibility factor at T and P	—	—

Once the tower diameter has been determined, the bed height may be calculated from Equation 19.9.

$$h_B = \frac{400\, m_w}{\pi\, x\, \rho_B\, d^2} \tag{19.9}$$

Where:			SI	FPS
	h_B =	bed height	m	ft
	m_w =	water loading/cycle	kg	lbm
	x =	useful desiccant capacity, %	—	—
	ρ_B =	bulk density of desiccant	kg/m^3	lbm/ft^3
	d =	bed diameter	m	ft

The bulk density, ρ_B, can be found in Table 19.1. Values for useful capacity, x, can be found from Equation 19.1 or from the values on page 409.

The actual tower height (SS, TT) will be the bed height plus the height of bed supports and sufficient space to ensure good flow distribution at the top of the bed. This additional height is typically 1-1.5 m [3.3-5 ft]. An example adsorption tower is shown in Figure 19.8.

The desired bed length to bed diameter ratio (h_B/d) should fall between about 2.5-6. A value less than 2.5 can result in lower useful desiccant capacity due to the relatively large MTZ/h_B ratio. A value greater than 6 can result in excessive ΔP. The total ΔP across an adsorbent tower should not exceed 55-70 kPa [8-10 psi].

If the bed is too short, the cycle time or number of beds should be increased. If the bed is too long the opposite is true.

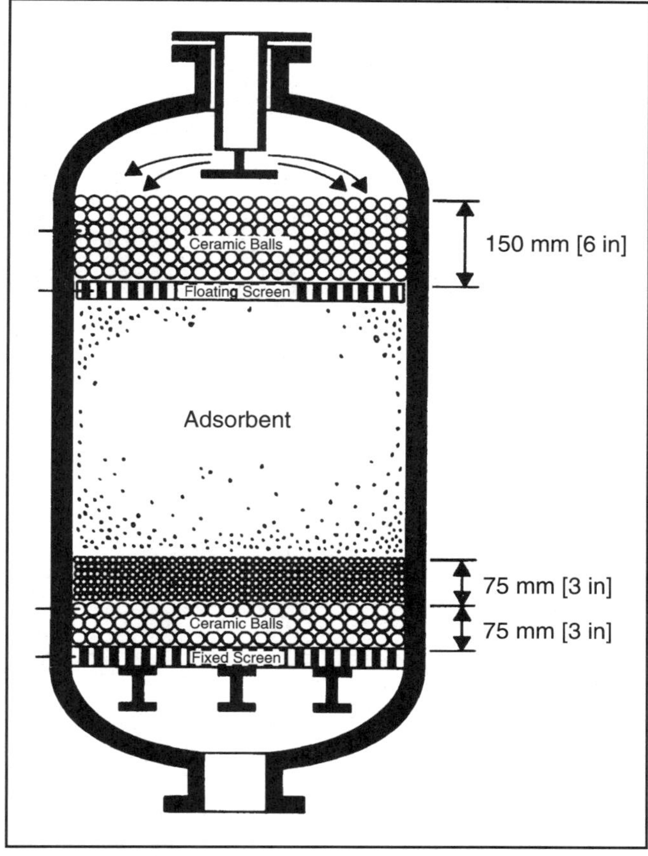

Figure 19.8 Adsorber with Floating Screen and Balls for Improved Gas Distribution

Example 19.1: Perform a preliminary sizing calculation on a molecular sieve dehydration system processing the gas shown below. The desiccant is 3 mm [1/8″] beads.

Feed Gas:
- flow = 2.0×10^6 std m^3/d [71 MMscfd]
- T = 30°C [86°F]
- P = 6000 kPa [870 psia]
- $z = 0.86$
- $\gamma = 0.65$
- $\mu = 0.014$ cp
- W = 720 kg/10^6 std m^3 [45 lbm/MMscf]

Assume a 3 bed system with 2 beds in parallel and a 12 hour adsorption cycle.

SI Solution:

Step 1: Calculate the superficial gas velocity, v_g

$$\rho = \frac{(P)(MW)}{zRT} = \frac{(6000)(0.65)(28.97)}{(0.86)(8.314)(303)} = 52.1\,kg/m^3$$

$$v_g = \frac{67}{\sqrt{52.1}} = 9.3\,m/min \qquad \text{Eq. 19.6}$$

Example 19.1 (Cont'd.):

Step 2: Calculate the bed diameter, d

$$q_a = \left(\frac{1000000}{1440}\right)\left(\frac{101.3}{6000}\right)\left(\frac{303}{288}\right)(0.86) = 10.6\ m^3/min \qquad \text{Eq. 19.8}$$

$$d = \sqrt{\frac{4q_a}{\pi v_g}} = \sqrt{\frac{(4)(10.6)}{(\pi)(9.3)}} = 1.2\ m \qquad \text{Eq. 19.7}$$

Step 3: Calculate the bed height for a useful loading, x, of 10%. The water loading per cycle, m_W is:

$$\left(\frac{1\times10^6\ std\ m^3}{d}\right)\left(\frac{12\ h}{24\ h}\right)\left(\frac{720\ kg}{10^6\ std\ m^3}\right) = \frac{360\ kg}{cycle} H_2O$$

$$h_B = \frac{(400)(360)}{(\pi)(10)(705)(1.2)^2} = 4.5\ m \qquad \text{Eq. 19.9}$$

length to diameter ratio = 4.5/1.2 = 3.8 OK!

Check ΔP:

$$\frac{\Delta P}{L} = B\mu\, v_g + C\rho_g\, v_g^2 \qquad \text{Eq. 19.4}$$

$$= (4.16)(0.014)(9.3) + (0.00135)(52.1)(9.3)^2 = 6.6\ kPa/m$$

Total ΔP = (6.6)(4.5) = 30 kPa OK!

Step 4: Estimate the useful capacity of the desiccant from Equations 19.1 and 19.2.

$$h_z = 0.155\left[\frac{(1)^{0.2389}(720)^{0.7895}(100)^{0.5249}}{(1.2)^{0.4778}}\right]\left[\frac{6000}{(303)(0.86)}\right]^{0.5506}$$

$$= 1615\ mm\ \text{for silica gel} \qquad \text{Eq. 19.2}$$

$$= (0.6)(1615) = 970\ mm\ \text{for mol sieve}$$

$$(x) = \frac{(x_s)(h_B) - (0.45)(h_Z)(x_s)}{(h_B)} \qquad \text{Eq. 19.1}$$

assume x_s = 14%

$$x = \frac{(14)(4.5) - (0.45)(0.97)(14)}{4.5} = \underline{\underline{12.6\ wt\%}}$$

a value of x = 10% is reasonable when considering the aging and deterioration of the dessicant.

Example 19.1 (Cont'd.):

FPS Solution:

Step 1: Calculate the superficial gas velocity, v_g

$$\rho = \frac{(P)(MW)}{zRT} = \frac{(870)(0.65)(28.97)}{(0.86)(10.73)(546)} = 3.25 \text{ lbm/ft}^3$$

$$v_g = \frac{55}{\sqrt{3.25}} = 30.5 \text{ ft/min} \qquad \text{Eq. 19.6}$$

Step 2: Calculate the bed diameter, d

$$q_a = \left(\frac{35\,500\,000}{1440}\right)\left(\frac{14.7}{870}\right)\left(\frac{545}{520}\right)(0.86) = 375 \text{ ft}^3/\text{min} \qquad \text{Eq. 19.8}$$

$$d = \sqrt{\frac{4q_a}{\pi v_g}} = \sqrt{\frac{(4)(375)}{(\pi)(30.5)}} = 4.0 \text{ ft} \qquad \text{Eq. 19.7}$$

Step 3: Calculate the bed height for a useful loading, x, of 10%. The water loading per cycle, m_W is:

$$\left(\frac{35.5 \text{ MMscf}}{\text{day}}\right)\left(\frac{12 \text{ hr}}{24 \text{ hr}}\right)\left(\frac{45 \text{ lbm}}{\text{MMscf}}\right) = \frac{799 \text{ lbm}}{\text{cycle}} H_2O$$

$$h_B = \frac{(400)(799)}{(\pi)(10)(43.8)(4.0)^2} = 14.5 \text{ ft} \qquad \text{Eq. 19.9}$$

length to diameter ratio = 14.5/4.0 = 3.6 OK!

Check ΔP:

$$\frac{\Delta P}{L} = B\mu v_g + C\rho_g v_g^2$$

$$= (0.0560)(0.014)(30.5) + (0.000\,088\,9)(3.25)(30.5)^2 \qquad \text{Eq. 19.9}$$

$$= 0.29 \text{ psia/ft}$$

Total ΔP = (0.29)(14.5) = 4.2 psi OK!

Step 4: Estimate the useful capacity of the desiccant from Equations 19.1 and 19.2.

$$h_z = 0.165\left[\frac{(35.5)^{0.2389}(45)^{0.7895}(100)^{0.5249}}{(4.0)^{0.4778}}\right]\left[\frac{870}{(545)(0.86)}\right]^{0.5506}$$

$$= 63.3 \text{ in} = 5.3 \text{ ft for silica gel} \qquad \text{Eq. 19.2}$$

$$= (0.6)(5.3) = 3.2 \text{ ft for mol sieve}$$

assume x_s = 14%

$$x = \frac{(14)(14.5) - (0.45)(3.2)(14)}{14.5} = \underline{\underline{12.6 \text{ wt\%}}} \qquad \text{Eq. 19.1}$$

a value of x = 10% is reasonable when considering the aging and deterioration of the dessicant.

REGENERATION DESIGN

Once the adsorber towers have been sized, the next step is to determine the amount of regeneration gas needed as well as the heating and cooling loads.

The heating load is most critical because this is the primary operating cost unless waste heat is available. Heating must accomplish all of the following:

1. Heat the desiccant to at least 204-288°C [400-550°F] (depends on desiccant).
2. Heat and then vaporize the adsorbed water.
3. Heat and then vaporize any hydrocarbons on the bed.
4. Heat the vessel shell and steel internals (if not insulated internally).
5. Heat the valves and piping in the line between the regeneration heater and the towers.
6. Supply heat lost through the insulation.

If the desiccant is a 4A or 5A molecular sieve, it is safe to assume that at the end of the heating cycle the bed temperature will be 260-288°C [500-550°F]. With 3A sieve, silica gel, and alumina these temperatures will be 180-220°C [350-425°F]. The actual regeneration temperature should be the minimum required to adequately regenerate the bed.

Figure 19.9 shows a regeneration temperature profile for a silica gel system. The temperature, T_H, is the hot regeneration gas into the bed. The temperature profile T_1 to T_4 is the outlet gas temperature leaving the bed. In this case, when the bed outlet temperature (T_4) reaches approximately 176°C [350°F], the heating cycle is finished and the cooling cycle begins. The temperature profile T_4 to T_5 shows the bed outlet temperature during the cooling cycle.

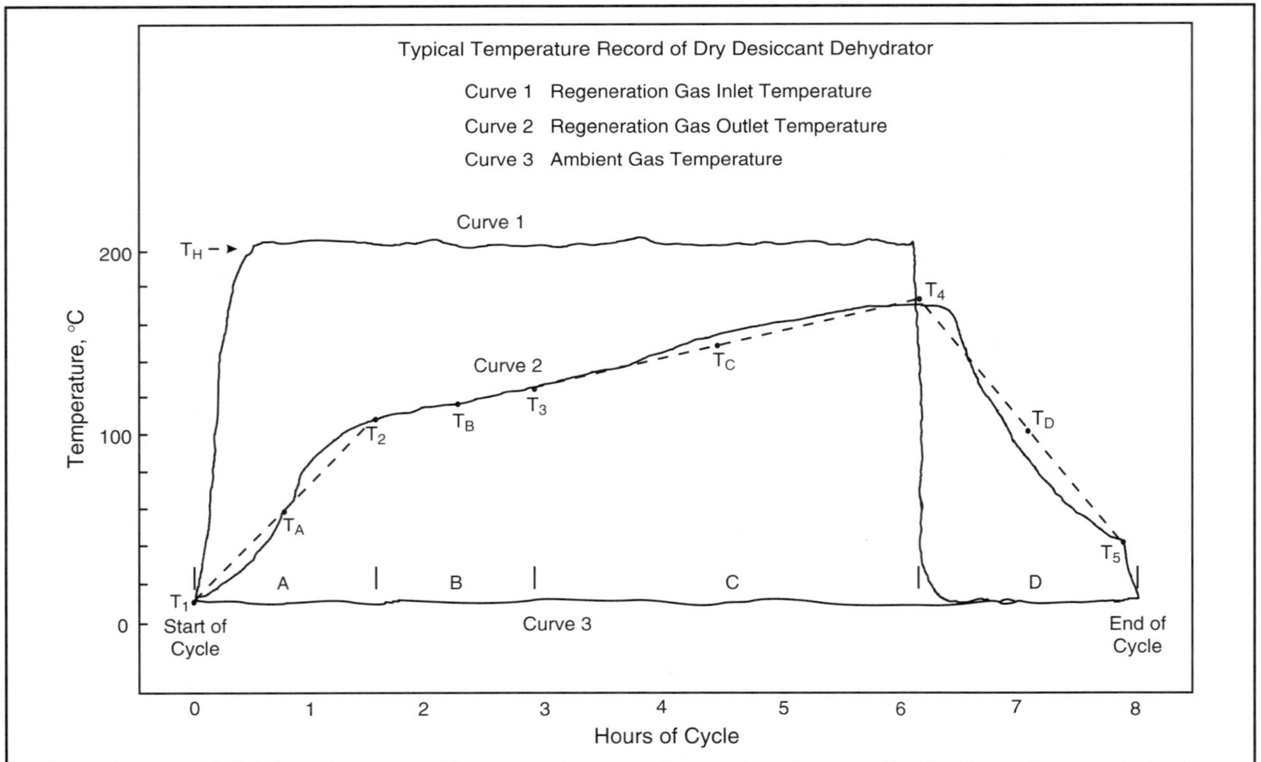

Figure 19.9 Temperature Curves for a Two-Tower Adsorber Plant

The entire regeneration cycle can be divided into four (4) specific time intervals. Interval A (Q_A) is virtually all sensible heat. It represents the time required to heat the bed, steel and adsorbed water from T_1 to T_2. At T_2, water begins to boil off the desiccant. For commercial desiccants, T_2 is about 110°C [230°F] and T_3 is about 127°C [260°F].

Interval B (Q_B) is where the most of the water is driven from the bed. This requires sufficient heat to not only revaporize the water but also to break the attractive forces which bind the water to the surface of the adsorbent. This is often called the heat of wetting. The sum of the latent heat and the heat of wetting is the heat of desorption. This value is approximately 4200 kJ/kg [1800 Btu/lbm] for sieves and 3260 kJ/kg [1400 Btu/lbm] for alumina and gels. The temperature profile for interval B is T_2 to T_3.

Once the bulk of the water has been driven from the bed, interval C (Q_C) represents the time required to remove heavy contaminants and residual water. This is sometime referred to as driving the "boot" off the bed. The temperature profile for Q_C is from T_3 to T_4.

The regeneration gas flowrate is established by heat balance. The regeneration gas rate must be adequate to deliver the required heat input in the time available. Likewise, it must also be sufficient to deliver the cooling in the time available. In equation form, this heat balance is summarized below.

$$Q_A = mC_p \, \Delta T_A \, \theta_A \tag{19.10a}$$

$$Q_B = mC_p \, \Delta T_B \, \theta_B \tag{19.10b}$$

$$Q_C = mC_p \, \Delta T_C \, \theta_C \tag{19.10c}$$

$$Q_D = mC_p \, \Delta T_D \, \theta_D \tag{19.10d}$$

Where:			SI	FPS
Q	=	heat load in a time interval	kJ	Btu
m	=	mass flow of regeneration gas	kg/h	lbm/hr
C_p	=	heat capacity of regeneration gas	kJ/kg·°C	Btu/lbm-°F
ΔT	=	effective temperature difference, e.g. ($T_H - T_B$) for a time interval	°C	°F
θ	=	length of time interval	h	hr

Equations 19.10(a-d) can be solved for m subject to the constraint

$$\theta_A + \theta_B + \theta_C + \theta_D < \text{time available for regeneration} \tag{19.11}$$

For a two tower system, the time available for regeneration is equal to the adsorption time. For the three tower system shown on page 401, the time available is one-half the adsorption time. For the three tower system in Figure 19.2, the time available is the adsorption time.

Equations 19.10 and 19.11 can only be solved by trial and error and are not suitable for manual calculations. One problem is the inability to easily determine the effective temperature differences ΔT_A, ΔT_B , etc.

An alternative is to calculate the total heating load, Q_H, ($Q_H = Q_A + Q_B + Q_C$) and cooling load, Q_D. This approach raises the problem of determining the overall temperature difference (ΔT) for the heating and cooling cycles. The temperature difference at any time is represented by the difference between curve 1 and curve 2 in Figure 19.9.

One method of estimating the average ΔT difference is to use a log-mean temperature difference.

$$\Delta T_H = \frac{(T_H - T_1) - (T_H - T_4)}{\ln\left(\dfrac{T_H - T_1}{T_H - T_4}\right)} \quad \text{heating} \quad (19.12)$$

$$\Delta T_D = \frac{(T_4 - T_1) - (T_5 - T_1)}{\ln\left(\dfrac{T_4 - T_1}{T_5 - T_1}\right)} \quad \text{cooling} \quad (19.13)$$

The calculation of the regeneration rate then follows:

$$m = \frac{Q_H}{C_p\, \Delta T_H\, \theta_H} = \frac{Q_D}{C_p\, \Delta T_D\, \theta_D} \quad (19.14)$$

To simplify the calculation further, many references suggest that Q_H be multiplied by 2.5 and T_H be taken as $T_H - T_1$. This eliminates the calculation of a mean temperature difference and gives results surprisingly close to the more complex methods. This is summarized below in Equation 19.15.

$$m = \frac{2.5\, Q_H}{C_p\, (T_H - T_1)\, \theta_H} \quad (19.15)$$

The calculation of the regeneration gas rate is still iterative since $\theta_H + \theta_D$ must be less than the time available. For first guess, the following guidelines can be used:

a. Assume m = 10% of the process gas rate
b. Allocate 60-70% of the total regeneration time to heating and 30-40% to cooling
c. θ_H should not be less than 1 hour or greater than 8 hours
d. The minimum heating gas velocity should exceed that which gives a ΔP = 0.23 kPa/m [0.01 psi/ft] to ensure good flow distribution and water removal. This velocity can be calculated from Equation 19.4.

The heat loads Q_H and Q_D can be calculated by heat balance. Q_H is the sum of the sensible heat required to heat the steel, desiccant, support balls, valves, piping, and water plus the heat of desorption.

Steel Shell

The heat required for the steel shell will depend on whether internal or external insulation is used. Internal insulation is of two types: (1) a steel "can" inside the shell that provides a stagnant gas space between the bed and shell or (2) cast or sprayed internal insulation. These are illustrated in Figure 19.10. With internal insulation, the bed diameter usually is about 15 cm [6 in] less than the shell I.D.

Internal insulation is usually required on towers when the regeneration time is limited. It may also be desirable on longer cycle units to save on fuel costs. For towers with internal insulation the shell sensible heat load is only about 25% that of an externally insulated tower.

The approximate wall thickness of adsorber is shown in Figure 19.11. This is based on ASME Section VIII Division 1 and A-516 Gr 70 plate steel. For Section VIII Division 2 and BS 5500 use a

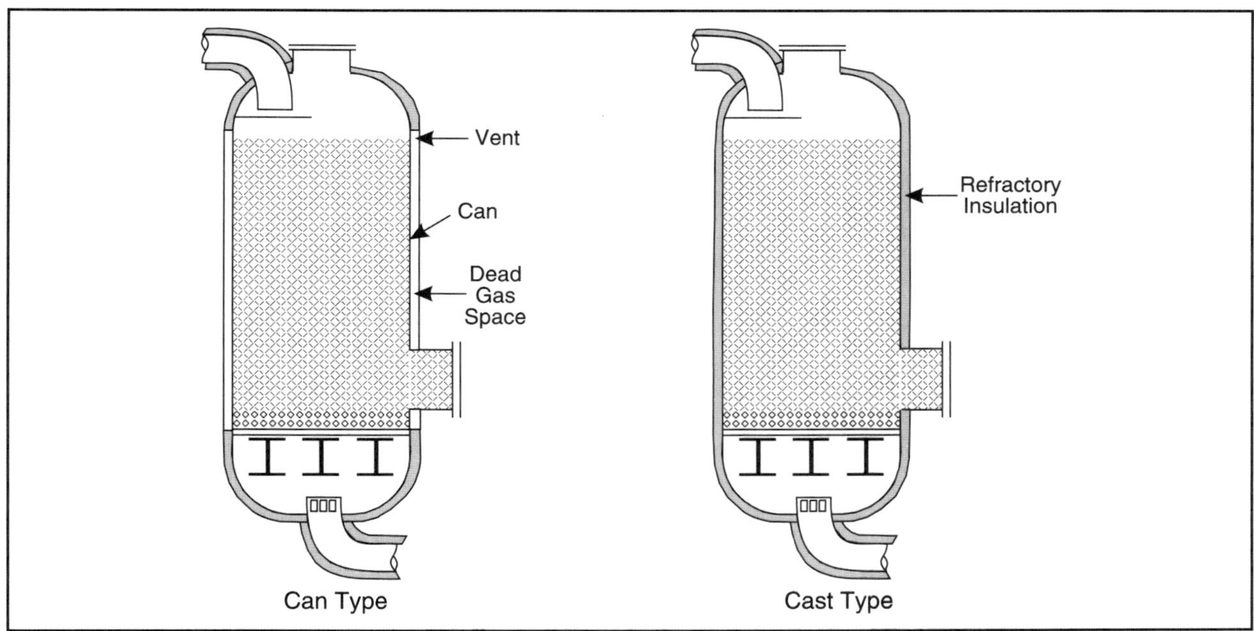

Figure 19.10 Types of Internal Insulation

wall thickness approximately 70% of that in Figure 19.11. The diameter should be the nearest commercial size available that will allow a desiccant bed diameter at least as large as that calculated. The mass of the shell and heads may be estimated from Equation 19.16.

$$m_s = A\,h\,d\,t \tag{19.16}$$

Where:			SI	FPS
	m_s =	mass of steel	kg	lbm
	h =	vessel length (HS-HS)	m	ft
	d =	vessel I.D.	mm	in
	t =	shell thickness	mm	in
	A =	weight factor	0.0347	15
	HS-HS =	head seam to head seam		

The following values are suitable for the heat balance calculation:

Heat Capacity, C_p	
Steel	0.50 kJ/kg·°C [0.12 Btu/lbm-°F]
Liquid Water	4.19 kJ/kg·°C [1.00 Btu/lbm-°F]
Desiccant	see Table 19.1
Heat of Desorption	
H_2O on sieves	4187 kJ/kg [1800 Btu/lbm]
H_2O on gel or alumina	3256 kJ/kg [1400 Btu/lbm]
Hydrocarbons	465 kJ/kg [200 Btu/lbm]

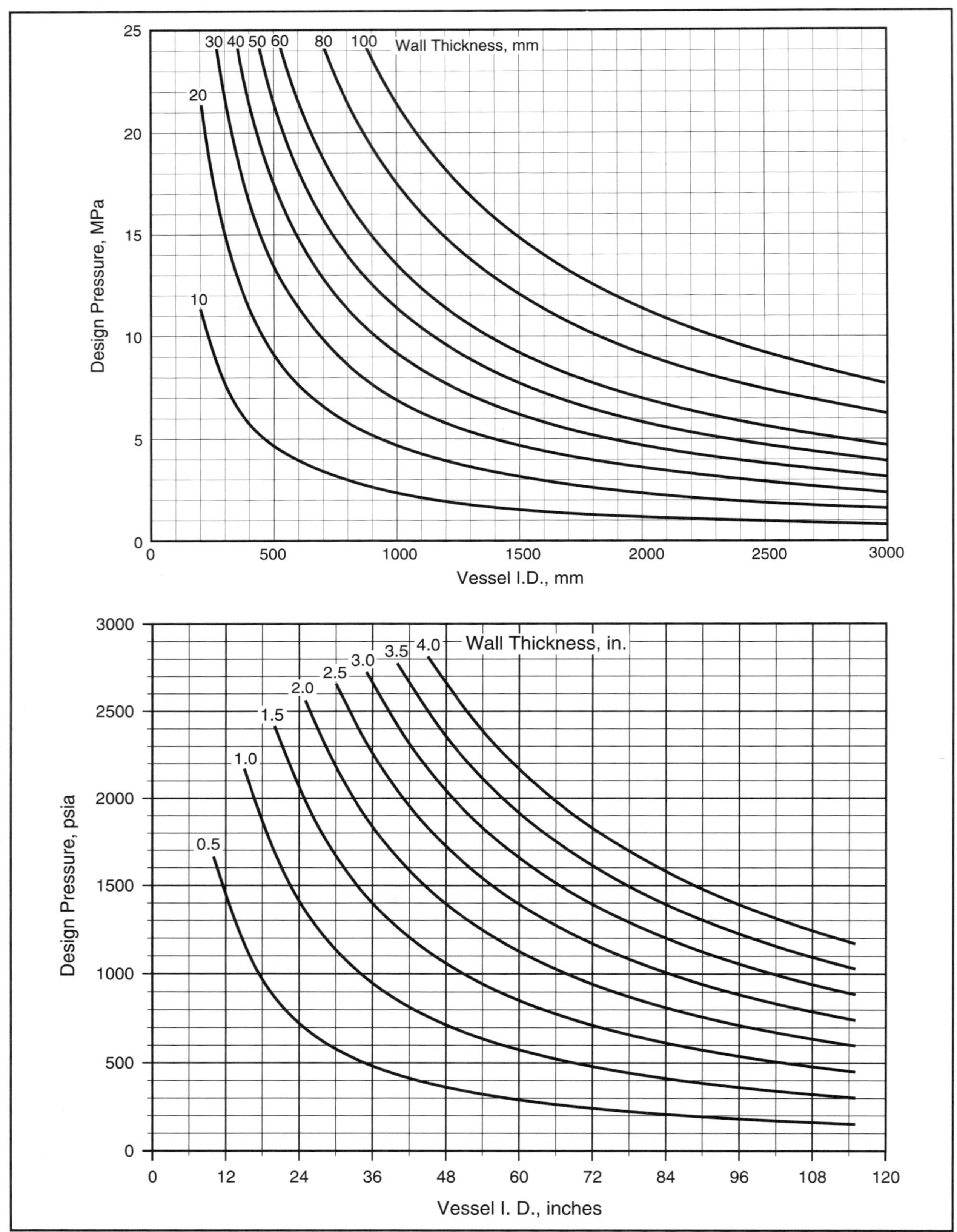

Figure 19.11 Approximate Pressure Vessel Wall Thickness

Calculation of Q_H

$$Q_H = (\text{mass of steel})(C_p)(T_4 - T_1) + (\text{mass of desiccant})(C_p)(T_4 - T_1) + (\text{mass of water})(C_p)(T_B - T_1) + (\text{mass of water})(\text{heat of desorption}) + (\text{heat losses and misc. heat load})(\text{piping, valves, support balls, etc.}) \quad (19.17)$$

The last item in Equation 19.17 is often estimated to be 25-30% of the calculated value for the steel, desiccant and water.

Calculation of Q_D

$$Q_D = (\text{mass of desiccant})(C_p)(T_4 - T_5) + (\text{mass of steel})(C_p)(T_4 - T_5) \quad (19.18)$$

Regeneration Gas Heater

$$Q_{RH} = mC_p(T_H - T_1) \quad (19.19)$$

Where:			SI	FPS
	Q_{RH} =	regeneration heater duty	kW	Btu/hr
	m =	regeneration gas mass flow	kg/s	lbm/hr
	C_p =	average regeneration gas heat capacity	kJ/kg·°C	Btu/lbm-°F
	T_H =	heater outlet temperature	°C	°F
	T_1 =	heater inlet temperature	°C	°F

Regeneration Gas Cooler

One needs to calculate the cooler load for all three intervals to find the highest load. It will normally occur in interval B for long cycle units. The latent heats of water and hydrocarbons must be known. Knowing the time for the interval and assuming the desorption is uniform during it, one can find the latent heat load. To this one must add the gas sensible heat load. The normal temperature approach will be 16-20°C [29-38°F] for air cooling and 8-10°C [15-18°F] for water cooling.

Example 19.2: Example Regeneration System Calculations. The regeneration gas properties for previous example are shown below.

Regen. Gas:
$z = 1.0$
$C_p = 2.3$ kJ/kg·°C [0.55 Btu/lbm-°F]
$\gamma = 0.59$
$P = 6000$ kPa [870 psia]
$T_H = 310$°C [590°F]
$T_4 = 290$°C [554°F]

SI Solution: Calculate mass of steel in tower – 1.2 m dia. × 6.0 m S-S

For a design pressure of 7000 kPa, t = 37 mm (use 40)

$$m_s = (0.0347)(6.0)(1200)(40) = 10\ 000 \text{ kg}$$

Example 19.2 (Cont'd.)

Calculate the heating load, Q_H

			MJ
Steel	(10 000)(0.50)(290 – 30)	=	1300
Desiccant	(3600)(0.96)(290 – 30)	=	900
Water (sensible)	(360)(4.19)(116 – 30)	=	130
Water (desorption)	(360)(4187)	=	1510
			3840

Add 25% to cover piping, valves, etc. + losses = 960

$$Q_H = 3840 + 960 = 4800 \text{ MJ}$$

The available regeneration time is 6 hours. Assume 4 hours for heating and 2 hours for cooling.

$$m = \frac{2.5\, Q_H}{C_p\, \theta_H (T_H - T_1)} = \frac{(2.5)(4\,800\,000)}{(2.3)(4)(310-30)} = 4660 \text{ kg/h}$$

$$q_s = \left(\frac{4660 \text{ kg}}{\text{h}}\right)\left(\frac{1 \text{ kmol}}{(0.59)(28.97) \text{ kg}}\right)\left(\frac{23.64 \text{ m}^3}{\text{kmol}}\right)\left(\frac{24 \text{ h}}{\text{d}}\right)$$

$$= 155\,000 \text{ m}^3/\text{d}$$

$$\approx 8\% \text{ of process rate}$$

Check cooling load, Q_D

			MJ
Steel	(10 000)(0.50)(290 – 45)	=	1225
Desiccant	(3600)(1.0)(290 – 45)	=	882
			2107

Estimate ΔT_c

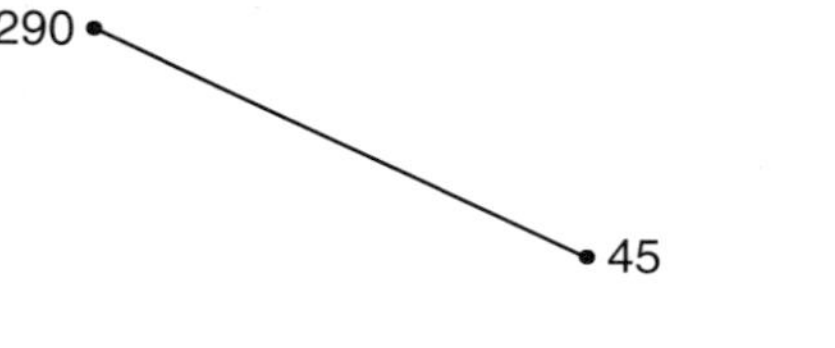

$$\Delta T_c = \frac{260 - 15}{\ln\left(\frac{260}{15}\right)} = 86°\text{C}$$

for regeneration gas flowrate of 4660 kg/h

$$Q = m C_p\, \Delta T_c\, \theta_c = (4660)(2.3)(86)(2) = 1\,840\,000 \text{ kJ}$$

Regeneration gas rate is about 15% too low to satisfy cooling duty. Regeneration gas rate should be increased, heating time shortened and cooling time lengthened.

Example 19.2 (Cont'd.)

Check to see if regeneration rate meets minimum velocity criteria using Eq. 19.5

for $\Delta P/L = 0.23$ kPa/m , $v_g = 1.8$ m/min

for d = 1.2 m   A = 1.3 m², so $q = Av_g = (1.13)(1.8) = 2.04$ m³/min

$$q_s = (1440)(2.04)\left(\frac{6000}{101.3}\right)\left(\frac{288}{433}\right)\left(\frac{1}{1.0}\right) = \underline{\underline{116\ 000 \text{ std m}^3/\text{d}}}$$

This is less than the 142 000 std m³/d required by heat balance.

Size regeneration gas heater

$$Q = mC_p(T_H - T_1)$$

$$= \left(\frac{4270 \text{ kg}}{\text{h}}\right)\left(\frac{1 \text{ h}}{3600 \text{ s}}\right)\left(\frac{2.3 \text{ kJ}}{\text{kg·°C}}\right)(310 - 30)°\text{C} = 764 \text{ kW}$$

FPS Solution: Calculate mass of steel in tower – 48 in dia. × 20.0 ft S-S

For a design pressure of 1000 psig, t = 1.4 in (use 1.5 in)

$m_s = (15)(20)(48)(1.5) = 21\ 600$ lbm

Calculate the heating load, Q_H

			MMBtu
Steel	(21 600)(0.12)(554 – 86)	=	1.21
Desiccant	(7990)(0.24)(554 – 86)	=	0.90
Water (sensible)	(799)(1.0)(240 – 86)	=	0.12
Water (desorption)	(799)(1800)	=	1.44
			3.67

Add 25% to cover piping, valves, etc. + losses = 0.92 MMBtu

$Q_H = 3.67 + 0.92 = 4.59$ MMBtu, use 4.6 MMBtu

The available regeneration time is 6 hours. Assume 4 hours for heating and 2 hours for cooling.

$$m = \frac{2.5\ Q_H}{C_p\ \theta_H(T_H - T_1)} = \frac{(2.5)(4\ 600\ 000)}{(0.55)(4)(590 - 86)} = 10\ 370 \text{ lbm/hr}$$

$$q_s = \left(\frac{10\ 370 \text{ lbm}}{\text{hr}}\right)\left(\frac{1 \text{ lbmol}}{(0.59)(28.97) \text{ lbm}}\right)\left(\frac{379.5 \text{ ft}^3}{\text{lbmol}}\right)\left(\frac{24 \text{ hr}}{\text{day}}\right)$$

$= 5.5$ MMscfd

$\approx 8\%$ of process rate

Example 19.2 (Cont'd.)

Check cooling load, Q_D

			MMBtu
Steel	(21 600)(0.12)(554 – 113)	=	1.14
Desiccant	(7990)(0.24)(554 – 113)	=	0.86
			2.00

Estimate ΔT_c

554 • → • 113

86 • → • 86

$$\Delta T_c = \frac{260 - 15}{\ln\left(\frac{260}{15}\right)} = 86°C$$

for regeneration gas flowrate of 10 370 lbm/hr

$$Q = m C_p \, \Delta T_c \, \theta_c = (10\ 370)(0.55)(155)(2) = 1.77 \ \text{MMBtu/hr}$$

Regeneration gas rate is about 15% too low to satisfy cooling duty. Regeneration gas rate should be increased, heating time shortened and cooling time lengthened.

Check to see if regeneration rate meets minimum velocity criteria using Eq. 19.5

for $\Delta P/L = 0.01$ psi/ft , $v_g = 5.9$ ft/min

for d = 48 in   A = 12.6 ft^2, so $q = Av_g = (12.6)(5.9) = 74 \ ft^3/min$

$$q_s = (1440)(74)\left(\frac{870}{14.7}\right)\left(\frac{520}{780}\right)\left(\frac{1}{1.0}\right) = \underline{\underline{4.2 \ \text{MMscfd}}}$$

This is less than the 5.5 MMscfd required by heat balance.

Size regeneration gas heater

$$Q = mC_p(T_H - T_1)$$

$$= \left(\frac{10\ 370 \ \text{lbm}}{\text{hr}}\right)\left(\frac{0.55 \ \text{Btu}}{\text{lbm}}\right)(590 - 86)°F = 2.9 \ \text{MMBtu/hr}$$

Regeneration Reflux

One cause of poor adsorber performance such as high pressure drop, dusting, shortened adsorbing time, or poor outlet dewpoint can be traced to an unusual phenomena which is often referred to as "refluxing." This is a situation where water is driven from the bed during regeneration, condenses on the cooler vessel walls and desiccant near the top of the bed and subsequently "rains" back down into the bed.

When refluxing occurs, the condensed water drains down the bed until it contacts the heat zone moving up the bed. At this point the water will boil. This rolling boil can grind the molecular sieve into a powder. Over time this powder is baked by the rising heat zone into a hard donut-shaped cake, sometimes weighing over 100 kg. Even in less severe cases this refluxing phenomenon will drastically reduce the molecular sieve capacity.

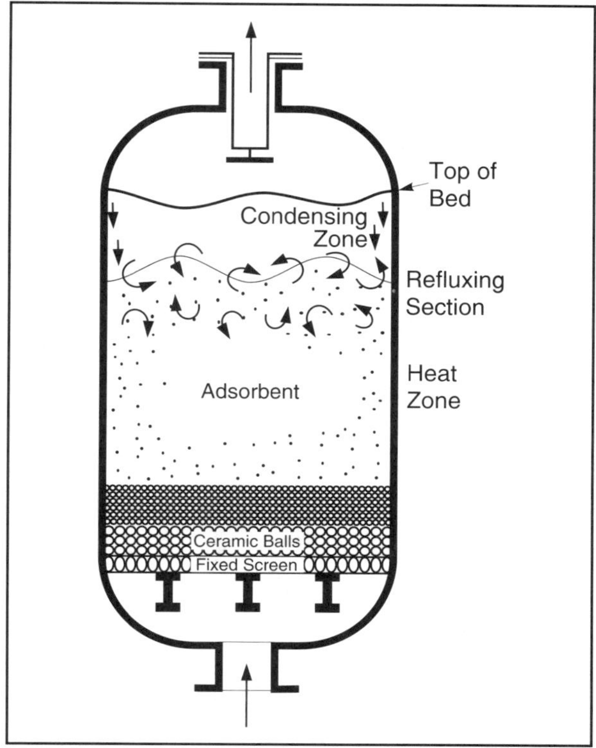

Figure 19.12 Regeneration Reflux

Refluxing is most common when the regeneration gas is at high pressure (> 4200 kPa [690 psia]) because of the high volumetric heat capacity and low water capacity of high pressure gas. One way to minimize this problem is to use a low pressure regeneration gas if feasible. Other suggestions to minimize the reflux effect are listed below.(19.5)

- A 150 mm [6 in] layer of 12-25 mm [1/2-1 in] inert support balls on top of the mol sieve bed may minimize the rolling boil but will not fix the problem.
- Slow down the initial rate of heat input. Normally the regeneration gas rate is just above the laminar flow region so the gas rate cannot be reduced but the rate of input heat can be reduced by ramping up the inlet gas temperature. One approach, if time is available, is to start the heating cycle with an ambient temperature purge for one hour at the design regeneration gas flowrate. The temperature should be increased by 55°C [100°F] increments each hour up to the design inlet temperature. Normally this heating should continue at full flow for one hour after the outlet temperature has leveled out (approaches the inlet temperature by 27-42°C [50-75°F]).
- Minimize heat loss by using internal insulation.
- Change the heating gas flow direction to downflow. This will push out any condensed water eliminating the refluxing boil. Co-current regeneration requires higher flowrates to thoroughly strip the bed. Also the downward flow will push heavy liquid contaminants on down through the bed possibly increasing the fouling rate.
- An alternate scheme is to reverse all flows and adsorb upward and heat down flow if the adsorbing flowrate is below the bed fluidization (lifting) velocity. The adsorbent supplier can easily check this.

MECHANICAL DESIGN CONSIDERATIONS

Once a unit configuration suitable to your conditions has been selected and the vendor has furnished a process design, mechanical design can proceed. It is then necessary to design the vessels including distribution devices, associated valving, the regeneration gas heater, and the necessary piping layout.

1. **Vessel Design.** As discussed in an earlier section, the vessel sizing is largely set once superficial gas velocity and mass of desiccant required have been determined. To this one must add the space required for support members, screens, inert balls, gas distribution devices, and velocity reduction. Typical spacing for these items is shown in Figure 19.8. Bed weights are substantial and, coupled with the P forces incurred during adsorption, adequate support must be provided.

   Additional length of vessel shell to ensure adequate room for these system elements and good flow distribution can significantly improve operation. Remember that the mol sieve must be loaded and unloaded several times during project life, so provide adequate openings for these operations.

   Traditionally, wire mesh and grating assemblies have been used to support desiccant beds. These assemblies have served well with the larger desiccant particles. To minimize some of the problems in the sealing and maintenance of the wire mesh and gratings, Johnson screens (UOP) have been used, particularly for the finer mol sieve grades. A 0.84 mm [0.033 in] slot opening will retain the common 4-8 mesh and 8-12 mesh mol sieve materials.

   The annular space between the vessel wall and the edge of the bed support screen must be sealed to prevent loss of mol sieve. Asbestos rope packing forced into this space will prevent sieve loss.

   Internal or external insulation of the dehydration vessels may be selected. Internal insulation reduces the regeneration energy requirements somewhat. Small cracks in internal insulation can contribute to bypassing gas around the sieve bed. It is estimated that only 0.1% bypassed gas can prevent successful expander or LNG plant operation. Internal insulation may also extend the dryout or curing time required for plant startup. "Can type" insulation is not recommended because cracks often develop in the seal welds and the can warps over time due to the temperature cycling.

   Gas channeling may result from several causes. One cause is poor flow distribution at either the inlet or outlet of the molecular sieve beds. Poor flow distribution may also contribute to impingement of high velocity gas on the bed surface. One of the simplest, lowest cost methods for control of gas flow distribution is to provide simple baffle plates over the inlet and outlet nozzles. A screen-wrapped slotted pipe with gas exiting radially into the vessel can also be used.

   A moisture sample probe should be located in the dehydrator desiccant bed about 0.5 m [1.6 ft] from the outlet end of the bed and extending to the center. This probe, when used in conjunction with the outlet gas moisture probe, offers a source of valuable information for troubleshooting dehydration problems — particularly if a possible channel exists down the wall of the vessel. It also permits capacity tests for optimizing time cycles to be run with reasonable safety because movement of the water front can be detected prior to breakthrough. The probe can be a long thermowell drilled with 0.8 mm [1/32 in] holes on the sides near the end of the probe.

2. **Heat Loss Management.** There are several items that contribute to the regeneration gas heater load, e.g., heat required to heat desiccant, to heat the water, to vaporize the water, to heat the steel, to heat the insulation, to heat piping and finally to make up heat losses from the vessels and piping. This last item can be of significance, especially for small units. At very low regeneration gas flowrates, it is conceivable that gas leaving the heater will not reach the bed at an adequate temperature or that heat losses around the vessel will be great enough to interfere with regeneration effectiveness. The regeneration gas system must be designed to ensure that heat losses are not excessive or that the heater and gas flowrates are large enough to provide for such losses.

Desiccant Loading

The loading of adsorption dehydration and treating vessels is an important step in assuring that the unit will perform as designed. All openings for liquid level switches or other openings in or near the bed in liquid dehydrators and liquid product treaters will have to be equipped with screens to prevent sieve entry. The following procedures are recommended during loading:

1. Inert balls (75-100 mm [3-4 in] in depth) are usually placed on top of the bottom support screen to minimize adsorbent nesting in the screen opening and to aid in gas distribution. A rule of thumb says that the inert balls can be two to four times the size of the sieve particles. If necessary two sizes of inert balls may be used. The larger balls near the screen must be retained by the screen. Where available materials permit, graded desiccant may be used instead of inert balls.
2. After the support balls have been loaded, the moisture probe should be inserted. Installation of probes is very difficult after the desiccant has been loaded.
3. The desiccant should be loaded through a plastic or cloth sock that reaches to the support balls from the top loading nozzle. The sock must be moved in a random manner to level the bed and prevent coning and is pulled out as the vessel fills. An equivalent method to sock loading would be to lower the desiccant into the vessel and place it at random locations using a dump bucket. If the desiccant is simply dumped from the top or at the center from a sock or bucket it will segregate (large particles on sides, small in center) and the bed will have more flow capacity at the sides, thus causing early adsorbent breakthrough.
4. A layer of inert balls (13 mm [0.5 in] diameter) should be placed on top of the adsorption bed. This layer of balls will prevent movement of desiccant on the surface caused by high local velocities and may aid in gas distribution. A floating stainless steel screen should be placed between the inert balls and the sieve to prevent migration of the denser balls down into the bed.

Feed Gas Conditioning

Without exception the most frequently encountered operating problem in adsorption systems is feed gas conditioning. The gas entering the bed should be free of all entrained hydrocarbons, treating chemicals (e.g., glycol, amine), free water, and solids. Although most desiccants are designed to withstand some carryover, significant or persistent carryover of entrained material will cause premature reduction of bed capacity and/or mechanical damage to the desiccant material.

A properly sized impingement separator followed by a coalescing filter separator should be installed upstream of any solid bed dehydration system. If the feed gas is at its retrograde dewpoint (pressure is above cricondentherm pressure) it is advisable to heat the feed gas slightly (~5°C [9°F]) to preclude retrograde condensation.

Some of the most common bed contaminants are listed below:[(19.6)]

1. **Hydrocarbons.** Heavy hydrocarbons such as crude oil or lube oil fractions are adsorbed by the binder in macropores which are much larger than the active sites for water. These high boiling point fractions can undergo a number of reactions during regeneration including cracking, polymerization, or coking. Each of these can leave behind a high MW non-volatile "coke" which blocks or slows the diffusion of water molecules to the active sites. This results in a loss of dynamic equilibrium and premature breakthrough.

   Light hydrocarbons, such as NGLs, can also be adsorbed in the macropores. Again this slows diffusion, hence the adsorption rate, and results in premature breakthrough. The difference is that light hydrocarbons are driven off the bed during regeneration and do not leave a non-volatile residue. Light hydrocarbon contamination can occur due to entrainment or retrograde condensation.

2. **Glycols.** Glycols behave similar to heavy oils in adsorption bed contamination. They are adsorbed in the macropores and decompose during regeneration. Again this can form a heavy non-volatile coke on the sieve or in some cases can result in the "cementing" of sieve particles together to form "chunks". This encourages gas channeling which in turn leads to premature breakthrough.

3. **Amines.** Like hydrocarbons and glycol, amines can also contribute to coking. In addition, ammonia is formed during regeneration. Ammonia can attack the binder and weaken the physical structure of the sieve. Two or three water "wash trays" at the top of the amine contactor are recommended to minimize the carryover of amine compounds.

4. **Salt.** Salt usually enters a desiccant bed dissolved in entrained water. Unfortunately, it does not leave when the water is vaporized and is not removed from the bed during regeneration. Thus, the solid accumulates and blocks pores, macropores – and in extreme cases – binds the beads together to form "chunks". Once sufficient salt has accumulated to reduce adsorbent capacity below the minimum level required to maintain cycle times, it is usually necessary to replace the adsorbent. Unfortunately, there is no dependable way to recover desiccant capacity from such a unit.

   Adsorbers subject to such contamination are those treating gas from salt water bearing formations, and those treating LPG, propylene, etc. from "salt dome" storage caverns.

5. **Oxygen.** If there is any oxygen in the system, or in the regeneration gas, it will react with H_2S and some other sulfur compounds on the surface of the sieves and deposit elemental sulfur. In extreme cases this will not only block macro and micropores but also the space between the sieve particles, resulting in one large "lump" which may have to be removed with pick and shovel.

   Complications resulting from oxygen in a hydrocarbon system are not limited to the production of sulfur. Reaction with the hydrocarbons present, especially during the high temperature portion of the regeneration cycle, can result in heavy coke lay down and fouling of the sieves.

   Since oxygen can enter a system by a number of routes, it's a good idea to request an oxygen determination during any routine stream analysis. If only trace amounts are detected early, and the source is discovered and cut off, it should be possible to prevent severe damage to the sieves.

6. **H_2S/Sulfur Compounds.** H_2S is adsorbed on 4A and 5A sieves. In fact molecular sieves are sometimes used for H_2S removal from natural gas. This application is discussed briefly

later in this text. When H_2S and CO_2 are present in the feed gas, special sieves which minimize the formation of COS should be specified.

Another issue which should be considered is the concentration of H_2S in the regeneration gas. Any H_2S adsorbed on the desiccant comes off the bed during regeneration in a "spike" lasting 5-15 minutes. In other words, all of the H_2S adsorbed on the bed during the adsorption cycle is removed in this short 5-15 minute interval. This can increase the concentration of H_2S in the regeneration gas to several hundred or in some cases several thousand ppm. Depending on the disposition of regeneration gas this may require temporary flaring or subsequent H_2S removal from the regeneration gas.

Unfortunately, all sulfur compounds are not designed for easy removal from the sieves. Heavy mercaptans and other large molecule, high boiling, sulfur compounds, are not efficiently removed during routine molecular sieve regeneration cycles. As a result, they tend to build in concentrations as a bed ages and produce a capacity reduction much as do the previously mentioned heavy oils.

7. **Methanol.** Methanol is frequently used for hydrate inhibition in the production and gathering systems. Methanol has a vapor pressure higher than water, so significant quantities of methanol can be present in the vapor phase in the feed gas. Methanol is adsorbed on 4A sieves and can reduce the capacity of desiccant to adsorb water. If methanol is present in the feed gas, the mol sieve supplier should be notified so that additional adsorption capacity can be included in the design. Some installations have successfully used 3A sieve to preclude the adsorption of methanol.

SPECIAL ISSUES/NEW DEVELOPMENTS

Adsorbents have been used to dehydrate, sweeten and remove NGLs from natural gas streams for over 50 years. Early units employed silica gel and/or alumina. Molecular sieves were introduced to the gas processing industry in the 1960s. Several improvements in the quality and versatility of these desiccants have resulted in new and innovative applications. A few of these are discussed below:

1. **Compound Beds.** Compound beds are desiccant beds which use more than one desiccant type or desiccant size. The purpose is to increase the useful capacity of the bed by increasing the equilibrium capacity or shortening the MTZ, or both. The most common example of a compound bed is the use of 8-12 mesh desiccant at the bottom of the bed and 4-8 mesh at the top. The equilibrium capacity of the two desiccants is the same, but the rate of adsorption in the smaller (8-12 mesh) desiccant is faster – hence a smaller MTZ.

   Since the 8-12 mesh desiccant is at the bottom of the bed, the useful capacity of the bed is increased. This can result in longer cycle times and/or shorter bed depths. The 4-8 mesh desiccant at the top of the bed minimizes pressure drop and the ΔP forces on the bed supports.

   Another compound bed application involves the use of alumina at the top of the bed and molecular sieves at the bottom. Alumina has a higher static equilibrium capacity for water as well as a lower heat of adsorption. This results in a higher useful capacity and lower regeneration heating requirements. Alumina will often not dry the gas to sufficiently low outlet dewpoints but the molecular sieve at the bottom of the bed will remove the residual water not picked up on the alumina.

2. **H_2S.** As previously mentioned, H_2S is adsorbed on 4A, 5A, 10X and 13X molecular sieves. Indeed, molecular sieves are sometimes used for H_2S removal. This application is

generally limited to those cases where the concentration of H_2S in the feed gas is less than 200-300 ppm. The advantage of using molecular sieves is the simultaneous dehydration and sweetening of natural gas. 5A sieve is commonly used for this application; however, 4A sieve also has some useful capacity for H_2S.

There are several important differences between H_2S and H_2O adsorption. First, the equilibrium capacity for H_2S is much lower than for water. In addition, the MTZ is longer for H_2S. These two factors mean much lower useful capacities for H_2S. The result is significantly shorter cycle times and/or longer beds are required for H_2S removal compared to water. Water will be adsorbed along with the H_2S, but the water loading on the bed will be much less than the full useful capacity. This is the case because the breakthrough occurs with H_2S not H_2O.

If CO_2 is present in the gas the mol sieve can catalyze the reaction of H_2S and CO_2 to form carbonyl sulfide (COS). COS is very difficult to remove from natural gas and NGL streams. In those instances where COS can form it is recommended to select a desiccant which minimizes the formation of COS.

A more detailed discussion of gas sweetening using adsorption processes can be found in Volume 4 of the Campbell "Gas Conditioning and Processing" series, *Gas Treating and Sulfur Recovery*.

3. **Adsorption vs. Glycol.** In general, when the outlet water dewpoint requirement is above –40°C [–40°F], glycol (TEG) dehydration or glycol (MEG) injection is generally preferred due to lower capital and operating costs. If H_2S is present in the gas, a fixed bed system may be chosen over glycol because of the coabsorption of H_2S in the glycol and subsequent environmental impact at the regenerator. In adsorption systems, the regeneration cycle can be "closed" so that no H_2S need be vented or incinerated.

   The heat of regeneration in glycol units is less than that required in molecular sieve systems. TEG systems average about 9200-11 600 kJ/kg H_2O [4000-5000 Btu/lb H_2O] while molecular sieve systems require approximately twice as much energy.

   Another factor which affects the dehydration decision is the presence of aromatic hydrocarbons in the feed gas. Aromatic hydrocarbons have become a significant environmental issue in glycol units as they can be vented from the regenerator. Many states/countries now require reduction of these emissions.

   Aromatic hydrocarbons are not adsorbed on alumina and molecular sieve and are recovered as a liquid product in silica gel units — hence no environmental problem. Some companies have used dry desiccant units (in lieu of glycol) for this reason.

   Another related topic which is often raised in the design of dry desiccant systems is whether to dehydrate the gas with glycol upstream of the adsorption. This may be difficult to justify economically, but there are a number of positive benefits to this configuration. First, the water content of the feed gas is significantly lower, hence longer adsorption cycle times and fewer regeneration cycles. This can increase the life of the desiccant and save several bed changeouts over the facility life.

   A second advantage is energy efficiency. Due to the lower regeneration heating requirements for glycol, significant energy savings can be realized by removing the bulk of the water with glycol upstream of the dry desiccant unit.

HYDROCARBON RECOVERY

The basic mechanism for hydrocarbon recovery is similar but more complex than dehydration. One is faced with describing multiple zone behavior.

In a dehydrator, the purpose of the condenser is to remove the desorbed liquids from the gas stream. In the short cycle plant, for hydrocarbon recovery, condenser operation has a critical effect on recovery. The adsorption bed simply serves to concentrate the recoverable components so that condensation is more efficient. The temperature and pressure of condensation is a critical parameter governing plant economics. Depending on the length of the adsorption cycle, a silica gel plant with ambient condensation is limited to minimal butane recovery and 75-90% of the pentanes and heavier. There are many alternatives in between.

Consider that the hydrocarbon deposited on the bed is desorbed by 10-15% as much regeneration gas. The net effect is to make the gas 6-10 times richer in condensable hydrocarbons, thus making recovery easier. A refrigerated condenser may be used to further enhance recovery but the condensation temperature should exceed the hydrate temperature.

Silica gel plants have been used primarily on lean gas streams where other methods of processing were not economically attractive. The most popular application has been the processing of gas produced from gas storage reservoirs. The untapped potential for hydrocarbon dewpoint control and as an alternative to refrigeration appears large particularly at higher pressures.

Process Characteristics

The capacity of most silica gel is about the same for hydrocarbons as for water. Activated carbon, of course, has no effective capacity for water.

Figure 19.13 shows the equilibrium capacity of silica gel for various hydrocarbons in a two component gas where the second component is methane. The figure shows both static equilibrium (from cell tests) and dynamic equilibrium (from flow tests).

Notice that all curves are approaching the monolayer capacity of the gel. The monolayer capacity is found by assuming that only one layer of molecules is held to the solid surface. Knowing both surface area and molecule size, one may compute the capacity. This same characteristic has been found for all gels, aluminas and molecular sieves. This means that ultimate capacity for any component is fixed by surface area – provided that the component is small enough to enter the interior of the adsorbing particle.

Actual capacity for any component is fixed by the zone movement previously described, bed geometry, equilibrium capacity and gas flowrate. The surface of the adsorbent is always occupied by some molecule. As the zone of a given component progresses down the bed it must displace the molecules already there. The rate of displacement depends on their relative wettability.

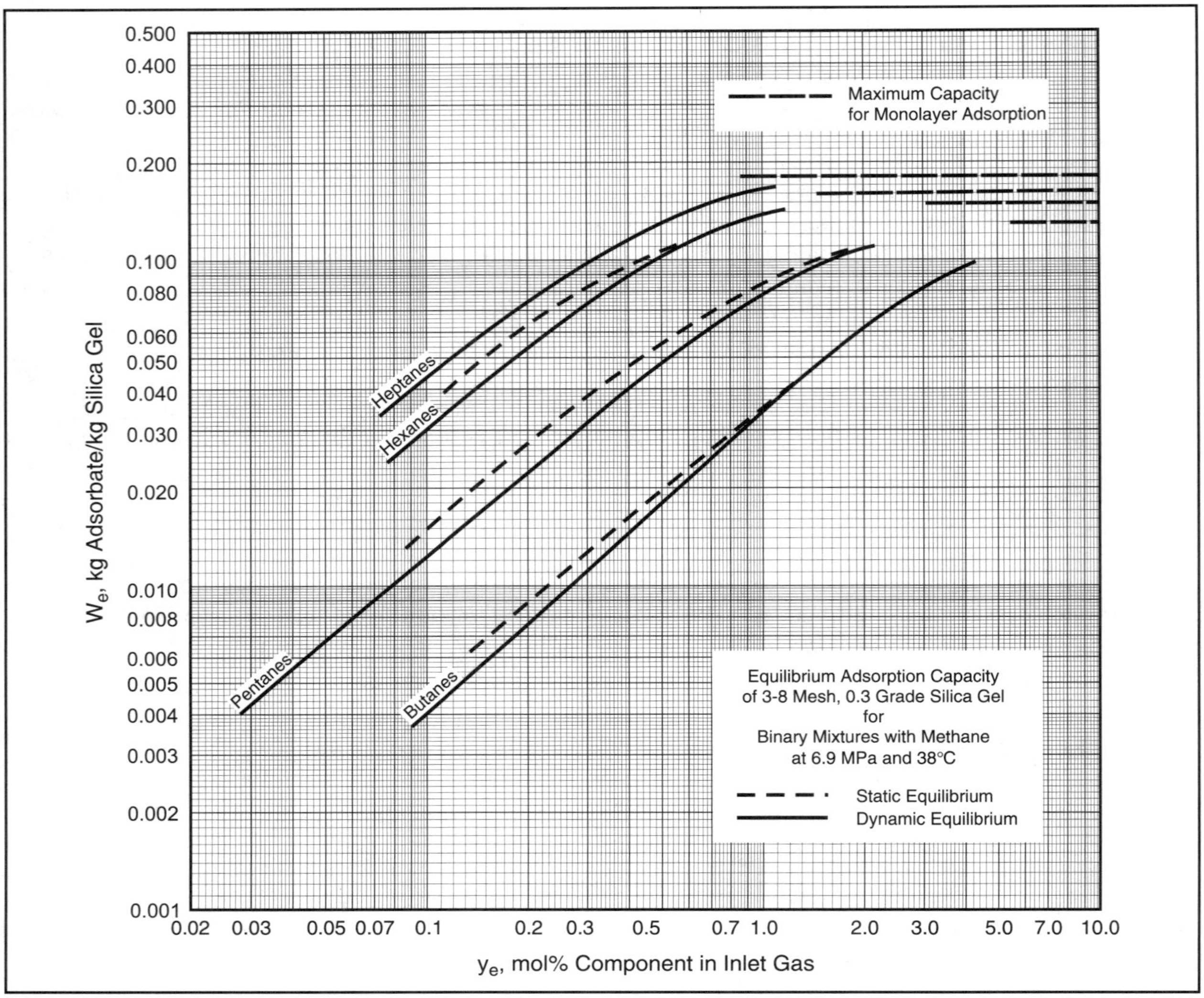

Figure 19.13 Hydrocarbon Equilibrium on Silica Gel(19.7)

Theoretically, the zone for any component cannot move any faster than it can completely displace the materials ahead of it. In actual practice, at commercial flowrates, the zone tends to "over-run" each other. Therefore, true chromatographic separation does not occur. This is illustrated in Figure 19.14.

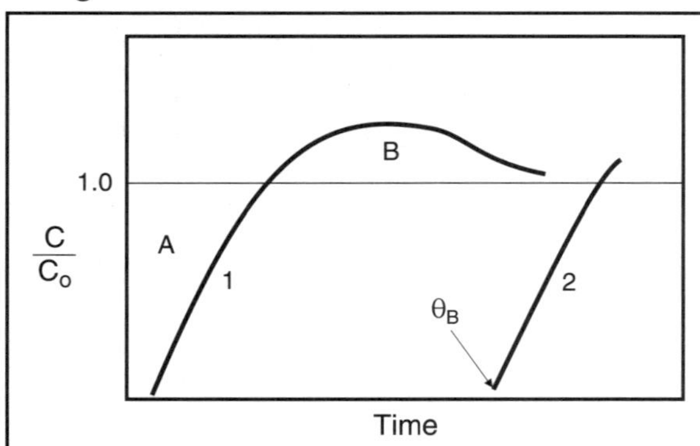

Figure 19.14 Illustration of the Adsorption/Desorption Process

Once the front of the mass transfer zone (MTZ) for a particular component reaches the outlet of the bed the ratio of outlet to inlet concentration (C/C_o) starts to increase. When this ratio reaches one, all primary adsorption ceases for that component (in this case, component 1). Desorption now begins because the MTZ for the less volatile component (component 2) behind it is displacing the component 1 previously adsorbed. The concentration ratio rises above one but again *approaches* one when this next zone starts breaking out the end of the bed. This process continues until the cycle is terminated and regeneration of the bed begins.

Area A is representative of the amount of Component 1 adsorbed and Area B of the amount desorbed *by the next zone*. The latter is smaller than the former. At time $_B$ for Component 2 (Area A – Area B)/Area A is generally about 0.35-0.40. Thus, immediate displacement has not occurred.

This is shown (at right) by the test data on a field unit.

Time, min	Inlet mol%, C		
	C_3	iC_5	nC_5
	1.020	0.120	0.085
	Outlet mol% (C_0)		
2	0.804	0.031	0.012
12	0.976	0.074	0.031
22	0.938	0.075	0.043
32	0.962	0.077	0.065
42	0.970	0.155	0.130
52	0.952	0.109	0.102

Even if one recognizes the 6-10% error in sampling and analysis, no sharp separation has occurred. The propane has broken through in less than two minutes. This data is not characteristic of an efficient process because even good pentanes recovery is not obtained early in the cycle. Beyond this point, the exit stream is being *enriched*.

If liquid recovery is the goal, some net amount of component is available even after its zone passes from the tower. The recovery will simply be less. If hydrocarbon dewpoint control is the goal, such enrichment is probably intolerable for components heavier than C_6. In such a case, C_6 breakthrough represents maximum cycle time. Liquid recovery plants tend to use shorter cycles than dewpoint control plants since the recovery of C_5 is often a process objective. In dewpoint control plants the objective is to meet a hydrocarbon dewpoint specification which may require efficient recovery of C_7 or C_8+ components only.

Figure 19.15 shows the breakthrough curves for several natural gas components on a silica gel bed.[19.8] No time values are shown on the x-axis because breakthrough times depend on gas velocity, bed length and inlet composition of C_5+ components. Component breakthrough typically occurs in or-

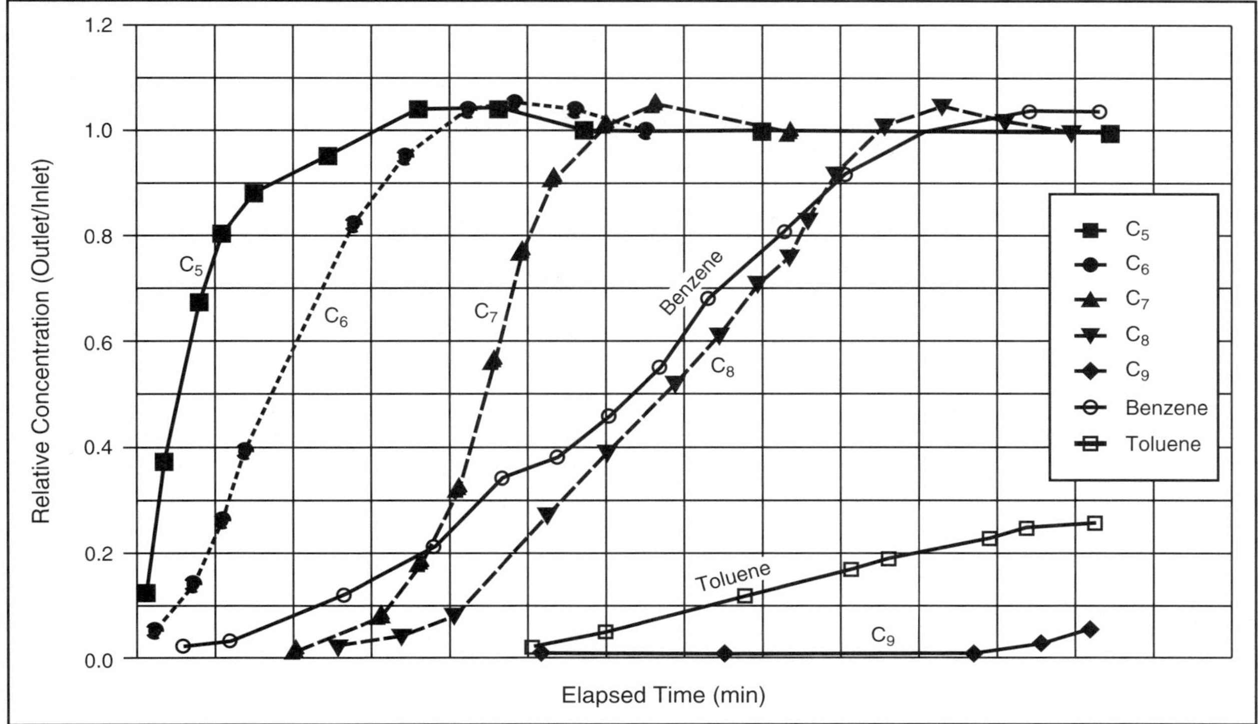

Figure 19.15 Breakthrough Profiles for Hydrocarbons on Silica Gel

der of increasing boiling point with the exception of aromatic hydrocarbons which are retained for a longer time due to their slight polarity.

If the process objective is to recover hydrocarbons for sale, the adsorption time will be short to maximize recovery of C_5+ hydrocarbons. If the process objective is hydrocarbon dewpoint control, longer cycle times might be acceptable since recovery of only C_7+ or C_8+ hydrocarbons may be required.

Figure 19.16 shows the breakthrough curves for C_5 and C_6 on silica gel at different feed gas compositions and superficial bed velocities. The gas in question contained methane, ethane, negligible propane and butanes, and the amounts of pentanes and hexanes shown. There was no heavier component present to displace the hexanes. Water content was negligible. Note breakthrough times are shorter for higher feed gas concentrations and superficial velocities.

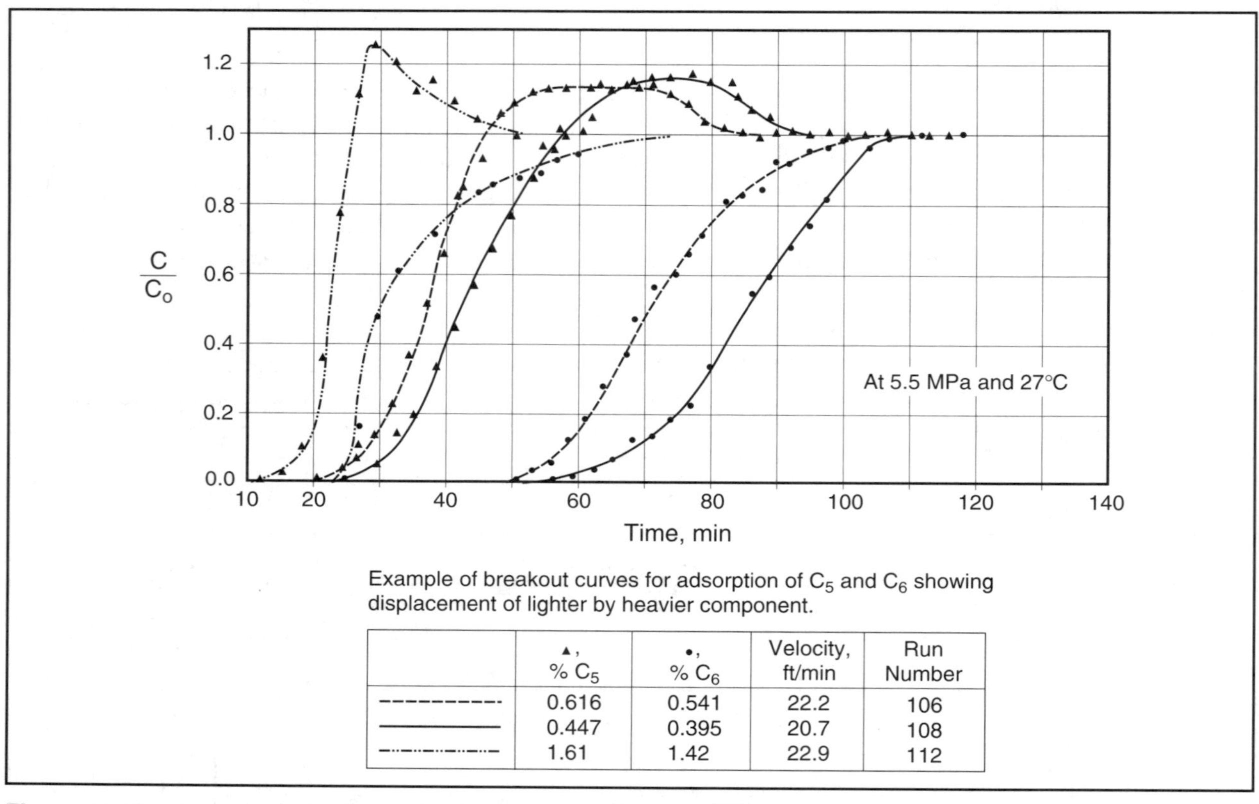

	▲, % C_5	•, % C_6	Velocity, ft/min	Run Number
- - - - - - - - -	0.616	0.541	22.2	106
————————	0.447	0.395	20.7	108
-··-··-··-··-	1.61	1.42	22.9	112

Figure 19.16 Breakthrough Curves for C_5 and C_6 on Silica Gel[(19.9)]

With activated carbon the zones tend to move slower. For one thing, water does not promote displacement. Basically though, the zone speed is lower because of the greater affinity for the lighter hydrocarbons.

Regeneration and Recovery

Figure 19.17 summarizes the regeneration behavior of a short cycle plant. Notice that the materials do not desorb at a constant rate. The pentanes and lighter start desorbing almost immediately. The hexanes and heavier concentration in the exit gas peak after a finite time.

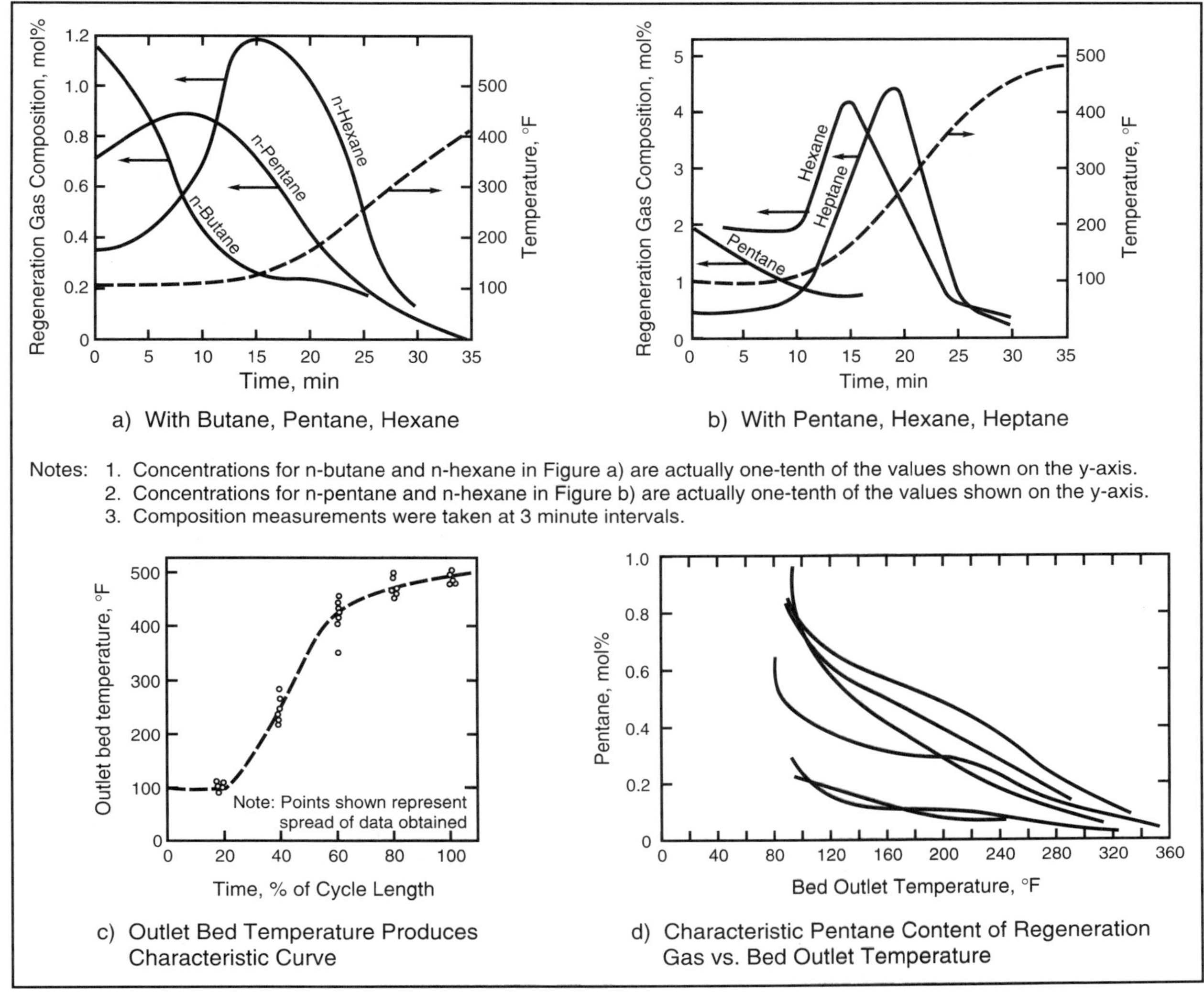

Figure 19.17 Typical Regeneration for Hydrocarbon Adsorbers(19.10)

Figures 19.17 a) & b) show the composition of various hydrocarbon components in the regeneration gas vs. time. Obviously, the composition of the stream to the condenser varies continually with time. For this reason a series of flash calculations must be made with time to accurately represent the liquid recovery to be expected. The minimum amount of regeneration gas necessary to increase the concentration of recoverable components in the condenser should be used. The use of refrigeration in the condenser will likewise enhance recovery. Even at temperatures as high as 16-20°C [60-68°F], a marked improvement recovery efficiency may be realized.

General Specification

Proper design of silica gel and activated carbon hydrocarbon recovery units is complicated enough that computer simulation is necessary. A detailed design method is beyond the scope of this text although the principles governing it have been presented here. It is sufficient to say that the general empirical methods to date fail to utilize the full potential of the adsorption process. References 19.8 and 19.11 summarize recent developments in the use of silica gel units for hydrocarbon dewpoint control.

Most obvious problems stem from improper specification. A liquid recovery guarantee is useful for planning purposes but is more likely to be a basis for controversy than a measure of plant performance.

The following minimum specifications are suggested:

1. Gas flow limited to the rate previously discussed for dehydration (to promote desiccant life).
2. Cycle length should be fixed by the following considerations:
    a. Not less than 15 minutes for gas containing pentanes and heavier.
    b. Adsorption time set by breakthrough time for the component for which recovery is desired or which must be removed for dewpoint control.
3. Bed length should be at least 5 m [16.5 ft].
4. Inlet regeneration gas temperature should be at least 230°C [446°F] and preferably 260°C [500°F] when processing gases containing pentanes and heavier.
5. The alternative of using refrigeration instead of ambient cooling in the condenser should be considered.

Items 1-4 are not independent; each affects the other. If the adsorption cycle length is less than 15 minutes it is almost impossible to properly regenerate the bed. If this 15 minutes is greater than the breakthrough time for key component, some compromise is needed. Breakthrough time depends on gas velocity and bed length (for a given gas composition and adsorbent). Economics and/or process needs will govern the compromise. As a matter of information – so one may make an intelligent decision – it is wise to also specify that the vendor furnish you with adsorption efficiency as well as condenser recovery. Adsorption efficiency is simply that fraction of the component entering during the proposed cycle length that is retained on the adsorbent. This enables you to not only compare the relative merit of competitive bids but to make necessary changes prior to purchase.

LIQUID DEHYDRATION

Silica gel, alumina and molecular sieves may be used to dry hydrocarbon liquids. The flow scheme is similar to that for gas. Some larger plants are designed so that the flow may be reversed to "loosen" the bed if it has been compacted or to free the retaining screens of sediment. This provision is seldom needed for fractionated liquids. If there is any possibility of free liquid water being present, a coalescer should be provided upstream of the desiccant beds.

One difference between gas and liquid dehydration is in the regeneration cycle. Two systems are commonly used to provide regeneration:

1. Gas
2. Closed Vapor (either gas or vaporized product)

Table 19.4 summarizes these processes. Natural gas is typically used for regeneration in liquid desiccant systems. The gas is heated to approximately 160°C [320°F] for alumina, higher temperatures are required for molecular sieves.

TABLE 19.4

Comparison of Regeneration Practices

Method	Advantage	Disadvantages	Common Usage
Natural Gas	1. Uses readily available material. 2. Low operating cost. 3. Simple construction. 4. Readily adaptable to automatic control.	1. Introduces some additional safety hazard. 2. Requires compressor if high pressure gas not available.	Field locations, Product Pipe lines, Non-Volatile Liquids.
Closed Vapor (vaporized product)	1. Simple operation. 2. Low cost of operation. 3. Minimizes loss of valuable volatile liquids. 4. No contamination of product.	1. Control of system more critical. 2. Requires pumping equipment. 3. Requires efficient condensation of exit regeneration gas.	With volatile liquids such as propane, butane, etc. Where composition of feed is substantially constant.

Design Considerations

The solubility of water in sweet hydrocarbons is shown in Figure 19.18. Notice that it is far more soluble in many unsaturated and aromatic hydrocarbons than in the normal paraffins. Knowledge of composition is thus very important. The presence of sulfur compounds enhances water solubility. If

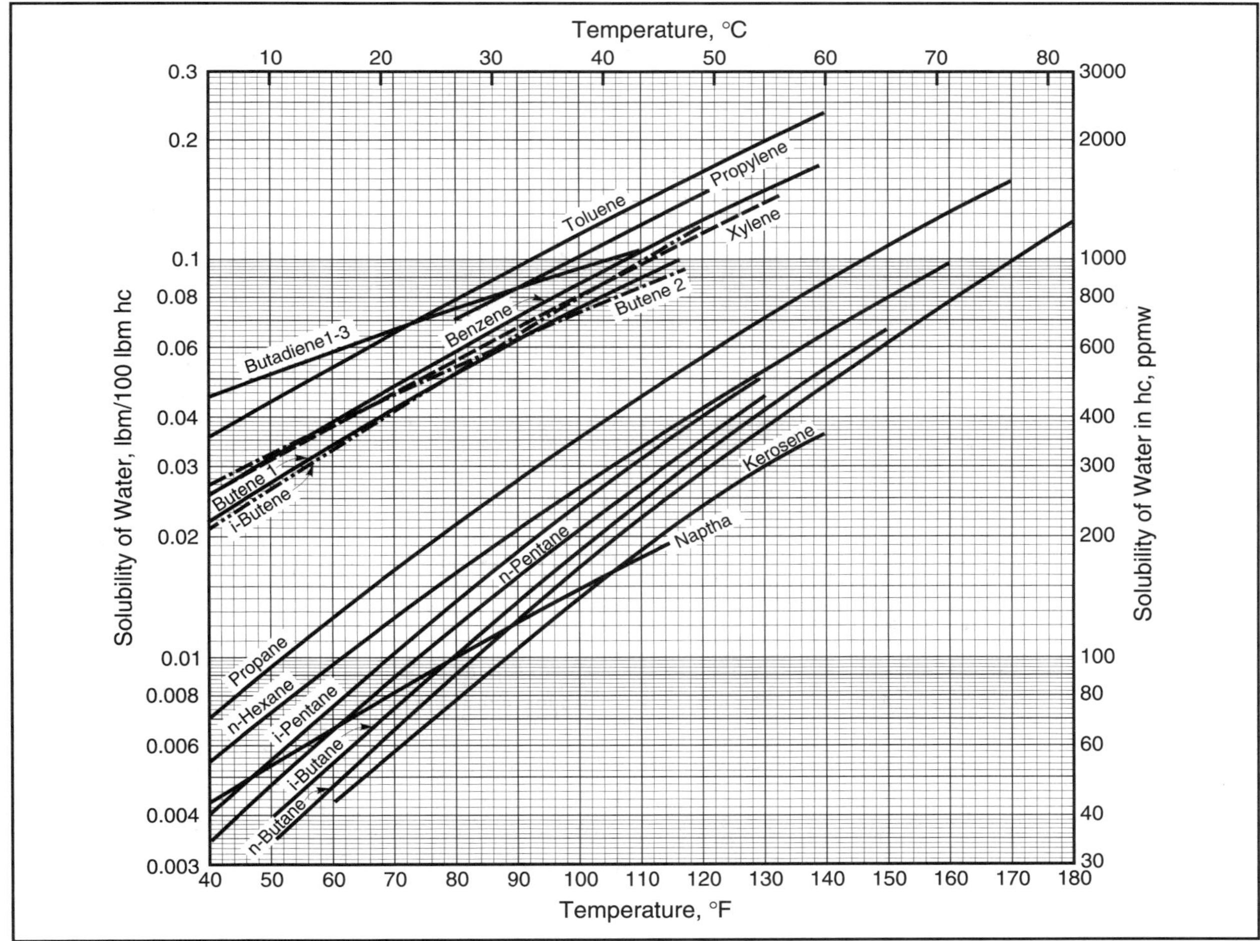

Figure 19.18 Solubility of Water In Hydrocarbons

no data are available for the specific liquid, a *weight fraction* relationship may be used to estimate liquid mixture water content.

Most contracts specify that the dried liquid show a negative result to the Cobalt Bromide test, which is equivalent to a water content of 15-30 ppmw. This is virtually bone dry for ppmw is weight percent times 10 000. In design, one assumes all incoming water is removed in the bed.

Liquid velocity should be 1-2 m/min. This will fix tower diameter. Tower length will usually be shorter than for gas. An (L/D) ratio of 2-3:1 is common. As little as three seconds contact time is commonly provided. Some operators require a minimum bed length of 1.5 meters.

Activated alumina has been used quite widely for liquid drying since it is relatively inexpensive and tower costs are minimized at low pressure. An effective capacity of 5-12 kg of water per 100 kg alumina is common, this is equivalent to that in gas service. The capacity of gels and sieves in liquid dehydration service is likewise similar to their gas drying capacity.

A 3A sieve is useful for drying liquids to minimize coadsorption of NGL components and heavier hydrocarbon molecules. The pore opening size is too small to admit these contaminants to the interior particle surface. Olefins, for example, may "tie-up" the available surface and reduce water capacity when using alumina.

Molecular sieves may be used also to both dry and sweeten liquids as discussed in Volume 4.

REFERENCES

19.1 Cummings, W. P., "Solid Bed Adsorbers," *Petr. Learn. Prog.*, Houston, Texas (1980).

19.2 Simpson, E. A. and W. P. Cummings, *Chem. Engr. Prog.*, Vol. 60, No. 4, (April 1964), p. 57.

19.3 Lee, Hanju and W. P. Cummings, *Chem. Engr. Prog. Symp. Series*, Vol. 63, No. 74, (1967), p. 42.

19.4 Ergun, S. "Fluid Flow Through Packed Columns," *Chem. Eng. Prog.*, Vol. 48, No. 2 (Feb. 1952), p. 89.

19.5 Tent, R. E., "Regeneration Reflux," Zeochem Product Bulletin.

19.6 Dewhirst, Mike, "Effect of Contaminants on Molecular Sieves," 14th Annual GPA European Chapter Meeting (Sept. 1997), Antwerp, Belgium.

19.7 Ashford, F., PhD. Thesis, Univ. of Oklahoma (1970).

19.8 Harris, D. M. and D. W. Ingram, "Adsorption: The Flexible Solution for Gas Processing for NAM's Underground Storage System," European GPA Continental Meeting, Budapest (1999).

19.9 Needham, R. B. et al., *I & E C Proc. Dev.*, Vol. 5, (1966), p. 122.

19.10 Campbell, J. M., *Oil and Gas J.*, (Feb. 21, 1966), p. 93.

19.11 Schultz, T., "Gas Conditioning for Underground Storage Applications," European GPA Continental Meeting, Zurich (Sept. 1998).

19.12 Maddox, R. N. and D. J. Morgan, "Gas Treating and Sulfur Recovery," Vol. 4, 4th Ed. (1998), John M. Campbell and Company, Norman, Oklahoma.

19.13 Turnock, P. H. and K. J. Gustafson, Gas Cond. Conf., Norman, Oklahoma (1972).

19.14 Kunkel, L. V. and J. W. Chobotuk, NGPA Meeting, Dallas, Texas (1973).

General Conversion Factors

Base SI metric units

Quantity	Name	Symbol
Length	meter	m
Mass	kilogram	kg
Time	second	s
Electric current	ampere	A
Thermodynamic temperature	kelvin	K
Luminous intensity	candela	cd
Amount of substance	mole	mol

Important derived and supplementary SI units

Quantity	Name	Symbol	Formula
Electric capacitance	farad	F	A-s/V
Electric charge	coulomb	C	A-s
Electric conductance	siemens	S	A/V
Electric inductance	henry	H	Wb/A
Electric potential	volt	V	W/A
Electric quantity	coulomb	C	A-s
Electric resistance	ohm	Ω	V/A
Electromotive force	volt	V	W/A
Energy	joule	J	N-m
Energy flux	watt	W	J/s
Force	newton	N	kg-m/s^2
Frequency	hertz	Hz	1/s
Illuminance	lux	1x	1m/m^2
Luminous flux	lumen	1m	cd-sr
Magnetic flux	weber	Wb	V-s
Magnetic flux density	tesla	T	Wb/m^2
Power	watt	W	J/s
Pressure	pascal	Pa	N/m^2
Quantity of heat	joule	J	N-m
Solid Angle	steradian	sr	---
Work	joule	J	N-m

SI prefixes and multiplication factors

Prefix	SI Symbol	Factor
tera	T	10^{12}
giga	G	10^{9}
mega	M	10^{6}
kilo	k	10^{3}
hecto	h	10^{2}
deka	da	10^{1}
deci	d	10^{-1}
centi	c	10^{-2}
milli	m	10^{-3}
micro	μ	10^{-6}
nano	n	10^{-9}
pico	p	10^{-12}
femto	f	10^{-15}
atto	a	10^{-18}

To convert from	to	Multiply by
acre	hectare	0.40469
acre	meter^2	4,046.9
acre-feet	meter^3	1,233.5
ampere-hour	coulomb	3,600.0
angstrom	nanometer	0.10000
astronomical unit	gigameter	149.598
atmosphere	bar	1.01325
atmosphere (normal)	kilopascal	101.325
atmosphere(normal)	pascal	101,325
atmosphere (1 kgf/cm^3)	pascal	98,067
bar	kilopascal	100.000
bar	pascal	100,000
barrel (42 US gal)	meter^3	0.15899
barrel/acre-foot	decimeter^3/meter^3	0.12889
barrel/day	decimeter^3/second	0.00184
barrel/day	meter^3/day	0.15899
barrel/day (1 kg/dm^3)	tonne/annum	58.0304
barrel/foot	meter^3/meter	0.52161
barrel/hour	decimeter^3/second	0.04416
barrel/hour	meter^3/hour	0.15899
barrel/US ton	meter^3/tonne	0.17525
board-foot	meter^3	0.00236
British thermal unit	joule	1,055.1
British thermal unit	kilojoule	1.05506
British thermal unit	watt-hour	0.29288
BTU/bhp-hour	watt/kilowatt	0.39301
BTU/foot^3	kilojoule/meter^3	37.2590
BTU/foot^3	kilowatt-hour/meter^3	0.01035
BTU/US gallon	kilojoule/meter^3	278.716
BTU/US gallon	kilowatt-hour/meter^3	0.07742
BTU/hour	watt	0.29307
BTU/minute	watt	17.5843
BTU/second	kilowatt	1.05506
BTU/hour-foot^2	watt/meter^2	3.15459
BTU/hour-foot^3	watt/meter^3	10.3497
BTU/hr-foot^2-°F	watt/meter^2-kelvin	5.67826
BTU/second-foot^3-°F	kilowatt/meter^2-kelvin	20.4418
BTU/hour-foot^3-°F	watt/meter^3-kelvin	18.6295
BTU/second-foot^3-°F	kilowatt/meter^3-kelvin	67.0661
BTU/hour-foot^2-°F/foot	watt/meter-kelvin	1.73074
BTU/hour-foot^2-°F/inch	watt/meter-kelvin	0.14428
BTU/second-foot^2-°F/in.	watt/meter-kelvin	519.220
BTU/pound mass	joules/gram	2.32600
BTU/pound mass	watt-hour/kilogram	0.64611
BTU/pound mol	joules/mol	2.32600
BTU/pound-°F	joules/gram-kelvin	4.18680
BTU/pound mol-°F	joules/mol-kelvin	4.18680
bushel	decimeter^3	35.2391
bushel	meter^3	0.03524
calorie	joule	4.18400
calorie/pound	joule/kilogram	9.22414
centipoise	newton-second/meter^2	0.00100
centipoise	pascal-second	0.00100
centistroke	millimeter^2/second	1.00000
chain	meter	20.1168
cycle/second	hertz	1.00000
darcy	micrometer^2	0.98692
degree (angle)	radian	0.01745
°F/100 foot	millikelvin/meter	18.2269
°F.-foot^2-hour/BTU	kelvin-meter^2/watt	0.17611
dyne	millinewton	0.01000
erg	microjoule	0.10000
erg	millijoule	0.00010
fathom	meter	1.82880
foot	centimeter	30.4800
foot	meter	0.30480
foot/day	meter/day	0.30480
foot/°F.	meter/kelvin	0.54864
foot/hour	meter/hour	0.30480
foot/hour	millimeter/second	0.08467
foot/minute	centimeter/second	0.50800
foot/minute	meter/minute	0.30480
foot/minute	meter/second	0.00508
foot/second	meter/second	0.30480
foot/second^2	meter/second^2	0.30480
foot^2	centimeter^2	929.030
foot^2	meter^2	0.09290
foot^2/hour	millimeter^2/second	25.8064
foot^2/second	millimeter^2/second	92.903
foot^3	decimeter^3	28.3169
foot^3	meter^3	0.02832
foot^3/day	decimeter^3/second	0.00033
foot^3/day	meter^3/day	0.02832
foot^3/day	meter^3/hour	0.00118

To convert from	to	Multiply by
foot^3/foot	meter^3/meter	0.09290
foot^3/hour	decimeter^3/second	0.00787
foot^3/hour	meter^3/hour	0.02832
foot^3/minute	decimeter^3/second	0.47195
foot^3/minute	meter^3/minute	0.02832
foot^3/pound	meter^3/kilogram	0.06243
foot^3/pound mol	meter^3/kilomol	0.06243
foot^3/second	meter^3/second	0.02832
footcandle	lux	10.0764
footcandle-second	lux-second	10.0764
foot-pound force	joule	1.35582
foot-pound/US gallon	kilojoule/meter^3	0.35817
foot-pound/minute	milliwatt	22.5970
foot-pound/minute	watt	0.02260
foot-pound/second	watt	1.35582
gallon US liquid	meter^3	0.00379
gallon US/foot	meter^3/meter	0.01242
gallon US/horsepower-hr.	decimeter^3/megajoule	1.41009
gallon US/horsepower-hr.	millimeter^3/joule	1.41009
gallon US/1,000 barrel	centimeter^3/meter^3	23.8095
gallon US/foot^3	decimeter^3/meter^3	133.681
gallon US/hour	decimeter^3/second	0.00105
gallon US/hour	meter^3/hour	0.00379
gallon US/mile	decimeter^3/100 km	235.215
gallon US/minute	decimeter^3/second	0.06309
gallon US/minute	meter^3/hour	0.22712
gallon US/ton US	decimeter^3/tonne	4.17270
grain	milligram	64.7989
grain/gallon US	gram/meter^3	17.1181
grain/100 foot^3	milligram/meter^3	22.8835
horsepower	kilowatt	0.74570
horsepower (boiler)	kilowatt	9.80950
horsepower/foot^3	kilowatt/meter^3	26.3341
horsepower-hour	kilowatt/hour	0.74570
horsepower-hour	megajoule	2.68452
inch	centimeter	2.54000
inch	millimeter	25.4000
inch	pica	6.02250
inch	point	72.2700
inch Hg (60°F)	kilopascal	3.37685
inch H_2O (60°F)	kilopascal	0.24884
inch/minute	centimeter/minute	2.54000
inch/second	centimeter/second	2.54000
inch/second	millimeter/second	25.4000
inch/year	millimeter/annum	25.4000
inch^2	centimeter^2	6.45160
inch^2	millimeter^2	645.160
inch^2/second	millimeter^2/second	645.160
inch^3	centimeter^3	16.3871
inch^4	centimeter^4	41.6231
joule	British thermal unit	0.00095
joule	foot-pound force	0.73756
kilogram	pound mass av.	2.20462
kilogram/meter^3	pound mass/foot^3	0.06243
kilometer/hour	mile/hour statute	0.62137
kilowatt-hour	kilojoule	3,600.0
kilowatt-hour	megajoule	3.60000
knot	kilometer/hour	1.85200
lambert	candela/meter^2	3,183.1
link	meter	0.20117
meter	foot	3.28084
meter^2	foot^2	10.7639
meter^3	barrel (42 US gal)	6.28981
meter^3	foot^3	35.3147
meter/minute	foot/minute	3.28084
meter^3/minute	gallon US/minute	264.172
micron	micrometer	1.00000
mil	micrometer	25.4000
mile,nautical	kilometer	1.85200
mile US statute	kilometer	1.60934
mile^2 US statute	kilometer^2	2.58999
mile (US stat.)/US gal	kilometer/decimeter^3	0.42514
mile (US stat.)/hour	kilometer/hour	1.60934
minute (angle)	radian	0.00029
newton	pound force av.	0.22481
ounce force av	newton	0.27801
ounce mass av	gram	28.3495
ounce mass av	kilogram	0.02835
ounce US fluid	centimeter^3	29.5735
ounce troy	gram	31.1035
part/million (volume)	centimeter^3/meter^3	1.00000
pascal	pound force/foot^2	0.02089
pascal	pound force/inch^2	0.00015

To convert from	to	Multiply by
pica	centimeter	0.42175
pica	inch	0.16604
pint US dry	decimeter^3	0.55061
pint US liquid	decimeter^3	0.47318
point	centimeter	0.03515
point	inch	0.01384
pound force	newton	4.44822
pound mass av	kilogram	0.45359
pound mass troy	kilogram	0.37324
pound mol	kilomol	0.45359
pound force/foot^2	pascal	47.8803
pound force/inch^2	kilopascal	6.89476
pound force/inch^2/foot	kilopascal/meter	22.6206
pound mass/barrel	gram/decimeter^3	2.85301
pound mass/barrel	kilogram/meter^3	2.85301
pound mass/foot	kilogram/meter	1.48816
pound mass/foot^2	kilogram/meter^2	4.88243
pound mass/foot^3	kilogram/meter^3	16.0185
pound mass/US gallon	gram/centimeter^3	0.11983
pound mass/US gallon	kilogram-decimeter^3	0.11983
pound mass/horsepower-hr.	kilogram/kilowatt-hour	0.60828
pound mass/horsepower-hr.	kilogram/megajoule	0.16897
pound mass/horsepower-hr.	milligram/joule	0.16897
pound mass/hour	tonne/day	0.45359
pound mass/minute	kilogram/minute	0.45359
pound mass/second	kilogram/second	0.45359
pound mass/hour-foot	gram/second-meter	0.41338
pound mass/second-foot	kilogram/second-meter	1.48816
pound mass/second-foot	pascal-second	1.48816
pound mass/second foot^2	kilogram/second-meter^2	4.88243
pound mol	kilomol	0.45359
pound mol/foot^3	kilomol/meter^3	16.0185
pound mol/US gallon	kilomol/meter^3	119.826
pound mol/hour	kilomol/hour	0.45359
pound mol/second	kilomol/second	0.45359
pound force-foot	joule	1.35582
pound force-foot	newton-meter	1.35582
pound force-inch	joule	0.11298
pound force-inch	newton-meter	0.11298
pound force-foot^2	kilogram-meter^2	0.04214
pound force-foot/inch	newton-meter/meter	53.3787
pound force-foot/inch	newton-meter/meter	4.44822
pound force-foot/inch^2	joule/centimeter	0.00210
pound force-foot/second	kilogram-meter/second	0.13826
pound force-second/foot^2	pascal-second	47.8803
quart US	decimeter^3	0.94635
radian (angle)	degree	57.2958
radian (angle)	revolution	0.15915
radian (angle)/minute	revolution/minute	0.15915
°Rankin	kelvin	0.55556
revolution	radian (angle)	6.28319
revolution/minute	radian/minute	6.28319
section	hectare	258.999
therm	megajoule	105.506
ton force US	kilonewton	8.89644
ton mass US short	kilogram	907.185
ton mass US short	toone	0.90718
tonne	kilogram	1,000.0
tonne	megagram	1.00000
ton force US/foot^2	megapascal	0.09576
ton force US/inch^2	megapascal	13.7895
ton force US-mile	megajoule	14.3174
ton mass US/day	tonne/hour	0.03780
ton mass US/hour	tonne/day	0.90718
ton mass US/day	kilogram/second	0.01050
ton mass US/hour	kilogram/second	0.25120
ton mass US/minute	kilogram/second	15.1197
ton mass US/year	tonne-annum	0.90718
ton mass US/foot^2	tonne/meter^2	9.76486
torr	pascal	133.322
watt	BTU/minute	0.05687
watt	joule/second	1.00000
watt-hour	joule	3,600.0
watt-second	joule	1.00000
yard	meter	0.91440
yard^2	meter^2	0.83613
yard^3	meter^3	0.76455
year	annum	1.00000

This	times this	gives this
Acre	43,560	sq. ft.
Acre-feet	7,757.8	bbl.(42 gal.)
	43,560	cu. ft.
Ampere-hours	3,600	Coulombs
Angstroms	0.0001	Microns
Atm. at 32*F.)	33.90	ft. of water
	1,034	g./sq.cm.
	29.92	in. Hg
	760.18	mm. Hg
	14.697	psi.
Bbl.(42 gal.)	5.6146	cu. ft.
	9,702.03	cu. in.
Bbl./day(42 gal.)	1.75	g.p.h.
	0.0292	g.p.m.
Bbl./hr.(42 gal.)	0.0936	cu. ft./min.
	0.7	g.p.m.
Board feet	144.0	cu. in.
B.t.u.	778	ft.-lb.
	252	gram-cal.
	0.0003927	hp.-hr.
	0.0002928	kw.-hr.
B.t.u./min.	46,681.68	ft.-lb./hr.
	778	ft.-lb./min.
	12.967	ft.-lb./sec.
	0.02356	hp.
	0.01758	kw.
B.t.u./sec.	46,681.68	ft.-lb./min.
	778	ft.-lb./sec.
	1.4146	hp.
	1.05487	kw.
Bushels(U.S. std.)	1.2445	cu. ft.
	2,150	cu. in.
Candles/sq. in.	452.39	ft.-Lamberts
	0.48695	Lamberts
Candlepower (sph.)	12.566	lumens
Carat (Intl.)	3.08647	grains
	0.2	g.
	200	mg.
Centimeters	0.03281	ft.
	0.3937	in.
	393.7	mils
Cm. of mercury	0.01316	atm.
	0.4461	ft. of water
	27.85	lb./sq. ft.
	0.1934	psi.
Cm. per second	1.9685	ft./min.
	0.03281	ft./sec.
	0.02237	m.p.h.
Centipoise	1.0000	g./cm.-sec.
	0.00067197	lb./ft.-sec.
Circular mils	0.7854	sq. mils
Coulombs	0.1	Abcoulombs
	0.0002778	amp.-hr.
Cubic centimeters	0.00003531	cu. ft.
	0.0002642	gal. (U.S. liq.)
	0.001	liters
Cubic feet	0.00002296	acre-ft.
	0.1781	bbl. (42 gal.)
	0.80356	bu.
	28,320	cc.
	1,728	cu. in.
	7,481	gal. (U.S. liq.)
	28.32	liters
Cu. ft., million	22.95	acre-ft.
Cu. ft. per min	10.686	bbl.(42 gal.)/hr.
	0.1781	bbl.(42 gal.)/min.
	1,440	cu. ft./day
	7,481	g.p.m. (U.S. liq.)
Cubic inches	0.000465	bu.
	16.39	cc.
	0.0005787	cu. ft.
	0.004329	gal.(U.S. liq.)
	0.016387	liters
Degrees of arc	60	minutes
	0.017453	radians
Dynes	0.0010197	g.
	0.000002248	lb.
Dyne-cm.	1	ergs
	0.00000007376	ft.-lb.
	0.0010197	g.-cm.
Ems	0.1666	in.
Ergs	1	dyne-cm.
Faradays	26.80	amp.-hr.
Fathoms	6	ft.
Feet	30.48	cm.
	0.16667	fathoms
	0.000164468	mile(U.S. naut.)
	0.000189394	mile(U.S. stat.)
Feet per minute	0.5080	cm./sec.
	0.00987	knots/hr.
	0.01136	m.p.h.
Feet per second	0.68121	m.p.h.
	0.01136	miles/min.
Feet of water	0.02950	atm.
	0.4335	psi.
Foot-candles	1	lumens/sq. ft.
Foot-Lamberts	0.0022105	candles/sq. in.
	1	lumens/sq. ft.
Foot-pounds	0.0012853	B.t.u.
	0.000000505	hp.-hr.
	1.35582	joules
	0.0003241	kg.-cal.
	0.0000003766	kw.-hr.
Foot-pounds/hour	0.001284	B.t.u./hr.
	0.00002141	B.t.u./min
	0.000000357	B.t.u./sec
	0.01666	ft.-lb./min.
	0.000000505	hp.
	0.0000003766	kw.
Foot-pounds/min.	0.077118	B.t.u./hr.
	0.001286	B.t.u./min.
	0.0000303	hp.
	0.00002256	kw.
Foot-pounds/sec.	0.001285	B.t.u./sec.
	0.001818	hp.
Furlong	660	ft.
Gal. (U.S. dry)	0.15556	cu. ft.
	1.1637	gal. (U.S. liq.)
Gal. (U.S. liq.)	0.02381	bbl. (42 gal.)
	0.133681	cu. ft.
	231	cu. in.
	0.859365	gal. (U.S. dry)
	3.7854	liters
Gal./hr. (U.S. liq.)	0.1337	cu. t./hr.
	0.002228	cu. ft./min.
	0.01666	g.p.m. (U.S. liq.)
G.p.m. (U.S. liq.)	34.2857	bbl./day (42 gal.)
	1.4286	bbl./hr. (42 gal.)
	0.02381	bbl./min. (42 gal.)
	192.49	cu. ft./day
	0.1337	cu. ft./min.
	1,440	gal./day
Grains	0.00014286	lb.,avoir.
Grains/gal. (U.S. liq.)	0.01714	g./l./
	17.118	p.p.m.
Grams	980.7	dynes
	0.035274	oz.,avoir.
	0.0022046	lb.,avoir.
Gram-calories	0.003968	B.t.u.
	3.0875	ft.-lb.
Grams per liter	58.418	grains/gal
	1,000	p.p.m.
	0.0624	lb./cu. ft.
Horsepower	2,544	B.t.u./hr.
	42.44	B.t.u./min.
	33,000	ft.-lb./min.
	550	ft.-lb./sec.
	745.70	joules/sec.
	0.74570	kw.
Horsepower-hours	2,544	B.t.u.
	1,980,000	ft.-lb.
	0.7457	kw.-hr.
Inches	2.54	cm.
	0.08333	ft.
	25,400.05	microns
	1,000	mils
Inches of mercury	0.03242	atm.
	0.49116	psi.
Joules	0.0009486	B.t.u.
	0.239	calories
	0.73756	ft.-lb.
Kilograms	2.205	lb.,avoir.
Kilogram-calories	3.9683	B.t.u.
	3,087.5	ft.-lb.
Kilowatts	56.92	B.t.u./min.
	0.948	B.t.u./sec.
	44,253	ft.-lb./min.
	1.341	hp.
Kilowatt-hours	3,415	B.t.u.
	2,655,199	ft.-lb.
	1.341	hp.-hr.
Knots (naut.mile)	6,080	ft.
	1.152	mile (U.S. stat.)
Knots	101.34	ft./min.
	1.152	m.p.h. (U.S. stat.)
Liters	1,000	cc.
	0.03531	cu. ft.
	61.0271	cu. in.
	0.26418	gal. (U.S. liq.)
	1.057	qt. (U.S. liq.)
Logarithm, com	2.3026	logarithm,Nap.
Logarithm, Nap	0.4343	logarithm, com
Meters	3.2808	ft.
	39.37	in.
	1,000,000	microns
Microns	10,000	angstroms
	0.00003937	in.
	0.001	mm.
Mils	0.00254	cm.
	0.001	in.
	25.4	microns
Miles (U.S. stat.)	5,280	ft.
	1.6093	km.
Miles (U.S. naut.)	6,080	ft.
Miles per hour	44.704	cm./sec.
	88	ft./min.
	1.4667	ft./sec.
	0.01666	miles/min.
Milliliters	0.061	cu. in.
Millimeters	0.03937	in.
	39.37	mils
Minutes of arc	0.016667	deg. of arc
Ounces (avoir.)	28.35	g.
Parts per million	0.05835	grains/gal.
	0.001	g./l.
Points	0.013889	in.
Pounds, avoir.	444,823	dynes
	7,000	grains
	453.59	g.
	1.215	lb., Troy, Apoth
Pounds per gallon	7.4805	lb./cu. ft.
Pounds per cu. ft.	0.016018	g./cc.
	0.13368	lb./gal. (U.S. liq.)
Pounds per sq. in.	0.06804	atm.
	5.1715	cm. Hg.
	2.307	ft. of water
	2.036	in. Hg.
	144	lb./sq. ft.
Quarts	0.946	liters
Radians	57.2958	deg. of arc
Radians per sec.	9.549	r.p.m.
Rods	16.5	ft.
	5.5	yd.
Sections	640	acres
	1	sq. miles
Square feet	0.00002296	acres
	144	sq. in.
Square mils	1.2732	circular mils
Township	23,040	acres
	36	sections
	36	sq. miles
Vara	2.7777	ft.

°C	Temp. °C or °F -459.69 to -4	°F
-273.16	-459.69	
-267.78	-450	
-262.22	-440	
-256.67	-430	
-251.11	-420	
-245.56	-410	
-240.00	-400	
-234.44	-390	
-228.89	-380	
-223.33	-370	
-217.78	-360	
-212.22	-350	
-206.67	-340	
-201.11	-330	
-195.56	-320	
-190.00	-310	
-184.44	-300	
-178.89	-290	
-173.33	-280	
-169.53	-273.16	-459.7
-168.89	-272	-457.6
-167.78	-270	-454.0
-162.22	-260	-436.0
-156.67	-250	-418.0
-151.11	-240	-400.0
-145.56	-230	-382.0
-140.00	-220	-364.0
-134.44	-210	-346.0
-128.89	-200	-328.0
-123.33	-190	-310.0
-117.78	-180	-292.0
-112.22	-170	-274.0
-106.67	-160	-256.0
-101.11	-150	-238.0
-95.56	-140	-220.0
-90.00	-130	-202.0
-84.44	-120	-184.0
-78.89	-110	-166.0
-73.33	-100	-148.0
-70.56	-95	-139.0
-67.78	-90	-130.0
-65.00	-85	-121.0
-62.22	-80	-112.0
-59.45	-75	-103.0
-56.67	-70	-94.0
-53.89	-65	-85.0
-51.11	-60	-76.0
-48.34	-55	-67.0
-45.56	-50	-58.0
-42.78	-45	-49.0
-40.00	-40	-40.0
-39.45	-39	-38.2
-38.89	-38	-36.4
-38.34	-37	-34.6
-37.78	-36	-32.8
-37.23	-35	-31.0
-36.67	-34	-29.2
-36.12	-33	-27.4
-35.56	-32	-25.6
-35.00	-31	-23.8
-34.44	-30	-22.0
-33.89	-29	-20.2
-33.33	-28	-18.4
-32.78	-27	-16.6
-32.22	-26	-14.8
-31.67	-25	-13.0
-31.11	-24	-11.2
-30.56	-23	-9.4
-30.00	-22	-7.6
-29.45	-21	-5.8
-28.89	-20	-4.0
-28.34	-19	-2.2
-27.78	-18	-0.4
-27.23	-17	1.4
-26.67	-16	3.2
-26.12	-15	5.0
-25.56	-14	6.8
-25.00	-13	8.6
-24.44	-12	10.4
-23.89	-11	12.2
-23.33	-10	14.0
-22.78	-9	15.8
-22.22	-8	17.6
-21.67	-7	19.4
-21.11	-6	21.2
-20.56	-5	23.0
-20.00	-4	24.8

°C	Temp. °C or °F -3 to 83	°F
-19.44	-3	26.6
-18.89	-2	28.4
-18.33	-1	30.2
-17.8	0	32.0
-17.2	1	33.8
-16.7	2	35.6
-16.1	3	37.4
-15.6	4	39.2
-15.0	5	41.0
-14.4	6	42.8
-13.9	7	44.6
-13.3	8	46.4
-12.8	9	48.2
-12.2	10	50.0
-11.7	11	51.8
-11.1	12	53.6
-10.6	13	55.4
-10.0	14	57.2
-9.44	15	59.0
-8.89	16	60.8
-8.33	17	62.6
-7.78	18	64.4
-7.22	19	66.2
-6.67	20	68.0
-6.11	21	69.8
-5.56	22	71.6
-5.00	23	73.4
-4.44	24	75.2
-3.89	25	77.0
-3.33	26	78.8
-2.78	27	80.6
-2.22	28	82.4
-1.67	29	84.2
-1.11	30	86.0
-0.56	31	87.8
0	32	89.6
0.56	33	91.4
1.11	34	93.2
1.67	35	95.0
2.22	36	96.8
2.78	37	98.6
3.33	38	100.4
3.89	39	102.2
4.44	40	104.0
5.00	41	105.8
5.56	42	107.6
6.11	43	109.4
6.67	44	111.2
7.22	45	113.0
7.78	46	114.8
8.33	47	116.6
8.89	48	118.4
9.44	49	120.2
10.0	50	122.0
10.6	51	123.8
11.1	52	125.6
11.7	53	127.4
12.2	54	129.2
12.8	55	131.0
13.3	56	132.8
13.9	57	134.6
14.4	58	136.4
15.0	59	138.2
15.6	60	140.0
16.1	61	141.8
16.7	62	143.6
17.2	63	145.4
17.8	64	147.2
18.3	65	149.0
18.9	66	150.8
19.4	67	152.6
20.0	68	154.4
20.6	69	156.2
21.1	70	158.0
21.7	71	159.8
22.2	72	161.6
22.8	73	163.4
23.3	74	165.2
23.9	75	167.0
24.4	76	168.8
25.0	77	170.6
25.6	78	172.4
26.1	79	174.2
26.7	80	176.0
27.2	81	177.8
27.8	82	179.6
28.3	83	181.4

°C	Temp. °C or °F 84 to 800	°F
28.9	84	183.2
29.4	85	185.0
30.0	86	186.8
30.6	87	188.6
31.1	88	190.4
31.7	89	192.2
32.2	90	194.0
32.8	91	195.8
33.3	92	197.6
33.9	93	199.4
34.4	94	201.2
35.0	95	203.0
35.6	96	204.8
36.1	97	206.6
36.7	98	208.4
37.2	99	210.2
37.8	100	212.0
43.3	110	230.0
48.9	120	248.0
54.4	130	266.0
60.0	140	284.0
65.6	150	302.0
71.1	160	320.0
76.7	170	338.0
82.2	180	356.0
87.8	190	374.0
93.3	200	392.0
98.9	210	410.0
104.4	220	428.0
110.0	230	446.0
115.6	240	464.0
121.1	250	482.0
126.7	260	500.0
132.2	270	518.0
137.8	280	536.0
143.3	290	554.0
148.9	300	572.0
154.4	310	590.0
160.0	320	608.0
165.6	330	626.0
171.1	340	644.0
176.7	350	662.0
182.2	360	680.0
187.8	370	698.0
193.3	380	716.0
198.9	390	734.0
204.4	400	752.0
210.0	410	770.0
215.6	420	788.0
221.1	430	806.0
226.7	440	824.6
232.2	450	842.0
237.8	460	860.0
243.3	470	878.0
248.9	480	896.0
254.4	490	914.0
260.0	500	932.0
265.6	510	950.0
271.1	520	968.0
276.7	530	986.0
282.2	540	1,004.0
287.8	550	1,022.0
293.3	560	1,040.0
298.9	570	1,058.0
304.4	580	1,076.0
310.0	590	1,094.0
315.6	600	1,112.0
321.1	610	1,130.0
326.7	620	1,148.0
332.2	630	1,166.0
337.8	640	1,184.0
343.3	650	1,202.0
248.9	660	1,220.0
354.4	670	1,238.0
360.0	680	1,256.0
365.6	690	1,274.0
371.1	700	1,292.0
376.7	710	1,310.0
382.2	720	1,328.0
387.8	730	1,346.0
393.3	740	1,364.0
398.9	750	1,382.0
404.4	760	1,400.0
410.0	770	1,418.0
415.6	780	1,436.0
421.1	790	1,454.0
426.7	800	1,472.0

°C	Temp. °C or °F 810 to 1,670	°F
432.2	810	1,490.0
437.8	820	1,508.0
443.3	830	1,526.0
448.9	840	1,544.0
454.4	850	1,562.0
460.0	860	1,580.0
465.6	870	1,598.0
471.1	880	1,616.0
476.7	890	1,634.0
482.2	900	1,652.0
487.8	910	1,670.0
493.3	920	1,688.0
498.9	930	1,706.0
504.4	940	1,724.0
510.0	950	1,742.0
515.6	960	1,760.0
521.1	970	1,778.0
526.7	980	1,796.0
532.2	990	1,814.0
537.8	1,000	1,832.0
543.3	1,010	1,850.0
548.9	1,020	1,868.0
554.4	1,030	1,886.0
560.0	1,040	1,904.0
565.6	1,050	1,922.0
571.1	1,060	1,940.0
576.7	1,070	1,958.0
582.2	1,080	1,976.0
587.8	1,090	1,994.0
593.3	1,100	2,012.0
598.9	1,110	2,030.0
604.4	1,120	2,048.0
610.0	1,130	2,066.0
615.6	1,140	2,084.0
621.1	1,150	2,102.0
626.7	1,160	2,120.0
632.2	1,170	2,138.0
637.8	1,180	2,156.0
643.3	1,190	2,174.0
648.9	1,200	2,192.0
654.4	1,210	2,210.0
660.0	1,220	2,228.0
665.6	1,230	2,246.0
671.1	1,240	2,264.0
676.7	1,250	2,282.0
682.2	1,260	2,300.0
687.8	1,270	2,318.0
693.3	1,280	2,336.0
698.9	1,290	2,354.0
704.4	1,300	2,372.0
710.0	1,310	2,390.0
715.6	1,320	2,408.0
721.1	1,330	2,426.0
726.7	1,340	2,444.0
732.2	1,350	2,462.0
737.8	1,360	2,480.0
743.3	1,370	2,498.0
748.9	1,380	2,516.0
754.4	1,390	2,534.0
760.0	1,400	2,552.0
765.6	1,410	2,570.0
771.1	1,420	2,588.0
776.7	1,430	2,606.0
782.2	1,440	2,624.0
787.8	1,450	2,642.0
793.3	1,460	2,660.0
798.9	1,470	2,678.0
804.4	1,480	2,696.0
810.0	1,490	2,714.0
815.6	1,500	2,732.0
821.1	1,510	2,750.0
826.7	1,520	2,768.0
832.2	1,530	2,786.0
837.8	1,540	2,804.0
843.3	1,550	2,822.0
848.9	1,560	2,840.0
854.4	1,570	2,858.0
860.0	1,580	2,876.0
865.6	1,590	2,894.0
871.1	1,600	2,912.0
876.7	1,610	2,930.0
882.2	1,620	2,948.0
887.8	1,630	2,966.0
893.3	1,640	2,984.0
898.9	1,650	3,002.0
904.4	1,660	3,020.0
910.0	1,670	3,038.0

°C	Temp. °C or °F 1,680 to 3,000	°F
915.6	1,680	3,056.0
921.1	1,690	3,074.0
926.7	1,700	3,092.0
932.2	1,710	3,110.0
937.8	1,720	3,128.0
943.3	1,730	3,146.0
948.9	1,740	3,164.0
954.4	1,750	3,182.0
960.0	1,760	3,200.0
965.6	1,770	3,218.0
971.1	1,780	3,236.0
976.7	1,790	3,254.0
982.2	1,800	3,272.0
987.8	1,810	3,290.0
993.3	1,820	3,308.0
998.9	1,830	3,326.0
1,004.4	1,840	3,344.0
1,010.0	1,850	3,362.0
1,015.6	1,860	3,380.0
1,021.1	1,870	3,398.0
1,026.7	1,880	3,416.0
1,032.2	1,890	3,434.0
1,037.8	1,900	3,452.0
1,043.3	1,910	3,470.0
1,048.9	1,920	3,488.0
1,054.4	1,930	3,506.0
1,060.0	1,940	3,524.0
1,065.6	1,950	3,542.0
1,071.1	1,960	3,560.0
1,076.7	1,970	3,578.0
1,082.2	1,980	3,596.0
1,087.8	1,990	3,614.0
1,093.3	2,000	3,632.0
1,098.9	2,010	3,650.0
1,104.4	2,020	3,668.0
1,110.0	2,030	3,686.0
1,115.6	2,040	3,704.0
1,121.1	2,050	3,722.0
1,126.7	2,060	3,740.0
1,132.2	2,070	3,758.0
1,137.8	2,080	3,776.0
1,143.3	2,090	3,794.0
1,148.9	2,100	3,812.0
1,154.4	2,110	3,830.0
1,160.0	2,120	3,848.0
1,165.6	2,130	3,866.0
1,171.1	2,140	3,884.0
1,176.7	2,150	3,902.0
1,182.2	2,160	3,920.0
1,187.8	2,170	3,938.0
1,193.3	2,180	3,956.0
1,198.9	2,190	3,974.0
1,204.4	2,200	3,992.0
1,210.0	2,210	4,010.0
1,215.6	2,220	4,028.0
1,221.1	2,230	4,046.0
1,226.7	2,240	4,064.0
1,232.2	2,250	4,082.0
1,237.8	2,260	4,100.0
1,243.3	2,270	4,118.0
1,248.9	2,280	4,136.0
1,254.4	2,290	4,154.0
1,260.0	2,300	4,172.0
1,265.6	2,310	4,190.0
1,271.1	2,320	4,208.0
1,276.7	2,330	4,226.0
1,282.2	2,340	4,244.0
1,287.8	2,350	4,262.0
1,293.3	2,360	4,280.0
1,298.9	2,370	4,298.0
1,304.4	2,380	4,316.0
1,310.0	2,390	4,334.0
1,343.3	2,450	4,442.0
1,371.1	2,500	4,532.0
1,398.9	2,550	4,622.0
1,426.7	2,600	4,712.0
1,454.4	2,650	4,802.0
1,482.2	2,700	4,892.0
1,510.0	2,750	4,982.0
1,537.8	2,800	5,072.0
1,565.1	2,850	5,162.0
1,593.3	2,900	5,252.0
1,621.1	2,950	5,342.0
1,648.9	3,000	5,432.0

THERMODYNAMIC DATA

Both SI and FPS unit data are shown. Most is from previous editions. Samples of data from other sources are included.

The H-S diagram for steam, the P-H diagram for Freon 12 and two tables are fromt he reference shown in the title block. The P-H diagram for Freon 22 is a sample of data from "Thermodynamic Properties in SI" by Reynolds, available from the Department of Mechanical Engineering, Stanford University, Stanford, California 94305.

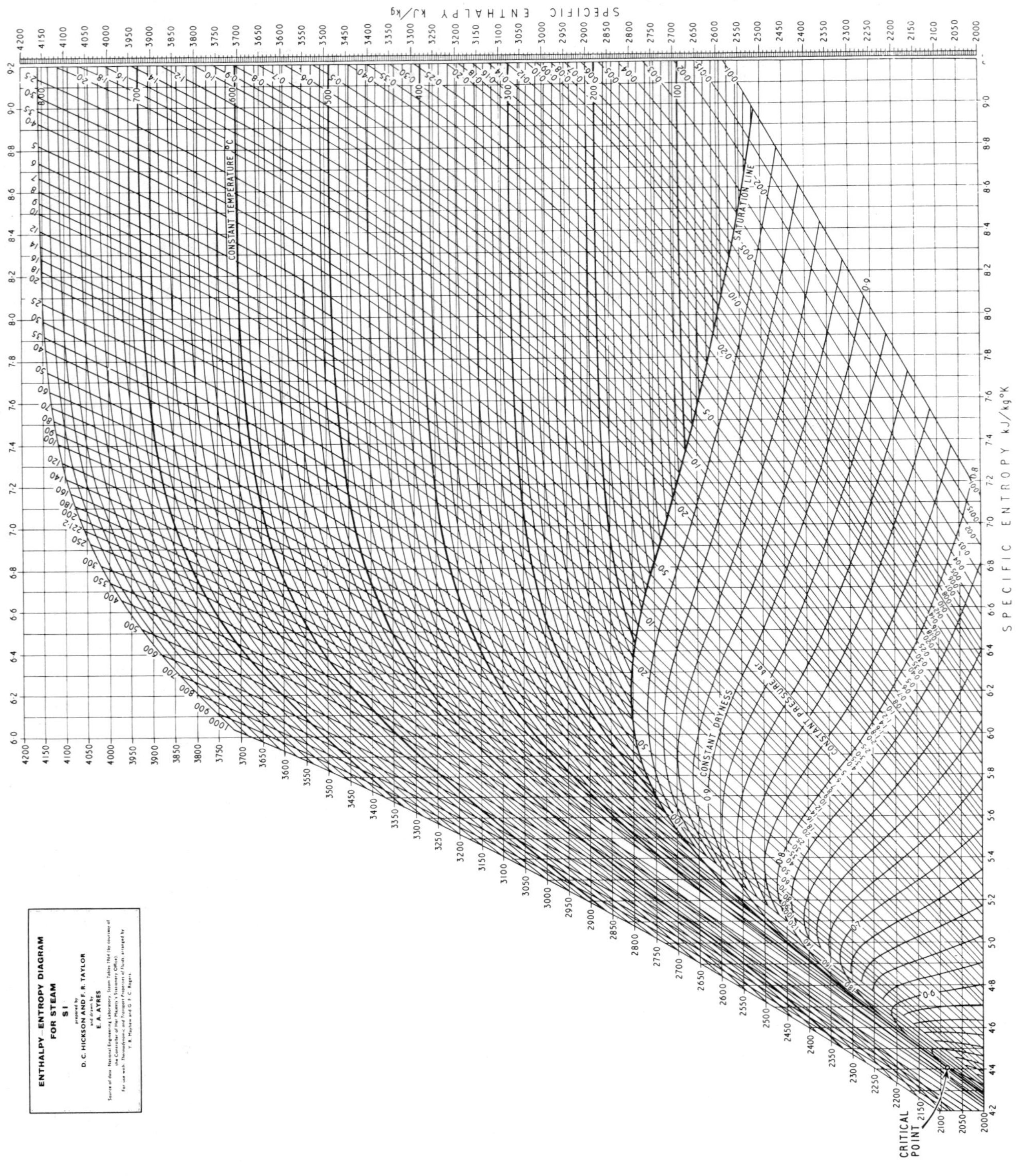
ENTHALPY—ENTROPY DIAGRAM
FOR STEAM
SI
D. C. HICKSON AND F. R. TAYLOR
E. A. AYRES
SPECIFIC ENTHALPY kJ/kg
SPECIFIC ENTROPY kJ/kg°K
CONSTANT TEMPERATURE °C
SATURATION LINE
CONSTANT DRYNESS
CONSTANT PRESSURE bar
CRITICAL POINT

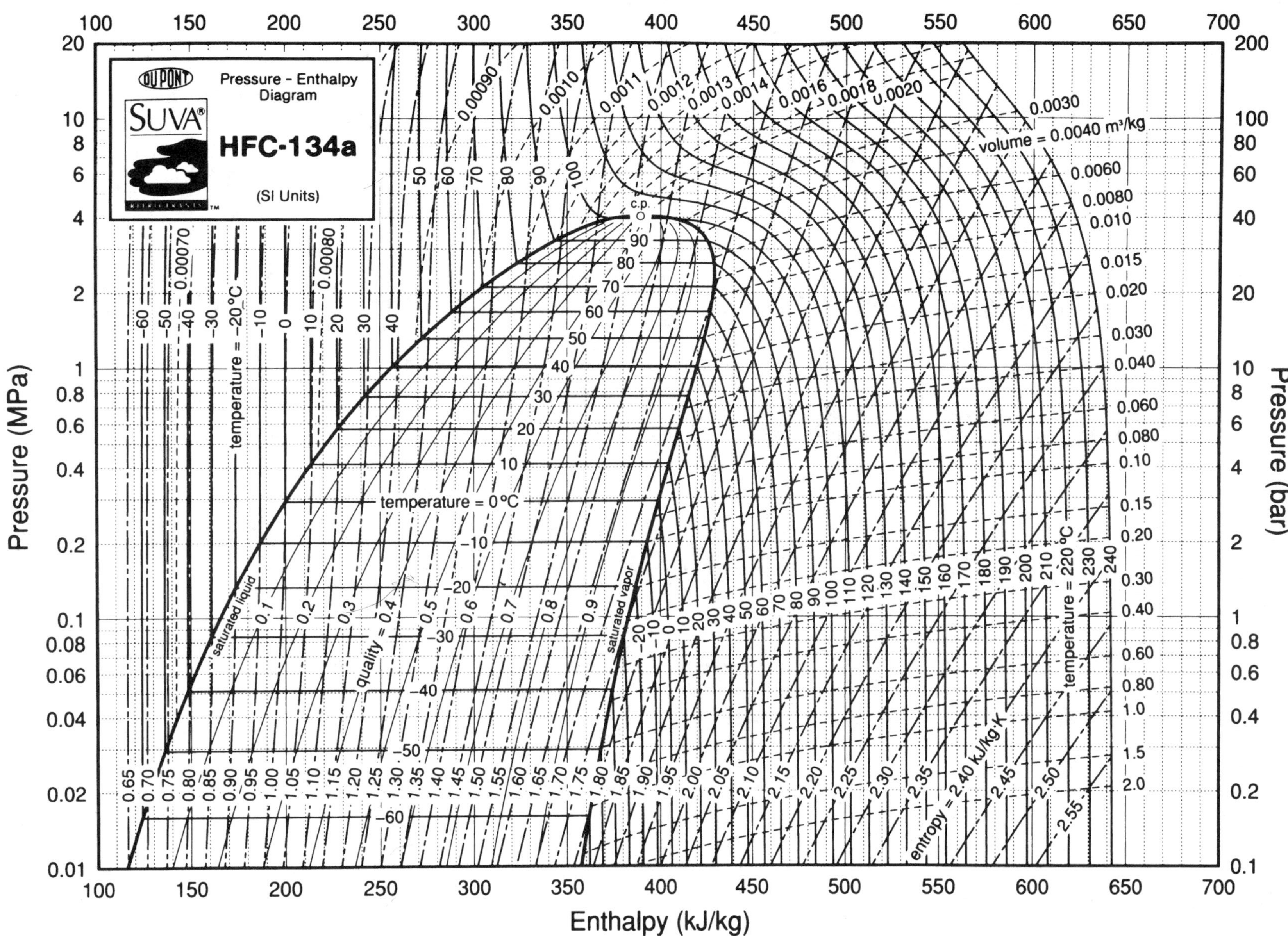

DUPONT
SUVA®
Pressure - Enthalpy Diagram
HFC-134a
(SI Units)
Pressure (MPa)
Pressure (bar)
Enthalpy (kJ/kg)
saturated liquid
saturated vapor
quality = 0.4
temperature = 0°C
temperature = -20°C
temperature = 220°C
entropy = 2.40 kJ/kg·K
volume = 0.0040 m³/kg
c.p.

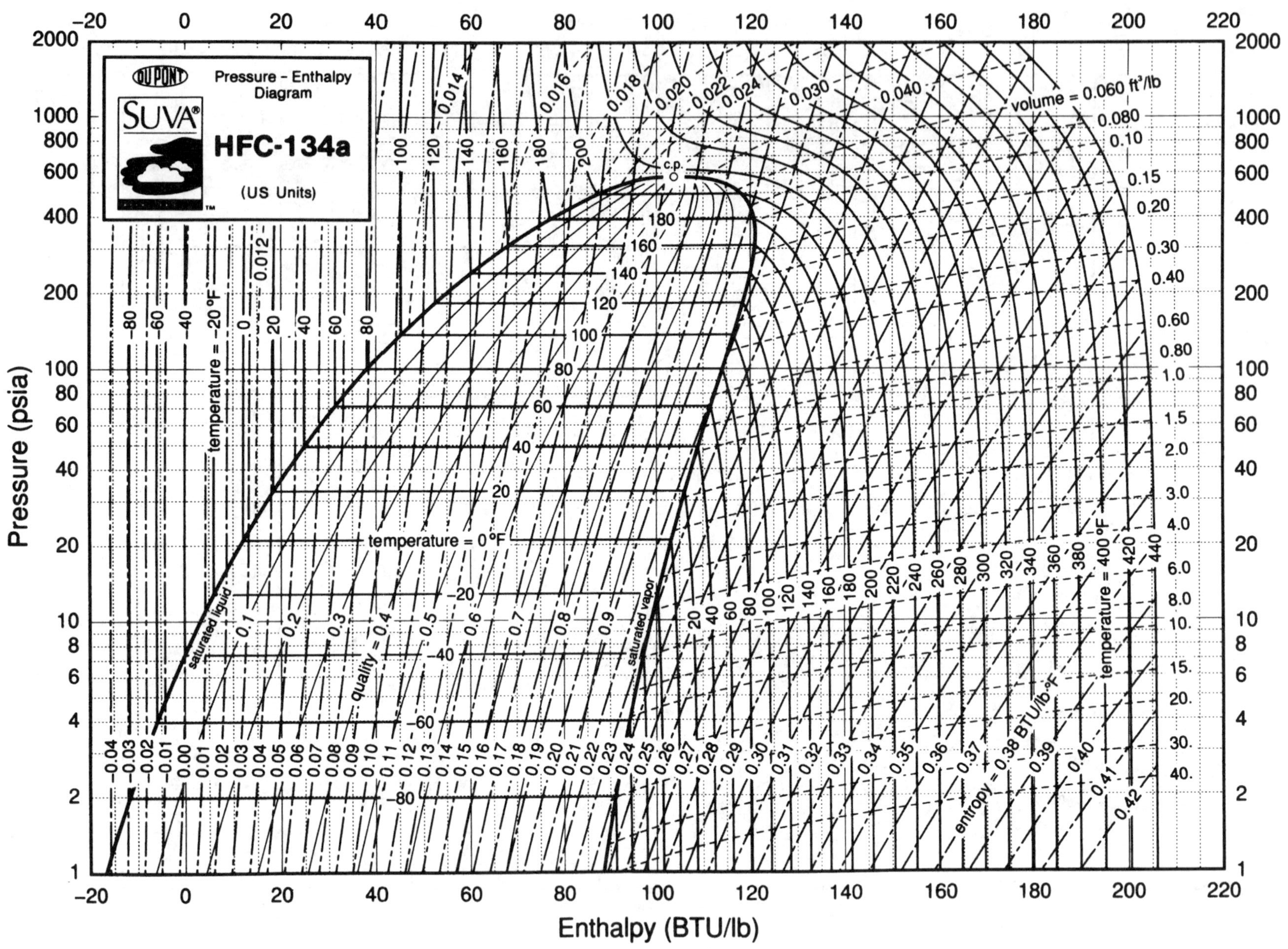

DUPONT
SUVA®
Pressure - Enthalpy Diagram
HFC-134a
(US Units)
Pressure (psia)
Enthalpy (BTU/lb)
c.p.
saturated liquid
saturated vapor
temperature = 0°F
temperature = −20 °F
temperature = 400°F
quality = 0.4
entropy = 0.38 BTU/lb·°F
volume = 0.060 ft³/lb

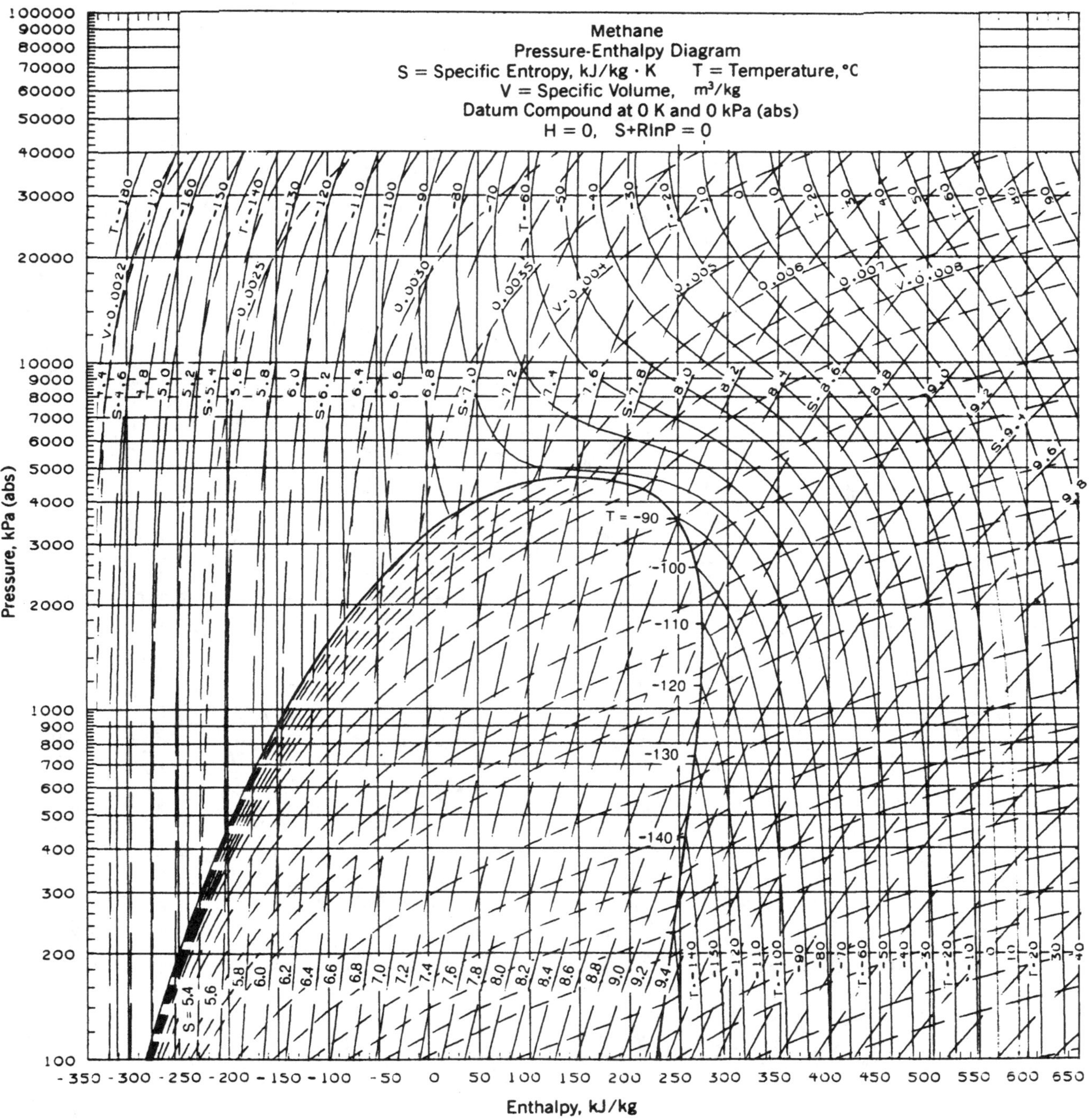
Methane
Pressure-Enthalpy Diagram
S = Specific Entropy, kJ/kg · K   T = Temperature, °C
V = Specific Volume, m³/kg
Datum Compound at 0 K and 0 kPa (abs)
H = 0, S+RlnP = 0
Pressure, kPa (abs)
Enthalpy, kJ/kg

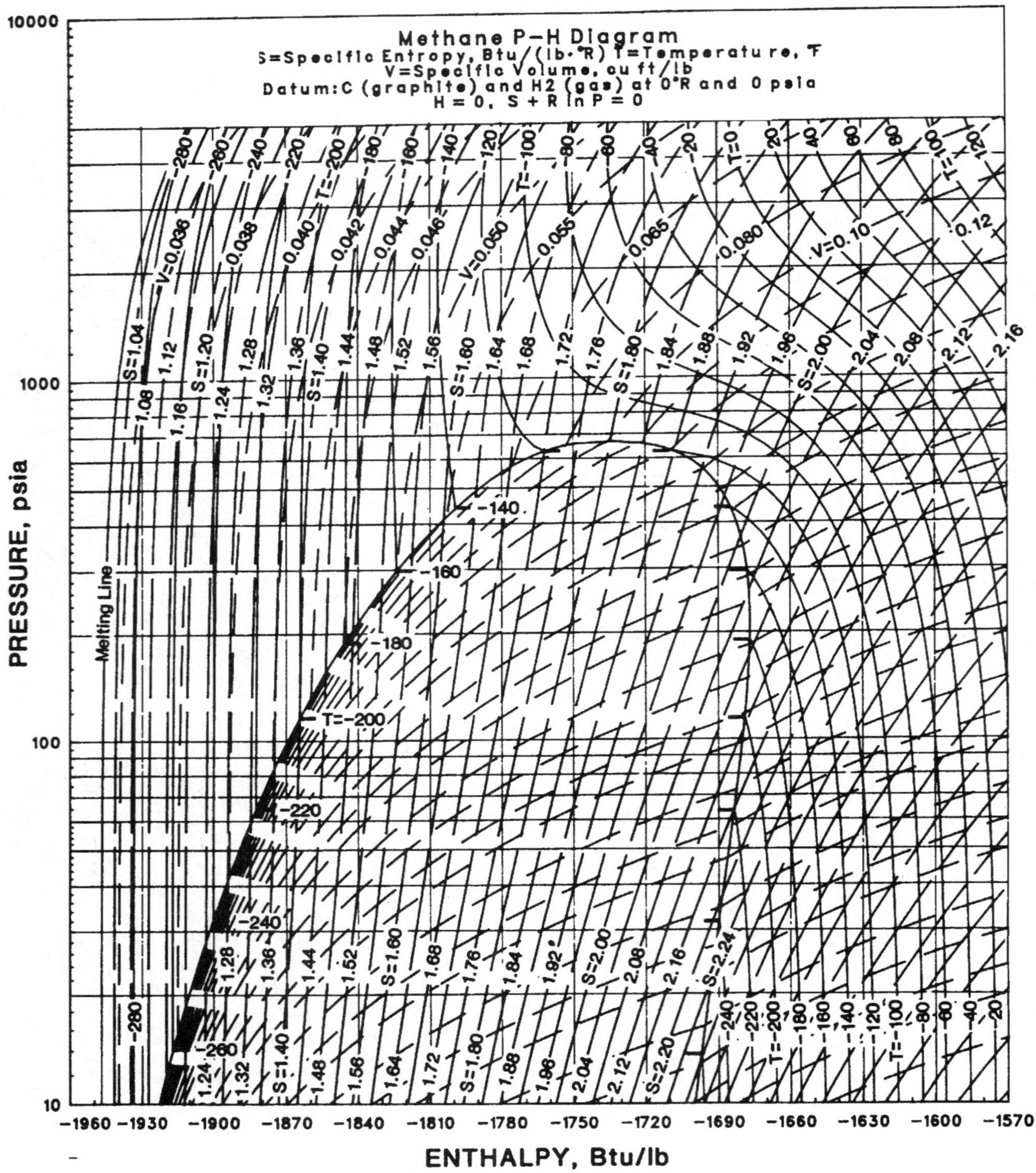
Methane P–H Diagram
S=Specific Entropy, Btu/(lb·°R) T=Temperature, °F
V=Specific Volume, cu ft/lb
Datum:C (graphite) and H2 (gas) at 0°R and 0 psia
H = 0, S + R ln P = 0
PRESSURE, psia
10000
1000
100
10
ENTHALPY, Btu/lb
−1960
−1930
−1900
−1870
−1840
−1810
−1780
−1750
−1720
−1690
−1660
−1630
−1600
−1570
Melting Line
T=−200
−140
−160
−180
−220
−240
−260
−280

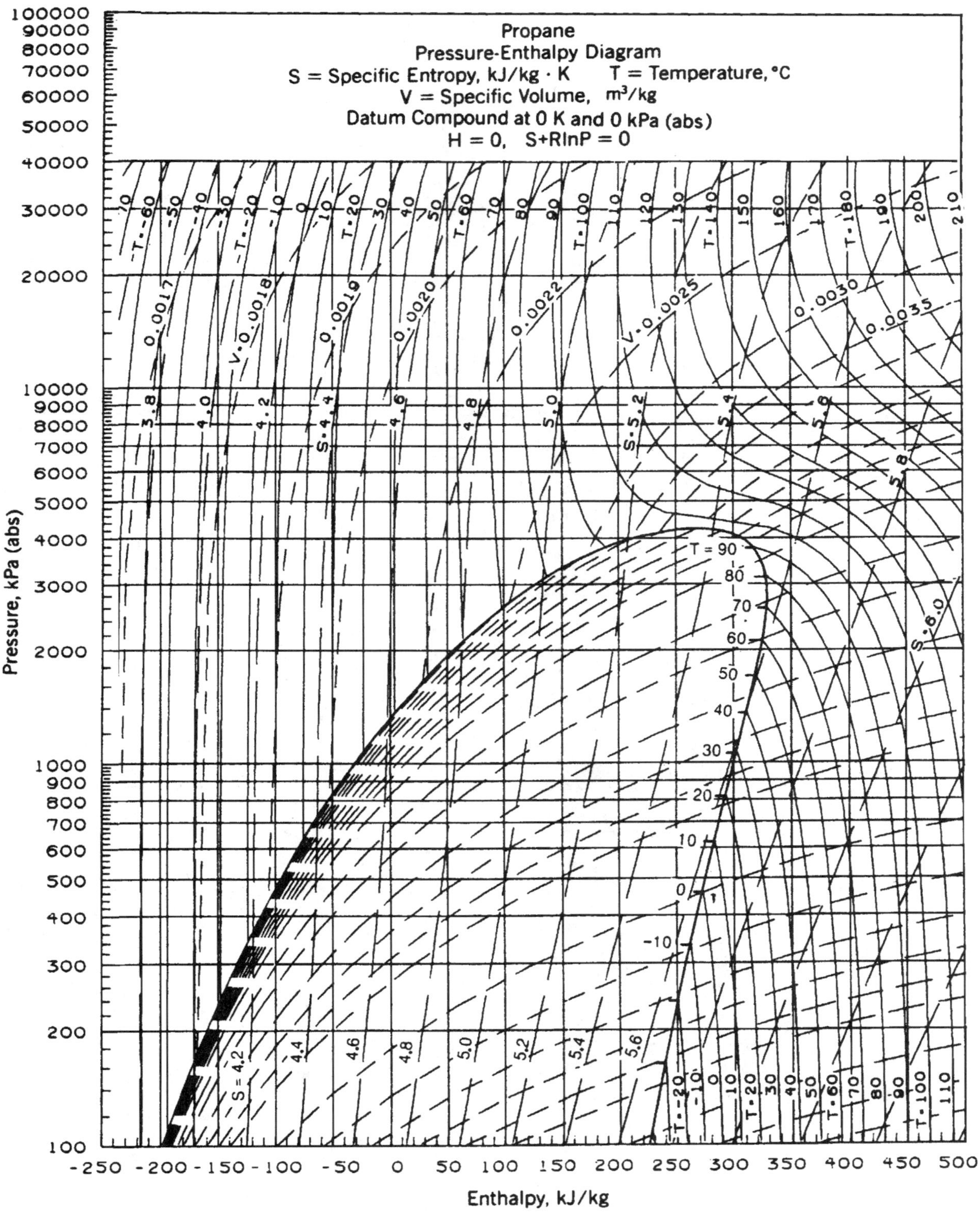
Propane
Pressure-Enthalpy Diagram
S = Specific Entropy, kJ/kg · K
T = Temperature, °C
V = Specific Volume, m³/kg
Datum Compound at 0 K and 0 kPa (abs)
H = 0, S+RlnP = 0
Pressure, kPa (abs)
Enthalpy, kJ/kg

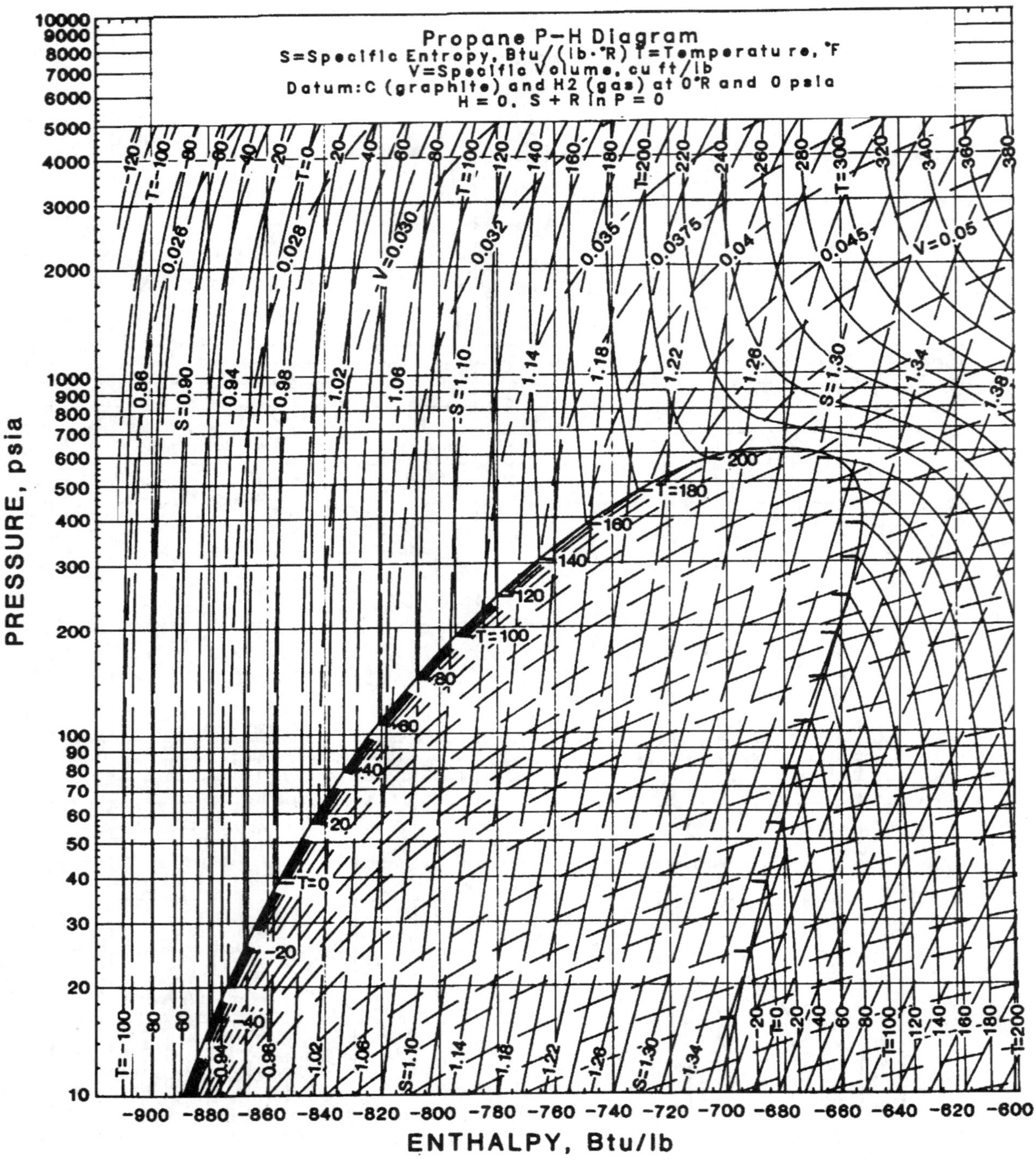
Propane P–H Diagram
S=Specific Entropy, Btu/(lb·°R) T=Temperature, °F
V=Specific Volume, cu ft/lb
Datum: C (graphite) and H2 (gas) at 0°R and 0 psia
H = 0, S + R ln P = 0
PRESSURE, psia
ENTHALPY, Btu/lb

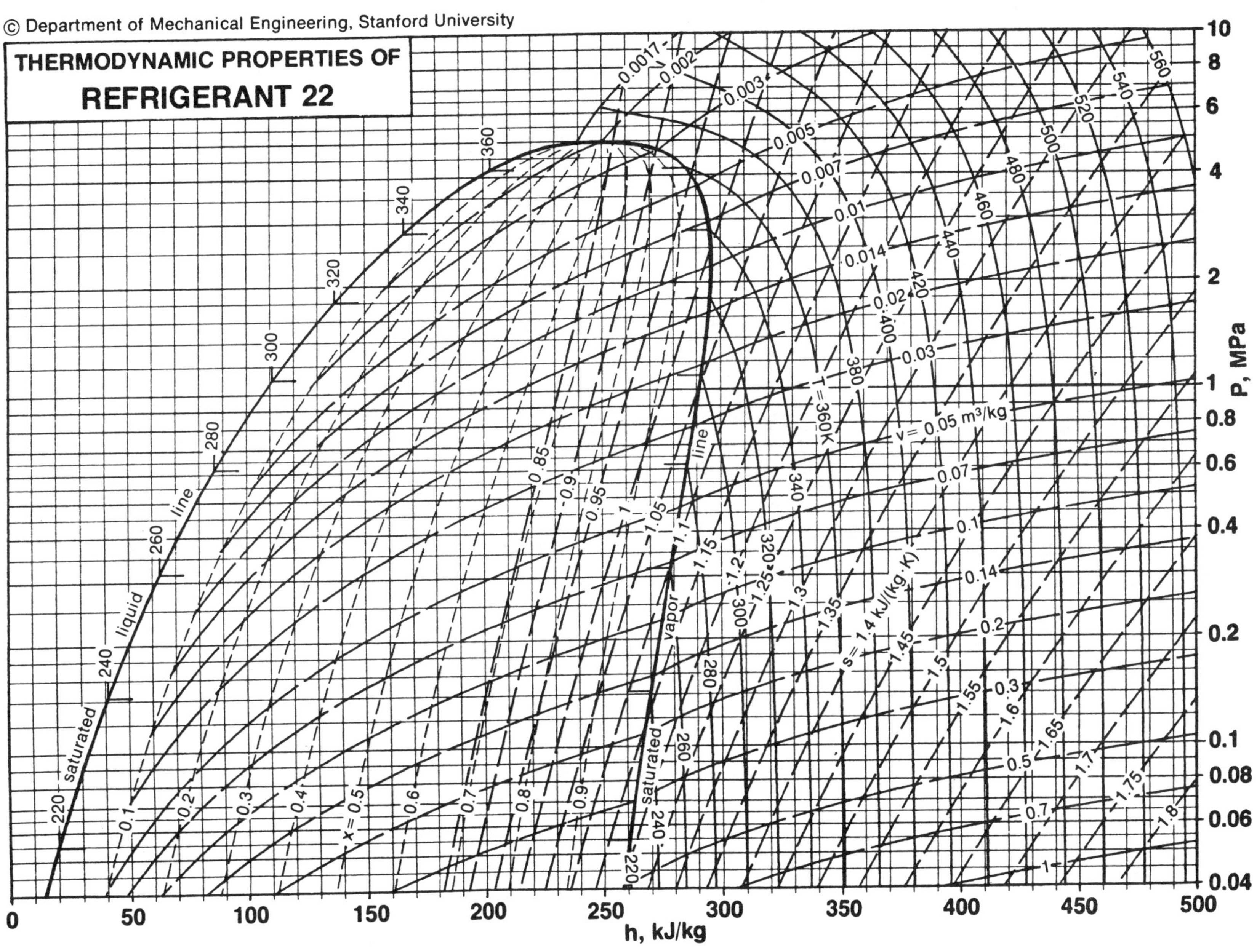

© Department of Mechanical Engineering, Stanford University
THERMODYNAMIC PROPERTIES OF
REFRIGERANT 22
P, MPa
h, kJ/kg
saturated liquid line
saturated vapor line
T = 360K
v = 0.05 m³/kg
s = 1.4 kJ/(kg · K)
x = 0.5

T = °C. P = kPa

THE PROPERTIES OF SATURATED STEAM

T	P	v_f	v_g	h_f	h_g	s_f	s_g
40.00	7.38	1.0092	19550.	167.37	2574.3	.5731	8.2586
50.00	12.33	1.0166	12042.	209.19	2591.4	.7054	8.0782
60.00	19.90	1.0161	7678.7	250.89	2609.4	.8306	7.9109
65.00	25.03	1.0234	6205.6	271.95	2617.7	.8880	7.8282
70.00	31.05	1.0191	5045.6	292.57	2626.0	.9625	7.7631
75.00	38.58	1.0271	4133.2	313.81	2635.1	1.0111	7.6791
80.00	47.32	1.0290	3409.3	334.66	2643.1	1.0753	7.6131
83.00	53.40	1.0311	3045.1	347.29	2648.2	1.1097	7.5705
86.00	60.05	1.0344	2728.0	359.83	2653.0	1.1460	7.5321
90.00	70.04	1.0378	2362.2	376.63	2659.4	1.1927	7.4798
100.00	101.28	1.0438	1673.0	418.77	2675.3	1.3063	7.3545
105.00	120.75	1.0482	1419.4	439.83	2683.1	1.3618	7.2958
110.00	143.21	1.0510	1210.0	460.98	2690.8	1.4178	7.2382
115.00	169.01	1.0553	1036.3	482.16	2698.2	1.4727	7.1824
120.00	198.42	1.0616	891.50	503.47	2705.5	1.5270	7.1281
125.00	232.07	1.0640	770.53	524.54	2712.0	1.5801	7.0764
130.00	269.98	1.0707	668.10	545.97	2719.5	1.6336	7.0250
135.00	312.94	1.0749	581.89	567.29	2726.0	1.6859	6.9758
137.00	331.39	1.0785	551.00	575.91	2729.5	1.7064	6.9555
140.00	361.40	1.0788	508.83	588.55	2731.1	1.7388	6.9279
145.00	415.34	1.0852	446.07	610.20	2738.7	1.7895	6.8807
147.00	438.68	1.0877	423.85	618.73	2740.9	1.8102	6.8620
150.00	475.94	1.0907	392.32	631.80	2745.1	1.8409	6.8355
153.00	515.36	1.0942	364.10	644.70	2748.4	1.8712	6.8083
155.00	543.24	1.0969	346.39	653.38	2751.2	1.8915	6.7907
157.00	572.06	1.0976	329.92	662.05	2752.2	1.9115	6.7729
160.00	617.90	1.1023	306.75	675.07	2756.4	1.9415	6.7471
163.00	666.51	1.1058	285.56	688.12	2759.4	1.9713	6.7213
165.00	700.56	1.1085	272.40	696.83	2761.5	1.9913	6.7039
168.00	754.28	1.1124	253.98	709.93	2764.6	2.0210	6.6787
170.00	792.94	1.1107	242.25	718.97	2767.7	2,0424	6.6620
175.00	892.13	1.1212	216.57	740.62	2771.2	2.0895	6.6217
178.00	957.29	1.1239	202.49	753.91	2774.2	2.1191	6.5971
180.00	1001.9	1.1315	194.23	762.62	2774.9	2.1373	6.5832
184.00	1098.1	1.1326	177.60	780.42	2779.2	2.1771	6.5495
186.00	1148.6	1.1367	170.24	789.31	2780.3	2.1961	6.5346
189.00	1227.5	1.1399	159.67	802.57	2783.2	2.2250	6.5107
191.00	1282.7	1.1427	153.09	811.56	2784.3	2.2438	6.4958
194.00	1368.7	1.1473	143.81	824.94	2786.8	2.2725	6.4724
205.00	1723.9	1.1643	115.04	874.80	2793.3	2.3763	6.3901
215.00	2105.4	1.1824	94.646	887.62	2797.5	2.4698	6.3170
220.00	2319.2	1.1926	86.098	586.19	2798.9	2.5157	6.2810
230.00	2796.7	1.2063	71.479	909.56	2801.2	2.6089	6.2095
235.00	3062.	1.2159	65.284	918.34	2801.4	2.6548	6.1745
240.00	3345.6	1.2262	59.609	1262.1	2801.1	2.6995	6.1361
245.00	3651.8	1.2373	54.647	1061.2	2800.6	2.7481	6.1076
250.00	3975.3	1.2518	50.094	1085.0	2798.4	2.7908	6.0655
255.00	4322.1	1.2499	45.859	1110.1	2797.7	2.8395	6.0339
260.00	4692.3	1.2903	42.150	1134.3	2794.0	2.8820	5.9982
265.00	5084.0	1.2818	38.695	1159.4	2791.8	2.9305	5.9619
270.00	5502.5	1.3048	35.579	1184.7	2787.5	2.9747	5.9265
275.00	5946.1	1.3141	32.723	1210.4	2782.9	3.0204	5.8902
277.00	6131.1	1.3205	31.652	1220.7	2780.7	3.0384	5.8757
280.00	6415.0	1.3334	30.113	1236.2	2777.7	3.0674	5.8534
285.00	6913.9	1.3499	27.718	1262.5	2771.4	3.1128	5.8168
290.00	7440.2	1.3719	25.511	1289.2	2764.4	3.1596	5.7795
300.00	8586.3	1.4036	21.618	1344.3	2747.1	3.2536	5.7012
305.00	9208.0	1.4262	19.898	1372.7	2736.6	3.3012	5.6609
310.00	9863.5	1.4647	18.278	1404.7	2721.4	3.3543	5.6139
315.00	10554.	1.4574	16.843	1428.0	2716.1	3.3936	5.5828
335.00	13704.	1.5943	11.833	1557.7	2643.3	3.6017	5.3869
350.00	16526.	1.8019	8.7090	1681.4	2548.0	3.7939	5.1863
370.00	21040.	2.3377	4.8901	1911.4	2320.7	4.1429	4.7793
375.00	22331.	3.3872	2.6537	2154.7	2033.3	4.5143	4.3262

THE PROPERTIES OF SUPERHEATED STEAM

T	P=6.89kPa(Sat.Temp.=38.74°C)			P=137.8kPa(Sat.Temp.=108.87°C)			P=34.5kPa(Sat.Temp.=72.36°C)			P=68.9kPa(Sat.Temp.=89.56°C)		
	v	h	s	v	h	s	v	h	s	v	h	s
95.0	24621.	2678.5	8.5966	1399.8	2773.4	7.4611	4902.1	2674.9	7.8456	2437.1	2669.9	7.5148
150.0	28311.	2783.0	8.8613	1589.4	2882.9	7.6043	5649.1	2781.2	8.1149	2816.8	2778.7	7.7905
205.0	32000.	2888.7	9.0960	1776.7	2992.1	7.9206	6391.4	2887.6	8.3516	3190.1	2886.2	8.0292
260.0	35684.	2996.1	9.3085	1962.7	3102.1	8.1171	7130.5	2995.4	8.5645	3561.5	2994.2	8.2434
315.0	39367.	3105.2	9.5030	2148.1	3213.8	8.2985	7868.4	3104.5	8.7598	3931.1	3103.8	8.4387
370.0	43050.	3215.9	9.6835	2333.0	3327.0	8.4670	8606.4	3215.4	8.9399	4300.7	3215.0	8.6196
425.0	46735.	3328.6	9.8516	2517.7	3442.0	8.6262	9343.7	3328.4	9.1085	4669.7	3327.9	8.7882
480.0	50411.	3443.5	10.009	2702.6	3559.2	8.7757	10080.	3443.2	9.2667	5038.6	3442.8	8.9464
535.0	54095.	3560.3	10.160	2886.7	3678.0	8.9185	10817.	3560.0	9.4167	5407.6	3559.6	9.0964
590.0	57778.	3679.3	10.301	3088.3	3810.4	9.0663	11554.	3679.0	9.5589	5776.0	3678.8	9.2386
650.0	61797.	3811.1	10.450	3457.0	4059.1	9.3211	12357.	3811.1	9.7068	6177.9	3810.8	9.3866
760.0	69158.	4059.8	10.704	3825.5	4316.4	9.5579	13834.	4059.8	9.9612	6915.1	4059.5	9.6414
870.0	76524.	4317.2	10.941				15305.	4316.9	10.198	7651.9	4316.7	9.8777

T	P=101.3kPa(Sat.Temp.=100°C)			P=275.7kPa(Sat.Temp.=130.69°C)			P=413.5kPa(Sat.Temp.-144.84°C)			P=551.3kPa(Sat.Temp.=155.57°C)		
	v	h	s	v	h	s	v	h	s	v	h	s
150.00	1911.1	2776.2	7.6085	691.23	2762.4	7.1207	454.59	2750.4	6.9108	388.81	2863.3	7.0325
205.00	2167.6	2884.6	7.8496	789.31	2876.7	7.3745	522.37	2870.0	7.1765	438.24	2979.3	7.2624
260.00	2421.0	2993.3	8.0642	884.48	2987.9	7.5949	587.01	2983.7	7.4014	486.27	3093.0	7.4656
315.00	2673.1	3103.1	8.2603	978.43	3099.1	7.7935	650.30	3096.1	7.6021	533.55	3206.8	7.6503
370.00	2925.3	3214.5	8.4413	1071.7	3211.4	7.9756	712.97	3209.1	7.7860	580.53	3321.4	7.8218
425.00	3176.2	3327.5	8.6098	1164.7	3325.1	8.1455	775.29	3323.3	7.9562	627.21	3437.4	7.9817
480.00	3427.8	3442.5	8.7685	1257.3	3440.4	8.3045	837.27	3439.0	8.1162	677.97	3566.0	8.1459
535.00	3676.7	3559.4	8.9185	1350.0	3557.7	8.4550	899.21	3556.6	8.2666	720.23	3675.0	8.2760
590.00	3929.0	3678.3	9.0610	1442.7	3677.1	8.5975	960.96	3676.0	8.4095	770.80	3807.6	8.4244
650.00	4203.3	3810.6	9.2090	1543.2	3809.5	8.7460	1028.2	3808.5	8.5580	63.38	4057.0	8.6755
760.00	4705.2	4059.3	9.4634	1728.0	4058.4	9.0008	1551.5	4057.7	8.8128	955.78	4314.8	8.9166
870.00	5206.4	4316.6	9.7002	1912.2	4316.0	9.2375	1274.8	4315.3	9.0500			

T	P=689.2kPa(Sat.Temp.=164.34°C)			P=827.0kPa(Sat.Temp.=171.81°C			P=964.8kPa(Sat.Temp.=178.34°C)			P=1102.7kPa(Sat.Temp.=184.18°C)		
	v	h	s	v	h	s	v	h	s	v	h	s
205.00	308.62	2856.1	6.9183	255.13	2848.7	6.8217	217.82	2841.1	6.7381	188.07	2833.0	6.6632
260.00	348.90	2974.7	7.1531	289.42	2970.2	7.0627	246.84	2965.6	6.9848	214.94	2960.7	6.9162
315.00	387.79	3089.8	7.3589	322.12	3086.5	7.2705	275.17	3083.5	7.1951	240.03	3080.2	7.1294
370.00	425.92	3204.4	7.5448	354.13	3201.9	7.4581	303.64	3199.5	7.3839	264.46	3197.0	7.3195
425.00	463.70	3319.5	7.7167	386.79	3317.6	7.6304	329.92	3315.8	7.5576	288.38	3313.7	7.4939
480.00	501.16	3436.0	7.8770	417.13	3434.3	7.7916	358.48	3432.9	7.7187	312.12	3431.3	7.6560
540.00	541.88	3564.8	8.0416	451.18	3563.4	7.9567	386.37	3562.3	7.8842	337.81	3560.9	7.8215
590.00	575.76	3673.9	8.1717	479.44	3672.7	8.0867	410.65	3671.5	8.0147	359.11	3670.3	7.9523
650.00	616.29	3806.4	8.3206	513.28	3805.5	8.2356	440.71	3804.6	8.1640	384.53	3803.7	8.1016
760.00	690.45	4056.3	8.5762	575.21	4055.6	8.4912	492.87	4054.7	8.4197	431.13	4054.0	8.3577
870.00	764.50	4314.1	8.8134	636.95	4313.7	8.7288	547.86	4312.9	8.6572	477.49	4312.5	8.5953

T	P=1240.5kPa(Sat.Temp.=189.48°C)			P=1378.3kPa(Sat.Temp.=194.33°C)			P=1516.2kPa(Sat.Temp.=198.81°C)			P=1654.0kPa(Sat.Temp.=202.98°C)		
	v	h	s	v	h	s	v	h	s	v	h	s
205.00	166.63	2824.7	6.5951	147.64	2816.1	6.5628	132.89	2807.3	6.4731	120.55	2798.1	6.4171
260.00	190.03	2955.8	6.8550	170.18	2950.9	6.7994	153.88	2945.8	6.7479	140.28	2940.7	6.7001
315.00	212.72	3076.7	7.0711	190.83	3073.4	6.8901	172.86	3070.2	6.9698	157.96	3066.7	6.9254
370.00	234.54	3194.6	7.2625	210.61	3192.0	7.4687	191.04	3189.7	7.1641	174.71	3187.1	7.1209
425.00	255.94	3311.8	7.4373	229.97	3309.9	4.7009	208.72	3308.1	7.3406	191.04	3306.2	7.2983
480.00	277.07	3429.6	7.5999	249.06	3428.2	7.5500	226.17	3426.6	7.5043	207.05	3425.0	7.4624
540.00	300.00	3559.7	7.7658	239.76	3558.3	7.7160	244.98	3557.2	7.6712	224.38	3555.8	7.6298
590.00	318.93	3669.4	7.8970	286.84	3668.2	7.8472	260.60	3667.0	7.8024	238.71	3666.1	7.7613
650.00	341.65	3802.7	8.0464	307.34	3801.8	7.9970	279.21	3800.9	7.9522	255.83	3799.9	7.9112
760.00	383.06	4053.3	8.3028	344.66	4052.6	8.2534	313.20	4051.9	8.2091	286.98	4051.2	8.1684
870.00	424.33	4311.8	8.5405	381.86	4311.3	8.4914	347.08	4310.6	8.4471	318.06	4310.2	8.4065

h = kJ/kg, v ≑ cm^3/g and S = kJ/kg·K

Datum - Sat. liquid h = 0 at 0°C.

THE PROPERTIES OF SUPERHEATED METHANE (continued)

T	P=2067.5kPa (Sat.Temp.=-106.39°C)			P=3445.90kPa (Sat.Temp.=-91.61°C)			P=5513.4kPa			P=6891.8kPa			P=10337.7kPa		
	v	h	s	v	h	s	v	h	s	v	h	s	v	h	s
-115.00															
-105.00															
-95.00	34.808	632.20	4.1547												
-85.00	38.765	660.60	4.3155	18.498	606.89	3.8389									
-80.00	40.512	674.07	4.3895	20.303	627.27	3.9425									
-75.00	42.261	687.44	4.4599	21.834	646.32	4.0379	10.103	576.99	3.5390						
-60.00	47.323	726.35	4.6481	25.659	693.70	4.2770	12.843	633.43	3.8147	8.4124	570.54	3.4708			
-50.00	50.413	751.35	4.7627	27.870	722.50	4.4103	14.929	673.60	4.0017	10.513	628.14	3.7376	5.5479	528.01	3.2672
-40.00	53.438	775.81	4.8680	30.028	750.00	4.5297	16.693	707.67	4.1537	12.217	672.56	3.9343	6.6735	589.07	3.5056
-30.00	56.379	799.87	4.9665	32.125	776.44	4.6399	18.275	739.25	4.2882	13.649	711.32	4.0910	7.8017	638.03	3.6888
-15.00	60.726	835.58	5.1106	34.898	815.56	4.7944	20.362	784.61	4.4672	15.530	762.66	4.2905	9.3276	704.23	3.9297
-5.00	63.618	859.71	5.2004	36.781	841.34	4.8915	21.710	813.66	4.5748	16.736	794.03	4.4067	10.249	744.36	4.0652
5.00	66.452	883.35	5.2864	38.617	866.97	4.9835	22.975	842.02	4.6740	17.834	825.10	4.5124	11.094	782.30	4.1859
15.00	69.211	907.02	5.3689	40.4 0	892.28	5.0708	24.209	869.66	4.7675	18.865	854.89	4.6107	11.893	817.65	4.2963
25.00	71.981	931.03	5.4492	42.180	917.61	5.1560	25.479	896.71	4.8580	19.890	882.26	4.7048	12.669	850.42	8.4023
40.00	76.198	966.53	5.5682	44.798	954.63	5.2786	27.241	936.29	4.9882	21.391	924.49	4.8400	13.775	896.23	4.5509
50.00	78.900	990.36	5.6437	46.526	979.21	5.3570	28.409	961.83	5.0699	22.407	950.48	4.9250	14.473	925.21	4.6446
60.00	81.593	1014.4	5.7158	48.257	1003.7	5.4299	29.528	987.44	5.1473	23.348	976.98	5.0049	15.164	953.49	4.7319
70.00	84.306	1038.7	5.7861	49.933	1028.5	5.5003	30.651	1013.3	5.2223	24.211	1003.8	5.0823	15.837	981.50	4.8152
80.00	86.912	1062.3	5.8572	51.640	1053.0	5.5731	31.751	1038.7	5.2964	25.246	1029.2	5.1602	16.483	1008.4	4.8988
95.00	90.921	1100.2	5.9602	54.084	1091.1	5.6768	33.377	1078.3	5.4049	26.502	1069.6	5.2711	17.462	1050.0	5.0154
105.00	93.548	1125.7	6.0277	55.714	1116.9	5.7446	34.457	1104.8	5.4752	27.386	1096.7	5.3434	18.089	1077.6	5.0911
115.00	96.247	1151.6	6.0933	57.344	1142.0	5.8109	35.579	1131.3	5.5438	28.281	1124.1	5.4131	18.707	1105.4	5.1638
125.00	98.865	1177.1	6.1581	58.960	1183.4	5.8771	36.668	1157.9	5.6113	29.220	1151.1	5.4818	19.325	1133.1	5.2348
140.00	102.82	1216.1	6.2553	61.396	1223.4	5.9749	38.181	1198.4	5.7118	30.470	1191.6	5.8535	20.242	1175.2	5.3394
150.00	105.41	1242.5	6.3181	63.050	1231.9	6.0385	39.264	1225.6	5.7770	31.375	1219.1	5.6489	20.828	1203.6	5.4067
160.00	108.00	1269.3	6.3803	64.675	1242.3	6.1014	40.328	1253.0	5.8410	32.212	1246.7	5.7133	21.407	1232.1	5.4730
170.00	110.58	1296.5	6.4421	66.247	1288.4	6.1644	41.340	1280.8	5.9047	32.999	1274.6	5.7778	21.996	1260.8	5.5390
180.00	113.22	1323.9	6.5036	67.865	1322.0	6.2264	42.343	1309.0	5.9678	33.829	1302.9	5.8413	22.592	1289.6	5.6040
190.00	115.85	1351.7	6.5643	69.494	1335.8	6.2874	43.370	1337.5	6.0297	34.699	1331.7	5.9034	23.169	1318.8	5.6680

h = kJ/kg, v = cm^3/g and s = kJ/kg·K

T = °C. P = kPa

THE PROPERTIES OF SATURATED PROPANE

T	P	v_f	v_g	h_f	h_g	s_f	s_g	T	P	v_f	v_g	h_f	h_g	s_f	s_g
-62.22	38.94	1.6543	1011.3	378.14	823.26	3.6819	5.7912	15.56	736.73	1.9665	61.429	557.21	912.09	4.4007	5.6275
-61.11	41.35	1.6606	955.15	380.47	824.65	3.6932	5.7870	21.11	856.65	2.0033	53.313	571.40	917.21	4.4481	5.6216
-60.00	43.56	1.6606	911.45	382.79	826.05	3.7041	5.7828	26.67	989.66	2.0408	46.509	585.81	921.86	4.4954	5.6158
-56.67	51.55	1.6731	780.35	389.77	830.23	3.7376	5.7698	32.22	1137.2	2.0782	40.141	600.47	926.28	4.5427	5.6103
-51.11	67.40	1.6874	609.92	401.63	837.21	3.7932	5.7527	37.78	1300.5	2.1163	34.835	615.35	930.70	4.5900	5.6053
-45.56	86.84	1.7062	482.57	413.49	843.72	3.8468	5.7368	43.33	1480.4	2.1550	30.402	630.47	934.65	4.6390	5.6011
-40.00	110.27	1.7249	384.56	425.58	850.47	3.9000	5.7234	48.89	1677.5	2.2049	26.594	646.51	939.07	4.6871	5.5969
-34.44	139.08	1.7442	313.39	438.14	857.21	3.9528	5.7108	54.44	1891.8	2.2549	23.098	663.26	942.79	4.7353	5.5919
-28.89	172.64	1.7642	253.46	450.70	863.95	4.0059	5.6982	60.00	2125.4	2.3111	19.977	680.70	946.51	4.7855	5.5881
-23.33	213.30	1.7848	207.88	463.72	870.70	4.0570	5.6865	65.56	2380.4	2.3829	17.355	698.14	949.30	4.8366	5.5793
-17.78	260.58	1.8060	171.05	476.74	877.21	4.1081	5.6752	71.11	2653.3	2.4734	14.983	717.21	950.70	4.8902	5.5697
-12.22	315.99	1.8291	143.58	490.00	883.72	4.1583	5.6652	76.67	2935.9	2.5795	12.985	738.37	950.23	4.9471	5.5567
-6.67	379.05	1.8541	120.49	503.72	889.77	4.2077	5.6564	82.22	3261.2	2.7262	11.237	761.63	947.91	5.0116	5.5362
-1.11	452.79	1.8797	99.885	516.98	895.58	4.2567	5.6484	86.67	3432.8	2.8985	9.6763	783.49	943.95	5.0685	5.5136
4.44	536.18	1.9072	83.029	530.00	901.16	4.3053	5.6409	87.78	3607.2	2.9416	9.3017	788.84	940.93	5.0828	5.5082
10.00	630.60	1.9359	71.168	543.72	906.74	4.3534	5.6338	93.33	3962.8	3.2525	7.0543	822.09	926.28	5.1749	5.4596

THE PROPERTIES OF SUPERHEATED PROPANE

T	P = 50.65kPa (Sat.Temp.=-57.06°C)			P = 84.36kPa (Sat.Tem.=-46.11°C)			P = 101.28kPa (Sat.Temp.=-42.06°C)			P = 137.84kPa (Sat.Temp.=-34.61°C)		
	v	h	s	v	h	s	v	h	s	v	h	s
-50.00	822.56	839.39	5.8130									
-40.00	860.88	853.02	5.8711	510.78	851.63	5.7761	422.95	850.70	5.7401			
-30.00	899.16	866.89	5.9293	533.79	865.81	5.8334	442.48	865.16	5.7988	322.20	863.95	5.7372
-15.00	955.54	888.34	6.0152	568.29	887.46	5.9194	471.80	887.06	5.8822	343.56	886.33	5.8241
0.00	1013.4	911.14	6.1011	602.63	910.37	6.0051	500.80	910.01	5.9900	365.08	908.96	5.9104
10.00	1050.6	927.09	6.1585	625.90	925.96	6.9625	520.19	925.82	5.9681	379.56	925.15	5.9669
20.00	1089.2	942.52	6.2154	648.64	941.65	6.1197	539.28	941.72	6.1909	393.79	940.92	6.0236
30.00	1126.8	961.12	6.2730	672.20	960.62	6.1755	558.49	958.89	6.0170	408.14	957.93	6.0800
40.00	1164.4	973.74	6.3294	694.70	974.31	6.2316	578.11	975.81	6.1281	422.44	974.90	6.1354
50.00	1202.9	994.70	6.3847	717.35	993.61	6.2872	596.88	993.13	6.2527	436.63	992.46	6.1904
60.00	1241.1	1012.1	6.4393	740.39	1013.7	6.3422	616.10	1011.2	6.3070	450.79	1010.7	6.2454
70.00	1278.7	1024.1	6.4936	763.43	1030.3	6.3968	635.36	1029.8	6.3634	464.95	1029.3	6.3005
80.00	1316.8	1046.8	6.5478	786.46	1048.6	6.4511	652.54	1048.9	6.3860	479.11	1048.3	6.3555
90.00	1355.0	1074.5	6.6021	809.52	1069.8	6.5053	669.51	1068.7	6.4262	493.27	1067.7	6.4105

T	P = 206.75kPa (Sat.Temp.=-24.18°C)			P = 275.67kPa (Sat.Temp.=-16.17°C)			P = 413.51kPa (Sat.Temp.=4.00°C)			P = 551.34kPa (Sat.Temp.=5.38°C)		
	v	h	s	v	h	s	v	h	s	v	h	s
-15.00	224.94	884.19	5.7428									
0.00	239.75	907.14	5.8281	176.96	905.01	5.7679						
10.00	249.65	923.21	5.8844	184.58	921.12	5.8257	119.32	916.82	5.7395			
20.00	259.52	939.23	5.9413	192.25	937.56	5.8826	124.65	933.37	5.8003	90.822	928.88	5.7316
30.00	269.22	956.20	5.9973	199.64	954.59	5.9388	129.92	950.76	5.8591	95.020	947.11	5.7910
40.00	278.80	973.74	6.0531	206.95	971.73	5.9956	135.03	968.71	5.9171	99.111	964.84	5.8494
50.00	288.35	991.26	6.1089	214.25	989.45	6.0520	140.09	986.63	5.9741	103.06	982.98	5.9073
60.00	297.91	1009.3	6.1646	221.56	1007.9	6.1077	145.14	1005.1	6.0307	107.00	1002.1	5.9654
70.00	307.46	1028.1	6.2204	228.86	1026.9	6.1635	150.20	1024.4	6.0872	110.94	1021.8	6.0234
80.00	317.01	1047.3	6.2762	236.16	1046.3	6.2193	155.26	1044.0	6.1434	114.77	1041.6	6.0808
90.00	326.56	1067.0	6.3319	243.47	1066.1	6.2746	160.25	1064.1	6.1993	118.57	1062.0	6.1375

T	P = 689.18kPa (Sat.Temp.=13.12°C)			P = 895.93kPa (Sat.Temp.=22.89°C)			P = 1102.7kPa (Sat.Temp.=30.95°C)			P = 1309.4kPa (Sat.Temp.=38.06°C)			P = 1516.2kPa (Sat.Temp.=44.36°C)		
	v	h	s	v	h	s	v	h	s	v	h	s	v	h	s
20.00	70.164	923.71	5.6734												
30.00	73.857	942.53	5.7417	53.892	934.03	5.6691									
40.00	77.305	960.77	5.7942	57.207	954.07	5.7322	44.155	945.57	5.6760						
50.00	80.897	979.58	5.8530	60.159	973.72	5.7939	46.886	966.59	5.7424	37.705	959.75	5.6934	30.924	951.05	5.6438
60.00	84.090	999.07	5.9159	62.802	993.72	5.8561	49.368	987.91	5.8058	40.047	981.63	5.7606	33.174	974.88	5.7150
70.00	87.050	1018.9	5.9748	65.386	1014.5	5.9176	51.683	1009.0	5.8676	42.173	1003.1	5.8251	35.205	997.00	5.7817
80.00	90.373	1039.2	6.0328	67.974	1035.0	5.9780	53.904	1030.4	5.9295	44.171	1025.3	5.8880	37.049	1019.7	5.8470
90.00	93.717	1059.9	6.0918	70.523	1055.9	6.0377	56.058	1051.9	5.9908	46.048	1047.2	5.9486	38.786	1042.0	5.9090

Datum - h = 0 and s = 0 for solid at absolute zero

T = °C. P = kPa

THE PROPERTIES OF SATURATED AMMONIA

T	P	v_f	v_g	h_f	h_g	s_f	s_g
-50.00	40.87	1.4246	2625.1	-44.42	1373.0	-0.1943	6.1600
-45.00	66.05	1.4362	1919.9	-22.14	1338.4	-0.3063	5.9975
-40.00	71.74	1.4489	1552.0	0.00	1389.8	-0.0000	5.9628
-35.00	92.43	1.5175	1222.2	18.23	1401.7	0.1105	5.8760
-30.00	119.54	1.4749	963.89	45.21	1405.3	0.1874	5.7856
-28.00	131.59	1.4821	880.53	53.50	1408.3	0.2246	5.7521
-26.00	144.65	1.4860	805.50	62.66	1411.2	0.2605	5.7191
-24.00	158.69	1.4923	738.66	71.77	1414.2	0.2970	5.6870
-22.00	173.82	1.4977	678.42	80.75	1417.0	0.3329	5.6554
-20.00	190.14	1.5033	623.72	89.77	1419.8	0.3684	5.6241
-18.00	207.64	1.5102	574.28	98.78	1422.5	0.4043	5.5936
-16.00	226.33	1.5151	529.57	107.75	1425.0	0.4397	5.5637
-14.00	246.50	1.5212	488.95	116.88	1427.7	0.4749	5.5343
-12.00	267.57	1.5274	452.08	126.24	1430.4	0.5095	5.5049
-10.00	291.20	1.5335	418.43	135.17	1432.6	0.5444	5.4766
-8.00	314.68	1.5397	387.86	144.32	1435.1	0.5788	5.4483

T	P	v_f	v_g	h_f	h_g	s_f	s_g
-6.00	340.80	1.5460	359.85	153.54	1437.7	0.6132	5.4210
-4.00	368.78	1.5521	334.31	162.74	1439.8	0.6473	5.3937
-2.00	398.11	1.5590	310.95	171.95	1441.9	0.6812	5.3666
0.00	429.29	1.5663	289.48	181.16	1444.2	0.7151	5.3403
2.00	462.31	1.5731	269.84	190.35	1446.3	0.7487	5.3142
4.00	497.26	1.5792	251.72	199.79	1448.4	0.7818	5.2886
6.00	534.32	1.5864	235.09	208.92	1450.2	0.8152	5.2632
10.00	614.68	1.6006	205.64	227.67	1454.0	0.8813	5.2138
15.00	728.03	1.6191	174.81	251.33	1458.4	0.9631	5.1538
20.00	856.87	1.6387	149.36	274.93	1462.2	1.0443	5.0955
25.00	1002.07	1.6586	128.28	298.96	1465.7	1.1245	5.0389
30.00	1165.55	1.6809	110.60	322.97	1468.6	1.2035	4.9837
35.00	1349.32	1.7000	95.775	347.24	1471.0	1.2820	4.9300
40.00	1553.84	1.7320	83.283	371.93	1472.8	1.3603	4.8774
45.00	1780.52	1.7672	72.592	396.60	1474.0	1.4373	4.8249
50.00	2031.70	1.7748	63.450	422.00	1474.4	1.5147	4.7728

THE PROPERTIES OF SUPERHEATED AMMONIA

T	P = 34.46kPa (Sat.Temp.=-52.84°C)			P = 48.24kPa (Sat.Temp.=-47.16°C)		
	v	h	s	v	h	s
-45.00	3195.1	1385.3	6.2959	2271.6	1382.6	6.1228
-40.00	3268.7	1396.1	6.3426	2325.4	1393.7	6.1709
-35.00	3342.3	1406.7	6.3879	2378.4	1404.6	6.2169
-30.00	3415.4	1417.2	6.4316	2430.7	1415.2	6.2620
-25.00	3488.8	1427.7	6.4753	2484.5	1426.3	6.3044
-20.00	3558.9	1438.1	6.5153	2535.8	1436.0	6.3490
-10.00	3705.0	1459.0	6.5975	2640.6	1457.8	6.4298
-5.00	3777.2	1469.6	6.6371	2692.2	1468.4	6.4696
0.00	3850.0	1480.0	6.6761	2744.0	1479.1	6.5086
5.00	3921.3	1490.5	6.7139	2795.5	1489.5	6.5469
10.00	3992.9	1501.2	6.7512	2847.3	1500.0	6.5846
15.00	4065.1	1511.7	6.7881	2889.3	1510.6	6.6215
20.00	4135.9	1521.9	6.8245	2949.9	1521.5	6.6578
25.00	4208.2	1532.9	6.8599	3012.0	1531.9	6.6940
30.00	4279.6	1543.4	6.8953	3053.1	1542.0	6.7283
35.00	4350.9	1553.1	6.9306	3104.2	1554.0	6.7656
40.00	4422.3	1565.7	6.9633	3155.6	1562.8	6.7978
50.00	4565.0	1586.0	7.0319	3257.7	1585.4	6.8663
55.00	4636.4	1596.8	7.0646	3308.6	1596.1	6.8992
60.00	4707.7	1607.4	7.0975	3359.9	1607.0	6.9321
65.00	4779.0	1618.3	7.1299	3411.0	1617.8	6.9645
70.00	4850.4	1629.2	7.1616	3461.6	1628.8	6.9963
75.00	4921.4	1640.1	7.1932	3512.5	1639.6	7.0279
80.00	4992.2	1651.1	7.2249	3564.0	1650.6	7.0595

T	P = 68.92kPa (Sat.Temp.=-40.74°C)			P = 96.49kPa (Sat.Temp.=-34.31°C)		
	v	h	s	v	h	s
-30.00	1693.5	1412.7	6.0784			
-25.00	1730.6	1423.6	6.1229	1228.3	1419.9	5.9474
-20.00	1767.8	1434.5	6.1662	1255.4	1431.4	5.9919
-15.00	1804.7	1445.2	6.2085	1282.1	1442.3	6.0354
-10.00	1842.0	1455.9	6.2499	1309.4	1453.2	6.0780
-5.00	1878.9	1466.6	6.2901	1336.0	1464.1	6.1185
0.00	1915.1	1477.3	6.3295	1362.5	1474.9	6.1584
5.00	1952.0	1487.9	6.3682	1388.8	1485.9	6.1979
10.00	1988.3	1498.6	6.4062	1415.2	1496.7	6.2367
15.00	2024.2	1509.3	6.4436	1441.5	1507.3	6.2745
20.00	2060.5	1519.8	6.4802	1468.0	1518.3	6.3111
25.00	2097.6	1531.1	6.5163	1492.9	1529.2	6.3481
30.00	2132.1	1540.6	6.5524	1519.9	1539.4	6.3843
35.00	2168.6	1553.1	6.5867	1545.7	1551.0	6.4185
40.00	2207.6	1561.8	6.6226	1571.8	1561.8	6.4543
50.00	2277.5	1584.4	6.6897	1623.3	1583.2	6.5222
55.00	2313.1	1595.2	6.7231	1649.0	1593.9	6.5556
60.00	2348.5	1606.1	6.7558	1674.9	1604.8	6.5888
65.00	2384.7	1616.8	6.7883	1700.9	1616.0	6.6217
70.00	2420.4	1628.2	6.8207	1726.6	1627.0	6.6538
75.00	2456.5	1638.9	6.8520	1751.6	1637.9	6.6854
80.00	2492.0	1650.0	6.8835	1777.9	1649.1	6.7168
85.00	2527.6	1661.0	6.9145	1803.2	1660.1	6.7481
90.00	2563.9	1672.2	6.9454	1828.7	1671.3	6.7793
100.00						

T	P = 124.05kPa (Sat.Temp.=-29.23°C)		
	v	h	s
-30.00			
-25.00	949.20	1416.4	5.8128
-20.00	970.69	1427.9	5.8591
-15.00	991.84	1439.4	5.9038
-10.00	1013.5	1450.5	5.9463
-5.00	1034.5	1461.6	5.9887
0.00	1055.4	1472.8	6.0295
5.00	1075.9	1483.8	6.0694
10.00	1096.9	1494.7	6.1085
15.00	1117.9	1505.6	6.1466
20.00	1137.6	1516.3	6.1842
25.00	1158.8	1527.5	6.2209
30.00	1179.0	1538.4	6.2565
35.00	1199.2	1549.6	6.2942
40.00	1219.7	1559.3	6.3259
50.00	1260.0	1581.4	6.3964
55.00	1280.0	1592.9	6.4300
60.00	1300.4	1604.0	6.4636
65.00	1320.7	1614.8	6.4966
70.00	1340.6	1626.0	6.5281
75.00	1360.3	1637.0	6.5607
80.00	1380.8	1648.1	6.5920
85.00	1400.6	1659.2	6.6233
90.00	1420.3	1670.3	6.6545
100.00	1462.0	1692.6	6.7151

T	P = 165.40kPa (Sat.Temp.=-23.10°C)			P = 206.75kPa (Sat.Temp.=-18.09°C)		
	v	h	s	v	h	s
-15.00	737.68	1434.7	5.7490	585.00	1430.0	5.6246
-10.00	753.63	1446.3	5.7937	598.60	1442.0	5.6717
-5.00	770.52	1457.8	5.8368	611.94	1454.0	5.7160
0.00	786.19	1469.3	5.8787	624.92	1465.5	5.7594
5.00	802.09	1480.5	5.9196	638.40	1477.1	5.8011
10.00	818.43	1491.6	5.9595	651.12	1488.6	5.8416
15.00	834.25	1502.7	5.9983	663.16	1499.8	5.8820
20.00	849.48	1514.0	6.0365	676.93	1511.4	5.9202
25.00	864.98	1524.6	6.0742	688.28	1522.1	5.8588
30.00	879.86	1536.1	6.1103	701.81	1533.9	5.9952
35.00	897.90	1547.5	6.1479	714.17	1544.2	6.0322
40.00	909.94	1557.1	6.1811	726.53	1556.8	6.0679
50.00	942.29	1580.1	6.2517	751.24	1578.1	6.1379
55.00	957.46	1591.0	6.2856	763.67	1589.4	6.1723
60.00	973.63	1602.1	6.3191	775.98	1600.5	6.2061
65.00	987.80	1613.1	6.3520	788.15	1611.5	6.2391
70.00	1003.0	1624.5	6.3844	800.90	1622.7	6.2726
75.00	1018.1	1635.1	6.4165	812.79	1633.7	6.3037
80.00	1033.7	1646.9	6.4492	824.25	1645.5	6.3370
90.00	1063.7	1669.0	6.5108	849.28	1667.6	6.3992
95.00	1078.7	1680.3	6.5416	861.09	1678.9	6.4303
105.00	1108.5	1702.9	6.6021	885.34	1701.5	6.4908
115.00	1139.0	1725.7	6.6617	909.51	1724.3	6.5503
125.00				933.39	1747.4	6.6091

T	P = 261.89kPa (Sat.Temp.=-12.54°C)		
	v	h	s
-10.00	467.30	1435.8	5.5374
-5.00	478.31	1448.5	5.5843
0.00	488.94	1460.5	5.6293
5.00	499.55	1472.5	5.6729
10.00	510.04	1484.4	5.7150
15.00	520.33	1496.0	5.7557
20.00	530.53	1507.4	5.7954
25.00	540.75	1519.0	5.8346
30.00	550.89	1530.3	5.8713
40.00	570.86	1552.9	5.9457
45.00	580.72	1564.2	5.9814
50.00	590.71	1575.5	6.0167
55.00	600.44	1586.9	6.0512
60.00	610.29	1598.1	6.0851
65.00	620.29	1609.4	6.1184
70.00	629.31	1620.7	6.1525
75.00	639.87	1632.1	6.1837
80.00	649.19	1642.8	6.2175
85.00	659.30	1654.8	6.2484
95.00	678.22	1677.3	6.3111
105.00	697.76	1700.1	6.3723
115.00	716.86	1723.9	6.4322
125.00	735.96	1745.9	6.4908
135.00	755.07	1769.1	6.5482
145.00			

h = kJ/kg, v = cm^3/g and s = kJ/kg·K

Datum - Sat. liquid h and s at -40°C.

THE PROPERTIES OF SUPERHEATED AMMONIA (continued)

T	P = 330.81kPa (Sat.Temp.=-6.78°C) v	h	s
-10.00			
-5.00	373.66	1441.2	5.4488
0.00	382.49	1453.9	5.4965
5.00	391.20	1466.6	5.5420
10.00	399.79	1478.8	5.5856
15.00	408.22	1490.7	5.6280
20.00	416.56	1502.8	5.6686
25.00	424.74	1514.3	5.7092
30.00	433.06	1526.3	5.7462
40.00	449.16	1549.3	5.8223
45.00	457.10	1560.7	5.8592
50.00	465.05	1572.5	5.8948
55.00	472.95	1583.7	5.9299
60.00	480.82	1595.1	5.9645
65.00	488.68	1606.8	5.9984
70.00	496.50	1617.8	6.0320
75.00	504.29	1629.6	6.0652
80.00	5.1199	1640.8	6.0972
85.00	519.79	1652.1	6.1301
95.00	535.19	1675.2	6.1926
105.00	550.53	1698.0	6.2543
115.00	565.81	1721.0	6.3146
125.00	581.04	1744.3	6.3735
135.00	596.26	1767.7	6.4314
145.00	611.45	1791.2	6.4881

T	P = 413.51kPa (Sat.Temp.=-0.99°C) v	h	s
5.00	308.77	1459.1	5.4117
10.00	315.89	1471.9	5.4575
15.00	322.86	1484.7	5.5014
20.00	329.75	1496.9	5.5450
25.00	336.69	1509.5	5.5849
30.00	343.20	1521.1	5.6265
40.00	356.54	1545.0	5.7031
45.00	363.08	1556.7	5.7402
50.00	369.57	1568.4	5.7767
55.00	376.04	1580.2	5.8122
60.00	382.43	1591.6	5.8473
65.00	388.77	1603.1	5.8820
70.00	395.17	1614.9	5.9160
75.00	401.41	1626.5	5.9496
80.00	407.72	1637.6	5.9819
90.00	420.22	1660.8	6.0469
95.00	426.47	1672.6	6.0784
100.00	432.69	1684.2	6.1096
105.00	438.87	1695.7	6.1404
115.00	451.17	1718.9	6.2011
125.00	463.48	1742.4	6.2604
135.00	475.74	1765.8	6.3186
145.00	487.94	1789.2	6.3759

T	P = 551.3kPa (Sat.Temp.=6.89°C) v	h	s	P = 689.2kPa (Sat.Temp.=13.36°C) v	h	s
10.00	231.73	1459.8	5.2833			
15.00	237.39	1473.5	5.3308	185.84	1461.9	5.1897
20.00	242.78	1487.1	5.3770	190.47	1476.0	5.2391
25.00	248.27	1499.4	5.4205	195.08	1489.8	5.2863
30.00	253.51	1512.8	5.4630	199.57	1503.6	5.3299
40.00	263.85	1537.4	5.5437	208.17	1529.5	5.4154
45.00	268.95	1549.7	5.5826	212.41	1542.1	5.4558
50.00	273.98	1561.7	5.6203	216.53	1554.7	5.4950
55.00	278.99	1573.7	5.6571	220.64	1567.2	5.5333
60.00	283.92	1585.6	5.6932	224.74	1579.5	5.5705
65.00	288.81	1597.5	5.7286	228.78	1591.7	5.6069
70.00	293.69	1609.5	5.7637	232.82	1603.7	5.6419
75.00	298.53	1621.3	5.7970	236.70	1616.0	5.6780
80.00	303.41	1632.5	5.8322	240.64	1628.0	5.7110
90.00	312.90	1656.3	5.8972	248.42	1651.8	5.7784
95.00	317.66	1668.2	5.9289	252.33	1663.7	5.8115
100.00	322.34	1680.0	5.9608	256.19	1675.7	5.8433
105.00	327.10	1691.8	5.9923	259.99	1687.6	5.8752
110.00	331.80	1703.5	6.0231	263.82	1699.5	5.9067
115.00	336.46	1715.2	6.0537	267.64	1711.5	5.9376
120.00	341.18	1727.1	6.0839	271.41	1723.4	5.9683
125.00	345.86	1738.9	6.1137	275.17	1735.3	5.9982
140.00	359.79	1774.5	6.2013	286.46	1771.4	6.0870
145.00	364.44	1786.7	6.2299	290.19	1783.5	6.1164

T	P = 964.9kPa (Sat.Temp.=23.77°C) v	h	s	P = 1102.7kPa Sat.Temp.=28.13°C) v	h	s	P = 1240.5kPa (Sat.Temp.=32.10°C) v	h	s
30.00	137.57	1483.9	5.1165						
35.00	140.96	1498.6	5.1641	121.21	1489.0	5.0748	105.76	1479.3	4.9922
40.00	144.33	1512.6	5.2105	124.27	1504.1	5.1210	108.66	1495.2	5.0430
45.00	147.59	1526.9	5.2539	127.30	1518.8	5.1840	111.38	1510.3	5.0917
50.00	150.78	1540.4	5.2965	130.16	1532.8	5.1798	114.11	1525.2	5.1375
55.00	153.91	1553.6	5.3377	132.97	1546.7	5.2875	116.70	1539.5	5.1814
60.00	157.01	1566.7	5.3771	135.78	1560.2	5.2971	119.24	1553.5	5.2239
65.00	160.06	1579.7	5.4153	138.54	1573.4	5.2417	121.77	1567.3	5.2649
70.00	162.99	1592.5	5.4538	141.23	1586.5	5.3752	124.22	1580.8	5.3051
75.00	166.03	1605.3	5.4897	143.88	1599.9	5.4133	126.74	1594.0	5.3427
80.00	169.01	1617.3	5.5255	146.57	1612.3	5.4486	128.95	1607.1	5.3816
90.00	174.80	1642.6	5.5952	151.69	1637.9	5.5201	133.73	1633.2	5.4534
95.00	177.59	1655.2	5.6291	154.21	1650.4	5.5550	136.06	1645.8	5.4882
100.00	180.42	1667.2	5.6625	156.77	1663.1	5.5887	138.35	1658.1	5.5224
105.00	183.31	1679.6	5.6951	159.28	1675.6	5.6220	140.57	1673.8	5.5563
110.00	186.10	1691.7	5.7271	161.75	1687.9	5.6547	142.83	1684.0	5.5894
115.00	188.81	1704.0	5.7589	164.23	1700.3	5.6867	145.10	1682.6	5.6219
120.00	191.74	1716.2	5.7897	166.67	1712.6	5.7183	147.29	1739.2	5.6536
125.00	194.33	1728.8	5.8217	169.19	1725.0	5.7500	149.40	1677.9	5.6862
130.00	197.27	1740.4	5.8511	171.60	1737.3	5.7797	152.16	1734.4	5.7162
140.00	202.65	1765.2	5.9115	176.43	1762.0	5.8409	156.08	1758.9	5.7776
145.00	205.35	1777.3	5.9407	178.85	1774.4	5.8705	158.20	1771.4	5.8074
150.00	208.05	1789.5	5.9697	181.26	1786.7	5.8998	160.41	1783.7	5.8371
160.00	213.5	1814.0	6.0269	186.03	1811.4	5.9574	166.87	1808.6	5.8954
170.00				190.75	1836.1	6.0139	170.29	1833.7	5.9518

T	P = 1378.4kPa (Sat.Temp.=35.74°C) v	h	s
45.00	98.650	1501.5	5.0186
50.00	101.22	1517.2	5.0669
55.00	103.63	1532.1	5.1129
60.00	106.00	1546.5	5.1569
65.00	108.38	1560.8	5.1991
70.00	110.63	1574.5	5.2401
75.00	112.82	1588.7	5.2800
80.00	115.17	1601.4	5.3171
85.00	117.18	1615.2	5.3568
95.00	121.50	1641.3	5.4276
100.00	123.56	1654.3	5.4626
105.00	125.65	1667.3	5.4969
110.00	127.73	1680.0	5.5303
115.00	129.77	1692.5	5.5632
120.00	131.71	1705.4	5.5958
125.00	133.89	1717.9	5.6272
130.00	135.70	1730.5	5.6601
135.00	137.82	1743.0	5.6894
145.00	141.74	1768.1	5.7509
150.00	143.71	1780.7	5.7806
160.00	147.58	1805.8	5.8393
170.00	151.40	1830.9	5.8967
180.00	155.22	1856.1	5.9527
190.00	159.05	1881.2	6.0075

T	P = 1516.2kPa (Sat.Temp=39.12°C) v	h	s	P = 1654.0kPa (Sat.Temp.=42.27°C) v	h	s	P = 1791.9kPa (Sat.Temp.=45.23°C) v	h	s	P = 2067.5kPa (Sat.Tem.=50.67°C) v	h	s
45.00	88.195	1492.6	4.9502	79.492	1483.3	4.8840						
50.00	90.633	1508.9	4.9999	81.778	1500.3	4.9368	74.257	1491.2	4.8759			
55.00	92.947	1524.5	5.0486	84.032	1516.4	4.9869	76.403	1508.4	4.9280	64.097	1490.6	4.8157
60.00	95.202	1539.5	5.0941	86.150	1532.1	5.0346	78.472	1524.7	4.9777	66.049	1508.6	4.8701
65.00	97.393	1554.2	5.1369	88.168	1547.5	5.0800	80.440	1540.4	5.0249	67.878	1525.8	4.9211
70.00	99.536	1568.6	5.1798	90.216	1562.0	5.1234	82.372	1555.7	5.0701	69.783	1542.5	4.9694
75.00	101.50	1582.3	5.2209	92.130	1576.7	5.1643	84.211	1570.7	5.1118	71.352	1557.6	5.0157
80.00	103.65	1596.7	5.2598	94.198	1590.9	5.2066	86.030	1584.9	5.1557	73.120	1573.9	5.0579
85.00	105.77	1610.1	5.2978	95.923	1604.3	5.2435	87.841	1600.3	5.1928	74.763	1588.6	5.1028
95.00	109.61	1636.8	5.3714	99.680	1631.9	5.3198	91.289	1627.3	5.2709	77.873	1617.7	5.1813
100.00	111.53	1649.9	5.4073	101.46	1645.3	5.3557	93.029	1640.8	5.3078	79.443	1631.6	5.2198
105.00	113.46	1663.8	5.4420	103.28	1658.7	5.3910	94.696	1654.3	5.3437	80.948	1645.5	5.2567
110.00	115.37	1675.8	5.4759	105.07	1671.9	5.4257	96.326	1667.7	5.3788	82.405	1659.3	5.2930
115.00	117.21	1688.8	5.5095	106.80	1684.9	5.4598	97.967	1680.9	5.4133	83.856	1672.9	5.3286
120.00	119.10	1701.8	5.5423	108.56	1697.6	5.4923	99.568	1693.9	5.4460	85.342	1686.5	5.3626
125.00	120.89	1714.0	5.5748	110.25	1711.1	5.5263	101.21	1707.5	5.4812	86.723	1700.0	5.3968
130.00	122.81	1727.6	5.6060	111.93	1723.3	5.5577	102.86	1719.86	5.5112	88.196	1713.2	5.4307
135.00	124.60	1739.4	5.6380	113.63	1736.7	5.5891	104.30	1733.8	5.5455	89.496	1726.5	5.4616
145.00	128.20	1765.1	5.6991	116.99	1762.2	5.6514	107.50	1759.0	5.6069	92.345	1752.8	5.5264
150.00	130.00	1777.7	5.7292	118.67	1774.9	5.6815	109.03	1771.9	5.6376	93.691	1765.9	5.5573
160.00	133.60	1803.0	5.7883	121.98	1800.2	5.7409	112.12	1797.4	5.6974	96.389	1791.9	5.6183
170.00	137.14	1828.4	5.8460	125.25	1825.8	5.7990	115.16	1824.0	5.7559	99.087	1817.8	5.6775
180.00	140.63	1853.7	5.9023	128.46	1851.2	5.8558	118.19	1848.8	5.8130	101.74	1843.8	5.7354
190.00	144.10	1879.0	5.9576	131.64	1876.4	5.9114	121.2	1874.2	5.8689	104.36	1869.7	5.7920

h = kJ/kg, v = cm^3/g and s = kJ/kg·K

T = °C. P = kPa

THE PROPERTIES OF SATURATED CARBON DIOXIDE

T	P	v_f	v_g	h_f	h_g	s_f	s_g	T	P	v_f	v_g	h_f	h_g	s_f	s_g
-95.00	23.42	0.6298	1444.6	-282.22	300.71	2.5426	5.8260	-30.00	1426.7	0.9308	27.002	18.76	322.28	3.9394	5.1878
-85.00	59.83	0.6393	602.03	-276.55	306.48	2.6091	5.7275	-25.00	1662.0	0.9503	22.884	29.52	322.26	3.9792	5.1587
-75.00	133.70	0.6412	268.80	-255.91	311.74	2.6683	5.5466	-20.00	1967.2	0.9708	19.470	38.07	322.33	4.0122	5.1328
-70.00	198.35	0.6479	185.64	-252.44	313.47	2.7049	5.4844	-15.00	2289.4	0.9906	16.599	50.82	324.25	4.0662	5.1246
-65.00	286.98	0.6511	128.85	-242.60	314.98	2.7425	5.4235	-5.00	3044.4	1.0478	12.144	72.56	321.53	4.1426	5.0713
-60.00	410.05	0.6576	91.451	-235.69	315.89	2.7864	5.3694	0.00	3483.9	1.0817	10.375	85.10	319.87	4.1864	5.0451
-56.61	517.57	0.8490	72.229	-31.86	316.05	3.7200	5.3273	5.00	3970.1	1.1321	8.847	99.37	317.11	4.2323	5.0168
-50.00	682.63	0.8667	55.411	-19.49	317.88	3.7767	5.2883	10.00	4504.5	1.1662	7.5225	112.56	313.95	4.2781	4.9894
-45.00	632.56	0.8815	45.813	-9.79	319.25	3.8187	5.2610	15.00	5090.7	1.2203	6.3318	127.16	308.13	4.3284	4.9550
-40.00	1004.2	0.8971	38.162	0.00	320.47	3.8594	5.2348	20.00	5731.7	1.3032	5.2465	144.45	298.89	4.3840	4.9171
-35.00	1201.8	0.9133	32.001	9.48	321.21	3.8996	5.2080	25.00	6428.7	1.3964	4.1991	162.56	284.83	4.4457	4.8688
								30.00	7202.0	1.9099	2.6387	207.36	242.38	4.5908	4.7116

THE PROPERTIES OF SUPERHEATED CARBON DIOXIDE

T	P = 6.89kPa			P = 68.92kPa			P = 137.84kPa			P = 275.67kPa			P = 551.34kPa		
	v	h	s	v	h	s	v	h	s	v	h	s	v	h	s
-55.00	5983.8	662.13	6.2013	591.68	660.86	5.7662	292.31	659.45	5.6341	143.00	656.63	5.4967			
-45.00	6258.0	670.24	6.2369	619.74	669.07	5.8019	307.20	667.92	5.6700	150.49	665.35	5.5331	72.273	660.48	5.3521
-15.00	7080.6	694.75	6.3359	703.78	693.61	5.9009	348.89	692.42	5.7695	173.10	690.20	5.6340	84.524	685.55	5.4723
10.00	7766.0	715.58	6.4125	772.85	714.65	5.9775	382.00	713.49	5.8464	190.59	711.40	5.7116	93.517	707.21	5.5617
35.00	8451.4	737.15	6.4843	841.87	736.44	6.0493	419.03	735.51	5.9183	207.74	733.52	5.7842	103.23	730.19	5.6397
65.00	9274.1	763.23	6.5651	924.91	762.76	6.1301	461.65	762.05	5.9991	229.86	760.89	5.8659	113.93	758.53	5.7256
90.00	9958.8	785.55	6.6286	993.42	785.31	6.1936	495.77	784.87	6.0625	246.91	783.88	5.9302	122.52	782.31	5.7922
150.00	11604.	841.52	6.7704	1158.6	841.28	6.3354	578.79	840.83	6.2044	288.86	840.36	6.0729	143.85	838.75	5.9381
205.00	13106.	895.21	6.8889	1310.0	894.98	6.4539	654.39	894.75	6.3228	326.90	894.29	6.1922	163.00	893.36	6.0586
315.00	16122.	1009.6	7.1036	1612.2	1009.6	6.6686	805.81	1009.4	6.5375	402.81	1009.2	6.4069	201.20	1008.7	6.2754
425.00	19132.	1130.9	7.2920	1913.9	1130.9	6.8569	956.60	1130.9	6.7259	477.43	1130.6	6.5953	239.10	1130.4	6.4642
535.00	22148.	1258.1	7.4606	2215.4	1258.1	7.0256	1107.4	1258.1	6.8946	553.81	1258.1	6.7639	276.72	1257.9	6.6329
650.00	25301.	1395.8	7.6202	2530.7	1395.8	7.1852	1265.1	1395.8	7.0541	632.55	1395.8	6.9235	316.27	1395.5	6.7925
760.00	28317.	1531.6	7.7594	2831.7	1531.6	7.3244	1415.9	1531.6	7.1933	709.80	1531.6	7.0627	353.97	1531.6	6.9317
870.00	31327.	1670.2	7.8859	3133.3	1670.2	7.4509	1566.0	1670.2	7.3198	783.37	1670.2	7.1892	391.73	1670.2	7.0581
980.00	34336.	1811.1	8.0040	3434.4	1811.1	7.5690	1716.8	1811.1	7.4380	858.55	1811.1	7.3074	429.20	1811.1	7.1763

T	P = 827.02kPa			P = 1102.6kPa			P = 1378.4kPa			P = 1654.0kPa			P = 2067.5kPa		
	v	h	s	v	h	s	v	h	s	v	h	s	v	h	s
-15.00	54.975	680.71	5.3848	40.121	676.15	5.3146	31.093	671.59	5.2548	25.152	666.75	5.2030			
10.00	61.173	702.79	5.4788	44.992	698.60	5.4127	35.284	694.42	5.3612	28.804	690.00	5.3147	22.243	683.72	5.2595
35.00	67.032	726.56	5.5588	49.406	722.94	5.4986	39.227	719.32	5.4497	32.177	715.69	5.4084	25.170	710.42	5.3546
65.00	75.289	756.19	5.6456	56.006	753.84	5.5878	44.407	751.49	5.5413	36.685	749.14	5.5020	28.892	745.62	5.4533
90.00	81.024	780.64	5.7128	60.300	779.02	5.6555	47.852	777.40	5.6101	39.554	775.74	5.5713	31.267	773.41	5.5253
150.00	95.465	837.59	5.8611	71.303	836.21	5.8042	56.813	834.82	5.7607	47.152	833.67	5.7263	37.476	831.58	5.6800
205.00	108.38	892.43	5.9820	81.068	891.51	5.9259	64.693	890.59	5.8832	53.786	889.65	5.8472	42.930	888.03	5.8049
315.00	133.97	1008.3	6.1988	100.41	1007.8	6.1431	80.267	1007.3	6.1008	66.802	1006.6	6.0657	53.363	1005..	6.0234
425.00	159.37	1130.2	6.3876	119.45	1129.7	6.3332	95.534	1129.5	6.2913	79.272	1129.3	6.2557	63.580	1128.8	6.2138
535.00	184.53	1257.9	6.5563	138.36	1257.7	6.5018	110.68	1257.7	6.4600	92.196	1257.4	6.4252	73.785	1257.2	6.3833
650.00	210.82	1395.5	6.7158	158.13	1395.5	6.6614	126.51	1395.5	6.6195	105.44	1395.5	6.5848	84.317	1395.3	6.5429
760.00	236.04	1531.6	6.8550	177.05	1531.6	6.8006	141.65	1531.4	6.7588	118.05	1531.4	6.7240	94.453	1531.4	6.6821
870.00	261.19	1670.2	6.9815	195.90	1670.2	6.9271	156.73	1670.2	6.8852	130.66	1670.2	6.8505	104.53	1670.0	6.8086
980.00	286.17	1811.1	7.0997	214.68	1811.1	7.0453	171.81	1811.1	7.0034	143.15	1811.1	6.9686	114.54	1811.1	6.9268

T	P = 2481.1kPa			P = 3032.4kPa			P = 3583.7kPa			P = 4135.1kPa			P = 5513.4kPa		
	v	h	s	v	h	s	v	h	s	v	h	s	v	h	s
10.00	17.842	677.21	5.2067	13.834	668.84	5.1422	11.062	660.23	5.0861	9.0645	651.86	5.0325			
35.00	20.544	704.81	5.3065	16.255	697.57	5.2473	13.313	690.60	5.1962	11.163	683.36	5.1475	7.2510	665.24	5.0576
65.00	23.551	742.10	5.4097	18.941	737.40	5.3560	15.655	732.71	5.3099	13.222	728.01	5.2663	9.2286	716.27	5.1857
90.00	25.706	770.79	5.4851	20.695	767.56	5.4352	17.213	764.26	5.3926	14.666	761.04	5.3520	10.514	752.81	5.2768
150.00	31.032	829.73	5.6444	25.185	826.96	5.6005	21.130	824.42	5.5633	18.150	821.66	5.5287	13.318	815.18	5.4631
205.00	35.600	886.64	5.7702	28.962	884.79	5.7292	24.385	882.93	5.6949	21.023	881.07	5.6639	15.561	876.21	5.6032
315.00	44.985	1005.2	5.9890	36.279	1004.3	5.9505	30.633	1003.3	5.9183	26.505	1001.9	5.8902	19.788	999.61	5.8332
425.00	52.950	1128.3	6.1795	43.279	1127.8	6.1414	36.625	1127.2	6.1095	31.716	1126.7	6.0824	23.739	1125.3	6.0279
535.00	61.425	1256.9	6.3490	50.261	1256.5	6.3109	42.502	1256.0	6.2791	36.835	1255.8	6.2519	27.596	1254.8	6.1975
650.00	70.317	1395.3	6.5086	57.510	1395.1	6.4705	48.660	1394.9	6.4387	42.172	1394.4	6.4115	31.627	1393.9	6.3570
760.00	78.722	1531.4	6.6478	64.426	1531.2	6.6097	54.518	1530.9	6.5779	47.264	1530.7	6.5507	35.459	1530.2	6.4962
870.00	87.127	1670.0	6.7743	71.224	1669.8	6.7362	60.322	1669.8	6.7019	52.295	1669.8	6.6771	39.242	1669.3	6.6227
980.00	95.534	1810.8	6.8924	78.198	1810.8	6.8543	66.178	1810.9	6.8223	57.342	1810.6	6.7953	43.035	1810.4	6.7409

T	P = 6891.8kPa			P = 8270.1kPa			P = 9648.5kPa			P = 12405kPa			P = 11026kPa		
	v	h	s	v	h	s	v	h	s	v	h	s	v	h	s
65.00	6.8454	704.53	5.1132												
90.00	8.0288	744.66	5.2097	6.3717	737.30	5.1552									
150.00	10.421	808.48	5.4055	8.4998	801.79	5.3587	7.1234	795.31	5.3220						
205.00	12.291	871.58	5.5522	10.135	866.94	5.5104	8.5983	862.07	5.4769	6.5474	847.28	5.4150	7.4461	855.60	5.4451
315.00	15.752	997.28	5.7884	13.071	994.95	5.7516	11.151	992.38	5.7214	8.5867	987.48	5.6674	9.7078	990.05	5.6938
425.00	18.982	1123.6	5.9852	15.761	1122.5	5.9492	13.452	1121.1	5.9194	10.430	1118.2	5.8683	11.821	1119.6	5.8937
535.00	22.065	1253.9	6.1556	18.369	1253.0	6.1204	15.730	1252.0	6.0911	12.216	1250.4	6.0418	13.753	1251.3	6.0654
650.00	25.308	1393.2	6.3152	21.089	1392.5	6.2800	18.076	1392.1	6.2511	14.058	1390.7	6.2030	15.814	1391.4	6.2259
760.00	28.373	1529.8	6.4544	23.654	1529.5	6.4192	20.283	1529.1	6.3903	15.782	1527.9	6.3422	17.748	1528.4	6.3610
870.00	31.414	1669.1	6.5808	26.188	1668.8	6.5457	22.460	1668.4	6.5168	17.488	1667.9	6.4686	19.664	1668.1	6.4916
980.00	34.437	1810.2	6.6990	28.722	1809.9	6.6638	24.025	1809.9	6.6350	19.187	1809.5	6.5868	21.568	1809.7	6.6102

Datum - Sat. liquid h = 0 at -40°C.

h = kJ/kg, v = cm^3/g and s = kJ/kg·K

Below a triple point temperature of -56.6°C the values shown are for solid and gaseous phases only

T = ^{o}C. P = kPa

THE PROPERTIES OF SATURATED ETHYLENE

T	P	v_f	v_g	h_f	h_g	s_f	s_g	T	P	v_f	v_g	h_f	h_g	s_f	s_g
-165.00	.23		135188.	-146.32	415.54	-1.0480	4.1683	-75.00	441.31	1.9511	98.020	72.22	504.26	0.3894	2.5701
-180.00	.51		66007.	-136.03	421.43	-0.9509	3.9762	-60.00	775.02	1.9994	90.801	111.17	512.61	0.6044	2.4879
-150.00	2.08		18236.	-111.60	432.72	-0.7678	3.6544	-50.00	1048.0	2.0749	51.537	137.64	516.28	0.7074	2.4044
-140.00	6.14		6436.3	-87.35	445.12	-0.5786	3.4206	-40.00	1422.0	2.1650	37.020	164.65	518.14	0.8101	2.3262
-130.00	15.61		2697.5	-83.42	462.69	-0.4113	3.2440	-30.00	1927.8	2.2914	26.996	193.68	516.76	0.9268	2.2564
-120.00	34.44		1299.8	-38.74	458.92	-0.2225	3.0578	-20.00	2528.0	2.3849	20.142	225.46	516.67	1.0497	2.1991
-110.00	69.13		682.32	-15.69	485.10	-0.1083	2.9446	-5.00	3639.0	2.7003	12.580	286.18	503.30	1.2714	2.0811
-105.00	93.50	1.7420	518.14	-2.90	482.78	-0.0082	2.8797	5.00	4570.5	3.2157	8.3607	350.43	474.99	1.4949	1.9427
-95.00	164.28	1.7905	308.77	21.48	490.06	0.1161	2.7459	10.00	5128.9	4.4060	4.2584	400.72	397.13	1.6636	1.6508
-85.00	268.28	1.8382	199.36	46.98	497.78	0.2672	2.6627								

THE PROPERTIES OF SUPERHEATED ETHYLENE

T	P = 101.3kPa			P = 202.7kPa			P = 405.2kPa			P = 607.8kPa		
	v	h	s	v	h	s	v	h	s	v	h	s
-95.00	504.97	494.26	2.9135									
-85.00	536.21	507.03	2.9851	260.66	501.41	2.7543						
-70.00	582.59	526.02	3.0854	284.97	521.67	2.8532	135.97	512.73	2.6324			
-60.00	613.38	539.24	3.1443	301.03	535.30	2.9157	144.65	527.65	2.6977	92.551	518.75	2.5577
-50.00	642.98	552.39	3.2058	316.82	549.01	2.9797	153.24	542.06	2.7616	98.353	534.57	2.6233
-40.00	673.85	565.81	3.2657	332.37	562.79	3.0396	161.56	556.51	2.8261	104.44	550.23	2.6879
-25.00	713.89	586.59	3.3496	355.56	583.95	3.1235	173.84	578.45	2.9136	115.70	573.03	2.7816
-15.00	754.00	600.70	3.4 78	370.88	598.52	3.1816	181.88	593.59	2.9727	111.35	588.59	2.8437
-5.00	783.36	614.92	3.4629	386.09	612.79	3.2370	189.79	608.48	3.0330	120.56	603.90	2.9025
5.00	808.47	629.69	3.5156	401.25	627.61	3.2894	197.64	623.66	3.0884	131.19	619.59	2.9590
15.00	838.18	644.97	3.5686	416.36	643.08	3.3425	205.39	639.36	3.1412	135.05	635.59	3.0154
35.00	897.64	676.08	3.6717	446.46	674.54	3.4480	220.82	671.41	3.2494	145.63	668.35	3.1249
60.00	971.38	716.98	3.8018	483.75	715.35	3.5797	239.91	713.02	3.3787	158.63	710.23	3.2531
80.00	1030.4	751.45	3.8974	513.73	750.12	3.6741	255.10	748.06	3.4757	168.91	745.58	3.3529
105.00	1104.1	796.36	4.0259	550.72	795.23	3.8081	273.85	793.14	3.6033	181.63	791.27	3.4816
125.00	1163.1	833.99	4.1200	580.63	833.29	3.9020	288.85	831.42	3.7013	191.89	629.53	3.5756
160.00	1266.0	903.26	4.2915	631.77	902.56	4.0779	315.01	900.93	3.8726	209.26	899.53	3.7514
195.00	1369.0	976.63	4.4540	683.56	976.18	4.2405	341.03	974.80	4.0397	226.82	973.40	3.9182
225.00	1457.2	1043.6	4.5896	727.96	1043.2	4.3799	363.22	1042.0	4.1792	241.71	1040.8	4.0577
200.00	1560.1	1123.02	4.7436	779.10	1122.6	4.5343	389.17	1121.6	4.3333	259.08	1120.5	4.2119

T	P = 1013kPa			P = 1520kPa			P = 2026kPa			P = 3039kPa		
	v	h	s	v	h	s	v	h	s	v	h	s
-50.00	54.194	519.22	2.4298									
-40.00	58.432	536.74	2.5079									
-25.00	64.263	561.80	2.6010	39.600	544.43	2.4293						
-15.00	68.135	577.93	2.6639	42.551	563.02	2.5023	29.335	546.57	2.3920			
-5.00	71.736	594.42	2.7281	45.262	581.42	2.5746	31.814	567.39	2.4301	17.958	532.32	2.2263
5.00	75.239	611.17	2.7872	47.904	599.88	2.6411	34.148	587.36	2.4817	19.960	558.34	2.3159
15.00	78.726	627.90	2.8437	50.487	618.09	2.7020	36.288	606.58	2.5845	21.766	581.93	2.3988
35.00	85.359	661.65	2.9595	55.354	653.30	2.8131	40.157	644.44	2.7032	24.809	625.44	2.5459
60.00	93.579	704.88	3.0940	61.054	697.91	2.9601	44.761	690.47	2.8554	28.342	674.88	2.7047
80.00	99.990	740.90	3.1938	65.484	734.75	3.0569	48.196	728.66	2.9570	30.796	715.34	2.8167
105.00	107.79	787.09	3.3269	70.909	782.00	3.1971	52.409	776.68	3.1007	33.967	765.54	2.9588
125.00	103.97	825.78	3.4248	75.122	821.32	3.2956	55.726	816.63	3.1984	36.330	806.53	3.0644
160.00	124.79	896.28	3.6006	82.530	892.56	3.4709	61.367	888.60	3.3787	40.266	880.47	3.2448
195.00	135.41	970.64	3.7674	89.802	967.40	3.6377	66.936	964.17	3.5456	44.129	957.49	3.4160
225.00	144.51	1038.5	3.9031	95.925	1035.7	3.7733	71.599	1032.6	3.6850	47.274	1027.0	3.5554
260.00	155.01	1118.4	4.0570	103.01	1116.1	3.9272	76.974	1113.5	3.8393	51.004	1108.6	3.7137

T	P = 4052kPa			P = 50.65kPa			P = 6078kPa			P = 8104kPa		
	v	h	s	v	h	s	v	h	s	v	h	s
5.00	12.120	516.28	2.1133									
15.00	14.159	550.16	2.2302	9.4991	523.84	2.0998	-5.5428	444.63	1.7917	3.3873	369.09	1.6150
35.00	16.959	603.05	2.4061	12.008	573.53	2.2794	8.5626	545.78	2.1329	4.7435	459.65	1.9794
60.00	20.164	659.07	2.5833	15.170	640.93	2.4702	11.611	621.88	2.3781	7.5538	577.44	2.1855
80.00	22.184	701.63	2.7066	16.860	686.87	2.8049	13.472	671.94	2.5263	9.1955	639.48	2.3830
105.00	24.710	754.19	2.8541	19.323	742.38	2.7621	15.525	730.60	2.6826	10.896	705.61	2.5491
125.00	26.565	798.67	2.9596	20.723	786.33	2.8714	16.933	776.26	2.7995	12.077	755.07	2.6735
160.00	29.716	872.33	3.1443	23.410	863.95	3.0606	19.165	855.35	2.9977	13.984	838.37	2.8805
195.00	32.662	950.80	3.3196	25.498	932.26	3.2156	21.024	925.55	3.1490	15.450	911.10	3.0434
225.00	35.134	1021.2	3.4591	27.853	1015.1	3.3836	23.067	1009.1	3.3202	17.089	997.73	3.2159
260.00	38.019	1103.5	3.6216	30.215	1098.6	3.5462	25.096	1093.7	3.4876	18.791	1084.0	3.3829

T	P = 10 130kPa			P = 15 195kPa			P = 20 260kPa			P = 25 325kPa			P = 30 390kPa		
	v	h	s	v	h	s	v	h	s	v	h	s	v	h	s
20.00	2.8209	345.44	1.4120	2.4768	337.25	1.3345	2.3893	333.90	1.2704	2.2657	331.97	1.2333	2.1977	331.25	1.1968
35.00	3.4254	408.79	1.6241	2.7603	382.83	1.4845	2.5910	373.84	1.4029	2.4385	368.96	1.3599	2.3682	368.24	1.3209
60.00	5.1191	530.93	2.0097	3.4960	464.65	1.7459	2.9341	444.42	1.6412	2.6844	436.05	1.5742	2.5595	433.26	1.5240
80.00	6.6400	604.69	2.2162	4.0985	538.74	1.9590	3.3034	507.90	1.8213	3.0150	493.54	1.7354	2.7963	487.09	1.6712
105.00	8.2067	679.03	2.3276	5.0804	623.24	2.1982	3.8781	588.54	2.0437	3.3753	568.89	1.9385	3.0879	556.65	1.8711
125.00	9.2195	733.13	2.5594	5.8136	684.71	2.3492	4.3235	650.17	2.2023	3.6532	628.36	2.0931	3.3495	613.54	2.0142
160.00	10.803	821.63	2.7842	6.9295	782.56	2.5916	5.3064	752.09	2.4493	4.3075	736.47	2.3488	3.8081	715.58	2.2651
190.00	12.158	897.56	2.9508	7.8882	865.25	2.7773	5.9874	838.60	2.6424	4.8135	817.76	2.5420	4.2556	802.86	2.4621
225.00	13.005	986.49	3.1355	8.8834	960.22	2.9722	6.7063	937.04	2.8463	5.4590	918.73	2.7495	4.7840	904.91	2.6701
260.00	14.920	1074.4	3.3034	9.9260	1052.6	3.1485	7.4289	1033.0	3.0312	6.1804	1017.4	2.9391	5.1815	1004.7	2.8638

h = kJ/kg, v = cm^3/g and s = kJ/kg·K

Datum - Sat. liquid h and s = 0 at 1 atmosphere pressure

Properties of Saturated Propane*

Temp., °F.	Abs. pressure, lb./sq.in.	Volume, cu. ft./lb.		Enthalpy, B.t.u./lb.		Entropy, B.t.u./(lb.)(°R.)		Temp., °F.	Abs. pressure, lb./sq. in.	Volume, cu. ft./lb.		Enthalpy, B.t.u./lb.		Entropy, B.t.u./(lb.)(°R.)	
		Liquid	Vapor	Liquid	Vapor	Liquid	Vapor			Liquid	Vapor	Liquid	Vapor	Liquid	Vapor
t	p	v_f	v_g	h_f	h_g	s_f	s_g	t	p	v_f	v_g	h_f	h_g	s_f	s_g
−80	5.65	0.0265	16.2	162.6	354.0	0.8794	1.3832	70	124.3	.03209	.854	245.7	394.4	1.0624	1.3427
−70	7.48	.0268	12.5	167.6	357.0	.8927	1.3781	80	143.6	.03269	.745	251.9	396.4	1.0737	1.3413
−60	9.78	.02703	9.77	172.7	360.0	.9060	1.3740	90	165.0	.03329	.643	258.2	398.3	1.0850	1.3400
−50	12.60	.02733	7.73	177.8	362.8	.9188	1.3702	100	188.7	.03390	.558	264.6	400.2	1.0963	1.3388
−40	16.00	.02763	6.16	183.0	365.7	.9315	1.3670	110	214.8	.03452	.487	271.1	401.9	1.1080	1.3378
−30	20.18	.02794	5.02	188.4	368.6	.9441	1.3640	120	243.4	.03532	.426	278.0	403.8	1.1195	1.3368
−20	25.05	.02826	4.06	193.8	371.5	.9568	1.3610	130	274.5	.03612	.370	285.2	405.4	1.1310	1.3356
−10	30.95	.02859	3.33	199.4	374.4	.9690	1.3582	140	308.4	.03702	.320	292.7	407.0	1.1430	1.3347
0	37.81	.02893	2.74	205.0	377.2	.9812	1.3555	150	345.4	.03817	.278	300.2	408.2	1.1552	1.3326
10	45.85	.02930	2.30	210.7	380.0	.9932	1.3531	160	385.0	.03962	.240	308.4	408.8	1.1680	1.3303
20	55.00	.02970	1.93	216.6	382.6	1.0050	1.3510	170	426.0	.04132	.208	317.5	408.6	1.1816	1.3272
30	65.70	.03011	1.60	222.3	385.1	1.0167	1.3491	180	473.2	.04367	.180	327.5	407.6	1.1970	1.3223
40	77.80	.03055	1.33	227.9	387.5	1.0283	1.3473	190	523.4	.04712	.149	339.2	404.6	1.2140	1.3156
50	91.50	.03101	1.14	233.8	389.9	1.0398	1.3456	200	575.0	.0521	.113	353.5	398.3	1.2360	1.3040
60	106.9	.03150	0.984	239.6	392.2	1.0511	1.3441								

* Stearns and George, *Ind. Eng. Chem.*, **35**, 602 (1943); with permission.

Properties of Superheated Propane*

v, volume, cu. ft./lb.; h, enthalpy, B.t.u./lb.; s, entropy, B.t.u./(lb.)(°R.); p, absolute pressure, lb./sq. in.
Parenthetic figures after pressures are saturation temperatures

Temp., °F.	7.35 (−70.7°F.)			12.24 (−51°F.)			14.696 (−43.708°F.)			20 (−30.30°F.)		
	v	h	s	v	h	s	v	h	s	v	h	s
Sat.	12.72	356.5	1.3785	7.90	362.75	1.3718	6.66	364.6	1.3681	5.050	368.4	1.3640
−60	13.11	360.3	1.3869									
−40	13.79	366.8	1.4023	8.182	366.2	1.3796	6.775	365.8	1.3710	.....	.....	......
−20	14.47	373.4	1.4177	8.591	373.0	1.3948	7.123	372.8	1.3866	5.186	372.2	1.3719
0	15.14	380.2	1.4329	9.000	379.9	1.4100	7.471	379.8	1.4020	5.439	379.3	1.3873
20	15.82	387.4	1.4481	9.409	387.0	1.4252	7.816	386.8	1.4772	5.695	386.5	1.4025
40	16.50	394.8	1.4633	9.818	394.4	1.4404	8.160	394.3	1.4324	5.951	393.9	1.4177
60	17.17	402.4	1.4785	10.23	402.0	1.4556	8.503	401.9	1.4474	6.207	401.6	1.4327
80	17.85	410.2	1.4937	10.64	410.0	1.4706	8.844	409.8	1.4623	6.461	409.4	1.4477
100	18.52	418.4	1.5088	11.050	418.1	1.4854	9.187	418.0	1.4772	6.716	417.6	1.4625
120	19.20	426.6	1.5235	11.45	426.2	1.5002	9.528	426.2	1.4918	6.969	425.9	1.4771
140	19.88	435.2	1.5380	11.86	435.9	1.5148	9.869	434.8	1.5064	7.221	434.6	1.4917
160	20.55	444.1	1.5524	12.27	443.8	1.5293	10.21	443.7	1.5210	7.473	443.5	1.5063
180	21.23	453.1	1.5668	12.68	452.9	1.5437	10.51	452.9	1.5256	7.725	452.6	1.5209
200	21.90	462.5	1.5812	13.09	462.4	1.5581	10.84	462.4	1.5500	7.977	461.9	1.5355

Temp., °F.	30 (−11.52°F.)			40 (+2.90°F.)			60 (24.80°F.)			80 (41.69°F.)		
	v	h	s	v	h	s	v	h	s	v	h	s
Sat.	3.30	374.00	1.3588	2.61	378.2	1.3548	1.76	383.8	1.3501	1.32	387.8	1.3470
0	3.559	378.3	1.3678	.....	.....	......	.....	.....	......	.....	.....	......
20	3.735	385.7	1.3830	2.753	384.7	1.3684	.....	.....	......	.....	.....	......
40	3.911	393.1	1.3980	2.889	392.2	1.3838	1.863	390.1	1.3630	.....	.....	......
60	4.087	400.8	1.4130	3.025	400.0	1.3990	1.959	398.2	1.3789	1.424	396.1	1.3624
80	4.261	408.7	1.4280	3.159	408.0	1.4140	2.053	406.3	1.3948	1.500	404.6	1.3785
100	4.432	417.0	1.4428	3.289	416.2	1.4290	2.145	414.8	1.4102	1.573	413.2	1.3940
120	4.602	425.4	1.4576	3.419	424.6	1.4440	2.235	423.4	1.4254	1.644	421.8	1.4094
140	4.772	434.0	1.4724	3.549	433.4	1.4588	2.325	432.2	1.4404	1.714	430.9	1.4248
160	4.942	443.0	1.4872	3.679	442.5	1.4736	2.415	441.4	1.4554	1.784	440.3	1.4402
180	5.112	452.2	1.5020	3.809	451.8	1.4884	2.505	450.8	1.4703	1.852	449.8	1.4554
200	5.282	461.8	1.5168	3.939	461.3	1.5030	2.593	460.5	1.4851	1.920	459.7	1.4704

Temp., °F.	100 (55.62°F.)			130 (73.20°F.)			160 (87.71°F.)		
	v	h	s	v	h	s	v	h	s
Sat.	1.06	391.2	1.3448	0.8165	395.0	1.3424	0.6685	397.9	1.3404
60	1.094	393.5	1.3488	......	.....	......	......	.....	......
80	1.164	402.6	1.3656	.8456	398.8	1.3486	......	.....	......
100	1.227	411.4	1.3816	.9045	408.3	1.3659	.6969	404.7	1.3521
120	1.289	420.3	1.3962	.9588	417.8	1.3822	.7464	414.6	1.3698
140	1.347	429.6	1.4130	1.006	427.3	1.3987	.7908	424.8	1.3867
160	1.400	439.1	1.4286	1.052	437.2	1.4150	.8319	434.9	1.4031
180	1.460	448.8	1.4440	1.098	447.0	1.4310	.8712	445.1	1.4195
200	1.516	458.8	1.4598	1.143	457.1	1.4468	.9093	455.4	1.4357

Temp., °F.	190 (100.50°F.)			220 (111.85°F.)			250 (122.12°F.)			300 (137.55°F.)		
	v	h	s	v	h	s	v	h	s	v	h	s
Sat.	0.5540	400.4	1.3388	0.4738	402.5	1.3375	0.4130	404.2	1.3355	0.3332	406.6	1.3345
120	.5995	411.6	1.3580	.4911	407.7	1.3460	......	.....	......	......	.....	......
140	.6415	422.1	1.3759	.5314	419.2	1.3650	.4473	415.9	1.3550	.3392	408.7	1.3372
160	.6792	432.4	1.3930	.5673	429.8	1.3827	.4816	427.1	1.3732	.3745	422.0	1.3580
180	.7144	443.0	1.4096	.5998	440.6	1.3999	.5121	438.0	1.3908	.4037	433.2	1.3765
200	.7472	453.4	1.4255	.6302	451.2	1.4161	.5408	448.9	1.4074	.4303	444.9	1.3944

* Stearns and George, *Ind. Eng. Chem.*, **25**, 602 (1943); with permission.

Properties of Saturated Dowtherm A*

(73.5 % diphenyloxide, 26.5 % diphenyl)

Temp., °F. t	Abs. pressure, lb./sq. in. p	Volume, cu. ft./lb. Liquid v_f	Volume, cu. ft./lb. Vapor v_g	Enthalpy, B.t.u./lb. Liquid h_f	Enthalpy, B.t.u./lb. Vapor h_g
53.6	......	......	......	0	164
100	......	......	......	18.0	176
150	......	......	......	38.4	192
200	......	0.0160	833	60.0	210
220	......	.0162	500	69.0	217
240	0.20	.0163	294	78.2	224
260	.29	.0165	179	87.7	232
280	.49	.0166	125	97.5	240
300	.74	.0168	83.3	108.0	250
320	1.1	.0170	52.6	118.0	258
340	1.6	.0171	38.5	128.0	266
360	2.2	.0173	27.8	138.0	275
380	3.0	.0175	20.0	150.0	286
400	4.1	.0176	14.7	162.0	296
420	5.4	.0178	11.4	174.0	306
440	6.9	.0180	8.40	186.0	316
460	8.8	.0181	6.54	198.0	326
480	12	.0183	5.13	210.0	334
500	15	.0185	4.08	222.0	345
520	19	.0188	3.23	234.0	354
540	24	.0190	2.44	247.0	365
560	30	.0193	2.00	260.0	375
580	36	.0195	1.67	274.0	386
600	43	.0198	1.47	288.0	398
620	51	.0201	1.20	302.0	409
640	62	.0204	0.980	316.0	421
660	74	.0207	.826	330.0	432
680	88	.0211	.694	344.0	443
700	103	.0213	.617	358.0	455
720	120	.0218	.526	372.0	465
750	150	.0225	.417	393.0	482

* Dow Chemical Co.

Properties of Saturated Ammonia*

Temp., °F. t	Abs. pressure, lb./sq. in. p	Volume, cu. ft./lb. Liquid v_f	Volume, cu. ft./lb. Vapor v_g	Enthalpy, B.t.u./lb. Liquid h_f	Enthalpy, B.t.u./lb. Vapor h_g	Entropy, B.t.u./(lb.)(°R.) Liquid s_f	Entropy, B.t.u./(lb.)(°R.) Vapor s_g	Temp., °F. t	Abs. pressure, lb./sq. in. p	Volume, cu. ft./lb. Liquid v_f	Volume, cu. ft./lb. Vapor v_g	Enthalpy, B.t.u./lb. Liquid h_f	Enthalpy, B.t.u./lb. Vapor h_g	Entropy, B.t.u./(lb.)(°R.) Liquid s_f	Entropy, B.t.u./(lb.)(°R.) Vapor s_g
−60	5.55	0.02278	44.73	−21.2	589.6	−0.0517	1.4769	24	52.59	....	5.443	69.1	618.9	.1528	1.2897
−50	7.67	.02299	33.08	−10.6	593.7	− .0256	1.4497	28	57.28	....	5.021	73.5	619.9	.1618	1.2825
−40	10.41	.02322	24.86	0.0	597.6	.0000	1.4242	32	62.29	....	4.637	77.9	621.0	.1708	1.2755
−30	13.90	....	18.97	10.7	601.4	.0250	1.4001	36	67.63	....	4.289	82.3	622.0	.1797	1.2686
−20	18.30	.02369	14.68	21.4	605.0	.0497	1.3774	40	73.32	.02533	3.971	86.8	623.0	.1885	1.2618
−16	20.34	....	13.29	25.6	606.4	.0594	1.3686	50	89.19	.02564	3.294	97.9	625.2	.2105	1.2453
−12	22.56	....	12.06	30.0	607.8	.0690	1.3600	60	107.6	.02597	2.751	109.2	627.3	.2322	1.2294
− 8	24.97	....	10.97	34.3	609.2	.0786	1.3516	70	128.8	.02632	2.312	120.5	629.1	.2537	1.2140
− 4	27.59	....	9.991	38.6	610.5	.0880	1.3433	80	153.0	.02668	1.955	132.0	630.7	.2749	1.1991
0	30.42	.02419	9.116	42.9	611.8	.0975	1.3352	90	180.6	.02707	1.661	143.5	632.0	.2958	1.1846
4	33.47	....	8.333	47.2	613.0	.1069	1.3273	100	211.9	.02747	1.419	155.2	633.0	.3166	1.1705
8	36.77	....	7.629	51.6	614.3	.1162	1.3195	110	247.0	.02790	1.217	167.0	633.7	.3372	1.1566
12	40.31	....	6.996	56.0	615.5	.1254	1.3118	120	286.4	.02836	1.047	179.0	634.0	.3576	1.1427
16	44.12	....	6.425	60.3	616.6	.1346	1.3043	125	307.8	.02860	0.973	183.9	634.0	.3659	1.1372
20	48.21	.02474	5.910	64.7	617.8	.1437	1.2969								

* *U. S. Bur. Standards Circ.* 142, 1923.

Table 204. Properties of Superheated Ammonia*

v, volume, cu. ft./lb.; h, enthalpy, B.t.u /lb.; s, entropy, B.t.u./(lb.)(°R.)
Absolute pressure, lb. per sq. in. (saturation temperature, °F., in parentheses)

Temp., °F.	5 (−63.11)			7 (−52.88°)			10 (−41.34)			14 (−29.76)			18 (−20.61)		
	v	h	s	v	h	s	v	h	s	v	h	s	v	h	s
Sat.	49.31	588.5	1.4857	36.01	592.5	1.4574	25.81	597.1	1.4276	18.85	601.4	1.3996	14.90	604.8	1.3787
−50	51.05	595.2	1.5025	36.29	594.0	1.4611									
−40	52.36	600.3	1.5149	37.25	599.3	1.4739									
−30	53.67	605.4	1.5269	38.19	604.5	1.4861	26.58	603.2	1.4420						
−20	54.97	610.4	1.5385	39.13	609.6	1.4979	27.26	608.5	1.4542	19.33	606.8	1.4119	14.93	605.1	1.3795
−10	56.26	615.4	1.5498	40.07	614.7	1.5094	27.92	613.7	1.4659	19.82	612.2	1.4241	15.32	610.7	1.3921
0	57.55	620.4	1.5608	41.00	619.8	1.5206	28.58	618.9	1.4773	20.30	617.6	1.4358	15.70	616.2	1.4042
10	58.84	625.4	1.5716	41.93	624.9	1.5314	29.24	624.0	1.4884	20.78	622.8	1.4472	16.08	621.6	1.4158
20	60.12	630.4	1.5821	42.85	629.9	1.5421	29.90	629.1	1.4992	21.26	628.0	1.4582	16.46	626.9	1.4270
30	61.41	635.4	1.5925	43.77	635.0	1.5525	30.55	634.2	1.5097	21.73	633.2	1.4688	16.83	632.2	1.4380
40	62.69	640.4	1.6026	44.69	640.0	1.5627	31.20	639.3	1.5200	22.20	638.4	1.4793	17.20	637.5	1.4486
50	63.96	645.5	1.6125	45.61	645.0	1.5727	31.85	644.4	1.5301	22.67	643.6	1.4896	17.57	642.7	1.4590
60	65.24	650.5	1.6223	46.53	650.1	1.5825	32.49	649.5	1.5400	23.14	648.7	1.4996	17.94	647.9	1.4691
70	66.51	655.5	1.6319	47.44	655.2	1.5921	33.14	654.6	1.5497	23.60	653.9	1.5094	18.30	653.1	1.4790
80	67.79	660.6	1.6413	48.36	660.2	1.6016	33.78	659.7	1.5593	24.06	659.0	1.5191	18.67	658.4	1.4887

* *U. S. Bur. Standards Circ.* 142, 1923.

Properties of Superheated Ammonia*— *(Continued)*

Temp. °F.	5 (−63.11)			7 (−52.88°)			10 (−41.34)			14 (−29.76)			18 (−20.61)		
	v	h	s	v	h	s	v	h	s	v	h	s	v	h	s
90	69.06	665.6	1.6506	49.27	665.3	1.6110	34.42	664.8	1.5687	24.53	664.2	1.5285	19.03	663.6	1.4983
100	70.33	670.7	1.6598	50.18	670.4	1.6202	35.07	670.0	1.5779	24.99	669.4	1.5378	19.39	668.8	1.5077
110	71.60	675.8	1.6689	51.09	675.5	1.6292	35.71	675.1	1.5870	25.45	674.5	1.5470	19.75	674.0	1.5169
120	72.87	680.9	1.6778	52.00	680.7	1.6382	36.35	680.3	1.5960	25.91	679.7	1.5560	20.11	679.2	1.5260
130	74.14	686.1	1.6865	52.91	685.8	1.6470	36.99	685.4	1.6049	26.37	684.9	1.5649	20.47	684.4	1.5349
140	75.41	691.2	1.6952	53.82	691.0	1.6557	37.62	690.6	1.6136	26.83	690.1	1.5737	20.83	689.7	1.5438
150	76.68	696.4	1.7038	54.73	696.2	1.6643	38.26	695.8	1.6222	27.29	695.4	1.5824	21.19	694.9	1.5525
160	77.95	701.6	1.7122	55.63	701.4	1.6727	38.90	701.1	1.6307	27.74	700.6	1.5909	21.54	700.2	1.5610
170	79.21	706.8	1.7206	56.54	706.6	1.6811	39.54	706.3	1.6391	28.20	705.9	1.5993	21.90	705.5	1.5695
180	80.48	712.1	1.7289	57.45	711.9	1.6894	40.17	711.6	1.6474	28.66	711.2	1.6076	22.26	710.8	1.5778
190	.....	.....	......	.....	.....	......	40.81	716.9	1.6556	29.11	716.5	1.6159	22.61	716.1	1.5861
200	.....	.....	......	.....	.....	......	41.45	722.2	1.6637	29.57	721.8	1.6240	22.97	721.4	1.5943
220	.....	.....	......	.....	.....	......	.....	.....	......	.....	.....	......	23.68	732.2	1.6103

Temp., °F.	24 (−9.58)			30 (−0.57)			38 (9.42)			48 (19.80)		
	v	h	s	v	h	s	v	h	s	v	h	s
Sat.	11.39	608.6	1.3549	9.236	611.6	1.3364	7.396	614.7	1.3168	5.934	617.7	1.2973
0	11.67	614.1	1.3670	9.250	611.9	1.3371						
10	11.96	619.7	1.3791	9.492	617.8	1.3497	7.407	615.0	1.3175			
20	12.25	625.2	1.3907	9.731	623.5	1.3618	7.603	621.0	1.3301	5.937	617.8	1.2976
30	12.54	630.7	1.4019	9.966	629.1	1.3733	7.795	626.9	1.3422	6.096	624.0	1.3103
40	12.82	636.1	1.4128	10.20	634.6	1.3845	7.983	632.6	1.3538	6.251	630.0	1.3225
50	13.11	641.4	1.4234	10.43	640.1	1.3953	8.170	638.3	1.3650	6.404	635.9	1.3341
60	13.39	646.7	1.4337	10.65	645.5	1.4059	8.353	643.8	1.3758	6.554	641.6	1.3453
70	13.66	652.0	1.4438	10.88	650.9	1.4161	8.535	649.3	1.3863	6.702	647.3	1.3561
80	13.94	657.3	1.4537	11.10	656.2	1.4261	8.716	654.8	1.3965	6.848	652.9	1.3666
90	14.22	662.6	1.4634	11.33	661.6	1.4359	8.895	660.2	1.4065	6.993	658.5	1.3768
100	14.49	667.8	1.4729	11.55	666.9	1.4456	9.073	665.6	1.4163	7.137	664.0	1.3868
110	14.76	673.1	1.4822	11.77	672.2	1.4550	9.250	671.0	1.4258	7.280	669.5	1.3965
120	15.04	678.4	1.4914	11.99	677.5	1.4642	9.426	676.4	1.4352	7.421	675.0	1.4061
130	15.31	683.6	1.5004	12.21	682.9	1.4733	9.602	681.8	1.4444	7.562	680.5	1.4154
140	15.58	688.9	1.5093	12.43	688.2	1.4823	9.776	687.2	1.4534	7.702	685.9	1.4246
150	15.85	694.2	1.5180	12.65	693.5	1.4911	9.950	692.6	1.4623	7.842	691.4	1.4336
160	16.12	699.5	1.5266	12.87	698.8	1.4998	10.12	698.0	1.4711	7.981	696.8	1.4425
170	16.39	704.8	1.5352	13.08	704.2	1.5083	10.30	703.3	1.4797	8.119	702.3	1.4512
180	16.66	710.2	1.5436	13.30	709.6	1.5168	10.47	708.7	1.4883	8.257	707.7	1.4598
190	16.93	715.5	1.5518	13.52	714.9	1.5251	10.64	714.2	1.4966	8.395	713.2	1.4683
200	17.20	720.9	1.5600	13.73	720.3	1.5334	10.81	719.6	1.5049	8.532	718.7	1.4766
220	17.73	731.7	1.5761	14.16	731.1	1.5495	11.16	730.5	1.5212	8.805	729.6	1.4930
240	18.27	742.6	1.5919	14.59	742.0	1.5653	11.50	741.4	1.5371	9.077	740.6	1.5090
260	.....	.....	......	15.02	753.0	1.5808	11.84	752.4	1.5526	9.348	751.7	1.5246
280	.....	.....	......	.....	.....	......	12.18	763.5	1.5678	9.619	762.9	1.5399
300	.....	.....	......	.....	.....	......	.....	.....	......	9.888	774.1	1.5548

Temp., °F.	60 (30.21)			80 (44.40)			100 (56.05)			120 (66.02)		
	v	h	s	v	h	s	v	h	s	v	h	s
Sat.	4.805	620.5	1.2787	3.655	624.0	1.2545	2.952	626.5	1.2356	2.476	628.4	1.2201
30												
40	4.933	626.8	1.2913	.....	.....	......	.....	.....	......	.....	.....	......
50	5.060	632.9	1.3035	3.712	627.7	1.2619	.....	.....	......	.....	.....	......
60	5.184	639.0	1.3152	3.812	634.3	1.2745	2.985	629.3	1.2409	.....	.....	......
70	5.307	644.9	1.3265	3.909	640.6	1.2866	3.068	636.0	1.2539	2.505	631.3	1.2255
80	5.428	650.7	1.3373	4.005	646.7	1.2981	3.149	642.6	1.2661	2.576	638.3	1.2386
90	5.547	656.4	1.3479	4.098	652.8	1.3092	3.227	649.0	1.2778	2.645	645.0	1.2510
100	5.665	662.1	1.3581	4.190	658.7	1.3199	3.304	655.2	1.2891	2.712	651.6	1.2628
110	5.781	667.7	1.3681	4.281	664.6	1.3303	3.380	661.3	1.2999	2.778	658.0	1.2741
120	5.897	673.3	1.3778	4.371	670.4	1.3404	3.454	667.3	1.3104	2.842	664.2	1.2850
130	6.012	678.9	1.3873	4.460	676.1	1.3502	3.527	673.3	1.3206	2.905	670.4	1.2956
140	6.126	684.4	1.3966	4.548	681.8	1.3598	3.600	679.2	1.3305	2.967	676.5	1.3058
150	6.239	689.9	1.4058	4.635	687.5	1.3692	3.672	685.0	1.3401	3.029	682.5	1.3157
160	6.352	695.5	1.4148	4.722	693.2	1.3784	3.743	690.8	1.3495	3.089	688.4	1.3254
170	6.464	701.0	1.4236	4.808	698.8	1.3874	3.813	696.6	1.3588	3.149	694.3	1.3348
180	6.576	706.5	1.4323	4.893	704.4	1.3963	3.883	702.3	1.3678	3.209	700.2	1.3441
190	6.687	712.0	1.4409	4.978	710.0	1.4050	3.952	708.0	1.3767	3.268	706.0	1.3531
200	6.798	717.5	1.4493	5.063	715.6	1.4136	4.021	713.7	1.3854	3.326	711.8	1.3620
210	6.909	723.1	1.4576	5.147	721.3	1.4220	4.090	719.4	1.3940	3.385	717.6	1.3707
220	7.019	728.6	1.4658	5.231	726.9	1.4304	4.158	725.1	1.4024	3.442	723.1	1.3793
230	.....	.....	......	5.315	732.5	1.4386	4.226	730.8	1.4108	3.500	729.2	1.3877
240	7.238	739.7	1.4819	5.398	738.1	1.4467	4.294	736.5	1.4190	3.557	734.9	1.3960
250	.....	.....	......	5.482	743.8	1.4547	4.361	742.2	1.4271	3.614	740.7	1.4042
260	7.457	750.9	1.4976	5.565	749.4	1.4626	4.428	747.9	1.4350	3.671	746.5	1.4123
270	.....	.....	......	.....	.....	......	.....	.....	......	3.727	752.2	1.4202
280	7.675	762.1	1.5130	5.730	760.7	1.4781	4.562	759.4	1.4507	3.783	758.0	1.4281
290	.....	.....	......	.....	.....	......	.....	.....	......	3.839	763.8	1.4359
300	7.892	773.3	1.5281	5.894	772.1	1.4933	4.695	770.8	1.4660	3.895	769.6	1.4435

Properties of Superheated Ammonia*—(*Concluded*)

Temp., °F.	140 (74.79)			160 (82.64)			180 (89.78)		
	v	h	s	v	h	s	v	h	s
Sat.	2.132	629.9	1.2068	1.872	631.1	1.1952	1.667	632.0	1.1850
80	2.166	633.8	1.2140						
90	2.228	640.9	1.2272	1.914	636.6	1.2055	1.668	632.2	1.1853
100	2.288	647.8	1.2396	1.969	643.9	1.2186	1.720	639.9	1.1992
110	2.347	654.5	1.2515	2.023	651.0	1.2311	1.770	647.3	1.2123
120	2.404	661.1	1.2628	2.075	657.8	1.2429	1.818	654.4	1.2247
130	2.460	667.4	1.2738	2.125	664.4	1.2542	1.865	661.3	1.2364
140	2.515	673.7	1.2843	2.175	670.9	1.2652	1.910	668.0	1.2477
150	2.569	679.9	1.2945	2.224	677.2	1.2575	1.955	674.6	1.2586
160	2.622	686.0	1.3045	2.272	683.5	1.2859	1.999	681.0	1.2691
170	2.675	692.0	1.3141	2.319	689.7	1.2958	2.042	687.3	1.2792
180	2.727	698.0	1.3236	2.365	695.8	1.3054	2.084	693.6	1.2891
190	2.779	704.0	1.3328	2.411	701.9	1.3148	2.126	699.8	1.2987
200	2.830	709.9	1.3418	2.457	707.9	1.3240	2.167	705.9	1.3081
210	2.880	715.8	1.3507	2.502	713.9	1.3331	2.208	712.0	1.3172
220	2.931	721.6	1.3594	2.547	719.9	1.3419	2.248	718.1	1.3262
230	2.981	727.5	1.3679	2.591	725.8	1.3506	2.288	724.1	1.3350
240	3.030	733.3	1.3763	2.635	731.7	1.3591	2.328	730.1	1.3436
250	3.080	739.2	1.3846	2.679	737.6	1.3675	2.367	736.1	1.3521
260	3.129	745.0	1.3928	2.723	743.5	1.3757	2.407	732.0	1.3605
270	3.179	750.8	1.4008	2.766	749.4	1.3838	2.446	748.0	1.3687
280	3.227	756.7	1.4088	2.809	755.3	1.3919	2.484	753.9	1.3768
290	3.275	762.5	1.4166	2.852	761.2	1.3998	2.523	759.9	1.3847
300	3.323	768.3	1.4243	2.895	767.1	1.4076	2.561	765.8	1.3926
320	3.420	780.0	1.4395	2.980	778.9	1.4229	2.637	777.7	1.4081
340	.....	.....	......	3.064	790.7	1.4379	2.713	789.6	1.4231

Temp., °F.	200 (96.34)			220 (102.42)			240 (108.09)			260 (113.42)			300 (123.21)		
	v	h	s	v	h	s	v	h	s	v	h	s	v	h	s
Sat.	1.502	632.7	1.1756	1.367	633.2	1.1671	1.253	633.6	1.1592	1.155	633.9	1.1518	0.999	634.0	1.1383
110	1.567	643.4	1.1947	1.400	639.4	1.1781	1.261	635.3	1.1621						
120	1.612	650.9	1.2077	1.443	647.3	1.1917	1.302	643.5	1.1764	1.182	639.5	1.1617			
130	1.656	658.1	1.2200	1.485	654.8	1.2045	1.342	651.3	1.1898	1.220	647.8	1.1757	1.023	640.1	1.1487
140	1.698	665.0	1.2317	1.525	662.0	1.2167	1.380	658.8	1.2025	1.257	655.6	1.1889	1.058	648.7	1.1632
150	1.740	671.8	1.2429	1.564	669.0	1.2281	1.416	666.1	1.2145	1.292	663.1	1.2014	1.091	656.9	1.1767
160	1.780	678.4	1.2537	1.601	675.8	1.2394	1.452	673.1	1.2259	1.326	670.4	1.2132	1.123	664.7	1.1894
170	1.820	684.9	1.2641	1.638	682.5	1.2501	1.487	680.0	1.2369	1.359	677.5	1.2245	1.153	672.2	1.2014
180	1.859	691.3	1.2742	1.675	689.1	1.2604	1.521	686.7	1.2475	1.391	684.4	1.2354	1.183	679.5	1.2129
190	1.897	697.7	1.2840	1.710	695.5	1.2704	1.554	693.3	1.2577	1.422	691.1	1.2458	1.211	686.5	1.2239
200	1.935	703.9	1.2935	1.745	701.9	1.2801	1.587	699.8	1.2677	1.453	697.7	1.2560	1.239	693.5	1.2344
210	1.972	710.1	1.3029	1.780	708.2	1.2896	1.619	706.2	1.2773	1.484	704.3	1.2658	1.267	700.3	1.2447
220	2.009	716.3	1.3120	1.814	714.4	1.2989	1.651	712.6	1.2867	1.514	710.7	1.2754	1.294	706.9	1.2546
230	2.046	722.4	1.3209	1.848	720.6	1.3079	1.683	718.9	1.2959	1.543	717.1	1.2847	1.320	713.5	1.2642
240	2.082	728.4	1.3296	1.881	726.8	1.3168	1.714	725.1	1.3049	1.572	723.4	1.2938	1.346	720.0	1.2736
250	2.118	734.5	1.3382	1.914	732.9	1.3255	1.745	731.3	1.3137	1.601	729.7	1.3027	1.372	726.5	1.2827
260	2.154	740.5	1.3467	1.947	739.0	1.3340	1.775	737.5	1.3224	1.630	736.0	1.3115	1.397	732.9	1.2917
270	2.189	746.5	1.3550	1.980	745.1	1.3424	1.805	743.6	1.3308	1.658	742.2	1.3200	1.422	739.2	1.3004
280	2.225	752.5	1.3631	2.012	751.1	1.3507	1.835	749.8	1.3392	1.686	748.4	1.3285	1.447	745.5	1.3090
290	2.260	758.5	1.3712	2.044	757.2	1.3588	1.865	755.9	1.3474	1.714	754.5	1.3367	1.472	751.8	1.3175
300	2.295	764.5	1.3791	2.076	763.2	1.3668	1.895	762.0	1.3554	1.741	760.7	1.3449	1.496	758.1	1.3257
320	2.364	776.5	1.3947	2.140	775.3	1.3825	1.954	774.1	1.3712	1.796	772.9	1.3608	1.544	770.5	1.3419
340	2.432	788.5	1.4099	2.203	787.4	1.3978	2.012	786.3	1.3866	1.850	785.2	1.3763	1.592	782.9	1.3576
360	2.500	800.5	1.4247	2.265	799.5	1.4127	2.069	798.4	1.4016	1.904	797.4	1.3914	1.639	795.3	1.3729
380	2.568	812.5	1.4392	2.327	811.6	1.4273	2.126	810.6	1.4163	1.957	809.6	1.4062	1.686	807.7	1.3878

Properties of Saturated Ethylene*

Temp., °F.	Abs. pressure, atm.	Volume, cu. ft./lb.		Enthalpy, B.t.u./lb.		Entropy, B.t.u./(lb.)(°R.)	
		Liquid	Vapor	Liquid	Vapor	Liquid	Vapor
t	p	v_f	v_g	h_f	h_g	s_f	s_g
−272.47	0.0012	....	4064.0	− 68.19	176.5	−0.2826	1.024
−260.0	.0037	....	1405.0	− 60.84	180.1	− .2351	0.972
−240.0	.0177	....	328.6	− 49.16	185.7	− .1887	.880
−220.0	.0606	....	103.1	− 37.56	191.4	− .1382	.817
−200.0	.169	....	39.74	− 26.03	197.1	− .0922	.767
−180.0	.402	....	17.92	− 14.57	202.6	− .0498	.727
−160.0	.837	....	9.047	− 3.12	207.2	− .0103	.692
−154.66	1.0000	0.02818	7.6712	0.0	207.9	.0000	.6814
−140.00	1.5775	.02877	5.005	+ 8.6	210.5	+ .0249	.6563
−120.00	2.7376	.02964	2.987	20.8	214.2	.0648	.6340
−100.00	4.4616	.03122	1.879	32.8	217.2	.0995	.6121
− 80.00	6.8697	.03179	1.732	45.2	219.9	.1374	.5974
− 60.00	10.099	.03308	0.857	57.9	221.9	.1666	.5769
− 40.00	14.0338	.03468	.593	70.8	222.8	.1935	.5556
− 20.00	19.722	.03662	.419	84.7	222.7	.2245	.5383
0.00	26.397	.03912	.301	100.3	221.0	.2577	.5202
+ 20.00	34.55	.04292	.212	119.6	216.6	.2968	.4991
+ 40.00	44.54	.05035	.139	148.7	206.2	.3533	.4683
+ 49.82	50.50	.070	.070	171.8	171.8	.3964	.3964

* York and White, *Trans. Am. Inst. Chem. Engrs.*, **40**, 227 (1944).

Properties of Superheated Ethylene*

v, volume, cu. ft./lb.; h, enthalpy, B.t.u./lb.; s, entropy, B.t.u./(lb.)(°R.)

Abs. pressure, atm.		Temperature, °F.																			
		−140°	−120°	−100°	−80°	−60°	−40°	−20°	0°	20°	40°	60°	100°	140°	180°	220°	260°	320°	380°	440°	500°
	v	8.061	8.617	9.168	9.714	10.256	10.794	11.329	11.862	12.497	12.924	13.453	14.51	15.56	16.61	17.66	18.71	20.28	21.85	23.42	24.99
1	h	212.2	218.3	224.3	230.6	236.9	243.3	249.9	256.6	263.4	270.4	277.7	292.6	308.3	324.8	342.0	360.0	388.4	418.4	450.4	482.9
	s	0.695	0.714	0.732	0.748	0.764	0.780	0.795	0.810	0.825	0.839	0.853	0.881	0.908	0.934	0.961	0.986	1.025	1.062	1.098	1.133
	v	.....	4.190	4.479	4.765	5.047	5.324	5.599	5.873	6.144	6.414	6.683	7.217	7.749	8.280	8.808	9.34	10.12	10.91	11.70	12.48
2	h	.....	215.9	222.3	228.9	235.4	242.0	248.7	255.6	262.5	269.5	276.9	291.9	307.6	324.2	341.5	359.7	388.1	418.2	450.2	482.7
	s	.....	0.659	0.677	0.693	0.710	0.726	0.741	0.756	0.771	0.785	0.799	0.828	0.855	0.881	0.909	0.934	0.974	1.011	1.048	1.083
	v	.....	.....	2.130	2.287	2.439	2.588	2.733	2.878	3.019	3.159	3.297	3.571	3.843	4.113	4.380	4.647	5.046	5.443	5.838	6.234
4	h	.....	.....	218.2	225.5	232.4	239.3	246.3	253.4	260.6	267.8	275.3	290.6	306.6	323.3	340.6	358.9	387.4	417.6	449.7	482.3
	s	.....	.....	0.623	0.641	0.658	0.675	0.691	0.706	0.722	0.737	0.751	0.780	0.807	0.833	0.860	0.886	0.925	0.963	1.000	1.035
	v	.....	.....	.....	1.456	1.566	1.673	1.776	1.833	1.877	2.099	2.168	2.356	2.541	2.724	2.905	3.084	3.352	3.620	3.885	4.150
6	h	.....	.....	.....	221.6	229.1	236.6	243.9	251.2	258.6	266.0	273.7	289.3	305.4	322.3	339.8	358.1	386.8	417.0	449.2	481.8
	s	.....	.....	.....	0.607	0.625	0.642	0.659	0.675	0.691	0.706	0.721	0.750	0.777	0.804	0.831	0.856	0.896	0.934	0.971	1.006
	v	.....	.....	.....	.....	0.860	0.936	1.006	1.074	1.140	1.202	1.264	1.382	1.499	1.613	1.724	1.834	1.999	2.161	2.323	2.483
10	h	.....	.....	.....	.....	222.4	230.8	238.8	246.6	254.4	262.4	270.4	286.5	303.1	320.3	338.0	356.5	385.4	415.8	448.2	480.9
	s	.....	.....	.....	.....	0.578	0.599	0.616	0.632	0.649	0.665	0.680	0.711	0.739	0.766	0.794	0.820	0.860	0.898	0.934	0.969
	v	.....	.....	.....	.....	.....	.....	0.614	0.669	0.718	0.765	0.811	0.897	0.978	1.057	1.134	1.209	1.322	1.433	1.542	1.650
15	h	.....	.....	.....	.....	.....	.....	230.9	239.9	248.7	257.5	266.2	283.0	300.1	317.7	335.8	354.6	383.8	414.4	447.0	479.9
	s	.....	.....	.....	.....	.....	.....	0.573	0.593	0.612	0.630	0.646	0.677	0.707	0.734	0.763	0.789	0.829	0.867	0.903	0.938
	v	.....	.....	.....	.....	.....	.....	.....	0.459	0.503	0.545*	0.583	0.652	0.717	0.778	0.838	0.897	0.983	1.068	1.151	1.233
20	h	.....	.....	.....	.....	.....	.....	.....	232.5	242.5	252.1	261.3	279.2	296.9	315.1	333.5	352.6	382.1	413.0	445.7	478.8
	s	.....	.....	.....	.....	.....	.....	.....	0.560	0.582	0.591	0.618	0.651	0.682	0.710	0.740	0.766	0.807	0.845	0.882	0.917
	v	.....	.....	.....	.....	.....	.....	.....	.....	0.282	0.318	0.350	0.405	0.454	0.498	0.543	0.585	0.645	0.704	0.760	0.817
30	h	.....	.....	.....	.....	.....	.....	.....	.....	226.9	239.5	250.8	271.2	290.2	309.5	328.7	348.3	378.6	410.1	443.3	476.7
	s	.....	.....	.....	.....	.....	.....	.....	.....	0.528	0.552	0.574	0.613	0.646	0.676	0.706	0.734	0.775	0.814	0.851	0.887
	v	.....	.....	.....	.....	.....	.....	.....	.....	.....	0.192	0.228	0.279	0.323	0.359	0.395	0.428	0.476	0.521	0.565	0.609
40	h	.....	.....	.....	.....	.....	.....	.....	.....	.....	221.1	237.3	262.3	283.4	303.7	323.8	344.1	375.1	407.2	440.8	474.5
	s	.....	.....	.....	.....	.....	.....	.....	.....	.....	0.503	0.534	0.580	0.617	0.650	0.681	0.709	0.751	0.791	0.828	0.865
	v	.....	.....	.....	.....	.....	.....	.....	.....	.....	.....	0.138	0.201	0.243	0.275	0.309	0.334	0.375	0.412	0.448	0.484
50	h	.....	.....	.....	.....	.....	.....	.....	.....	.....	.....	218.1	250.3	275.6	297.5	318.7	339.7	371.5	404.2	438.2	472.4
	s	.....	.....	.....	.....	.....	.....	.....	.....	.....	.....	0.484	0.550	0.590	0.626	0.659	0.688	0.731	0.772	0.810	0.847
	v	.....	.....	.....	.....	.....	.....	.....	.....	.....	.....	0.069	0.144	0.186	0.219	0.248	0.273	0.307	0.340	0.371	0.402
60	h	.....	.....	.....	.....	.....	.....	.....	.....	.....	.....	171.0	238.9	267.4	291.2	313.6	335.4	367.8	401.4	435.8	470.3
	s	.....	.....	.....	.....	.....	.....	.....	.....	.....	.....	0.390	0.517	0.568	0.607	0.640	0.671	0.716	0.756	0.795	0.833
	v	.....	.....	.....	.....	.....	.....	.....	.....	.....	.....	0.051	0.082	0.121	0.150	0.174	0.195	0.224	0.250	0.275	0.301
80	h	.....	.....	.....	.....	.....	.....	.....	.....	.....	.....	146.4	205.0	248.3	277.6	302.8	326.4	360.5	395.3	430.8	466.1
	s	.....	.....	.....	.....	.....	.....	.....	.....	.....	.....	0.340	0.477	0.522	0.569	0.608	0.641	0.688	0.731	0.770	0.808
	v	.....	.....	.....	.....	.....	.....	.....	.....	.....	.....	0.044	0.057	0.082	0.109	0.131	0.149	0.174	0.197	0.219	0.239
100	h	.....	.....	.....	.....	.....	.....	.....	.....	.....	.....	143.5	182.5	228.3	263.0	291.3	317.1	353.3	389.6	426.0	462.0
	s	.....	.....	.....	.....	.....	.....	.....	.....	.....	.....	0.327	0.401	0.480	0.535	0.579	0.614	0.665	0.709	0.751	0.789
	v	.....	.....	.....	.....	.....	.....	.....	.....	.....	.....	0.039	0.046	0.056	0.067	0.081	0.094	0.111	0.128	0.143	0.159
150	h	.....	.....	.....	.....	.....	.....	.....	.....	.....	.....	139.1	167.7	199.8	235.0	267.2	296.5	336.5	376.0	414.8	452.6
	s	.....	.....	.....	.....	.....	.....	.....	.....	.....	.....	0.308	0.361	0.417	0.474	0.524	0.564	0.619	0.668	0.712	0.752
	v	.....	.....	.....	.....	.....	.....	.....	.....	.....	.....	0.037	0.042	0.047	0.054	0.062	0.070	0.085	0.097	0.108	0.119
200	h	.....	.....	.....	.....	.....	.....	.....	.....	.....	.....	138.3	163.8	191.1	221.5	252.3	281.7	323.4	364.7	404.9	444.2
	s	.....	.....	.....	.....	.....	.....	.....	.....	.....	.....	0.295	0.342	0.392	0.440	0.487	0.529	0.585	0.636	0.682	0.724
	v	.....	.....	.....	.....	.....	.....	.....	.....	.....	.....	0.035	0.039	0.043	0.049	0.054	0.059	0.069	0.078	0.088	0.099
250	h	.....	.....	.....	.....	.....	.....	.....	.....	.....	.....	138.2	161.8	187.5	215.1	243.9	272.3	314.1	355.8	397.1	437.5
	s	.....	.....	.....	.....	.....	.....	.....	.....	.....	.....	0.286	0.331	0.376	0.419	0.462	0.503	0.561	0.612	0.659	0.702
	v	.....	.....	.....	.....	.....	.....	.....	.....	.....	.....	0.034	0.038	0.041	0.045	0.049	0.054	0.061	0.069	0.077	0.083
300	h	.....	.....	.....	.....	.....	.....	.....	.....	.....	.....	137.9	161.4	186.3	212.1	238.7	265.9	307.7	349.4	391.2	432.0
	s	.....	.....	.....	.....	.....	.....	.....	.....	.....	.....	0.277	0.322	0.364	0.404	0.446	0.484	0.541	0.593	0.640	0.684

* York and White, *Trans. Am. Inst. Chem. Engrs.*, **40**, 227 (1944).

Properties of Saturated Methane*

Temp., °F.	Abs. pressure, lb./sq. in.	Volume, cu. ft./lb.		Enthalpy, B.t.u./lb.		Entropy B.t.u./(lb.)(°R.)		Temp., °F.	Abs. pressure, lb./sq. in.	Volume, cu. ft./lb.		Enthalpy, B.t.u./lb.		Entropy B.t.u./(lb.)(°R.)	
		Liquid	Vapor	Liquid	Vapor	Liquid	Vapor			Liquid	Vapor	Liquid	Vapor	Liquid	Vapor
t	p	v_f	v_g	h_f	h_g	s_f	s_g	t	p	v_f	v_g	h_f	h_g	s_f	s_g
−280	4.90	0.03635	24.04	0	228.2	0	1.2699	−180	191.5	.04575	.773	87.8	257.0	.3767	.9816
−270	8.44	.03698	14.61	8.2	232.3	0.0423	1.2236	−170	240.0	.04745	.610	98.0	257.2	.4127	.9622
−260	13.80	.03766	9.31	16.6	236.4	.0823	1.1830	−160	297.0	.04944	.483	108.7	256.5	.4476	.9411
−250	21.71	.03839	6.13	25.0	240.3	.1201	1.1468	−150	364	.05197	.381	120.3	254.5	.4839	.9169
−240	32.4	.03915	4.24	33.3	243.9	.1578	1.1164	−140	440	.05224	.3008	133.2	251.2	.5214	.8905
−230	46.4	.03999	3.04	42.0	247.3	.1962	1.0900	−130	527	.05999	.2318	148.1	245.9	.5656	.8622
−220	64.5	.04092	2.23	50.6	250.2	.2333	1.0660	−120	627	.06961	.1613	171.8	231.4	.6329	.8083
−210	87.6	.04193	1.67	59.5	252.8	.2693	1.0434	−115.8	673	.0983	.0983	203.4	203.4	.7232	.7232
−200	115.7	.04306	1.281	68.8	254.8	.3062	1.0224								
−190	150.0	.04431	0.990	78.2	256.2	.3419	1.0019								

* Matthews and Hurd, *Trans. Am. Inst. Chem. Engrs.* **42**, 55 (1946).

Properties of Superheated Methane*

v, volume, cu. ft./lb.; h, enthalpy, B.t.u./lb.; s, entropy, B.t.u./(lb.)(°R.)
Parenthetic figures after pressures are saturation temperatures

Temp., °F.	10 lb./sq. in. abs. (−266.6°F.)			20 lb./sq. in. abs. (−251.8°F.)			30 lb./sq. in. abs. (−242°F.)			40 lb./sq. in. abs. (−234.3°F.)			60 lb./sq. in. abs. (−222.2°F.)			80 lb./sq. in. abs. (−213.0°F.)			100 lb./sq. in. abs. (−205.5°F.)		
t	v	h	s	v	h	s	v	h	s	i	h	s	v	h	s	v	h	s	v	h	s
−260	12.98	237.0	1.2262	.....	.....	.....	.....	.....	.....	.....	.....	.....	.....	.....	.....	.....	.....	.....	.....	.....	.....
−240	14.39	247.3	1.2750	7.04	245.8	1.1830	4.60	244.2	1.1273	.....	.....	.....	.....	.....	.....	.....	.....	.....	.....	.....	.....
−220	15.78	257.5	1.3214	7.76	256.3	1.2313	5.09	254.9	1.1767	3.75	253.6	1.1365	2.421	251.1	1.0775	.....	.....	.....	.....	.....	.....
−200	17.15	267.7	1.3644	8.47	266.6	1.2752	5.57	265.4	1.2217	4.12	264.3	1.1826	2.678	262.1	1.1258	1.954	259.5	1.0832	1.518	256.9	1.0476
−180	18.52	277.9	1.4032	9.17	276.9	1.3148	6.06	275.9	1.2618	4.49	274.9	1.2237	2.934	273.0	1.1681	2.153	270.7	1.1269	1.684	268.4	1.0935
−160	19.87	287.9	1.4386	9.86	287.0	1.3507	6.53	286.1	1.2982	4.85	285.2	1.2607	3.184	283.5	1.2062	2.348	281.5	1.1662	1.847	279.6	1.1339
−140	21.28	297.9	1.4710	10.56	297.1	1.3836	6.99	296.3	1.3315	5.21	295.5	1.2944	3.429	293.9	1.2407	2.539	292.3	1.2015	2.002	290.6	1.1701
−120	22.62	308.0	1.5009	11.25	307.2	1.4138	7.45	306.4	1.3621	5.56	305.8	1.3252	3.670	304.4	1.2721	2.725	302.9	1.2336	2.155	301.4	1.2029
−115.8	22.91	310.0	1.5067	11.40	309.2	1.4197	7.56	308.3	1.3681	5.64	307.9	1.3312	3.72	306.5	1.2782	2.761	305.0	1.2398	2.183	303.6	1.2092
−100	23.97	318.1	1.5290	11.94	317.4	1.4422	7.91	316.7	1.3908	5.91	316.1	1.3541	3.91	314.9	1.3015	2.903	313.5	1.2635	2.301	312.2	1.2332
− 80	25.33	328.1	1.5561	12.61	327.6	1.4693	8.37	327.0	1.4182	6.25	326.4	1.3816	4.14	325.2	1.3293	3.080	324.1	1.2918	2.444	322.9	1.2619
− 60	26.69	338.2	1.5820	13.29	337.6	1.4953	8.83	337.0	1.4443	6.60	236.6	1.4079	4.37	335.6	1.3558	3.255	334.5	1.3186	2.588	333.4	1.2891
− 40	28.02	348.3	1.6066	13.97	347.8	1.5200	9.28	347.3	1.4691	6.94	346.9	1.4328	4.60	345.9	1.3811	3.432	345.0	1.3440	2.729	344.0	1.3148
− 20	29.36	358.5	1.6303	14.64	358.0	1.5437	9.74	357.5	1.4929	7.29	357.1	1.4567	4.83	356.3	1.4052	3.607	355.4	1.3683	2.871	354.5	1.3393
0	30.72	368.7	1.6531	15.32	368.3	1.5667	10.19	367.9	1.5159	7.63	367.5	1.4798	5.06	366.7	1.4284	3.78	365.9	1.3917	3.014	365.2	1.3628
20	32.06	378.9	1.6752	16.00	378.6	1.5888	10.64	378.2	1.5380	7.97	377.8	1.5020	5.30	377.0	1.4508	3.96	376.4	1.4142	3.155	375.7	1.3854
40	33.40	389.4	1.6964	16.66	389.0	1.6101	11.10	388.6	1.5594	8.31	388.2	1.5234	5.52	387.6	1.4723	4.13	387.0	1.4358	3.293	386.3	1.4072
60	34.73	399.9	1.7169	17.34	399.5	1.6306	11.54	399.2	1.5800	8.65	398.9	1.5440	5.75	398.3	1.4930	4.30	397.7	1.4566	3.431	397.0	1.4280
80	36.10	410.5	1.7370	18.02	410.2	1.6507	12.00	409.9	1.6002	8.99	409.6	1.5642	5.98	409.1	1.5133	4.47	408.4	1.4770	3.569	407.9	1.4485
100	37.44	421.4	1.7570	18.70	421.1	1.6707	12.44	420.8	1.6202	9.33	420.5	1.5843	6.21	420.1	1.5334	4.65	419.4	1.4971	3.71	418.9	1.4687
120	38.78	432.2	1.7763	19.37	431.9	1.6900	12.90	431.6	1.6396	9.66	431.4	1.6036	6.44	431.0	1.5529	4.82	430.3	1.5166	3.86	429.8	1.4882
140	40.12	443.2	1.7946	20.05	442.9	1.7084	13.35	442.7	1.6580	10.01	442.4	1.6221	6.66	442.0	1.5714	4.99	441.4	1.5351	3.98	440.9	1.5068
160	41.46	454.4	1.8127	20.73	454.2	1.7265	13.79	453.9	1.6760	10.34	453.7	1.6402	6.89	453.3	1.5895	5.16	452.7	1.5533	4.12	452.2	1.5250
180	42.80	465.7	1.8307	21.40	465.5	1.7445	14.25	465.2	1.6940	10.69	465.0	1.6582	7.12	464.6	1.6075	5.33	464.1	1.5714	4.26	463.7	1.5432
200	44.13	477.3	1.8484	22.07	477.1	1.7622	14.70	476.9	1.7118	11.03	476.7	1.6760	7.34	476.3	1.6254	5.50	475.8	1.5893	4.40	475.4	1.5611
220	45.47	488.9	1.8659	22.74	488.7	1.7798	15.16	488.5	1.7294	11.36	488.3	1.6936	7.56	487.9	1.6430	5.66	487.4	1.6069	4.52	487.1	1.5788
240	46.81	500.9	1.8829	23.41	500.7	1.7968	15.61	500.5	1.7464	11.70	500.3	1.7106	7.77	499.9	1.6600	5.84	499.5	1.6241	4.66	499.2	1.5960
260	48.15	512.9	1.8998	24.08	512.8	1.8137	16.06	512.6	1.7633	12.04	512.4	1.7276	8.02	512.0	1.6670	6.01	511.6	1.6411	4.80	511.3	1.6130
280	49.49	525.1	1.9166	24.75	525.0	1.8305	16.50	524.8	1.7801	12.37	524.6	1.7444	8.24	524.2	1.6938	6.17	523.9	1.6579	4.93	523.5	1.6299
300	50.83	537.6	1.9331	25.42	537.4	1.8470	16.94	537.2	1.7966	12.71	537.0	1.7609	8.46	536.7	1.7103	6.35	536.3	1.6744	5.07	536.0	1.6464
320	52.16	550.2	1.9493	26.09	550.0	1.8633	17.39	549.8	1.8129	13.04	549.6	1.7771	8.69	549.3	1.7266	6.52	549.0	1.6907	5.21	548.6	1.6627
340	53.50	563.0	1.9655	26.75	562.9	1.8795	17.83	562.7	1.8291	13.38	562.5	1.7934	8.91	562.2	1.7429	6.68	561.9	1.7070	5.35	561.5	1.6790
360	54.84	576.0	1.9815	27.43	575.9	1.8955	18.28	575.7	1.8451	13.72	575.5	1.8096	9.14	575.2	1.7589	6.86	574.9	1.7230	5.49	574.6	1.6951
380	56.18	589.2	1.9973	28.10	589.1	1.9113	18.73	588.9	1.8609	14.05	588.7	1.8252	9.36	588.5	1.7747	7.03	588.2	1.7389	5.61	587.8	1.7110
400	57.51	602.6	2.0132	28.76	602.5	1.9271	19.17	602.4	1.8768	14.38	602.2	1.8411	9.58	602.0	1.7906	7.19	601.7	1.7548	5.75	601.3	1.7269
420	58.85	616.1	2.0290	29.43	616.0	1.9429	19.62	615.8	1.8926	14.72	615.7	1.8569	9.81	615.5	1.8064	7.36	615.3	1.7706	5.89	614.9	1.7427
440	60.19	629.9	2.0446	30.10	629.8	1.9585	20.07	629.7	1.9082	15.05	629.6	1.8725	10.04	629.4	1.8221	7.53	629.1	1.7862	6.02	628.8	1.7583
460	61.53	643.9	2.0601	30.77	643.8	1.9741	20.51	643.7	1.9237	15.38	643.6	1.8880	10.26	643.4	1.8376	7.69	643.2	1.8018	6.15	642.9	1.7739
480	62.87	658.0	2.0755	31.44	657.9	1.9895	20.96	657.8	1.9391	15.72	657.7	1.9034	10.48	657.5	1.8530	7.87	657.4	1.8172	6.28	657.0	1.7893
500	64.20	672.4	2.0907	32.10	672.3	2.0047	21.40	672.3	1.9543	16.05	672.2	1.9186	10.70	672.0	1.8682	8.03	671.8	1.8325	6.42	671.5	1.8046

Temp., °F.	150 lb./sq. in. abs. (−190.0°F.)			200 lb./sq. in. abs. (−178.2°F.)			300 lb./sq. in. abs. (−159.5°F.)			500 lb./sq. in. abs. (−132.9°F.)			800 lb./sq. in. abs.			1000 lb./sq. in. abs.			1500 lb./sq. in. abs.		
t	v	h	s	v	h	s	v	h	s	v	h	s	v	h	s	v	h	s	v	h	s
−180	1.052	262.4	1.0267	.....	.....	.....	.....	.....	.....	.....	.....	.....	.....	.....	.....	.....	.....	.....	.....	.....	.....
−160	1.172	274.5	1.0716	0.830	269.4	1.0240	.....	.....	.....	.....	.....	.....	.....	.....	.....	.....	.....	.....	.....	.....	.....
−140	1.283	286.3	1.1106	.923	281.6	1.0655	0.553	271.1	0.9899	.....	.....	.....	.....	.....	.....	.....	.....	.....	.....	.....	.....
−120	1.391	297.7	1.1452	1.010	293.5	1.1016	.624	284.7	1.0327	0.3000	261.9	0.9197	.....	.....	.....	.....	.....	.....	.....	.....	.....
−115.8	1.414	299.9	1.1518	1.027	295.8	1.1085	.637	287.4	1.0409	.3142	265.9	0.9313	.....	.....	.....	.....	.....	.....	.....	.....	.....
−100	1.495	308.9	1.1769	1.092	305.2	1.1346	.687	297.5	1.0706	.3566	280.4	0.9715	0.1441	236.9	0.8141	.....	.....	.....	.....	.....	.....
− 80	1.597	319.9	1.2065	1.172	316.6	1.1654	.747	309.9	1.1040	.402	295.5	1.0139	.1969	267.9	.8992	0.1262	238.3	0.8118	.....	.....	.....
− 60	1.695	330.7	1.2344	1.247	327.9	1.1941	.802	321.9	1.1345	.443	309.3	1.0500	.2359	287.9	.9514	.1650	267.8	.8865	0.0870	223.5	0.7723
− 40	1.793	341.5	1.2606	1.324	339.0	1.2211	.856	333.6	1.1627	.481	322.5	1.0819	.2674	304.3	.9921	.1957	289.2	.9397	.1069	253.3	.8373
− 20	1.890	352.3	1.2856	1.399	350.0	1.2467	.909	345.1	1.1889	.518	335.2	1.1111	.2953	319.4	1.0276	.2212	307.6	.9809	.1270	276.6	.8860
0	1.989	363.0	1.3096	1.475	360.9	1.2711	.961	356.5	1.2145	.551	347.7	1.1386	.3202	333.9	1.0595	.2438	324.0	1.0165	.1453	297.8	.9292
20	2.084	373.6	1.3326	1.548	371.8	1.2944	1.011	367.9	1.2386	.584	359.9	1.1645	.3441	347.8	1.0885	.2647	339.3	1.0481	.1617	317.3	.9657
40	2.177	384.5	1.3546	1.620	382.8	1.3167	1.062	379.3	1.2615	.617	372.2	1.1891	.367	361.4	1.1151	.2848	354.0	1.0764	.1770	335.5	.9983
60	2.269	395.4	1.3757	1.691	393.8	1.3380	1.111	390.6	1.2834	.649	384.3	1.2123	.389	374.6	1.1399	.3031	368.2	1.1025	.1912	352.4	1.0276
80	2.364	406.3	1.3964	1.762	404.9	1.3589	1.161	402.1	1.3047	.681	396.3	1.2347	.411	387.5	1.1638	.3217	381.6	1.1273	.2051	367.9	1.0554
100	2.459	417.5	1.4168	1.835	416.1	1.3795	1.211	413.4	1.3258	.712	408.1	1.2565	.432	400.1	1.1869	.3396	395.0	1.1513	.2182	382.5	1.0818
120	2.554	428.6	1.4366	1.906	427.3	1.3994	1.259	424.7	1.3460	.742	419.9	1.2774	.453	412.4	1.2088	.3568	407.5	1.1741	.2306	396.5	1.1069
140	2.648	439.8	1.4552	1.976	438.6	1.4182	1.307	436.2	1.3652	.773	431.6	1.2969	.473	424.6	1.2294	.374	420.1	1.1954	.2429	410.0	1.1302
160	2.739	451.2	1.4736	2.046	450.0	1.4367	1.355	447.7	1.3839	.803	443.4	1.3157	.493	436.9	1.2493	.390	432.8	1.2160	.2548	423.3	1.1524
180	2.833	462.7	1.4919	2.116	461.6	1.4551	1.403	459.4	1.4025	.833	455.3	1.3347	.513	449.2	1.2689	.407	445.3	1.2363	.2664	436.4	1.1741
200	2.924	474.4	1.5099	2.187	473.4	1.4732	1.450	471.4	1.4208	.862	467.4	1.3531	.532	461.8	1.2881	.422	458.1	1.2560	.2780	449.6	1.1948
220	3.012	486.1	1.5276	2.253	485.2	1.4911	1.496	483.4	1.4388	.891	479.7	1.3712	.551	474.4	1.3068	.438	470.9	1.2753	.2892	462.7	1.2150
240	3.106	498.4	1.5448	2.323	497.4	1.5084	1.544	495.8	1.4562	.920	492.3	1.3888	.571	487.1	1.3250	.454	484.0	1.2938	.3002	476.0	1.2343
260	3.197	510.6	1.5619	2.392	509.7	1.5255	1.591	508.0	1.4735	.949	504.8	1.4063	.590	499.8	1.3429	.470	496.9	1.3120	.3112	489.2	1.2531
280	3.287	523.0	1.5788	2.461	522.1	1.5425	1.638	520.4	1.4907	.978	517.4	1.4236	.608	512.7	1.3607	.485	509.8	1.3300	.3221	502.6	1.2716
300	3.378	535.5	1.5954	2.531	534.7	1.5592	1.684	533.0	1.5074	1.007	530.2	1.4406	.627	525.7	1.3781	.501	522.9	1.3475	.3326	516.2	1.2896
320	3.459	548.2	1.6117	2.599	547.4	1.5756	1.730	545.8	1.5239	1.036	534.2	1.4573	.646	538.8	1.3951	.516	536.1	1.3646	.3429	529.8	1.3072
340	3.559	561.1	1.6281	2.667	560.3	1.5919	1.776	558.8	1.5403	1.064	556.3	1.4740	.664	552.1	1.4120	.530	549.4	1.3817	.3534	543.5	1.3247
360	3.66	574.2	1.6442	2.736	573.4	1.6080	1.823	571.9	1.5566	1.093	569.6	1.4904	.682	565.6	1.4287	.545	563.0	1.3985	.364	557.3	1.3419
380	3.75	587.5	1.6601	2.805	586.7	1.6239	1.869	585.3	1.5726	1.121	583.1	1.5066	.700	579.2	1.4450	.560	576.7	1.4149	.374	571.3	1.3589

* Matthews and Hurd, *Trans. Am. Inst. Chem. Engrs.* **42**, 55 (1946).

Properties of Saturated Carbon Dioxide*†‡

Temp., °F. t	Abs. pressure, lb./sq. in. p	Volume, cu. ft./lb. Condensed phase* v_f	Volume, cu. ft./lb. Vapor v_g	Enthalpy, B.t.u./lb. Condensed phase* h_f	Enthalpy, B.t.u./lb. Vapor h_g	Entropy B.t.u./(lb.)(°R.) Condensed phase* s_f	Entropy B.t.u./(lb.)(°R.) Vapor s_g
−140	3.18	0.01008	24.320	−121.5	129.2	0.6065	1.3908
−120	8.90	.01018	9.179	−116.0	132.0	.6232	1.3636
−100	22.22	.01032	3.804	−110.1	134.3	.6403	1.3199
− 90	33.98	.01040	2.525	−106.7	135.1	.6499	1.3033
− 80	50.85	.01048	1.700	−102.5	135.7	.6607	1.2881
− 70	74.82	.01059	1.162	− 98.0	135.9	.6724	1.2726
− 69.9	75.10	.01059	1.157	− 97.9	135.9	.6725	1.2724
− 69.9	75.10	.01360	1.1570	− 13.7	135.9	.8885	1.2724
− 60	94.7	.01384	0.9270	− 9.2	136.6	.8997	1.2647
− 50	118.2	.01409	.7492	− 4.7	137.2	.9110	1.2572
− 40	145.8	.01437	.6113	.00	137.8	.9218	1.2503
− 30	177.8	.01466	.5029	4.5	138.2	.9325	1.2436
− 20	214.9	.01498	.4168	9.1	138.5	.9430	1.2372
− 10	257.3	.01532	.3472	13.9	138.7	.9532	1.2303
0	305.5	.01570	.2904	18.8	138.9	.9636	1.2247
10	360.2	.01614	.2437	24.0	138.7	.9744	1.2188
20	421.8	.01663	.2049	29.4	138.3	.9856	1.2127
30	490.8	.01719	.1722	35.4	137.8	.9976	1.2067
40	567.8	.01787	.1444	41.7	136.7	1.0092	1.1994
50	653.6	.01868	.1205	48.4	135.0	1.0218	1.1917
60	748.6	.01970	.0994	55.5	132.1	1.0353	1.1826
70	853.4	.02112	.08040	63.7	127.5	1.0500	1.1724
80	968.7	.02370	.06064	73.9	118.7	1.0694	1.1555
87.8	1069.4	.03454	.03454	97.0	97.0	1.1098	1.1098

* Above the solid line the condensed phase is solid; below the line it is liquid.
† "Refrigerating Data Book," 5th ed., American Society of Refrigerating Engineers, New York, 1942.
‡ s_f = 1.0 at 32°F.
h_f = 36.7 at 32°F.

Properties of Superheated Carbon Dioxide*,†

v, volume, cu. ft./lb.; h, enthalpy, B.t.u./lb.; s, entropy, B.t.u/(lb.)(°R.); p, absolute pressure, lb./sq. in.

p		−75°F.	−50°F.	0°F.	50°F.	100°F.	150°F.	200°F.	300°F.	400°F.	600°F.	800°F.	1000°F.	1200°F.	1400°F.	1600°F.	1800°F.
1.00	v	93.90	100.0	112.2	124.4	136.6	148.8	161.0	185.4	209.7	258.5	307.2	356.0	404.8	453.6	502.3	551.0
	h	283.2	288.0	297.8	307.7	318.0	328.4	339.1	361.4	384.7	434.4	487.1	542.4	599.6	658.6	718.8	780.0
	s	1.4772	1.4892	1.5112	1.5316	1.5506	1.5684	1.5852	1.6165	1.6451	1.6969	1.7423	1.7829	1.8197	1.8533	1.8838	1.9123
10.0	v	9.280	9.902	11.15	12.38	13.61	14.84	16.06	18.51	20.96	25.85	30.73	35.61	40.49	45.36	50.24	55.11
	h	282.6	287.5	297.3	307.3	317.7	328.2	339.0	361.3	384.6	434.4	487.1	542.4	599.6	658.6	718.8	780.0
	s	1.3733	1.3853	1.4073	1.4277	1.4467	1.4645	1.4813	1.5126	1.5412	1.5930	1.6384	1.6790	1.7158	1.7494	1.7799	1.8084
20.0	v	4.586	4.904	5.542	6.119	6.778	7.407	8.016	9.247	10.47	12.92	15.36	17.80	20.24	22.68	25.11	27.55
	h	281.9	287.0	296.8	306.8	317.3	327.9	338.8	361.1	384.5	434.3	487.1	542.4	599.6	658.6	718.8	780.0
	s	1.3417	1.3538	1.3759	1.3964	1.4154	1.4332	1.4500	1.4813	1.5099	1.5617	1.6071	1.6477	1.6845	1.7181	1.7486	1.7771
40.0	v	2.239	2.404	2.738	3.053	3.363	3.688	3.993	4.615	5.230	6.458	7.688	8.901	10.12	11.37	12.56	13.78
	h	280.6	285.9	295.8	305.9	316.5	327.4	338.4	360.9	384.3	434.2	487.0	542.4	599.6	658.6	718.8	780.0
	s	1.3088	1.3211	1.3435	1.3642	1.3834	1.4014	1.4184	1.4499	1.4787	1.5305	1.5759	1.6165	1.6533	1.6869	1.7174	1.7459
80.0	v	......	1.154	1.335	1.498	1.657	1.828	1.982	2.298	2.608	3.226	3.839	4.448	5.060	5.670	6.281	6.887
	h	......	283.8	293.8	304.1	315.1	326.4	337.7	360.2	383.9	434.0	486.9	542.3	599.5	658.6	718.8	780.0
	s	......	1.2778	1.3044	1.3284	1.3490	1.3679	1.3855	1.4177	1.4468	1.4991	1.5446	1.5852	1.6220	1.6556	1.6861	1.7146
120	v	......	......	0.8665	0.9799	1.088	1.208	1.311	1.525	1.734	2.148	2.559	2.966	3.373	3.781	4.188	4.592
	h	......	......	291.7	302.2	313.6	325.4	337.0	359.7	383.5	433.8	486.8	542.3	599.5	658.6	718.8	780.0
	s	......	......	1.2833	1.3086	1.3297	1.3488	1.3666	1.3993	1.4285	1.4808	1.5263	1.5669	1.6037	1.6373	1.6678	1.6963
160	v	......	......	0.6305	0.7207	0.8033	0.8986	0.9760	1.139	1.297	1.610	1.918	2.224	2.530	2.836	3.141	3.445
	h	......	......	289.7	300.4	312.1	324.4	336.3	359.1	383.1	433.6	486.6	542.2	599.5	658.6	718.8	780.0
	s	......	......	1.2666	1.2928	1.3154	1.3350	1.3529	1.3857	1.4151	1.4675	1.5133	1.5539	1.5907	1.6243	1.6548	1.6833
200	v	......	......	0.4891	0.5652	0.6376	0.7125	0.7748	0.9075	1.035	1.287	1.534	1.779	2.024	2.269	2.513	2.757
	h	......	......	287.7	298.6	310.6	323.4	335.6	358.5	382.7	433.4	486.5	542.2	599.5	658.5	718.8	780.0
	s	......	......	1.2519	1.2805	1.3038	1.3239	1.3421	1.3753	1.4049	1.4574	1.5033	1.5439	1.5807	1.6143	1.6448	1.6733
240	v	......	......	0.3948	0.4614	0.5237	0.5886	0.6407	0.7532	0.8604	1.071	1.273	1.482	1.687	1.891	2.095	2.297
	h	......	......	285.6	296.7	309.1	322.4	334.9	358.0	382.3	433.1	486.4	542.1	599.5	658.5	718.8	780.0
	s	......	......	1.2395	1.2694	1.2940	1.3145	1.3330	1.3671	1.3963	1.4490	1.4948	1.5356	1.5724	1.6060	1.6365	1.6650
300	v	......	......	......	0.3563	0.4100	0.4636	0.5065	0.5985	0.6868	0.8556	1.021	1.186	1.349	1.513	1.676	1.838
	h	......	......	......	294.0	306.9	320.9	333.9	357.1	381.6	432.8	486.2	542.0	599.4	658.5	718.7	780.0
	s	......	......	......	1.2562	1.2813	1.3029	1.3219	1.3560	1.3862	1.4389	1.4848	1.5256	1.5624	1.5960	1.6265	1.6550
360	v	......	......	......	0.2858	0.3341	0.3780	0.4171	0.4958	0.5693	0.7212	0.8502	0.9874	1.125	1.261	1.397	1.533
	h	......	......	......	291.2	304.6	319.4	332.8	356.3	381.0	432.5	486.0	541.9	599.4	658.5	718.7	779.9
	s	......	......	......	1.2436	1.2699	1.2925	1.3124	1.3475	1.3779	1.4307	1.4766	1.5174	1.5542	1.5878	1.6183	1.6468
440	v	......	......	......	0.2216	0.2652	0.3040	0.3358	0.4022	0.4633	0.5817	0.6950	0.8079	0.9201	1.032	1.142	1.255
	h	......	......	......	287.6	301.6	317.4	331.4	355.1	380.2	432.1	485.8	541.7	599.3	658.4	718.6	779.9
	s	......	......	......	1.2282	1.2559	1.2797	1.3006	1.3370	1.3681	1.4215	1.4675	1.5083	1.5451	1.5787	1.6092	1.6377
520	v	......	......	......	0.1772	0.2174	0.2513	0.2795	0.3374	0.3901	0.4912	0.5881	0.6832	0.7785	0.8733	0.9672	1.062
	h	......	......	......	283.9	298.7	315.4	330.0	354.0	379.4	431.7	485.5	541.5	599.2	658.3	718.6	779.9
	s	......	......	......	1.2148	1.2438	1.2687	1.2905	1.3281	1.3599	1.4138	1.4599	1.5007	1.5375	1.5711	1.6010	1.6301
600	v	......	......	......	0.1452	0.1823	0.2123	0.2383	0.2898	0.3363	0.4250	0.5093	0.5921	0.6747	0.7571	0.8385	0.9202
	h	......	......	......	280.3	295.7	313.4	328.6	352.8	378.6	431.1	485.3	541.4	599.0	658.2	718.6	779.8
	s	......	......	......	1.2020	1.2323	1.2583	1.2809	1.3198	1.3525	1.4071	1.4534	1.4942	1.5310	1.5646	1.5951	1.6236
800	v	......	......	......	......	0.1196	0.1483	0.1712	0.2126	0.2489	0.3173	0.3812	0.4436	0.5060	0.5680	0.6292	0.6906
	h	......	......	......	......	288.2	308.4	325.1	350.0	376.5	430.1	484.7	541.0	598.8	658.0	718.4	779.7
	s	......	......	......	......	1.2111	1.2391	1.2631	1.3041	1.3380	1.3935	1.4404	1.4812	1.5180	1.5516	1.5821	1.6106
1000	v	......	......	......	......	......	0.1101	0.1310	0.1663	0.1966	0.2526	0.3048	0.3547	0.4049	0.4545	0.5037	0.5526
	h	......	......	......	......	......	303.4	321.6	347.1	374.5	429.1	484.0	540.6	598.5	657.8	718.3	779.6
	s	......	......	......	......	......	1.2218	1.2472	1.2903	1.3258	1.3828	1.4302	1.4712	1.5080	1.5416	1.5721	1.6006
1200	v	......	......	......	......	......	......	0.1042	0.1356	0.1621	0.2096	0.2531	0.2953	0.3374	0.3789	0.4199	0.4609
	h	......	......	......	......	......	......	318.4	344.2	372.5	428.1	483.5	540.2	598.2	657.7	718.2	779.5
	s	......	......	......	......	......	......	1.2343	1.2791	1.3158	1.3740	1.4216	1.4628	1.4996	1.5332	1.5637	1.5922
1400	v	......	......	......	......	......	......	......	0.1136	0.1375	0.1788	0.2160	0.2529	0.2892	0.3249	0.3601	0.3551
	h	......	......	......	......	......	......	......	341.4	370.4	427.0	482.9	539.8	598.0	657.5	718.0	779.5
	s	......	......	......	......	......	......	......	1.2703	1.3078	1.3668	1.4145	1.4558	1.4927	1.5263	1.5568	1.5853
1600	v	......	......	......	......	......	......	......	......	0.1191	0.1557	0.1898	0.2211	0.2530	0.2843	0.3153	0.3461
	h	......	......	......	......	......	......	......	......	367.6	426.0	482.3	539.5	597.7	657.2	717.9	779.4
	s	......	......	......	......	......	......	......	......	1.3002	1.3602	1.4083	1.4497	1.4867	1.5193	1.5508	1.5793
1800	v	......	......	......	......	......	......	......	......	0.1047	0.1377	0.1675	0.1964	0.2249	0.2528	0.2804	0.3079
	h	......	......	......	......	......	......	......	......	364.0	424.9	481.7	539.1	597.4	657.0	717.8	779.3
	s	......	......	......	......	......	......	......	......	1.2930	1.3539	1.4023	1.4440	1.4812	1.5148	1.5453	1.5738

* Sweigert, Weber, and Allen, *Ind. Eng. Chem.*, **38**, 185 (1946); with permission.
† s_f = 1.0 at 32°F. h_f = 180 at 32°F. Therefore, according to the bases of Table 208, the entropies of these two CO_2 tables are consistent, but (180 − 36.7) or 143.3 B.t.u./lb. must be added to the enthalpies of saturated CO_2 to make them consistent with those of superheated CO_2.

Properties of Saturated Steam: Pressure Table*

Abs. pressure lb./sq. in. p	Temp., °F. t	Volume, cu. ft./lb.		Enthalpy, B.t.u./lb.		Entropy, B.t.u./(lb.)(°R.)		Internal energy, B.t.u./lb.	
		Liquid v_f	Vapor v_g	Liquid h_f	Vapor h_g	Liquid s_f	Vapor s_g	Liquid u_f	Vapor u_g
1.0	101.74	0.01614	333.6	69.70	1106.0	0.1326	1.9782	69.70	1044.3
2.0	126.08	.01623	173.73	93.99	1116.3	.1749	1.9200	93.98	1051.9
3.0	141.48	.01630	118.71	109.37	1122.6	.2008	1.8863	109.36	1056.7
4.0	152.97	.01636	90.63	120.86	1127.3	.2198	1.8625	120.85	1060.2
5.0	162.24	.01640	73.52	130.13	1131.1	.2347	1.8441	130.12	1063.1
6.0	170.06	.01645	61.98	137.96	1134.2	.2472	1.8292	137.94	1065.4
7.0	176.85	.01649	53.64	144.76	1136.9	.2581	1.8167	144.74	1067.4
8.0	182.86	.01653	47.34	150.79	1139.3	.2674	1.8057	150.77	1069.2
9.0	188.28	.01656	42.40	156.22	1141.4	.2759	1.7962	156.19	1070.8
10	193.21	.01659	38.42	161.17	1143.3	.2835	1.7876	161.14	1072.2
14.696	212.00	.01672	26.80	180.07	1150.4	.3120	1.7566	180.02	1077.5
15	213.03	.01672	26.29	181.11	1150.8	.3135	1.7549	181.06	1077.8
20	227.96	.01683	20.089	196.16	1156.3	.3356	1.7319	196.10	1081.9
25	240.07	.01692	16.303	208.42	1160.6	.3533	1.7139	208.34	1085.1
30	250.33	.01701	13.746	218.82	1164.1	.3680	1.6993	218.73	1087.8
35	259.28	.01708	11.898	227.91	1167.1	.3807	1.6870	227.80	1090.1
40	267.25	.01715	10.498	236.03	1169.7	.3919	1.6763	235.90	1092.0
45	274.44	.01721	9.401	243.36	1172.0	.4019	1.6669	243.22	1093.7
50	281.01	.01727	8.515	250.09	1174.1	.4110	1.6585	249.93	1095.3
55	287.07	.01732	7.787	256.30	1175.9	.4193	1.6509	256.12	1096.7
60	292.71	.01738	7.175	262.09	1177.6	.4270	1.6438	261.90	1097.9
65	297.97	.01743	6.655	267.50	1179.1	.4342	1.6374	267.29	1099.1
70	302.92	.01748	6.206	272.61	1180.6	.4409	1.6315	272.38	1100.2
75	307.60	.01753	5.816	277.43	1181.9	.4472	1.6259	277.19	1101.2
80	312.03	.01757	5.472	282.02	1183.1	.4531	1.6207	281.76	1102.1
85	316.25	.01761	5.168	286.39	1184.2	.4587	1.6158	286.11	1102.9
90	320.27	.01766	4.896	290.56	1185.3	.4641	1.6112	290.27	1103.7
95	324.12	.01770	4.652	294.56	1186.2	.4692	1.6068	294.25	1104.5
100	327.81	.01774	4.432	298.40	1187.2	.4740	1.6026	298.08	1105.2
110	334.77	.01782	4.049	305.66	1188.9	.4832	1.5948	305.30	1106.5
120	341.25	.01789	3.728	312.44	1190.4	.4916	1.5878	312.05	1107.6
130	347.32	.01796	3.455	318.81	1191.7	.4995	1.5812	318.38	1108.6
140	353.02	.01802	3.220	324.82	1193.0	.5069	1.5751	324.35	1109.6
150	358.42	.01809	3.015	330.51	1194.1	.5138	1.5694	330.01	1110.5
160	363.53	.01815	2.834	335.93	1195.1	.5204	1.5640	335.39	1111.2
170	368.41	.01822	2.675	341.09	1196.0	.5266	1.5590	340.52	1111.9
180	373.06	.01827	2.532	346.03	1196.9	.5325	1.5542	345.42	1112.5
190	377.51	.01833	2.404	350.79	1197.6	.5381	1.5497	350.15	1113.1
200	381.79	.01839	2.288	355.36	1198.4	.5435	1.5453	354.68	1113.7
250	400.95	.01865	1.8438	376.00	1201.1	.5675	1.5263	375.14	1115.8
300	417.33	.01890	1.5433	393.84	1202.8	.5879	1.5104	392.79	1117.1
350	431.72	.01913	1.3260	409.69	1203.9	.6056	1.4966	408.45	1118.0
400	444.59	.0193	1.1613	424.0	1204.5	.6214	1.4844	422.6	1118.5
450	456.28	.0195	1.0320	437.2	1204.6	.6356	1.4734	435.5	1118.7
500	467.01	.0197	0.9278	499.4	1204.4	.6487	1.4634	447.6	1118.6
550	476.94	.0199	.8424	460.8	1203.9	.6608	1.4542	458.8	1118.2
600	486.21	.0201	.7698	471.6	1203.2	.6720	1.4454	469.4	1117.7
650	494.90	.0203	.7083	481.8	1202.3	.6826	1.4374	479.4	1117.1
700	503.10	.0205	.6554	491.5	1201.2	.6925	1.4296	488.8	1116.3
750	510.86	.0207	.6092	500.8	1200.0	.7019	1.4223	598.0	1115.4
800	518.23	.0209	.5687	509.7	1198.6	.7108	1.4153	506.6	1114.4
850	525.26	.0210	.5327	518.3	1197.1	.7194	1.4085	515.0	1113.3
900	531.98	.0212	.5006	526.6	1195.4	.7275	1.4020	523.1	1112.1
950	538.43	.0214	.4717	534.6	1193.7	.7355	1.3957	530.9	1110.8
1000	544.61	.0216	.4456	542.4	1191.8	.7430	1.3897	538.4	1109.4
1100	556.31	.0220	.4001	557.4	1187.8	.7575	1.3780	552.9	1106.4
1200	567.22	.0223	.3619	571.7	1183.4	.7711	1.3667	566.7	1103.0
1300	577.46	.0227	.3293	585.4	1178.6	.7840	1.3559	580.0	1099.4
1400	587.10	.0231	.3012	598.7	1173.4	.7963	1.3454	592.7	1095.4
1500	596.23	.0235	.2765	611.6	1167.9	.8082	1.3351	605.1	1091.2
2000	635.82	.0257	.1878	671.7	1135.1	.8619	1.2849	662.2	1065.6
2500	668.13	.0287	.1307	730.6	1091.1	.9126	1.2322	717.3	1030.6
3000	695.36	.0346	.0858	802.5	1020.3	.9731	1.1615	783.4	972.7
3206.2	705.40	.0503	.0503	902.7	902.7	1.0580	1.0580	872.9	872.9

* Abridged from Keenan and Keyes, "Thermodynamic Properties of Steam," Wiley, New York, 1936. Copyright, 1937, by Joseph H. Keenan and Frederick G. Keyes.

Properties of Superheated Steam*

v, volume, cu. ft./lb.; *h*, enthalpy, B.t.u./lb.; *s*, entropy, B.t.u./(lb.)(°R.)

Abs. pressure, lb./sq. in. (sat. temp.)		Temp., °F.												
		200	300	400	500	600	700	800	900	1000	1100	1200	1400	1600
1	v	392.6	452.3	512.0	571.6	631.2	690.8	750.4	809.9	869.5	929.1	988.7	1107.8	1227.0
	h	1150.4	1195.8	1241.7	1288.3	1335.7	1383.8	1432.8	1482.7	1533.5	1585.2	1637.7	1745.7	1857.5
(101.74)	s	2.0512	2.1153	2.1720	2.2233	2.2702	2.3137	2.3542	2.3923	2.4283	2.4625	2.4952	2.5566	2.6137
5	v	78.16	90.25	102.26	114.22	126.16	138.10	150.03	161.95	173.87	185.79	197.71	221.6	245.4
	h	1148.8	1195.0	1241.2	1288.0	1335.4	1383.6	1432.7	1482.6	1533.4	1585.1	1637.7	1745.7	1857.4
(162.24)	s	1.8718	1.9370	1.9942	2.0456	2.0927	2.1361	2.1767	2.2148	2.2509	2.2851	2.3178	2.3792	2.4363
10	v	38.85	45.00	51.04	57.05	63.03	69.01	74.98	80.95	86.92	92.88	98.84	110.77	122.69
	h	1146.6	1193.9	1240.6	1287.5	1335.1	1383.4	1432.5	1482.4	1533.2	1585.0	1637.6	1745.6	1857.3
(193.21)	s	1.7927	1.8595	1.9172	1.9689	2.0160	2.0596	2.1002	2.1383	2.1744	2.2086	2.2413	2.3028	2.3598
14.696	v	......	30.53	34.68	38.78	42.86	46.94	51.00	55.07	59.13	63.19	67.25	75.37	83.48
	h	......	1192.8	1239.9	1287.1	1334.8	1383.2	1432.3	1482.3	1533.1	1584.8	1637.5	1745.5	1857.3
(212.00)	s	......	1.8160	1.8743	1.9261	1.9734	2.0170	2.0576	2.0958	2.1319	2.1662	2.1989	2.2603	2.3174
20	v	......	22.36	25.43	28.46	31.47	34.47	37.46	40.45	43.44	46.42	49.41	55.37	61.34
	h	......	1191.6	1239.2	1286.6	1334.4	1382.9	1432.1	1482.1	1533.0	1584.7	1637.4	1745.4	1857.2
(227.96)	s	......	1.7808	1.8396	1.8918	1.9392	1.9829	2.0235	2.0618	2.0978	2.1321	2.1648	2.2263	2.2834
40	v	......	11.040	12.628	14.168	15.688	17.198	18.702	20.20	21.70	23.20	24.69	27.68	30.66
	h	......	1186.8	1236.5	1284.8	1333.1	1381.9	1431.3	1481.4	1532.4	1584.3	1637.0	1745.1	1857.0
(267.25)	s	......	1.6994	1.7608	1.8140	1.8619	1.9058	1.9467	1.9850	2.0212	2.0555	2.0883	2.1498	2.2069
60	v	......	7.259	8.357	9.403	10.427	11.441	12.449	13.452	14.454	15.453	16.451	18.446	20.44
	h	......	1181.6	1233.6	1283.0	1331.8	1380.9	1430.5	1480.8	1531.9	1583.8	1636.6	1744.8	1856.7
(292.71)	s	......	1.6492	1.7135	1.7678	1.8162	1.8605	1.9015	1.9400	1.9762	2.0106	2.0434	2.1049	2.1621
80	v	......	......	6.220	7.020	7.797	8.562	9.322	10.077	10.830	11.582	12.332	13.830	15.325
	h	......	......	1230.7	1281.1	1330.5	1379.9	1429.7	1480.1	1531.3	1583.4	1636.2	1744.5	1856.5
(312.03)	s	......	......	1.6791	1.7346	1.7836	1.8281	1.8694	1.9079	1.9442	1.9787	2.0115	2.0721	2.1303
100	v	......	......	4.937	5.589	6.218	6.835	7.446	8.052	8.656	9.259	9.860	11.060	12.258
	h	......	......	1227.6	1279.1	1329.1	1378.9	1428.9	1479.5	1530.8	1582.9	1635.7	1744.2	1856.2
(327.81)	s	......	......	1.6518	1.7085	1.7581	1.8029	1.8443	1.8829	1.9193	1.9538	1.9867	2.0484	2.1056
120	v	......	......	4.081	4.636	5.165	5.683	6.195	6.702	7.207	7.710	8.212	9.214	10.213
	h	......	......	1224.4	1277.2	1327.7	1377.8	1428.1	1478.8	1530.2	1582.4	1635.3	1743.9	1856.0
(341.25)	s	......	......	1.6287	1.6869	1.7370	1.7822	1.8237	1.8625	1.8990	1.9335	1.9664	2.0281	2.0854
140	v	......	......	3.468	3.954	4.413	4.861	5.301	5.738	6.172	6.604	7.035	7.895	8.752
	h	......	......	1221.1	1275.2	1326.4	1376.8	1427.3	1478.2	1529.7	1581.9	1634.9	1743.5	1855.7
(353.02)	s	......	......	1.6087	1.6683	1.7190	1.7645	1.8063	1.8451	1.8817	1.9163	1.9493	2.0110	2.0683
160	v	......	......	3.008	3.443	3.849	4.244	4.631	5.015	5.396	5.775	6.152	6.906	7.656
	h	......	......	1217.6	1273.1	1325.0	1375.7	1426.4	1477.5	1529.1	1581.4	1634.5	1743.2	1855.5
(363.53)	s	......	......	1.5908	1.6519	1.7033	1.7491	1.7911	1.8301	1.8667	1.9014	1.9344	1.9962	2.0535
180	v	......	......	2.649	3.044	3.411	3.764	4.110	4.452	4.792	5.129	5.466	6.136	6.804
	h	......	......	1214.0	1271.0	1323.5	1374.7	1425.6	1476.8	1528.6	1581.0	1634.1	1742.9	1855.2
(373.06)	s	......	......	1.5745	1.6373	1.6894	1.7355	1.7776	1.8167	1.8534	1.8882	1.9212	1.9831	2.0404
200	v	......	......	2.361	2.726	3.060	3.380	3.693	4.002	4.309	4.613	4.917	5.521	6.123
	h	......	......	1210.3	1268.9	1322.1	1373.6	1424.8	1476.2	1528.0	1580.5	1633.7	1742.6	1855.0
(381.79)	s	......	......	1.5594	1.6240	1.6767	1.7232	1.7655	1.8048	1.8415	1.8763	1.9094	1.9713	2.0287
220	v	......	......	2.125	2.465	2.772	3.066	3.352	3.634	3.913	4.191	4.467	5.017	5.565
	h	......	......	1206.5	1266.7	1320.7	1372.6	1424.0	1475.5	1527.5	1580.0	1633.3	1742.3	1854.7
(389.86)	s	......	......	1.5453	1.6117	1.6652	1.7120	1.7545	1.7939	1.8308	1.8656	1.8987	1.9607	2.0181
240	v	......	......	1.9276	2.247	2.533	2.804	3.068	3.327	3.584	3.839	4.093	4.597	5.100
	h	......	......	1202.5	1264.5	1319.2	1371.5	1423.2	1474.8	1526.9	1579.6	1632.9	1742.0	1854.5
(397.37)	s	......	......	1.5319	1.6003	1.6546	1.7017	1.7444	1.7839	1.8209	1.8558	1.8889	1.9510	2.0084
260	v	......	......	......	2.063	2.330	2.582	2.827	3.067	3.305	3.541	3.776	4.242	4.707
	h	......	......	......	1262.3	1317.7	1370.4	1422.3	1474.2	1526.3	1579.1	1632.5	1741.7	1854.2
(404.42)	s	......	......	......	1.5897	1.6447	1.6922	1.7352	1.7748	1.8118	1.8467	1.8799	1.9420	1.9995
280	v	......	......	......	1.9047	2.156	2.392	2.621	2.845	3.066	3.286	3.504	3.938	4.370
	h	......	......	......	1260.0	1316.2	1369.4	1421.5	1473.5	1525.8	1578.6	1632.1	1741.4	1854.0
(411.05)	s	......	......	......	1.5796	1.6354	1.6834	1.7265	1.7662	1.8033	1.8383	1.8716	1.9337	1.9912
300	v	......	......	......	1.7675	2.005	2.227	2.442	2.652	2.859	3.065	3.269	3.674	4.078
	h	......	......	......	1257.6	1314.7	1368.3	1420.6	1472.8	1525.2	1578.1	1631.7	1741.0	1853.7
(417.33)	s	......	......	......	1.5701	1.6268	1.6751	1.7184	1.7582	1.7954	1.8305	1.8638	1.9260	1.9835
350	v	......	......	......	1.4923	1.7036	1.8980	2.084	2.266	2.445	2.622	2.798	3.147	3.493
	h	......	......	......	1251.5	1310.9	1365.5	1418.5	1471.1	1523.8	1577.0	1630.7	1740.3	1853.1
(431.72)	s	......	......	......	1.5481	1.6070	1.6563	1.7002	1.7403	1.7777	1.8130	1.8463	1.9086	1.9663
400	v	......	......	......	1.2851	1.4770	1.6508	1.8161	1.9767	2.134	2.290	2.445	2.751	3.055
	h	......	......	......	1245.1	1306.9	1362.7	1416.4	1469.4	1522.4	1575.8	1629.6	1739.5	1852.5
(444.59)	s	......	......	......	1.5281	1.5894	1.6398	1.6842	1.7247	1.7623	1.7977	1.8311	1.8936	1.9513

* Abridged from Keenan and Keyes, "Thermodynamic Properties of Steam," Wiley, New York, 1936. Copyright, 1937, by Joseph H. Keenan and Frederick G. Keyes.

INDEX

In searching for a given item in this book, please also refer to the Table of Contents, List of Figures, and List of Tables in the front of this book. The latter two are particularly useful since they show the content of the accompanying material. The reader is advised that this index only covers Volume 2 of the "Gas Conditioning and Processing" series. Volume 1 has a separate index.

A

B

C

D

E

F

G

H

I

O

P

R

S

T

U

V

W

Z

NOTES

NOTES

NOTES

NOTES

NOTES

NOTES